Contributing Authors

Gerald Smith
University of Michigan
*Chapter 3: Introduction to
Classification*

Anthony Gharrett
University of Alaska
Chapter 27: Genetics

Michael Barton
Centre College
Chapter 28: Behavior
Chapter 29: Ecology

Camm Swift
Loyola Marymount University
Chapter 30: Distribution and Migration

Reviewer List

Dr. Steven Bortone, *University of West Florida*
Dr. John Briggs, *University of Georgia–Athens*
Dr. Donald Buth, *University of California–Los Angeles*
Dr. Robert Cashner, *University of New Orleans*
Dr. Tom Coon, *Michigan State University*
Dr. James Dooley, *Adelphi University*
Dr. Shelby D. Gerking, *Arizona State University*
Dr. David Heins, *Tulane University*
Dr. Dean Hendrickson, *Texas Memorial Museum*
Dr. William Hershberger, *University of Washington*
Dr. Karel F. Liem, *Harvard University*
Dr. John Lundberg, *University of Arizona*
Dr. Mark Luttenton, *Grand Valley State University*
Dr. Scott Schaefer, *Academy of Natural Sciences*
Dr. C. Lovett Smith, *American Museum of Natural History–New York City*
Dr. Jamie E. Thomerson, *Southern Illinois University*
Dr. Jackie Webb, *Villanova University*
Dr. Mark Westneat, *Field Museum of Natural History*

Biology of Fishes

...is to be retu... ...
...last date st...

Biology of Fishes

Second Edition

Carl E. Bond

Professor Emeritus of Fisheries and Wildlife

Oregon State University

Corvallis, Oregon

SAUNDERS COLLEGE PUBLISHING

Harcourt Brace College Publishers
Fort Worth   Philadelphia   San Diego
New York   Orlando   Austin   San Antonio
Toronto   Montreal   London   Sydney   Tokyo

Text Typeface: Times Roman
Compositor: Monotype Composition
Executive Editor: Julie Levin Alexander
Assistant Editor: Jane Sanders Wood
Project Editor: Linda Boyle
Copy Editor: Patricia Daly
Art Director: Joan Wendt
Cover Designer: Chazz Bjanes
Text Designer: Kathryn Needle
Text Artwork: John Norton
Production Manager: Alicia Jackson

Printed in the United States of America

BIOLOGY OF FISHES

ISBN 0-03-070342-5

Library of Congress Catalog Card Number: 95-067414

67890123 039 10 98765432

(w) 597 B

Dedication

This book is dedicated to the memory of
Professor Roland E. Dimick, who introduced many
to the study of fishes.

Preface

From time to time I have been asked why and when I chose a career in ichthyology. The seeds were probably sown by an elementary school teacher who spent her Saturdays taking her students with her to the trout streams and sunfish lakes so numerous in western Oregon. Although she was a marvelous flyfisher, she didn't mind the hazel wood poles, kirby hooks, and gobs of worms that her companions used. She showed us the difference between the fishes we encountered—bluegills, pumpkinseeds, largemouth bass, crappies, bullheads, cutthroat trout, carp, and even sculpins and the sandrollers, which we called troutperch.

The fishing habit stayed with me when I began teaching elementary school, and when I was transported to West Africa with the Navy "seabees" in World War II. I became familiar with herrings, mudskippers, sleepers, mullets, barracudas, and other fishes. Assignments on Oahu and Midway, where I was fortunate enough to participate in a topographic survey of the reef, led to a familiarity with many tropical species of fishes.

When I returned home at war's end, I decided to go to Oregon State College in order to study Forestry, but Ben Hur Lampman, Poet Laureate of Oregon and fishing friend of my oldest brother, told me that the forests were only the background for my interests. He suggested that I visit Professor Roland Dimick at the Department of Fish and Game Management when I went to Corvallis. At the college, a friend in the registrar's office told me where to find the School of Forestry but suggested I could visit Professor Dimick on the way. I had an interview with Mr. Dimick and by the time I finally entered the Forestry building I was already an Associate Professor of Fisheries.

Research and teaching in fisheries and ichthyology gave me an opportunity to build the teaching collection of fishes into a research collection and to influence some graduate students to study ichthyology instead of fisheries. While on sabbatical leave I had the opportunity to study at The University of Michigan, where I received a Ph.D. degree.

Except for students and my family (all of whom could haul seines and clip fins well enough to please me) the study of fishes has consumed my life. I am still learning about fishes, even though the fields of fisheries and ichthyology have advanced in technology in ways that are like magic compared to what we had 30 years ago. There is much to be discovered with new methods that are being applied to many aspects of the fields, including conservation, systematics, genetics and life history studies and more. Although the scientists who study fishes have steadily increased in numbers over the years, I believe there is still room for more.

I hope this book will influence some students to turn to ichthyology and will allow others to know about fishes in ways that will enrich their lives as much as mine has been enriched. The goal is to present as complete an ichthyology text as possible under the constraints of time and space.

CHANGES IN THIS EDITION

We believe there are many improvements to the second edition. Chapter outlines initiate each subject, and because this edition is intended as an upper division text, many literature citations are available. The book is now in four parts: One, Introduction to the Fishes; Two, Evolution and Diversity of Fishes; Three, Organismic Biology of Fishes; and Four, Populations, Species, and Communities. This text is arranged to allow the reader to begin with a section that will give an understanding of the fishes and the scientists who developed ichthyology. In the following sections of Part One they will learn much of the anatomy on which classification is based and the ideas on how and why fishes are classified.

Part Two includes eleven chapters devoted to the evolution and classification of fishes. At the request of several reviewers the classification system used is that of Dr. Joseph S. Nelson, except the Chondrichthyes, in which the classification of Dr. L. J. V. Compagno is used. Life history as well as classification is addressed in these chapters.

The chapters in Part Three combine both structure and function of organ systems. This combination was arranged at the request of various professors who used the first edition. Section Four presents genetics, behavior, ecology, distribution, and human endeavor.

NEW TO THIS EDITION

New material in this edition includes three chapters on subjects not addressed in the first edition: Introduction to Classification, written by Dr. Gerald R. Smith of the University of Michigan; Locomotion, written by the senior author; and a growing field, Genetics of Fishes, written by Dr. Anthony Gharrett of the University of Alaska. These additions allow this edition to be a more complete ichthyology text.

Some other chapters under new authorship include Chapter 28, Behavior: Doing What Fishes Do Best, and Chapter 29, Ecology: Environments, Habitats, and Adaptations, written by Dr. Michael Barton of Centre College, and Chapter 30, Distribution and Migrations, written by Camm Swift of Loyola Marymount University.

The illustration program, comprising new and revised figures, has been prepared by John Norton. A list of useful Greek and Latin terms is included for reference, and there is an expanded glossary prepared by Drs. Bond, Gharrett, and Smith.

ACKNOWLEDGMENTS

Several persons have been important to the completion of this book. Gerald Smith, Anthony Gharrett, Michael Barton, and Camm Swift interrupted research and teaching to write important chapters. Department Heads of the Department of Fisheries and Wildlife at Oregon State University, Richard Tubb, Erik Fritzell, and, interim Department Head, Lawrence Curtis allowed me access to departmental facilities and staff. Dr. Curtis also helped with advice on physiology. Dr. Richard Strauss helped by writing the prospectus and outline for this edition.

Many of my departmental colleagues aided with advice and encouragement, particularly Douglas Markle, Carl Schreck, James Hall, B. J. Verts, and Leslie Carraway. Other staff members who have aided in many ways include Jan Mosely,

LaVon Mauer, Charlotte Vickers, and Melanie Bonnichsen. Paulo Petry, Phillip Harris, and Chris Donohoe are graduate students who have spent helpful time with me.

The Saunders College Publishing editorial staff who had patience and lent encouragement through often difficult times were Julie Levin Alexander, Jane Sanders Wood, Dena Digilio-Betz, Linda Boyle, and Joan Wendt. John Norton, who drew the new illustrations for the book and patiently worked with me on several "picky" points of some strange fishes, has become a good friend.

My wife, Lenora, has been more important than anyone to this book. She has helped with typing from time to time, but mainly she has helped me through the illnesses and difficulties of the past three years.

We all celebrate the publication of this book.

Contents

Part Three   Organismic Biology of Fishes   **239**

Part One

Introduction to the Fishes

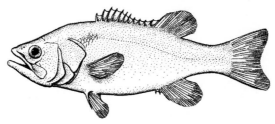

A Teleost

Chapter

1

Introduction

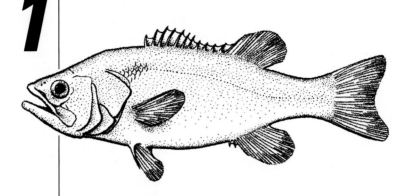

Fishes: What They Are

This book is intended to provide general knowledge of fish and fishlike vertebrates and will cover animals that have a cranium (subphylum Myxini and subphylum Vertebrata), live in water, possess gills that are used throughout the life span, and possess fins. Hagfishes (Myxini) and lampreys lack paired fins, either because of primitive absence or secondary loss. General groupings of the living animals that fall into this assemblage are the hagfishes, lampreys, sharks, rays (including skates), chimaeras, and bony fishes (Fig. 1–1). The first two are jawless and represent levels of vertebrate organization that are widely separated from living vertebrates with jaws. However, their inclusion in a study of fishes is traditional, justified by their aquatic habitat and general similarity in structure to the remainder of the vertebrates. Through their study, we might better appreciate the lineages of aquatic vertebrates.

Fishes are of great economic value for food, recreation, and aquarium use (see Chapter 31) and are also important to the ecology of water systems. Diminishing stocks of commercially important species and the extinction of species, mainly from freshwater habitats, because of human activity are current realities.

The first known fishlike vertebrates occur in the fossil record of the Ordovician Period, nearly 500 million years ago (Table 1–1). Table 1–2 lists the higher categories of fishes. These were jawless pteraspidomorphs with shields on the fins (†order Arandaspidimorphi[1]) known as "ostracoderms" because of their armored skin. By the end of the Silurian Period, over 400 million years ago, the bony fishes coexisted with several lines of both jawless and jaw-bearing fishes (gnathostomes), many of which were destined to become extinct. All major groups of jawed fishes presently living were in existence by the middle of the Devonian, more than 350 million years ago. Although probably of great antiquity, the soft-bodied hagfishes and lampreys have left a meager fossil record; the only fossils of these two groups are known from the Carboniferous (Bardack, 1991; Carroll, 1988).

The animals called fishes here are an assemblage of chordates that includes diverse forms, but all have important features in common. This is a paraphyletic grouping, one that includes what is considered to be the most recent common ancestor without including all the groups that have descended from it. This is in contrast to a monophyletic group, which includes not only the most recent common ancestor but all the descendants (see Nelson, 1994).

Major Groups of Living Fishes

Jawless "Fishes"

Hagfishes and lampreys are placed into the superclass Agnatha (without jaws), but in different classes. Living hagfishes (class Myxini, order Myxiniformes) have nonfunctional eyes and are elongate, marine predators and scavengers that lack paired fins and vertebrae. They have cartilaginous skeletons. There are about 25 species, found mainly in temperate seas, and some are subjected to a fishery because of the use of their skins for small leather items. They are characterized by barbels (slender

---

[1]Throughout the text, extinct taxa will be designated by a dagger.

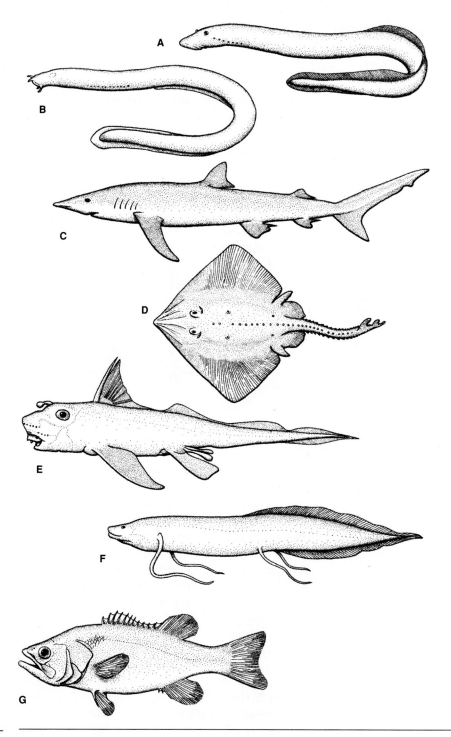

FIGURE 1–1 Examples of groups of living fishes. **A,** Lamprey (class Cephalaspidomorphi, order Petromy-zontiformes); **B,** hagfish (class Myxini, order Myxiniformes); **C,** shark (class Chondrichthyes, subclass Elasmobranchii); **D,** ray (skate) (class Chondrichthyes, subclass Elasmobranchii); **E,** chimaera (class Chondrichthyes, subclass Holocephali); **F,** lungfish (class Sarcopterygii, superorder Ceratodontimorpha); **G,** teleost (class Actinopterygii, subclass Neopterygii).

TABLE 1–1

Range of Some Major Groups of Fishes Through Time[1]

Era	*Period*	*Epoch*	*Dur[2]*	*BP[3]*
Cenozoic	Quaternary	Recent		
		Pleistocene	1.8	1.8
	Tertiary	Pliocene	3.2	5
		Miocene	19	24
		Oligocene	14	38
		Eocene	20	58
		Palaeocene	7	65
Mesozoic	Cretaceous		79	144
	Jurassic		64	208
	Triassic		37	245
Palaeozoic	Permian		41	286
	Carboniferous		74	360
	Devonian		48	408
	Silurian		30	438
	Ordovician		67	505
	Cambrian		65	570

Major Group

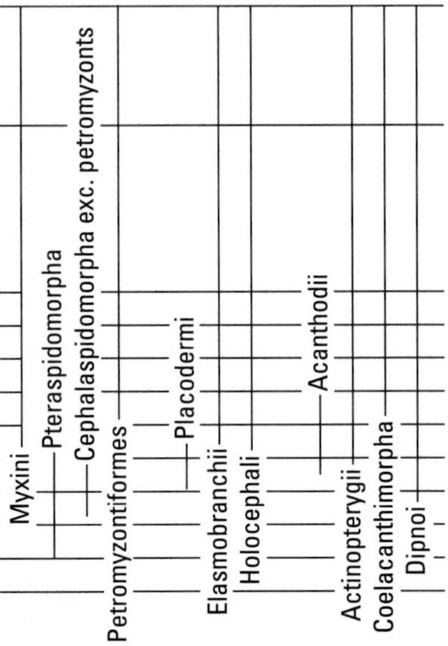
Myxini
Pteraspidomorpha
Cephalaspidomorpha exc. petromyzonts
Petromyzontiformes
Placodermi
Elasmobranchii
Holocephali
Acanthodii
Actinopterygii
Coelacanthimorpha
Dipnoi

[1]Geological time scale based on Stanley (1989), existence of groups based on Moy-Thomas and Miles (1971), Carroll (1988), and Nelson (1994).
[2]Approximate duration in millions of years.
[3]Approximate millions of years before present at beginning of each period or epoch.

sensory structures) around the mouth and the nasal opening. Species range in length from about 300 mm to a meter. Some researchers (Janvier, 1981) doubt that hagfishes should be considered vertebrates.

Lampreys are the living representatives of the class Cephalaspidomorphi, so named because most of the known fossil members had armored heads. The lampreys (order Petromyzontiformes) comprise about 35 species that range in length from 75 mm to more than a meter. All undergo a long larval period buried in bottom materials and so are seldom seen except during migrations. Larger species (some anadromous) are parasitic and feed on fishes and, at times, on marine mammals. The smaller kinds are usually nonparasitic and filter feed only during the larval stage. Although all spawn in fresh waters, some of the parasitic forms grow to maturity in the sea. Lampreys are eel shaped and have a cartilaginous skeleton. They lack paired fins, and the adults have an oral suctorial disc set with keratinous (hornlike) teeth. They are found in cold and temperate waters.

Sharks and Rays

Sharks and rays of the subclass Elasmobranchii (plate-gill) of the class Chondrichthyes (cartilage fish) are widely distributed in marine waters and are of interest for many reasons. Many species are the objects of fisheries, not only for the harvest

TABLE 1–2	**Higher Categories of Fishes to Level of Division (after Nelson, 1994)**

Superclass Agnatha
    Class Myxini (hagfishes)
    †Class Pteraspidomorphi (extinct armored fishes)
    Class Cephalaspidomorphi (lampreys, extinct relatives)
Superclass Gnathostomata
   †Grade Placodermomorphi
    †Class Placodermi (placoderms)
   Grade Chondrichthiomorphi
    Class Chondrichthyes
      Subclass Holocephali (chimaeras)
      Subclass Elasmobranchii (sharks and rays)
   Grade Teleostomi (bony fishes)
    †Class Acanthodii (acanthodians)
    Class Sarcopterygii (lobe-finned fishes)
      Subclass Coelacanthimorpha (coelacanths)
      Subclass (unnamed)
        †Infraclass Porolepimorpha
      Infraclass Dipnoi (lungfishes)
    Class Actinopterygii (ray-finned fishes)
      Subclass Chondrostei (sturgeons and bichirs)
      Subclass Neopterygii
        Division Teleostei (most bony fishes)

of their flesh but also for other materials such as their hides for leather and their jaws and teeth for souvenirs. A few sharks and rays are dangerous to humans. Stingrays and others with stinging spines can injure swimmers or fishermen, and both small and large predatory sharks have been known to attack human beings. Sharks are generally large animals; the various species average about 2 m long, but the whale shark is the largest fish known. This titan reaches 12 m or more. The dwarf shark (*Squaliolus*) matures at about 200 mm. There are, according to Compagno (1991), at least 376 and perhaps as many as 480 species of sharks. These are all marine, although a few species, including the bull shark, which has been implicated in many shark attacks, enter fresh water.

There are 500 or more species of skates and rays, and, like the sharks, there are many species in tropical areas. There are true freshwater species in two families of stingrays, representatives of which live in tropical rivers. Rays range in width from about 0.1 m to a "wingspan" of over 7 m in the largest manta.

Chimaeras

Chimaeras of the subclass Holocephali are related to the sharks and rays and share certain features of anatomy. (The name of the subclass emphasizes the fusion of elements of the skull in many of the group.) Their bizarre appearance has earned them such names as rabbitfish, ratfish, and ghost sharks, among others. Despite their looks, some chimaeras are used as food. Their oily livers produce a fine lubricant that once was used commonly by fishermen. There are fewer than 40 species; except for the few that live on the continental shelf, most live in deep marine habitat and are seldom seen or caught.

Bony Fishes

The remainder of the living fishes differ in that they have skeletons of bone, or mostly bone instead of cartilage entirely. These are the bony fishes, grade Teleostomi (formerly class Osteichthyes), which are numerous and varied. Tallies of known species indicate that there are about 25,000 species (Nelson, 1994). Eschmeyer (1990) estimates that there are about 55,000 named species and subspecies combined. New species are being described constantly. Bony fishes represent about 40 percent of all living vertebrates. Some of the extinct relatives of lungfishes (Dipnoi) and coelacanths (actinistians) are the closest known relatives of primitive amphibians. Lungfishes and coelacanths (so called because of their hollow notochord) are classified as separate subclasses in the class Sarcopterygii (fleshy finned), which includes tetrapods (Nelson, 1994). *Latimeria,* the living coelacanth, belongs to a subclass thought to have been extinct for millions of years until a specimen was captured near the coast of South Africa in 1938. This remarkable animal retains many unique features that had been deduced accurately from fossils of its extinct relatives, although the living species shows specialization for its particular deep water mode of life—for instance, in its osmoregulation (see Chapter 24) and in its gas bladder.

The lungfishes are placed in the infraclass Dipnoi (two-breathing). They have well-developed lungs that allow them to live in waters of low oxygen content. There is one living species of lungfish in South America, one in Australia, and four in Africa. All live in fresh water. Most reach sizes between 600 mm and 1 m in length, but an African species and the Australian species reach about 1.8 m.

Other bony fishes of the class Actinopterygii (ray-fin) proceeded along evolutionary lines distinct from those trends that produced terrestrial vertebrates, although some of the members of the subclass Chondrostei, the Polypteridae (many fins), have some characteristics reminiscent of lungfishes. Polypteridae were formerly placed in the infraclass Cladistia. This group contains a dozen or so species of African bichirs and reedfishes. In addition, the Chondrostei includes about 21 sturgeons and two paddlefishes.

The subclass Neopterygii (new fins) contains the gars and the bowfin of North America and the division Teleostei (complete or perfect bone), which includes about 20,000 bony fish species, of which diverse types are found in all oceans and in most fresh waters.

Range of Habitats. About 41 percent of these species are confined to fresh water, and the remainder are marine or salt tolerant. They are found from the abyss to the supratidal zone of the ocean, from thick swamps to the rushing torrents of the Andes and Himalayas, and from hot springs to freezing bogs and marine waters so cold that antifreeze is required in the blood (see Chapter 23). They swim by many methods, walk and wriggle both in and out of water, leap, glide, and even fly. They range in size from minuscule 1-cm gobies to giant tunas, marlins, swordfishes, catfishes, and the arapaima of South America, all of which reach lengths of about 3.5 m or more. Their colors rival the butterflies and birds, their shapes and postures are amazing, and their modes of life and some of their anatomical and behavioral adaptations for feeding and breeding approach (or reach) the fantastic.

Evolutionary Trends. Teleostomes include soft-rayed fishes that have retained some anatomical characteristics that are considered primitive relative to the spiny-rayed forms, which show more derived features. Some of the characteristics of soft-rayed groups include ganoid scales in the gars and bichirs, cycloid scales in most, abdominal pelvic fins, and the presence in many of orbitosphenoid and mesocoracoid bones (see Chapter 2). Soft-rayed fishes include the bony-tongues or osteoglossomorphs, which are possibly the most primitive living teleosts (Lauder and Liem, 1983b), tarpons, herrings, and trouts. Eels, carps, catfishes, and their relatives share in many of the primitive characteristics. Fishes with more derived characteristics include lanternfishes, flyingfishes, cods, and the spiny-rayed groups such as perches, basses, scorpionfishes, and related forms. Many of these have spines in the fins, ctenoid scales, and pelvic fins in a thoracic position.

The Study of Fishes

Phylogeny of fishes is intricate and difficult to trace. The groups of fishes, living and fossil, are diverse; the fossil record is far from complete; and anatomical structures are subject to differing interpretations. Nonetheless, ichthyologists are having success in understanding the phylogenetic relationships. Cladistic analyses, aimed at using shared, derived characters (synapomorphies) to show relationships among the fishes, have aided in our understanding; but in spite of the successes of the past decade, a consensus on the treatment of the evolutionary hypotheses has not yet been reached. There are differing opinions among ichthyologists regarding the classification of some fishes. The names, rank, and placement of groups discussed in this text will be based on, but not identical to, that of Nelson (1994).

As might be expected, the bony fishes will receive the most attention in this book. This is due not only to their numbers but also to their importance from many standpoints. Living species will receive most coverage, but the important extinct groups will not be ignored. Of the vast numbers of fishes that have become extinct—that evolved, perhaps flourished, and then failed to make some adjustment along the way—some are known in fossil form. Many have been studied by those clever detectives, the paleontologists, whose painstaking work has provided a better evaluation of the relationships of fishes than ever could be gained by attention only to living species. The often fragmentary, accidentally preserved remains of a random fraction of the extinct forms give a tantalizing glimpse of waypoints and endpoints in fish evolution.

Fishes are of interest for diverse reasons. They afford sport for the angler, provide food for millions all over the world, and have other commercial uses as animal food and as raw materials. Many species are kept as pets and have made advanced amateur ichthyologists out of an army of hobbyists. In the academic field, some scientists concentrate on comparative morphology of fishes, some study their biology, and others study their systematic relationships and great evolutionary significance. Many believe that fish behavior is the most interesting aspect of the study of fishes. The behavior and activity of the individual fish is at once directed by its environment and influences its environment. The place of a particular species in an environment depends on its structural and physiological capabilities for taking advantage of food sources and for finding suitable habitat for reproduction, however far from the feeding grounds.

A fish, active in its habitat and interacting with members of its own species or with other species, is a complex of structures, organ systems, and behavioral patterns that have evolved over the millions of years that fishes or their ancestors have been swimming in the abundant waters of this planet. Ichthyologists face many interconnected problems in studying fish behavior, because it is necessary to know with what biological equipment the fish works, how this equipment evolved, and (perhaps of prime importance) the reason for differing behavior in closely related forms in spite of a similarity in structure. In studying these matters, ichthyologists generate more than knowledge for its own sake because fishes are used for many purposes; and many aspects of fish biology, such as ecology, behavior, reproduction, and population dynamics, must be studied to manage wild stocks or to culture captive or domesticated stocks.

Our required knowledge concerning fishes is not complete, although laypeople, naturalists, scientists, and others have been recording information for centuries. Because of the difficulties inherent in the collection, preservation, and transportation of specimens from the deep sea and little-known land areas, thousands of species are known mainly as a few specimens in jars on the shelves of museums, with little or no information available on the living organisms. Museum collections are of great importance to the study of fish evolution. For many species that enter fisheries all over the world, only the basic features of their life histories are known. Others that have been studied in detail continue to present investigators with mysteries, as expanding knowledge allows the peculiarities of physiology and behavior to be evaluated and appreciated. So there is much yet to be done, much to be discovered, analyzed, reported, and understood. The labors that remain for ichthyologists are great and compelling.

Almost all ichthyologists are concerned with more than pure science and are involved to some extent in practical aspects of fishery science, even if this goes no farther than studying the systematics of economically significant groups. Those involved in the science of ichthyology for science's sake often see the results of their investigations put to use by others.

In the sequence of subjects in this book, the authors present first the comparative morphology necessary to study the systematics of fishes. The second section describes the systematic diversity of fishes, their evolution and relationships, their world and ecological distribution, and their general importance to fisheries. The third section explores the structure and function of organ systems. The final section deals with subjects treated at the level of populations, species, and communities, including ecology, distribution, and the relations of fishes to human beings.

Historical Figures in the Development of Ichthyology

To supplement this cursory coverage of ichthyologists from Aristotle to the middle of the 20th century, please refer to the following, on which this treatment is based: Jordan (1905), Dymond (1964), Myers (1964), and Hubbs (1964a, 1964b). To extend the knowledge of ichthyologists beyond this coverage, one should become familiar with the journal *Copeia* and others that deal with fishes.

Aristotle (383–322 B.C.) observed and recorded facts on 118 species of Greek fishes, and a few other Greeks and Romans added to the knowledge of Mediterranean fishes somewhat later, but for centuries very little original information was added to Aristotle's work. Not until the 16th century was there any significant effort to increase the knowledge of Mediterranean fishes. A few new books with figures and some attempts at classification appeared about 1555. One of the best was *De Piscibus Marinum* by Guillaume Rondelet, which included 244 species.

About 100 years later, exploration in Brazil led to the posthumous publication of Georg Markgraf's *Naturalis Brasilae,* which added nearly 100 new species to the known fishes. John Ray and Francis Willughby discussed classification in *Historia Piscium,* which covered over 400 nominal species and was published in 1686. In the mid-1700s, naturalists who had the opportunity to travel brought out works on fishes of many areas, from China to North America to the West Indies.

Carolus Linnaeus (Karl von Linné), professor at the University of Upsala, had students who were interested in fishes and wrote on non-European fishes, but one of his fellow students and associates, Peter Artedi (Artedius) (1705–1735), became known as the father of ichthyology. Artedi recognized five orders of "fishes": Malacopterygii, soft-rayed fishes; Acanthopterygii, spiny-rayed fishes; Branchiostegi, trunkfishes, anglerfishes, etc.; Chondropterygii, sharks, rays, lampreys, and sturgeons; and Plagiuri, cetaceans. He subdivided the orders into genera and the genera into species, to which he applied names consisting of a descriptive phrase. Artedi drowned accidentally in an Amsterdam canal, and his manuscripts were published by Linnaeus.

In his tenth and twelfth editions of *Systemae Naturae,* Linnaeus converted the names of all the animals to binomials and added several genera of fishes. He revised the orders of Artedi mainly on the basis of the position of the pelvic fins. Fishes without pelvic fins were called Apodes; those with abdominal, thoracic, and jugular pelvics were named, respectively, Abdominales, Thoracici, and Jugulares. He divided the Branchiostegi and Chondropterygii into two sections, Spiraculus compositis and Spiraculus solitariis.

During the late 18th and early 19th centuries, museums were receiving collections of fishes from many parts of the world, from explorers too numerous to mention. Some who dealt with American fishes were Georg Steller, who accompanied Bering to Siberia and Alaska; Johann Schöpf, who wrote on fishes of New York; and the eccentric Rafinesque, author of *Ichthyologia Ohiensis.* Myers (1964) recounts a story about Rafinesque using Audubon's Cremona violin to knock down bats. Audubon retaliated by providing Rafinesque with drawings of imaginary fishes, which Rafinesque then described. A compiler of note was Johann Walbaum, who in 1792 applied binomial names to fishes described by Steller, Schöpf, and Pennant. Several North American species were included. Other early students of North American fishes were Charles Alexandre LeSueur, a student of Cuvier; Samuel L. Mitchill, who published on the fishes of New York; and Sir John Richardson, who was also a student of the fishes of Australia and China and who covered northern fishes of North America in his *Fauna Boreali-Americana,* published 1831 to 1837.

M. E. Bloch of Germany published, from 1782 to 1785, a work called *Ichthyologia,* which was a series of volumes of plates. Following his death, his associate, Johann Schneider, published the great work containing 1519 species, *M. E. Blochii Systemae Ichthyologia.*

In France, between 1798 and 1803, Bernard German de Lacépède published the five-volume *Histoire Naturelle des Poissons.* According to Jordan (1905), the publication (1817 to 1830) by Georges Léopold Cuvier of *Règne Animal arrangé aprés son Organization* initiated a new era in ichthyology. Jordan states that the publication "is, in the history of ichthyology, not less important than the 'Systemae Naturae' itself, and from it dates practically our knowledge of families of fishes and the interrelationships of the different groups" (p. 400). Cuvier worked with his student, Achille Valenciennes, on the great 22-volume *Histoire Naturelle des Poissons,* much of which was written and published by Valenciennes after Cuvier's death in 1832. This work, which was not completed, contained accounts of 4514 species.

Albert Günther (1830–1914) of the British Museum published, from 1859 to 1870, a *Catalogue of the Fishes of the British Museum.* This is an eight-volume work with descriptions of over 6800 species and mention of nearly 1700 others. George Boulenger (1858–1937) of the British Museum published an excellent volume on the percoid fishes in 1898.

Of great importance to the development of ichthyology in the United States was Louis Agassiz (1807–1873), a Swiss scientist who had carried out important research on freshwater fishes and fish fossils before moving to the United States in 1846. He continued his work as a professor at Harvard University. His students included David Starr Jordan, Charles Girard, and Samuel Garman.

Space allows only brief mention of selected ichthyologists active in the 1800s:

Francis Hamilton (1762–1829), Scotland, described numerous species from India.

Felipe Poey y Aloy (1799–1891), Cuba, studied fishes of Cuba.

Johannes Müller (1808–1858), Germany, comparative anatomist, revised some of Cuvier's work and defined several groups of fishes.

William O. Ayres (1817–1891), United States, wrote on Pacific Coast fishes of the United States. He was California's first ichthyologist.

Pieter von Bleeker (1819–1878), Holland, provided much knowledge of the fishes of the East Indies.

Francis Day (1829–1899), England, published a two-volume work on the fishes of India.

Charles Girard (1822–1895), United States (from France), made a tremendous contribution to the knowledge of fishes of the western United States.

Spencer Fullerton Baird (1823–1887), United States, first U.S. Commissioner of Fisheries, sponsored and participated in much ichthyological work by the United States Fish Commission and the United States National Museum.

Edward Drinker Cope (1840–1897), United States, accomplished zoologist and paleontologist, studied freshwater fishes and fossil fishes in both the eastern and western United States. The journal *Copeia* is named in his honor.

George Brown Goode (1851–1899), United States, coauthor with Tarleton Bean (1846–1915) of *Oceanic Ichthyology* subtitled *A Treatise on the Deep-Sea and Pelagic Fishes of the World.*

Of those whose work spanned the late 19th and early 20th centuries, the greatest, in the opinions of many, was David Starr Jordan (1851–1931), who is characterized by Hubbs (1964a) as "a man of massive frame and even vaster intellect" (p. 50). Jordan was an educator as well as an ichthyologist and served as the president of Indiana University and as the first president of Stanford University. His writings encompass about 650 articles and books on ichthyology and nearly 1400 general works on education and politics, including a two-volume autobiography (*Days of a Man,* 1922). Under Jordan's influence, Stanford University became a center for ichthyological research.

Jordan's collaborators included Charles Henry Gilbert (1859–1928), a meticulous worker who contributed tremendously to our knowledge of Pacific Coast fishes, and Barton Warren Evermann, coauthor of *Fishes of North and Middle America.* Some others of the Stanford group included John O. Snyder, who studied the fishes of Mexico and Hawaii and the freshwater fishes of the Pacific Coast. Another was Edwin C. Starks (1867–1932), who published on fish osteology and on the fishes of Panama and Puget Sound.

Other selected students of fishes who completed their careers in the first part of the 20th century are as follows:

Franz Steindachner (1834–1919), Germany, studied fishes from many countries, including Mexico, the United States, and Japan.

Theodore Nicholas Gill (1837–1914), United States, who Jordan (1905) called "the keenest interpreter of taxonomic facts yet known in the history of ichthyology" (p. 405), defined several orders of fishes.

Samuel Garman (1843–1927), United States, a student of Agassiz, worked in ichthyology at Harvard University on, among others, elasmobranchs and deep sea fishes.

Seth Meek (1859–1914), United States, published on freshwater fishes of Central America and the United States.

Carl H. Eigenmann (1863–1927), United States (from Germany), was appointed professor of zoology at Indiana University by Jordan; he later became Curator of Fishes at the Carnegie Museum. His researches on such subjects as Pacific Coast fishes, embryology, blind cave fishes, and the freshwater fishes of South America led to his being referred to as one of the foremost ichthyologists of the time.

Bashford Dean (1867–1928), United States, studied primitive fishes and published the outstanding *Bibliography of Fishes.*

C. Tate Regan (1878–1943), England, a taxonomist at the British Museum of Natural History, studied several orders in detail and published a classification of fishes in the 14th edition of the *Encyclopaedia Brittania.*

John Richardson Norman (1899–1944), England, studied flatfishes and produced a draft synopsis of orders, families, and genera of fishes that was unpublished but was distributed after his death by the British Museum of Natural History, where he had worked.

John Treadwell Nicholls (1883–1958), United States, studied the fishes of China and initiated publication of *Copeia,* the journal of the American Society of Ichthyologists and Herpetologists.

Those who have been active beyond the middle of the 20th century are too numerous to discuss here. The American Society of Ichthyologists and Herpetologists currently has over 1100 members who are listed with an interest in ichthyology. However, two recent scientists will be mentioned here because of their influence on the development of ichthyology in the United States.

Donn E. Rosen (1929–1986) spent most of his career at the American Museum of Natural History in New York. His work encompassed systematics of the live-bearers (Poeciliidae) as well as other groups, fish reproduction, and studies on the higher relationships of fishes. He was the recipient of many honors. Carl L. Hubbs (1894–1979), a student of Charles Henry Gilbert, was, in most opinions, the greatest American ichthyologist since David Starr Jordan. A tireless, vigorous worker, he collected and classified fishes from many areas, but mainly the United States. In both the field and laboratory, his wife Laura was on hand to take notes and help in other ways. He was at the University of Michigan in his early professional career but later was at the Scripps Institution of Oceanography. His work on fishes of the arid western United States is especially well known. (For details of his accomplishments, see Norris, 1974. For a list of the ichthyologists who have contributed to studies of world fishes through the late 1960s, see the 148-page bibliography in Lindberg, 1974.)

In the last three decades, due to a general acceptance of phylogenetic systematics, there have been great changes in our views of the relationships of fishes. We are still in a period of intense study and change. There are lingering differences of opinion on the classification of fishes, but as new knowledge is generated, a consensus must be closer.

Chapter *Chapter* 2

General Morphology

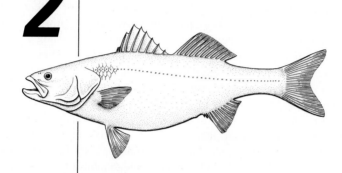

Challenges of the Aquatic Environment

During their long evolution, fishes have adapted to numerous physical and chemical aspects of the aquatic environment. Compared to air, water is an extremely viscous and dense medium, so organisms with a specific gravity close to that of water sink slowly and must exert little effort to maintain a given depth. However, rapid movement through water requires significant energy. Water is surprisingly dense; a standard 10-gallon (37.85 = l) aquarium holds over 83 pounds (37.85 kg) of water, and a moving body of water can thus exert considerable force. Water can hold at most a few parts per million of dissolved oxygen, so aquatic organisms must get along with little oxygen, develop means to supply enough oxygen to support active metabolism, or breathe air. The broad ability of water to dissolve various substances is well known, and these dissolved substances could govern the ability of the fish to maintain the proper osmoconcentration of its body fluids (i.e., NaCl and other metal salts). Other solutes of significance could be pollutants with harmful effects, significant biological odors, or nutrients involved in primary production by plants. The reflective air-water interface limits the light that can pass through. Once light has penetrated, it is absorbed rapidly and differentially by particles in the water. On the other hand, sound is propagated well in water. Pressure increase with depth can limit vertical movement of many organisms because of limited tolerance.

The aquatic environment is not just open water. Its boundaries include sandy slopes, rocky crags and cliffs, flat muddy plains, and coral reefs. The ability of organisms to adapt to life on and near those surfaces has resulted in remarkable biological diversity and highly complex aquatic ecosystems.

Fishes have evolved adaptations for lives both active and inactive, for swift swimming as well as slow, for efficient gathering of food and oxygen, for awareness of their environment, and for many successful modes of reproduction.

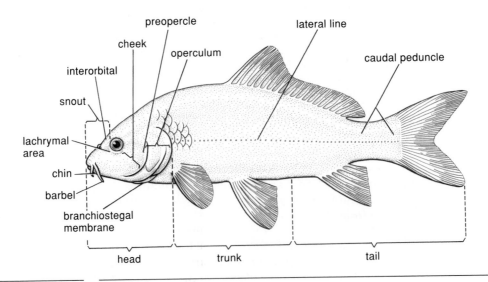

FIGURE 2–1   Diagram of bony fish, showing external features.

This chapter will introduce the morphological features that adapt fishes to life in water, discuss the morphology that ichthyologists use in studying the relationships of fishes, and lay the groundwork for more detailed information, in later chapters, on how different groups of fishes meet the various challenges of life in water.

External Body Form and Fins

The majority of fishes have a more or less streamlined body necessary for locomotion in water. Usually a head, trunk, and tail can be distinguished (Fig. 2–1). In most bony fishes, the head extends from the tip of the snout to the posterior edge of the gill cover. Because the posterior boundary of the head may be obscure in the lampreys, hagfishes, sharks, rays, and bony fishes, such as eels, that have the opercula covered, systematists and morphologists studying these groups use other points of reference for the three body regions. The trunk or abdominal region includes the internal organs. It begins at the posterior margin of the gill cavity and extends, in most fishes, to approximately the origin of the anal fin, but the true distinction between the trunk and tail is internal (see Vertebral Column, page 48).

The Head

Anatomical regions of the head of a bony fish (Fig. 2–1) include the snout, from the eye to the anterior tip of the upper jaw; the operculum, or gill cover; the cheek, between the eye and the angle of the preopercle; the branchiostegal membrane, below the operculum; the chin, or mentum; and the interorbital. The lachrymal region is below the anterior edge of the eye. Many fishes have special or unique features of the head, and knowledge of some of these can aid in recognizing species, in studying systematics, or in assessing the mode of life of the fish.

The mouth is located near the anterior of the fish, but its exact position may vary. It may be inferior, as in many sharks, rays, and the sturgeons; subterminal, as in dace; terminal, as in trout; or oblique or even superior, as in sandfish (Fig. 2–2). The lower jaw or mandibles and the paired premaxillae and maxillae are usually visible externally on bony fishes. Some species have supramaxillary bones attached to the maxillae. Most derived bony fishes have protrusible mouths. In most of the spiny-rayed fishes and their relatives, ascending processes of the premaxillae keep those bones oriented as the processes slide down and forward from their resting place in the nasal region.Other types of jaw protrusion are seen in the carplike fishes and sturgeons (see Lauder, 1983). Many species have their lips or jaws bound to the snout or chin by a continuous bridge of skin, or frenum, so the mouth is nonprotractile.

Barbels (Fig. 2–1), fleshy, elongate structures that carry tactile and chemosensory receptors, may be present around the nostrils and mouth and on chins. These may be minute and simple or conspicuous and sometimes branched, as in some catfishes. They take the designation of the structure that bears them, such as maxillary, mandibular, nasal, rostral, and mental (on chin). Similar to barbels but usually without special sensory functions are cirri, or various flaps of skin on the lips or other parts of the head. Many of these projections seem to blend the fish with its surroundings or otherwise make it less conspicuous.

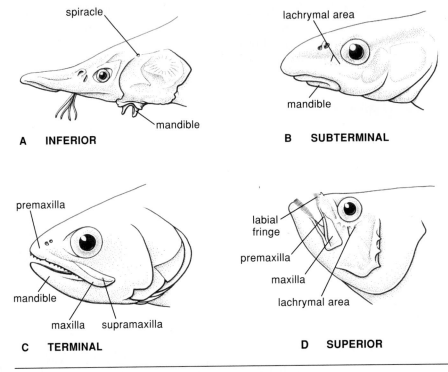

spiracle

mandible

A   INFERIOR

lachrymal area

mandible

B   SUBTERMINAL

premaxilla

mandible

maxilla   supramaxilla

C   TERMINAL

labial fringe

premaxilla

maxilla

lachrymal area

D   SUPERIOR

FIGURE 2–2   Examples of mouth positions in fishes. **A,** Inferior (sturgeon); **B,** subterminal (dace); **C,** terminal (trout); **D,** superior (sandfish). (**D** based on Jordan and Evermann, 1900.)

Lampreys have a series of fleshy fimbria surrounding the mouth, which is a jawless sucking disc (Fig. 2–3**A**). The jawless mouth and single nasal opening of hagfish have four barbels each. Sharks and rays may have oronasal grooves and labial folds in the mouth region (Fig. 2–3**B**). These aid in maintaining a flow of water through the nostrils to smell or taste as the animal moves or pumps water through the gills (Bell, 1993) (see Chapter 22).

Some other prominent features of the head of bony fishes are spines on various bones (Fig. 2–4**A**). These are commonly found on the preopercle or opercle and make some common fishes, such as yellow perch (*Perca flavescens*) and various species of sculpins (*Cottus*), hard to handle. Head spines usually take their names from the bones that bear them, such as opercular or parietal, but are sometimes named from their location (e.g., preocular or nuchal). Sensory canals on the head (part of the lateral line sensory system) can be recognized by rows of pores or open grooves in the skin (Fig. 2–4**B**). Sensory organs in the canals respond to water movement and other mechanical stimuli (see Chapter 21).

Nostrils of living fishes, except for hagfishes, lungfishes, and some specialized bony fishes, have no internal openings to the oral cavity. There may be a blind sac on each side, with its single opening separated into incurrent and excurrent portions. Usually the sac has anterior and posterior nares barely separated from each other. In some fishes, such as eels, the olfactory organ is a greatly expanded, tubelike cavity with the nares widely separated (Fig. 2–5).

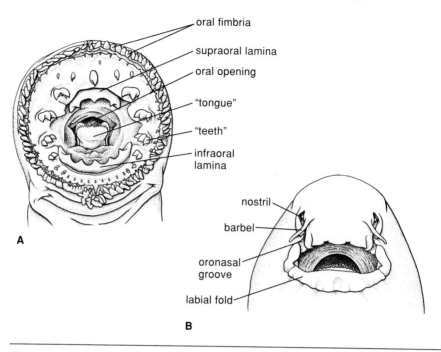

oral fimbria

supraoral lamina

oral opening

"tongue"

"teeth"

infraoral lamina

nostril

barbel

oronasal groove

labial fold

A

B

FIGURE 2–3   **A,** Oral disc of lamprey (*Lampreta minima*); **B,** ventral view of mouth and rostrum of shark (*Chiloscyllium indicum*) showing grooves and labial folds.

In rays, many sharks, and some primitive bony fishes such as the bichirs (*Polypterus*), sturgeons (*Acipenser*), and paddlefishes (*Polyodon*), an opening called the spiracle is found behind the eye (Fig. 2–2). This aperture is the remnant of a full gill slit—originally between the mandibular and hyoid arches (Romer, 1970)—that has been reduced in size as a result of the modification of dorsal hyoid arch elements to serve as a suspensory apparatus for the mandibular arch. In bottom-dwelling rays and some benthic sharks, the respiratory current is brought to the gills through the spiracles.

Body Form

The body form (Fig. 2–6) of a fish can be used in quick appraisal of its way of life. A common body form of fast-swimming, open-water fishes is exemplified by the tunas and mackerels (Scombridae). This streamlined configuration, with an elliptical to round cross section and narrow caudal peduncle just in front of the caudal fin, is called fusiform. This term is often applied to the body shapes of fishes that are considerably more laterally compressed (compressiform) than the tunas, such as Pacific salmons (*Oncorhynchus*).

Many fishes that are not constantly in motion but that may be capable of quick bursts of speed are markedly compressed laterally and are called compressiform. Familiar fishes of this shape are sunfishes (genus *Lepomis* of the family Centrarchidae), snappers (Lutjanidae), porgies (Sparidae), and flounders (Pleuronectidae). Fish that are flattened dorsoventrally are termed depressiform. Depressiform fishes include

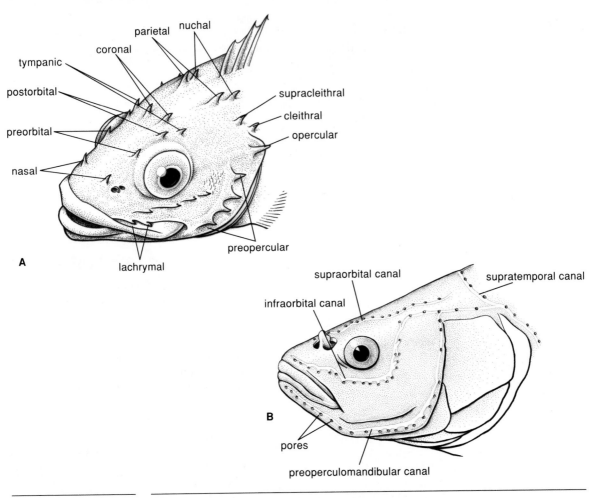

FIGURE 2–4   **A,** Diagram of head of rockfish (Scorpaenidae), showing spines (Adapted from Hart, 1973); **B,** diagram of head of tui chub (*Gila bicolor*), Cyprinidae, showing position of cephalic sensory canals and pores.

skates (Rajidae) and their relatives, angel sharks (Squatinidae), toadfishes (Batrachoididae), and goosefishes (Lophiidae). Obviously this shape suits the fish for life on the bottom, but the greatly flattened mantas (Mobulidae) and eagle rays (Myliobatididae), derived from bottom-living forms, have adapted to swimming above the bottom.

Eel-shaped fishes are called anguilliform, from *Anguilla,* the genus that includes the American eel. Other descriptive terms used in connection with body form are filiform, for thread-shaped fishes such as snipe eels (Nemichthyidae); taeniform, for the ribbon-like shape of such fishes as gunnels (Pholidae), pricklebacks (Stichaeidae), hairtails, and cutlassfishes (Trichiuridae); sagittiform, for the somewhat arrow-like shape of pikes (Esocidae), gars (Lepisosteidae), and others; and globiform, exemplified by the rotund lumpsuckers (Cyclopteridae).

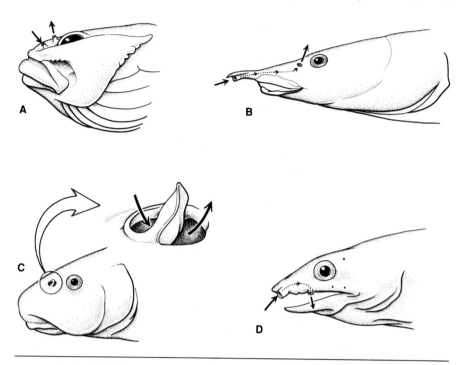

FIGURE 2–5   Representative nostrils; arrows show incurrent and excurrent apertures. **A,** Sculpin (Cottidae); **B,** spinyback (Mastacembelidae); **C,** typical bony fish nostrils divided by flap of skin (Catostomidae); **D,** worm eel (*Myrophis*).

Of course, not all fishes have body forms that can be described by these convenient terms. Boxfishes and cowfishes (Ostraciidae), seahorses (Syngnathidae), and sea moths (Pegasidae) (Fig. 2–7) are some examples of odd-shaped fishes. A familiar freshwater fish, the brown bullhead (*Ameiurus nebulosus*), is an example of a fish with a combination of shapes; it has a depressed head, a body of round cross section, and a laterally compressed caudal peduncle.

A body form often encountered in marine fishes, many from considerable depths, is that exemplified by the chimaeras (Fig. 1–1) and grenadiers (Macrouridae), which have a large head and forebody with a tapering afterbody and tail. This "chimaeriform" body, or one resembling it, can be seen in some poachers (Agonidae), spiny eels (Halosauridae), and a few others. Some of these fishes hold the body still and swim by undulating the pectoral fins, but others swim by undulations of the body.

Topography of the Body

Some regions of the body are described by terms that aid in locating identifying features (Fig. 2–8). The dorsal surface just behind the occiput (the posterior terminus of the skull) is the nuchal region, sometimes characterized by a hump. The most anterioventral part of the body is usually the narrow isthmus that extends far forward below and between the gill openings. Posterior to this is the breast, and posterior to

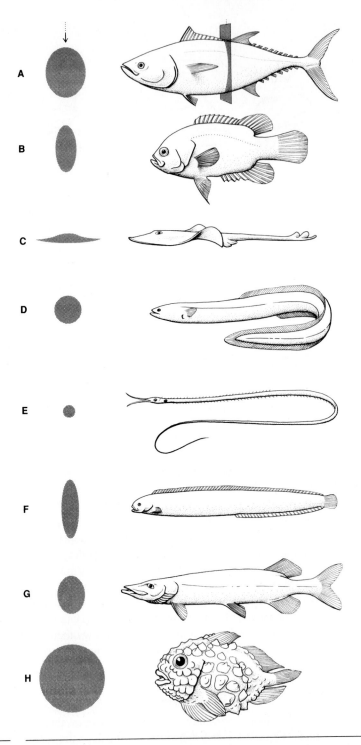

FIGURE 2–6 Representative body shapes in fishes, with typical cross sections. **A,** Fusiform (tuna, Scombridae); **B,** compressiform (sunfish, Centrarchidae); **C,** depressiform (skate, Rajidae) dorsal view; **D,** anguilliform (eel, Anguillidae); **E,** filiform (snipe eel, Nemichthyidae); **F,** taeniform (gunnel, Pholidae); **G,** sagittiform (pike, Esocidae); **H,** globiform (lumpsucker, Cyclopteridae). (**H** based on Jordan and Evermann, 1900.)

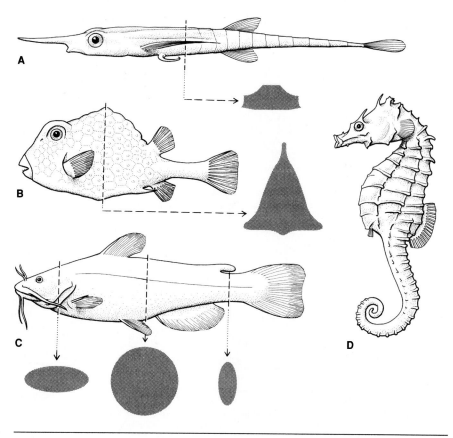

FIGURE 2–7   Examples of body shapes in fishes. **A,** Sea moth (Pegasidae); **B,** cowfish (Ostraciidae); **C,** bull-head (Ictaluridae); **D,** seahorse (Syngnathidae). (**B** and **C** based on Jordan and Evermann, 1900.)

that is the belly. The narrow part of the body of the fish just anterior to the caudal fin is the caudal peduncle.

Conspicuous along each side of many fishes is the trunk canal of the lateral line system, a continuation of the network of sensory canals on the head. The usually single line may be an open groove in the skin, as in some chimaeras, or a row of pores in the skin or scales. Lines may be multiple, as in the greenlings (Hexagrammidae), or reduced in various ways (see Coombs et al., 1988). Herrings (Clupeidae), for instance, lack an extended lateral line; it often appears only on a few anterior scales.

The gill openings of lampreys are in a lateral position and appear as a row of seven nearly circular apertures. Hagfishes, depending on species, may have from 1 to 16 circular gill openings placed well behind the head. In sharks and rays, five to seven individual gill openings occur in a series. Those of sharks are mostly lateral, whereas those of rays are mostly ventral. Gill openings of bony fishes and chimaeras are typically in a lateral position just anterior to the pectoral girdle and covered by the operculum, but they may be placed well behind the operculum, as in eels, or in a ventral position, as in the swamp eels (Synbranchidae).

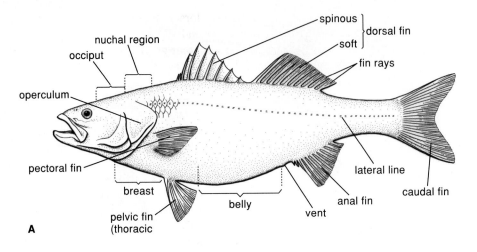

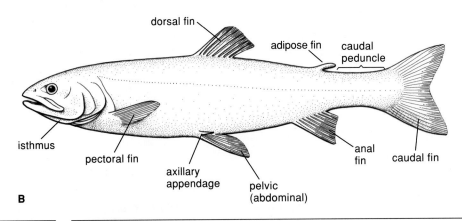

FIGURE 2–8   Body regions and fins. **A,** spiny-rayed fish; **B,** soft-rayed fish.

Fins

Fins, of course, are conspicuous features on the fish body. Fins are stiffened by structures called rays, which may be soft and flexible or spinous in the advanced fishes (see page 55). The median or unpaired fins are the dorsal, anal, and caudal fins. These are in line with the axial skeleton and are supported by median elements associated with the vertebral column. The paired fins are the pectoral and pelvic (ventral) fins (Fig. 2–8). They are supported, respectively, by the pectoral and pelvic girdles.

Median or Unpaired Fins. These are the dorsal fin(s), along the back; the caudal or tail fin; and the anal fin(s), located ventrally just behind the anus in most fishes (Fig. 2–8).

The dorsal fin may extend the length of the back, be divided into two or three separate fins, or be single and small. The dorsal fin is lacking in some families, such as the gymnotid knifefishes of South America. In the derived bony fishes, the anterior part of the dorsal (or the first dorsal, if there are two) is supported by spines, which can be stiff and sharp or secondarily modified into flexible structures, as in the freshwater sculpins (*Cottus*). In some groups—for example, salmons and trouts (Salmonidae), various catfishes, most characoids, and the lanternfishes (Myctophidae)—there is a small, fleshy, usually rayless adipose dorsal fin on the caudal peduncle.

The dorsal fins function in stabilization and in achieving quick changes in direction, but they can be used in conjunction with the caudal and anal fins in braking. Many species that have dorsal fins extending the length of the back can move by undulating the fin, but some species with short dorsals, such as pipefishes and seahorses, also use the dorsal for locomotion (see Chapter 18).

Modified dorsal fins (Fig. 2–9) include the sucking disc atop the head of the remoras (Echeneidae), which allows them to cling to sharks or to other large fishes and be carried along as hitchhikers. The angling apparatus of anglerfishes (Lophiiformes) is a modified dorsal fin spine. Some species use a showy dorsal fin in displays for intraspecific communication; size and color of dorsal fins are sexually dimorphic in many species.

In the bichirs (Polypteridae) of Africa (Fig. 2–10A), the dorsal fin is divided into a unique series of finlets, each consisting of a spine with a few soft rays attached along the length of the spine. Most scombrids have a series of finlets posterior to the dorsal fin (Fig. 2–6A). These consist of detached soft rays, usually branched and set in tough skin. Several species of the cod family (Gadidae) have three dorsal fins that have soft rays only.

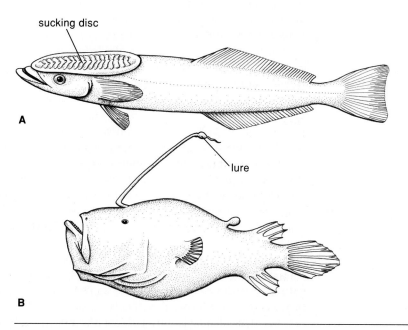

sucking disc

A

lure

B

FIGURE 2–9   Modified dorsal fins. **A**, Sucking disc (remora, Echeneidae); **B**, fishing rod and lure (anglerfish, Ceratiidae). (**B** based on Jordan and Evermann, 1890.)

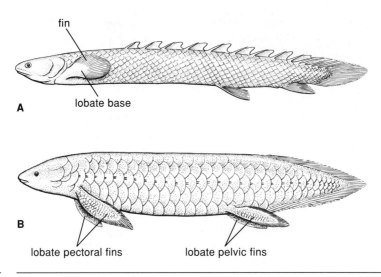

FIGURE 2–10   Example (**A**) of fish with modified dorsal fin, and (**A**), (**B**) fishes with lobate pectoral fins. **A,** Bichir (*Polypterus*); **B,** Australian lungfish (*Neoceratodus*). (**B** based on Goodrich, in Lankester, 1909.)

Most fishes have anal fins located anterior to the anus (Fig. 1–1). Fishes that lack an anal fin include lampreys, chimaeras, skates, rays, some sharks (*Squalus, Somniosus,* etc.), and a few bony fishes, including the king-of-the-salmon (*Trachipterus*) and male pipefishes (*Syngnathus*).

The anal fin is generally short based, but there are many species that have anal fins that exceed the dorsal fin in length. Some have long anal bases, so the anal fin stretches from the anus to the caudal fin even when the anus is located nearly under the chin, as in the knifefishes (Gymnotidae and Rhamphichthyidae) of South America. Flounders (Pleuronectidae) and gouramies (Osphronemidae) are compressiform fish that have long-based anal fins. Only a relatively few fishes, such as cods (Gadidae), have more than one anal fin. Some, such as the jack mackerel (*Trachurus*) (Fig. 2–15**C**), have the anal fin spines separate from the soft rays, forming a small spinous anal fin. Usually, however, if anal fin spines are present, they are located at the anterior of the single anal fin. Finlets posterior to the anal fin are present in the sauries (Scomberesocidae) and tunas, mackerels, and allied fishes (Scombridae). Males of some species (notably Poeciliidae) have the anal fin modified into an intromittent organ called a gonopodium. Some of the opisthoproctids (spookfishes) have the anal fin at the posterior terminus of the body, displacing the caudal fin upward.

Caudal fins appear in a variety of shapes, sizes, and kinds, and their internal structure often reflects phylogenetic relationships more than the other fins. Swimming habitats may be deduced to some extent from the caudal fin (Fig. 2–11). Those fishes that have a crescent (lunate) caudal fin and a narrow caudal peduncle are generally among the speediest of fishes and are capable of rapid, sustained motion. Many pelagic species have forked tails and are constantly on the move. Species with truncate, rounded, or emarginate caudal fins may be strong swimmers but somewhat slower than those mentioned earlier. Fishes with small caudal fins or caudal fins

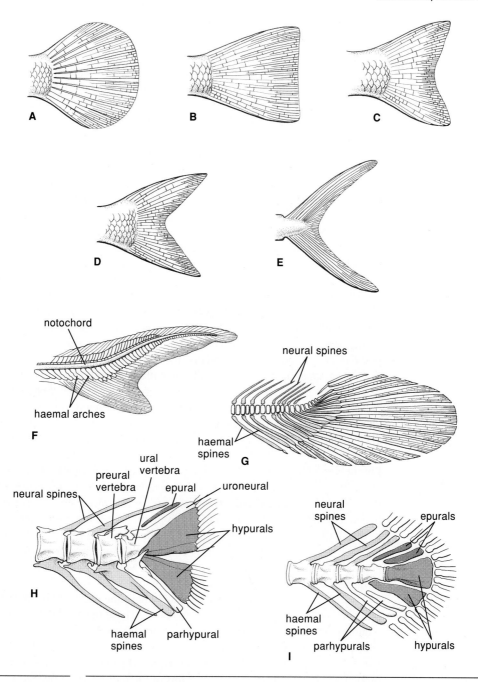

FIGURE 2–11   Representative shapes of caudal fins. **A,** Rounded; **B,** truncate; **C,** emarginate; **D,** forked; **E,** lu-
nate. Types of caudal fins, showing structure. **F,** heterocercal (sturgeon, Acipenseridae); **G,**
abbreviate heterocercal (bowfin, Amiidae); **H,** homocercal (striped bass, Moronidae); **I,**
isocercal (cod, Gadidae). (**A** based on Goodrich, 1930; **B** based on Jordan and Evermann,
1900.)

continuous with the dorsal and anal fins tend to be weak swimmers or may move by wriggling along the bottom.

Most familiar bony fishes have homocercal caudal fins—the type that appear externally to be symmetrical but that are asymmetrical internally, because the tip of the vertebral axis turns upward with most of the fin attached below it. Evidently, this is an evolutionary stage derived from the markedly asymmetrical heterocercal (epicercal) structure of the sharks, sturgeons, and several extinct groups, in which the body axis turns upward posteriorly and almost all of the caudal fin is borne ventrally near the end of the tail. Intermediate stages can be seen in the gars (Lepisosteidae) and bowfin (*Amia*), in which the caudal fins are called "abbreviate" heterocercal because they are only slightly asymmetrical externally. On the upper edge of the caudal fin of gars, sturgeons, and many extinct fishes, there are a series of modified, elongate scales or fulcra. These are called "fringing fulcral scales" by Moy-Thomas and Miles (1970). (They are also present on the leading edges of paired fins in some primitive bony fishes.) The homocercal tail of some of the more primitive teleosts differs from that seen in the advanced forms in that, like the gar and bowfin, more than one vertebra is included in the upturned portion (Fig. 2–11**G**) (Patterson, 1968).

The symmetrical tail of the cods and hakes (Gadidae) is called isocercal. The true caudal fin of cods is small and is borne on a symmetrical plate at the end of a tapering series of vertebrae. Most of what appears to be caudal fin is actually composed of dorsal and anal-fin elements (Fig. 2–11**I**).

Long, tapering, or whiplike tails are called leptocercal. Other symmetrical tails that come to a more abrupt point, as in lungfishes, are called diphycercal. In a few fishes, the caudal portion of the body is absorbed during development so the dorsal and anal fins bridge over the posterior terminus of the body in what is called a gephyrocercal tail. This is seen in the molas or ocean sunfishes (Molidae).

Paired Fins.[1] The pectoral fins of bony fishes, composed of soft rays only, are borne by the pectoral (shoulder) girdle, which in most fishes forms the posterior border of the gill cavity. The right and left halves of the girdle meet ventrally at the body midline and diverge dorsally, each half fastening more or less firmly to the posterior part of the skull. Pectoral fins are usually prominent; only a relatively few groups lack them or have them reduced in size. Groups with less prominent pectoral fins are generally elongate, eel shaped, or taeniform. Many of such fishes are adapted for wriggling along the bottom or through vegetation.

The more primitive bony fishes usually have the pectoral fins placed low in the body below the centers of buoyancy and mass. These fins tend to have oblique bases (slanting downward and posteriorly), in part due to the presence of the mesocoracoid bone in the pectoral girdle. Although suitable for trimming the balance of the fish either at rest or in motion, these low-placed pectoral fins have less use in locomotion and braking than the pectoral fins of the more derived fishes, which lack the mesocoracoid bone and in which the pectoral fins are generally placed higher on the body and have vertical bases. These fins, which are closer to the centers of mass and buoyancy, are versatile and more efficient in use for locomotion, maneuvering, braking to sudden stops, aggressive displays, and other purposes (see Chapter 18). In

---

[1]See Nursall, 1962, for origin.

some deep-bodied species, a shift in position of the gas bladder has accompanied the evolutionary change in the position of the pectoral fins. In some, the gas bladder is above the center of gravity, so maintenance of balance and locomotion by means of the pectoral fins are facilitated.

The low position of the pectoral fin in primitive fishes (and, to a lesser extent, some advanced fishes) allows the fins to touch bottom and provide support in species that live close to the substrate. These fins often bear numerous taste buds and touch receptors. The bichirs (*Polypterus*) bear the pectoral fins on a lobate base, with a skeleton peculiar to the group, and use them somewhat as arms. The skeleton of the fin lobe consists of two rodlike bones with a cartilaginous plate between them, connected proximally to the scapula and coracoid (see Fig. 2–25, pages 50–51). Distally, the rods and plate attach to ossified radials that bear the fin rays (Berg, 1947). This armlike pectoral fin is called a brachiopterygium (Fig. 2–10**A**). The crossopterygian (*Latimeria*) and the Australian lungfish (*Neoceratodus*) (Fig. 2–10**B**) have lobate pectoral fins of a different structure. Highly modified pectoral fins supported by a central joined axis with no radials (fin-ray supports) or fin rays are seen in the African and South American lungfishes, *Protopterus* and *Lepidosiren*.

Some species have pectoral fins that aid them in remaining at the bottom or even on a vertical surface. Sisorid catfishes apparently use the pectoral fins as well as a pad between the fins in clinging to surfaces (Fig. 2–12**A**). The algae eaters (Gyrinocheilidae) have pectoral fins (and mouths) modified for use in maintaining suction.

The fishes of an African genus, *Pantodon,* are called butterflyfishes because of their large expanded pectoral fins. These 12-cm fishes can make 2-m leaps over the water surface. Unique among fishes is the pectoral apparatus of the South American family Gastropelecidae of the order Characiformes. These are known as "freshwater flying fishes." Bones of the pectoral girdle are expanded ventrally to give a somewhat hatchet-like appearance and to provide a wide area for attachment of the massive pectoral musculature. Because the powerful pectoral fins can be used to propel the fishes out of the water, gasteropelicids are skilled at jumping.

Catfishes have developed spinelike structures consisting of consolidated soft rays at the leading edge of the pectoral fins. Anyone who has made a bad grab while sorting a seineful of mixed warm-water fishes and has embedded the spine of a bullhead or madtom in a thumb knows that this is a potent defensive weapon. A few catfishes make use of these spines in terrestrial locomotion; the best known of these are species of *Clarias,* the so-called walking catfishes. *Clarias batrachus* has been accidentally released in Florida, where conditions for life and reproduction are suitable enough that it is spreading (in part by overland locomotion) and is reported as becoming a pest.

Many derived fishes are structurally specialized for certain habitats and ways of life, and the pectoral fins are often involved in the specialization (Fig. 2–12). Threadfins (Polynemidae) have pectoral fins that are each divided into two parts; the lower one consists of several filaments that reach great lengths in some species. These filaments are thought to function as tactile organs and, when extended and fanned out, can apprise the foraging fish of edibles over a wide area. Others with detached pectoral rays are the flying gurnards (Dactylopteridae), the searobins (Triglidae), and some of the stonefishes (Synanceiidae). These families have finger-like rays that are probably tactile in function as well as useful in crawling along the bottom (see Chapter 21). Other kinds of fishes that walk over the bottom using the pectoral fins include the batfish, *Ogcocephalus*. Development of the pectoral fins into "wings" for gliding is found in the flyingfishes (Exocoetidae). Flyingfishes can remain airborne for as long as 20 seconds and can glide a distance of 150 m or more.

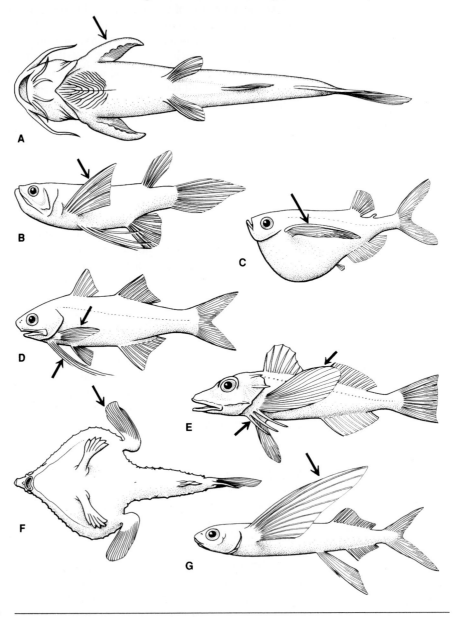

FIGURE 2–12 Examples of fish with modified pectoral fins. **A,** Ventral view of sisorid catfish (*Glyptothorax*); **B,** freshwater butterflyfish (*Pantodon*); **C,** hatchetfish (*Gastropelecus*); **D,** threadfin (Polynemidae); **E,** gurnard (Triglidae); **F,** ventral view of batfish (Ogcocephalidae) with armlike pectorals well behind pelvics; **G,** flying fish (Exocoetidae). (**B** based on Herald, 1961; **D, E,** and **G** based on Jordan and Evermann, 1900.)

Sexual dimorphism of the pectoral fins (and other fins) is fairly common among fishes. Particularly in fishes that have specialized reproductive behaviors, coloration and size as well as other features may differ between the sexes. Fishes with sexually dimorphic paired fins include the dragonets (Callionymidae), killifishes and pupfishes (Cyprinodontidae), gobies (Gobiidae), and suckers (Catostomidae).

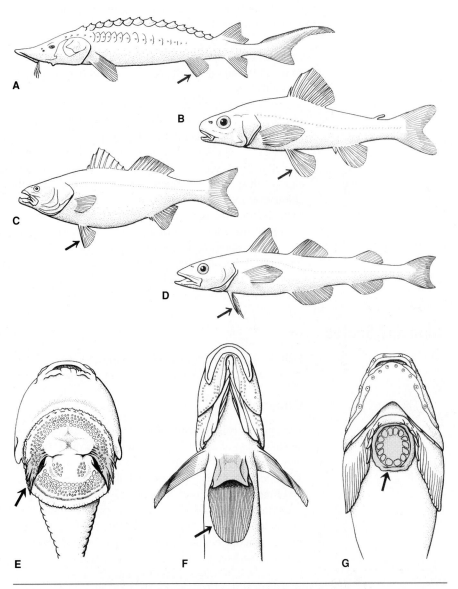

FIGURE 2–13   Examples of pelvic fin placement, pelvic fins circled. **A,** Abdominal (sturgeon, Acipenseridae); **B,** subabdominal (sand roller, Percopsidae); **C,** thoracic (bass, Moronidae); **D,** jugular (pollock, Gadidae). (Based on Jordan and Evermann, 1900.) Pelvic fins modified as sucking devices. **E,** Clingfish (Gobiesocidae); **F,** goby (Gobiidae); **G,** snailfish (Liparidae).

Pelvic appendages of fishes are often smaller than the pectorals, more restricted in function, and subject to greater variation in placement (Figs. 2–11, 2–12, 2–13). Primitive bony fishes are characterized by abdominal pelvic girdles and fins, as are the sharks. The pelvic girdle is embedded in the flesh of the belly and has no internal connection with other skeletal elements. A few groups have the pelvic girdle moved forward toward the pectoral girdle but lack contact between the bones. Derived bony fishes usually have thoracic pelvic fins, placed below or a little behind the pectoral fins, with a more or less firm connection between the pelvic and pectoral girdles. Pelvic fins of derived bony fishes usually have a spine and a few soft rays. The more primitive bony fishes tend to have many-rayed pelvics. In the cods and relatives (Gadidae), some of the blennies (Blennioidei), toadfishes (Batrachoididae), and others, the pelvic fins are placed in a jugular position anterior to the pectorals.

Pelvic fins are reduced or lost in many forms, especially in elongate fishes that wriggle along the bottom. Pelvic fins usually function in stabilizing and braking; they tend to be of less use than the pectoral fins in locomotion, except in certain flyingfishes, in which the pelvic fins are used in gliding. Batfishes employ pelvic fins as well as pectorals in walking on the bottom. Several groups of fishes show modification and specialized function of the pelvic fins. In the males of sharks, rays, and chimaeras, the pelvic fins are modified for use as intromittent organs in copulation. Many benthic species such as sculpins, which live on hard surfaces, use pelvic fins to help hold them in place. In gobies (Gobiidae), clingfishes (Gobiesocidae), lumpfishes (Cyclopteridae), and algae eaters (Gyrinocheilidae), pelvic fins have evolved into or have become incorporated into ventral sucking structures that aid the fish in holding to the substrate (Fig. 2–13). Use of pelvic fins as tactile organs may be exemplified by the gouramies, such as *Osphronemus*.

Skin and Scales

Fish skin (see Van Oosten, 1957; Whitear 1986a, 1986b) is made up of the usual two layers, an outer epidermis and an inner dermis (Fig. 2–14).

Epidermis

The epidermis is typically thin, composed of from 10 to 30 layers of cells (an average thickness of about 250 µm) in most familiar fishes. Seahorses and their relatives may have only two or three layers of epidermal cells on the surface of their armor. The thickness of the epidermis on these fishes (about 20 µm) contrasts greatly with that found on the lips of sturgeons, which may be up to 3 mm thick. Exteriorly, the epidermis consists of squamous cells produced in a columnar germinative layer next to the dermis; these move outward, where they are sloughed off eventually. Except for the mucous covering, live cells of the epidermis are essentially in contact with the medium because no cornified layer is present. However, in many species, a nonliving secretion of the epidermis, called cuticle, covers the cells. A layer of cuticular secretion is usually associated with structures that have contact with the bottom, such as the sucking disc of the clingfishes or the detached pectoral rays of gurnards, but it occurs on others. Some bony fishes—notably minnows (Cyprinidae) and their relatives—and certain salmoniforms secrete horny nuptial tubercles or pearl organs

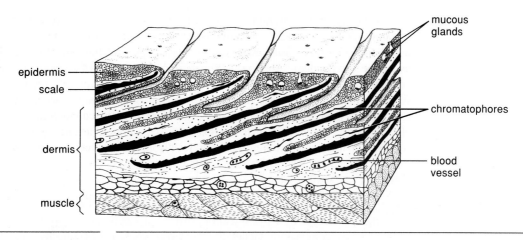

epidermis

scale

dermis

muscle

mucous glands

chromatophores

blood vessel

FIGURE 2–14   Section of fish skin. (Based on Wunder, 1936.)

that cover part of the skin (Fig. 26–5). These tubercles roughen the skin and provide friction during contact by breeding fishes.

Some epidermal cells are unicellular mucous glands that discharge the mucus that forms the slimy outer coating of fishes. There appear to be two types of mucous cells, one that discharges abruptly and refills and another that produces slime gradually over a longer period. Mucus consists largely of glycoproteins that can absorb great amounts of water. The champions among mucus producers are the hagfishes, which have mucous glands with ducts that discharge slime and mucoprotein threads of considerable length (Whitear, 1986a). There are many stories of hagfishes turning buckets of water into a thick jelly and one of a hagfish that turned the water of a display tank so viscous than an octopus could not pass it through its respiratory apparatus.

The slime of fishes is largely protective in function. In addition to protecting the epidermis and making fishes difficult to grasp, slime can tie up particulate irritants and some heavy metal salts and slough them off. Bacteria may be kept from the live epidermal cells by the mucus. Although mucus performs a lubricating function by helping fishes slip through the water, it appears to give only a slight advantage to most fishes and may function mainly in fast starts (Hoyt, 1975; Rosen and Cornford, 1971). Mucus also can precipitate certain suspended solids in muddy water. Special functions of mucus from the skin include use as nest-building material in gouramies (Osphronemidae) and as cocoon material in the African lungfishes (Protopteridae), which require an air-tight covering on their burrow during estivation. When parrotfishes (Scaridae) rest at night, they form a mucous envelope around themselves. The mucus is secreted from the oral cavity. The special, thick mucus of the clownfishes or anemonefishes (*Amphiprion, Premnas*) protects them from the stings of the anemones around which they live (Lubbock, 1980). The mucus of snakeheads (Channidae = Ophiocephalidae) is used in western India to produce an extra-strong building mortar (Antony, 1952).

Although the thin epidermis of a few cell layers thickness is relatively simple in structure, thicker epidermis may contain nerve endings and pigment cells. Blood vessels usually are absent; nutritive materials diffuse through the intercellular matrix that holds the cells together. In some fishes, epidermal cells specialized for venom production are found associated with fin or head spines. Chimaeras (Holocephali), stingrays (Dasyatidae), stonefishes (Synanceiidae), weevers (Trachinidae), and certain catfishes are examples of fishes that possess various kinds of stinging spines. The venoms are painful, and some can be fatal to human beings (see Halstead, 1978).

The thick epidermis of the sucking discs of clingfishes (Gobiesocidae) contains alveolar cells, which form cushions that help shape the surface of the organ to the substrate. In addition, photophores are derived from epidermis (Whitear, 1986a).

Dermis

The dermis is much thicker and much more complex than the epidermis. It usually is made up of two layers: a stratum spongiosum, just beneath the epidermis, and a deeper stratum compactum. Generally, the dermis consists of connective tissue with a paucity of cells. A subcutis of connective tissue lies next to the musculature. The dermis contains pigment cells, blood vessels, nerves, and the dermal skeleton, which may consist of plates or scales of various types. Fish dermis is characterized by two sets of collagen fibers that are disposed around the body in opposing spirals in such a manner that the fish can bend without causing wrinkles in the skin (Whitear, 1986b).

In some bony fishes, especially those lacking scales, the skin may be thick and tough. Such fishes may be skinned and the skin made into leather. Shark skin, with its small, rough placoid scales, has a variety of specialized uses. Hagfish skin is prized for making a thin, tough leather, usually sold as "eelskin," that is used in the manufacture of wallets and other small items.

Scales

Most fishes have a covering of scales, which are dermal in origin (Orvig, 1968a). These may be lacking, as in catfishes; embedded in the skin, as in eels; or modified into bony plates or scutes (Fig. 2–15), as in sturgeons (Acipenseridae), sticklebacks (Gasterosteidae), armored catfishes (Callichthyidae, Loricariidae, etc.), and poachers (Agonidae). Placoid scales, or dermal denticles, are typical of the sharks and their relatives. This type of scale consists of a basal plate, containing some bone cells, that is buried in the skin with a raised portion exposed (Fig. 2–15). The overall structure is similar to that of a tooth, with which these scales are homologous, and has a pulp cavity and tubules leading into the dentine (Nelson, 1970). These denticles, with their hard outer layer of vitrodentine (which is similar to tooth enamel), make possible the use of dried shark skin as an abrasive similar to fine sandpaper.

Many extinct lobe-finned fishes have cosmoid scales with a layer of noncellular cosmine that lies beneath a thin outer layer of vitrodentine slightly different from that of the placoid scale. Below the cosmine is a layer of vascularized bone, called isopedine, and an inner layer of laminate bone. Although the scales of the living coelacanth (*Latimeria*) are said to be simplified cosmoid scales (Lagler et al., 1977), Meinke (1982) concludes that coelacanths did not develop cosmine. The scales of extant lungfishes are of the elasmoid type (Whitear, 1986b). Ganoid scales are encountered on reedfishes (*Erpetoichthys*), bichirs (*Polypterus*), gars (*Lepisosteus*), and in modified form on the caudal fin of sturgeons (Acipenseridae) and paddle-

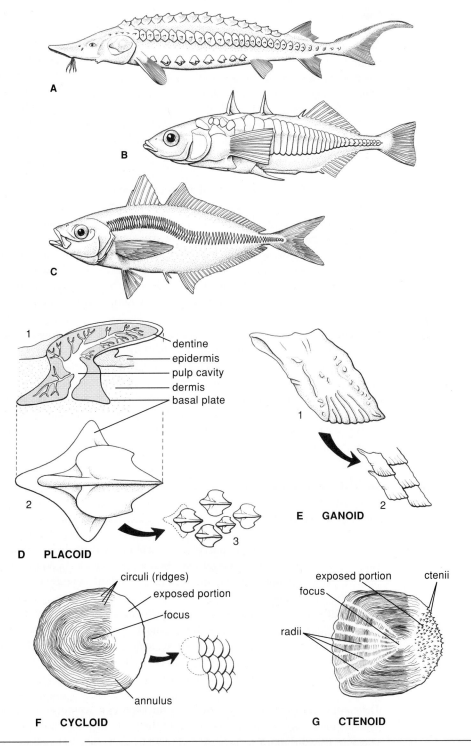

Labels in the figure:

- dentine
- epidermis
- pulp cavity
- dermis
- basal plate

D   PLACOID

1

2

3

E   GANOID

1

2

circuli (ridges)

exposed portion

focus

annulus

F   CYCLOID

exposed portion

focus

ctenii

radii

G   CTENOID

A

B

C

FIGURE 2–15 Examples of fishes with scutes. **A,** Sturgeon (Acipenseridae); **B,** stickleback (Gasterosteidae); **C,** jack (Carangidae); examples of types of scales (anterior to left). **D,** placoid—1, sagittal section, 2, top view, 3, disposition on skin; **E,** ganoid—1, single scale, 2, disposition on fish; **F,** cycloid; **G,** ctenoid. (**A** and **B** based on Jordan and Evermann, 1900.)

fishes (Polyodontidae). In these, the ganoid scale has a typical rhomboid shape, with an anterior, peglike extension of each overlapped by the scale in front. The outer layer of this scale is acellular ganoin, with a cosmine-like dentine layer beneath it. Lamellar bone forms the basal plate (Whitear, 1986b). In extinct †Palaeonisciformes and the living bichirs and reedfishes, the cosmine layer is perforated by tubules similar to those present in cosmoid scales and is underlain by a vascular area of transverse canals. This type of scale is sometimes termed "palaeoniscoid." The cosmine and tubules are reduced in gars, sturgeons, and paddlefishes.

Scales of bony fishes are relatively simple, consisting of a mineralized surface layer and a thin deeper layer of collagenous tissue (Whitear, 1986b). These are thin as compared with ganoid or cosmoid scales and lie in pockets of the dermis, usually overlapping the scale behind in an imbricated manner. Scales of soft-rayed actinopterygians are generally ovoid to nearly circular (subcircular) and lack spines or projections on the surface or posterior margin; this type of smooth-rimmed scale is termed cycloid. Derived bony fishes tend to have ctenoid scales, with minute spines on the exposed portions of the scales or in a comblike row on the posterior margin. Other terms applied to the scales of actinopterygians are *bony, bony ridge,* and *elasmoid.*

The bony layer of the scale is usually characterized by concentric ridges that represent growth increments during the life of the fish (Figs. 2–15**F** and **G**). Spacing and other characteristics of the ridges (circuli) give biologists clues to the life history of the individual fish. Year marks (annuli), spawning marks, and signs of other developmental events may be interpreted by a skilled scale reader. The innermost plate of the scale is called the focus. Lines called radii often lead outward from the focus toward the edge of the scale. In most bony fishes, there are lateral lines of sensory pores through scales.

Internal Support: The Skeletal System

The Axial Skeleton

The Skull. The skull (syncranium) of all vertebrates may be divided into two discrete functional units: (1) the neurocranium (braincase), which supports, surrounds, and protects the brain and sense organs; and (2) the splanchnocranium or branchiocranium (the visceral, branchial, or gill arches), which supports the gills and from which the mandibular (jaws) and hyoid arches were derived. (Note that many vertebrate morphology texts and most laboratory manuals refer to the elasmobranch neurocranium as the chondrocranium. It is more accurately termed the chondroneurocranium or endocranium [Compagno, 1988].) These units differ not only in function but in embryological origin as well.

Development of the Skull or Syncranium. There is a general similarity among vertebrates in the embryonic development of the skull, although different groups show characteristic differences. Because there will be some discussion of the development of fish skulls later in this chapter, a generalized and abbreviated account of the process will be outlined here (Fig. 2–16). (Extensive coverage of the subject is provided in Hanken and Hall, 1993.)

The neurocranium, derived mainly from the neural crest (Gans, 1993), forms at the anterior end of the notochord, beginning with the formation of two parachordal cartilages. These cartilages, one on each side of the notochord and posterior to the forebrain, enlarge and form a structure, the basal plate, by fusing around the notochord. The basal plate enlarges and fuses with paired occipital arch cartilages that develop over the hindbrain, thus forming the rear wall of the neurocranium. Also uniting with the basal plate and occipital arch cartilages are paired otic capsules that form around the inner ears. The synotic tectum, a cartilage that forms over the posterior part of the brain, joins with the occipital arch cartilages. Anteriorly, two prechordal cartilages called *trabeculae* form, anteriorly from the neural crest, the posterior part from mesoderm (Balinsky, 1970). As the trabeculae grow, they fuse at their anterior ends to form the ethmoid plate and unite with the developing basal plate posteriorly.

The splanchnocranium, like the trabeculae described earlier, is derived from neural crest mesenchyme. This mesenchyme flows ventrally between the future gill slits to form the precursors of what will become a series of skeletal bars that run between the gill slits and support and (in conjunction with the branchiomeric musculature) operate the gills. The ancestral number of these visceral arches is not known, but many systematists believe that there were eight and that most or all of them functioned primitively in the support of gills.

In some fishes, a pair of polar cartilages forms between the trabeculae and joins them and the parachordals, so three paired elements together contribute to the formation of the cranial floor. Nasal capsules form anteriorly from their respective

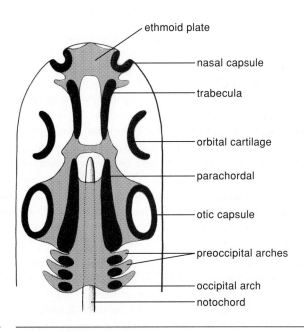

FIGURE 2–16   Generalized diagram of early stages of development of chondrocranium. *In black,* separate cartilages; *in gray,* the formation of ethmoid and basal plates.

cartilages, and orbital cartilages extend forward from the otic capsules. Antorbital processes develop from prechordal mesoderm. Enlargement and dorsal growth of these structures result in the cartilaginous cranium, which will be covered partially or replaced by bone in the bony fishes.

The Splanchnocranium and Jaws. The series of branchial arches that form around the pharyngeal region form the splanchnocranium. A primitive, undifferentiated branchial arch consists of four lateral paired elements (from dorsal to ventral, the pharyngobranchial, epibranchial, ceratobranchial, and hypobranchial elements) and a single, median, ventral element (the basibranchial) to which the paired hypobranchials articulate. In most fishes, most of the branchiocranium is associated with the support of gills, but in gnathostomes the mandibular and hyoid arches have been modified and have acquired other functions.

The first or anteriormost arch in living fishes is called the mandibular arch because of its contribution to the formation of the primary (i.e., primitive) upper and lower jaws (see Schultze, 1993). As in the other arches, there are major upper and lower elements on each side. In the course of evolution, the upper elements (palatoquadrate cartilages) of the mandibular arch have slanted forward along the underside of the neurocranium and have joined anteriomedially to complete the primitive upper jaw. Essentially, this remains as the functional tooth-bearing upper jaw of sharks and rays but does not form the border of the mouth in bony fishes, in which jaws of dermal bone surround and supplement the primary upper jaw.

The lower elements of the mandibular arch are called the mandibular or Meckel's cartilages. These represent the lower elements of the primary lower jaw, articulating with the palatoquadrate cartilage posteriorly and joining one another anteriomedially. In sharks and rays, the mandibular cartilage bears teeth and forms a complete mandible. In bony fishes, it becomes ossified and forms the jaw joint, but it is hidden within a secondary, lower jaw, which is formed by dermal bone. The posterior ossification of the palatoquadrate cartilage is the quadrate bone; the ossification of Meckel's cartilage is the articular bone.

The second arch in the series is the hyoid arch. The upper element on each side, the hyomandibular cartilage, has evolved in fishes into an important suspensory structure that contributes to the suspension of the jaws and hyoid bar. The spiracle, which is the remnant of the upper part of a full primitive gill slit, is located between the mandibular and hyoid arches. The lower element of the hyoid arch is the ceratohyal cartilage, which becomes the supporting structure of the median basihyoid cartilage. Remaining visceral arches typically support the gills and are called branchial arches. The ceratobranchial element of the fifth arch of bony fishes is usually modified to bear pharyngeal teeth.

Dermatocranium. The skulls of ostracoderms and placoderms and the ancestors of living fishes were surrounded by a dense envelope of dermal bone, whereas the neurocranium and splanchnocranium remained cartilaginous. The skulls of living elasmobranch fishes provide an excellent introduction to the structural and functional relations between the braincase and branchial arches, because in these fishes the confusing envelope of dermal bone (the dermatocranium) has been lost. As bony fishes evolved, the dermal bone became somewhat reduced and more closely associated with the neurocranium and splanchnocranium. Gradually the latter two functional units also acquired endochondral ossifications. These dermal and endochondral bones will be discussed later in the section on the bony fish skull (see page 42).

Skull of Living Agnathans. Hagfishes, considered to be the sister group[2] of the rest of the craniates, have a cranium that appears to have been arrested in an early stage of development (Fig. 2–17**A**). The skull "is made up of sinuous cartilaginous arches and plates to which muscles attach, and which strengthen the walls of ducts" (Janvier, 1993, p. 135). Trabeculae are present, and there is a cartilaginous floor consisting of the fused parachordals. The cartilaginous otic capsules are fused to the parachordals, but the remainder of the sides and top of the structure are membranous. Visceral arches are not well developed in hagfishes. A rudimentary framework of cartilage external to the pharynx and gill pouches serves as a branchial skeleton, with a series of cartilages supporting the rasping organ (lingual apparatus) and its muscles. The velum, which pumps water through the pharynx, has a complicated skeleton of cartilaginous rods.

The cranium of lampreys is more complete than that of hagfishes (Fig. 2–17**B**). A partial roof for the brain is formed anteriorly by extensions of the trabeculae. Posteriorly there are sidewalls, but only connective tissue covers most of the brain. This rudimentary neurocranium, homologous with that of other vertebrates, constitutes a minor part of the total skull of lampreys. There are also dorsal cartilages supporting the region anterior to the nasal opening, a series of cartilages around the circular mouth, and a long lingual cartilage ventrally located. Support of the gill region is provided by an intricate branchial basket of cartilage that develops just beneath the skin. Posterior extensions of the branchial basket also cover and protect the heart. This basket is not homologous with the branchial skeleton of gnathostomes.

Skull of Elasmobranchs. In sharks, the chondroneurocranium usually forms a complete box, pierced in places by foramina and fenestrae for the passage of nerves and blood vessels (Fig. 2–18). The olfactory and otic capsules are integral parts of this box. Anterior to the olfactory capsule is the rostrum, which may be extremely elongated in some sharks (e.g., the saw shark, *Pristiophorus;* Fig. 5–7**D**).

Rays may have large dorsal fontanelles both in the rostral region and more posteriorly, so the cranial roof appears to be incomplete. The rostrum in rays is often long. The extremes of length are seen in the eagle ray (*Myliobatis*), in which the rostrum is undeveloped, and in the sawfish (*Pristis*), in which the rostrum may comprise nearly one third of the entire length of the fish.

In most elasmobranchs, there are prominent processes anterior and posterior to the orbit (Fig. 2–18). Some species show various crests and other sculpturing on the roof of the cranium. Posteriorly, occipital condyles form a surface for articulation with the vertebral column, and the foramen magnum allows passage of the spinal cord.

---

[2]*Sister group* is defined by Wiley (1981, p. 7) as "a species or a higher taxon that is hypothesized to be the closest genealogical relative of a given taxon exclusive of the ancestral species of both taxa. Thus, when we say that two taxa are sister groups, we mean that they are hypothesized to share an ancestral species not shared by any other taxon."

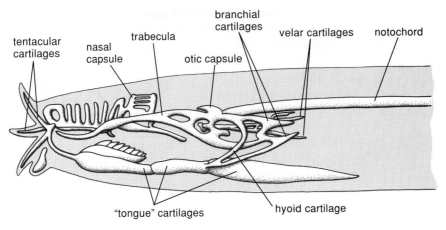

A   HAGFISH

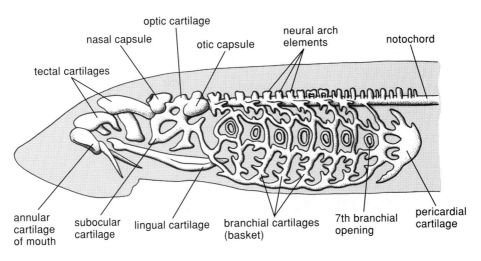

B   LAMPREY

FIGURE 2–17   Diagrams of skulls of living agnaths. **A,** Hagfish; **B,** lamprey. (Based on Goodrich, in Lankester, 1909.)

Although modern elasmobranchs have virtually no bone in their skeletons, the cartilage may be variously calcified to some extent, even to the point of being as hard as bone. Prismatic calcifications of hydroapatite may strengthen major skeletal structures of adult elasmobranchs (Compagno, 1988), so a mosaic appearance is given to the surface of a dried chondrocranium. Vertebral centra are often heavily calcified.

The jaws of elasmobranchs consist of the modified first visceral arch, which is greatly enlarged in some species. The palatoquadrate cartilage extends forward along the underside of the neurocranium, remaining free and movable in most species. It is usually attached to the cranium by means of the upper part of the hyoid arch, the hyomandibular cartilage, which articulates with the otic region. The lower

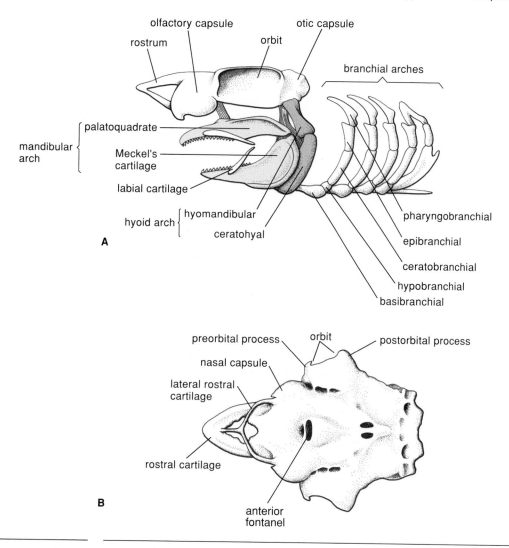

FIGURE 2–18   **A,** Diagram of elasmobranch skull (left side only of visceral skeleton). (Based on Bridge and Boulenger, 1910.) **B,** Dorsal view of shark skull.

jaw, or Meckel's cartilage, articulates with the posterior part of the palatoquadrate. This type of jaw suspension is called *hyostylic* (Fig. 2–19**A**). In a few sharks, the palatoquadrate is attached to the neurocranium as well as to the hyomandibular in what is termed *amphistylic* suspension (Fig. 2–19**B**). Maisey (1980) points out that amphistyly is not qualitatively different from hyostyly and perhaps should be considered a type of hyostyly, because the hyomandibular aids in suspension of both types.

There is a graded transition among species from small hyomandibular cartilages, with little suspensory function, to large hyomandibulas, with major suspensory function. Rays are only hyostylic, but most sharks at least have ligamentous

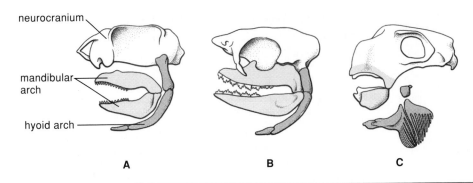

neurocranium

mandibular arch

hyoid arch

A   B   C

FIGURE 2–19 Schematic diagrams showing relationship of mandibular arch to neurocranium in three types of jaw suspensions. (For clarity, the mandibular arch has been lowered in **A**; the hyoid arch has been flexed downward in **A** and **B** and has been lowered in **C**.) **A**, Hyostylic (upper jaw not firmly attached to neurocranium, with ligamentous attachment to hyomandibular); **B**, amphistylic (upper jaw attaches anteriorly to basal angle of neurocranium and posteriorly to postorbital process); **C**, holostylic (upper jaw fused to neurocranium).

connections between the palatoquadrate cartilage and the neurocranium. Connections, either ligamentous or firm, range in location in various species from rostral to postorbital positions. Some kinds of connections restrict movement of the palatoquadrate, whereas others allow ventral and forward movement.

The sharks that possess an orbital process on the palatoquadrate and show an attachment to the orbit include *Notorynchus, Clamydoselachus, Hexanchus, Heptranchias,* the squaloids, pristiophoroids (saw sharks), and *Squatina.* Maisey (1980) calls this kind of jaw suspension "*orbitostylic*" and suggests that sharks that have such a suspension are more closely related to each other than to other sharks. All of these have ethmoidal and/or postorbital connections to the neurocranium as well as the hyomandibular and orbital connections.

The lower part of the hyoid arch is called the anterohyal (ceratohyal) cartilage, and both it and the hyomandibular may bear gill rays, slender cartilaginous rods that strengthen the interbranchial septa from which gill tissue projects as lamellae. A basihyal cartilage connects the anterohyal cartilages ventrally.

There are five branchial arches in most sharks and rays, but a few species have six or seven. These arches typically consist of pharyngobranchials, epibranchials, ceratobranchials, and hypobranchials. Connective tissue might or might not hold pharyngobranchials to the roof of the pharynx, depending on the species. Basibranchial cartilages connect the hypobranchials on the ventral midline. Gill rays on the epibranchials and ceratobranchials support the gills on all but the last arch.

The chimaeras (holocephalans) have the palatoquadrate cartilage entirely fused to a cartilaginous neurocranium, a type of autostylic jaw suspension generally called *holostylic* (Fig. 2–19C). The hyomandibular cartilage serves no sensory purpose for either the upper or lower jaw. The hyoid arch is slightly more modified than the branchial arches and bears gill rays.

Skull of Bony Fishes. Figures 2–20, 2–21, and 2–22 show the structures mentioned in this section. Frequent referral to these figures will be helpful. Good coverage of the ontogenetic development of the bony fish skull can be found in Weisel (1967) and Morris and Gaudin (1975). See also Schultze (1993), DeBeer (1937), and Gregory (1933). Excellent treatment of many phases of skull development diversity and structure can be found in the three-volume work edited by Hankin and Hall (1933).

Some primitive bony fishes, such as paddlefishes and sturgeons, retain much of the cartilaginous neurocranium with few ossifications. Others, such as the bowfin and salmonids, also retain much cartilage in the cranium, but in the majority of bony fishes the cartilage is mostly replaced by bone. Ossifications forming around and replacing cartilage are respectively called perichondral and endochondral cartilage bones, whereas membrane or dermal bones are formed in the dermis and are not preceded by a cartilage model. In teleosts, endochondral bones are especially prominent in the posterior region of the neurocranium. The ventral unpaired basioccipital bone usually forms the occipital condyle, which articulates with the vertebral column. In perciform fishes and many other spiny-rayed groups, the lateral paired exoccipital bones contribute to the occipital condyle and form, along with the basioccipital, a tripartite structure that articulates with the first vertebra. In most teleosts, the exoccipitals surround the spinal cord completely as it enters the skull through the foramen magnum. The dorsal, median, supraoccipital bone, in addition to forming part of the cranial roof, furnishes an anterior attachment surface for the epaxial trunk muscles. This bone may be extended into a crest to which the muscles attach. The crest reaches a large size in deep-bodied fishes with a large mass of muscle extending onto the skull, as in many advanced teleosts.

Much of the posterior part of the teleost skull consists of five endochondral bones that form in each otic capsule, protecting the membranous labyrinth. The largest of the complex are the paired prootic bones, which are anterior to the basioccipital and comprise a considerable portion of the lateral floor of the cranium in many species. Part of the posterior boundary of the orbit consists of the sphenotic bone, dorsal to the anterior part of the prootic. The sphenotic in part forms around the anterior semicircular canal. The hyomandibular bone, which supports the jaws, articulates with the sphenotic. The prominent ridges that usually mark the widest part of the cranium are formed by the pterotic bones, which ossify around the lateral semicircular canals and usually combine with a dermal element to produce a compound bone. The epiotic bone, which ossifies in part around the posterior semicircular canal, usually can be recognized as a process between the pterotic and the supraoccipital. The epiotic bone is the site of attachment of the pectoral girdle to the cranium. An intercalar bone, an ossification of a ligament, appears in the back wall of the cranium between the pterotic bone and the exoccipitals. This has replaced an endochondral bone (opisthotic) that has been lost by living fishes (Rojo, 1991).

Cartilage bones of the trabecular section of the cranium include the following: the paired pterosphenoid bones, which make up part of the posterior wall of the orbit and connect with the prootic bones posteriorly; the median orbitosphenoid bone, which forms a bony interorbital septum through which the olfactory nerves pass in clupeiforms, salmoniforms, cypriniforms, and other more primitive bony fishes (this bone is absent in derived teleosts); and the median basisphenoid bone in the posterior part of the orbit. The arms of the Y-shaped basisphenoid articulate with the prootics, and the shaft connects ventrally to the parasphenoid bone. Anterior to the orbit are paired lateral ethmoid bones, which are of endochondral origin. The lateral

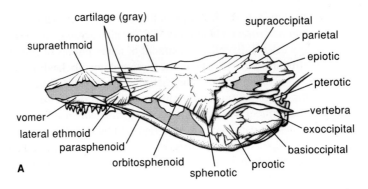

A

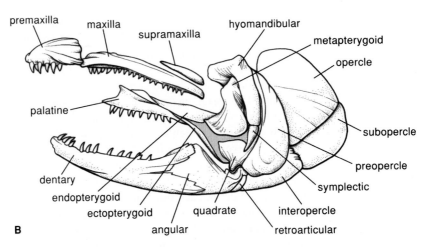

B

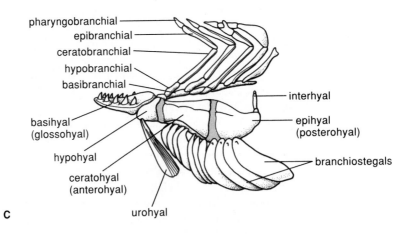

C

FIGURE 2–20   Skull bones of steelhead trout (*Oncorhynchus mykiss*). **A,** Neurocranium; **B,** jaws, suspensorium, and operculum (left side only); **C,** branchiohyoid apparatus (left side only). (Pharyngobranchials attach to ventral midline of neurocranium; interhyal attaches to hyomandibular posterior to symplectic.)

ethmoids form a complex with dermal elements (prefrontals) in some fishes and are sites of attachment of the paired dermal lachrymal bones.

The cartilage (endochondral) bones comprise the primary neurocranium, to which dermal bones are added. The dermal bones, also called investing bones, may include plates originating from scales, plates formed from coalescing tooth bases, and bones that form directly from membranes. The most posterior dermal bones of the skull are the paired parietal bones, which usually flank the median supraoccipital bone and constitute part of the roof of the cranium. Anterior to these are the large frontal bones, which make up most of the cranial roof. Parietals and frontals are both

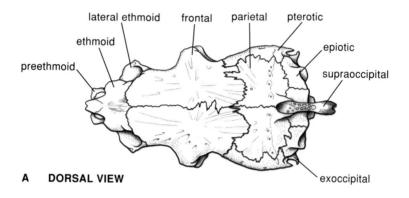

A DORSAL VIEW

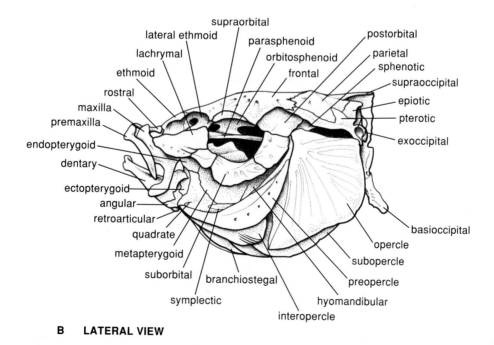

B LATERAL VIEW

FIGURE 2–21 Skull of carp (*Cyprinus carpio*). **A,** Dorsal view of neurocranium; **B,** lateral view of skull.

traversed by cephalic lateral line canals in many fishes. Typically an unpaired ethmoid bone (supraethmoid, dermal mesethmoid) roofs the snout in front of the frontals. Paired nasal bones, which develop around cephalic sensory canals, are located on each side of the ethmoid. Ventrally the vomer usually forms the anterior point of the neurocranium; it is often attached to the ethmoid in higher teleosts but is separate in the lower bony fishes. The vomer forms part of the roof of the mouth and

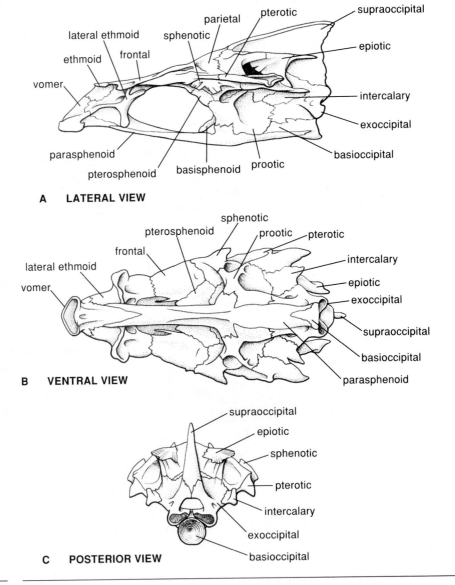

A   LATERAL VIEW

B   VENTRAL VIEW

C   POSTERIOR VIEW

FIGURE 2–22   Neurocranium of percoid (based on *Morone saxatilis*). **A,** Lateral view; **B,** ventral view; **C,** posterior view.

often bears teeth. The long parasphenoid forms the ventral midline of the cranium, extending between the vomer and the basioccipital. The parasphenoid bears teeth in some of the teleosts.

A series of bones (infraorbitals or circumorbitals) partially surround the orbit, although the number and extent of these are reduced in many fishes. Infraorbitals enclose cephalic sensory canals and cover the head musculature. Usually the anterior bone of the group, the lachrymal, is larger than the others in the series. Others, such as the suborbital and postorbital bones, are named according to their positions with respect to the eye.

The suspension of the primary upper jaw and the primary and secondary lower jaws is by bones that form in the palatoquadrate cartilage. That cartilage is ossified in part as the quadrate bone at the posterior end. The quadrate is shaped somewhat like a quadrant of a circle, with the point downward. The lower jaw articulates with the point of the quadrate. Dorsally the quadrate is attached to the metapterygoid, another ossification of the palatoquadrate cartilage. This bone is instrumental in suspending the remainder of the primary upper jaw from the hyomandibular. Anterior to the quadrate and metapterygoid are two dermal bones that form along the lower edge of the palatoquadrate cartilage. One of these, the endopterygoid, stiffens the roof of the mouth. The other, the ectopterygoid, connects the quadrate and palatine, which is in the anterior part of the roof of the mouth, just behind and lateral to the head of the vomer. The palatine bone may have both endochondral and dermal components. If the dermal component is lacking, the bone is called the autopalatine.

In the upper jaw the vomer and palatines, and sometimes the ectopterygoids, endopterygoids, and the parasphenoid, may bear teeth. However, the so-called secondary upper jaw, composed of the dermal premaxillary and maxillary bones, usually constitutes the main tooth-bearing (dentigerous) surface.

Within the lower jaw, Meckel's cartilage remains largely unossified in bony fishes except for an anterior mentomeckelian element and a posterior articular element, both of which may form complexes with the dermal elements of the lower jaw. The major tooth-bearing bone of the lower jaw is the dentary. Between the dentary and the quadrate, from which the lower jaw is suspended, is the angular or anguloarticular bone, called the "articular" in most older literature. The retroarticular bone, consisting of endochondral and dermal elements, is on the posterior lower corner of the angular; it was often called the "angular" in older literature. A sensory canal runs through the angular and dentary bones.

In many nonteleost bony fishes, the lower jaw contains many more bones than the lower jaw of teleosts. Included are prearticulars on the inner surface; tooth-bearing coronoids on the upper edge; splenials and postsplenials, which bear sensory canals; and supraangulars in the posterior part of the jaw.

The hyoid arch is ossified in multiple centers of ossification. The uppermost is the hyomandibular, which articulates with the otic region of the cranium and acts as a suspension for the primary upper jaw, the lower jaw, the hyoid apparatus, and the operculum. The metapterygoid articulates with the anterior face of the hyomandibular. A peglike bone, the symplectic, extends from the bottom of the hyomandibular to the quadrate. The interhyal attaches to the hyomandibular just behind the symplectic and suspends the remainder of the hyoid arch, which consists of the paired posterohyal (epihyal), anterohyal (ceratohyal) (see Nelson, 1969, and Rojo, 1991), upper hypohyal, and lower hypohyal bones and the unpaired basihyal (glossohyal) bone. The latter bears teeth in many fishes.

An unpaired bone, the urohyal, extends backward from the basihyals into the isthmus and constitutes the firm ventral connection between the head and trunk. The urohyal is a cartilage bone in sarcopterygians but an ossification of a tendon in teleosts (Arratia and Schultze, 1990). Important dermal bones that connect with the posterohyal and anterohyal are the branchiostegals, which in some fishes protect the gills ventrally. In others, the branchiostegals stiffen a membrane that can be of greater importance than the operculum in pumping water over the gills. The operculum, which serves as a shield for the gills and as part of the branchial pump, is composed of four pairs of dermal bones. These are usually platelike and are associated with the hyoid apparatus. The largest bone in the operculum is usually the opercle, which attaches by its anterodorsal corner to a condyle on the posterior edge of the hyomandibular. The preopercle, which carries a sensory canal, usually attaches along the hyomandibular for much of its length. The interopercle is below the preopercle, and the subopercle lies ventral to the opercle.

The branchial arches of teleosts are composed of a series of endochondral bones plus dermal tooth plates and gill rakers. The first three arches consist of a pharyngobranchial and an epibranchial in the upper section and a ceratobranchial and a hypobranchial in the lower part. The pharyngobranchial of the fourth arch is typically fused to that of the third arch or reduced so only the epibranchial and ceratobranchial are evident. The fifth arch is reduced further to one bone that may represent a ceratobranchial, and it is generally modified to bear pharyngeal teeth. A series of basibranchials is set between the left and right halves of the arches. These sometimes bear teeth, which appear just behind the teeth on the basihyal.

The Vertebral Column. In hagfishes and lampreys, the notochord persists without constriction (Goodrich, 1930), and the only rudiments of vertebrae in hagfishes are small cartilages resembling neuropophyses in the caudal region. Lampreys possess neural elements along the full length of the notochord (see Bridge, 1910, and Fig. 40 in Grassé, 1958 [vol. 13, part 1]).

The notochord of elasmobranchs is constricted by cartilaginous vertebral centra so, if extracted from the body intact, it would resemble a string of beads; the constricted portions would contrast markedly in diameter with the unconstricted portions that fit into the concavities of the amphicelous (biconcave) centra. Some species have a single calcified cylinder formed within the centrum (the cyclospondylous condition), whereas others may possess two or more concentric cylinders (tectospondylous). In some, calcified radiating lamellae extend from the calcified cylinder, giving a somewhat star-shaped pattern in cross section (asterospondylous). Each centrum in the trunk of the elasmobranch has ventrolateral transverse processes (basapophyses), which bear the cartilaginous ribs. Dorsally there is a neural spine surrounding the neural canal, through which the spinal cord runs. These neural arches consist of dorsal and ventral intercalary plates that alternate with basal dorsal plates (Daniel, 1934). In the tail region, the caudal vertebrae bear ventral hemal arches and spines.

Some primitive bony fishes have vertebrae that differ from most others. For instance the bowfin, *Amia calva,* has two vertebrae in each body segment in the posterior section of the vertebral column, a condition called diplospondyly. Gars

(Lepisosteidae) have opisthocelous vertebrae, so called because they are concave posteriorly and convex anteriorly.

Although in some eels the front and back surfaces of the centra are flat and the blenniid *Andamia* has centra that are convex anteriorly, the typical teleost has ossified amphicelous centra (concave anteriorly and posteriorly) (Fig. 2–23) with the notochord filling the concavities. Basapophyses (parapophyses) are present but might not be fused to the centra. Neural arches and spines are present, and the caudal vertebrae have hemal arches and spines. Zygapophyses can occur both anteriorly and posteriorly on the centra. The zygapophyses are generally small in fishes, with those of adjacent vertebrae not making contact, but they may be large and interlock in powerful swimmers such as the tunas and mackerels. The interlock of the zygapophyses prevents excessive rotation of vertebrae, keeping them lined up in true dorsoventral position.

Ventral ribs (pleural ribs) usually attach to the vertebral basapophyses. Intermuscular bones that extend into the horizontal skeletogenous septum are often called dorsal ribs, regardless of whether they are borne on the centrum or on the pleural rib. Usually, the bones that lie in the myosepta (intermuscular bones) take their names from the structures that bear them; for example, those borne on the neural arch are called epineurals, those borne on the centra are called epicentrals, and those on the ribs are called epipleurals. However, the terminology for these bones varies somewhat (see, for example, Forey, 1973; Goodrich, 1930; Grassé, 1958). Patterson and Johnson (1995) point out that intermuscular bones are subject to much modification and that in spiny-rayed fishes the epineurals may be displaced ventrally into the horizontal septum or onto the ribs, so in such fishes the bones that appear to be epipleurals are homologous with epineurals. Epineurals develop as bony growths on the neural arches but in most fishes lose the bony connection and are attached by ligaments. Epipleurals and epicentrals are usually attached by ligaments or may be represented by ligaments (see Patterson and Johnson, 1995).

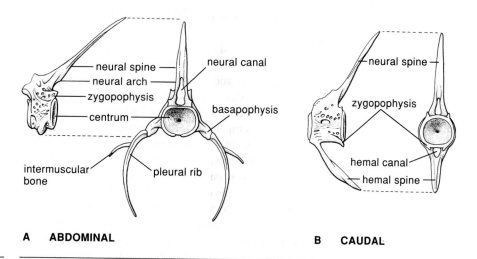

A   ABDOMINAL

B   CAUDAL

FIGURE 2–23   Vertebrae of teleost. **A,** Lateral and posterior views of abdominal vertebrae; **B,** lateral and posterior views of caudal vertebrae.

Intermuscular bones that are attached to the vertebrae by flexible or ligamentous joints may bend with the flexure of the musculature. A fossil fish (*Chongichthys:* Chongichthyidae) has sturdy, firmly attached epineural bones on the neural spines in the region of the dorsal fin (Arratia, 1982). This presumably derived condition must have limited flexion of the body somewhat. Intermuscular bones are more common in the more primitive bony fishes than in the higher groups.

The number of vertebrae tends to be higher in the less advanced bony fishes than in the advanced forms; for instance, the salmons and trouts have around 60 and many of the perchlike fishes have from 25 to 35. Some elongate specialists, however, may have numerous vertebrae. Crestfishes (Lophotidae) have up to 200, and some species of snipe eels (Nemichthyidae) have up to 750 (Nelson, 1994). Lindsey (1975) noted that among related species, those that grow to a larger adult body size tend to have more vertebrae that those with a smaller adult size. He called this phenomenon "pleomerism," indicating many divisions.

There is much modification of the vertebral column in the region of the caudal fin (Fig. 2–11**B**). In some of the less advanced teleosts, the column may be upturned, with three or more progressively smaller centra involved (the upturned portion is called the urostyle). Below these there is a supporting structure made up of about six hypural bones that appear to be modified hemal spines, whereas above the vertebral column there are one or more uroneurals. The hypural complex supports the rays of the caudal fin. In derived teleosts, the vertebral column ends in a urostyle, which is a turned-up portion of the last vertebral centrum. The hypurals of higher teleosts are usually fused into larger plates, sometimes with one supporting the upper lobe of the caudal fin and one supporting the lower. Patterson (1968, p. 234) defined the teleosts as follows: "Actinopterygian fishes in which the vertebral centra are perichordally ossified, the lower lobe of the caudal fin is primitively supported by two hypurals articulating with a single centrum, and in which the neural arches are modified into elongate uroneurals, the anterior uroneurals extending forwards on to the preural centra."

Median Fins and the Appendicular Skeleton

The Fins. The median fins of elasmobranchs are supported by basal cartilages that are often segmented into proximal, middle, and distal elements (Fig. 2–24). In some sharks, the proximal elements may fuse into a single plate. In some flattened species, the basals may join with the neural spines. Median fins of *Latimeria* are subject to different interpretations. Uyeno (1991) views the "three-lobed" caudal fin (Nelson, 1994) as consisting of a small caudal fin (with rays not supported by pterygiophores) at the posterior terminus of the fish between the second anal fin and third dorsal fin, in which the rays are borne on pterygiophores. The first anal fin and the second dorsal fin are lobate and extremely flexible. The median fins of lungfishes are supported by a series of basal elements that bear radials, which in turn bear fin rays. The fin rays are more numerous than the radials.

In teleosts, each ray of the median fins is typically supported by two ossified and one cartilaginous pterygiophores. Proximal pterygiophores are elongate tapered bones set deeply into the medial skeletogenous septum, usually between the neural or hemal spines. Because of this, they are often called interspinous bones; those supporting the dorsal fin may be called interneurals and those of the anal fin may be

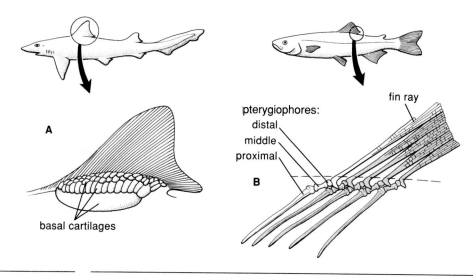

FIGURE 2–24 Skeletal supports of dorsal fin. **A,** Shark; **B,** bony fish (dashed lines show approximate body contour). (**A** based on Goodrich, 1930.)

called interhemals. The middle pterygiophores are ossified and jointed flexibly to the proximal elements on one end and to the distal pterygiophores, if present, on the outer end. In soft-rayed fins, the cartilaginous distal pterygiophores usually fit between the bases of the two halves of the rays. In spinous fins, the distal pterygiophores may be lost or fused to the middle element. In extreme cases, fusion of the three elements is complete, and fin spines attach to single supporting structures. The pectoral fins of elasmobranchs are supported by a cartilaginous girdle consisting of an upper scapular section and a lower coracoid element (Figs. 2–25A and **B**). A small suprascapular cartilage may be present. In sharks, the two halves of the girdle are separate from each other dorsally and do not attach to the vertebral column. In rays, the two halves join each other or the vertebral column.

The pectoral fin articulates with the coracoid. There are three basal cartilages: an anterior propterygium, a middle mesopterygium, and a posterior metapterygium. In rays, the articulation is horizontal and the propterygium and metapterygium extend far forward and backward, respectively. Series of jointed radials, which attach to the basal cartilages, bear the fin rays at their distal ends.

The pelvic girdle of elasmobranchs is rather simple and consists of a bar of cartilage crossing the ventral midline. Extending posteriorly from the ends of the cartilage are the elongate basipterygia (one in each fin in some species) that bear the jointed radials of the fin.

In the pectoral girdle of typical teleosts, the scapula and coracoid are ossified as endochondral bones, and part of their outer edges forms the articular surface for the radials (actinosts) of the pectoral fin (Figs. 2–25C, **D,** and **E**). This complex is applied to the inner surface of a secondary pectoral girdle consisting of a series of dermal bones. Actual attachment is to the cleithrum, usually the largest of the series. Cleithra meet at the ventral midline and extend upward toward the cranium. A supracleithrum attaches to the cleithrum and extends forward, where it attaches to

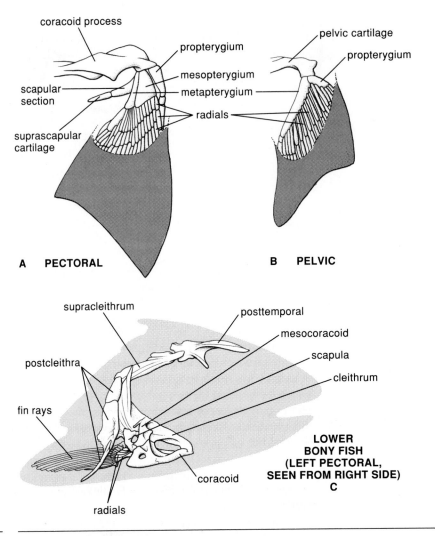

FIGURE 2–25 Skeletal supports of paired fins of shark. **A,** Ventral view of left pectoral; **B,** ventral view of left pelvic. Skeleton of paired fins of teleosts. **C,** Medial aspect of left pectoral bones of *Oncorhynchus mykiss*. (**A** and **B** based on drawings by John McKern.)

the posttemporal bone, which is usually forked. If forked, the upper branch of the posttemporal attaches to the epiotic and the lower branch to the pterotic or the intercalar. A series of postcleithra is present in most teleosts.

Most teleosts in which the pectorals are low and have oblique bases possess a mesocoracoid bone that forms a brace between the coracoid and the cleithrum. This is typical of soft-rayed fishes such as herrings, salmons, carps, and catfishes but is absent in higher teleosts—perches and basses, among others—that have vertical-based pectorals set higher on the sides.

The pelvic fin skeleton in teleosts is made up of platelike basipterygia, one for each fin. These bones usually are joined to each other posteriorly and may meet an-

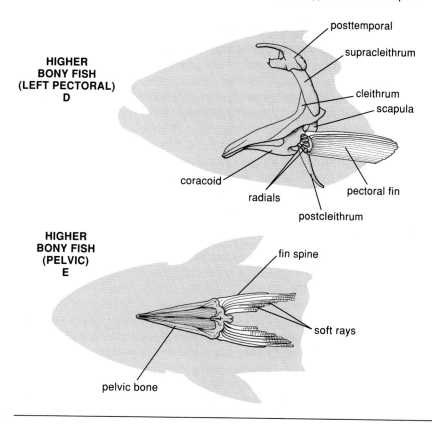

HIGHER
BONY FISH
(LEFT PECTORAL)
D

posttemporal

supracleithrum

cleithrum

scapula

coracoid

radials

pectoral fin

postcleithrum

HIGHER
BONY FISH
(PELVIC)
E

fin spine

soft rays

pelvic bone

**FIGURE 2–25
continued**   Skeletal supports of paired fins of teleost. **D**, Lateral view of left pectoral bones of *Morone saxatilis;* **E**, ventral view of pelvic skeleton of *M. saxatilis.*

teriorly. Remnants of pterygiophores may be present where the fin rays join the basipterygium.

Fin Rays. Fin rays of various groups of fishes differ in structure and origin. Those of lampreys and hagfishes are simply rods of cartilage extending from the notochord, whereas elasmobranchs have fibrous, horny rays arising from the dermis. These structures, usually unbranched, are called ceratotrichia because they are composed mostly of keratin, a scleroprotein that is characteristic of horn, feathers, and hair. Other rays of a horny nature, composed of elastoidin fibers, are actinotrichia (rodhairs) found in adipose fins and in the embryos of bony fishes. The actinotrichia of the coelacanth (*Latimeria*) are similar to those of bony fishes. Lungfishes have fin rays called camptotrichia that are different from the rays of other fishes. They are composed of an outer layer of flexible, fibrous bone over an uncalcified interior.

The dermal fin rays of bony fishes are called lepidotrichia (Whitear, 1986b) because they are believed to have arisen from rows of scales covering the horny primitive fin rays. The lepidotrichia of the soft fins of most bony fishes are usually made up of several elements placed end to end in two closely apposed bilateral rows, so the rays have a jointed appearance and are composed of two sides (Fig. 2–26).

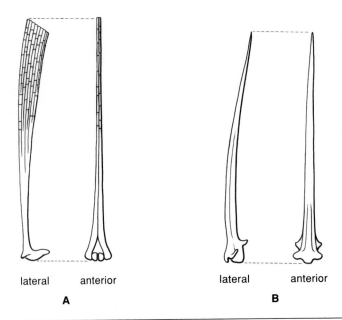

lateral   anterior   lateral   anterior

A   **B**

FIGURE 2–26 Comparison of soft and spinous rays. **A,** Lateral and anterior views of soft ray; **B,** same views of fin spine. Note branching, segmentation, and double construction of soft ray.

The typical unmodified soft ray is not only jointed and double but may be branched. Some "soft" rays, such as those at the leading edge of the dorsal fin of a carp, are not really soft to the touch but have been modified by loss of the branched and jointed characteristics and have hardened into spinelike structures. True spines—such as those in the dorsal, anal, and pelvic fins of typical acanthopterygian fishes—are undivided, unbranched, typically hard, sharp structures, although these may be modified secondarily toward flexibility in some groups, such as sculpins. Lagler et al. (1962) state that the fin spines of acanthopterygians are formed from actinotrichs.

General Relationships of Internal Organs

Alimentary Canal and Associated Structures[3]

The alimentary canal and associated structures of fishes follow the generalized vertebrate plan, with some notable exceptions. The mouth cavity is usually equipped anteriorly with oral valves that aid in pumping water over the gills by preventing backflow through the mouth. There can be teeth within the mouth and pharyngeal cavities on several bones other than the jawbones, as described earlier. Possession of a pharynx perforated by gill slits is a difference between fishes and adult tetrapods.

---

[3]Although details of internal organ systems will be given in subsequent chapters (e.g., Chapter 25), a general description of the placement of viscera is provided here for the sake of orientation. This orientation will be aided by referral to Chapter 25, especially Figures 25–4 and 25–5.

Internally, the pharynx leads to a short gullet or esophagus, which is separated from the variously shaped stomach by a sphincter. There are certain groups of fishes that lack a true stomach, and in these the esophagus communicates directly with the intestine. Beyond the distal end of the stomach, just past the pyloric sphincter, are blind sacs (pyloric caeca) opening off the intestine. These are found in most families of bony fishes, but not in the cartilaginous fishes.

The gut may be relatively straight and short, S-shaped, or variously elongated and coiled or folded, depending on the food habits of the species. Usually the lumen of the posterior part of the gut is larger than that of the anterior part. Elasmobranchs have greatly different small and large intestines; the latter contain an internal coil or helix, the "spiral valve," that increases the internal surface area and slows passage of food for more efficient extraction of nutrients. Some lower bony fishes, such as bichirs, lungfishes, sturgeons, gars, and bowfin, have a spiral valve. In bony fishes, with few exceptions, the gut reaches the exterior at the anus or vent, but in elasmobranchs the gut empties into a cloaca, which also receives the ducts from the urogenital system.

The liver and spleen, both of which are usually located near the stomach, are associated with the alimentary tract. Elasmobranchs and a few bony fishes have a discrete pancreas; otherwise it is associated with the liver to form a "hepatopancreas." Another organ associated with the alimentary canal is the gas bladder or swimbladder, which is derived from the digestive system and remains attached to it by a tube (pneumatic duct) in most soft-rayed fishes. Such fishes are called physostomous, from the Greek roots *physo* (bladder) and *stom* (mouth). Most of the spinyrayed fishes and their allies have lost the open connection except in the larval stages, when in many species air is gulped into the bladder. These are called physoclistic (clist = closed). The gas bladder is typically a torpedo-shaped, thin-walled sac in the upper part of the body cavity immediately below the kidney. Gas bladders vary considerably among the fishes in both shape and function.

Urogenital Organs

The dark red kidney tissue of bony fishes is typically located above the body cavity along the vertebrae, separated from the other viscera by membranes. Appearance and organization of the kidney material vary among fishes; the prominent large, pulpy kidneys of trout contrast greatly with the thinner organs of the actinopterygian fishes. The anterior part of the bony fish kidney (head kidney or pronephros) is specialized for blood cell formation and is usually more expanded than the posterior part (opisthonephros). Elasmobranchs have more compact kidneys, usually located in the posterior part of the body cavity. Hagfish and lamprey kidneys are long and straplike.

Urinary ducts lead to a recognizable bladder in few fishes. Typically, the urinary and reproductive ducts join to form the urogenital sinus in elasmobranchs. The sinus empties to the cloaca. In bony fishes, the ducts may join to form a single urogenital duct prior to reaching the exterior just behind the anus or, depending on the species, may be arranged so the left and right ducts of each system join to form single tubes (one urinary and one reproductive) that open to the exterior separately, the urinary duct anterior to the reproductive duct.

Although there is much variation in shape, size, and structure, gonads of fishes are located along the dorsal part of the body cavity beneath the kidneys. Testes are usually much thinner than ovaries and lighter in color. Developing ovaries take on a yellowish coloration. Sperm ducts and oviducts may be more or less evident depending on species, age, and stage of development.

The Heart

The heart is placed at the lower, anterior end of the body cavity, more or less below the gills and between the lower muscles and bones of the pectoral girdle. Elasmobranch hearts consist of a sinus venosus, atrium, ventricle, and contractile conus arteriosus. Bony fish hearts have the conus reduced to a small valve-bearing structure associated with an expansible basal section of the aorta called the bulbus arteriosus. The sinus venosus receives blood from the ducts of Cuvier and the hepatic veins. (See Chapter 23 for a discussion of the circulatory system.)

Part Two

Evolution and Diversity of Fishes

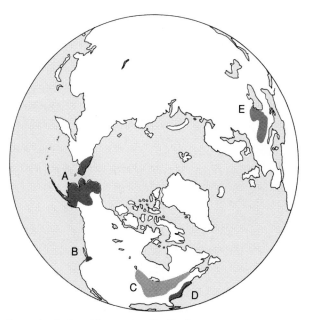

Example of allopatric species distribution.

Introduction to Classification

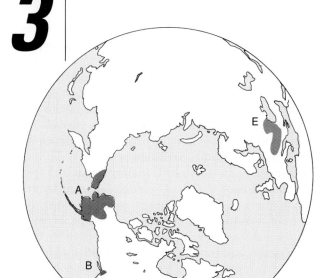

Purposes of Classification

Communicating about Fishes

Effective communication about biology and management of fishes requires consistent nomenclature. Use of fish names has reached enormous importance with the increase in conservation, natural history, scientific, and aquarist interest in fishes in the late 20th century. Fishes are important commercial and cultural resources that now require constant legal attention because of the crisis in their numbers. Consider how much interest there is in steelhead, pupfish, groundfish, salmon, tuna, *Tilapia,* white shark, snail darter, lamprey, orange roughy, barracuda, chinook, sockeye, swordfish, or *Salmo.* Whether we use fishes as food, pets, or educational resources, reliability of species names and classification is essential.

Fishes are formally classified by Latinized genus and species names in an internationally agreed-upon system of zoological nomenclature (e.g., *Salvelinus fontinalis,* brook trout). Scientific names of genera and species are standardized worldwide to avoid confusion. This classification system not only supplies a framework for international knowledge and conservation of fishes but allows scientists and managers to predict unknown attributes when confronted by a new fish problem.

Predictive Classification

The predictive value of a classification serves ecological and evolutionary hypotheses testing and resource management. Not all observations or experiments will ever be made for all species, and it is often necessary to infer responses from knowledge based on the study of closely related species. This is never foolproof, but prediction is more reliable with an accurate classification.

Knowing that a fish is in the family Salmonidae tells us that it is adapted to high latitudes, breeds in fresh water, grows to medium size, and has unique bones, including a diagnostic fan-shaped bone, the stegural, in the tail. The family Salmonidae includes all of the species that share a common ancestry; these also then share a considerable evolutionary history (Fig. 3–1). Therefore, we can sometimes solve a species management problem by predicting unknown attributes, such as life history, sensory ability, behavior, or physiology, from information gathered from research on other species of the family. Shared evolutionary history permits a generic name (e.g., *Abudefduf,* which is a damselfish, Pomacentridae) to supply more detailed information (e.g., they live in inshore marine waters near reefs and are territorial, aggressive, and small, with only one pair of nostrils).

A classification system has predictive qualities if it represents phylogenetic relationships of fishes. Related species share more characteristics with each other than with unrelated species. Members of a species (e.g., *Poecilia reticulata,* guppies) share genetic, ecological, mating, and behavioral systems and usually differ in these features from related species. Members of a genus (e.g., *Thunnus,* tuna) share more general—less specific—features. Members of a family (e.g., Centrarchidae, sunfish and bass) share fewer but more general kinds of attributes. Members of an order (e.g., Anguilliformes, eels) share basic characters with each other and differ from other orders (e.g., Pleuronectiformes, flatfishes, or Siluriformes, catfishes) quite

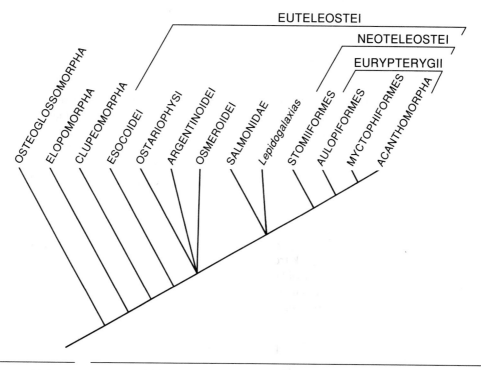

FIGURE 3–1 Teleostean phylogeny (from Fink, 1984), illustrating hierarchical relationships of groups and their implied shared ancestry.

broadly. This information system is based on the hierarchical sharing of traits among successive branches of related lineages.

Similarity and Special Similarity

There has been a debate about the best way to achieve a predictive classification. Some (Sneath and Sokal, 1973) have argued that a system based on general similarity has the best predictive value because many diverse similarities will, by definition, predict many similar responses. An alternative view is that shared phylogenetic history is the best predictor of unknown qualities because so much of an organism's morphology, physiology, and ecology derives from historically shared genetic pathways. The most powerful method for discovering genealogically related groups of organisms is the presence of shared specializations (Hennig, 1967). Unrelated groups may share primitive traits (e.g., sharks and sturgeons share heterocercal caudal fins), but primitive traits are uninformative about shared or predictive attributes. The special similarity of derived features (e.g., the specialized caudal fins of the cods and their relatives) provides solid evidence for relationship and, in combination with other derived characters, enables systematists to estimate the basic hierarchy of relationships (see the section titled "Phylogenetic Methods" later in this chapter).

Even an optimum classification system has its limitations. Species have converged as well as diverged, and some traits are especially influenced by environment.

These factors interfere with the predictive value of any classification. A landlocked Pacific salmon might feed and grow like landlocked Atlantic salmon, although in reproduction they are more like their sister species. Genetics and morphological adaptations are often correlated with history, whereas some aspects of growth, behavior, and life history are often correlated with environment. But all phenotypic traits result from the interaction of genetics and environment, and both historical and environmental information are necessary for prediction in science and management.

Names, Groups, and Categories

Fish Names and Rules of Nomenclature

The Linnaean system of classification and names dates back to 1758 and has been built up, refined, and broadened to the present. The most important building blocks of the system are the genus and species names—for example, *Lepomis gibbosus* (Linnaei, 1758), the formal name of the pumpkinseed sunfish. The generic name is capitalized but the species name is not; both names are printed in a type font (usually italics) different from the surrounding text; the endings are governed by rules of Latin grammar regarding gender and case. The author and date of the name follow the name; parentheses indicate that the author originally named the species in a different genus. The date establishes the priority of the name. (This is important because most species were named more than once as fish were brought back to museums by different expeditions. The oldest available name is the one we are required to use, with few exceptions.) To be formally available, a name must have been proposed in a genus plus species context, using a generic name with a specific name not used before in that genus, and based on a type specimen (or, in older literature, specimens). The words chosen for names may be Latin or Greek or Latinized words of any language representable in the Latin alphabet. For example, the scientific name of the chinook salmon, *Oncorhynchus tshawytscha,* is composed of a Latinized Greek generic name meaning hooknose and a Latinized Tungusic or Chukchi Siberian salmon name. There are many other rules governing application of the rules of priority, grammar, endings, and special problems. A small book of rules, *The International Code of Zoological Nomenclature* (International Trust for Zoological Nomenclature, British Museum [Natural History], London, 1985), governs the formal conventions, so we will not go into them in detail here. The International Committee on Zoological Nomenclature oversees the rules and keeps them up to date. This committee has the power to establish permanently a name for the sake of stability, when requested. A committee of the American Fisheries Society has worked for years to ensure that the names of our 2428 species of North American fishes are based on the best interpretation of the rules (Robins et al., 1991). Occasionally, nomenclatural rules make it necessary to change a name in use.

Common Names

Common names for North American fishes have been gleaned from tradition or made up by a committee (Robins et al., 1991) for the convenience of those uncomfortable with Latin names and also to provide a stable continuity around the occasional changes in formal classification. Some have argued that common names must reflect natural groups, the way formal names must. But that would destroy the traditional value, contribution to stability, and ecological uses of common names. For example, the terms *trout* and *salmon* have been used to refer roughly to life history patterns: Trouts usually do not migrate to the sea (except for anadromous rainbow and cutthroat trouts) and salmon usually do (except for kokanee and landlocked Atlantic salmon). These names are useful because of their widespread recognition; their inconsistent application can be mitigated by use of scientific names.

Higher Classification of Fish Groups

The higher classification of fishes utilizes a hierarchy of levels, but two—families and orders—are traditionally held to be important. Here we must distinguish between fish groups and categories. The group names (e.g., Cottidae, Sphyrnidae) are intended to refer to real, evolved, and related entities in nature; the categories or levels (family, order) are not. As more is learned about the natural groups of fishes (e.g., Cichlidae, Otophysi, Elasmobranchii, Teleostei), the categories become more arbitrary and less relevant to scientific discussion. Nevertheless, they are used to tabulate biodiversity in a general way. (In descending order, some categories are Kingdom, Phylum, Class, Division, Order, Family, Tribe, Genus, Species; plus intermediate categories identified by the prefixes Super-, as in Superfamily, or sub-, as in Suborder). Group names for the family level in a zoological hierarchy always end in -idae (e.g., Sparidae), subfamilies are indicated by -inae (e.g., Gobiinae), and tribes are indicated by -ini (e.g., Etheostomatini). Many workers prefer to end ordinal names with a standard ending, -iformes, and suborders with -oidea, but the practice is becoming less standardized.

Stability Versus Phylogenetic Accuracy

Two important goals of classification are in conflict: the need for stability for communication versus the need to reflect correctly our best knowledge about relationships. Currently, the major changes in classification are caused by re-evaluation of the monophyly of fish groups. The actual entities in nature, which share evolutionary history, are called **clades** (e.g., named groups in Figs. 3–1, 3–2**B** and **D,** and 3–5). Clades are **monophyletic**—that is, they include all of the descendants of a common ancestor and exclude any look-alikes that evolved from a different ancestor. Clades are **diagnosed** by their member's possession of unique specializations. For example, the Otophysi (Fink and Fink, 1981) is a natural group diagnosed by the presence of uniquely elaborate modifications of the anterior vertebrae, for hearing. The group includes all relatives of minnows, characins, catfishes, and South American electric fishes. The Otophysi is a subgroup of the Ostariophysi, which is diagnosed by less specialized anterior vertebrae and the presence of an alarm substance recognized by other members of the group. The hierarchy of the groups is a function only of distribution of character evidence; the categorical ranks order, suborder, etc., are not a part of the analytical process.

It is scientifically crucial that a phylogenetic group be monophyletic, for any subsequent evolutionary study based on false information about the group's real membership is likely to be false correspondingly. For example, recognition of a group "Agnatha" for lampreys (Petromyzontiformes) and hagfishes (Myxiniformes) implies that the Gnathostomata has an evolutionary history equally separated from the two jawless groups. But the diagnostic evidence shows that Petromyzontiformes is the sister group to all other Vertebrata, not to Myxiniformes (Forey and Janvier, 1994). Therefore, it is useful to recognize a monophyletic group, Vertebrata, including Petromyzontiformes and Gnathostomata but not hagfishes. Myxiniformes is the sister to Gnathostomata; together they form the Craniata (Fig. 3–2**A**). The paraphyletic "Agnatha" is thus defined by a shared primitive character, lack of jaws, so it is not a natural group (Fig. 3–2**B**). It is an assemblage without the special shared evolution that gives meaning to a monophyletic group.

Monophyletic groups are discovered analytically by cladistic or phylogenetic studies (described in the section titled "Phylogenetic Methods"). These groups,

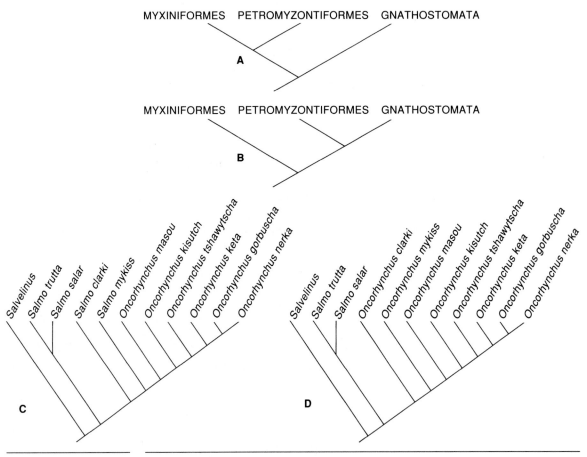

FIGURE 3–2   Alternate phylogenies of "agnathans" and gnathostomes. **A,** Implies that Petromyzontiformes share a significant evolutionary history with Myxiniformes rather than with Gnathostomata; **B,** implies that Petromyzontiformes share history with Gnathostomata, separate from Myxiniformes. Strong evidence (Forey and Janvier, 1994) supports hypothesis **B,** making the "Agnatha" paraphyletic. **C, D,** Classifications of trouts and salmons: **C,** inclusion of *mykiss* and *clarki* in *Salmo* made *Salmo* paraphyletic and misrepresented historical relationships; **D,** a solution to the problem.

called clades, exist in hierarchical relationship to each other. Each clade of fishes has a unique evolutionary existence in time and space, with a historical beginning and end and clearly defined spatial distribution. Polyphyletic or paraphyletic groups are arbitrarily **defined** to contain members of other clades or to lack entities that belong in the clade, respectively; these are not clades because their position in time and space is arbitrary, depending on subjective definitions. Paraphyletic groups can be subjectively defined out of existence or into existence, but this criticism does not apply to monophyletic groups (Patterson and Smith, 1987).

If a study concludes that the real membership of a group differs from that represented in the traditional classification, the author suggests a new, corrected classification. Generally, the new classification is adopted or rejected in a few years, depending on whether the evidence is confirmed by further, independent study. For example, Pacific trouts were recently called *Salmo gairdneri* and *Salmo clarki* (Fig. 3–2**C**) but were removed from the genus of Atlantic salmon and trouts, *Salmo,* and placed in the genus *Oncorhynchus* with Pacific salmon (Fig. 3–2**D**). Four different research groups, studying diverse morphological, fossil, biochemical, and molecular evidence, independently found that Pacific salmon and trout were part of the same monophyletic clade, separate from the Atlantic trout and salmon clade. (Robert Behnke also discovered that the Kamchatka rainbow trout, *O. mykiss,* and the American rainbow trout, *O. gairdneri,* were one and the same species; we must therefore use the name of the Asian population, which was named first.) The name changes, however annoying at the time, enable more accurate predictions to be made in ecology and evolution as well as resource management.

Name changes aimed at adjusting the arbitrary levels in classification, or adjusting ambiguous priority, are less justified. Stability of nomenclature is extremely important and should be sacrificed only for accuracy, not for artificial reasons. Difficult choices arise because it is often discovered that a group can be made monophyletic by adding or subtracting a taxon or changing the level at which a name is applied.

Species

One category enjoys special importance in our classification system. The species is held to be more real and less arbitrary than other categories. (But here, too, the species **category** is arbitrary, existing mostly in theory, in contrast to **individual species,** which are real in nature.) Fish species are reproductively independent lineages, diagnosed by different genetically based character states and ecological roles (Smith, Rosenfield, and Porterfield, 1995). A proposed new species is more readily accepted if it can be shown to possess several unique characters drawn from several lines of evidence—morphological, behavioral, chromosomal, biochemical, molecular, or ecological. A named species that differs in only one feature is suspect until someone demonstrates that it is not an ecophenotypic or polymorphic variation of another species (see Chapter 27). A lineage supported by evidence for a distinct mating system is more convincing. Congruence of diverse characters is the key to recognizing the difference between species and partly differentiated populations within a species.

An entire spectrum of species—well differentiated to barely different—is encountered in the study of fishes. This is because we are observers of many different stages in the speciation process; we are not placed here as the final stage of creation.

Some species began to diverge 10,000 years ago, while many began more than 1 million years ago. Some have changed in obvious shapes and colors, some only in small, hidden features. Not all species have been discovered yet, so we still have an inadequate assessment of fish diversity. The genetics, reproduction, and ecology of all fish species need more intensive study by interested students. Important new systematic information depends especially on the study of species with a long history of intensive scientific work. For example, current research on life history and genetics of trout, salmon, and bass is changing our general understanding of all fishes. It is also true, however, that discovery of a new species may reveal startling new aspects of fish biology (e.g., all-female and hermaphroditic species, or protandrous and protogynous species).

Processes That Cause Fish Species Origins

Formation of new species is the consequence of natural selection, usually in geographically isolated populations, as demonstrated especially by Charles Darwin (1859), Ernst Mayr (1963), and others in the past 50 years. Yet exactly how natural selection makes species is not well understood. We will examine some of the peculiarities of fish species to demonstrate part of the natural history that is already understood and the exciting potential for further study.

Most fish species began the first part of the speciation process when populations of their ancestral species became isolated by natural barriers, thousands or millions of years ago. The second stage in species evolution involves the accumulation of different genetic pathways, selected by changing environments. This **allopatric** (different countries) or **vicariant** (replacement) model fits most of the data we see in the study of North American fishes. Allopatric sister populations are found on opposite sides of geographic or ecological barriers. Examples of the allopatric model include the Olympic, central, eastern, and European mudminnows and the Alaska blackfish (*Umbra* and *Dallia;* Fig. 3–3**A**), four species of squawfishes (*Ptychocheilus;* Fig. 3–3**B**), the lake suckers and their relatives (*Chasmistes, Deltistes,* and *Xyrauchen;* Fig. 3–3**C**), and the darters of the *Etheostoma variatum* group (Fig. 3–3**D**). An example of speciation in progress may be represented by two subspecies of *Erimyzon oblongus, E. o. oblongus* in eastern North America and *E. o. claviformis* in midwestern United States (Fig. 3–4), which differ morphologically, except in southeastern United States, where the characters overlap. Populations divided by natural barriers are often exposed to different environmental conditions. When this is the case, differential survival and reproduction of genetic variants changes the genetic makeup of the separated populations through time.

The most famous examples of this speciation model are the species pairs on the Atlantic and Pacific sides of Panama. The isthmus emerged as land 3 to 5 million years ago, dividing all of the local marine fishes into pairs of populations (Jackson et al., 1995). Many of these have changed sufficiently to enable systematists to discriminate each member of one population from each member of the other, so they are recognized as distinct species (Collins, 1995).

When dispersal or hydrographic changes bring sister populations into contact after a period of isolation, the different forms either merge or diverge genetically, depending on the amount of genetic change during separation and the environmental context at the time and place of secondary contact. A freshwater example of species newly sympatric after long separation by glaciation is *Prosopium cylindraceum* (round whitefish) of northern and eastern North America and *P. williamsoni* (mountain whitefish) of western North America, which are sympatric and distinct in northern British Columbia and adjacent Northwest Territories (Scott and Crossman, 1972). In the Mississippi and Great Lakes drainages, a northern species *Luxilus cornutus* (common shiner) and a southern species *L. chrysocephalus* (striped shiner) established a zone of secondary contact near the southern margin of glaciation, following the withdrawal of ice. They provide an example of two kinds of fishes that engage in frequent hybridization, with different genetic results at different localities.

A variation of the allopatric speciation model may apply to colonists of different environments within a lake. They may evolve by divergent resource use and reproduction, despite limited exchange of genes in intermediate depths and habitats. Examples of this form of speciation can be inferred from the depth distributions of Great Lakes ciscoes (Koelz, 1929) and Lake Baikal sculpins (Kozhov, 1963). Individual species living in these lakes are each adapted to different depths, temperatures, light, pressure, and food resources, and they are genetically separated from each other by different places and times of spawning.

New morphological or genetic traits which diagnose diverging species are molded by differential survival and reproduction in diverse ecological contexts. Changing circumstances also provide new contexts in which mate choice affects reproduction. This occurs when mate choice involves being at the best place at the optimal time for spawning, or when a female's choice of the male with the most mature appearance leads to evolution of more extreme secondary sexual characteristics. The ecological contexts most likely to be involved in fish speciation are those affecting fertilization of eggs and survival of young. But natural selection may also result from competition for mates among females or among males. In all cases, a behavioral change in the mate-getting system becomes, at some time or other, part of the process of speciation. The search to understand species and speciation of fishes requires study of responses to reproductive ecology and competition for mates.

Speciation that begins with a change in the mate recognition system seems to characterize many North American freshwater fishes, such as small, colorful darters, minnows, pupfish, and live-bearers. In many of these forms, the species differences are mainly in male breeding colors, implying that female choice of mates is an important part of the speciation process (Houde and Endler, 1990; Smith et al., 1995). A similar possibility has been suggested for the diverse flocks of closely related cichlid species in the African rift lakes (Dominey, 1984). This process invokes the action of sexual selection—competition for mates among males (see Chapter 26) and female choice of mates showing indicators of maturity and genetic success (e.g., large body size, bright colors, vigor, large territories, or good nests). One of the forces behind sexual selection (and consequent differentiation) is the struggle between the sexes to control reproduction.

Sometimes, when populations come into contact after a period of genetic separation, interbreeding occurs. If the intermediate individuals have lower fitness than diverged parental types, selection may favor mate recognition systems that promote assortative (therefore more successful) mating (i.e., preference for mates belonging to the same genetic type). This process, theoretically important but largely

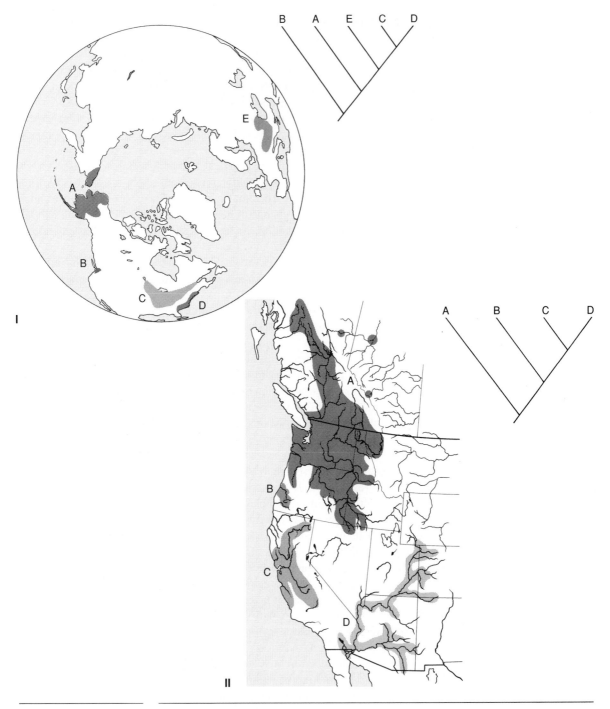

FIGURE 3–3 Examples of allopatric species distributions suggestive of vicariant speciation. **I,** A, Alaska blackfish; B, Olympic; C, central; D, eastern; and E, European mudminnows (*Dallia pectoralis, Novumbra hubbsi, Umbra limi, Umbra pygmaea, and Umbra krameri*); **II,** four species of squawfishes (*Ptychocheilus* A, *oregonensis;* B, *umpquae;* C, *grandis;* D, *lucius).*

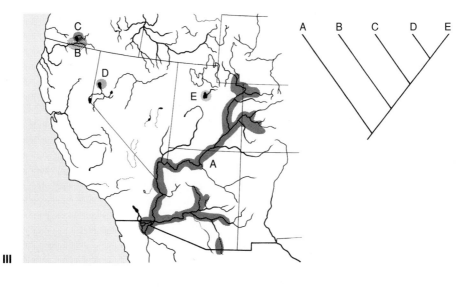

III

IV

FIGURE 3–3
(Continued)

III, River and lake suckers of the *Chasmistes* group. A, razorback sucker; B, Lost River sucker; C, shortnose sucker; D, cui-ui; E, June sucker (*Xyrauchen texanus, Deltistes luxatus, C. cujus, C. liorus*); and **IV,** five darters of the subgenus *Poecilichthys*. A, variegate darter; B, Arkansas saddled darter; C, Missouri saddled darter; D, Kanawha darter; E, candy darter (*Etheostoma variatum, E. euzonum, tetrazonum, E. kanawhae, E. osburni*).

undemonstrated in fishes, is called reproductive character displacement. It is recognized by the presence of greater distinctiveness between sympatric members of two populations than between allopatric members of the same two species. Divergent recognition cues in fishes may include body shapes, color patterns, courtship

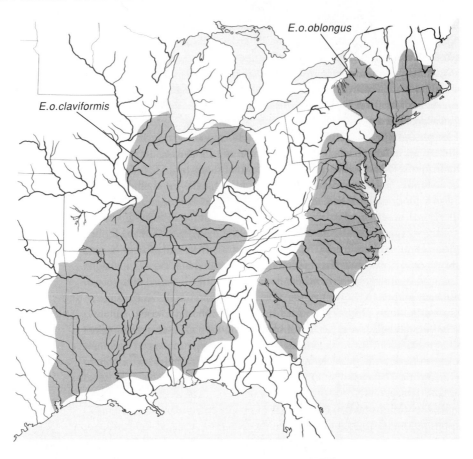

FIGURE 3–4 Distribution of eastern and central subspecies of creek chubsuckers, which have not diverged to species status, judging from the lack of unique diagnostic characters for the two populations.

behaviors, courtship sounds, electrical signals, luminescent displays, olfactory cues, nest characteristics, and time and place of spawning.

Ecological Character Displacement

The ecological differences and associated morphological distinctions between fish species are often exaggerated by enhanced growth and survival of individuals utilizing a part of the resource base not shared by the competing species. That is, lower fitness of overlapping and competing individuals may select for divergence. This process is called ecological character displacement (Robinson and Wilson, 1995) and is responsible for species differences in size, jaws, teeth, gill rakers, habitat choice, and feeding behaviors.

Speciation, although of crucial importance, is not a primary "process" in evolution. It is simply a byproduct of descent with modification in different populations. Most or all significant evolutionary changes occur when one variant in a population has more success than the others (e.g., long and finely spaced gill rakers catch more food in the pelagic zone; short and broad gill rakers catch more food from the benthic zone). The genes responsible for the phenotypes that have more successful survival and reproduction—higher fitness—increase in proportion to their alternatives during subsequent generations, thus increasing the observable phenotypes to which they contribute. Speciation is an effect of these changes, not a procedure driven by nature for some advantage.

The diversification of lineages provides the marvelous adaptations we see in fishes. The most spectacular consequences are those involving breakthroughs in the match-up of morphology and behavior to ecology. Examples are the coral reef percoids, South American characins, South American catfishes, cichlids, Holarctic minnows, North Pacific scorpaeniforms, inshore blennies and gobies, and Antarctic icefishes, to name just a few. These "adaptive radiations" are great flowerings of fish diversity that occur when new opportunities for trophic interactions are set in motion by unusually diverse spatial habitats. Coral reefs provide complex three-dimensional structure; African lakes provide depth gradients along fluctuating shorelines and islands; South American, North American, and Eurasian rivers provide millions of miles of branching fluvial gradients; inshore marine habitats are linear and heterogeneous; and the Antarctic shelf provides inshore gradients at a stable temperature that excludes most competitors. What each of these provides is a spatially varying foundation on which the evolving fishes, among themselves, increase the trophic opportunities for increasing specialization and proliferation.

Phylogenetic Methods

The pattern of branching of fish species and clades is inferred from shared specializations among a sample of the organisms. The result of a phylogenetic analysis is a tree diagram of genealogical relationships, on which the evidence for monophyletic taxa is indicated by notation of homologous character states on the stems representing common ancestry (Fig. 3–5). The shared character states (e.g., the numbers of spines, size of scales, colors of fins, or connections among bones) are thus the homologies that diagnose monophyletic groups.

The method hinges on correct identification of advanced versus primitive states. Advanced character states are called **apomorphies.** Specializations unique to only one taxon are called **autapomorphies.** Shared specializations, called **synapomorphies,** are distinguished from the primitive states (**plesiomorphies**) by comparison to several **outgroups**—nearby relatives outside the study group. This step, called polarization of characters, designates states as advanced if they differ from the state most widely distributed among diverse outgroups. For example, anal fin spines are an advanced state; the absence of anal spines is more general among fishes. The presence of pelvic, dorsal, and anal fin spines of Acanthomorpha is evidence that acanthomorph families share a common ancestry. States determined to be apomorphies by this criterion are sometimes inconsistent (e.g., internal fertilization and viviparity appears to be advanced in livebearing poeciliids and primitive in some outgroups, such as elasmobranchs and coelacanths. A hypothesis of relationship is tested by the **congruence** (consistency) among numerous independent characters in a cladistic analysis of all possible related groups (Fig. 3–5). For example,

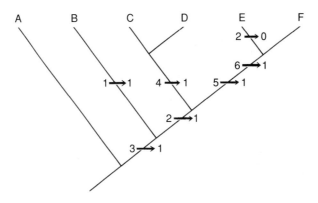

Characters	Taxa					
	A	B	C	D	E	F
1) spines: 0 = no, 1 = yes	0	1	0	0	0	0
2) fins: 0 = long, 1 = short	0	0	1	1	0	1
3) electrosensory: 0 = no, 1 = yes	0	1	1	1	1	1
4) color: 0 = black, 1 = red	0	0	1	1	0	0
5) scales: 0 = large, 1 = small	0	0	0	0	1	1
6) eyes: 0 = small, 1 = large	0	0	0	0	1	1

FIGURE 3–5 Matrix of six characters and six taxa, with resulting tree. Taxa A and B are outgroups; C, D, E, F, are the study group. The tree hypothesis is discovered by minimizing the number of ad hoc hypotheses about homoplasy, i.e., finding the resolved tree requiring the minimum number of evolutionary steps. Characters 5 and 6 are correlated; all of the characters except 2 are congruent.

the conclusion that internal fertilization and viviparity are primitive relative to egg laying is mostly congruent with the classification in which electroreception and possession of lungs are primitive states that are lost in teleosts and then sometimes regained in certain specialized groups.

The principle of parsimony is the basis of phylogenetic hypothesis testing. Parsimony dictates that we accept the hypothesis of relationship that requires the fewest ad hoc hypotheses about character change. This is logically extended to the principle that the shortest cladistic tree that can be calculated from a complete data set is the best estimate of phylogenetic relationships (Fig. 3–5). Acceptance of the parsimony principle does not require a belief that evolution is parsimonious; rather it requires that in a large set of independent data describing character changes, the most consistent agreement of character changes over a tree is likely to represent the history of branching lineages because random changes are unlikely to yield a strong signal of shared specializations. Computer programs to search data sets for shared changes and calculate the shortest trees were first developed by fish students. HENNIG86 (by J. S. Farris) and PAUP (Phylogenetic Analysis Using Parsimony, by D. Swofford) are now used by systematists worldwide.

Character Evidence, Phylogeny, and Homology

Discrete characters of fish samples are discovered by careful study, and the character states are then assigned integer code values (0, 1, 2, etc.) for phylogenetic analyses

(Fig. 3–5). (Code values are entered into the computer programs as nonadditive or "any-directional" if character-state changes are not constrained to be sequential, e.g., $0 \to 1 \to 2 \to 3$.) Characters may be drawn from morphology, DNA sequences, proteins, chromosomes, behavior, or ecology. The only requirements are that characters and character states be homologous (i.e., representative of equivalent structures, judged to be similar due to inheritance from common ancestry), independent (i.e., correlated by history, not by interdependent function or environmental influence), and genetically based. The obvious problem is that homology, thus defined, is what we are trying to discover. Indeed, we only establish strong evidence of homology by phylogenetic study. This problem is solved in practice by using the results of previous studies as a basis for recognizing the homology of, for example, fins, skull bones, or gene sequences in the study group. These decisions are made with confidence limited by the nature of the previous studies. In the course of a study of fishes, we find support for homology of, for example, thoracic pelvic fins of percomorphs, Weberian bones of Otophysi, gill rakers of Osteichthyes, specific DNA restriction sites or proteins of a genus or family, and specialized behaviors or life history traits of certain groups. (Note that homologies are always traits *of a group;* they are homologies because they are shared apomorphies of specimens or subgroups within that group.) States are established as homologous if their distribution on the cladogram is found to be consistent with a single origin. Cladograms frequently show that similar character states had separate origins; therefore their similarity is convergent, i.e., homoplasious. The electroreceptive ability of gymnotoid fishes is a homology, demonstrated by congruence with other characters. The congruent characters show it to be analogous, not homologous, to electroreception in mormyrid fishes.

Current expectations are that morphological and molecular characters are more conservative in their origins than life history and behavioral characters. But we have a long history of morphological studies and powerful molecular technologies to guide us in the appropriate choices among those characters. This background enables us to focus on conservative and consistent sources of evidence and avoid features caused by environmental perturbations or molecular noise. As life histories and behavior are better studied, it is likely that additional informative characters will be discovered. In all cases, the ultimate arbiter is consistency on the cladogram. Congruence of independent characters in an analysis is evidence of shared ancestry and shared history. The power of this method to identify trends in evolution of life histories, behaviors, and morphological adaptations is increasing rapidly.

Adaptations

Changes that make a fish population more fit (e.g., the wondrous ways fishes sense their environment and each other, swim, capture food, avoid becoming food, and attract mates) are called adaptations. (Adaptations that are clearly modified from previous, different functions may be called exaptations.) These offer the primary challenges to fish systematists, whose task it is to understand evolutionary changes by testing evolutionary hypotheses concerning the origin and modification of new structures and behaviors for new functions.

Adaptations have traditionally been studied by demonstrating ecological reasons why observed states of structures or responses are more fit. Because alternate explanations and alternate histories are usually imaginable, adaptations suffered a credibility crisis among skeptical workers. The opportunity to follow the origin, descent, and modification of adaptive systems and their precursors within the framework of a cladistic hypothesis (i.e., the causal progressions of character changes on a phylogenetic tree) is leading to a renewed interest in testing hypotheses about adaptations. New statistical methods are being developed to enable confident tests of alternative historical hypotheses of character change.

Fossils and Extinction

Paleontology provides time-related evidence of history for those systems that have left an adequate fossil record. In a few fortunate cases, we can actually study the likely ancestors and observe primitive characters and observe their descent and modification in their own ecological context (Bell, 1994). This is possible when hydrographic evidence enables us to reject hypothetical dispersal events that could cause lineage mixing or substitution. Isolated Late Cenozoic lake and river systems provide examples of long-term transitions into the recent fish faunas.

Fossil fishes tell us that evolution is usually slow, sometimes reversible, and highly dependent on ecological conditions. For example, most lakes are ephemeral in geologic time, with histories shorter than usually required for significant adaptation of stream fishes to lake conditions. Exceptions are lakes in tectonic basins that subside long enough to rejuvenate the deep habitats. In such cases, related groups of lacustrine fishes branch repeatedly from initial colonists, resulting in species flocks (e.g., African rift lake cichlids, Baikal sculpins, Lake Titicaca orestias, Lake Lanao cyprinids, Laurentian Great Lakes ciscoes, and others). These examples and fossil fishes from rift lakes, such as the Connecticut valley semionotids (McCune, 1984) and the Lake Idaho sculpins (Smith, 1987), show that intralucustrine diversification may occur in tens of thousands of years or less.

Because fossil fishes are in the long run limited to depositional environments, such as lowland rivers, lakes, and seas, the evolution of upland fishes, which inhabit eroding systems, can be known better from studies that use geographic factors as proxies for geologic time.

In the course of history, as lineages change in response to changing physical and biotic conditions, populations and species expand and contract. Populations shrink to extinction when individual fishes can no longer produce enough surviving eggs and young in the habitat to replace themselves because of competition, predation, or environmental change. The fossil record of North American fishes provides us with a background check on the frequency of extinction over the past 5 million years. In western North American fresh waters, where the amount of water for aquatic habitats has been unstable, species may go extinct at the rate of about 3 percent per million years. In eastern North America, where climate has slowly fluctuated between cool glacial and warm interglacial temperatures, the extinction rate has been much lower, perhaps 1 percent per million years.

These considerations suggest why the most ancient groups persist in the great rivers, with their elevational gradients, diverse habitats, and permanent lowland connections. Examples are paddlefish, sturgeons, bowfins, gars, and hiodontids of the Mississippi River; osteoglossids, characins, and catfishes of the Amazon River; and cyprinoid diversity of the Mekong River.

In the past 100 years, due primarily to culturally induced environmental change, about 4 percent of the species and subspecies of freshwater fishes have gone extinct in North America, including Mexico. Southwestern North American fishes comprise only one fourth the species density on the continent but have suffered more than double the number of extinctions because of the more drastic habitat changes associated with land and water use in the American Southwest.

Systematics and Conservation

Systematists are biologists with the tools to resolve questions about the distribution and irreplaceability of populations or species being considered for protection. Using morphological, behavioral, ecological, genetic, protein, and molecular data, systematists resolve questions about the limits of species. Because species are historical individuals (Ghiselin, 1974), they are subject to extinction and are irreplaceable. A fragment of a population that is defined by possession of a certain trait is a class of organisms (whose existence is a function of the definition), and is replaceable if other fishes may acquire the trait ecophenotypically. For example, a uniquely recognizable population of salmon that is genetically constrained to spawn at a certain time and place is a lineage subject to extinction; but a part of a population that spawns at a certain time and place because of nongenetic, ecological, and behavioral interactions is replaceable. The first is an individual, diagnosed by genetic characters; the second is a class, defined by its ecophenotype. The distinction is similar to the distinction between fish species and subspecific fragments. To evaluate species status, fish systematists use diverse evidence to estimate the evolutionary independence of lineages, which is the same information necessary to evaluate irreplaceability.

Jawless Fishes

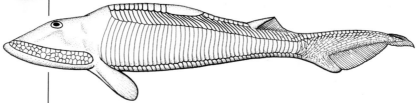

Origin of the Fishes

Far back in the Cambrian or pre-Cambrian Period, there were selective pressures that resulted in the evolution of fishlike vertebrates. The typical scenario of what could have occurred involves invertebrate chordates with free-swimming larvae that had the chordate characteristics of a notochord; a hollow, dorsal nerve tube; a ventral heart; and clefts in the pharynx. One such group is the subphylum Urochordata, the tunicates, which for the most part settle down to a sessile existence after the larval stage and live a life unlike most vertebrates. There must have been, as there are now, groups that changed from the settled life to a free-living existence in the plankton.

The present-day class Larvacea comprises free-living tunicates that retain many larval characteristics in the adult stage, although they enclose themselves in a mucous "house" and live as plankton as adults. One can imagine that there could have been ascidians or related invertebrates that remained in the tadpole larval stage, reproduced, and formed the evolutionary basis for the more complex early vertebrates. (Ability to reproduce as larvae is part of the condition known as neoteny.)

The subphylum Cephalochordata, typified by the lancelets (*Branchiostoma*), is a reasonable model for what the forerunner of the fishlike vertebrates could have been like. It has a notochord stiff enough to allow swimming by contractions of bilateral trunk muscles, a perforated pharynx, and metamerism (segmental body plan). In the fossil record, there are no species that are considered intermediate between the lancelets and the earliest known vertebrates.

There are suggestions (see Northcutt and Gans, 1983) that vertebrate development was made possible by the evolution of certain embryonic features—such as neural crest cells, ectodermal placodes, and hypomere, a mesodermal plate that can provide muscle in the gut and associated organs. Some of the important features of vertebrates are (1) the formation of sensory structures in the head region from ectodermal placodes (including the lens of the eye, the olfactory epithelium, the lateral line system, and the inner ear), and (2) derivation from neural crest cells of the dermal skeleton, skeletal material of jaws and branchial arches, trabeculae, parts of sensory capsules, pigment cells, and some cardiac and aortic-arch motor neurons as well as some sensory cranial nerves. In addition, the well-equipped vertebrate has other features, such as a brain, cranium, perforated and muscular pharynx, body cavity, muscular gut, heart, eventually vertebrae and calcium phosphate in the skeleton, and a tail (usually postanal).

Agnatha: Jawless Fishes

The superclass Agnatha comprises fishlike craniates without jaws. These are the oldest of the fishlike animals in terms of their fossil history. These early fishes were generally small (15 cm or less in length, although some reached nearly 2 m; Radinsky, 1987) and must have fed by sucking in or scooping up organic matter through their jawless mouths. Although not all are closely related, the known forms of agnathans, both living and fossil, inherited certain common characteristics from the lines that began in the Cambrian or earlier.

There are fossils of very small (6 mm) animals called conodonts, which resemble hagfishes, known from deposits from the Cambrian to the Triassic. They are known mainly from teeth, which Krejsa et al. (1987) and Krejsa and Slavkin (1987) compared to those of embryonic and juvenile hagfish and suggested an ancestral relationship of conodonts to hagfish. Aldridge et al. (1993) reported on conodont specimens that provide the most complete knowledge of the anatomy of the animals and concluded that conodonts are vertebrates and probably the primitive sister group of Heterostraci. Forey and Janvier (1994) point out that the conodont fossils show no evidence of gill openings and that their hard tissues are unlike the enamel and bone of vertebrates, so they do not fit the group into their vertebrate phylogeny.

Some of the most important agnathan features indicating a development different from that of the jaw-bearing vertebrates are the entire arrangement of the gills and the branchial skeleton. All the gill tissue (which is endodermal in origin) as well as the branchial arteries and nerves are internal to the branchial skeleton, which is a cartilaginous "basket" external to the gill pouches in the living forms. The branchial skeleton is fused to the neurocranium. Gill openings are pores, not slits. In the agnathans, vertebrae do not replace the notochord; there are two semicircular canals in each ear; and pelvic fins are absent, although some forms have pectoral fins of a primitive nature.

Agnathans are most abundant in the Silurian and Lower Devonian. Their numbers decreased through the Middle and Upper Devonian, and only two atypical groups, the lampreys and hagfishes, have survived into modern times. The jawless fishes, living and fossil, are a polyphyletic (with more than one ancestral line) assemblage of diverse lineages. Nelson (1994), Forey and Janvier (1993, 1995), Janvier (1981, 1984, 1986), Janvier and Lund (1983), Halstead (1982), and Yalden (1985) are among those who have proposed various hypotheses of relationships among the agnathans. The following classification is that of Nelson (1994), with comments on the arrangement of the agnathans by Forey and Janvier (1993).

Class Myxini

Order Myxiniformes. The hagfishes (Fig. 1–1) have at least one fossil representative from the Carboniferous (Pennsylvanian) that resembles modern species (Bardack, 1991; Carroll, 1988). The eyes are degenerate and completely covered by skin. Large tentacles surround the terminal nasal opening and the mouth, which is not developed as a sucking disc. The feeding action consists of rapid eversion and retraction of teeth situated on each side of the mouth, so the effect is that of jaws operating laterally as opposed to the up-and-down motion of true jaws. The similarity in the musculature and other structures of this lingual biting system has led Yalden (1985) to postulate that Myxinoidea could be included in Cephalaspidomorphi. Although Janvier (1981) doubted that hagfishes were vertebrates, largely because they lack rudiments (arcualia) of vertebrae, Forey and Janvier (1994) state that hagfishes are the most primitive of vertebrates. Hagfishes differ in several ways from the lampreys. They are only marine and probably have no larval stage (Hardisty, 1979) as they develop from a large meroblastic egg, and no larvae have ever been collected. Their eggs (Fig. 4–1) may be over 25 mm long and equipped with hooklike tendrils that can hold them together in a string or clump. Developing eggs have seldom been found, so knowledge of hagfish embryology is limited (Gorbman and Tamarin, 1985). Walvig (1963) reports that one investigator found only 151 hagfish eggs over a 20-year period. The largest cluster contained 21 eggs.

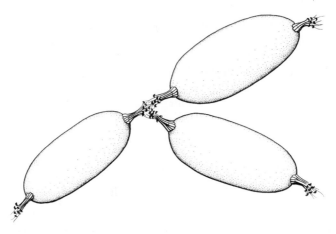

FIGURE 4–1   Eggs of hagfish, class Myxini (actual length ca. 30 mm).

As few as 5 to as many as 15 pairs of gills, depending on species, may open from the pharynx. In some genera, the efferent branchial ducts are collected into a tube with a single external opening on each side. On the left side a peculiar duct (pharyngocutaneous duct), apparently homologous with the gills, originates at the posterior portion of the pharynx and communicates either with the left branchial duct or, in the South American *Neomyxine tridentiger,* directly to the exterior.

The cartilaginous skeleton is not well developed, with no rudiments of vertebrae and only a membranous roof in the skull. The dorsal roots of the spinal nerves are united with the ventral roots. The two semicircular canals are connected in such a manner that they appear as one (Berg, 1947).

Hagfishes have been given considerable notice by medical researchers because of the array of contractile structures in their vascular system. In addition to the usual heart, considered to be a primitive one, there are a caudal heart, a cardinal vein heart, and a portal vein heart that pumps blood from an intestinal vein and the right jugular vein to the liver.

Hagfishes live on soft bottoms of mud, silt, or clay at depths between 25 and 600 m, although a Japanese species may be found at 5 m and others have been taken as deep as 1000 m. They burrow in the bottom and perhaps feed on worms and other soft-bodied animals. They act both as scavengers and predators and may attack fishes in nets close to the bottom.

Six genera are known, most from temperate waters. These are usually placed in the family Myxinidae, but those with multiple branchial openings are sometimes treated as a separate family, Eptatretidae, or as a subfamily, Eptatretinae (see Nelson, 1994). There are over 40 species; two of the best known are the Atlantic hagfish (*Myxine glutinosa*) and the Pacific hagfish (*Eptatretus stouti*) of the eastern Pacific Ocean. Economic importance is centered in the Orient, where hagfish leather is made. Recently there has been concern that some hagfish stocks are being overexploited by that fishery. There is some use as food in Japan. From a negative standpoint, hagfish are known to mutilate the catches of commercial fishermen. This often occurs when fish are tangled in gill nets or hooked on long lines.

†Class Pteraspidomorphi

These are extinct agnathans that appear to have had paired nasal sacs with separate openings and no nasohypophyseal canal. Because of this, they have been called Diplorhina by some authors. Usually there is only one pair of branchial openings, although there are several gill pouches, but arandaspidiforms have several individual gill openings. Bony armor lacking true bone cells is present.

†**Order Arandaspidiformes.** These may be the oldest known fishes. Most genera are known from marine Ordovician deposits in the Southern Hemisphere. The several branchial openings have individual covers of bone (Nelson, 1994). †*Anatolepis* is a very small animal, known from small fragments of fossil armor, that lived from the Upper Cambrian to the Lower Ordovician and appeared about 520 million years ago (Radinsky, 1987). It has armor of apatite, a bonelike material, which features sculpturing resembling that of Ordovician vertebrates. Some paleontologists consider †*Anatolepis* to be the earliest vertebrate (Carroll, 1988), but others state that it cannot be identified definitely as a vertebrate and might be an arthropod (Colbert and Morales, 1991; Janvier, 1991). More complete fossil records of this order of early vertebrates are found in Middle Ordovician deposits of the Southern Hemisphere. †*Sacabambaspis,* from Bolivia, is the earliest well-known agnath, and †*Arandaspis* and †*Porophoraspis,* from Australia, are of nearly the same age (Colbert and Morales, 1991).

†**Order Pteraspidiformes.** This group, known from the Lower Silurian to the Upper Devonian, is also known as Heterostraci. Most are covered by bony plates over the anterior part of the body and scales on the posterior body, including the tail. Paired fins are lacking. Caudal fin shape is from hypocercal to heterocercal.

Pteraspidiforms reach their greatest development in the Upper Silurian and the Lower Devonian. They have bony plates that form a shield on the head and forebody (Fig. 4–2A). Typically the head is flat, the eyes lateral, and the mouth subterminal to slightly superior. The caudal is hypocercal in most. Paired fins are not present. Length is usually less than 300 mm, but some reach 1.5 m. One of the most primitive of the heterostracans is †*Anglaspis,* which is almost completely covered by large bony plates anteriorly and large bony scales posteriorly (these scales are smaller on the almost symmetrical caudal fin, probably the most flexible section of the body).

Some genera have part of the head shield produced laterally to form what appear to be underwater "gliding" surfaces (Fig. 4–2B). One (†*Doryaspis*) has an anterior extension of pseudorostrum similar in appearance to the rostrum of the saw shark (*Pristiophorus*). Specializations such as tubular mouths, dorsally placed mouths, and stabilizing keels have led paleontologists to believe that the pteraspidimorphs, in their adaptive radiation, took advantage of a wider range of habitats than did the cephalaspidomorphs, which were primarily bottom feeders.

†**Order Thelodontiformes.** The thelodonts (Fig. 4–3A) are small fishes usually 100 to 200 mm long (†*Thelodus parvidens* fossils are up to 1 m long [Turner, 1986]) that lived mainly during the Silurian and Devonian. They differ from other agnathans in

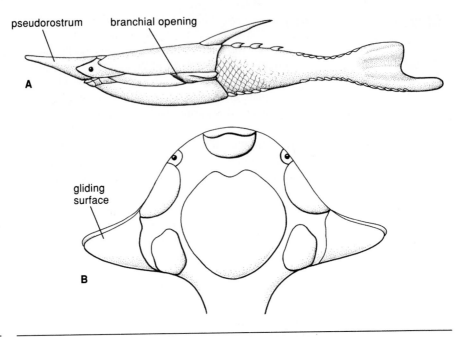

pseudorostrum

branchial opening

A

gliding
surface

B

FIGURE 4–2   Representatives of class †Pteraspidomorphi. **A**, *Pteraspis*, lateral view (actual size ca. 22 cm); **B**, dorsal view of hypothetical cephalaspid, showing laterally produced plates (actual size ca. 80 cm). (**A** after Moy-Thomas and Miles, 1971.)

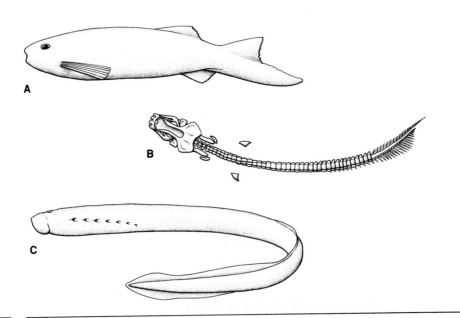

A

B

C

FIGURE 4–3   Representative of **A**, Order †Thelodontiformes (actual size ca. 7.5 cm); **B**, †*Palaeospondylus* (actual size ca. 3 cm); **C**, larva of lamprey, order Petromyzontiformes (actual size 13 cm). (**A** and **B** after Moy-Thomas and Miles, 1971.)

having a covering of small, denticle-like scales instead of plates or solid armor. Dorsal, anal, asymmetrical to hypocercal caudal, and lateral fins are present, and there appear to be eight or nine branchial sacs opening separately to the exterior. Evidence of monophyly of thelodonts has been the body form and the scales (Wilson and Caldwell, 1993; see also Janvier, 1986).

Fossils of fishes called "fork-tailed thelodonts" have been discovered in Silurian and Devonian deposits in northwest Canada. These are compressiform and contrast in body shape with the often depressiform thelodonts. The caudal fins are huge and supported by 8 to 14 scaled lobes connected by a web. The web is scaled in the Devonian specimens. The fossils are interpreted as possessing stomachs, indicating that stomachs evolved before jaws (Wilson and Caldwell, 1993).

†*Palaeospondylus.* A fish of uncertain rank and placement, the enigmatic *Palaeospondylus gunni* (Fig. 4–4) is known from many fossil specimens from the Middle

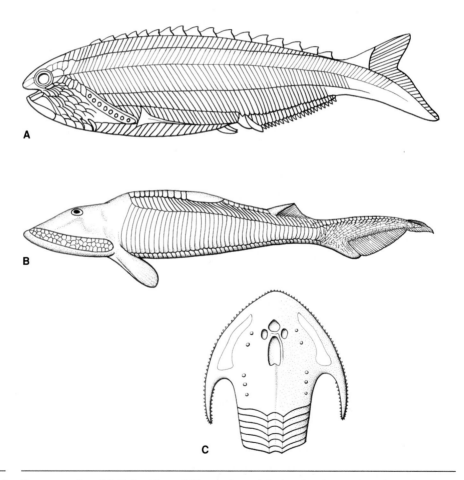

FIGURE 4–4   Representative of **A,** Order †Anaspidiformes (actual size ca. 6 cm); representatives of order †Cephalaspidiformes. **B,** †*Hemicyclaspis,* lateral view (actual size ca. 18 cm); **C,** †*Thyestes,* dorsal view of head (actual size ca. 2 cm wide). (**A** and **C** after Berg, 1947.)

Devonian of Scotland (Moy-Thomas and Miles, 1971). It ranges from 12 mm to about 50 mm in length and has well developed calcified vertebrae, paired fins, and a rather intricate skull. The caudal is heterocercal.

The small size and the state of preservation have not allowed conclusive work to be done on the genus, and consequently many different interpretations have been presented. Some investigators consider it to be an agnathan related to the hagfish, others think that it is an adult gnathostome, and still others believe it represents a larval gnathostome. Moy-Thomas and Miles (1971) and Nelson (1994) present it as an agnathan. Certainly the possession of vertebrae seems to be an advancement over the agnathans, but the head structure does not appear to lead to a firm interpretation of true jaws, so the animal remains a mystery.

Class Cephalaspidomorphi

This group is called Monorhina by some authors because of the single median opening leading to the nasal sac and the hypophyseal area beneath the brain. Another important characteristic is the number of gill openings, as many as 15 on each side. True bone cells and pectoral fins are present in some of the extinct genera. This group flourished in the Silurian and Lower Devonian, but only the lampreys survive today.

Order Petromyzontiformes. Lampreys (Petromyzontiformes) have two known fossil representatives, †*Mayomyzon* and †*Hardistiella,* both from the Carboniferous. The latter differs from the extant species in having an anal fin with fin rays and a hypocercal caudal (Janvier and Lund, 1983). †*Mayomyzon* is similar to the remainder of the living species but lacks teeth. The scarcity of fossil representatives may not reflect an ancient rarity of lampreys and relatives but rather a lack of readily fossilized hard parts.

Lampreys have no paired fins or scales (Fig. 1–1). They are eel-like, with lateral eyes and a ventral mouth consisting of a circular disc set with horny teeth. The nasohypophyseal opening is between the eyes. The skeleton is cartilaginous and not well developed except for the skull and branchial region. No vertebral centra are developed, and the neural arches are rudimentary. Dorsal and caudal fins are present. Myotomes are not divided horizontally into epaxial and hypaxial muscles, as in jawed fishes.

Lampreys possess some peculiar internal features. The internal labyrinth has a ciliated epithelium, the left duct of Cuvier is absent, and the dorsal and ventral roots of the spinal nerves are not connected with each other. In adults, the gills open into a respiratory tube that begins at the mouth, extends under the esophagus, and ends at the seventh pair of gills. In the eyeless, toothless larvae (called ammocoetes), the gills open to the long pharynx, but this is cut off posteriorly during metamorphosis and a new esophagus forms. The gallbladder and bile ducts disappear in adults.

Some relationship to bony fishes may be indicated by the embryonic formation of a neural keel instead of a neural tube, as in sharks and rays. In addition, a bulbus arteriosus is present instead of a conus arteriosus. Many researchers believe lampreys to be more closely related to gnathostomes than to hagfishes (Forey and Janvier, 1994).

All lampreys have a long larval life. Very small eggs are deposited in nests made in gravel bottoms of streams. The incubation period varies with temperature but is usually two to four weeks. When the tiny larvae hatch, they drift to soft bottoms in pools and eddies and begin a life of filtering plankton and detritus at the mud-water interface. They may remain as larvae for about five years, after which metamorphosis takes place and a new type of existence is begun.

With regard to feeding mode, there are two types of lampreys, parasitic and nonparasitic. After metamorphosis, the parasitic types feed by attacking fishes with their sucking mouths, rasping holes in the skin with their piston-like tongues, and pumping out blood and body fluids, a process aided by secretion of an anticoagulant from paired buccal or "salivary" glands. A few species ingest small fragments of flesh and viscera as well as scales and small bones (Beamish, 1980; Bond et al., 1983). This removal of tissue from prey may be facilitated by the longitudinal laminae of "teeth" on the tongue closing together in a lateral biting motion (Yalden, 1985). Some parasitic lampreys are anadromous, spending their postmetamorphic growth period in salt water before returning to streams to spawn and die. These may reach a meter in length. Other parasitic lampreys remain in fresh water and may grow to a half meter or more, as do the landlocked populations of sea lamprey (*Petromyzon marinus*) in the Great Lakes; or some strictly freshwater species may reach adult size at 15 cm or even smaller (Kan and Bond, 1981).

Nonparasitic lampreys are usually called "brook" lampreys. These confine feeding to the larval stage, and after metamorphosis they spend a few months in hiding while the gonads mature. They then spawn and die, usually at lengths of less than 20 cm. In many instances, species of brook lampreys are paired with a sister species of parasitic lamprey living in the same drainages. There is a general acceptance of the concept that the parasitic forms have given rise to the nonparasitic species. In some instances, more than one nonparasitic species has evolved from a parasitic species (Hardisty, 1979; Potter, 1980), so Vladykov and Kott (1979) have suggested using the term *satellite species* in place of *paired species*. Paired or satellite species have been reported in all lamprey genera except *Petromyzon, Caspiomyzon,* and *Geotria* (Hardisty and Potter, 1971).

On the Pacific Coast of North America, there are the river lamprey, *Lampetra ayresi* (parasitic), and the closely related western brook lamprey, *L. richardsoni* (nonparasitic). The western brook lamprey has a wider distribution than the river lamprey. It is found in many small creeks, both coastal and inland, whereas the river lamprey lives in larger streams generally close to marine waters, where it feeds on herring and other fishes. Another Pacific Coast series consists of the widely distributed, parasitic Pacific lamprey, *Lampetra tridentata,* and its assumed derivatives, the nonparasitic Pit-Klamath brook lamprey, *L. lethophaga,* which is restricted to creeks in the upper reaches of the Klamath and Pit river systems, and the nonparasitic Kern brook lamprey, of the Kern river system.

The Arctic lamprey, *Lampetra japonica,* distributed from Siberia to Japan and Alaska, has at least five satellite species, two found in Asia and three in North America (Potter, 1980). There are three pairs of species in the North American genus *Ichthyomyzon,* according to Potter (1980). They are listed here with the ancestral species first, followed by the nonparasitic derivative: *I. unicuspis,* silver lamprey; *I. fossor,* northern brook lamprey; *I. castaneus,* chestnut lamprey; *I. gagei,* southern brook lamprey; *I. bdellium*, Ohio lamprey; and *I. greeleyi*, mountain brook lamprey. Other examples of satellite species include the following: *Lampetra fluviatilis* and the nonparasitic *L. planeri* of Europe; *Eudontomyzon danfordi* and the nonparasitic *E. vladykovi* and other brook lampreys of eastern Europe; and *Mordacia mordax* and the nonparasitic *M. praecox* of Australia.

Lampreys inhabit temperate seas and streams and occur in both the Northern and Southern Hemispheres. All are in the family Petromyzontidae. The genus *Geotria* from the Falkland Islands, South America, New Zealand, and Australia is placed in the subfamily Geotriinae, and the genus *Mordacia* of Australia is in the subfamily Mordaciinae. Lampreys of the Northern Hemisphere are in the subfamily Petromyzontinae.

The economic value of lampreys is slight, even though they are used as food in some areas and they have been used as a source for a light oil. Their negative economic impact can be great locally if they attack and scar or kill fishes of sport or commercial value. The chronicle of the invasion of the upper Great Lakes by the sea lamprey is a sad one, for the valuable lake trout (*Salvelinus namaycush*) and lake whitefish (*Coregonus clupeaformis*) virtually disappeared as commercial species when mortality due to the lamprey was superimposed on fishing mortality. In many other instances the effect has not been as severe, but smaller lampreys, such as the chestnut lamprey (*Ichthyomyzon castaneus*) or landlocked Pacific lampreys (*Lampetra tridentata*), are known to attack and injure, if not kill, various freshwater game fishes. The smallest known parasitic lamprey (*L. minima*), made extinct by a fish control operation, was capable, apparently because of strength of numbers, of killing fingerling trout and tui chub (*Gila bicolor*). Most adults were under 80 mm long (Bond and Kan, 1973).

Larval lampreys are eaten by a variety of fishes and are sometimes used as bait. Adult lampreys are excellent bait for sturgeon and are found in the stomachs of other fishes, including sharks. Seasonally they may form a great portion of the food of the California sea lions, which live near the mouths of rivers that sustain runs of the Pacific lamprey (*Lampetra tridentatus).

†Order Anaspidiformes. Anaspidiforms are small fishes, usually less than 150 mm long, that lived from the Lower Silurian to the Devonian Period. They are characterized by a terminal mouth, 6 to 15 gill ports, and lateral eyes (Fig. 4–4A). If armor is present, it consists of overlapping plates. Some species have a ridge of scutes down the back, some show lateral fin folds, and all have hypocercal tails. The body is rounded, the mouth terminal, and the nasohypophyseal and pineal openings dorsal. The pectoral fins are represented by spines, and the anal fin is small or lacking.

Typical genera are †*Lasanius* and †*Birkenia*. Some genera (e.g., †*Jaymoytius*) have little ossification. That genus has a branchial basket similar to that of lampreys, and some contend that anaspidans are the close relatives of lampreys (Janvier, 1986).

†Order Galeaspidiformes. This is an order of armored fishes known from the Lower Silurian to the Middle Devonian of China. They have a bony carapace, two semicircular canals, and a braincase of endochondral bone (bone formed by the replacement of cartilage). The nasal sacs are paired but are associated with an opening to the pharynx. The dorsal shield is characterized by an anterior opening that may have been associated with a sensory organ. There is no pineal opening in the carapace. The tail appears to be hypocercal. There are no paired fins. Genera include †*Sangiaspis,* †*Galeaspis,* and †*Polybranchiaspis.*

†Order Cephalaspidiformes. This order contains at least five extinct (Upper Silurian to Upper Devonian) families of small fishes usually under 600 mm long, with a shield of bone covering the head and some of the anterior portion of the body, as in †*Hemicyclapsis* (Fig. 4–4**B**). The mouth is ventral. Eyes, nasohypophyseal opening, and pineal opening are dorsal. Bony overlapping scales are present in some orders, as are pectoral fins; the caudal fin is heterocercal. The head shield of †Tremataspidae is exceptionally long and covers much of the body. The head is rounded in cross section. Genera include †*Tremataspis* and †*Dartmuthia.*

In families †Ateleaspidae and †Cephalaspidae, the head is flattened and shows evidence of a "lateral sensory field" that could have been sensitive to pressure or electricity (Berg, 1947) (Fig. 4–4**B**). The head shield of many specimens shows impressions of the brain, with well-delineated nerves and canals associated with the sensory fields. The brain resembles that of lampreys (Carroll, 1988). Typical genera of cephalaspids are †*Cephalaspis* and †*Thyestes,* which have posteriorly directed "horns" at the lateral edges of the head shield (Fig. 4–4**C**).

Early Gnathostomes, Elasmobranchs, and Relatives

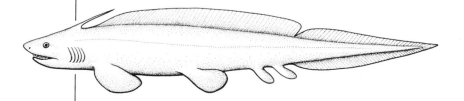

Origin of Jaws: Introduction to the Gnathostomata
Jaws
The Gnathostomes
†Class Placodermi
Class Chondrichthyes
Subclass Holocephali
  †Superorder Paraselachimorpha
  Superorder Holocephalimorpha
Extinct Holocephalan Groups
Subclass Elasmobranchii
  Cohort †Palaeoselachii
  Cohort Euselachii
  Subcohort Neoselachii
  Superorder Squalomorphii
  Superorder Squatinomorphii
  Superorder Rajomorphii
  Superorder Galeomorphii

Origin of Jaws: Introduction to the Gnathostomata

Jaws

The evolution of jaws allowed fishes to use a wider selection of food and to have greater efficiency in taking it in competition with jawless fishes. A great proportion of agnathans lived on or close to the bottom, where they could consume benthic organic material or prey. The origin of jaws gave fishes the ability to grasp and hold prey to be swallowed or to grasp and tear loose portions from attached prey or from soft-bodied organisms too large to swallow. This may have freed the early gnathostomes from the bottom and allowed them to evolve the ability to range farther and faster to exploit food and other resources.

The structure of some primitive jaws was similar to that of gill arches in certain extinct groups of fishes, and the hypothesis that jaws evolved from branchial arches has long been considered plausible. There is, however, little or no evidence that gill tissue was associated with the mandibular arch. Further, the jaws are innervated by cranial nerve V (trigeminal), which in agnathans innervates an area anterior to the pharynx not associated with the gills. At present, the jaws are thought to have arisen from a visceral arch that may have formed, independent of gill tissue, as supporting elements at the border of the mouth (see Chapter 2 and Carroll, 1988).

The Gnathostomes

The vertebrates that bear jaws constitute the superclass Gnathostomata. In the fossil record, gnathostomes are known from the Upper Silurian, more than 400 million years before present (mybp). They flourished in the Devonian, and several lineages survived into the Carboniferous and beyond whereas most clades of jawless vertebrates dwindled. Gnathostomes include not only the modern fishes but amphibians, reptiles, and their derivatives.

Early gnathostome fossils share derived characteristics that are not present in agnathans. Jaws are present; gill tissue, branchial arteries, and branchial nerves are external to the gill arches; pectoral and pelvic fins are present; three semicircular canals are present; and the branchial skeleton is not fused to the neurocranium. The notochord, persistent in early forms, is partially or completely replaced by vertebrae in modern gnathostomes.

Gnathostomes include two main groups of fishes, the cartilaginous fishes and the bony fishes. The cartilaginous fishes—sharks and rays (the elasmobranchs) and chimaeras—are placed in the class Chondrichthyes. The bony fishes are considered (Nelson, 1994) to constitute the taxon Euteleostomi. That taxon is intermediate between a grade and a class and replaces the term *Osteichthyes,* formerly used as a class name.

There is not complete agreement on how some groups are associated, however. For instance, although these views may not be acceptable, Jarvik (1980) stated that lungfishes should be aligned with the elasmobranchs, and Lagios (1979, 1982) argued that certain anatomical and physiological factors indicate that *Latimeria* should be considered as closely related to the sharks. In addition, there are differing views on the phylogenetic affinities of the two major extinct lines of gnathostomes, the acanthodians and placoderms. Some paleontologists interpret acanthodians as being sharklike, whereas others consider them to be closely related to the bony fishes (Miles, 1973); the justification of the association depends on the emphasis given certain features of anatomy. In this book, following Nelson (1994), Acanthodii will be considered a class closely aligned with the bony fishes. Extinct groups are indicated by a dagger (†).

†Class Placodermi

The bony-plated placoderms share features with bony fishes, sharks, rays, and chimaeras and are most often aligned with the Elasmobranchii. Schaeffer and Williams (1977) perceive no close relationship with either major group and express the opinion that the placoderms are the sister group of the remainder of the gnathostomes. Maisey (1986) considers the placoderms a problematic group and does not include them with Chondrichthyes. In the following treatment, Placodermi will be treated as a class without placement in either of the major groups. Placoderms are rather diverse in structure and body form (some are sharklike and others flattened like rays), but all are typified by an armor of bony plates on the head and forebody (Fig. 5–1). Usually there is a "neck joint" between the armor of head and body so that, apparently, the head could be raised. Scales or small tessellated plates or tesserae are present in many species. The endoskeleton is at least partially ossified. Pectoral and pelvic fins are present, and the caudal fin is heterocercal in most. The eyes are typically rather far forward. Most placoderms are known from the Devonian, but some appeared in the Upper Silurian and some persisted into the Carboniferous.

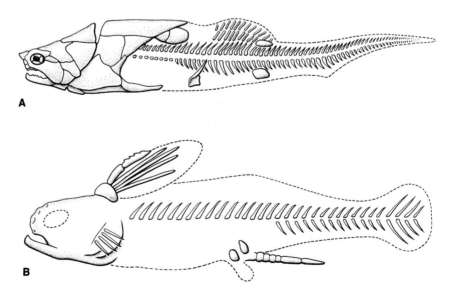

A

B

FIGURE 5–1 Representative of **A,** Class †Placodermi, order Arthrodiriformes (actual size ca. 35 cm). Representative of **B,** †Iniopterygiformes (actual size ca. 30 cm). (**A** after Dean, 1895; **B** after Zangerl and Case, 1973.)

The following groups of placoderms (referred to here as orders) are given different taxonomic ranks by various authors; some appear as subclasses or classes. Authorities usually recognize from 7 (Nelson, 1994) to 11 (Denison, 1978) orders, some of which will be mentioned here.

†**Order Rhenaniformes.** This order is composed of depressiform fishes with terminal mouths that lack gnathal plates. Instead of toothlike structures on plates, these have tubercles on the palatoquadrate, which is connected to the cranium by a hyomandibular, much like the jaw linkage in elasmobranchs. The mouth was apparently protrusible. The overall appearance of the Rhenanida is much like that of rays, with broad, flat pectoral fins that border the rather robust body from the eyes to the pelvic fins. The tail is dyphycercal. Eyes and nostrils are dorsal, and the latter are set almost between the eyes.

There are a few large plates on the head, but much of the head and all the body and fins are covered by a mosaic of small, scalelike plates. Part of the cranium is ossified. Vertebral centra appear to be present but are actually fused neural and hemal arches (Carroll, 1988). These are mostly small fishes, but some reached a meter or more. They lived in the Devonian (see Long, 1995).

†**Order Antiarchiformes (Pterichthyes).** This group has such peculiar pectoral appendages that some systematists have considered them to constitute a separate class of vertebrates. The pectorals are large and covered with plates of bone. They articulate with the large body shield, are jointed in the middle, and have an ossified or calcified endoskeleton. The head is relatively small. The body of some species is covered by overlapping scales. Some had diverticula extending from the gill cavity into the body that may have been used in air breathing (Denison, 1978).

†**Order Acanthothoraciformes (Palaeacanthaspidoidei).** These fishes are unusual in that the head armor may be represented within a species by large plates or by smaller plates resembling scales. There is no neck joint. These are small depressiform fishes from the Lower Devonian.

†**Order Ptyctodontiformes.** This order contains small placoderms usually less than 200 mm long. Head and body armor is not as extensive or as heavy as in the arthrodires; there is armor only on the anterior part of the body and none on the snout. A plate on the cheek apparently covers the gill opening. In body form and in several other characteristics, including the tooth plates, the ptyctodontiformes resemble holocephalans. Pectoral fins are large, as are the pelvics. Claspers tipped with bony plates are associated with the latter. As in holocephalans, prepelvic claspers are present.

†**Order Petalichthyiformes.** These are characterized by numerous head plates; scales on the body, on the pectoral fins, and in the snout region; and large lateral spines in front of the pectorals. The eyes are dorsal but anterior. The caudal fin is thought to be diphycercal.

†**Order Phyllolepiformes.** This order is represented by one depressiform genus with a reduced number of armor plates. The nuchal plate on the head and the median dor-

sal body plate are enlarged. The neck joint does not appear to have been movable. The snout region seems to be unarmored.

†**Order Arthrodiriformes.** This order, also referred to as Coccostei, contains most of the known placoderms. These are characterized by a heavily armored head and fore-body, and some have impressive, tusklike gnathal plates that form biting surfaces. The gills open between the head and body armor, and the slit usually is covered by a bony plate. A few genera contain species that reached more than 6 m, but most are considerably smaller.

Class Chondrichthyes (Elasmobranchimorphi)

Subclass Holocephali

These are elasmobranchiomorph fishes with the palatoquadrate completely fused to the cranium (holostylic jaw suspension). The hyomandibular is little modified and plays no part in suspension of the jaws. The skeleton is cartilaginous and the noto-chord persistent. Branchial arches are all placed below the neurocranium, and the gill openings are covered by fleshy opercula. Except in a few fossil lines, the teeth are grinding plates with no enamel. In living forms there is no spiracle and no cloaca, and the oviducts open to the exterior separately. There is usually a strong spine at the leading edge of the first dorsal fin, supported by a synarcual plate formed from neural arch elements.

Holocephali are known from the Upper Devonian to the present. Extinct forms known as bradyodonts (so called because the teeth are thought to have slow growth and replacement) have been treated variously by modern authors, depending on evaluation of certain aspects of structure; some are placed closer to sharks than to chimaeras. There are several extinct taxa that resemble holocephalans but are placed variously in and out of the group by paleoichthyologists. These fishes will be treated briefly following the coverage of the generally accepted Holocephali. Holocepha-lans show many remarkable resemblances to the ptyctodontid placoderms and might be related to them.

The ranking of groups with a classification scheme differs among students of the subclass. For instance, Patterson (1965) includes many of the extinct groups as suborders under Chimaeriformes, whereas Lund (1986) considers most of those groups as orders. Because the extinct holocephalans are extremely diverse, we will follow the latter plan to emphasize the diversity, using the classification of Nelson (1994). New forms are still being described and evaluated, and there seems to be no consensus on their relationships.

†*Superorder Paraselachimorpha*

Another group that resembles holocephalans and has at times been aligned with them is the iniopterygians, which may be more closely related to sharks.

†**Order Iniopterygiformes (Iniopterygia).** This order contains Paleozoic (Carbonifer-ous) fishes with the pectoral fin attached to the nuchal region (Fig. 5–1**B**) and is characterized by prominent spines often armed with hooks. Unlike modern

chimaeras, the teeth are rather sharklike; the dentition consists of denticles arranged in rows. This is considered a primitive characteristic for the subclass. Some paleo-ichthyologists believe that the iniopterygians belong in a separate group from the chimaeras and place them closer to, or with, the sharks.

†**Order Petalodontiformes (Petalodontida).** These are known from the Lower Carboniferous to the Upper Permian. Some authors have placed them with Holocephali. †*Janassa* is depressiform with a prominent rostrum and large, horizontally oriented, paired fins. †*Belantsea* (Fig. 5–2A) (Lund, 1989) is compressiform with a deep body and large head with high, rounded dorsals, the first of which is large and originates on the head above the orbits. The rounded pelvic fins are set near the caudal, there is no anal fin, and the caudal is structurally heterocercal but externally resembles a homocercal tail. Teeth are large and serrate. The gills are placed posterior to the neurocranium.

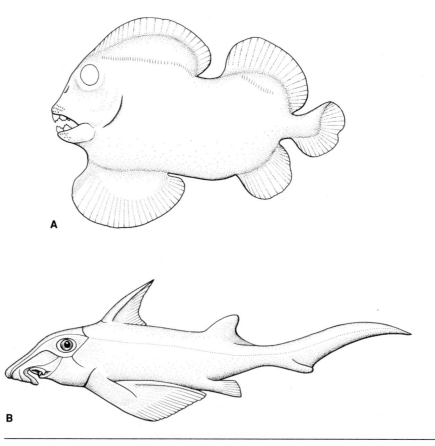

A

B

FIGURE 5–2   Representative of **A**, †Petalodontiformes, *Belantsea* (actual size ca. 29 cm). Representative of **B**, Chimaeriformes (*Callorhynchus*). (**A** after Lund, 1989.)

Additional orders of Paraselachimorpha

†Order Eugeneodontiformes   †Order Desmiodontiformes

†Order Orodontiformes   †Order Helodontiformes

Superorder Holocephalimorpha

These are modern holocephali. Nelson (1984) places living chimaeras into this super-order, a group distinguished by holostylic jaw suspension and grinding plates as teeth.

Order Chimaeriformes. This order contains all the modern chimaeras, which are known from the Lower Jurassic to Recent. These living chimaeroids are among the most bizarre of fishes, as indicated by their various common names. These names are based on anatomical features or general appearance of the species and include spookfish, ghost shark, chimaera, ratfish, rabbitfish, and elephantfish. Modern species are usually placed in three families, all marine. Species are usually between 60 cm and 2 m long.

Chimaeridae (ratfishes or shortnose chimaeras) contains about 20 species with a short snout, diphycercal or leptocercal tail, and a long second dorsal fin (Fig. 1–1). The first dorsal has a long, sharp, venomous spine at the leading edge. Males are equipped with a frontal clasper on the top of the head. It is somewhat finger-like, with a patch of denticles on the ventral aspect of the tip. In addition, males have a set of abdominal claspers or tenacula situated in pockets just in front of the pelvic fins, as well as pelvic claspers of bifid or trifid construction.

Members of this widespread family are found in mid-depths or in shallow water. A North Pacific species, the spotted ratfish, *Hydrolagus colliei,* is common in Puget Sound and is sometimes found intertidally. *H. colliei* is of little commercial significance, although fishermen at one time extracted a fine oil from the livers. *Hydrolagus* has about 15 species; *Chimaera* has at least six. Fossils of the latter genus are found in Cretaceous deposits.

Rhinochimaeridae share the sharp dorsal spine and diphycercal tail of the chimaerids but have a long, depressed, and pointed snout from which they take the name "longnose chimaeras." There are well-developed frontal claspers and reduced paired claspers. This family contains deep water forms of wide distribution with three genera, *Neoharriota, Harriota,* and *Rhinochimaera,* each with two species. Members of this family are known from the Cretaceous and Jurassic.

Callorhynchidae, the plownose chimaeras, have heterocercal tails, a dorsal spine, a short second dorsal fin, and a peculiar snout that turns back on itself ventrally, forming a flattened appendage just in front of the mouth (Fig. 5–2**B**). Frontal claspers are present. There is one genus, *Callorhynchus,* with about four species in shallow to moderately deep waters around the Southern Hemisphere ranging north to Argentina, Peru, Australia, and South Africa. Fossils of representatives of this family are known from the Upper Cretaceous.

The habitat and comparative rarity of most chimaeroids preclude detailed studies of their habits and life histories. Egg cases of the deep sea species are sometimes obtained, so some details of the embryology are known. The shallow water *Hydrolagus colliei* is somewhat better known. This species appears to have wide tastes in food and eats fishes, crustaceans, and molluscs. Young emerge from the spindle-like egg cases in the fall at about 140 mm and appear to grow to around 300 mm in the first year; much of this length is made up of the tail and caudal filament. Sexual maturity seems to be reached in the fourth year of life.

Extinct Holocephalan Groups

†**Order Cochliodontiformes.** This order is known from the Upper Devonian to the Permian and has occipital and mandibular spines and characteristics of dentition that set it apart. Lund (1986) includes the myriacanthoids and menaspoids as part of this order. The latter are characterized by large mandibular and frontal spines and large, platelike scales.

†**Order Squalorajiformes.** These fishes from the Jurassic are depressiform, with a long snout and a long frontal clasper in the males. Abdominal claspers are absent.

†**Order Chondrenchelyiformes.** This order, from the Lower Carboniferous, is part of a group of taxa that Nelson (1984) set apart in the superorder Paraselachimorpha under Holocephali. The superorder was proposed by Lund (1977) for holocephalans that show sharklike characters of dentition and do not all have the holostylic jaw condition. The bradyodont chondrenchelyiforms have a chimaeroid-like skull but an elongate body and long dorsal fin resembling the extinct sharks of the family Xenacanthidae.

†**Order Helodontiformes.** Known from the Upper Carboniferous, this is another somewhat sharklike holocephalan that Nelson (1984) considered to be a paraselachian. It has a heterocercal caudal fin and is a bradyodont group. Other holocephalan-like groups known mainly from their teeth are †Order Copodontiformes and †Order Psammodontiformes. Lund (1984, 1986) and Carroll (1988) include the extinct genus †*Echinochimaera* from the Upper Mississippian in the Chimaeriformes. This is a strange fish with occipital spines, a high first dorsal fin with a strong spine, and denticles (some large and spinous) on the body.

Subclass Elasmobranchii

Elasmobranchs are the modern sharks and rays and their fossil relatives. These fishes have cartilaginous endoskeletons, but the cartilage in many species is calcified. These calcifications may appear superficially on the endocranium in prismatic or granular form or within the endoskeletal cartilage, as in the vertebrae of many species. Most forms have dermal denticles in the form of placoid scales in the skin, but some have no scales. There is no operculum; there are five to seven separate gill openings on each side. Males have pelvic claspers. Fins are stiffened by horny rays called ceratotrichia. Jaw suspension is hyostylic or amphistylic. The branchial skeleton is posterior to the neurocranium. There is no gas bladder. A cloaca is present.

Elasmobranchs arose in the Upper Devonian and are a diverse group that has been classified in several different ways. Schaeffer (1967) pointed out that there are three general branches evident in the evolution of the elasmobranchs: a primitive cladodontoid branch (extinct); an intermediate and related hybodontoid group (extinct); and a modern branch with a few living forms occupying transitional positions between the hybodonts and the modern elasmobranchs. Maisey (1984) thought the living elasmobranchs were a closely related group and examined their possible monophyly.

Cladodonts are characterized by such features as teeth bearing an enlarged central cusp flanked by smaller cusps of the same conical shape (Figure 5–3**A**), a persistent notochord, a short rostral section of the brain case, palatoquadrate articulation with the enlarged postorbital processes of the cranium, and long jaws reaching from the snout to behind the skull. Examples of these sharks are discussed later in this chapter under †Palaeoselachii (see page 99).

The teeth of hybodonts are variable. Some genera are similar to cladodonts in dentition and others have more flattened teeth suitable for grinding (Fig. 5–3**B**). Hybodonts share most of the characteristics mentioned for the cladodonts but differ in the structure of the pectoral fin skeleton, possession of an anal fin, and reduction of caudal fin radials. In some, the rostral portion of the skull is enlarged. Modern elasmobranchs have vertebral centra that replace the notochord and shorter jaws, which in some may be protrusible. The jaws have hyostylic or "amphistylic" suspension. The modern elasmobranchs have larger neural and hemal elements of the vertebrae than hybodonts. Schaeffer (1967) indicated that the living sharks, except for the transitional Chlamydoselachidae, Heterodontidae, and Hexanchidae, represent two main phyletic lines, the galeoids and the squaloids. The rays, or batoids, represent another line of modern elasmobranchs.

As more is learned about paleozoic sharks and rays because of new fossil discoveries and subsequent research, there will be further alterations in our views of relationships among elasmobranchs. The recent discovery of several hitherto unknown sharks in Lower Carboniferous deposits of Montana should add to our understanding (Lund, 1990, 1991). Some investigators consider that paleozoic sharks represent a single premodern level of evolutionary organization, although there may be several types or designs (see Zangerl, 1973).

Nelson (1994) arranges the sharks into the following superorders (Table 5–1): (1) †Cladoselachimorpha, including the cladodonts; (2) †Xenacanthimorpha, including

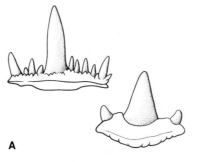
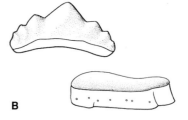

A

B

FIGURE 5–3   **A,** Cladodont teeth; **B,** hybodont teeth.

TABLE 5–1	Classification of Sharks and Rays

Subclass Elasmobranchii
    †Superorder Cladoselachimorpha
    †Superorder Xenacanthimorphs
    Superorder Euselachii. modern sharks and rays
        †Order Ctenacathiformes
        †Order Hybodontiformes
        Order Orectolobiformes. carpet sharks
        Order Carcharhiniformes. ground sharks
        Order Lamniformes. mackerel sharks
        Order Hexanchiformes. frill and cow sharks
        Order Squaliformes. dogfish, bramble and sleeper sharks
        Order Squatiniformes. angel sharks
        Order Pristiophoriformes. saw sharks
        Order Rajiformes. rays

After Nelson (1994).

TABLE 5–2	Classification of Sharks and Rays (Extinct Orders Not Listed)

Subclass Elasmobranchii
    †Cohort Palaeoselachii. 8 orders of extinct Paleozoic sharks
    Cohort Euselachii
        †Subcohort Protoselachii. 3 orders of extinct sharks
        Subcohort Neoselachii
            †Superorder Palaeospinacomorphii. extinct neoselachians
            Superorder Squalomorphii
                Order Hexanchiformes. frill sharks, cow sharks
                Order Squaliformes. dogfish sharks, bramble sharks, etc.
                Order Pristiophoriformes. saw sharks
            Superorder Squatinomorpha
                Order Squatiniformes. angel sharks
            Superorder Rajomorphii (Batoidea)
                Order Pristiformes. sawfishes
                Order Rhinobatiformes. guitar fishes
                Order Torpediniformes. electric rays
                Order Rajiformes. skates
                Order Myliobatiformes. stingrays, eagle rays, mantas
            Superorder Galeomorphii
                Order Heterodontiformes. bullhead sharks
                Order Orectolobiformes. carpet sharks, whale sharks
                Order Lamniformes. mackerel sharks
                Order Carcharhiniformes. requiem sharks

After Compagno (1991).

the family †Xenacanthidae; and (3) Euselachii, comprising the fossil †Ctenacanthiformes and †Hybodontiformes and seven orders of living sharks and rays. Unlike the classification of sharks and rays by Compagno (1977, 1991), Nelson (1994) does not separate those two groups into separate superorders. Other views of shark phylogeny and classification can be seen in Maisey (1984, 1986), Seret (1986), Cappetta (1987), and Carroll (1988).

Compagno (1973, 1977, 1991) places the sharks and rays into the subclass Elasmobranchii along with their Paleozoic relatives. Paleozoic sharks are placed in the cohort †Palaeoselachii, which Compagno (1991) sees as comprising eight orders. Modern sharks and rays and their extinct near relatives (xenacanths, ctenacanths, and hybodonts) are included in the cohort Euselachii. Compagno does not follow the practice of splitting the living Euselachii into sharks and rays as lines of equal rank. Instead he recognizes four evolutionary lines, presented as three superorders of sharks and one superorder of rays (Table 5–2). In the following treatment, Compagno's 1991 arrangement will be followed.

Cohort †*Palaeoselachii*

This group contains a number of forms that not only show primitive features, as compared with the modern level of selachians, but show some remarkable adaptations as well. The order †Cladoselachiformes (Fig. 5–4A) contains Devonian species with broad-based paired fins, heavy dorsal fin spines, and branched teeth (cladodont). The heterocercal caudal fin is lunate. The order †Symmoriiformes, from the Carboniferous, has cladodont teeth and lacks fin spines of the usual type. One

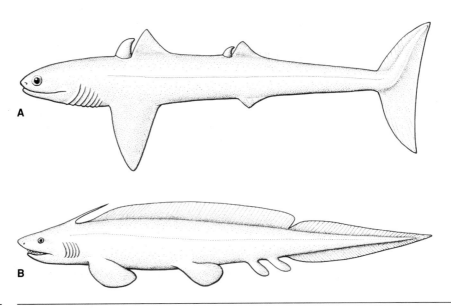

FIGURE 5–4   Representative of **A,** †Cladoselachiformes (actual size ca. 50 cm). Representative of **B,** †Xenacanthiformes (actual size ca. 75 cm). (**A** after Romer, 1970; **B** after Schaeffer and Williams, 1977.)

family shows strange sexual dimorphism in that the males of one species have an odd, spinelike structure above the pectoral girdle. The appendage curves forward to form a flattened denticle-covered blade over the top of the head, which is covered by similar spinous denticles. Zangerl (1984) stated that the complex could have served to mimic a large mouth.

Cohort Euselachii

Under this group, Compagno (1991) lists two subcohorts. One, the †Protoselachii, includes extinct groups that are close to the modern level. Orders included are †Ctenacanthiformes, †Xenacanthiformes (Fig. 5–4**B**), and †Hybodontiformes. The ctenacanths are known from the Devonian into the Triassic and, although they have broad-based paired fins, have developed mesopterygia and propterygia in the pectoral fins. The xenacanthids existed from the Lower Devonian to the Upper Triassic, mostly in freshwater habitats. They have a long archipterygial axis supporting the pectoral fins, which are broad based in some species. †*Xenacanthus* of the Permian was rather eel shaped with a diphycercal tail and a prominent occipital spine. The hybodonts appear to be closely related to the ctenacanths but have some modern features. The pectoral fins are supported by propterygia, mesopterygia, and metapterygia (Fig. 2–26**A**), and the fin radials, as in modern sharks, do not extend to the edges of the fins but bear ceratotrichia. Hybodonts have a long history, extending from the Upper Devonian to the Tertiary.

Subcohort Neoselachii

This comprises the modern sharks and rays as well as closely related extinct sharks of the superorder †Palaeospinacomorphii, which existed from the Lower Jurassic to the Early Tertiary. These sharks had long jaws and essentially cladodont teeth.

Superorder Squalomorphii

This superorder is based on similarities of cranial and pectoral anatomy and includes three orders of diverse appearance and dentition, one of which, the Hexanchiformes, usually has not been grouped with the others. All three orders include species with more than the usual five gill openings. Specializations include barbels and a tooth-studded rostrum in the saw shark, "cookie-cutting" dentition in *Isistius,* luminosity in that genus and in *Etmopterus,* small body size in some genera, and large body size in the Greenland shark (see also Shirai, 1992).

Order Hexanchiformes. These sharks have six or seven gill arches and slits. There is an anal fin and a single dorsal without a spine. The suborder Chlamydoselachoidea has one family, Chlamydoselachidae, and one species, *Chlamydoselachus anguineus,* the frill shark (Fig. 5–5**A**). This almost eel-shaped shark takes its name from the frilly extensions of the interbranchial septa, which overlap succeeding gill slits. Its notochord is unconstricted over most of the length of the trunk. The mouth is nearly terminal and is set with teeth that have a broad base bearing three sharp cusps. These characteristics are similar to those of the extinct "cladodont" sharks and suggest that the frill shark might represent a transitional form between primitive and modern sharks. This is a deep water species known from several localities and reaches almost 2 m long.

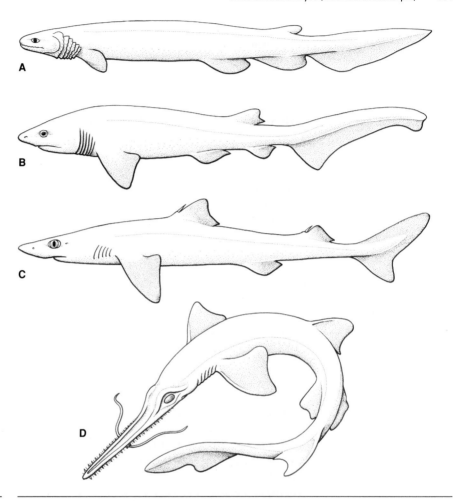

FIGURE 5–5   Representatives of superorder Sqaulomorphii. **A,** Frill shark (*Chlamydoselachus*); **B,** sixgill shark (*Hexanchus*); **C,** spiny dogfish (*Squalus*); **D,** saw shark (*Pliotrema*).

Suborder Hexanchoidei (Fig. 5–5**B**) contains the family Hexanchidae (the sixgill sharks of *Hexanchus* and the broad-headed sevengill shark *Notorhynchus*) and the family Heptranchidae (the "point-headed" sevengill sharks). Their mouth is subterminal and the teeth are mainly multicuspid, although more than one kind of tooth (heterodont condition) can be found in all species. The sixgill shark, *Hexanchus griseus,* reaches nearly 8 m and is widespread in temperate seas. Its food is usually herrings and other small fishes. It has been used in the manufacture of oil and meal.

Order Squaliformes. These sharks have five or six gill openings; two dorsal fins, often with a spine; and usually no anal fin. Families included are Echinorhinidae, the bramble sharks, and Squalidae, the dogfish sharks (Fig. 5–5**C**) and allies. Squalidae is usually considered to consist of several subfamilies, some of which, such as Dalatiinae and Somniosinae, are occasionally regarded as separate families. The dogfish shark, Squalus acanthias, has been found to contain a "broad-spectrum steroidal antibiotic" called squalamine (Moore et al., 1993). This compound is bactericidal, fungicidal, and damages protozoa. It might prove useful in the field of medicine.

The bramble shark or alligator dogfish, *Echinorhinus brucus,* is a robust shark that reaches 3 m. It is known mainly from warm seas, where it has been taken from depths as great as 900 meters. The smallest shark known is *Squaliolus laticaudus,* from the eastern Pacific near Japan and the Philippines. This midget is known to reach a length of only 15 cm. A closely related species from the Atlantic has been measured at 22 cm. *Squalus acanthias,* the spiny dogfish, is probably the best known of the family. It is widespread in temperate seas and is familiar to millions of comparative anatomy students. This abundant fish reaches about 2 m long. Its flesh is edible, but its commercial value is not as high now as formerly, when it was sought for the vitamin-rich oils of the liver. Development of synthetic vitamins caused the decline of the dogfish fishery. The genus *Somniosus* contains fishes known as sleeper and Greenland sharks that reach over 7 m. These are sluggish, cold water animals that act both as predators and scavengers but feed primarily on fishes.

The small, luminous sharks of the genus *Isistius,* including the Gulf dogfish, *I. plutodus,* show a remarkable specialization for cutting round plugs of flesh out of larger organisms. They have enlarged lower teeth and lips that aid in maintaining suction. They attack moving prey head on and allow the momentum of the larger animal to swivel them around, thus aiding in detaching the round piece of flesh. Because of this habit, they are called cookie-cutter sharks.

Order Pristiophoriformes. There is but one family in this order—Pristiophoridae, the saw sharks (Fig 5–5**D**). The two genera, *Pristiophorus* (with five gill openings) and *Pliotrema* (with six), have the rostrum extended into a long, flat blade armed on each edge with teeth. There are two large barbels on the undersurface of the rostrum. These sharks are mainly found in the warm Indo-Pacific, but a rare species is known from the Bahamas. Fossils of the family are known from the Cretaceous.

†**Order Protospinaciformes.** This is based on †*Protospinax* from the Jurassic, a kind of "shark-ray" which has some characteristics of both sharks and rays (see Cappetta, 1987).

Superorder Squatinomorphii

This group contains one genus of depressiform fishes, with pectorals expanded forward but not fused with the head. The gill openings are mainly lateral and the spiracles are large, as in most batoids. These fishes have two spineless dorsal fins set on the tail, no anal, and an essentially hypocercal caudal. They are the only living fishes to retain this ancient tail-fin design.

Order Squatiniformes. This order contains the single family Squatinidae and the genus *Squatina,* the monkfishes or angel sharks. These fishes are tropical to temperate in distribution and are usually found in shallow water. Despite their raylike appearance, their locomotion is sharklike and is accomplished by movements of the tail. The largest species reaches about 2.4 m and a weight of 72 kg.

Superorder Rajomorphii (Batoidea)

The skates and rays of the superorder Rajomorphii (Fig. 5–6) are recognized by a depressiform body, with the pectoral fins extending forward and fusing to the head so the five pairs of gill openings are ventral. There are, in addition, several skeletal characteristics that distinguish them from sharks. Some of these (see Maisey, 1984) are as follows. The suprascapulae are joined to each other over the vertebral column and either articulate with the column or fuse to a synarcual formed from the fusion of the anterior vertebrae. The posterior hypobranchial is in contact with the shoulder girdle, either articulating with or fusing to it. The palatoquadrate does not articulate with the neurocranium.

The batoids are benthic predators except for the mantas, which feed on plankton and small fishes in the open water. Feeding specializations in the group include the cephalic fins of the mantas, the crushing and grinding teeth of the eagle rays and others, and the great rostrum of the sawfish, with strong teeth set into sockets. Other specializations of note are the electric capabilities of the torpedoes, the stinging spines of stingrays, and the physiological modification of two families of stingrays that allows invasion of fresh water. The transition of myliobatiforms from the benthic to the pelagic habitat is a notable evolutionary trend.

Order Pristiformes. The sawfishes, Pristidae, make up this order. These are covergent with the saw sharks in general appearance and have the rostrum formed into a

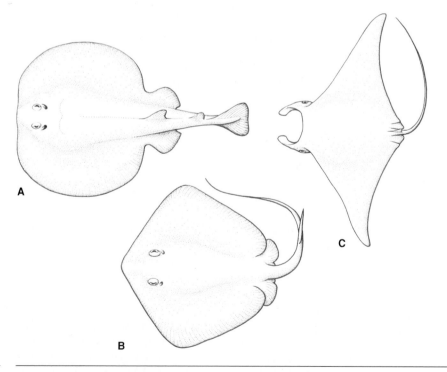

FIGURE 5–6   Representatives of superorder Rajomorphii. **A,** Electric ray (*Torpedo*); **B,** stingray (*Dasyatis*); **C,** manta (*Manta*).

long, flat blade armed with teeth set in sockets. These are shallow-water fishes of warm seas, bays, and tropical rivers. There appears to be a resident population in Lake Nicaragua. Some species are said to reach nearly 11 m long and a weight of about 2400 kg.

The saw is used in feeding; while the sawfish moves through a school of fishes, lateral movements of this weapon kill and injure individuals, which can be subsequently eaten. Fish impaled on the teeth of the saw are scraped off on the bottom and eaten. There are reports of serious injury and death resulting from bathers being in the way of startled sawfish in the Ganges. Sawfishes are ovoviviparous, but the rostrum and its teeth are soft until after birth.

Order Rhinobatiformes. This order contains four families of rather sharklike fishes—Rhinidae, Rhynchobatidae, Rhinobatidae, and Platyrhinidae—all generally referred to as guitarfishes, a name apparently based on the appearance given by the flattened head, pectoral fins, and snout followed by the tapering body and tail. These are fishes of shallow tropical and subtropical waters. They live on the bottom, feeding mainly on a variety of invertebrates. Most species reach about 1 m long. Commercial value is low, although they are edible.

Order Torpediniformes. This order (Fig. 5–6A) consists of four families of electric rays—Torpedinidae, Hypnidae, Narcinidae, and Narkidae. These have generally rounded discs, with the propterygia of the pectoral fins contacting the cranium anterior to the eyes. All have large electric organs in the disc on each side of the head. This organ is composed of columns of modified muscle tissue and allows the fish to deliver powerful shocks that can stun prey or possibly discourage predators, although these fish are known to be eaten by sharks. Up to 200 volts have been recorded from large specimens. The shock may be powerful enough to render humans unconscious, as evidenced by the word roots of the family names, which all refer to sleep or numbness. Mediterranean species were used by the ancients as a form of electrotherapy for ailments such as arthritis and gout. Electric rays are found in tropical and temperate waters over a considerable depth range. Some of the deep water forms, such as *Typhlonarke,* are blind. The largest species is thought to be *Torpedo nobiliana,* an Atlantic species that reaches 1.8 m in length.

Order Rajiformes. The order of skates contains batoids that lack some specialized features of the ensuing orders. They have no stinging spines on the tail, no sawlike rostrum, and no large electric organs between the pectoral fin and the skull, although small electric organs are present in the tail of skates.

The suborder Rajoidei contains fishes with the head, body, and pectoral fins combined into a flat disc, with a slender tail bearing very small dorsal and caudal fins. Most species are called skates. The four families included are Arhynchobatidae, Rajidae, Pseudorajidae, and Anacanthobatidae, although these are regarded by some authors to constitute the single family Rajidae. Most familiar are the skates of Rajidae, a nearly cosmopolitan marine group found from estuaries to the depths in warm and cold seas alike. Species range up to more than 2 m long, but most do not grow more than 75 cm long. Unlike most of the other rays, the skates are oviparous. Their eggs are enclosed in horny cases, which are called "mermaid purses" in some localities. Some of the larger species are used commercially in Europe. The thickest parts of the pectoral fins provide white, palatable flesh.

Order Myliobatiformes. Rays of this order have large pectoral fins that combine with the head to form a broad disc, with a slender tail that usually has strong stinging spines. The caudal and dorsal fins are reduced or absent. Eight families are included —Dasyatidae, the stingrays (Fig. 5–6**B**); Potamotrygonidae, river stingrays; Urolophidae, round stingrays; Gymnuridae, butterfly rays; Hexatrygonidae, sixgill rays; Myliobatidae, eagle and bat rays; Rhinopteridae, cownose rays (sometimes considered to be part of the preceding family); and Mobulidae, mantas or devil rays.

These are all warm water fishes that seldom enter cold waters and are usually found close to shore. Potamotrygonidae is found in rivers of South America. Some dasyatids of the genera *Dasyatis* and *Himantura* appear to be permanent residents of fresh water in Africa, Asia, and New Guinea. The various stingrays and the butterfly rays live on the bottom, often concealing themselves in sand or other fine materials. Their food is shellfishes and bottom-living fishes. The tail spines are typically barbed and grooved along the edges. The venom produced in the groove can make a wound caused by the spine to be both painful and dangerous. The largest stingrays may reach a width of 2 m.

The eagle rays and bat rays (*Aetobatis* and *Myliobatis*) and cownose rays (*Rhinoptera*) feed on the bottom, often dislodging bottom materials through the hydraulic action of powerful movements of their large pectorals. Clams, oysters, and other invertebrates make up most of their food. The teeth of these rays are in the form of broad grinding plates. Locomotion is by "flying" movements of the winglike pectorals. The long, whiplike tail is usually held straight behind. Some species reach 1.2 m in width.

Mantas or devil rays (*Manta, Mobula*) (Fig. 5–6**C**) have adapted to feeding on plankton and small schooling fishes. They swim through the water by means of the wide, slender-tipped pectoral "wings," holding the mouth open. The peculiar cephalic fins, positioned on either side of the mouth, are used to guide food into the mouth. These fins, when curled into the spiral resting position, give the impression of horns (hence the name devil rays). Although some species reach less than 1 m in width, others may reach several meters. A specimen of *Manta birostris* was measured at 6.6 m and is thought to have weighed over 1600 kg. Many of the mobulids, rhinopterids, myliobatids, and some dasyatids have a habit of leaping clear of the water and landing with a loud noise. Cartwheeling is another interesting behavior of these rays. Cownose rays are occasionally seen in more or less regularly oriented schools of up to 6000 individuals.

Superorder Galeomorphii

Four orders of sharks are brought together under this name because of similarities in the cranial skeleton and in the structure of the pectoral fins. All have an anal fin and five gill openings. Most have dorsal fins, but only the Heterodontiformes have dorsal fin spines. Most of the familiar genera of sharks belong to this superorder. There are only three known large sharks that are adapted to take plankton and small nekton: the whale shark, the basking shark, and the "megamouth" (*Megachasma pelagios*) from the Pacific. These three have numerous small teeth in as many as 70 rows

(Maisey, 1985) and are specialized for straining plankton, although the whale shark has been observed to take advantage of dense schools of small fishes. All others have dental arrangements for grasping or cutting prey. The members of this superorder range from camouflaged, lurking benthic predators to the swiftest and most rapacious of pelagic sharks. Some of the specializations for pelagic life include buoyancy control, acute acousticolateralis sense, and control of internal temperature. In addition, these sharks have remarkable abilities in detecting electrical stimuli.

Order Heterodontiformes. Only one genus, *Heterodontus* (Fig. 5–7**A**), of the family Heterodontidae is included. Species are referred to as horn sharks because of the strong spines at the front of each dorsal fin. The generic name alludes to the condition of the teeth in the short and modified mouth. Anterior teeth are small and sharp, whereas those in the back of the jaws are molariform. The horn sharks are found in the Indo-Pacific in tropical to warm temperate waters. They may reach 1.5 m in length.

Heterodontus has often been considered to be closely related to such primitive sharks as hybodonts and ctenacanthids. Some evidence for this involves the suspension of the upper jaw, which might be said to be structurally hyostylic but functionally amphistylic, according to Schaeffer (1967). Heterodontid teeth are known from the Jurassic.

Order Orectolobiformes. The sharks in this group are usually placed into two families: the Orectolobidae, containing the carpet sharks, nurse sharks, zebra sharks, and

FIGURE 5–7 Representatives of superorder Galeomorphii. **A,** Horn shark (*Heterodontus*); **B,** whale shark (*Rhincodon*).

wobbegongs; and the Rhincodontidae, or whale sharks (Fig. 5–7**B**). In Compagno's arrangement (1991), the Orectolobidae is restricted to a few genera and the remaining fishes are placed in the following families: Parascyllidae, Brachaeluridae, Hemiscyllidae, Stegostomatidae, and Ginglymostomatidae. These sharks are found mainly in the tropical parts of the Indo-Pacific, and most of the genera are present in Australian waters. One genus, *Gingylmostoma,* occurs in the Atlantic. For the most part, these are small sharks that reach less than a meter long, but the Atlantic nurse shark has a maximum length of about 4.2 m. Others, such as the zebra shark, *Stegostoma,* and some species of *Orectolobus,* may exceed 3 m long. The latter genus contains the strikingly marked wobbegongs of Australia. Some of the carpet sharks in the western Pacific are known as cat sharks, a name usually reserved for fishes of another order and family.

Rhincodon (= *Rhiniodon*) *typus,* the whale shark, is the largest living fish and attains a length of 15 m. One specimen of 11.5 m was estimated to weigh about 12,000 kg. This sluggish giant is present in all tropical seas. Its mouth is terminal and broad. It feeds on great quantities of small schooling fishes, such as herring, and on squid and planktonic crustacea. The mouth is equipped with numerous rows of small teeth, but most food items appear to be captured by straining through the fine gill rakers. The gill slits are very long and set rather high on the sides, partially above the pectoral fin. The body has a humpbacked appearance, and the caudal fin is large. The color pattern is striking and consists of yellowish or white spots on a gray to brown background. These huge animals are oviparous; the eggs are oval and surprisingly small (300×90 mm) considering the size of the sharks.

Order Lamniformes. Usually these are large and active sharks of shallow waters, with a few exceptions. Families included are Odontaspididae (= Carchariidae), the sand tigers, which are distributed in temperate to tropical waters and reach lengths of 3 m or more; Pseudocarchariidae; Mitsukurinidae (= Scapanorhynchidae), the goblin shark, a strange primitive shark with a long, flat rostrum; Alopiidae, the thresher sharks; Cetorhinidae, the basking shark and the megamouth (Maisey, 1985); and Lamnidae (= Isuridae), the mackerel sharks. The lamnids are among the best known sharks because of the large size and great appetite of some. One, the white shark, *Carcharodon carcharias,* has been implicated in many fatal attacks on human beings. This is a giant among sharks, but its maximum size is not reliably known. An Australian specimen was reported to have been slightly over 11 m long, and one taken in the Azores in 1981 is said to have been measured at 8.2 m. Weights for given lengths are somewhat variable. Specimens of 4 m taken in various places have been reported to weigh from 450 to nearly 1100 kg. One of nearly 7 m was weighed at 3300 kg. The white shark is found in tropical to temperate seas. In the eastern Pacific, it ranges as far north as 60°N to Petersburg, Alaska, and has attacked surfers in Oregon at temperatures around 13°C. It ranges at least to Newfoundland in the western North Atlantic. Lamnidae includes the mako sharks of the genus *Isurus* and the salmon and mackerel sharks of *Lamna.* The mako is often sought as a big game fish and is taken commercially in some areas; *L. nasus,* the porbeagle, has some importance as a commercial fish in Norway and Iceland.

The goblin shark, *Mitsukurina owstoni,* is a deep water fish that has been found in the Indian Ocean, the western Pacific off Japan, and the Atlantic near Portugal. It is characterized by a flat, elongate snout and protrusible jaws set with slender, sharp teeth. Some of these teeth were found in a malfunctioning communications cable brought up from 1300 m in the Indian Ocean. Whether the shark actually attacked the cable or was feeding on animals growing on it is an interesting point to ponder. The thresher sharks, *Alopias,* have the upper caudal lobe elongated so it comprises about one half of the total length of the fish. This great tail is used to herd the small schooling fishes on which the sharks prey (see Gubanov, 1972). When in shallow water, the tail slaps and splashes the surface. The several species of the genus are found in warm seas, and some are harvested commercially.

Unlike its swift and ferocious relatives, the gigantic basking shark (*Cetorhinus maxima*) is a slow-moving plankton feeder. It is close to the whale shark in reported maximum length, reaching 13.5 m. Its teeth, although numerous, are small and would appear to have limited function. The gill rakers, however, are long and slender and constitute an excellent sieve for the small crustaceans on which the shark feeds. The gill rakers are shed during the winter when plankton is scarce, and the shark passes the time in a type of hibernation. Basking sharks are distributed in cold and temperate seas and have often been the object of harpoon fisheries for their liver oil. *Megachasma pelagios,* the megamouth, was discovered off Hawaii when the first specimen entangled itself in a parachute-like sea anchor being dragged by a research vessel. The first specimen, about 4.25 m long, has a mouth about 1.2 m wide. Its teeth are small and the gill rakers are adapted to strain plankton, The fish was described and named in 1983 (Taylor et al., 1983).

Order Carcharhiniformes. This is a large assemblage that includes eight families and about 40 genera and encompasses many of the more familiar species of sharks. The families included are Scyliorhinidae, the cat sharks and spotted dogfishes; Proscyllidae, also called cat sharks; Pseudotriakidae, false cat sharks; Leptochariidae; Triakidae, the smooth hounds or smooth dogfishes; Hemigaleidae; Carcharhinidae, the requiem or "typical" sharks; and Sphyrnidae, the hammerhead sharks.

Cat sharks are small, often with striking color patterns, and are found in warm seas in many parts of the globe. Some are found at considerable depths and, like the brown shark of the eastern North Pacific, are of drab coloration. The swell sharks of the genus *Cephaloscyllium* are capable of swallowing air and inflating their stomachs when brought out of the water. The inflated specimens then float until they can deflate themselves, a task that appears easy for some but harder for others, taking hours or days to accomplish. Smooth dogfishes are widespread shallow water forms, some of which reach 2 m long. Familiar North American species are *Mustelus canis* of the Atlantic coast and *M. henlei* of the Pacific.

The requiem shark family (Carcharhinidae) contains several well-known, medium to large species from tropical and temperate waters. Blacktip and whitetip sharks, so named because of their fin coloration, are of the genus *Carcharhinus* and are found in warm seas. One member of the group, the bull shark of the western Atlantic, *C. leucas,* is found in the fresh waters of Lake Nicaragua, where it is known to make fatal attacks on bathers. The species is known to enter the Mississippi River. This or similar species of Africa and India also enter fresh water. Another of the family is the tiger shark, *Galeocerdo cuvieri.* This is a circumtropical species that reaches about 5.5 m and has some fame as a sport fish. The blue shark *Prionace glauca* (Fig.

1–1**C**) is found in most warm and temperate waters. It has a slender body, a remarkable blue coloration, and is an active feeder, often attacking hooked salmon to the dismay of the angler. The topes, *Galeorhinus* spp., are found in the Indo-Pacific and the eastern Atlantic. The soupfin shark, *G. zyopterus,* was once the target of a valuable fishery on the west coast of North America because of the vitamin content of its liver oil. Production of synthetic vitamin A lowered the price, so the fishery was abandoned.

The hammerhead sharks were also sought for the vitamin-rich liver oil. These medium to large sharks are characterized by flat, lateral expansions of the head, so from above or below the outline is that of the letter T. The eyes and nostrils are borne on the outer edge of the structure. These sharks are confined to warm waters, and the family is circumtropical. *Sphyrna mokarran* has been measured at 5.4 m and *S. zygaena* at over 4 m. They are reported to have hearty appetites and to feed on a variety of animals, including other hammerheads and the formidable stingrays. Several hypotheses have been advanced to explain the function of their expanded head. Some researchers have suggested that the flattened surface aids the swift animals in making tighter turns. Others have postulated that because the nostrils are out on the forward corners of the expansion, ranging in on sources of odors is facilitated.

Chapter

6 Teleostomi: Acanthodians and Sarcopterygians

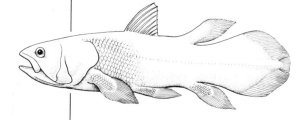

General Characteristics of the Teleostomi

Nelson (1994) chooses to abandon the use of the name Osteichthyes as a class and to adopt the use of Teleostomi as the grade embracing the remainder of the vertebrates, including the †Acanthodii, the Sarcopterygii (including the tetrapods), and the ray-finned fishes. Primitive members of this grade share certain characteristics with the Elasmobranchii and †Placodermi in that the caudal fin may be heterocercal and various groups may have spiracles, an intestinal spiral valve, valvular conus arteriosus, cloaca or the primitive placement of the anus between the bases of the pelvic fin. For the most part, however, the caudal fin is homocercal, the spiracles, conus arteriosus, and cloaca are absent, and the anus is variously placed, usually just anterior to the anal fin. The endoskeleton of bony fishes is typically ossified to some extent, with dermal bones in the head region. There is usually a gas bladder or lung, but that might be lost in secondarily specialized species. An operculum covers the gills, and the gill septa are reduced.

There are some resemblances between the †Acanthodii and primitive actinopterygians in the structure of the ventral part of the neurocranium and in the development of an operculum. Several opinions exist on the closeness of the evolutionary relationships between actinopterygians and acanthodians. Relationships of the †Acanthodii are uncertain. Widely divergent interpretations have been made of their affinities by paleoichthyologists, who considered them to be allied variously to placoderms, elasmobranchs, or bony fishes. Denison (1979, p. 19) stated that "it is necessary that these interpretations be made, but it is important that they be distinguished from facts." He provisionally accepts the interpretation by Miles (1973) that acanthodians are aligned with the Osteichthyes. Nelson (1994) places them with Teleostomi (see also Maisey, 1985; Carroll, 1988).

†Class Acanthodii

These primitive, bony fishes are elongate, with heterocercal tails, large eyes set far forward in the head, and prominent spines at the leading edges of all fins except the caudal. In some, there are spines in a row between the pectorals and pelvics. Gill clefts are covered by gill covers borne on the hyoid and branchial arches and by smaller plates subsidiary to those larger plates, but in advanced forms the hyoid gill cover is enlarged and appears to cover all the slits (Denison, 1979). The endocranium is ossified, and dermal bone is present on the head. Some show ossifications in the vertebral column. Nonimbricating small scales, with bony bases covered by dentine, are present. These fishes are known from the Upper Silurian to the Lower Permian and are the oldest known jawed vertebrates. Earliest remains are mainly from marine deposits, but by the beginning of the Devonian, the habitats included fresh water (Denison, 1979).

Acanthodians (Fig. 6–1**A**) were apparently fairly active swimmers and not primarily adapted to life on the bottom, as were the contemporaneous agnaths and placoderms. However, there is some evidence that species with long, heavy pectoral spines lived on the bottom (Denison, 1979). Whether they possessed gas bladders is not known (Carroll, 1988). Some had the teeth and gill rakers of predators; others lacked teeth and possessed long gill rakers that must have been effective in collecting small invertebrates. Fossilized contents of guts reveal that †*Acanthodes* fed on

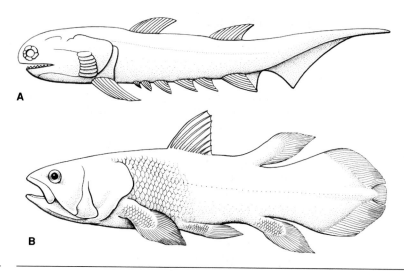

FIGURE 6–1   Representative of **A**, †Acanthodii (actual size ca. 15 cm). Representative of **B**, order Coelacanthiformes (*Latimeria chalumnae*). (**A** after Watson, 1937.)

small invertebrates but took fishes occasionally. Fish remains have also been found
in climatiids (Denison, 1979). The teeth are unlike those of both the elasmobranchs
and the other teleostomes, differing in microscopic structure and in apparent lack of
method of replacement. Tooth-bearing whorls apparently added new teeth toward
the medial aspect of the whorl, whereas new teeth were added to the front of the jaw
as it elongated with growth. Some species had nonreplaceable, isolated teeth on the
mouth lining (see Carroll, 1988).

Acanthodians are sometimes divided into as many as seven orders, but Moy-
Thomas and Miles (1971) and Nelson (1994) have recognized three: †Climati-
iformes, †Ischnacanthiformes, and †Acanthodiformes. Most of these were small
fishes, but some exceptional fossils are 2 m long.

Class Sarcopterygii: Lobe-Finned Fishes

Although Teleostomi includes the Tetrapoda within the class Sarcopterygii, the treatment herein will concentrate on the fishes. The lobe-finned fishes, which presumably
gave rise to the tetrapods, comprise the coelacanths (actinistians), lungfishes, the extinct porolepiforms, the extinct rhizodontiforms, and the extinct osteolepiforms
(rhipidistians). These fishes are characterized by fleshy lobate fins, usually with a
characteristic flexible skeleton. In addition, they have an epichordal lobe in the
caudal fin, which is variously shaped to be heterocercal, heterodiphycercal, or diphycercal. The latter two shapes often show an axial lobe, as in the living coelacanth,
Latimeria. Many ancestral lobe-fins had two dorsal fins, as does *Latimeria,* but the
living lungfishes do not. Cosmine is typical of the exoskeletons of many of the extinct lobe-fins, and the cosmoid scales and dermal bones of the fossil forms are characterized by pores that lead to a canal system, which researchers believe may have
been electrosensitive.

Subclass Coelacanthimorpha

The living coelacanth and its extinct relatives differ from lungfishes in having the palatoquadrate separate from the neurocranium and having the skull divided into anterior (ethmoid) and posterior (otico-occipital) sections at a joint between the frontals and the parietals, so a subcephalic muscle connecting the two parts can pull the anterior part ventrally in a biting motion (Carroll, 1988). Jaw suspension is hyostylic in most, but the hyomandibular is reduced in some representatives. The paired fins are paddle-like; some have a median axis with a proximal bone articulating by means of a ball-and-socket-like joint. A cloaca is absent. Choanae are present in some of the groups, and cosmoid scales are present in the extinct osteolepiforms and porolepiforms.

Order Coelacanthiformes. This order includes four extinct families that existed from the Upper Devonian to the Upper Cretaceous as well as the family Coelacanthidae, with several extinct genera and the only living member of the subclass, which is *Latimeria chalumnae* (Fig. 6–1**B**). Until a specimen was captured off southeast Africa in 1938, the order was thought to have become extinct in the Cretaceous. The South African ichthyologist Dr. J. L. B. Smith described the Recent coelacanth from the museum mount to which it had been converted and initiated a search for further specimens. Not until 1952 did an additional specimen come to his attention, from the Comoro Islands northwest of Madagascar. Since 1952, several additional specimens have been obtained and observed from submersibles in the Comoro region, mostly at depths of from 150 to 400 m. The discovery and subsequent study of *Latimeria* were of great significance to paleontologists, who had a rare opportunity to check their interpretation of fossil coelacanths against a living species. These interpretations involved type of reproduction, use of various fins in locomotion, structure of scales and gas bladder, and feeding habits.

Other than the extant *Latimeria,* the coelacanths are known as a comparatively homogeneous group from the Middle Devonian to the Upper Cretaceous. The group had achieved worldwide distribution by the Late Mesozoic. The fossils are found in both marine and freshwater deposits. There are over 120 species in 47 genera, and the fossils are found over a time span of 380 million years (Cloutier and Forey, 1991). They are thought by some to be derived from porolepiform stock (Andrews, 1973), but there is no consensus on that. There is evidence that the intracranial joint of the coelacanths is not homologous with that of the rhipidistians (Bjerring, 1973; Thomson, 1967).

These fishes usually have two external nostrils per side, but no choanae. Cosmine is lacking. There are two dorsal fins; the anterior one is placed in the front half of the body in contrast to the usual situation in rhipidistians. The caudal fin is diphycercal in all except one genus and typically has an axial lobe. The gas bladder or lung is calcified in extinct species.

Latimeria is large, typically about 2 m long, with reported lengths of more than 2.5 m. Weights of around 80 kg have been reported. The smallest free-living specimen captured was 42 cm long, barely longer than the estimated size at birth. Specimens of around 1.8 m have been estimated to be 11 years old. Life colors are brownish to bluish, with lighter spots because of unpigmented areas on scales. The scales are modified from the cosmoid type and consist of a dense isopedine plate and an outer mineralized layer to which denticles are fused. These denticles have a pulp cavity and dentine plus an outer layer that closely resembles enamel. There are reports of the Comoro Islands people using the scales like sandpaper to roughen inner tubes that are being repaired.

The pectoral, pelvic, second dorsal, and anal fins are all lobate and flexible. The pectoral is capable of 180-degree rotation. The larger cartilaginous fin supports, or "spines," are hollow, as in the fossil relatives (hence the name coelacanth, or hollow spine). Unlike extinct species, the gas bladder of *Latimeria* is not ossified; instead, it is filled with fat—a modification for buoyancy control. It retains a closed connection with the ventral part of the pharynx. Because of the prominence of the gas bladder, the kidney has been displaced to a ventral and posterior position.

The spiracular canal is closed, and internal nares are lacking. There is a large "rostral organ" of possible electroreceptive function beneath the skin of the snout (Bemis and Heatherington, 1982). Pleural ribs are lacking. The notochord is essentially a tube filled with fluid with the thin layer of notochordal cells surrounded by heavier connective tissue. The intercranial joint allows the anterior part of the cranium to be flexed up and down during feeding or breathing.

Although the reproductive biology is not known fully, the species is ovoviviparous even though the males do not have obvious intromittent organs. The 8- to 9-cm-diameter eggs are said to be the largest shell-less eggs among all the fishes; large females have been found to carry up to 20 (Thomson, 1991). The gestation period appears to be about 13 months, and the young are born at a length of about 32 cm.

Latimeria is a lurking predator that lives on steep rock slopes. It appears to be slow moving and spends much time hovering and drifting instead of pursuing prey rapidly. Locomotion of the species has been studied by Fricke et al. (1987) and Uyeno (1991). The second dorsal fin and the second anal fin are used in sculling motions, flexing to the same side at the same time and rotating as if to maintain trim and balance. The pectoral and pelvic fins seem to be used in an alternating sequence in slow swimming; the pelvic swings forward as the pectoral swings back. The first dorsal fin is kept erect, as if used as a rudder. Uyeno (1991) proposes a change in fin terminology, pointing out that the bulk of the tail complex is made up of the third dorsal fin and the second anal fin and that the so-called terminal lobe is truly the caudal fin.

Latimeria is thought to be a nocturnal predator on fishes and other nekton near the bottom (Uyeno and Tsutsumi, 1991). Stomach contents have included such deep water fishes as lanternfish, cardinalfish, snappers (*Symphysanodon*), deep sea witch-eel, a deep water alfonsino (*Beryx decadactylus*) and a swell shark, as well as a cuttlefish and parts of a cephalopod.

Several volumes and many journal articles have been written on *Latimeria* and cover the initial discovery (Smith, 1939) and many aspects of biology, including behavior and ecology (McCosker and Lagios, 1979; Musick et al., 1991; Thomson, 1991b).

There has been a contention, on grounds of physiology (especially osmoregulation and some anatomical characteristics), that the coelacanth is more closely related to the chondrichthyans than to the other bony fishes (Lagios, 1979, 1982), but this has not received wide acceptance and remains an open question.

Subclass Unnamed

Characteristics shared by the lungfishes and the extinct porolepiforms include the structure of the pectoral fins, which have a jointed axis, and the reduction of the ability to move the intracranial joint (Chang, 1991). In addition, neither group appears to have true choanae (see Nelson, 1993).

†*Infraclass Porolepimorpha*

†**Order Porolepiformes.** Members of this order lived from the Lower to Upper Devonian and were suggested by Jarvik (1968) to have given rise to the urodele amphibians, but this does not appear to have been widely accepted. Some had rhomboid scales containing cosmine, whereas others had thin, round scales (Nelson, 1994).

Infraclass Dipnoi

These fishes (lungfishes) have received various placement in different systems of classification. They have been considered a class, a subclass, an infraclass, and a series under the infraclass Choanata. They differ from other bony fishes in having autostylic suspension of the upper jaw (the palatoquadrate is fused to the neurocranium). In addition, the teeth are fused into crushing plates. Maxillae, premaxillae, and coronoids are lacking. The connection to the esophagus is ventral. Internal nares are present, as is a cloaca. The roofing bones of the skull seem not to be homologous with those of other bony fishes. They form a mosaic of numerous small bones in some extinct forms, usually with one or two relatively enlarged medial plates at the posterior part of the brain case and other paired plates surrounding these. Students of the group have not applied names to the bones but instead have designated them with letters of the alphabet. Figure 6–2**A** shows the skull roof of a modern lungfish, which has reduced numbers of plates with two elongate plates capping the median of the skull.

The paired fins of lungfishes consist of a long central axis with, in the Australian species, the fin rays disposed along it (Fig. 2–10**B**). In the South American and African lungfishes, the fin rays are lacking and the fins are produced as filaments. A special pulmonary blood circulation is developed in living species, and the atrium of the heart is divided into left and right chambers by an incomplete septum. A spiral valve is present in the intestine.

Fossil lungfishes are found on all continents (Marshall, 1987), some from marine deposits. Nearly 60 nominal genera of extinct lungfishes are known, and as many as five extinct orders are recognized by some authorities. About half of the genera data back to the Devonian. This discussion will concern only the two living orders.

Order Ceratodontiformes. This order contains the living Australian lungfish *Neoceratodus forsteri* and an extinct genus †*Ceratodus,* which was distributed more generally over the continents. Both are placed in the family Ceratodontidae, which is known from the Lower Triassic. *N. forsteri* is known from the Lower Cretaceous (Marshall, 1987).

Neoceratodus reaches nearly 2 m long and is a heavy-bodied fish, with large scales and paddle-like paired fins. The caudal fin is diphycercal. The species differs

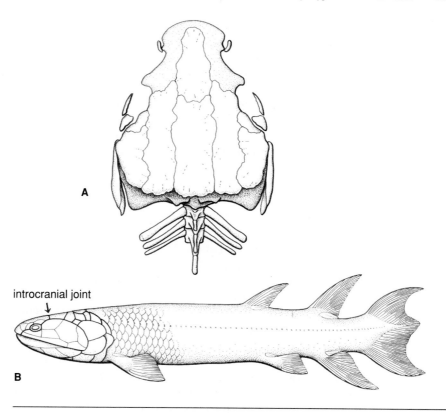

introcranial joint

A

B

FIGURE 6–2   **A,** Dorsal view of skull of Neoceratodus; **B,** reconstruction of †Eusthenopteron, with position of introcranial joint shown.

structurally from other living lungfishes by having an unpaired lung and a cartilaginous endocranium as well as four pairs of gills. There are many differences in life history and habit between this and other species. The Australian species can depend on the oxygen in the water unless conditions become stagnant or the fishes become extremely active. No special nest is made, the eggs are laid among vegetation, and the young have no special external gills. *Neoceratodus* apparently is omnivorous, feeding on vegetation and the many small forms of animal life that live among the plants, but it does not digest most plant material efficiently (Kemp, 1987). The species frequents permanent bodies of water and is incapable of estivation, but it can survive several weeks to a few months if kept moist in mud or vegetation (Kemp, 1987).

Order Lepidosireniformes. This order (Fig. 1–1**F**) contains two families: Lepidosirenidae, from South America, and Protopteridae, from Africa. Both have paired lungs, filamentous paired fins, and a membranous endocranium. Both families are known from the Upper Cretaceous.

Both families contain elongate fishes with fairly small scales. The habitat of these species is generally swampy and often contains low concentrations of dissolved oxygen, so the species have evolved the ability to utilize atmospheric oxygen.

The gills are reduced and relatively ineffective as compared to those of *Neocerato-dus*. If the swamps dry up, both the African and South American species can burrow into the muddy bottom and remain for several months in an inactive state, waiting for the next rainy season.

Protopteridae contains at least four species, all in the genus *Protopterus*. *Protopterus aethiopicus* reaches a length of 1.8 m; *P dolloi* reaches 1.3 m; *P. annectans* reaches slightly less than 1 m; and *P. amphibius* is not known to reach more than 443 mm (Greenwood, 1987). *Protopterus annectans,* an African species, is known for the formation of an effective mucous cocoon, in which it can estivate for seven or eight months (Greenwood, 1987). *Protopterus aethiopicus* has been maintained in its cocoon in the laboratory for as long as four years, far exceeding the normal estivation period (Coates, 1937). Other species may estivate in moist or water-filled burrows or, as in the case of the female of *P. dolloi,* seek out open water during the dry season (Greenwood, 1987). There is fossil evidence of estivation by lungfishes belonging to other families from far back in the Permian (Carroll, 1988).

When these fishes return to full activity after estivation, nests are constructed and breeding begins. The African species of *Protopterus* makes simple holes near the edge of the swamp, whereas the South American *Lepidosiren* constructs a burrow. In both genera, the eggs and larvae are guarded by the male. Larvae are held in place by a secretion from a cement organ on the breast region and have feathery external gills similar to those of salamanders.

There is apparently only one species in Lepidosirenidae, *Lepidosiren paradoxa*. It is characterized by reduced paired fins; five gill arches; and the development, in breeding males, of feathery, gill-like structures on the pelvic fins. These seem to act as gills in reverse, releasing oxygen in the vicinity of the young. *Lepidosiren* feeds on animals, especially snails, but also consumes algae. Maximum length is about 1.25 m. The species appears to be tenacious. One that arrived at Oregon State University with several inches of the tail missing not only healed the raw wound but regenerated the section almost perfectly within two years.

The paired fins of these species have radials along the axis, and there are six gill arches. The males do not develop gill-like structures on the pelvic fins, and the young retain vestiges of the external gills for some time after metamorphosis. *Protopterus* species are carnivorous and are said to be destructive to other fishes. They often attack others of their own species if held in the same tank.

†Subclass Rhizodontimorpha

The family †Rhizodontidae (Upper Devonian to Pennsylvanian) of the order †Rhizodontiformes is related to the osteolepiforms but lacks choanae (see Long, 1986).

†Subclass Osteolepimorpha (Rhipidistia)

These fishes existed during the time from the Lower Devonian to the Lower Permian. They are fishes with choanae, branched fin rays, and an intracranial joint somewhat different from that of the coelacanths in that the fifth cranial nerve passes through an opening posterior to the joint (Carroll, 1988).

†Order Osteolepiformes. This order is a well-known group of Devonian and Carboniferous fishes comprising at least five families. These differ from the porolepi-

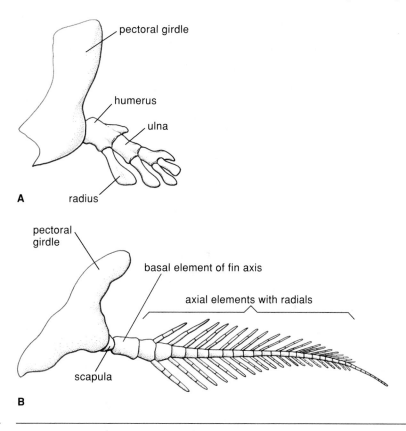

A   radius

B

FIGURE 6–3   Pectoral skeleton of **A,** †*Eusthenopteron,* compared with **B,** pectoral skeleton of modern lungfish.

forms in that they have a pineal foramen and the pectorals are inserted low on the body. Two families, †Panderichthyidae and †Eusthenopteridae (Fig. 6–2**B**), have received much attention as close relatives to the ancestors of primitive amphibians, especially because the skeleton of their paired fins (Fig. 6–3) can be compared almost directly with those amphibians (Carroll, 1988; Colbert, 1980). Other families are †Osteolepidae, †Rhizodopsidae, and †Tristichopteridae. Some members of this order reached about 4 m.

Views on the Origins of Tetrapods

The earliest known tetrapods (subclass Tetrapoda of class Sarcopterygii) are species of the genera †*Acanthostega* and †*Ichthyostega* from the late Devonian (Coates and Clack, 1991; Thomson, 1991). Tetrapods must have originated much earlier because footprints are known from the early Devonian of Australia (Long, 1990). Because of that chronological gap between the evidence that vertebrates had walked on land and the appearance of the amphibians in the fossil record, key parts of the puzzle of the fish-amphibian transition have not yet been found.

The desire to bridge the gap in knowledge of the origins of tetrapods has led to much research and speculation. Comparisons of early amphibian fossils with both fossil and extant sarcopterygians has resulted in contentions that tetrapods have arisen from lungfish-like ancestors or, in the opposite viewpoint, from osteolepiforms or close relatives. The latter view appears to have the best support in current literature (Carroll, 1988; Moy-Thomas and Miles, 1971), although there have been strong arguments in favor of close relationships to lungfishes. On the basis of similarity of β-hemoglobin chains, Gorr and Klienschmidt (1993) show that *Latimeria* appears closer to frogs than do other fishes.

Maisey (1986) presents a sister group relationship between recent lungfishes and tetrapods but mentions that many osteological synapomorphies (shared derived characteristics) are refutable and that there is the possibility of a closer relationship between osteolepiforms and tetrapods. Rosen et al. (1981) were strong proponents of lungfishes as the sister group of early tetrapods, but Holmes (1985) refuted many of their arguments. Coates and Clack (1991) reported on the branchial skeleton of †*Acanthostega,* which shows closest resemblance to the gill structure of the Devonian lungfish, †*Chirodipterus,* but is also similar to *Polypterus* and the extinct osteolepiform †*Eusthenopteron.* In addition, the shoulder girdle has points of resemblance to both lungfishes and †*Eusthenopteron.* Jamieson (1991) mentions that the spermatozoa of lungfishes indicate a relationship to both *Latimeria* and amphibians.

Long (1990) reports that juveniles of †*Eusthenopteron* show many similarities to the early tetrapods, more than any other fish group. He points out that paedomorphosis could have had a part in the transition from fish to tetrapods. The osteolepiforms of the family †Panderichthyidae from the Upper Devonian of Russia contain other fishes close in structure to early amphibians (Schultze and Arsenault, 1985; Thomson, 1991a). Schultze and Arsenault (1985) point out great similarity between the structure of the skull roof of the genus †*Elpistostege* and that of early amphibians.

The answer to the problem of tetrapod origins must wait for discovery and study of additional transitional fossils as well as for further research on the living representatives.

Primitive Actinopterygians

Class Actinopterygii
Subclass Chondrostei
    †Order Cheirolepiformes
    Order Polypteriformes
    †Order Palaeonisciformes
    †Order Tarrasiiformes
    †Order Phanerorhynchiformes
    †Order Saurichthyiformes
    Order Acipenseriformes
        †Suborder Chondrosteoidei
        Suborder Acipenseroidei
Subclass Neopterygii
    Order Semionotiformes
    Order Amiiformes

Class Actinopterygii

Actinopterygii is divided into two subclasses: Chondrostei, including the living bichirs, sturgeons, and paddlefishes as well as several extinct groups; and Neopterygii, including the living gars and bowfin plus extinct groups as well as the typical bony fishes from bonytongues, herrings, and tarpons to perches and their derivatives, plus extinct relatives.

Subclass Chondrostei

The chondrosteans share some ancestral characteristics, such as spiracles, heterocercal or abbreviate heterocercal tails, and more fin rays than ray supports in the dorsal and anal fins. There is evidence that the Chondrostei is not monophyletic (see McCune and Schaffer, 1986, and Nelson, 1994).

†**Order Cheirolepiformes.** This order has a single family (Cheirolepidae) known only from the Devonian. The species had large mouths and strongly heterocercal caudal fins, dorsal fins set far back over the anal fin, and small scales. †*Cheirolepis* had species up to about 0.5 m long.

Order Polypteriformes. There is some difficulty in fitting the bichirs and reedfish of Africa into the chondrostean framework. Their pectoral and dorsal fins are unlike those of any other fishes, they have a lung attached to the ventral part of the esophagus, and there are several other peculiarities of skeletal and soft anatomy that add to the belief that they should be placed in their own subgroup, outside or inside the Actinopterygii (Jarvik, 1983; Rosen et al., 1981). They are often placed into taxa called Brachiopterygii or Cladistia. Depending on the authors of the classification systems, they are placed variously as subclass, infraclass, or order (see Rosen et al., 1981; Lauder and Leim, 1983; Nelson, 1984, 1994; Carroll, 1988; and Benton, 1990). Nelson (1994) lists the group as an order of the subclass Chondrostei.

The order Polypteriformes contains the single family Polypteridae, the unusual bichirs (*Polypterus*) (Fig. 2–12), and reedfish (*Erpetoichthys* [= *Calamoichthys*]). This group is so well differentiated from other actinopterygians that it is sometimes placed by itself in a separate subclass Brachiopterygii (Nelson, 1984). The order, which is known from the Eocene of Egypt, was even formerly placed with the Crossopterygii (Jordan, 1923).

The dorsal fin of these fishes is different from that of any other group; it consists of a series of separate small fins (each with one large, spinelike ray) and one or more soft branches from the posterior edge of the spine (Fig. 2–10**A**). The spines are said to make handling these fishes dangerous.

The pectoral fin has a unique structure as well. It is lobate and is supported by two bony elements, the propterygium and metapterygium, with a cartilaginous, platelike mesopterygium between. Scales of polypteriforms are ganoid, but with three layers—ganoin, cosmine, and isopedine—as in the palaeonisciforms. Despite the hard scales, their bodies are flexible. The air bladder is lunglike and functions in respiration. A spiracle and spiral valve are present, and the heart has a conus arteriosus.

Polypterus contains about 10 species, all of African fresh waters. The species are medium- to large-sized and are predators. *P. bichir,* which reaches a length of about 1 m, is sought as food and is generally roasted in coals with the scales left on. The larvae have a rather amphibian-like appearance due to the large, feathery external gills (Fig. 7–1**A**). The single species of reedfish, *Erpetoichthys calabaricus* of West Africa (ropefish of the aquarium trade), reaches about 90 cm long. It lacks pelvic fins. It is somewhat amphibious in habit and stays out of the water from several minutes to more than an hour at a time in laboratory studies.

†**Order Palaeonisciformes.** This order is represented by a number of extinct actinopterygian families that Nelson (1994) places into four suborders: †Palaeoniscoidei,

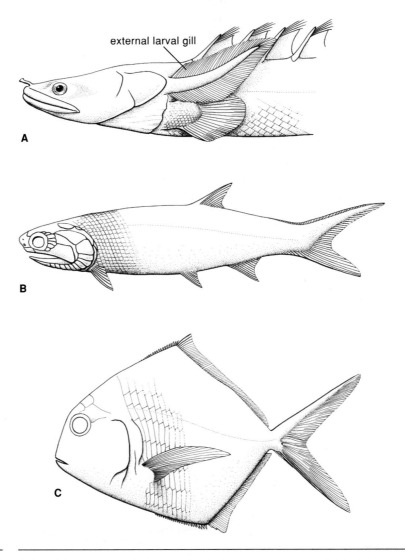

external larval gill

A

B

C

FIGURE 7–1   **A,** Anterior portion of *Polypterus,* showing external gills of larvae. Representatives of order †Palaeonisciformes, **B,** †palaeoniscoid (actual size ca. 15 cm); **C,** †platysomoid (actual size ca. 15 cm). (**B** after Romer, 1970; **C** after Moy-Thomas and Miles, 1971.)

†Redfieldioidei, †Platysomoidei, and †Dorypteroidei. These fishes had an evolutionary history from the Devonian into the Cretaceous and developed along several lines. Body forms range from the typical fusiform, as in palaeoniscoids (Fig. 7–2**A**), to the compressiform platysomoids, such as the †Chirondontidae (Fig. 7–2**B**), which have superficially symmetrical caudal fins and lack pelvic fins. For the most part, these are covered by ganoid scales and have large, bony heads. The mouths are generally long, with conical teeth, and seem to be designed mainly for catching and holding prey. Maxillaries are large and probably not movable. The platysomoids have more specialized mouths that are reduced in size and provided, in some instances, with crushing teeth.

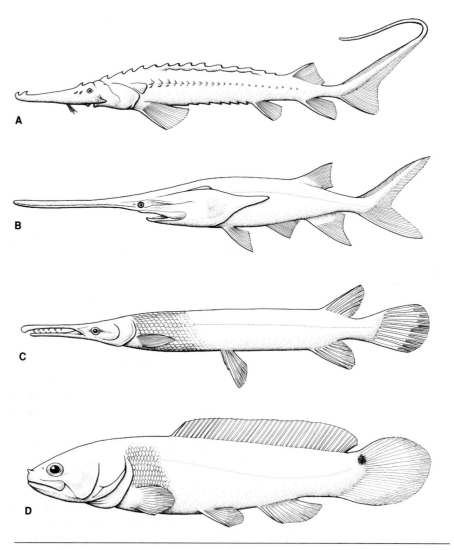

FIGURE 7–2 Representatives of order Acipenseriformes. **A,** Shovelnose sturgeon (*Pseudoscaphirhynchus*); **B,** paddlefish (*Polyodon*). Representatives of **C,** order Lepisosteiformes, gar (*Lepisosteus*); **D,** order Amiiformes, bowfin (*Amia*).

†**Order Tarrasiiformes.** Tarrasiidae differ in having a diphycercal, rather than hetero-cercal, caudal fin and long dorsal and anal fins continuous with the caudal. These are known from the Carboniferous (Mississippian).

†**Order Phanerorhynchiformes.** A single genus that resembles sturgeons is known from the Upper Carboniferous (Pennsylvanian). The fins are supported by a reduced number of fin rays.

†**Order Saurichthyiformes.** These fishes, from the Triassic and Jurassic, had an elon-gate body with the dorsal and anal fins set posteriorly, as in the gars. Their heads and mouths were elongate. Some species were about a meter long.

Order Acipenseriformes. This order includes the sturgeons and paddlefishes plus some more or less related extinct forms.

†**Suborder Chondrosteoidei.** Chondrosteoid fossils are known from the Lower Triassic to the Lower Cretaceous. One family, the †Chondrosteidae, resembled sturgeons in body shape, heterocercal tail, and inferior mouth. The probably related Errolichthyidae had an almost terminal mouth.

Suborder Acipenseroidei. Characteristics of the sturgeons and paddlefishes in-clude a cartilaginous endoskeleton, lack of vertebral centra, strongly heterocercal caudal fin, and radials supporting the rays of the pelvic fin. These creatures differ from most other actinopterygians in that they have the anus and urogenital opening at the base of the pelvics. Ganoid scales are present on the upper portion of the cau-dal fin, and some members retain a spiracle. The conus arteriosus has more than one set of valves, and a spiral valve is present in the intestine. A septate air bladder, which has some utility in breathing air, is present. Dermal bone is prominent on the head of sturgeons.

The sturgeons (Fig. 7–2**A**) of the family Acipenseridae are superficially distin-guishable from paddlefishes by having bony scutes along the sides and back and four barbels on the underside of the rostrum, which is shorter than in the paddle-fishes. The family contains four genera and over 20 species, which are distributed around the northern part of the Northern Hemisphere (holarctic distribution) and have marine, freshwater, and anadromous members. The waters of the former Soviet Union (Russia, Ukraine, etc.) are especially noted for the numbers of species of stur-geon found there. Sturgeons are sought by fishermen for their flesh and their roe, from which caviar is made. North American sturgeon fisheries are now relatively minor, but considerable tonnages are taken in the former Soviet Union and Iran.

The largest sturgeon is the beluga (*Huso*) of the Caspian and Black seas. It is reported to reach a length of about 9 m and a weight of 1500 kg. Large specimens may be 100 years old and carry over 7 million eggs. One of the most important com-mercial sturgeons is the Russian sturgeon (*Acipenser guldenstadti*) of the Caspian and the Sea of Azov, which reaches about 2.3 m long. The largest North American species is the white sturgeon (*Acipenser transmontanus*) of the Pacific Coast. Lengths of 6 m and weights of about 850 kg have been reported. This giant is found in both fresh and salt water from southern California to Cook Inlet of Alaska. It ap-pears to be at home in large rivers such as the Snake, Columbia, Sacramento, and Fraser.

The smallest members of the family appear to be the strange-looking shovel-nose sturgeons. *Scaphirhynchus platorhynchus* of the Mississippi reaches about 1 m long, whereas the small shovelnose of the Amu Darya River of the former Soviet Union (*Pseudoscaphirhynchus hermanni*) is reported to reach a maximum length of only 27 cm.

Some North American sturgeons are listed by the American Fisheries Society as rare or threatened in some way. The pallid sturgeon, *Scaphirhynchus albus,* and *S. suttkusi,* the Alabama sturgeon, are listed as endangered. *Acipenser oxyrhynchus,* the Atlantic sturgeon, is listed as of special concern, and a subspecies, the gulf sturgeon, *A.o. desotoi,* is considered threatened; also listed as threatened are the lake sturgeon, *A. fulvescens,* found in the Mississippi drainage and north through the Great Lakes, the Red River of the North, the St. Lawrence River, and Hudson Bay; and the shortnose sturgeon, *A. brevirostrum,* of fresh and marine waters of the Atlantic Coast. *A. medirostris,* the green sturgeon, is primarily a marine species that ranges from southern California to Alaska and westward to Asia. The species enters bays and rivers to spawn.

Food of sturgeons includes worms, crustaceans, and small fishes that can be sucked up by the greatly protrusible mouth. Beluga are reported to eat larger fishes and occasionally the young of the Caspian seal. A typical sturgeon life history includes a migration from feeding grounds to breeding grounds, usually to a river. Spawning takes place over gravel in fairly swift water. The demersal, adhesive eggs hatch after three to five days and the larvae—about 1 cm long—drift downstream to suitable rearing areas in the river or the sea. Growth is slow, and many species reach only about 1.2 m long after 10 years. Males reach sexual maturity earlier than the females. Medium-sized species may become mature at 8 to 12 years for the males and 10 to 15 years for the females.

The requirement for clean water, their slow growth and later maturity, and the fact that the eggs are the most valuable product from these animals pose problems for fishery managers. The white sturgeon was once a prized commercial fish in California, Oregon, and Washington, but almost unregulated fishing—especially for the large females—has diminished the stock, and replacement is slow. The species constitutes a minor fishery at this time but is the object of aquaculture in California. Some difficulty is being experienced in maintaining the Caspian stocks of sturgeon, and large hatcheries have been erected in both the former Soviet Union and Iran to supplement the yield of naturally spawned fish.

The paddlefishes of the family Polyodontidae are characterized by an extremely long snout with two minute barbels. They lack the bony scutes of the sturgeons. There are two living species, *Polyodon spathula* of eastern North America and *Psephurus gladius* of the Yangtze River in China. Both are found only in fresh water. *Psephurus* has a swordlike snout, relatively short gill rakers, and a protrusible mouth. Its food appears to be other fishes. Some specimens of *Psephurus* have been reported to 7 m long, but authentic records go only to 4 m. *Polyodon* has a broad, paddle-shaped snout, long gill rakers, and a nonprotrusible mouth. Crustaceans and other plankton are strained from the water by means of the long gill rakers. The paddle does not seem to function in food gathering, unless its role is sensory. *Polyodon* reaches a length of about 2 m, a weight of 76 kg, and spawns in swift rivers over gravel bars in the spring. Growth is more rapid than that of sturgeons. Although the range of the paddlefishes has been reduced by human activities, the species has become numerous in several reservoirs in the Mississippi drainage and forms the basis of a popular fishery based on snagging with treble hooks.

Some additional orders of the subclass Chondrostei are as follows:

†Order Ptycholepiformes. Fossils are found in North American Triassic and Jurassic deposits.

†Order Pholidopleuriformes. These are known from fossils from Middle to Late Triassic.

†Order Perleidiformes. Several families are known from the Triassic and Lower Jurassic.

†Order Luganoiiformes. This order is known from the Triassic.

Subclass Neopterygii

The gars, bowfin, and teleosts do not retain the full heterocercal tail and the spiracle typical of many chondrosteans, and only one living family has ganoid scales. The rays of the dorsal and anal fin correspond in number with the fin ray supports. These fishes are considered to belong to the subclass Neopterygii (Nelson, 1994). The gars and bowfin share some structures not seen in the Division Teleostomi (see Chapter 8). They have a close correspondence between many caudal fin rays and their supports. The track of the ventral branches of the spinal nerve roots penetrates the lateral musculature and courses ventrally outside the musculature. However, other evidence has led many systematists to conclude that the bowfin, although not a teleost, is more closely related to teleosts than either the bowfin or the teleosts are to the gars (see Wiley, 1976). The bowfin, gars, and fossil relatives were formerly placed in a group called Holostei, which was proposed to represent a group intermediate between the Chondrostei and the Neopterygii. Although this usage has been abandoned, a study of mitochondrial DNA from gars, bowfin, and teleosts by Normark et al. (1991) led those authors to conclude that *Amia* and *Lepisosteus* form a clade and that the earlier workers, who considered the gars and bowfin to be closely related, may have been correct. Another view of the relationships of these groups was the placement of gars into the Division Ginglymodi of Neopterygii and the placement of bowfin, their extinct relatives, and teleosts into the Division Halecostomi (the bowfin and closely related fossil genera in Subdivision Halecomorphi and the teleosts in Subdivision Teleostei). Those names and categories will be encountered even in fairly recent literature. Further information can be found in Patterson and Rosen (1977), Lauder and Liem (1983a,b), and Nelson (1984, 1994). Nelson (1994) chooses to place gars, the bowfin, and extinct relatives into Neopterygii but to sequence them as an unnamed series before the Division Teleostei.

Order Semionotiformes. This order includes (depending on the authority) two or three extinct families comprising several genera and the living gars, family Lepisosteidae. Extinct families are †Dapediidae, †Semionotidae, and †Macrosemiidae. These are known mainly from the Triassic to the Devonian. Gars are found in eastern North America from the Great Lakes region to Costa Rica, and one species reaches Cuba. These are elongate fishes with the body covered by heavy ganoid scales and the head covered by equally hard bone. Both jaws are elongate and are armed with several rows of strong, sharp teeth. Some of the upper teeth are borne on a series of infraorbital bones. The dorsal and anal fins are set far back, and the caudal is abbreviate heterocercal. Gars differ from other living fishes (except the blennioid genus *Andamia*) in that they have opisthocoelous vertebrae; these vertebrae are concave posteriorly and convex anteriorly. A spiral valve is present in the intestine, and the gas bladder is divided internally into interconnected chambers. Fossil gars date from the Upper Cretaceous of Europe and have been found in India, Africa, and North America.

Gars live in quiet, often weedy, waters and usually can be observed lying almost motionless near the surface. They can utilize atmospheric oxygen in addition to that obtained through the gills. Prey is captured by means of a rapid lateral strike with the jaws, often after waiting for it to come into range. According to Wiley (1976), there are two genera of gars: *Lepisosteus,* with four species, and *Atractosteus,* with three. Most live in fresh water, but the alligator gar, *Atractosteus spatula,* and its close relatives may enter salt water. These are the largest of the gars and reach 3 m long. Gars have little economic value—they are used as food to some extent in the southern United States and Mexico, and the ganoid scales are sometimes used as souvenirs and as ornaments. The eggs are reported to be poisonous.

Order Amiiformes. Although this order includes a diverse group of fossil fishes known from the Triassic to the Cretaceous, it has but one living representative, the North American bowfin (*Amia calva*) of the family Amiidae. The order has fossil representatives in Europe, Asia, and North America. *Amia* is found in eastern North America from the Great Lakes south and appears to prefer warm, shallow water. Males guard the nest—a depression made in aquatic vegetation—and the young, sometimes harboring the young in the mouth. The gas bladder is divided by internal septa and can function in aerial respiration, which allows the species to inhabit stagnant waters that have a low oxygen concentration. The bowfin is found only in fresh water. The sexes differ in maximum adult size. Females may reach 1 m long, but 60 cm is a more common length, with males being somewhat smaller than the females. An additional expression of sexual dimorphism is the ocellus, a dark spot at the base of the male's caudal fin. Despite its interesting habits and the scientific interest in the species as a "living fossil," the mediocre flesh and predatory habits of the bowfin have led many conservation agencies to regard them as unwanted. The eggs are used in some places as a source of caviar.

Division Teleostei: Fossil Teleosts, Bonytongues, Tarpons, Eels, and Herrings

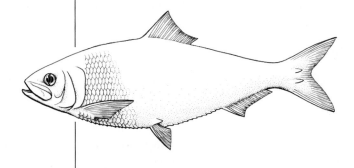

Division Teleostei

The Teleostei is a large assemblage of many diverse groups and includes 38 orders, about 425 families, and nearly 4070 genera of living fishes (Nelson, 1994), and it also includes extinct groups, some of which formerly were considered to represent a "preteleostean" level. Teleosts are defined by Patterson (1968, p. 234) as actinopterygians with "the lower lobe of the caudal fin primitively supported by two hypurals articulating with a single centrum" and have elongate uroneurals that are modified from neural arches (see Chapter 2). In most teleosts the caudal fin, though not symmetrical internally, appears to be symmetrical externally. The urohyal of teleosts is formed by ossification of the sternohyoideus tendon (Arratia and Schultze, 1990). Monophyly of the teleosts is discussed by Patterson and Rosen (1977).

Primitive Fossil Teleosts

†**Order Pholidophoriformes.** This order, which is known from the Triassic to the Cretaceous (Berg, 1940), features ganoid scales on most of its members, enamel on skull bones, and a lower jaw resembling that of the gars and *Amia*. However, these and two other extinct groups, the leptolepiforms and ichthyodectiforms, share many features with living teleosts, including those of the caudal fin, which is internally asymmetrical but externally homocercal and in which hypurals each bear more than one fin ray. Uroneurals are developed. Other teleost characteristics in these fishes, according to Patterson and Rosen (1977), are two supramaxillae, a myodome that extends into the basioccipital, fused vomers, and the division of the premaxilla into a lateral, moveable portion and a more medial lateral dermethmoid.

The pholidophoroids are small fishes (up to 40 cm) whose fossils have vertebrae in various stages of evolutionary development, some rudimentary. The cranial morphology of these fishes shares several specializations with that of teleosts (Patterson, 1975). The pholidophoroids may have given rise to the leptolepidiforms and other primitive teleosts in the Triassic or Jurassic (Carroll, 1988; Nelson, 1994).

†**Order Leptolepidiformes.** These fishes, known from the Triassic and Devonian, are generally considered to have attained the teleost level, although they show affinity to more primitive groups. Patterson (1975) shows that they are advanced over the pholidophoroids and share many specializations of skull anatomy with teleosts. Leptolepidids have better ossified vertebrae than pholidophorids and have cycloid scales (Carroll, 1988). Some fossils of adults do not exceed 5 cm long. The relationships of fishes usually assigned to this group were discussed by Patterson and Rosen (1977).

†**Order Tselfatoidiformes.** This order seems to be close to such groups as the herrings and tarpons and especially to the bonytongues (Osteoglossomorpha), with which it has been placed by some authorities (Patterson, 1967). These fishes are known from the Cretaceous.

Diversity of Living Teleosts

The living Teleostei lack ganoid scales and the spiral valve, and there are three or four lower jaw bones per side (fewer than in most fossil groups). There are four interrelated lines of teleosts. The more primitive representatives of all groups share some common features, and the body plans of many of the more advanced orders and suborders have features in common. The evolutionary trends are away from the generalized vertebrate body plan toward morphology that makes locomotion and food gathering more efficient or otherwise fits the species for exploitation of particular niches (see Gosline, 1971).

The more primitive teleosts usually have elongate bodies tending to modified fusiform shape, with 50 to 60 vertebrae. Although some groups, such as eels and other specialized fishes, have many vertebrae and a long, flexible body, the trend in derived fishes is toward shorter, deeper bodies and 20 to 30 vertebrae. The primitive fishes tend to have single, short-based dorsal fins near the middle of the back and only soft rays in all fins; whereas more advanced fishes have more than one dorsal fin or even long-based dorsals originating rather far forward. Spines usually add support for the anterior parts of dorsal, anal, and pelvic fins. The deepening of the body is usually accompanied by the encroachment of body musculature onto the top of the head, and the supraoccipital expands into a large crest as an attachment point for these muscles.

Pectoral fins of primitive fishes are typically set low on the body, with the base slanting downward and backward and being nearly horizontal in some. There are certain limitations of movement inherent in this positioning. These fins, although useful in guiding and braking, are not as versatile as the pectorals of more derived fishes, which are set higher on the sides with a base tending more to the vertical. The latter position might be more suitable for locomotion as well as for maneuvering and braking. Pectoral fins of many of these fishes allow them to hover or swim backward with little or no use of other fins. Several modifications of the pectoral girdle are involved in the positioning of the fins, including changes in relative sizes of bones and twisting of the axis of the girdle, but the consistent difference between the primitive and advanced groups is the presence of the mesocoracoid bone in the more primitive groups. This bone forms an arch with the scapula and coracoid and braces the fin base at an angle. The loss of the mesocoracoid bone in higher groups appears to allow the base of the pectoral to be aligned with the vertical axis of the girdle.

Pelvic fins in more primitive bony fishes are placed rather far back on the belly and are said to be "abdominal." The supporting bones, a reduced pelvic girdle, are not firmly connected to any other bony structure but are situated in the musculature of the body wall. More derived fishes have pelvics placed far forward on the breast region below the pectorals in the thoracic position. The pelvic girdle is attached to the lower portion of the pectoral girdle. A few groups of fishes have pelvics in an intermediate position with no connection to the pectoral girdle, and some maintain the intermediate position and a ligamentous connection. These are variously called subabdominal or subthoracic pelvic fins. In a few instances the pelvics are forward of the pectoral base in a jugular position, and this is usually considered an advanced characteristic. In many highly maneuverable fishes, the vertical alignment of the origin of the spinous dorsal, the pectorals, and the thoracic pelvics adds greatly to their ability to stop quickly or make tight turns. There is generally a change in the center of balance concomitant with the change in position of the paired fins because the position of the gas bladder changes. It is farther forward and higher in the body in most advanced fishes and aids more in maintaining balance.

Ancestral fishes differ from more derived forms in the structure of the upper jaw. The outer edge of the jaw consists of both the premaxilla and the maxilla in the primitive fishes. The premaxillae are seldom protractile, and both they and the maxillae may bear teeth. On the other hand, the premaxillae form the upper border of the mouth in the derived forms. The maxillae are excluded from the actual border of the gape, are situated above the premaxillae, and do not bear teeth. Advanced fishes may have a protrusible upper jaw with long, ascending processes of the premaxillae that slide along the anterodorsal part of the skull. Protrusibility leads to greater versatility in prey capture.

The more primitive teleosts retain an open pneumatic duct from the alimentary canal to the gas bladder (physostomous condition). This is in contrast to the physoclistous condition of higher teleosts, in which the duct is absent or closed. The pancreas is often a discrete gland in the more primitive forms but is usually diffuse in more advanced fishes and may be associated with the liver as a hepatopancreas in spiny-rayed fishes.

The orbitosphenoid bone forms a large part of the interorbital septum in many lower teleosts but is lacking in most of the middle and higher fishes. Scales of lower teleosts are usually cycloid; those of the higher teleosts are usually ctenoid. The hard, rough margins of most ctenoid scales may function in protection against injury and may serve a hydrodynamic function by holding a "boundary layer" of water next to the fish (see Chapter 18).

The foregoing are some of the generally recognized differences between primitive and derived groups of teleosts. (Others will be mentioned later in the discussions of orders.) These characteristics will occur in various combinations in the following material, and in some instances the difficulty in following well-defined lines of relationships will be evident.

The four major monophyletic lines (subdivisions) of living teleosts are as follows:

1. The osteoglossomorphs, including bonytongues, elephantfishes (mormyrids), featherbacks, mooneyes, and allies. This group is the Osteoglossomorpha, a name that has ranked from superorder to supercohort in various classifications. Another name that has been used for the group is Archaeophylaces. There is a single living order, Osteoglossiformes, and one extinct order, †Ichthyodectiformes.

2. The tarpons, tenpounders, eels, and relatives. These are the Elopomorpha (a name that has ranked as superorder or cohort) and comprise four extant orders: Elopiformes, Notacanthiformes, Anguilliformes, and Saccopharyngiformes. One extinct order, †Crossognathiformes, is considered to be closely related. This subdivision is also called Taeniopaedia.

3. The herrings and allies. The subdivisional name for this group is Clupeomorpha. There is a single living order, Clupeiformes, and an extinct order, †Ellimmichthyiformes.

4. All the remainder of the teleosts. These are the subdivision Euteleostei, which comprises, according to Nelson (1994), 32 orders, 391 families, 3795 genera, and 22,262 species.

In some former classification systems, various representatives of all four of the aforementioned groups were placed in a large inclusive order called Clupeiformes or

Isospondyli, mainly because of ancestral characteristics held in common by many soft-rayed teleosts (symplesiomorphies). These characteristics were discussed earlier in this chapter.

Subdivision Osteoglossomorpha

†Order Ichthyodectiformes. This order, known from the Jurassic and Cretaceous, consists of early teleosts that show relationships to the primitive living teleost groups and is placed with the osteoglossomorphs by Nelson (1994). These fishes are unusual in that they have large uroneurals that extend laterally over the centra of the preural vertebrae. In addition, they have an endoskeletal bone (ethmo-palatine) on the floor of the nasal cavity (Carroll, 1988; Nelson, 1994). Authorities recognize three to five families, all of which tend to be large, predaceous fishes. One genus, †*Xiphactinus* (*Portheus*), had species that reached 4 m long.

Order Osteoglossiformes. The order Osteoglossiformes contains some of the most primitive living teleosts, including the bonytongues, mooneyes, and mormyrids (elephantfishes and relatives), which have advanced electrical capability. Basically, the anatomy of the Osteoglossiformes is similar to that of other primitive bony fishes, and they were formerly included in a more inclusive interpretation of Clupeiformes or Isospondyli. Osteoglossiformes are similar to many Mesozoic bony fishes in the arrangement of the bite, which, instead of being primarily between the dentaries and the upper jaw, is between the toothed tongue and the toothed bones of the roof of the mouth. Usually the parasphenoid is involved, but in some the entopterygoids bear teeth as well. An extension from the parasphenoid forms a support for the entopterygoid. This structure is rare in teleosts and is found only in osteoglossiforms and alepocephaloids (slickheads), according to Gosline (1971). There is a pair of rodlike tendon bones associated with the base of the second gill arch. Usually some soft connection between the gas bladder and ear is present. The order Osteoglossiformes is thought to be an ancient group. Some extinct families dating from the Jurassic are placed in the order, and some close relatives of extant families are known from the Cretaceous.

Suborder Osteoglossoidei. This order includes the families Osteoglossidae and Pantodontidae. Osteoglossids are found in Africa, South America, southeast Asia, and Australia. The subfamily Heterotidinae includes the arapaima or pirarucu of South America and the genus *Heterotus* of Africa. *Arapaima gigas,* one of the longest freshwater fishes, reaches some 2.5 m long (Fig. 8–1**A**). It is prized as a food fish. *Heterotis* differs from the other genera in the family by its small mouth and dependence on smaller food items. Both *Heterotis* and *Arapaima* have air bladders with numerous small divisions that increase the area available for absorption of oxygen, which apparently makes the use of atmospheric oxygen for respiration possible. *Heterotis* has a special suprabranchial respiratory organ that aids in that function. *Heterotis* and *Arapaima* build large nests in shallow water and give protection to their young after hatching. The young of *Heterotis* are equipped with external gills. The genera *Osteoglossum* and *Scleropages* are placed in the subfamily Osteoglossinae. Two species of *Osteoglossum, O. ferreirai,* the black arawana, and *O. bicirrhosum,* the silver arawana, are often imported from South America as aquarium fishes.

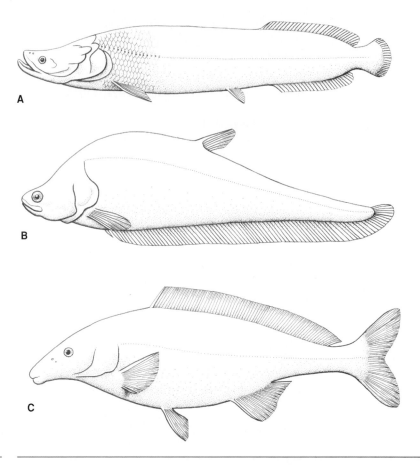

FIGURE 8–1   Representatives of **A,** order Osteoglossiformes, arapaima (*Arapaima*); **B,** featherfin knifefish (*Notopterus*); **C,** order Mormyriformes, elephantfish (*Mormyrus*).

Scleropages jardini and *S. leicharti* are regarded as the only truly freshwater teleosts native to Australia. They are primary freshwater fishes, with no relatives known from salt water. The other fishes found in the fresh waters of Australia belong to families that are marine, anadromous, or catadromous or that have fossil relatives from marine deposits. *Scleropages formosus* of Thailand and the Malay region is apparently a mouth brooder, as are the *Osteoglossum* species.

The family Pantodontidae contains one species, *Pantodon buchholzi* of Africa, which reaches a maximum size of less than 15 cm (Fig. 2–12**B**). This peculiar little fish has subthoracic pelvic fins that are made up of long, separate, rather filamentous rays. The pectorals are relatively large and expanded and are characterized by expanded cleithra that support large pectoral muscles. These apparently give the fish the ability to flap the pectorals like wings. The flying habits of the species have not been well studied, but the habit of making long leaps (or flights?) is well known. The common names butterflyfish and freshwater flyingfish are derived from this habit. The male has a modified anal fin that is thought to function in internal fertilization. The eggs float at the surface, as do the fry after hatching.

Suborder Notopteroidei. This suborder includes the following families: Hiodonti-dae (the mooneye and goldeye of North America); Notopteridae (the featherbacks of southeast Asia, the Indo-Malayan Archipelago, and Africa); and Mormyridae and Gymnarchidae ("electric" fishes of Africa). The superfamily Hiodontoidea includes Hiodontidae and †Lycopteridae, from the Jurassic and Cretaceous of Asia. The mooneye, *Hiodon tergisius,* and the goldeye, *H. alosoides,* are silvery, herring-like fishes of the northern section of eastern and central North America, extending west-ward to Alberta and into British Columbia. These have a connection between the gas bladder and the ear (otophysic connection) that consists of diverticula of the gas bladder that extend into the skull. They rarely reach as much as 45 cm long and have limited use as food and sport fishes.

The superfamily Notopteroidea, the featherbacks of Notopteridae, are strange-appearing, elongate, compressed fishes with a long anal fin beginning in the anterior third of the body and extending to the caudal, with which it is confluent (Fig. 8–1**B**). A very small dorsal fin is placed about midway down the back, except in the African genus, *Xenomystis,* which lacks a dorsal. Some have a humped back and a concave dorsal head profile. Several species are used as food, and some enter the aquarium trade as "knifefishes." Nelson (1994) lists four genera, *Notopterus* and *Chitala* of Asia and *Xenomystus* and *Papyrocranus* of Africa. *Papyrocranus* lacks pelvic fins and has diverticula of the gas bladder entering the cranium. Some species of *Chitala* reach 1.5 m long.

The superfamily Mormyroidei includes Mormyridae (the mormyrids or ele-phantfishes) and the closely related Gymnarchidae. All members of these families are from the fresh waters of Africa. The mormyrids were formerly placed in their own order by some ichthyologists because of their remarkable development of elec-trogenic and electrolocation capabilities and their exceptionally large cerebellum. In this family the operculum is small and covered by skin. Weak electric organs are present in the caudal region. The skull has a lateral foramen covered by a flat supratemporal bone. The cavity inside the skull at this foramen is occupied by a vesicle that originates as part of the gas bladder in the early developmental stages. The cerebellum of the mormyrids is enlarged as part of the adaptation for electro-reception and is, proportionately, the largest among the lower vertebrates. The premaxillaries are fused. They lack the opisthotic, angular, entopterygoid, and sym-plectic bones.

The family Mormyridae contains about ten genera, among which are some of the oddest-looking fishes known from fresh water, the so-called elephantnose fishes of the genus *Gnathonemus.* These have a snout that is greatly produced and curved downward with a very small mouth at the tip, often equipped with a thick barbel. These fishes were known to the ancient Egyptians, who depicted them in paintings and sculpture. Many species have a more normal shape, but all are characterized by a narrow caudal peduncle (Fig. 8–1**C**). These fishes are generally sought as food. The mormyrids have the sacculus and lagena of the inner ear separated from the utriculus and the semicircular canals, a characteristic they share with the Notopteri-dae. There are no pores in the cephalic lateral line complex.

The single species of Gymnarchidae, *Gymnarchus niloticus,* has been studied extensively because of its ability to locate objects electrically, an ability shared with the mormyrids. These fish differ from mormyrids in appearance; they are elongate and lack the pelvic, anal, and caudal fins. The dorsal fin, which runs most of the length of the back, is used for locomotion forward or backward by means of undulations that pass down the fin.

Subdivision Elopomorpha: Tarpons, Ladyfishes, Eels, and Relatives

These fishes, mostly marine and diverse in body form, constitute the subdivision Elopomorpha (Taeniopaedia) (see Forey, 1973a, 1973b; Lauder and Liem, 1983a, 1983b; Nelson, 1983). The group contains fishes ranging from the tarpons and bonefishes to some bizarre deep sea eels. These fishes are alike in that they all have the leptocephalus type of larva (Fig. 8–2**A**), have fused retroarticulars and angulars, and have prenasal and rostral ossicles. Their morphology is so varied that relationships within the cohort are subject to different interpretations (Smith, 1984). There are about 800 species in about 25 families, variously arranged by students of the group.

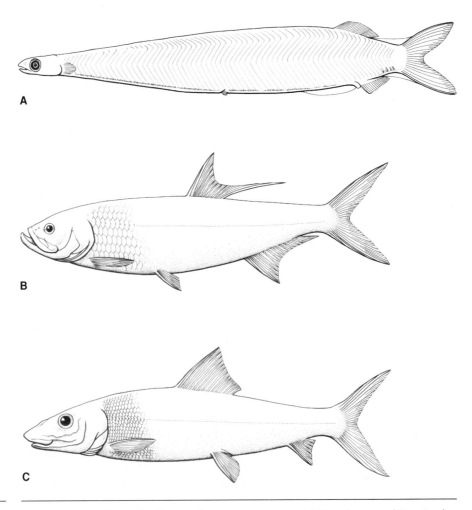

FIGURE 8–2   Representatives of order Elopiformes. **A,** leptocephalus larva of *Elops;* **B,** tarpon (*Megalops*); **C,** bonefish (*Albula*).

Order Elopiformes. This order includes the tarpons and ladyfishes. They are known from the Jurassic to the present and were especially abundant in the Cretaceous. These fishes show some relationships to the herring-like fishes and were formerly sometimes placed with them in an order called Isospondyli. However, they share the larval stage known as the leptocephalus with the eel-like fishes and are considered to be part of an evolutionary line that includes the eels, notacanths, and halosaurs. The leptocephalus larvae of various fishes may be ribbon-like or may have the general shape of a willow leaf. The flesh of the larvae is almost colorless and translucent, and the teeth are prominent. Larvae of elopiforms differ from those of eels in that they have forked caudal fins (Richards, 1984) (Fig. 8–2**A**).

In addition to the general primitive features of a lower teleost, the elopiforms have a gular plate (a bony plate between the branches of the lower jaw) and a commissure of the cephalic sensory canal system in the ethmoidal region. Some members have a conus arteriosus with two rows of valves. The number of branchiostegals ranges from 6 to 36, and the pelvics have from 10 to 18 rays.

The ladyfishes (tenpounders and machetes) of the family Elopidae and the tarpons of the family Megalopidae belong to this order. These are large, big-scaled fishes of warm seas. Not prized as food, they are sought mainly for sport. The ladyfish or tenpounder, *Elops saurus,* the best-known fish of its genus, reaches somewhat less than 1 m in length, reaches a weight of 14 kg, and is circumtropical. The tarpon, *Megalops atlanticus* (Fig. 8–2**B**), may reach 2.6 m in length and 150 kg, but the oxeye of the Indo-West Pacific, *M. cyprinoides,* usually does not exceed 1 m in length. The tarpons contrast somewhat with *Elops* because they possess a conus arteriosus, lack pseudobranchiae, and have a connection of the gas bladder to the otic region of the skull.

Order Albuliformes. There are some very dissimilar fishes in this order, but they share characteristics of the snout and inferior mouth, the sensory canals on the snout, and rostral ossicles. Lateral sensory canals can be large and cavernous (see Forey, 1973b).

Suborder Albuloidei. The families Albulidae (bonefishes) and Pterothrissidae (the deep sea bonefish) are in this suborder. These are fishes with abdominal pelvic fins, low-based pectoral fins, and forked caudal fins. They are placed in the single family Albulidae by some. The bonefish, *Albula vulpes* (Fig. 8–2**C**), is a prized game fish of tropical and subtropical waters around the world. *Albula* (= *Dixonina*) *nemoptera* is found in the Caribbean area and on the Pacific coast of Central America. *Pterothrissus* (= *Isteus*) has at least two species; one lives in deep water off Japan and the other off West Africa. The albuloids are placed in the Anguilliformes by some authorities (Lauder and Liem, 1983a).

Suborder Notacanthoidei. Notacanthoids have high-based pectoral fins, abdominal pelvic fins, and thin tapering tails with small or no caudal fins. The leptocephalus larvae of some of these may exceed the adults in length. The family Halosauridae contains widespread deep sea fishes that are known as fossils as far back as the Upper Cretaceous. They resemble true eels in that they are elongate, lack the mesocoracoid and orbitosphenoid, and have leptocephalus larvae. The pectoral girdle is not firmly connected to the skull. Although the pelvic fins are abdominal and the scales cycloid, these are physoclistous fishes that lack the orbitosphenoid. The tail tapers to a point, without a caudal fin, but individuals may possess regenerated tail tips that form a pseudocaudal fin. These fishes are soft rayed and have the upper border of the mouth made up of premaxillaries and maxillaries (McDowall, 1973). They may range as deep as 2000 m and may reach to about 500 cm in length. Genera include *Halosauropsis, Halosaurus,* and *Aldrovania.*

Notacanthidae, the spiny eels (Fig. 8–3**A**), have a series of spines down the back and in the anal fin, and the upper jaw is bordered by the premaxillaries only. These spiny-backed eels live at depths to 3000 m and may have a worldwide distribution. Adults reach about 1.5 m. Known genera are *Notacanthus, Polacanthonotus,* and *Lipgenys.*

Order Anguilliformes (Apodes). This is a large order of marine and catadromous species comprising about 20 families. (Catadromous fishes are hatched in marine water and enter fresh water for growth before returning to marine water to spawn.) These true eels are thought to have their origin in an albuloid-like ancestor, but they are so specialized that the relationships are obscure. The main link between the elopiformes and eels is the leptocephalus larvae, known from both groups. The eels are greatly elongate with large numbers of vertebrae, and the pectoral girdle is not connected with the skull. The dorsal and anal fins are confluent with the caudal fin. The gas bladder is present and usually physostomous. Pyloric caeca are absent, as are oviducts.

Osteological characteristics include the lack of mesocoracoid, opisthotic, symplectic, and posttemporals. Scales are absent in many families. Gill openings are usually small and placed back from the edge of the concealed operculum. Modern eels all lack the pelvic fins and girdle, and many have no pectorals.

Suborder Anguilloidei. Most of the eel families are in this suborder. These have the premaxillaries fused with the mesethmoid to form a tooth-bearing bone at the anterior point of the upper jaw. In some the lateral ethmoids and vomer may be involved. The gas bladder is present, as is the pneumatic duct.

Perhaps the best-known eels are the so-called freshwater eels of the family Anguillidae (Fig. 2–6**D**). These are widely used as food, especially in Europe and Asia, and are excellent examples of the catadromous life history, in which growth to maturity occurs in fresh water and spawning takes place in the ocean. The life history of these eels was not understood until early in this century. Although leptocephali were known to science, it was not until 1896 that two Italian scientists learned that they metamorphosed into young eels (elvers). Following this discovery, Johan Schmidt of Denmark began research to find where the leptocephali originated. After many years of study, he found that the closely related European eel, *Anguilla anguilla,* and American eel, *A. rostrata,* spawn in adjacent or overlapping areas of the Sargasso Sea. The fragile leptocephalus larvae then drift for months or years before they become transparent "glass eels" ready to enter fresh water. (The journey of the American eel larvae is about one year, and that of the European larvae is up to three years.) They metamorphose into pigmented elvers in coastal water before entering estuaries or fresh water, where they feed and grow.

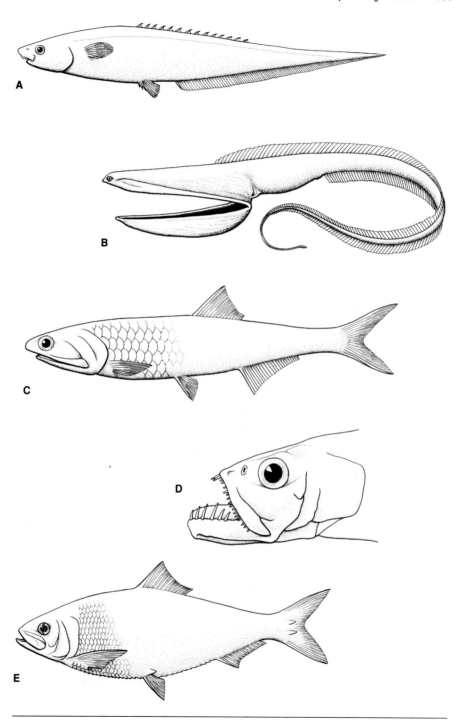

FIGURE 8–3 Representatives of order Anguilliformes. **A,** spiny eel (*Notacanthus*); **B,** gulper (*Eurypharynx*); representatives of order Clupeiformes. **C,** anchovy (*Engraulis*); **D,** head of wolf herring (*Chiro-centrus*); **E,** shad (*Alosa*).

The European eel is distributed in fresh water and estuaries from the Arctic Circle to the coast of Africa, through the Mediterranean Sea to the Black Sea, and in islands from Iceland to the Azores and Madeira. The distribution of the American eel is from northern South America to Labrador and (rarely) Greenland. Males of these eels spend about four to eight years in fresh water before returning to the spawning area at a length of about 60 cm. Females of the European eel may not migrate until they have spent about 12 years in fresh water and have reached about 1.5 m long. American eel females migrate at about 1 m long. Before they start the migration, they undergo a change in color from yellowish brown to silver and increase the size of the eyes. European eels require about a year to reach the Sargasso Sea.

The two species are distinguished mainly by the number of vertebrae. The American eel has about 107 vertebrae and the European eel about 114. Some biologists consider them to be a single species (see Robins et al., 1991b). The European eel has considerable value as a commercial fish, but the demand for the American eel is lower.

The spawning area of the Japanese eel, *A. japonica,* was not known until 1991. The species spawns in the summer west of the Mariana Islands at about latitude 15° N and longitude 140° E (Tsukamoto, 1992; Tsukamoto et al., 1989). This area is in the influence of the North Equatorial Current, so the larvae drift passively to the east at depths of 70 to 150 m until, at a length of about 20 mm, they begin daily vertical migrations that allow them to transfer to the north-flowing Kurashio Current. The transfer is aided by the trade wind (Kimura et al., 1993). In the Orient, especially Taiwan and Japan, the glass eels of the Japanese eel are collected by the millions and transferred to culture stations, where they are fed heavily until they reach marketable size.

Another anguilloid family is Heterencheylidae, a small group of slender eels without pectoral fins. They live mainly in the Mediterranean and tropical Atlantic, although at least one species lives in Panama Bay of the eastern Pacific. They are adapted to living in mud or sand. Genera are *Heterenchelys* (= *Pythonichthys*) and *Panturichthys.*

The family Moringuidae contains the worm or spaghetti eels, most species of which live in the tropical Indo-Pacific. Some are known from the Caribbean. These are filiform with low fins, and as juveniles they burrow in sand or mud. There are two genera, *Moringua* and *Neoconger.*

Suborder Muraenoidei. Eels of this suborder have no scales, and the lateral line is reduced. Members of the family Chlopsidae (Xenocongridae), the false morays, are found in tropical and subtropical waters. These are small, and some lack pectoral fins. Lateral line pores are found only on the head. *Kaupichthys nuchalis* of the Caribbean area is reported to hide inside sponges. There are about eight genera.

The family Myrocongridae has pectoral fins, a short lateral line, and a compressed body. There are two species: *Myroconger compressus,* a pale, deep bodied eel from the eastern tropical Atlantic, and *M. gracilis,* of which but three specimens are known.

Large eels are found among the morays of the family Muraenidae. These are widespread in tropical and subtropical waters and probably reach their greatest development around coral reefs. Although most species do not reach more than 1 m long, some species reach at least 2 m, and there are reports of individuals reaching 3 m. Many species are characterized by bold colors and patterns, and most have strong jaws and many sharp teeth. They have dorsal, anal, and caudal fins but lack pectoral fins and scales. Fortunately, large morays seldom attack divers, but when they do they can inflict grievous wounds. The flesh of some morays has been known to carry ciguatera poisoning (see Chapter 31). Genera, of which there are about 15, include *Muraena, Gymnothorax, Echidna,* and *Enchelycore.* There are nearly 400 species in this family. Note Box 8–1 concerning food of eel larvae.

BOX 8–1

Food of Eel Larvae

Food of eel leptocephali has long been a mystery because all guts examined for many years appeared to be empty (Moser, 1981). Mochioka and Iwazumu (1993) report finding material in the guts of leptocephali of species of eels of the families Congridae, Muraenidae, Muraenesocidae, Nettastomidae, and Ophichthyidae. In all instances, the gut contents were fecal pellets and remains of the gelatinous "houses" of larvaceans, which are tunicates that secrete coverings in which they live or to which they attach. Apparently, the large teeth of the leptocephali are used to grasp and hold the larvacean while the liquid in the house is sucked out. The liquid contains fecal pellets and other organic material that can be used as food.

Suborder Congroidei. Fishes of this suborder have the frontal bones fused together. Not all members have pectoral fins, and some lack scales. The family Congridae, congers and garden eels, has some large members, such as *Conger conger,* which reaches 2.7 m and is an excellent food fish. Congers are marine and are found in tropical and temperate waters. The pike congers of Muraenesocidae have narrow, long jaws with large teeth. Some species that reach more than 2 m are sought as food. They are found in the Indian and Pacific oceans, some in deep water. The suborder includes the following families: The Synaphobranchidae (neck or cutthroat eels), of tropical to temperate waters, have the gill openings confluent mid-ventrally. The family contains the subfamily Simenchelinae, to which belongs the parasitic or snubnose eel, *Simenchelys parasiticus.* This reportedly has the ability to burrow into and feed on larger fishes, a habit similar to that of the hagfishes (see Robins and Robins, 1976). Ophichthidae (snake eels), found in tropical to warm, temperate waters, has nearly 250 species. Colocongridae, of the Atlantic and Indo-West Pacific, has species that are characterized as "the least elongate anguilliform" (Nelson, 1994, p. 111). The family Derichthyidae, the longneck eels, has species that are mesopelagic to bathypelagic.

The snipe eels (Fig. 2–6E) (Nemichthyidae) are characterized by elongate jaws that curve away from each other toward the tips. The genera *Avocettina, Labichthys,* and *Nemichthys* are filiform and greatly elongate, with up to 750 vertebrae. The long dorsal and anal fins are confluent with the small caudal fin. These are deep water eels that have been caught at depths from about 500 m to over 7000 m. They range in length from around 0.5 m to 2 m, with most of the length in the tail. Nettastomidae (duckbill eels) are elongate, with up to 280 vertebrae. They are found in the Indian, Pacific, and Atlantic oceans. Serrivomeridae (sawtooth eels) are pelagic in warm to temperate waters.

Order Saccopharyngiformes. The gulpers and swallowers are an order of strange, eel-like fishes of deep marine waters. They lack the maxilla, premaxilla, opercular bones, and branchiostegals. The hyomandibular and the rest of the suspensorium project backward, so in some species the mouth can be much longer than the head. The gills are small and plume-like. Some species are elongate and reach a length of 1.8 m. Leptocephali are deep bodied and have V-shaped myotomes (Böhlke, 1966; Robins, 1989).

Suborder Cyematoidei. *Cyema* and *Neocyema,* the bobtail snipe eels or arrow eels, are much shorter than the snipe eels of Nemichthyidae, as if they have had the posterior part of the tail chopped off. These eels reach about 15 cm long. The dorsal and anal fins begin in the posterior half of the body and extend to the truncated end of the tail, where they extend backward like the feathers on a dart. As in the snipe eels, the long mouth cannot be fully closed. There are two species in the single family Cyematidae, of deep water in all warm seas. *Cyema atrum* has been found as deep as 5100 m.

Suborder Saccopharyngoidei. The suborder of the gulper eels contains some of the strangest known vertebrates. The three families that make up this order are all deep sea in habitat and are greatly specialized, in part by degeneration. They lack gas bladder, scales, and ribs. The caudal and pelvic fins are reduced or absent. The hyoid arch is reduced to only the hyomandibular, and the gill arches are not only degenerate but are situated well behind the head, with no bony or ligamentous connection to it. These eels lack the cleithrum, supracleithrum, lateral ethmoids, supraoccipital, orbitosphenoid, basisphenoid, and alisphenoids. The mouth is from three to seven times as long as the neurocranium (Fig. 8–3**B**).

The Saccopharyngidae, the swallowers or whiptail gulpers, contains nine species with large mouths and large, recurved teeth. These eels have extremely capacious stomachs and feed on fishes. Bioluminescent organs are present on the tail portion of the swallowers, which reaches a length of 2 m. Some, with an attenuated tail, have as many as 300 vertebrae.

The gulper of Eurypharyngidae has a tremendous mouth and small teeth. It reaches a length of about 60 cm. Juveniles have caudal fins, but the adult loses the caudal. These eels have five complete gills and six gill clefts inside the "opercular" cavity. The stomach is nondistensible, and the food is mainly bathypelagic crustacea. The remaining family of the suborder, Monognathidae, lacks the typical upper jaw bones—maxillaries and premaxillaries—but its members nonetheless are predators, probably feeding on shrimp. The seven or so known species of the family are known from about 70 specimens from deep water. They are among the most modified of the fishes. A bifid spine is developed from the front part of the neurocranium and protrudes anteriorly in the roof of the mouth as a rostral fang (Bertelsen et al., 1989). This "ethmoid tooth" is hollow and equipped with a gland with a duct to the

tubes. The males appear to change considerably at maturity, resorbing the mandible, losing the typical abdominal pouch, and greatly increasing the size of the olfactory organs (Bertelsen et al., 1989). Length is up to 11 cm.

The fossil order †Crossognathiformes may be aligned with the eels.

Subdivision Clupeomorpha

The herrings and anchovies make up the cohort Clupeomorpha, the living members of which are in the order Clupeiformes. The gas bladder is extended forward in two branches that enter the skull and terminate in small vesicles within ossified bullae (small, somewhat globular structures), forming an otophysic connection with the utriculus of the ear. In the caudal fin, the urostyle is made up of the centrum of the terminal vertebrae and the first uroneural. The first hypural is separated from the urostyle (see Lauder and Liem, 1983b). Many have scutes that project posteriorly along the ventral mid-line (those that also have scutes along the dorsal mid-line are termed "double armored").

Order Clupeiformes. This order is considered to embrace only the herring-like fishes instead of the larger assemblages formerly placed under this name. These herring-like fishes have been thought to be related to the elopiforms but lack some of the primitive characteristics, such as the gular plate and conus arteriosus, encountered in those fishes. Clupeiforms show the usual plesiomorphies (ancestral characteristics) of the lower bony fishes—soft rays, abdominal pelvics, cycloid scales, etc. Part of the sensory canal system spreads over the opercle and subopercle. The clupeiformes usually have silvery deciduous scales. Many have compressed, keel-like bellies, often with specialized, posteriorly directed scales, called scutes, on the mid-line. They are distributed in time from the Upper Jurassic. All except the extinct order †Ellimmichthyiformes and suborder Denticipitoidei have a recessus lateralis, which is a chamber in the pterotic bone with which some of the cephalic lateral line canals connect.

Suborder Denticipitoidei. This suborder contains one family, Denticipitidae, the tooth-head or denticle herring, in which a complete lateral line is present and the skull bones bear dermal denticles. The only species, *Denticeps clupeoides,* is confined to West Africa, mainly Nigeria.

Suborder Clupeoidei. Fishes of this suborder lack the lateral line along the sides of the body. Connection of the gas bladder with the gut is various. The caudal fin has 19 principal rays, contrasting with 16 in Denticipitoidei.

The family Engraulidae contains the anchovies, small silvery fishes with a rather rounded body cross section and a long snout and maxillary. The snout projects well beyond the lower jaw (Fig. 8–3C). Like herrings, anchovies are the objects of great fisheries. One of the greatest is the anchovetta fishery off Peru in the region of the Peru Current, where tremendous biological production occurs. The anchovetta, *Engraulis ringens,* forms the basis for a fishery that produces a harvest of up to 13 million metric tons in some years, in addition to the tremendous tonnage eaten by cormorants and other guano birds. Production and availability of the anchovetta depend on the ocean currents. In years when the wind patterns are wrong, the fishermen (and the birds) go nearly fishless. The *El Niño* wind blows warm water onshore perhaps every seven or eight years and causes population shifts in both birds and fishes. The currents caused by the winds may advect warm water north to the Pacific shores of North America. Anchovies are usually less than 25 cm long, and most species live in tropical or subtropical waters. There are 15 to 20 genera and over 100 species.

The ilishas, mostly of tropical and subtropical waters, have usually been considered a subfamily of Clupeidae, but recent works have treated them as a separate family, Pristigasteridae (Grande, 1982; Nelson, 1994; Whitehead, 1985). They have long-based anal fins and strong scutes on the ventral mid-line. Some species lack pelvic fins, and the genus *Raconda* has no dorsal fin. There are nine genera and about 35 species, some of which, such as *Ilisha africanus* and *I. elongatus,* have significant importance in commercial fisheries.

The family Chirocentridae contains the wolf herrings or dorabs, *Chirocentrus dorab* and *C. nudus,* which differ in several respects from typical herrings. *C. dorab* may reach a length of 3.5 m and has numerous large canine teeth (Fig. 8–3**D**). The gas bladder is divided internally by septa, and there is a bony appendage in the axilla of the pectoral fin. Wolf herrings are found in the Indo-Pacific and are used as food in some areas. Large specimens are said to be dangerous to fishermen.

The family Clupeidae includes the herrings, pilchards, shads, sardines, and similar fishes and ranks as one of the most important groups of fishes. Many species occur in dense schools, so they are subject to mass capture, and their oily flesh makes them the object of fisheries all over the world. The uses to which they are put are numerous—from food for humans and domestic animals to fertilizers and oils. The annual world harvest of clupeids has been estimated at 20 million metric tons.

Some important members of the family are herring, *Clupea harengus,* which has subspecies in both the north Atlantic and north Pacific; menhaden, *Brevoortia* spp., of the western Atlantic; and the American shad, *Alosa sapidissima* (Fig. 8–3**E**), which is sought for both food and sport. Others are *Sardinops sagax,* the Pacific sardine, once extremely important on the California coast but now depleted; and the sprat, *Sprattus sprattus,* of the North and Baltic seas. In all there are about 186 species of herrings in about 56 genera (Whitehead, 1985). Most are small fishes, seldom exceeding 0.5 m long. However, *Palonia castelnaui,* a freshwater herring of South America, reaches at least 1.5 m (Dr. Barry Chernoff, personal communication). The smallest known clupeomorph is also from South America. *Amazonsprattus scintilla,* described as a clupeid, reaches about 20 mm long and is mature at about 15 mm (Roberts, 1984).

From the standpoint of ecology, herrings are important as converters of plankton into fish flesh, so they form a great food resource for the large pelagic predators. Their life histories vary. Some spawn pelagically, and others spawn on seaweeds or other substrate. Some are anadromous, and some live in fresh water.

Chapter 9

Euteleostei: Carps, Catfishes, and Gymnotoids

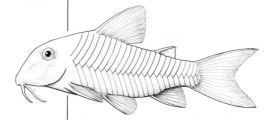

Subdivision Euteleostei

All the remaining orders of fishes covered in this chapter through Chapter 16 constitute the lineage (cohort, subdivision, or infradivision) called Euteleostei (Lauder and Liem, 1983a; Nelson, 1984, 1994; Rosen, 1973). Euteleostei probably is not a monophyletic group. It includes many diverse orders but seems to be at least provisionally accepted by many ichthyologists. Lauder and Liem (1983a, 1983b) point out that adipose fins and nuptial tubercles are known only from fishes of Euteleostei, but Nelson (1994, p. 125) points out that apparently there is no "unique character" common to all the species of the subdivision.

Superorder Ostariophysi

The fishes of this superorder are the dominant freshwater fishes of the world and include a few marine representatives as well. According to Nelson (1994), the more than 6500 species of this suborder make up about 27 percent of the known species of fishes in the world.

Ostariophysan fishes have modified anterior vertebrae and ribs, and most have a bony connection between the swimbladder and the ear—the Weberian apparatus—that aids in the reception of sound (see Chapter 21). In addition, these fishes, except gymnotoids (Fink and Fink, 1981), possess a peculiar alarm substance (a pheromone) (Pfeiffer, 1977), which is released into the water by injury to the skin and produces a fright reaction in other members of the superorder. Reaction to skin extracts of related fishes is not confined to fishes of this group (Smith, 1992). Percoid darters and certain gobies also have alarm substances.

Many of the ostariophysans have structures called unculi or "horny projections arising from single cells" (Roberts, 1982, p. 55). No other fishes are known to bear the structures, which are concentrated in areas of contact, such as lips, ventral surfaces, and sucking discs. They differ from the multicelled contact organs and breeding tubercles seen in these and other fishes (see Wiley and Collette, 1970).

The classification of the fishes now considered to belong in this superorder has been subject to different interpretations over the years. Classifications by some authors of the major taxa are shown in Table 9–1. The Ostariophysi are divided into two series (Fink and Fink, 1981): the Anotophysi, in which the anterior vertebrae are only slightly modified; and the Otophysi, in which a Weberian apparatus (Fig. 21–6) is present and in contact with the gas bladder and inner ear.

Series Anotophysi

Order Gonorynchiformes (Gonorhynchiformes). Anotophysi includes only the order Gonorynchiformes, ostariophysans with slightly modified anterior vertebrae and ribs. In some species, the peritoneum that covers the anterior of the gas bladder is thickened and in contact with the anterior ribs.

Suborder Chanoidei. This is a monotypic suborder containing only the well-known milkfish, *Chanos chanos* (Fig. 9–1A), family Chanidae, which is the object of pond culture in many areas of southeast Asia. It reaches a length of about 1.8 m (Wheeler, 1975). A streamlined fish of marine or brackish water, it has a large, forked caudal fin and is capable of great leaps. Milkfish grow rapidly on a diet of mainly planktonic algae. Young are collected for stocking in ponds by a variety of means, often

TABLE 9–1

**Changes in Classification of Ostariophysine Fishes
(selected systems since 1940)**

I. Berg (1940)
    Order Clupeiformes (in part)
        Suborder Chanoidei
        Suborder Gonorhynchoidei
    Order Cypriniformes
        Division Cyprini
        Suborder Caracinoidei
        Suborder Gymnotoidei
            Superfamily Sternarchoidae
            Superfamily Gymnotoidae
        Suborder Cyrinoidei
        Division Siluri
        Suborder Siluroidei
            Superfamily Diplomistoidae
            Superfamily Siluroidae
II. Greenwood et al. (1966)
    Superorder Protacanthopterygii (in part)
        Order Gonorynchiformes
        Suborder Gonorynchoidei
        Suborder Chanoidei
    Superorder Ostariophysi
        Order Cypriniformes
        Suborder Characoidei
        Suborder Gymnotoidei
        Suborder Cyprinoidei
        Order Siluriformes

III. Fink and Fink (1981)
     Superorder Ostariophysi
        Series Anotophysi
            Order Gonorynchiformes
        Series Otophysi
            Subseries Cypriniphysi
                Order Cypriniformes
            Subseries Characiphysi
                Order Characiformes
                Order Siluriformes
                    Suborder Siluroidei
                    Suborder Gymnotoidei
IV. Nelson (1984, 1994)
     Superorder Ostariophysi
        Series Anotophysi
            Order Gonorhynchiformes
                Suborder Chanoidei
                Suborder Gonorhynchoidei
                Suborder Kneroidei
        Series Otophysi
            Order Cypriniformes
            Order Characiformes
            Order Siluriformes
            Order Gymnotiformes
                Suborder Sternopygoidei
                Suborder Gymnotoidei

by providing shade along the beach in shallow water and then using a small seine or dip net to capture the fry that gather there.

Culture methods differ from country to country but usually involve a complex of ponds, sometimes interconnected to provide proper conditions for fry, fingerlings, and larger fishes being grown for the market. In Taiwan, special wintering ponds must be provided for protection, but in spite of the generally cooler climate there compared to some other places where milkfishes are cultured, yields of up to 2000 kg per hectare have been achieved.

Suborder Gonorynchoidei (Gonorhynchoidei). There is probably only one species in this suborder. It is the beaked salmon, or beaked sandfish, of the family Gonorynchidae. It is a small marine fish of the Indo-Pacific (Fig. 9–1B). It is slender, has a pointed snout and a single barbel, and burrows into sandy bottoms. It lacks orbitosphenoid, basisphenoid, urohyal, and gas bladder. The maxillaries are small, and the upper jaw is bordered only by the premaxillaries. Scales are ctenoid. A suprabranchial organ is present.

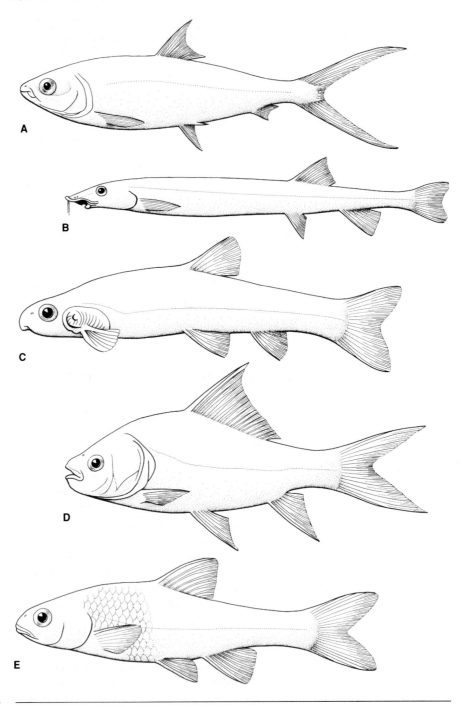

FIGURE 9–1 Representatives of order Gonorynchiformes, **A,** milkfish (*Chanos*); **B,** sandfish (*Gonorynchus*); **C,** shellear (*Kneria*). Representatives of order Cypriniformes, family Cyprinidae, **D,** catla (*Gibelion*); **E,** shiner (*Notropus*).

Suborder Knerioidei. This suborder includes the Kneriidae and Phractolaemidae, very small African freshwater fishes. Kneriidae has four genera of tropical Africa and the Nile drainage. *Kneria* has 12 species, one of which, *K. auriculata,* which is rarely more than 75 mm long, is called "shell-ear" because of the peculiar cup-shaped structure on the opercula of the males (Fig. 9–1C). These fishes are air-breathers that can remain out of the water for some hours and can wriggle up damp vertical structures to ascend streams. The shell-ear is said to estivate. The genus *Parakneria* has 13 species, one of which is the longest species in the family (15 cm). There are two monotypic (having only one species) genera, *Cromeria* and *Grasse-ichthys*. The family Phractolaemidae is monotypic. *Phractolaemus ansorgei* is a small fish (16 cm) that has an extremely small head and a peculiar small, protrusible mouth that opens upward. It is able to utilize atmospheric oxygen.

Series Otophysi

The otophysans are the dominant freshwater fishes in most of the world. This is a diverse group with more than 6000 species. All have a well-developed Weberian apparatus between the swimbladder and the ear that aids in sound reception. The otophysans are varied in body form, size, habits, and habitat. Although recent authors recognize two lineages based on structure of the otophysic connection, there seems to be no consensus on the relationships or classification. Fink and Fink (1981) recognize Cyprinophysi (which includes the minnows, carps, and suckers of the order Cypriniformes) and Characiphysi (which includes the characins, catfishes, and electric eels or gymnotoids). The following classification is that of Nelson (1994).

Order Cypriniformes. Cypriniformes contains the minnows, carps, loaches, suckers, and allied forms, all of which have toothless mouths and sickle-shaped, lower pharyngeal bones with teeth that usually bite upward against a cartilaginous pad borne on a posterior extension of the basioccipital bone. There is no adipose fin (except in the loach *Adiposia*), a dorsal fin is present, and the anus is in a normal position, not displaced far forward. A kinethmoid bone lies between the mesethmoid and the ascending processes of the premaxillaries. The orbitosphenoid, parietals, symplectics, and subopercular are present. The third and fourth vertebrae of the Weberian apparatus are not fused. Many members of the order have prominent intermuscular bones. Most of the cypriniforms are fully scaled, physostomous, and typically lack the true fin spines of the acanthopterygians. However, many do have enlarged and hardened rays, which are referred to as spines, at the front of the dorsal and anal fins. The cypriniforms were naturally distributed in Africa, Asia, Europe, and North America, but there were none in South America or Australia before they were introduced by humans. They are fishes of great importance, both ecologically and economically.

The Cyprinidae (carps and minnows) seems to have originated in southeast Asia but now is distributed in the fresh waters of Europe, Asia, Africa, and North America, with about 2000 species. About half the known species are native to Asia. Many important ornamental and aquarium fishes belong to this family, ranging from barbs, danios, and rasboras to the goldfish, *Carassius auratus,* and the carp, *Cyprinus carpio,* which has ornamental varieties. Because of its importance as a cultured food fish, the carp has been distributed almost worldwide from its original range in Eurasia. It has been the subject of selective breeding, which has produced high-backed, deep-bodied individuals with few scales, rapid growth, and great efficiency in food utilization. The carp is well accepted as food in Europe and Asia and is used to some extent elsewhere. In parts of North America, it is considered a pest that destroys the habitat of game fish while foraging by stirring up the bottom and uprooting aquatic vegetation that native species use for nesting and cover.

Some of the cyprinids reach a large size, weighing up to 130 kg or more. The mahseer of India, *Tor tor*, is one of the largest. Others are *Catlocarpio siamensis* of Thailand and *Gibelion catla* of India, which is a favored food fish (Fig. 9–1**D**). Other food fishes cultured in India are *Cirrhinus mrigala*, the mrigal; *Labeo rohita*, the rohu; and *L. calbasu*. Carps cultured in China include *Hypophthalmichthys nobilis*, the bighead carp; *H. molotrix*, the silver carp; and *Ctenopharyngodon idella*, the grass carp. The grass carp has been introduced to Europe and North America for use in control of aquatic vegetation. The grass carp is now established in the Mississippi drainage and is used in commercial fisheries in some places. The division of the family into subfamilies is subject to different interpretations of the relationships of the fishes. The subfamilies presented next are those accepted by Nelson (1994) following Howes (1991).

The subfamily Cypriniae includes about 700 species (many with two pairs of barbels) mostly from Africa, southeast Asia, and India, but there are many from other parts of Asia and Europe. Examples of fishes included are *Barbus*, the barbs; *Carassius*, the goldfish; *Cyprinus*, the common carp; *Cirrhinus*, the mrigal; *Gibelion*, the catla; and *Ctenopharyngodon*, the grass carp.

The subfamily Gobioninae is mostly distributed in eastern Asia, but *Gobio*, the gudgeon, is found from Europe to the Far East. Examples of genera are *Gnathopogon*, *Gobiobotia*, and *Sarcocheilichthys*.

The subfamily Rasborinae, found mainly in southeast Asia (Pakistan to China, south to Borneo) and Africa, provides many colorful fishes for the aquarium trade. Genera include *Barilius*, redfins and "barbs"; *Danio*, danios; *Engraulocypris*, the mukene; *Eosomus*, flying barbs; and *Rasbora*, rasboras.

The subfamily Acheilognathinae is found mainly in Europe and eastern Asia (not in central Asia or Siberia) to the Amur basin and Japan. *Rhodeus* is perhaps the most familiar genus. In this subfamily, a long ovipositor develops on the females as they approach spawning condition, and in most species it is used to deposit the eggs in the mantle cavity of molluscs. Other genera are *Acheilognathus* and *Tanakia*. The three genera mentioned here all occur in Japan.

The subfamily Leuciscinae is distributed in North America and northern Eurasia. Eurasian genera include *Abramis*, bream; *Blicca*, silver bream; *Elopichthys*, kanyu; *Leuciscus*, ide; *Hypophalmichthys*, silver and bighead carps; *Rutilus*, roach; and *Tribolodon*, "red belly dace" of Japan.

The genus *Phoxinus* is considered to be holarctic, occurring in the northern parts of both the eastern and western hemispheres. The genus includes the minnow of Eurasia and the blackside and redbelly daces of North America.

There are about 270 native species of minnows in North America, all of this subfamily. Most are distributed east of the continental divide. The majority are small fishes that can serve as forage for larger predators. After the revision of the North

American minnows by Mayden (1989), the genus with the largest number of species (about 70) is *Notropis* (Fig. 9–1**E**). Another species-rich genera is *Cyprinella* (about 25). Both of those are found mainly east of the Rocky Mountains.

The largest of the North American minnows are the squawfishes of the genus *Ptychocheilus*. They are found in the Colorado, Sacramento, Columbia, and contiguous drainages and are noted as voracious predators. The Colorado squawfish was reported by early ichthyologists to reach 1.5 m long. The northern squawfish now reaches about 75 cm, though Pliocene-age ancestors in Oregon and Idaho reached 1.5 m.

There are about 40 genera of leuciscines native to North America. Most have common names, including "shiner," "minnow," and "dace." Additional examples of those found mostly in the eastern drainages are *Campostoma*, stonerollers; *Couesius* (reaches west to the Fraser and Yukon rivers); *Dionda; Erimystax; Luxilus; Nocomus; Notimegonus; Phenacobius;* and *Pimephales*. Western minnows include *Gila* (with about 20 species), *Hesperoleucus, Lavinia, Pogonichthys, Richardsonius,* and *Orthodon. Rhinichthys* is a genus that is widespread and is found from coast to coast.

The family Gyrinocheilidae, the algae eaters, comprises fishes of the mountainous areas of southeast Asia. They hold themselves in place in swift water by sucking onto the substrate with the mouth. They have two gill apertures on each side, the inhalant opening above the exhalant opening. They have numerous gill rakers (up to 140) and feed only on algae.

The family Catostomidae, the suckers, is closely related to the minnows. Its members are thought to have had their origin in Asia, even though only one representative of the ancient suckers, *Myxocyprinus,* is found there now. Except for this species and *Catostomus catostomus* (Fig. 9–2**A**), which invaded Siberia from Alaska during the Pleistocene interglacial, the suckers are North American. Suckers have a single row of 16 or more pharyngeal teeth. Many of the members of the family have ventral mouths with thick papillose lips, exemplified by *Catostomus,* which has species in most major drainages of North America. The largest of the genus is *C. luxatus,* the Lost River sucker of the Klamath drainage in Oregon and California, an endangered species that reaches about 1 m.

Buffalofishes, *Ictiobus* spp. (Fig. 9–2**B**), and the quillbacks and carpsuckers, *Carpiodes* spp., are large, carplike catostomids of the Mississippi and contiguous river systems. They are used as food and occasionally are cultured in the southern United States.

Cobitidae (= Cobitididae), the loaches, are small, slender fishes of Eurasia and Africa (Fig. 9–2**C**). They have three to six pairs of barbels around the mouth, and the gas bladder is encapsulated in bone. There are around 18 genera and more than 100 species. Some (*Misgurnus* spp.) are known as "weatherfishes" in Europe because of their reactions to changing barometric pressure. Others—*Botia* spp., for example—are kept in aquaria because of their striking color patterns and vigorous activity. Many species swallow air to supplement their oxygen supply. Those that pass the air through the anus are called "squeakers."

The family Balitoridae, formerly known as Homalopteridae, contains the freshwater river loaches of Eurasia. These have at least three pairs of barbels close to the mouth and a Weberian apparatus that differs in structure from that of the cobitids. Nelson (1994) recognizes two subfamilies.

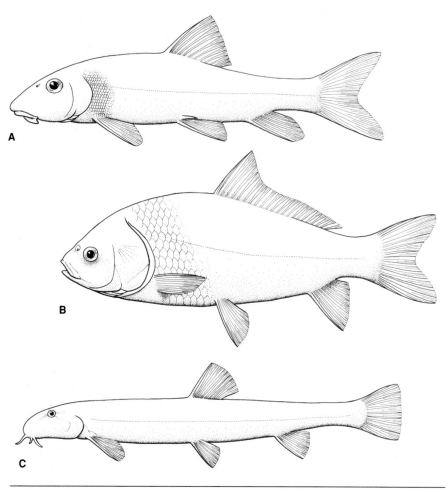

FIGURE 9–2   Representatives of order Cypriniformes, **A,** sucker (Catostomidae, *Catostomus*); **B,** buffalofish (Catostomidae, *Ictiobus*); **C,** loach (Cobitidae, *Cobitus*).

The subfamily Nemacheilinae includes several genera (up to 31, according to some authorities) of small fishes from Eurasia. Hundreds of nominal species have been described from this subfamily. Some, especially of the genus *Nemacheilus*, are exported from southeast Asia for use as aquarium fishes. A blind loach, *Nemacheilus smithi,* is known from caves in Iran. The genus *Yunnanilus* has numerous species in Yunnan province, China, and in Inlé Lake, Burma. One species lives in hot springs at 5200 m elevation, the highest fish habitat known (Kottelat and Chu, 1988). Another species lives in caves to 400 m below the surface.

The subfamily Balitorinae (flat loaches) contains fishes that are flattened on the ventral surface and generally have large pectoral and pelvic fins modified as adhesive organs. Most live in swift hill streams in southern Asia, including parts of India and China, Taiwan, and Borneo. Some have the gill openings reduced in size, and all have ventral mouths. There are usually three or four pairs of barbels around the mouth. These fishes are often called hill stream loaches, but some live in other habitats. Nelson (1994) mentions a species of *Homalopterus* from caves of Thailand that has no eyes or scales. There are nearly 30 genera and at least 120 species in the subfamily.

Order Characiformes. Species of this otophysan order usually have opposing pharyngeal teeth, an adipose fin, and separation of the first hypural from the compound ural centrum. The orbitosphenoid, parietal, symplectic, and subopercular bones are present. Jaw teeth are present, and barbels are absent.

This tropical order contains characins (Characidae) and their relatives. These are found in fresh water in both Africa and South America north through Central America and Mexico to the Rio Grande in Texas. Their continental distribution is no doubt due to dispersal prior to the separation of Africa and South America, but no genera are common to the two continents. About 25 genera and 200 species occur in Africa and over 200 genera and over 1300 species in the Americas.

Most authorities recognize several families—usually from 10 to 16 (e.g., Eschmeyer, 1990; Greenwood et al., 1966; Lauder and Liem, 1983a)—but others prefer to regard most of these as subfamilies (Gosline, 1971; Robins et al., 1991). The families listed here are those accepted by Nelson (1994).

The family Citharinidae of Africa is thought to be the most primitive of the characins (Fink and Fink, 1981). The fourth neural arch is not fused to the vertebra, there is a synchondral joint between the third and fourth vertebrae, and the pelvic bones are bifurcate anteriorly. Most have ctenoid scales, and some have slightly protrusible upper jaws. There are 20 genera and about 98 species (Nelson, 1994).

The subfamily Distichodontinae includes elongate predators and fin-eaters as well as compressiform herbivores and predators on insects and small crustacea. Members of the subfamily Citharinidae, the moonfishes (Fig. 9–3A), are deep bodied and have no teeth on the small maxillae. Moonfishes are good food fishes. They reach about 85 cm in length and 2.5 kg in weight.

Members of the family Hemiodontidae, of South America north to Panama, have no teeth in the lower jaw except in one genus (*Micromischodus*). The subfamily Parodontinae comprises algae-scraping, benthic fishes. The subfamily Hemiodontinae is found in open waters. The genus *Bivibranchia* and relatives have protrusible upper jaws. *Anodus* has no teeth in the jaws and has numerous gill rakers and other adaptations for feeding on plankton. There are about 50 species in the subfamily.

The family Curimatidae comprises detritus eaters of South America to Costa Rica. They have modifications of the gill chamber, such as loss of teeth on some pharyngeal tooth plates and an epibranchial organ. The subfamily Curimatinae lacks jaw teeth. The subfamily Prochilodontinae has teeth in the jaws and a protractile mouth with large lips. There are about 50 species in the family.

The family Anostomidae of South America comprises the headstanders of the aquarium trade. Many species habitually keep the head down while swimming. There are two subfamilies, Anostominae and Chilodontinae. The family has about 110 species.

The family Erythrinidae of South America includes some large-mouthed predators that can reach a length of 1 m. Some have lunglike gas bladders and are said to be able to move between waters over land.

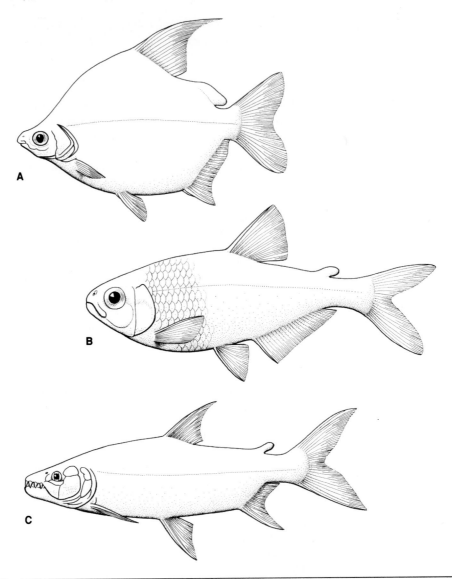

FIGURE 9–3 Representatives of order Characiformes, **A,** moonfish (Citharinidae); **B,** tetra (Characidae); **C,** tigerfish (Characidae, *Hydrocynus*).

The family Lebiasinidae of South America contains some aquarium fishes, such as *Lebiasina bimaculatus* of the subfamily Lebiasininae and the pencilfishes (*Nannostomus*) of subfamily Pyrrulininae.

The family Ctenoluciidae, the pike-characins, ranges from South America to Panama. These are large (to 1 m), predaceous fishes resembling pikes.

The family Hepsetidae of Africa is monotypic. *Hepsetus odoe* is a predaceous fish known as the Kafue pike. It reaches about 40 cm long. According to Géry (1977), this species spawns in a floating foam nest.

TABLE 9–2	**Subfamilies of Characidae**
Alestiinae	African tetras, Africa
Characinae	"true characins," South America
Tetragonopterinae	South American tetras, squarefins, South America
Iguanodectinae	South America
Glandulocaudinae	croaking tetras, South America
Serrasalminae	piranhas, pacus, South America
Rhoadsiinae	Ecuador to Costa Rica
Crenuchinae	sailfin characins, South America
Characidiinae	South American darters, South America

From Nelson (1994).

The family Gasteropelecidae of South America and Panama includes three genera, *Gasteropelecus, Carnegiella,* and *Thoracocharax.* Robins et al. (1991) list nine species that are exported as aquarium fishes. These are known as hatchetfishes or flying characins because of their ability to leap clear of the water with great force, aided by their winglike pectoral fins and powerful pectoral muscles.

The family Characidae contains numerous genera of diverse habits in both Africa and the Americas. There is little agreement among ichthyologists on the classification of the family. For instance, Géry (1977) and Nelson (1994) both list 11 subfamilies, but only four coincide.

Many small and colorful species of this family are kept in home aquaria and form the basis for a remarkable import-export trade. The tetras (Fig. 9–3**B**), such as *Hyphessobrycon* and *Hemigrammus,* are examples. The piranhas are notorious characins of the subfamily Serrasalminae, famed for their sharp teeth, strong jaws, and voracious feeding habits. Examples of genera of piranhas are *Serrasalmus* and *Pygocentrus.* Some other serrasalmines feed on small organisms, seeds, and fruit. The tigerfish *Hydrocynus* of Africa (Alestiinae) is another fierce predator (Fig. 9–3**C**), as is *Cynodon,* of South America. Table 9–2 lists the subfamilies of Characidae presented in Nelson (1994).

Order Siluriformes (Nematognathi). This is the order of the catfishes. They have no true scales, and the skin is bare or covered with bony plates, which may bear dermal denticles. The subopercle, symplectic, and parietal bones are absent. The Weberian apparatus may involve five or more vertebrae, with the second, third, fourth, and, in some, fifth vertebrae fused to one another. The premaxillary bears teeth, but the maxillary does not (except in the most primitive family and one higher genus) and is modified or is the basal skeletal unit of the maxillary barbel. Intermuscular bones are absent.

The pectoral and dorsal fins usually have large spines at the leading edges. These are modified soft rays and not homologous with the spines of acanthopterygians. Catfish spines are provided with a locking mechanism that holds them erect. Many species have venom glands associated with the spines, and wounds caused by the spines of certain species can be extremely painful or even fatal. All catfishes have barbels around the mouth, and an adipose fin is usually present. Fossil catfishes are known from the Paleocene and were diverse by the Eocene.

Various authorities recognize from 20 to over 30 families of catfishes and estimate that the total number of species is nearly 2500, of which over half are found in South America. Catfishes are known from all continents except Antarctica, where they occur as fossils. Those found in Australian fresh waters belong to the marine families Ariidae and Plotosidae (the former are widespread in warm seas, the latter in the Indo-Pacific). Most of the other catfish families are strictly freshwater fishes and are rarely found in water of more than a few parts per thousand salinity.

The family Diplomystidae (velvet catfishes) of South America is considered to contain the most primitive living catfishes. Their maxillae are normally developed and bear teeth. The fifth vertebra is not fused or firmly connected to the fourth, as in the remaining families. There are perhaps four species, of which *Diplomystes (Diplomyste) chilensis* and *Olivaichthys viedmensis* may be the best known. The South American Trichomycteridae (pencil catfishes) includes a number of genera that parasitically attack the gill cavities of larger fishes. One, *Vandellia,* the candiru, is attracted to nitrogenous wastes such as are normally produced by fish gills (Forster and Goldstein, 1969). They have been known to enter the urinary opening of mammals, including humans.

There are 15 families of catfishes in South America (Nelson, 1994). These include Aspredinidae (= Bunocephalidae), the banjo catfishes, some of which are used as aquarium fishes; Cetopsidae, the "whalelike" catfishes; Doradidae, the thorny catfishes, including armored forms that are used as aquarium fishes; Pimelodidae, the long-whiskered catfishes, including giant species to a length of 3 m; Helogenidae, marsh catfishes; and Hypophthalmidae, the loweye catfishes, with eyes set to look down. There is a group of South American catfish families that are considered monophyletic because they share gas bladders encapsulated in bone and teeth somewhat resembling placoid scales on the skin, armor, or fin rays. Unarmored families in this group are Trichomycteridae, the pencil catfishes; Nematogenyidae; and Astroblepidae, the climbing catfishes. Armored families are Callichthyidae (Fig. 9–4**A**), the plated catfishes; Scoloplacidae, the spiny dwarf catfishes; and Loricariidae, the suckermouth catfishes, which may encompass 550 species (Nelson, 1994). Many of the plated catfishes, especially of the genus *Corydoras,* are good aquarium fishes.

One of the best-known families of Old World catfishes is the Siluridae, the sheatfishes, which contains the giant wels of the Danube, *Silurus glanis.* This species may reach a weight of 130 kg or more (Fig. 9–4**B**). Many interesting silurids occur in Asia. One, *Wallagonia attu,* may reach 2 m long and, although slender, is a fierce predator; others are small and innocuous. In this family, the dorsal fin may be small or absent, and there is no adipose fin.

Another catfish family of interest is the Pangasiidae of Asia, which contains *Pangasianodon gigas.* This is one of the largest of freshwater fishes and reaches a length of 2.5 m. The closely related Schilbeidae occurs in both Asia and Africa, as does Bagridae. Clariidae of Asia and Africa and Heteropneustidae (Saccobranchidae) of Asia both contain air-breathing fishes. The former is equipped with an arborescent accessory breathing organ in the gill chamber, and the gill cavity of the latter communicates with large paired air sacs in the musculature of the body. These families are placed together in Clariidae by some ichthyologists because the two groups have a similar structure, except for the air-breathing apparatus.

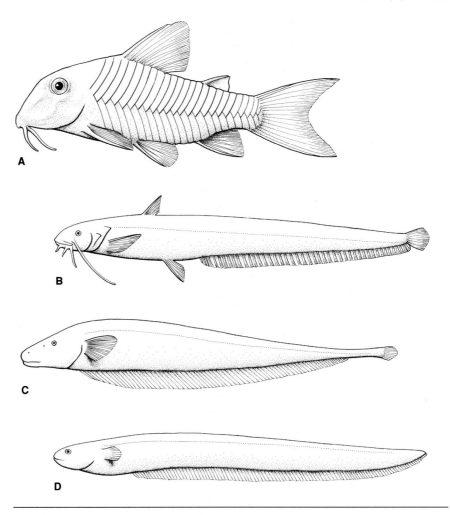

FIGURE 9–4   Representatives of order Siluriformes, **A,** armored catfish (Callichthyidae); **B,** wels (*Silurus*). Representatives of order Gymnotiformes, **C,** knifefish (*Apteronotus*); **D,** electric eel (*Electrophorus*).

Two families found mainly in swift mountain streams are Sisoridae, the sisorid catfishes (Turkey to China and Borneo), and Amblycipitidae, the torrent catfishes (Pakistan to Japan and Malaysia). The family Chacidae is known as squarehead or angler catfishes because of the large, flat, truncate head and the habit of wiggling maxillary barbels as a lure for other fishes. Chacidae is found from India to Borneo. Other Asian catfish families are Olyridae, Cranoglanididae (armorhead catfishes), Akysidae (stream catfishes), and Parakysidae.

The Malapteruridae of Africa contains electrogenic species, one of which, *Malapterurus electricus,* can deliver a severe shock. Another African family, Mochokidae, contains some species that habitually swim upside down. Some of these have reverse countershading, with dark bellies and lighter dorsal surfaces. The upsidedown swimming may aid in feeding on plankton in the surface layer and may help in obtaining oxygen from well-oxygenated water near the air-water interface

(Chapman et al., 1994). Amphiliidae, the loach catfishes, is an African family. The members are usually inhabitants of swift mountain streams.

The North American catfishes, Ictaluridae, which were native to western North America from Eocene to Pliocene, are now native only east of the Rocky Mountains but have been introduced and are established in most parts of the continent with suitably warm climate. The blue catfish, *Ictalurus furcatus,* is the giant of the group, weighing up to 70 kg. The flathead catfish, *Pylodictis olivaris,* is nearly as large, and the channel catfish, *I. punctatus,* may reach over 20 kg. The channel catfish is of great economic importance. It has long been a favored food fish and the object of a commercial fishery. It is cultured extensively throughout the warmer parts of the United States. Modern methods of fish culture, the availability of dry pelleted fish food, and consumer demand have made commercial rearing of channel catfish a profitable business.

The bullhead catfishes, *Ameiurus* (formerly in *Ictalurus*), are favored panfish and are sought by anglers across the continent. The small madtoms, *Noturus,* have venom glands associated with the pectoral spines and can cause painful wounds. There are three genera with blind species: *Trogloglanis* and *Satan* from Texas and *Preitella* from northern Mexico.

Order Gymnotiformes. This order has some characteristics in common with the catfishes and is placed into the Siluriformes by Fink and Fink (1981). Gymnotiforms are elongate fishes of South and Central America (Fig. 9–4C). They lack a dorsal fin but may have an elongate, fleshy filament on the back that resembles an adipose fin. The anus and urogenital apertures are placed far forward, usually in advance of the pectorals; the anal fin usually begins just behind the anus and stretches to the end of the tail. The caudal fin is usually lacking. There are no pelvic girdles or pelvic fins. The palatines are unossified; there are no ectopterygoids and no claustrum in the Weberian apparatus. These fishes are producers of electricity. Although Eschmeyer (1990) lists all the genera in this order under the family Gymnotidae, we follow Nelson (1994) in recognizing two suborders.

Suborder Sternopygoidei. This suborder contains weakly electric fishes that are capable of electric communication and electrolocation. Families usually recognized are Sternopygidae, Rhamphichthyidae (knifefishes), Hypopomidae, and Apteronotidae. Many of these have a bizarre appearance, and some resemble *Gymnarchus* or mormyrids, electric fishes of Africa—except for the dorsal and anal fins, of course.

Suborder Gymnotoidei. This suborder includes the families Gymnotidae and Electrophoridae. The latter contains the electric eel, which is a large, elongate, but heavy-bodied fish with about 250 vertebrae. The electric eel has about 80 percent of the body in the caudal region, which is largely made up of the electric organs. This species, *Electrophorus electricus* (Fig. 9–4**D**), is one of the strongest electrical fishes and is capable of producing 650 volts, although the average is about 350 volts. The electric eel has the buccopharyngeal cavity modified for air breathing. Gymnotiformes have electric organs and receptors that function in electrolocation (see Chapter 19).

Euteleostei: Pikes, Smelts, and Salmons

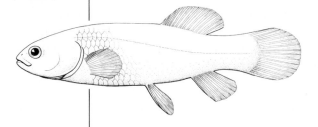

Superorder Protacanthopterygii

This taxon has been the object of many studies and has been the subject of several different interpretations, but Nelson (1994) considers that it includes the orders Esociformes, Osmeriformes, and Salmoniformes. This scheme will be followed in this discussion. Information on the superorder can be found in the following references: Begle (1991), Fink (1984), Greenwood et al. (1966), and Nelson (1984).

Order Esociformes. These fishes (Figs. 2–6**G** and 10–1**A**) constitute a well-recognized order that Lauder and Liem (1983b) suggested may be the most primitive of the euteleosts. Although usually accorded ordinal rank, they have been treated variously from suborder (Berg, 1947) to superorder (Jamieson, 1991). The order has

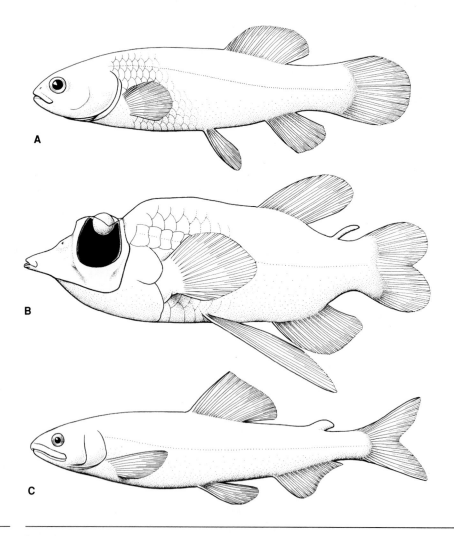

FIGURE 10–1   Representative of order Esociformes, **A,** mudminnow (Umbridae, *Umbra*). Representatives of order Osmeriformes, **B,** barreleye (Opisthoproctidae, *Macropinna*); **C,** ayu (Osmeridae, *Plecoglossus*).

some character states that are present in some more advanced groups, such as loss of the orbitosphenoid and mesocoracoid and loss of teeth from the maxillaries, which form a relatively small part of the mouth border. These fishes lack an adipose fin. Other characteristics include the presence of preethmoids (a primitive characteristic) and the representation of the mesethmoid by paired proethmoids. The esociforms are soft-rayed physostomes.

The family Esocidae contains only the pikes (*Esox* spp.), which are medium-sized to large, carnivorous fishes of lakes and slow rivers. Pikes are rather elongate fishes, with snout and jaws extended into a long, flattened mouth set with sharp teeth. They are known for their voracious feeding habits. The dorsal and anal fins are situated far back on the body. The northern pike, *E. lucius,* found in both North America and Eurasia, is a large fish that can weigh over 25 kg. Even larger, but not as widely distributed, is the muskellunge, *E. masquinongy,* found in the upper Mississippi drainage, the Great Lakes, and some contiguous drainages. The "muskie" is one of North America's greatest trophies for game fishermen. The Amur pike, *E. reicherti,* is found in Siberia. Some smaller species in North America are *E. niger,* the chain pickerel, which may reach about 2.25 kg; and *E. americanus,* which is smaller.

Pikes usually spawn in the early spring on flooded vegetation. Hatching occurs after about two weeks. The young initially feed on small organisms, such as water fleas (*Daphnia*) and insects, but turn to a fish diet after a few weeks. Pikes are generally sought as game fishes, but in Europe and Asia they support commercial fisheries as well.

Mudminnows of the family Umbridae are found in both North America and Europe. Unlike the pikes, these are small fishes, usually less than 15 cm long. There are four species: *Umbra krameri* of central Europe, *U. limi* of central North America, *U. pygmaea* of the Atlantic Coast of North America, and *Novumbra hubbsi* of the Olympic Peninsula in the state of Washington. All appear to prefer very slow water—usually bogs, stagnant ditches, and streams of low gradient. They will hide in the mud when disturbed, and *U. limi* is reported to survive dry periods by burrowing into the bottom sediments. The disjunct distribution of the mudminnows indicates that they were more widely distributed prior to glaciation.

The family Dalliidae, represented by the Alaskan blackfish, *Dallia pectoralis,* is usually included in the Umbridae, but it has many characteristics that appear to distinguish it from that family. The scapula, coracoid, and the two platelike radials of the pectoral fin are unossified. *Dallia pectoralis* lacks postcleithra and inframandibulars, and its pelvics have three rays in contrast to the six or seven of Umbridae. It is found on the Chukot Peninsula of Siberia and in Alaska, where it inhabits slow streams, lakes, and bogs. The winters are long and cold in these areas, so the blackfish must be inactive for a great part of the year and usually passes the coldest part of the winter buried in the bottom materials. Some sphagnum bogs freeze to the bottom, and the blackfish are sometimes immobilized in ice. They can be frozen externally, but if their internal body temperature does not drop low enough to crystallize the body fluids, they can survive. They appear to withstand oxygen-deficient periods in summer by utilizing atmospheric oxygen. Blackett (1962) reported rather slow growth for the species—165 mm at age three. Spawning is from May to August.

Individuals of the Alaskan blackfish were once kept at Oregon State University for a few years in a pan of water in a refrigerator and were used in several different experiments on their oxygen consumption. Although blackfish do not usually exceed 20 cm, they are useful as food for sled dogs and are occasionally used as food by humans.

Order Osmeriformes. In his treatment of the phylogeny of smelts and related groups, Begle (1991, p. 46) indicated that "it might be expedient in the future to elevate" the clade Osmerae to ordinal status. This has been done in Nelson (1994) and is followed here. The order Osmeriformes contains fishes that have been subject to many interpretations but have usually been considered to belong in Salmoniformes. The order includes argentines and allies, smelts, salangids, galaxiids, and related groups.

According to Begle (1991), the smelts and argentines form a monophyletic group because of the following characteristics: loss of the orbitosphenoid and basisphenoid; loss of nuptial tubercles; and the presence of a cartilaginous vane ventrally on the first basibranchial. Patterson and Johnson (1995) believe Osmeridae is closer to Salangidae and plecoglossidae. Many of these fishes have an adipose fin, which usually contains a small cartilage (Matsuoka and Iwai, 1983). A few emit an odor similar to that of cucumbers—which is caused by trans-2-cys-6-nonadiol and has been noted in smelts, argentinids, and a few other fishes (McDowall et al., 1993)—and thus must have originated independently more than once (Begle, 1991).

Suborder Argentinoidei. Argentinoidei includes about 160 species of deep sea fishes, variously arranged in up to nine families but usually five to seven (Begle, 1992; Eschmeyer, 1990; Nelson, 1984, 1994). In these fishes, the caudal fin is forked, the gas bladder is physoclistic or absent, and an epibranchial organ (crumenal organ), apparently for consolidation of small prey, is present. There are some modifications of the posterior gill arches. Photophores are present in some species.

Superfamily Argentinoidea. This superfamily contains four families: the Argentinidae or argentines, Bathylagidae or deep sea smelts, Opisthoproctidae or barreleyes (Fig. 10–1B), and Microstomatidae or pencilsmelts, which are listed in Argentinidae by Robins et al., 1991. Argentinoids are characterized by a silvery color and no teeth on the maxilla, an adipose fin is usually present, and the dorsal fin is close to the middle of the body. Some species of the argentines are harvested commercially.

Superfamily Alepocephaloidea. These are dark-colored fishes of the deep sea. They have the dorsal fin set over the anal fin, and they lack an adipose fin and gas bladder. Families usually included in this superfamily are the slickheads, Alepocephalidae; the apparently primitive Leptochilichthyidae; and the tubeshoulders, Platytroctidae (= Searsidae). The common name *tubeshoulder* is derived from the opening of a gland that lies beneath the cleithrum and produces a luminous substance. Begle (1992) considers all the Alepocephaloidea to belong in the single monophyletic family Alepocephalidae without any subfamilies.

Suborder Osmeroidei. This group contains diadromous or freshwater fishes of both the northern and southern hemispheres. They are mostly small, usually less than 30 cm long. Osmeroids have reduced mesopterygoid teeth, reduced articular bones, a short shaft of the vomer, and a ventral condyle on the pelvic plate (Begle, 1991).

Superfamily Osmeroidea. The family Osmeridae contains the smelts, small fishes found in both fresh and salt water of temperate and cold parts of the northern hemisphere. There are seven genera and eleven species of smelts. They are slender, silver-sided fishes of delicate flavor and are popular as food fishes. Most species prefer to spawn on sand or small gravel. Some, such as the eulachon (*Thaleichthys pacificus*), are anadromous, whereas others spawn on ocean beaches at high tide, usually in areas of some freshwater seepage. Large congregations of spawners make it easy to capture them with various dip nets. The holarctic capelin (*Mallotus villosus*) supports a commercial fishery in the North Atlantic, where the annual catch formerly exceeded a million metric tons.

The ayu of Japan, China, and Korea usually has been placed in the monotypic family Plecoglossidae but is closely related to the smelts, especially *Osmerus,* and is now placed with them in the Osmeridae (Begle, 1991). This species (*Plecoglossus altivelis*) (Fig. 10–1**C**) has a row of large, rather square-cut, chisel-like teeth on the maxillaries and dentaries (Howes and Sanford, 1987). The ayu feeds on diatoms and associated organisms growing on the rocks in river bottoms.

The ayu is an annual fish, without overlapping generations. It spawns in lower parts of rivers in the fall and early winter. The young are carried to sea and return to streams in late winter and early spring. Their growth is rapid, so they are large enough to sustain a fishery by early summer. Box 10–1 mentions some unusual types of fisheries for ayu.

BOX 10–1

Fishery for Ayu

The ayu is also known as "sweetfish" and is sought as a delicacy in Japan, where it formerly was the object of the cormorant fishery. In that fishery, tethered cormorants with rings around their throats to prevent them from swallowing prey were allowed to catch small fish. The birds were then pulled back into the boat, and the captured fish were taken from their mouths. Now the ayu is usually captured by snagging. One of the fishing methods is called "Tomozuri," which involves the use of a previously captured live ayu that is introduced, along with snagging hooks, to the territory of a free fish. When the resident fish comes to protect its territory, it is snagged.

The family Salangidae (icefishes, noodlefishes, and glassfishes), which many ichthyologists (including Nelson, 1994) consider to be closely related to Osmeridae, is aligned with the galaxioids by Begle (1991). These fishes are small (10 cm), slender, transparent fishes with a flattened head. Partly because of their resemblance to larval smelt, they are thought to be neotenic (Nelson, 1994). They are distributed in marine and fresh waters along the Asian coast from the greater Sunda Islands to the Amur River. During their spawning runs into fresh water, they are taken in commercial quantities. They may have only a one-year life span. The genus *Sundasalanx* is sometimes placed in its own family, Sundasalangidae. It lacks an adipose fin and has some osteological peculiarities, and the females of one species can mature at less than 15 cm (Roberts, 1981, 1984).

Superfamily Galaxiodea. These are fishes of temperate waters of the southern hemisphere. They lack a mesocoracoid and have 18 or fewer caudal fin rays (see Begle, 1991; Nelson, 1984). Three families are currently recognized: Retropinnidae, Lepidogalaxiidae, and Galaxiidae. In the family Retropinnidae, only the right gonad is present and a short horny keel is developed in front of the anus (McDowall, 1980). The family includes the southern hemisphere "graylings" of the genus *Prototroctes* (Fig. 10–2A) of Australia, Tasmania, and New Zealand (subfamily Prototroctinae); the southern smelts of the genera *Retropinna* of Australia, Tasmania, and New Zealand; and *Stokellia* of New Zealand (subfamily Retropinninae).

The southern graylings (*Prototroctes*) are not closely related to the salmonoid graylings (*Thymallus*) of the northern hemisphere but were given the name by early settlers because of a resemblance. These fishes have declined in numbers in Australia and Tasmania and are recently extinct in New Zealand. Changes in habitat due to land use and the introduction of exotic species are thought to have been responsible for the decline, but McDowall (1980) indicates that the species has declined in areas that have not been so affected.

The Australian grayling, *P. maraena,* deposits numerous (30,000 to 60,000 per female) small, demersal eggs in fresh water during the southern autumn. The larvae apparently drift to the sea, where they remain until spring, when they return to fresh water. Individuals are usually less than 30 cm long, but lengths over 45 cm have been noted. The southern smelts superficially resemble the osmerids but have the dorsal fin

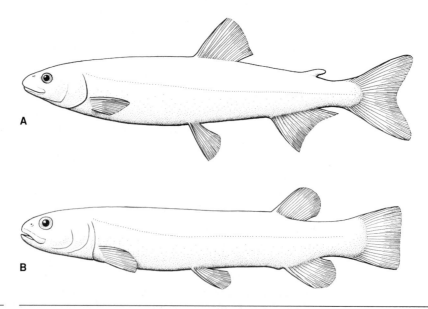

FIGURE 10–2   Representatives of order Osmeriformes, superfamily Galaxioidea, **A,** "southern grayling" (*Prototroctes*); **B,** kokopu (*Galaxias*).

fin far back near the adipose fin. They are translucent fishes usually less than 13 cm long and are often called cucumberfish because of their peculiar odor. Some species are apparently confined to fresh water, but others may be anadromous, ascending streams from the sea during the southern spring and summer (McDowall, 1980). Spawning takes place mainly in summer. Newly hatched fish drift to sea and return upriver when 5 to 6 cm long. On the upstream migration, these colorless young are often captured, along with other species, and used as food. These larval fishes are called "whitebait" and are cooked in patties. The spawning males have numerous pearl organs or nuptial tubercles. These species reach maturity in one year.

The monotypic family Lepidogalaxiidae has a restricted distribution in southwestern Australia. The salamanderfish, *Lepidogalaxias salamandroides,* is sexually mature at 40 mm. It is related to the osmeroid fishes (Begle, 1991), although Rosen (1974) placed it with the the Esocoidei. The dorsal fin is placed posteriorly as in pikes, mudminnows, and galaxiids. Caudal fin rays are reduced in number (nine) and are unbranched. The anal fin of the male has highly modified skeletal supports and fin rays and is covered by large scales set in thick skin (Rosen, 1974). This species resembles the galaxiids but differs in having fused frontal bones and greatly modified cephalic lateral line canal pores. This species is unusual in that it has no oculomotor muscles. The eye itself is immobile and is attached to the socket. The head, unlike the heads of most fishes, can be moved downward and laterally (McDowall and Pusey, 1983).

Lepidogalaxias can estivate in dry soil during droughts, although it does not appear to have accessory air-breathing organs (Berra et al., 1989). Pusey (1989) notes that it loses very little water during the first several weeks of estivation but does not accumulate much urea. Waters may dry for up to five months, during which there is significant mortality among estivating individuals. Reproduction follows estivation, and apparently females in which lipid reserves have dropped below minimum do not spawn. Males die after spawning at the age of about one year. Some females die following their first spawning, but others survive to spawn again the next year (Pusey, 1990).

The family Galaxiidae was once considered a separate order (Berg, 1940), but it has been variously placed into a suborder Galaxoidei or retained in the Salmonoidei. In this book it is included in the Osmeroidei, as treated by Begle (1991). The galaxiids lack the mesocoracoid and supramaxillae. These are mostly small fish less than 30 cm long.

The family Galaxiidae is viewed as comprising three subfamilies: Aplochitoninae, Lovettiinae, and Galaxiinae (Begle, 1991). Fishes representative of these groups include the peladillo, *Aplichiton zebra,* of southern South America and the Falkland Islands; the genus *Lovettia* of Australia and New Zealand; and several species of *Galaxias* and related genera.

Some of these species are diadromous. *Aplochiton marinus* of South America is said to spawn in the sea. Members of *Lovettia* are anadromous; the young drift to sea soon after hatching and return in a year. They are slender and transparent as they enter fresh water but darken as they mature. In mature males, the anus and urogenital opening are located in the anterior part of the abdomen.

Galaxiinae (Fig. 10–2**B**) are small freshwater or catadromous fishes of Australia, New Zealand, Tasmania, and the southern tips of Africa and South America.

The most widespread and speciose genus is *Galaxias*. *Galaxias zebratus* of South Africa is a translucent, scaleless fish of about 75 mm maximum length. Both *Galaxias* and *Brachygalaxias* occur in South America; *Galaxias* occurs in New Zealand, with *Neochanna,* and in Australia, with *Paragalaxias* and *Galaxiella.*

Galaxias maculatus spawns in vegetation along the shore at high tide and leaves the eggs to incubate above the level of the sea until the next extreme high tide two weeks later. Eggs hatch when they are again covered by water, and the larvae swim into the ocean, later to ascend the rivers. Landlocked populations are reported to spawn in tributaries to lakes on freshets, which subside and leave the eggs on shore. The eggs hatch during a subsequent freshet. Landlocked *Galaxias maculatus,* locally called "puye," occur in great numbers in Chilean lakes. The author once saw a spawning migration in Lake Puyehue that involved fish streaming along the shore in a school about 2.5 m wide and 1 m deep, moving at about average walking speed. The length of the school was not determined after 2 hours of observation, which included walking a kilometer along the shore opposite the direction of movement. Also unknown was when the migration began; it was proceeding at dawn. The migration was over by late afternoon, when the area was inspected again, but nearby streams were full of the small fish.

Order Salmoniformes. This order was formerly considered the basal group of a more inclusive Protacanthopterygii (Greenwood et al., 1966), but more recent phylogenetic studies raised serious doubts about the composition of the superorder and many former members were placed elsewhere (see Fink and Weitzman, 1982; Rosen, 1973). Fink (1984a) opined that the superordinal name *Protacanthopterygii* was no longer useful. He (Fink, 1984b) considered Salmoniformes to be coextensive with the family Salmonidae. Nelson (1994) also restricts the order to the single family. Begle (1991) avoids the term Salmoniformes.

Salmoniformes are soft rayed and mostly physostomous, although the swimbladder may be absent in some. There are no connections of the swimbladder with the ear (i.e., no otophysic connection). An adipose fin is often present. The order is often considered a basal one from which several of the higher groups could have evolved. Salmoniformes are known from the Cretaceous.

The family Salmonidae includes the trouts and salmons of the subfamily Salmoninae, the whitefishes of Coregoninae, and the graylings of Thymallinae (Fig. 10–3), all native to the northern hemisphere (Stearley and Smith, 1993). All have an adipose fin. Oviducts are reduced or absent. All retain a large proportion of cartilage in the cranium. The vertebral column attaches to a condyle made of both the basioccipital and the exoccipitals, a characteristic shared with higher fishes. Many of these are prized as food or sport fishes.

The genus *Salmo* contains the Atlantic salmon (*S. salar*) and the brown trout (*S. trutta*). The Pacific salmons and trouts make up the genus *Oncorhynchus* (Smith and Stearley, 1989). The six Pacific salmons are found in the North Pacific, five of these along the coast of North America and all six in Asia. The rainbow or steelhead trout (*Q. mykiss*), of the Pacific coasts of North America and Siberia, and the brown trout of Europe are currently sought primarily for sport, although both are used in aquaculture. Both have been transplanted to temperate parts of the southern hemisphere. The Atlantic salmon, a prized sport fish, is still fished commercially in some

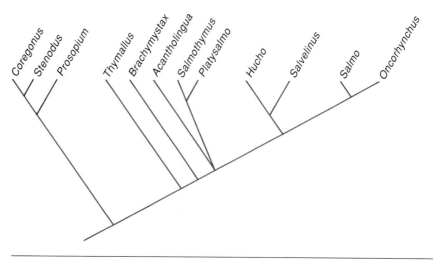

FIGURE 10–3   Cladogram illustrating relationship of genera of the family Salmonidae. (Adapted from Stearley and Smith, 1993.)

parts of the North Atlantic and is the basis of a remarkably successful aquaculture industry, especially in Scandinavia.

Various species of trouts and salmons or races within species may be anadromous, feeding and growing in the ocean but spawning and spending a portion of their early life in fresh water. Individuals of Pacific salmons die soon after spawning. However, in the rainbow trout and the cutthroat trout (*O. clarki*), the hormonal changes that cause mortality in the Pacific salmons are usually not as severe and are typically reversed. Members of the genus *Salmo* generally do not die following spawning.

The chinook salmon, *O. tshawytscha,* is the largest of the Pacific salmon, reaching a maximum weight of about 57 kg (Fig. 10–4). It is of great value as a commercial fish and supports a tremendous sport fishery. Once it ascended the Columbia River and its tributaries for a thousand miles or more, but now hydroelectric and irrigation dams have cut off its spawning areas in the upper reaches of the river system. Hatcheries have been used with moderate success to maintain certain stocks.

The coho salmon, *O. kisutch,* is also an excellent commercial fish. It is more numerous than the chinook in most areas, so it supports a larger sport fishery. The coho is well adapted to short coastal streams and lower tributaries of large river systems for spawning purposes and thus has not been affected as severely as the chinook by dam construction. However, it has been affected by changes in habitat partly due to deforestation, which has changed the stream flow patterns and has allowed siltation of spawning areas. In addition, there have been deleterious effects on the fitness of certain races by hybridization with hatchery stocks. The ease with which the species can be reared in modern hatcheries has made possible the support of coho runs in depleted streams, even though the genetic stability of the native stocks may be at risk. The introduction of the coho into the Great Lakes has provided a remarkable sport fishery, especially in Michigan. The salmon find ample

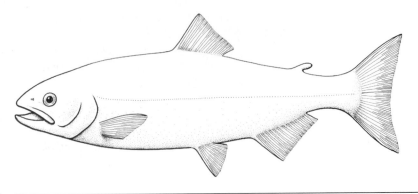

FIGURE 10–4   Representative of order Salmoniformes, Chinook salmon (*Oncorhynchus*).

food in the large stocks of alewife, *Alosa pseudoharengus,* which are abundant in the lakes.

The Pacific salmon with the most prized flesh is the sockeye salmon, *Oncorhynchus nerka.* The sockeye, unlike the piscivorous coho and chinook, depends heavily on pelagic crustaceans for food. It forms the basis for a short yearly season of intensive commercial fishing in Alaskan waters, especially in Bristol Bay, and good stocks are found in various Canadian waters. Because this salmon normally spends its early life in lakes, the best runs are into rivers with large lake systems. The great runs that entered the Fraser River of Canada were once endangered and diminished by a human-caused rock slide at Hells Gate in British Columbia. The torrential current at Hells Gate was a tremendous challenge to migrating fish, even without the increase in gradient caused by the slide. Construction of fish passage facilities and excellent management have done much to restore the sockeye runs in Canada. The landlocked form of the sockeye, known as kokanee, is valuable as a game fish in lakes of the Pacific Northwest.

Chum salmon, *O. keta,* and pink salmon, *O. gorbuscha,* are sought as commercial fishes and have only minor importance as game fishes. In most parts of their ranges, these two species ascend rivers only a short distance to spawn, and the young drift to the ocean immediately after emerging from their gravel nests, or redds. In the other species, the young spend from a few months to two years in fresh water. The pink salmon is remarkable in that the life cycle is almost invariably two years, and some streams have heavy runs in even-numbered years. The odd-year and even-year populations have evolved some morphological distinctions during thousands of years of chronological isolation. Hatchery production of pink salmon is carried out successfully in Siberia and southeast Alaska. In some years, the Alaska operations have produced so many fish that prices have been depressed.

The chars of the genus *Salvelinus* differ in coloration from the trout and salmon in that they have light spots on a darker background instead of dark spots on a light background. There are differences in cranial osteology as well. A biological difference is the season of spawning. Trout tend to spawn in late winter, spring, or early summer, whereas chars spawn mainly in autumn. Chars are freshwater and anadromous fishes found in cold waters of the northern hemisphere. In common usage they are usually called trout. Included in the genus are the Dolly Varden, *S. malma;* the brook trout, *S. fontinalis;* and the lake trout, *S. namaycush,* which supported a large commercial fishery in the Great Lakes prior to the spread of the sea lamprey, which feed on them. The bull trout, *S. confluentus,* formerly occupied many of the cold waters of the Columbia River system and ranged from northern California north through western Canada to Alaska but now is disappearing from parts of its range because of land use practices and introduction of the brook trout.

Other genera in Salmoninae are *Hucho,* the huchens, which are large, voracious fishes of cold waters of Europe and Asia; *Acantholingua* and *Salmothymus* (including *Platysalmo*) of eastern Europe; and the lenok (*Brachymystax*) of Asia (Stearley and Smith, 1993).

Within the Coregoninae, genera include *Coregonus,* the whitefishes, which are holarctic food fishes once abundant in the Great Lakes; *Stenodus,* the inconnu, a large predatory whitefish of arctic Asia and North America; and *Prosopium,* a holarctic genus that includes the Rocky Mountain whitefish *P. williamsoni.* The graylings (*Thymallus*) constitute the only genus in Thymallinae. Graylings are attractive, holarctic fishes with long, colorful dorsal fins and are sought over much of their range by anglers. The populations of Michigan and Montana have been diminished greatly.

Chapter 11

Euteleostei: Lightfishes Through Beardfishes

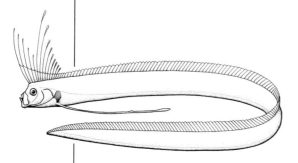

Neoteleostei

The remaining teleosts make up the monophyletic Neoteleostei, a group that is not given a formal rank by Nelson (1994). Nelson (1994) considers Neoteleostei to comprise seven superorders: Stenopterygii, Cyclosquamata, Scopelomorpha, Lampridomorpha, Polymyxiiformes, Paracanthopterygii, and Acanthopterygii. This arrangement will be followed here, even though the group names may be used for various levels of classification in other works. All these, except for Stenopterygii, make up the clade Eurypterygii (Johnson, 1992; Rosen, 1973).

Neoteleosts have developed a retractor dorsalis muscle (= retractor arcus branchialium) that originates on the vertebral column and inserts on the pharyngobranchials. This muscle allows these fishes to exercise great control over the manipulation of food by the pharyngeal jaws (Lauder and Liem, 1983a). Neoteleosts have a rostral cartilage that lies between the neurocranium and premaxillaries, and many have hinged teeth representing tooth attachment type 4 of Fink (1981).

Superorder Stenopterygii

Order Stomiiformes. The relationships of this widely distributed, deep sea group are discussed by Morrow (1964), Fink and Weitzman (1982), Fink (1984, 1985), and Fink and Fink (1986). Stomiiformes were formerly placed as a suborder in Salmoniformes (Protacanthopterygii) but have been separated because of their neoteleostean characters. They are presented by Rosen (1973) as the primitive sister group of the rest of the neoteleosts. They have, in addition to the retractor dorsalis muscle, an internal segment of the adductor mandibulae muscle of an advanced type (called $A_1\beta$) and ascending and articular premaxillary processes (Rosen, 1973).

These are soft-rayed physostomes, although some lack a gas bladder. The mouth is bordered by both the premaxilla and maxilla. Scales, when present, are cycloid. Some species have a small adipose fin rostral to the anal fin. Some genera include species with prominent chin barbels. Within the order, the body shapes range from the deep-bodied hatchetfishes to the anguilliform dragonfishes. In bioluminescent species, photophores are arranged in various patterns, and most have photophores that shine into the eye (Marshall, 1979).

Suborder Gonostomatoidei. This suborder contains the lightfishes and bristlemouths, Gonostomatidae, and the marine hatchetfishes, Sternoptychidae. Gill rakers are present (Nelson, 1994), and all have photophores in series. Usually each photophore has a duct or opening. The relationships of these families are discussed by Ahlstrom et al. (1984).

The gonostomatids are elongate fishes, with ventral series of photophores that extend into the isthmus and branchiostegals. Their mouths are large, with the maxillary extending far back beyond the eye. There are between 25 and 30 species in six genera. They live from the mesopelagic to the bathypelagic zone and have adaptations in coloration, feeding apparatus, and buoyancy mechanisms to suit the specific habitat. Members of the genus *Cyclothone* (Fig. 11–1**A**) are the most widespread of the deep pelagic fishes (Marshall, 1979) and are among the most abundant vertebrates (Ahlstrom et al., 1984).

The ten genera of Sternoptychidae are divided into two subfamilies: the Maurolicinae (pearlsides and relatives) and the marine hatchetfishes of Sternoptychinae

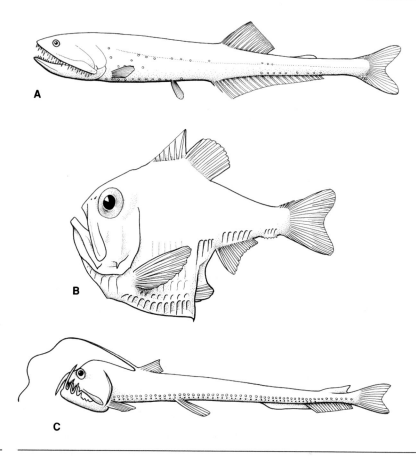

FIGURE 11–1   Representatives of order Stomiiformes, **A,** bristlemouth (Gonostomatidae, *Cyclothone*); **B,** marine hatchetfish (Sternoptychidae, *Argyropelecus*); **C,** viperfish (Stomiidae, *Chauliodus*).

(Nelson, 1984). The maurolicines, which were formerly placed in the Gonostomatidae, comprise seven genera and 14 species. They have a robust body shape but are not highly compressed. They have small ventral photophores and a small mouth. Marine hatchetfishes are bizarre in appearance in that they are extremely compressed and have large, usually oblique mouths (Fig. 11–1**B**). Scales are represented by elongate, narrow plates arranged vertically on the sides. Photophores are comparatively large, distributed in series on the lower lateral surfaces and directed ventrally. These are pelagic or mesopelagic fishes and, like the pearlsides, seldom exceed 75 mm in length. There are more than 30 species in three genera.

Suborder Phosichthyoidei. This group includes the Phosichthyidae (= Photichthyidae; see Eschmeyer and Bailey, 1990) and the Stomiidae, in which Fink (1985) included the following six families: Chauliodontidae, viperfishes (Fig. 11–1**C**); Stomiidae, scaly dragonfishes or barbeled dragonfishes; Astronesthidae, snaggletooths; Melanostomiidae, scaleless dragonfishes; Idiacanthidae, stalkeyes; and Malacosteidae, loosejaws. Nelson (1994) follows that arrangement.

There are seven genera and about 20 species of phosichthyids known from the Pacific, Indian, and Atlantic oceans. The "lighthousefishes," as they are sometimes called, are mesopelagic and migrate at night from water of about 200 to 300 m to shallower depths. They are rather small fishes without chin barbels and have two rows of photophores along their lower sides to the origin of the anal fin, and then a single row from there to the caudal fin. Fishes of the genus *Vincinguerria* are, along with *Cyclothone,* some of the most abundant vertebrates (Ahlstrom et al., 1984).

The fishes of the family Stomiidae, as defined by Fink, share the following characteristics (Fink, 1985): a chin barbel, lack of gill rakers in adults, lack of hypural 6, absence of scales that attach firmly in the skin, a geniohyoideus muscle in two main bodies, and the insertion of part of the adductor mandibulae on the postorbital photophore. Although the stomiids are mostly small fishes, reaching about 150 mm, they are predatory in habit. Some have extremely large teeth for their size and are bizarre in appearance. All are well specialized for life in the deep sea (see Marshall, 1979).

Order Ateleopodiformes. The family Ateleopodidae contains four genera and about 12 species of fishes from moderate depths of tropical and warm, temperate seas. Their relationships have been subject to different interpretations (Berg, 1940; Eschmeyer, 1990; Nelson, 1984; Olney et al., 1993). Nelson (1994) elevates them to ordinal status.

These are elongate fishes with a short head and trunk and a long tail, so they resemble chimaeras or grenadiers in general body shape. There is a short, sometimes high dorsal fin and a long anal fin that is confluent with the small caudal fin. The jugular pelvic fins are filamentous in some species.

Ateleopodids, or tadpolefishes, have no gas bladders, no fin spines, no pseudobranchiae, and few teeth. They lack several cranial bones, including the orbitosphenoid and basisphenoid. Genera include *Ijimaia, Ateleopus, Parateleopus,* and *Guentherus. Ijimaia dofleini* of Japan reaches about 1.7 m.

Superorder Cyclosquamata

The Aulopiformes are included in this group of marine fishes. These are soft rayed and are either physoclistic or lack a gas bladder. Maxillaries are excluded from the border of a nonprotrusible mouth. An adipose fin is generally present, and many have photophores. Most have abdominal pelvics.

Order Aulopiformes. This order contains marine fishes formerly referred to the Myctophiformes. They are distinguished by modifications of the upper elements of the second and third branchial arches; the pharyngobranchial of the second arch is at an angle to that of the third, leaving a space that is partially filled by a process of the second epibranchial. Some features of musculature also are used as distinguishing characteristics (Lauder and Liem, 1983a). The placement of these fishes in a separate order is accepted by many (e.g., Eschmeyer, 1990; Fink, 1984; Nelson, 1994), but some regard it as an open question (Johnson, 1982). Okiyama (1984) treats them as myctophiformes. Aulopiformes are known from the Cretaceous.

Suborder Aulopoidei. The Aulopidae (= Aulopodidae) are the threadsails or flagfins of warm marine waters. Some of the dozen or so members of the genus

Aulopus live in shallow water and one, *A. purpurissatus* (the "Sergeant Baker") of Australia, is caught by inshore anglers and is a food fish of minor importance. Nelson (1984) considers the family to be the most primitive of the order.

Suborder Chlorophthalmoidei. This suborder contains Chlorophthalmidae, greeneyes; Scopelarchidae, pearleyes; Notosudidae (Scopelosauridae), paperbones; and Ipnopidae, including *Ipnops,* "grideyes," and *Bathypterois,* tripodfishes, which were placed within the Chlorophthalmidae as the subfamily Ipnopinae by Eschmeyer (1990b). However, Nelson (1994) retains Ipnopidae. *Ipnops* has strange "eyes" consisting of broad retinas that lie under thin, transparent bones of the roof of the skull. The retinas are yellow and contain mainly rod cells. The tripodfishes (*Bathypterois*) have elongated, modified rays in the paired fins and in the anal and caudal fins. They have been observed to support themselves on the bottom at great depths using the pelvic and caudal fins as a tripod.

Suborder Alepisauroidei. This suborder comprises the Synodontidae, lizard fishes, including Harpadontinae and the Bombay ducks; the Giganturidae, giganturas and telescopefishes; the Paralepididae, barracudinas; the Anotopteridae, daggertooths; the Evermannellidae, sabertooth fishes; the Omosudidae, omosudids; and the Alepisauridae, lancetfishes. According to Johnson (1982), the Pseudotrichonotidae, which is sometimes placed in this suborder, is not a natural member of the group.

The alepisauroids are predatory fishes, and most of the deep sea forms have distensible stomachs and lack scales. The daggertooths and lancetfishes (Fig. 11–2**A**) are elongate fishes with large mouths and fearsome teeth. Lancetfish up to 1.25 m long are often found on or near the beaches of the Pacific Northwest during the spring months. In at least two known instances, startled anglers have hooked and landed lancetfish, one from the beach, the other from a jetty.

The lizardfishes (Fig. 11–2**B**) are mostly shore fishes of warm waters, but some are pelagic. Many of them have a strong, somewhat musky odor. The related *Harpadon nehereus,* or Bombay duck, is a colorless, translucent schooling fish that has a strong taste. It is common in the northern Indian Ocean and is harvested, dried, and used in the making of sauces.

Superorder Scopelomorpha

The remaining fishes constitute the clade Ctenosquamata of Rosen (1973). Although there may be doubts concerning the monophyly of this clade, Johnson (1992) believes that it is monophyletic because of the structure of the dorsal gill arches.

Order Myctophiformes. Other names that have been used for this, or a more inclusive group, are Scopeliformes and Iniomi. The current definition of this order is that it contains only the Myctophidae, lanternfishes, and the Neoscopelidae, blackchins. These are soft-rayed fishes that lack a mesocoracoid and in which the premaxillaries border the mouth to the exclusion of the maxillaries. An adipose fin is present, and if a gas bladder is present it is physoclistic. Many species invest the gas bladder with fatty substances for control of buoyancy. These fishes are distinguished by a large, toothed, third upper pharyngobranchial that is about double the size of the fourth, which is movably attached to the third. In addition, the retractor dorsalis muscle is di-

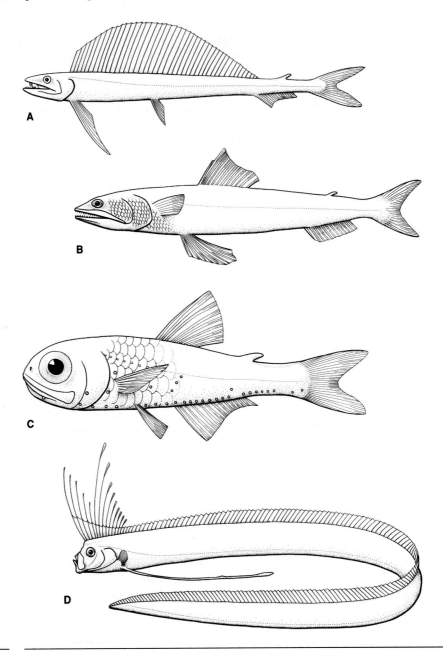

FIGURE 11–2 Representatives of order Aulopiformes, **A,** lancetfish (Alepisauridae, *Alepisaurus*); **B,** lizard-fish (Synodontidae, *Synodus*). Representative of **C,** Myctophiformes, lanternfish (Myctophidae). Representative of Lampridiformes, **D,** oarfish (Regalecidae, *Regalecus*).

vided into a small medial head, which attaches to the third pharyngobranchial, and a larger lateral head, which inserts on the fourth (Lauder and Liem, 1983a; Rosen, 1973). Diverse fossil myctophiformes are known from as long ago as the Cretaceous.

Myctophids (Fig. 11–2**C**) are small fishes found in all oceans from the surface to the depths, although most of the nearly 250 species live shallower than 1000 m. They have relatively large eyes and numerous photophores, each species apparently with its own patterns. Within a species, photophore patterns are sexually dimorphic.

Important genera are *Diaphus, Lampanyctus, Myctophum,* and *Tarletonbeania.* These fishes are noted for their daily vertical migrations, moving toward, or actually to, the surface at night and back into the aphotic zone by day. Although they have no direct commercial importance, they collectively have a tremendous biomass and contribute indirectly to our fisheries by converting plankton to food for commercially important fishes higher on the food chain. These small fishes, almost all much shorter than 30 cm, have considerable importance in the trophic ecology of the ocean.

The Neoscopelidae are somewhat larger than lanternfishes and generally have smaller eyes. They are black, and most have fewer photophores than lanternfishes. They live at depths ranging from 700 to 2000 m in most seas. There are five or six species, some of which are widespread in deep water.

Superorder Lampridomorpha

Order Lampridiformes (Allotriognathi). Lampridiformes are marine; some live in deep water. Fossils are known from the Eocene. This order includes the opah and the greatly elongate and ribbon-like oarfishes and ribbonfishes plus some smaller fishes with uncertain relationships. They are mostly soft rayed, although one or two modified fin spines may be present in the dorsal and anal fins of the family Veliferidae (Olney et al., 1993). Pelvic fins are thoracic and may have from 0 to 17 rays. The orbitosphenoid is present in some, but the mesocoracoid and opisthotic are absent. The protractile maxillaries have an outer blade and an inner process that meshes with a similar structure on the premaxillaries. Scales are cycloid or lacking. These fishes are physoclistic.

At one time this order was believed to be part of the series Percomorpha, but due to the study by Olney et al. (1993), they are now considered basal acanthomorphs (the sister group of all other acanthomorphs).

Suborder Lamproidei. This suborder includes the widely distributed opah, *Lampris guttatus,* the only member of the family Lampridae. Because of its greatly compressed, ovate body, the opah is also known as the moonfish. It is noted for its color pattern of blue or blue-gray on the back, silver on the sides, reddish silver on the belly, and red jaws and fins, all with an overlay of silver or whitish spots. Opahs may reach about 2 m in length and may weigh up to 270 kg.

Suborder Veliferoidei. This group includes only the family Veliferidae, called sailbearers because of the large dorsal and anal fins. The body is compressed as much as in the opah.

Suborder Trachipteroidei. This group includes fishes that are greatly compressed but are also elongate (taeniform; Fig. 11–2**D**). Many have bizarre coloration or fin shapes and so are thought to be the basis of sea serpent stories. The Lophotidae, the crestfishes, have a crest that extends forward on the head and bears the anterior part of the long dorsal fin. The Radiicephalidae contains one species that is similar to the crestfishes. The ribbonfishes, Trachipteridae, also have an elongate dorsal, as do the oarfishes, Regalecidae; both families lack an anal fin as adults (Olney, 1984). In the regalecids, the first several dorsal rays may be very high and the pectorals are elongate and oarlike.

Trachipterus altivelus, the king-of-the-salmon, occurs on the Pacific Coast of North America. *T. arcticus,* the dealfish, is found in the Atlantic. *Regalecus glesne,* the oarfish, is widespread and was apparently the basis of many of the sea serpent stories. This fish reaches 8 m long and has a thin, compressed body and a dorsal fin that might look like the mane of a horse. Its long pelvic fins, which earned the name "oarfish," could be called arms.

Suborder Stylephoroidei. This group contains the single family Stylephoridae, the tube-eyes, which live at greater depths than do the other members of the order Lampriformes. They derive their common name from the telescopic eyes, which may point upward or forward (Nelson, 1994). The lower lobe of their caudal fin is elongate, as are the first two rays of the dorsal fin. The mouth opening is small and at the end of a tubular snout, but the oral cavity can increase in volume rapidly (up to nearly 40-fold) so the fish can suction feed on plankton (Pietsch, 1978a). *Stylephorus chordatus* reaches about 30 cm.

Superorder Polymixomorpha

Order Polymixiiformes. These are tropical marine fishes called barbudos or beardfishes because of their long barbels. They are placed in one family, the Polymixiidae, the taxonomic position of which is not certain. It has been placed into the Paracanthopterygii, the Perciformes, or in an order of its own. Robins et al. (1991) and Eschmeyer (1990) retain it in the Beryciformes. Stiassny (1986, 1990) places it in an uncertain position as the sister group of the remainder of the Acanthomorpha. The family is distinguished by truncated posterior supramaxillae, modification of the anterior branchiostegals to serve as support for the hyoid barbel, and a palatovomerine ligament, which passes between the lateral maxillary processes (Stiassny, 1986). The three species of *Polymixia* live in mid-depths in the western Pacific and the Atlantic.

Euteleostei: Paracanthopterygii

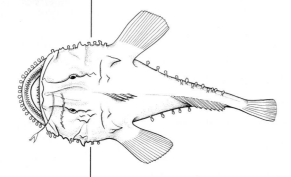

Is Paracanthopterygii Monophyletic?

The following five orders represent the superorder Paracanthopterygii, which was defined by Greenwood et al. (1966) and has been subjected to various changes and interpretations by several students of the group (Fraser, 1972; Patterson and Rosen, 1989; Rosen, 1985; Rosen and Patterson, 1969; Stiassny, 1986). The group is difficult to define by means of shared derived characteristics, but most species have no more than two epurals and have a complete neural spine on the second preural centrum. Some have a single supraneural (supraneural bones are positioned on the midline between the supraoccipital and the pterygiophores of the dorsal fin). They have a large intercalar (opisthotic) that forms part of the cranial wall and is pierced by the glossopharyngeal foramen (Nelson, 1994).

Rosen (1985) expressed doubts that the Paracanthopterygii could be accepted as a natural group. Patterson and Rosen (1989) present a cladogram that characterizes the superorder and its included lineages and eliminates the Gobiesociformes and Zoarcoidei, which were originally included by Greenwood et al. (1966); the Polymixiiformes, which Rosen and Patterson (1969) added; and other groups (e.g., Indostomiformes, Gobioidei), which other authors had proposed.

The orders presented here as belonging to the superorder are those that appear to represent something of a consensus among ichthyologists, even though there are unresolved questions regarding the relationships of the fishes included in the Percopsiformes and Ophidiiformes. Further study will no doubt result in additional rearrangements.

The Paracanthopterygii

Order Percopsiformes (Salmopercae). This is a small order of North American freshwater fishes of questionable monophyly (see Rosen, 1985, and Patterson and Rosen, 1989) characterized by a single supraneural behind the first or second neural spine, subabdominal or subthoracic pelvic fins, six branchiostegals, a caudal with 16 branched rays, and the maxillary excluded from the border of the mouth (gape). Some members are unusual in that they possess an adipose fin and have spines in the dorsal and anal fins. A subocular shelf, orbitosphenoid, and basisphenoid are absent (Nelson, 1994; Rosen and Patterson, 1969). The gas bladder is physoclistic. The caudal fin is supported by two plates that are the result of fusion of hypurals (Lauder and Liem, 1983a). Both ctenoid and cycloid scales are developed in this order. An extinct suborder, †Sphenocephaloidei, is known from marine deposits from the Cretaceous.

Suborder Percopsoidei. These fishes have true spines in the dorsal and anal, an adipose fin, ctenoid scales, and the anus in a normal position. The single family Percopsidae contains two species, *Percopsis omiscomaycus,* the trout-perch of eastern North American drainages (Fig. 12–1**A**), and *P. transmontana,* the sand roller of the Columbia River system. These are small fish of still or slow streams. *Percopsis* seems to be nocturnal in habit, remaining in deep water or hiding during the day and moving into shallow water at night.

Suborder Aphredoderoidei. This group lacks the adipose fin, has ctenoid scales, and has the anus in a jugular position. Both the anal and dorsal fins bear spines.

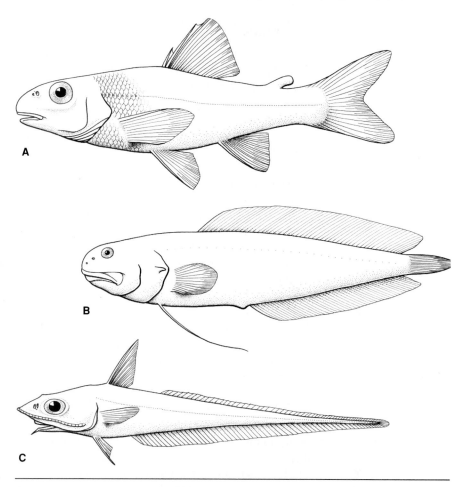

FIGURE 12–1 Representative of **A**, Order Percopsiformes, trout-perch (*Percopsis*). Representative of **B**, order Ophidiiformes, livebearing brotula (*Brotulina*). Representative of **C**, order Gadiformes, grenadier (*Macrourus*).

There are two families, Aphredoderidae and Amblyopsidae. The latter is regarded as constituting a separate suborder in some classifications.

Aphredoderus sayanus, the pirate perch, has ctenoid scales and subthoracic pelvic fins. It is a freshwater species that lives in streams along the Gulf and Atlantic coasts and in the Mississippi drainage of the United States. Although it ranges north to New York and the Lake states, it is more common in the southern United States. The amblyopsids are the cavefishes and swampfish of the southern part of the United States from Oklahoma to Kentucky. The eyes are reduced but not covered by skin in *Chologaster cornuta,* the swampfish, but members of the genera *Amblyopsis, Speoplatyrhinus,* and *Typhlichthys* are blind. The lateralis system of the cavefishes is well developed and includes lines of superficial neuromasts on the head and body. They have cycloid scales, and, in those species with pelvic fins, the fins are abdominal.

Order Ophidiiformes. These fishes (Fig. 12–1**B**) were once placed with the Perciformes, then considered to be a suborder of Gadiformes. Most recent treatments accord them ordinal status within the Paracanthopterygii (Cohen and Nielsen, 1978; Gordon et al., 1984; Nelson, 1994). The order contains a variety of fishes, some elongate, with tapering bodies and long dorsal and anal fins that are confluent with the caudal in many genera. Dorsal and anal pterygiophores are more numerous than the vertebrae adjacent to the fins. Pelvic fins have small spines in some species and are usually jugular to mental so that they resemble elongate barbels. Ophidiiformes are mostly benthic or benthopelagic, with few freshwater representatives. There may be as many as 400 species, some of which live at great depths.

Suborder Ophidioidei. Fishes of this suborder are oviparous. The anterior nostril is positioned some distance from the lip. The cusk-eels and brotulas of the family Ophidiidae range from the shallows to great depths in most seas. There are about 50 genera and over 165 species, usually arranged in four subfamilies. The subfamily Brotulinae contains only the benthic, circumtropical genus *Brotula,* which has about five species. Brotulotaeniinae is also a circumtropical group, with one genus, *Brotulotaenia,* and four species that are pelagic in mid-water. The subfamily Ophidiinae is widespread and includes both pelagic and benthic species, and some burrow in the substrate. Most are small, but members of the genus *Genypterus* found in Australia, South America, and South Africa may grow to over a meter long. *Genypterus reedi* and *G. chilensis,* both found in Chilean waters, are prized as food fishes.

The subfamily Neobythitinae contains at least 135 species in 38 genera (Gordon et al., 1984). The members are well distributed both geographically and bathymetrically. Some are rather brightly colored littoral fishes, but others are dark, deep water forms. *Abyssobrotula galatheae* is known as the deepest living fish and has been captured at 8470 m in the Puerto Rico trench. *Acanthonus armatus* is a benthopelagic species with a huge head that contains a large, fluid-filled cavity. The low specific gravity of the fluid provides buoyancy.

The pearlfishes of the family Carapidae lack caudal fins, the dorsal and anal fins are elongate, and the body tapers posteriorly to a point. The anus is jugular, and this is probably an adaptation for living with the posterior part of the body inside invertebrate hosts. Larvae are greatly elongate and bear a banner-like elongated dorsal ray (vexillum) just behind the head. There are two subfamilies and seven genera, with 31 species (Markle and Olney, 1990). Members of the subfamily Carapinae, which consists of thin-bodied little fishes, have no pelvic fins and have members that live commensally (some parasitically) in sea cucumbers, molluscs, and other invertebrates. The subfamily Pyramodontinae includes species that are deeper bodied and have pelvic fins. They are predatory fishes with large anterior teeth and a protractile upper jaw. They are nearly circumtropical.

Suborder Bythitoidei. In this group, the anterior nostril is close to the upper lip, and the males have an intromittent organ. In some species, the caudal fin is separated from the dorsal and anal fins. The family Bythitidae contains the viviparous (livebearing) brotulas, two subfamilies that are widely distributed in marine waters. There are about 80 species in 28 genera (Gordon et al., 1984). The Bythitinae have the dorsal and anal fins confluent with the caudal fin. *Stygnobrotula,* the black brotula of the Caribbean, frequents caves and crevices in coral reefs. Other genera include *Cataetyx* and *Oligopus,* which tend to inhabit reef-caves.

In the subfamily Brosmophysinae, the caudal fin is separate from the anal and dorsal fins. The red brotula, *Brosmophysis marginata,* of the west coast of North America lives at from 50 to 180 m and reaches a length of about 50 cm. The genus *Lucifuga* contains blind species that live in freshwater caves in Cuba. Others of the genus (with functional eyes) are found in limestone sinks in the Bahamas. Another genus, *Ogilbia,* has species in freshwater caves of Yucatan and in brackish waters of the Galapagos Islands (Nelson, 1984).

The family Aphyonidae is represented in the deep waters of all oceans. These are viviparous pelagic fishes with small, weakly developed eyes.

Order Gadiformes (Anacanthini). Patterson and Rosen (1989, p. 13) refer to the fishes of this order as "core paracanthopterygians." The relationships within the order have been subject to much study and interpretation in recent years, and various arrangements of suborders and families have been published. Important references include Cohen (1984, 1989), Cohen et al. (1990), Fahay and Markle (1984), and Siebert (1990). This order contains the cods and allies, including the grenadiers, hakes, and burbots. With few exceptions, these are marine. Gadiformes are known from the fossil record beginning in the Paleocene.

The cods and allied groups are soft-rayed, physoclistic fishes that usually have cycloid scales and thoracic or jugular pelvic fins. They have characteristics considered both primitive and derived among the teleosts. An additional advanced feature is the lack of the orbitosphenoid and mesocoracoid. The premaxillaries exclude the maxillaries from the gape. Although many gadiformes have tapering tails that are leptocercal in shape, a caudal fin is present in about two thirds of the members of the group (Fahay and Markle, 1984). The caudal fin is usually isocercal and has a distinctive internal skeleton, with the posterior vertebrae reduced in size. The neural and haemal spines of these vertebrae assist in the support of the caudal fin. The hypurals in many genera are reduced in number and variously fused. Many genera have accessory bones ("X" and "Y" bones) between the neural spines and haemal spines of preural vertebrae (Fahay and Markle, 1984; Patterson and Rosen, 1989). The first neural spine is in close contact or joined to the crest of the supraoccipital, except in *Muraenolepis* (Cohen, 1984).

Although there is active ongoing work on the phylogeny of gadiforms, there seems to be no general agreement about their relationships. Markle (1989) uses a classification that has four suborders (Ranicipitoidei, Melanonoidei, Macrouroidei, and Gadoidei), whereas Nelson (1984) listed Muraenolepidoidei, Gadoidei, and Macrouroidei. Eschmeyer (1990b) and Nelson (1994) list no suborders in their classifications, as will be done in this book.

The family Ranicipitidae contains *Raniceps raninus,* the tadpole fish, of the northeast Atlantic Ocean. The species is thought to be basal among the gadiformes (Dunn and Matarese, 1984; Markle, 1989). Characteristics include one branchiostegal on the epihyal, and the neural spine on the preural centrum is rounded (both of these are primitive, according to Dunn and Matarese, 1984). This fish has two postcleithra, six hypurals that fuse into three ontogenetically, six primary caudal rays (as does *Merluccius*), and three rays in the first dorsal fin.

The family Euclichthyidae has a single species, *Euclichthys polynemus,* from the waters around Australia and New Zealand. This has a small, asymmetrical, many-rayed caudal fin; four hypurals; and no otophysic connection. The Y bone is absent, and the pelvics have three elongate rays each.

The rattails and grenadiers of the family Macrouridae are elongate fishes with a large head, often with large eyes. The trunk of the body is short, with 10 to 16 abdominal vertebrae. The long, tapering body is fringed by long dorsal and anal fins (Fig. 12–1C), but there is no caudal fin. Some may have a spinous second dorsal ray and ctenoid scales. Photophores containing luminous symbiotic bacteria are present in many species, and drumming muscles are associated with the gas bladder of the males of certain species. Macrourids are benthopelagic fishes, usually found at depths of from 200 to 2000 m. (See Marshall and Iwamoto, 1973.) Howes (1989) suggests that certain genera—*Bathygadus, Gadomus,* and *Trachyrhincus*—usually placed in this family should be considered more closely related to gadids.

Steindachneriidae contains a single species, the luminous hake (*Steindachneria argentea*), from the Gulf of Mexico and adjacent areas of the tropical Atlantic. It lacks a caudal fin, and the anus (but not the genital papilla) is displaced forward between the pelvic fins. There are photophores on the body and head.

The Moridae, or codlings, are deep sea fishes with an otophysic connection consisting of very large diverticula from the gas bladder that pass through wide foramina in the exoccipitals. Melanonidae are confined to the southern parts of the Atlantic and Pacific oceans and have no otophysic connection.

Macruronidae, the straptails, consists of species from temperate seas of the southern hemisphere. These have the long-based second dorsal and anal fins confluent with the reduced caudal fin.

The family Bregmacerotidae contains one genus, *Bregmaceros,* the codlets of warm marine waters. There are about seven species, mostly pelagic. They are distinctive both in appearance and characteristics (Cohen, 1984). The first dorsal fin consists of a single elongate ray set just behind the head, and the second dorsal fin extends along most of the posterior two thirds of the body. The caudal fin is emarginate. There is a single anal fin that matches the second dorsal in length. The pelvic fins have five rays, three of which are separate and elongate. Cohen (1984) remarks that the codlets are not closely related to the other gadiforms.

The family Muraenolepididae is found in cold marine waters of the southern hemisphere. It is characterized by 10 to 13 pectoral radials (actinosts), a narrow gill opening entirely below the base of the pectoral, and the confluence of the second dorsal, caudal, and anal fins. The structure of the caudal fin is thought to be primitive for the gadiformes. Fahay and Markle (1984) present the hypothesis that X and Y bones in gadiformes are represented in *Muraenolepis* as the radials of the next to last dorsal and anal rays. The first dorsal is made up of a single ray. The first neural spine is free from the supraoccipital but has winglike structures that extend to each side of the crest of that bone (Cohen, 1984).

The family name Phycidae has been used recently for fishes traditionally included in Gadidae that take the common name "hake," notably those of the genera *Phycis* and *Urophycis.*

Closely related to the cods is the family Merlucciidae, called merlucciid or silver hakes. Robins et al. (1991a) include these fishes in Gadidae. These are commercial fishes of the Atlantic, eastern Pacific, and the seas around southern New Zealand and South America. The Pacific hake, *Merluccius productus,* has become a major commercial species off the Pacific Northwest. It originally supported moderate fisheries until the fleet of the former Soviet Union began exploiting it. The species is called "Pacific whiting" in the fishing industry.

The family Gadidae includes fishes with four or five actinosts and normally placed gill openings. There are two or three dorsal fins. The caudal fin is of the "pseudocaudal" type in that it consists mostly of dorsal and anal elements, with a very small true caudal making up the central part. Many cods have diverticula of the gas bladder in connection with the inner ear, which apparently increase the acuity of hearing. Members of the Gadidae are found in temperate and cold waters of both the northern and southern hemispheres.

By far, the most important family in the order from an economical standpoint is Gadidae (Fig. 2–13**D**), which contains the cods, haddocks, pollock, pollack, lings, and whitings. These all have commercial value as food or as a source of high-quality fish meal. Landings of the walleye pollock, *Theragra chalcogramma,* from the Pacific Ocean and Bering Sea usually exceed 6 million metric tons annually, the highest for the cod family.

The well-known cod fisheries of the western North Atlantic, off New England and the Maritime Provinces, have been operated by many nations for more than three centuries. The object of most of the effort has been the Atlantic cod, *Gadus morhua,* which may reach 1.2 m long and a weight of 45 kg. Longlining has been the primary method employed over the years, but trawling, especially since the advent of steam-powered vessels, forms the basis of the fisheries. In addition, gill nets are employed in the cod fisheries of New England. Landings of the species in recent years have amounted to nearly 2 million metric tons per year.

Another important commercial gadid in the North Atlantic is the haddock, *Melanogrammus aeglefinus,* which may be suffering from overfishing by trawlers. The species reaches about 1.1 m long and 16 kg. The pollock, *Pollachius virens,* and the blue whiting, *Micromesistius poutassou,* are additional important commercial species. The Pacific Ocean counterpart of the Atlantic cod is the Pacific cod, *G. macrocephalus,* which was once fished in Canadian and Alaskan waters by sailing vessels from San Francisco. The rocklings, some species called hakes, and the burbot traditionally were placed in Gadidae as the subfamily Lotinae but are now considered by some authors to constitute a separate family, Lotidae. (See classifications by Cohen, 1984, and Eschmeyer, 1990.) Some of the genera include *Brosme,* the cusks; *Molva,* the European ling; and *Ciliata,* rocklings. *Lota lota,* the burbot, is a holarctic fish of fresh waters.

Order Batrachoidiformes (Haplodoci). This order contains the toadfishes and midshipmen, usually broad-headed, big-mouthed, bottom fishes of tropical and temperate seas. They are related to the Lophiiformes, and Patterson and Rosen (1989) consider the two orders to constitute the clade Pediculati. In both orders, the cranium is flattened and the parasphenoid joins the frontals (Patterson and Rosen, 1989).

Members of this order lack ribs. The pelvic fins are jugular and have one spine and two or three soft rays. The pectoral radials are elongate. Both the premaxillaries and maxillaries form the border of the mouth, but only the premaxillaries bear teeth. There are three pairs of gills. A short spinous dorsal fin, bearing two to four spines, is rostral to a long, soft dorsal fin. Venom is produced in glands at the bases of the dorsal spines in certain genera, and some (*Thalassophryne* and *Daector*) have hollow spines through which the painful venom can flow.

Many species can produce a variety of sounds with special muscles that attach to, and can vibrate, the gas bladder. These sounds have been described as hoots, grunts, and boat whistle blasts (see Chapter 19). There is only one family, Batrachoididae, divided into three subfamilies (Collette, 1966b; Smith, 1952). There are about 19 genera and 64 species.

The subfamily Batrachoidinae contains the toadfishes, which are shallow water marine fishes of most warm and warm-temperate seas. They have no venom glands and no photophores. There are three spines in the first dorsal fin. Most have barbels and skin flaps around the mouth, and members of some genera, such as *Batrachomoeus,* have remarkable camouflage that suits them for life on coral. The genus *Opsanus* is represented on the Gulf and Atlantic coasts of North America.

Members of the subfamily Porichthyinae are found on the continental shelf in North and South America. The first dorsal fin consists of two spines, there are no scales, and some species (for instance, *Porichthys plectrodon,* the Atlantic midshipman), have venom glands (Hoese and Moore, 1977). They have multiple lateral lines (Hart, 1973; Nelson, 1984). The genus *Porichthys* takes the name midshipman from the several rows of photophores that course along the body like buttons (Fig. 12–2**A**). These are among the few shallow water fishes that produce light. The genus *Aphos* of the Pacific coast of South America lacks photophores.

The subfamily Thalassophyrninae contains venomous toadfishes. In contrast to the solid dorsal and opercular spines of the other batrachiformes, these have two sharp, hollow spines in the first dorsal fin and one on the operculum. Venom glands in association with the spines allow injection of venom. There are no photophores, and the lateral line is single or lacking. These are fishes mainly of warm waters of Central and South America, with a few freshwater species, especially in the Amazon.

Order Lophiiformes. This is the order of the anglerfishes and frogfishes. These fishes are characterized by a spinous dorsal fin that is composed of one to a few flexible rays, the first one to three of which may be cephalic and modified into a fishing apparatus or illicium, which often bears a flap or a bulbous bait or lure (esca) at the tip. These fishes are similar to the toadfishes and clingfishes in that they have no ribs and have reduced gill openings, which are small and tubelike (Pietsch, 1984). Pectoral radials are long and narrow, and the caudal fin is borne on a single hypural plate that is the product of fusion of the two ural centra and the first preural centrum (Rosen and Patterson, 1969). The order is divided into five suborders (Pietsch, 1987).

Suborder Lophioidei. This suborder consists of one family, Lophiidae, the anglers or goosefishes (Fig. 12–2**B**). These fishes have a large, flat head and a wide mouth set with sharp, depressible teeth. They have jugular pelvics, with one spine and five soft rays.

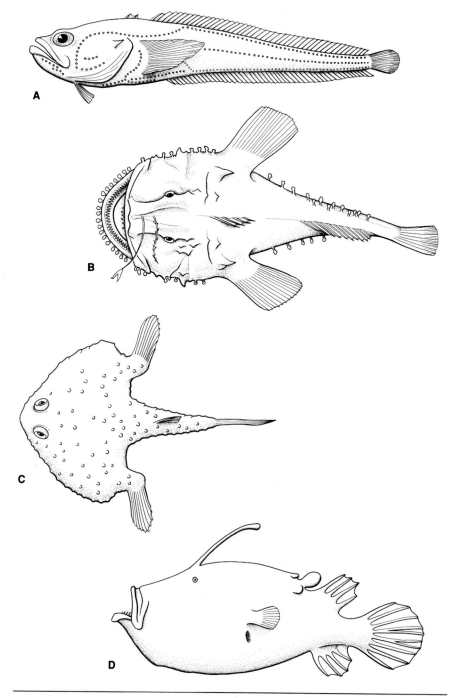

FIGURE 12–2   Representative of **A,** Order Batrachoidiformes, midshipman (*Porichthys*). Representatives of **B,** order Lophiiformes, goosefish (*Lophius*), dorsal view; **C,** batfish (*Ogcocephalus*), dorsal view; **D,** Lophiiformes, seadevil (Ceratioidei).

Goosefishes are widely distributed from cold waters (e.g., Barents Sea) to the tropics. *Lophius americanus* is the common Atlantic Coast species in the Americas; *L. piscatorius* and *L. budegassa* are European species. Members of the genus *Lophius* may reach 1.3 m. Lophiids lie concealed on the bottom and attract prey by moving their angling apparatus. *Lophius americanus* deposits up to 2.6 million eggs in a "veil" of a gelatinous material secreted by the ovaries. The veils can be up to 10 m long and nearly a meter wide (Breder and Rosen, 1966). *Lophius* has eight species. Other genera are *Lophiomus* (one species), *Lophioides* (13 species), and *Sladenia* (three species).

Goosefishes are good food fishes, and they are utilized in the United States, Europe, and Asia. Because of their usual unattractive common names and bizarre appearance, they are marketed in the United States using the British common name "monkfish" (not to be confused with the angel sharks).

Suborder Antennarioidei. Included here are the small frogfishes, batfishes, and allies. The pelvic fin is set well in advance of the pectoral. Scales are absent, but the skin bears spinules or denticles. The body is laterally compressed.

The family Antennariidae, the frogfishes, is a circumtropical group of 12 genera and 41 species, with a concentration of species in the Indo-Australian Archipelago (Pietsch and Grobecker, 1987). They are primarily benthic, shallow water fishes, but a few species are found deeper than 100 m, and one (*Antennarius nummifer*) has been taken as deep as 293 m (Pietsch and Grobecker, 1987).

One of the best-known frogfishes is the sargassum fish, *Histrio histrio*. This species may live far from land in floating mats of sargassum, depositing its eggs among the weeds in gelatinous veils or rafts, much like those of the goosefishes but about a third as large. *Histrio* and some species of *Antennarius* are known to inflate themselves by swallowing air, but Pietsch and Grobecker (1987, p. 343) remark that inflation is usually in response to much "poking and manipulation."

The monotypic Tetrabrachiidae (Pietsch, 1981) is found from the Moluccas Islands to northern Australia. *Tetrabrachium ocellatum* has highly modified pectoral fins that are divided into upper and lower sections, with the lower part bound to the body by a membrane. Lophichthyidae contains but one species, *Lophichthys boschmai* of New Guinea.

The Brachionichthyidae of southern Australia has a spinous dorsal fin set forward on the head, with the three spines connected by membranes.

Suborder Chaunacoidei. This suborder contains the sea toads or coffinfishes of the family Chaunacidae, which are found in warm and temperate seas at depths from 90 to over 2000 m (Caruso, 1989). They have a very short illicium with a cirrated esca. The head is large and globose, and the gill openings are placed well behind the pectorals. There are two genera: *Chaunax,* which has about 12 species, and *Bathychaunax,* which has two species (Caruso, 1989). These fishes are placed with the following two suborders in a single suborder by Nelson (1994).

Suborder Ogcocephaloidei. This suborder contains the single family Ogcocephalidae, the batfishes (Fig. 12–2C), which Pietsch and Grobecker (1987) indicate are closely related to the ceratioids. These are flattened, have scales that are modified into tubercles or bucklers (Bradbury, 1967), and have a retractable illicium. They can walk over the bottom on their pectoral and pelvic fins (Nelson, 1994). According to Bradbury (1967), there are nine genera and about 60 species of frogfishes, distributed in tropical oceans.

Suborder Ceratioidei. This large group is made up of 11 families of deep sea anglerfishes that live usually below 300 m (Figs. 2–9**B** and 12–2**D**). These are small fishes with large mouths. The females are larger than the males. Pelvic fins are absent. The illicium usually carries a lure containing luminous bacteria. Some species bear barbels that have autogenic luminous systems (Marshall, 1979). Bertelsen (1951) remains the major reference on the suborder. There are about 135 species in 34 genera (Bertelsen, 1984).

In some families, the males are parasitic on the females and attach firmly with their jaws, becoming, in some cases, completely dependent on the female for their blood supply. Obligatory parasitic males are known in Ceratiidae, Neoceratiidae, and Linophrynidae (including Aceratiidae and Photocarynidae). Males of Caulophrynidae may be facultative parasites. In Melanocetidae, Himantolophidae, and Oneirodidae, the males are known to be free swimming but have a remarkably large olfactory apparatus and jaws suited to clamping onto the female. Other families are Diceratiidae, Gigantactinidae, Thaumatichthyidae, and Centrophrynidae.

Euteleostei: Acanthopterygii, Mugilomorpha, and Atherinomorpha

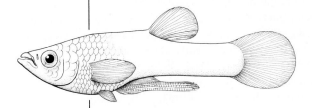

Superorder Acanthopterygii

The remainder of the neoteleosts constitute the superorder Acanthopterygii. The dorsal, anal, and pelvic fins of most of the orders of this group are characterized by hard, sharp spines. Most orders have ctenoid scales. The upper jaw is protrusible in most families, and the retractor dorsalis muscle (= retractor arcuum branchialum) is generally inserted on the third pharyngobranchial (Rosen, 1973).

Series Mugilomorpha

The relationships of the Mugilidae and related families are not clear, have been subject to several interpretations, and have often been placed as a suborder of Perciformes (see Nelson, 1984). A short historical review of the classification of mugilids is given by de Sylva (1984a). Stiassny (1990) proposed that the mugiloids should be a series (a taxon between superorder and order) on the level of Atherinomorpha and Percomorpha but notes that the character of united pelvic bones, which she believes shows the monophyly of the percomorphs, cannot be distinguished between the mugiloids and percomorphs. Stiassny (1993, p. 217) states that "the precise phylogenetic relationships of the mullets remain enigmatic," but she shows them as the sister group of the atherinomorphs. Johnson and Patterson (1993) include mullets in their proposed clade Smegmamorpha along with sticklebacks, silversides, pygmy sunfishes, and synbranchiformes. Nelson (1994) places mullets as a preperciform series and order near Atheriniformes.

Order Mugiliformes. Mullets, family Mugilidae, are characterized by subabdominal pelvic fins, a ligamentous connection of the pelvic bones to the cleithra, and well-separated spinous and soft dorsal fins. The pectoral fins are placed high, and the lateral line on the trunk may be poorly developed or absent. Mullets have feeble teeth in the jaws, and their gill rakers and pharyngeal teeth are adapted for processing small particles. They eat microscopic plants and other organic material. The anterior part of the stomach is modified into a gizzard-like structure.

The mullets are found in marine to fresh waters in warm climates. There are nearly 100 species in about a dozen genera, including *Agonostomus, Liza, Rhinomugil,* and *Mugil.* One species, *Mugil cephalus,* the striped mullet, is nearly circumtropical. Mullets are regarded as excellent food in some areas and are not only captured but are cultured in ponds. In most places where mullet culture is practiced, including Hong Kong and Taiwan, the young fish or "seed" are gathered from natural spawning areas or are simply allowed to flow with the tide into ponds, which are then closed off. Taiwanese mullet culturists have learned how to produce mullet juveniles through artificial propagation.

Series Atherinomorpha

Atherinomorphs have similarities that may be due to convergent adaptations to life near the water's surface, including the placement of pectoral, dorsal, and pelvic fins; dorsal position of mouth; and similarity in disposition of olfactory lamella. Rosen and Parenti (1981) point out that the members have relatively large demersal eggs with long adhesive filaments and many oil droplets, protrusible upper jaws with crossed palatomaxillary ligaments, and no third to fifth infraorbitals.

The included fishes are the orders of the flyingfishes and relatives (Beloniformes), the killifishes and topminnows (Cyprinodontiformes), and the silversides and rainbowfishes (Atheriniformes or "atherinoids"). Rosen and Parenti (1981) presented the silversides and relatives as polyphyletic. Lauder and Liem (1983a, p. 160) state that they "cannot be characterized." Parenti (1993) diagnosed the series Atherinomorpha as monophyletic using characteristics including features of osteology, musculature, anatomy of the circulatory system, testes, and the olfactory organ.

The series Atherinomorpha comprises about 170 genera and 1100 species, in up to at least 20 generally recognized families.

Order Atheriniformes. The silversides, priapiumfishes, and relatives have been shifted considerably in their placement by ichthyologists. There is no consensus on their relationships, and subordinal and familial placement appears to be subject to much differing opinion. Silversides have some features in common with the mullets and formerly were placed with them in a suborder of Perciformes (Regan, 1929) or in the separate order Mugiliformes (Berg, 1940; Lagler et al., 1977). Priapiumfishes have been placed variously with Cyprinodontiformes, Mugiliformes, or in their own order, Phallostethiformes.

This order is characterized by two dorsal fins, the first weak but composed of true spines. Pelvic fins are small, subthoracic in silversides, but in an anterior position, if present, in the priapiumfishes. In silversides the pelvics consist of a spine and five soft rays. Both ctenoid and cycloid scales are present in the group. The lateral line is poorly developed or absent.

Suborder Bedotioidei. This suborder contains the single family Bedotiidae, which consists of the genera *Bedotia* and *Rheocles* (Stiassny, 1990). Species are found in the inland waters of Madagascar. They are small, about 8 cm, and colorful. They are believed to be plesiomorphic atheriniformes (Parenti, 1993; Stiassny, 1990) and have one lachrymal bone per side plus one long bone posterior to it instead of two, as in others of the order. The posterior seven or so vertebrae show remarkable thickening of centra and neural and haemal spines. The first and second dorsal fins are close together in *Bedotia* (Parenti, 1993).

Suborder Melanotaenioidei. The Melanotaeniidae (rainbowfishes) is a family of freshwater fishes of Australia, with a few species in New Guinea. These are often highly colored species that are in demand as aquarium fishes. The family Pseudomugilidae is sometimes separated from the Melanotaeniidae. Some of these can enter salt water and have a fairly wide coastwise distribution in Australia.

Suborder Atherinoidei. The silversides, Atherinidae (Fig. 13–1**A**), are small fishes, mostly marine, with some freshwater representatives. Most have a prominent silver band along each side and lack the lateral line. They are found along most warm and warm-temperate coasts and often invade fresh water in the absence of varied freshwater faunas. The genus *Chirostoma* contains the charal and pejerreys from plateaus of Mexico and parts of South America, where they are important food fishes. *Labidesthes sicculus* is a freshwater species of North America.

Leuresthes tenuis, the grunion of California, is famous for spawning at night on beaches during spring and summer high tides. Females swim ashore with the highest waves at high tide during the full moon. They partially bury themselves in the sand and deposit eggs between waves. The eggs are fertilized by males, which curl

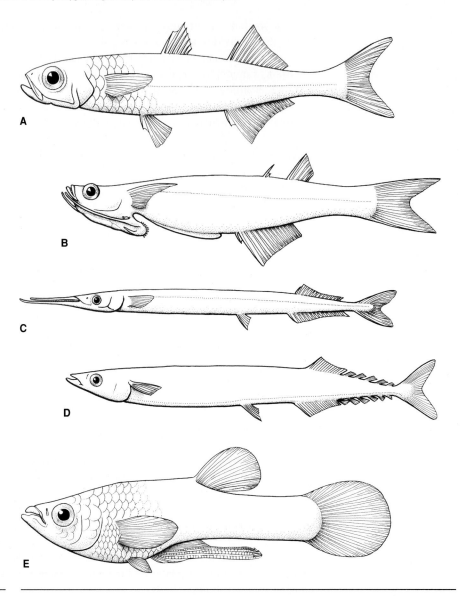

FIGURE 13–1 Representatives of Atheriniformes, **A**, Silverside (Atherinidae, *Atherina*); **B**, priapiumfish (Phallostethidae, *Neostethus*). Representatives of Beloniformes, **C**, needlefish (*Belone*); **D**, saury (*Scomberesox*). Representative of Cyprinodontiformes, **E**, guppy (*Poecilia*).

around the half-buried female. Eggs remain in the sand above the water line until the next series of high tides, two weeks later, and then hatch when washed out by the waves (Breder and Rosen, 1966; Walker, 1959).

Atherinopsis californiensis is one of the largest silversides, reaching a length of more than 30 cm. Its spawning is more typical of the family as a whole. The eggs have filaments that attach to vegetation or to other objects.

Members of the family Notocheiridae (Isonidae) are called "surf sardines." These nearly transparent little fishes of the Indian and Pacific oceans, including

Japan, are relatively deep-bodied and compressiform, with a keel along the ventral mid-line.

Telmatherinidae contains the Celebes rainbowfish of the Celebes Islands. *Telmatherina ladigesi* is a beautifully colored fish with elongate dorsal and anal fins and is prized as an aquarium fish.

The Phallostethidae (including Neostethidae) of Thailand, the Philippines, and the Indo-Malayan Archipelago are called priapiumfishes because of the strange intromittent organ of the male (Fig. 13–1**B**). This device is situated in the region of the isthmus and is supported by a skeletal structure made up of elements from the pelvic and pectoral girdles plus the first pair of ribs. The anus and genital pore are jugular in position.

The genus *Dentatherina* from the western Pacific, which Eschmeyer (1990b) considers to be in Atherinidae, is sometimes placed in its own family (Dentatherinidae). Parenti (1984) places it close to Phallostethidae.

Order Beloniformes (Synentognathi). This and the order Cyprinodontiformes comprise the atherinomorph Division II of Rosen and Parenti (1981). This order includes the ricefishes and the mostly marine flyingfishes, halfbeaks, and needlefishes. Beloniforms are soft-rayed, physoclistic, and have abdominal pelvic fins with six rays. The lower pharyngeal bones are fused to each other. There is no orbitosphenoid or mesocoracoid, and the maxillae are excluded from the gape.

This group has several characteristics that are related to their life near the water surface. The dorsal fin is set far back over the anal fin, and the caudal fin usually has only 13 branched rays, with the lower lobe longer and with more principal rays than the upper (Rosen and Parenti, 1981). The lateral line of the trunk, if present, is situated very low on the body. The pectorals are set high on the sides. The nasal organs are variously reduced.

These are active fishes, mostly of warm seas, although some enter fresh water and some are found in temperate marine waters. Fossil Beloniformes are known from the Eocene. There are nearly 40 genera and 180 living species in the order.

Suborder Adrianichthyoidei. This suborder was defined by Rosen and Parenti (1981) and contains only the family Adrianichthyidae, including Horaichthyidae and Oryziidae. Nelson (1984) retains the two latter groups as valid families, but Collette et al. (1984) and Robins et al. (1991) follow Rosen and Parenti. In this group there is no vomer, metapterygoid, ectopterygoid, or supracleithrum. There are two nasal openings per side, and there is no lateral line canal on the body. The articular surface of the fourth epibranchial is expanded (see Nelson, 1984; Rosen and Parenti, 1981). Nelson (1994) lists Oryziinae, Adrianichthyinae, and Horaichthyinae as subfamilies.

Adrianichthyidae contains fishes of Asia and the East Indies. The ricefishes and medakas of the genus *Oryzias* are found in fresh and brackish waters of Southeast Asia, the East Indies to Timor and Celebes, Luzon, and north to Japan. These are popular aquarium and experimental fishes. The medakas have been referred to as the "white rats" of the fishes. *Adrianichthys* and *Xenopoecilus* of the fresh waters of Celebes are large-jawed fishes with internal fertilization and eggs that hatch upon extrusion. In *Horaichthys* of India, the gonopodium of the males is deflected to the right and the genital orifice of the females is deflected to the left.

Suborder Exocoetoidei. This suborder, according to Rosen and Parenti (1981), is defined by a platelike process on the posterior, ventral part of the basioccipital, one nasal opening on each side, a median lower pharyngeal tooth plate, and more than three anterior branchiostegals. At some stage of the life history, there is an elongate lower jaw. The eggs of many species bear long tendrils or filaments. Collette et al. (1984) add that the eggs of the fishes in this group have no (or minute) oil droplets and that the preanal length of the larvae is 66 percent of the total length.

The family Belonidae contains the needlefishes, which are slender, elongate fishes with the jaws produced into a long, beaklike mouth set with sharp, fine teeth (Fig. 13–1**C**). The largest members of the family may reach over 1.5 m long. Needlefishes are noted for their jumping ability. Schools may scatter with great leaps at the swift approach of a boat, or they may jump over objects as if in play.

The family Scomberesocidae contains the sauries (Fig. 13–1**D**), which usually have a short beak and a series of mackerel-like finlets behind the dorsal and anal fins. The leaping habit is somewhat developed. *Cololabis saira* of the North Pacific, a plankton feeder, occurs in large schools and is the object of a commercial fishery.

The Hemiramphidae, the halfbeaks, also are noted as being remarkable jumpers. In this family, the lower jaw is greatly extended but the upper jaw is fairly short. The lower lobe of the caudal is larger than the upper. Their food is plankton.

The Exocoetidae is the best-known family in the order. It contains the flying-fishes, which are seen gliding over the water's surface in many warm seas (Fig. 2–12**G**). The pectoral and pelvic fins are both enlarged in some species and serve as gliding planes (four-wing flyingfishes). In others, the pectorals may be greatly enlarged and are the gliding planes (two-winged flyingfishes). The lower lobe of the caudal is usually elongate. Careful observation of a flyingfish on takeoff or in the process of gaining additional gliding time when near the crest of a wave will reveal the lower part of the caudal employed in an extremely rapid sculling motion (it is the only part of the fish in contact with the water). According to Rayner (1991), flights of 50 m and a speed of 90 km/hr are not unusual. The California flyingfish, *Cypselurus californicus,* is one of the largest of the family and reaches a length of about 45 cm.

Order Cyprinodontiformes (Microcyprini). This is the order of killifishes, topminnows, and relatives. They are soft-rayed physoclists with abdominal pelvic fins. The single dorsal is set behind the middle of the body. The caudal fin has a varying number of rays and is never forked. The orbitosphenoid, mesocoracoids, and basisphenoids are lacking, and the maxillaries are excluded from the border of the mouth by the premaxillaries. The lateral line canal system is incomplete or not well formed on the head and body. These fishes show some similarities to the Beloniformes and are sometimes placed with them in a single order (Nelson, 1984).

Cyprinodontiformes are widely distributed and occur naturally in all continents except Australia and Antarctica. They are all small and usually found in estuarine or marine environments. There are about 670 species of cyprinodontiformes, arranged in from 47 (Parenti, 1981) to 88 (Nelson, 1994) genera.

Suborder Aplocheiloidei. These are small freshwater fishes of the Americas from Florida to Argentina, as well as tropical Africa and the southern part of Asia. Parenti (1981) arranges them in two families, the Aplocheilidae and Rivulidae; but Nelson (1984), Robins et al. (1991), and Eschmeyer (1990) list the latter as a subfamily of

Aplocheilidae. The suborder is characterized by having the first two dorsal radials each supporting a dorsal fin ray, three basibranchials, and the pelvic bones close to each other.

The Aplocheilidae contains numerous species that enter the aquarium trade. The aplocheilines are fishes of Africa and southern Asia and include panchaxes and lyretails of the genus *Aphiosemion;* panchaxes of *Epiplatys* and *Pachypanchax;* and the nothos of *Nothobranchius.* The latter are of interest because of their life histories; they are "annual" fishes that deposit eggs in the mud of temporary ponds and then die when the ponds dry. The eggs delay full development until the ensuing rainy season, when they hatch and repopulate the ponds.

Rivulins are distributed from Florida to Argentina and include such aquarium fishes as the genera *Rivulus, Pterolebias,* and *Cynolebias.* Species of the latter genus are called "pearlfishes" in the aquarium trade, although the family Carapidae is known by the same name.

Suborder Cyprinodontoidei. These fishes have a metapterygoid, the first dorsal fin ray is borne on two radials, and three basibranchials are present. Upper and lower hypurals are fused to the last vertebral centrum. (The lower hypurals of *Xenopoecilius* are not fused; Rosen, 1964). Species are distributed in fresh waters and inshore marine habitats throughout much of the temperate and tropical parts of the world but do not occur natively in Australia and New Guinea. Many members of the suborder are brightly colored and enter the aquarium fish trade.

According to Parenti (1981), the suborder contains seven families, some of which are given a subfamilial rank by Eschmeyer (1990b). Robins et al. (1991) do not recognize some of the familial designations of genera proposed by Parenti, as will be noted.

The family Profundulidae, which is not recognized by Eschmeyer (1990b) or Robins et al. (1991), has one genus (*Profundulus*) with five species in Mexico and Central America. *P. hildebrandi* of Mexico may be endangered.

The familiar killifishes of the genus *Fundulus,* which are often used as experimental animals, are placed in the family Fundulidae by Parenti (1981) but are retained in Cyprinodontidae by Robins et al. (1991a). Other genera include *Adinia, Leptolucania,* and *Lucania.* The rainwater killifish, *Lucania parva* of the Atlantic Coast and the Gulf of Mexico, has, for some unknown reason, been introduced to saline lakes in Utah and to parts of the Pacific Coast.

The family Valenciidae was erected for the genus *Valencia* of the northern coast of the Mediterranean. There are two species, both of which are used in the aquarium trade. *V. hispanica,* which ranges from Spain to Italy, is considered to be endangered because of drainage of marshes and the introduction of the competitive and predatory *Gambusia affinis.*

The foureye fishes of the family Anablepidae actually have only two eyes, but each pupil is divided into an upper section and a lower one. The eyes protrude from the top of the head so that the upper half can be above the surface, allowing light from above water to reach the retina. Simultaneously, the lower section of the pupil allows light from below the surface to enter the eye.

Another peculiarity of the family is the asymmetrical orientation of the intromittent organ of the male and the genital opening of the female. About 60 percent of the males have the intromittent organ oriented to the right; the remainder are sinistral.

Fortunately, the reverse is true for the orientation of the genital orifice of the females. Foureye fishes are found from southern Mexico to northern South America.

Jenynsia, a genus of southern Brazil and northern Argentina, is placed with Anablepidae by Parenti (1981) but is retained as an *Anableps.* The gonopodium and female genital aperture are oriented either to right or left.

The family Poeciliidae contains the livebearing tooth carps—guppies, mollies, and swordtails. (Parenti, 1981, includes certain egglayers in the family and considers the traditional Poeciliidae to constitute a subfamily; see also Parenti and Rauchenberger, 1989.) Livebearers are found primarily in the warmer parts of North and South America and are among the greatest favorites of aquarists. The guppy, *Poecilia reticulata* (Fig. 13–1**E**), is one of the easiest livebearers to maintain and breed. Some members of the family have reputations as "mosquito fish" because of their habit of feeding close to the water's surface on insect larvae. *Gambusia affinis* and *G. holbrooki* of the United States have been introduced to many areas of the world to aid in insect control, but the insect control value of these introductions is slight at best. The *Gambusia* species compete with, and are predators on, various native species.

The genera *Poecilia* and *Poeciliopsis* are noted for the existence of unisexual (female) "species" produced by hybridization or by activation of eggs by sperm that does not contribute genetic material to the developing eggs (see Schultz, 1989, and Wetherington et al., 1989).

Goodeidae includes the subfamilies Empatrichthyinae and Goodeinae (Nelson, 1994; Parenti, 1981). Robins et al. (1991) retain the oviparous springfishes (*Crenichthys*) and poolfishes (*Empetrichthys*) of Empetrichthyinae in the family Cyprinodontidae, as will be done here.

The splitfins of Goodeinae are viviparous topminnows endemic to the plateau of central Mexico. Developing embryos absorb nutrients through branched filaments called trophotaeniae, which are attached at the anal region. The common name refers to the anal fin of the male because of a slightly separated anterior lobe that may serve as a gonopodium. Various members in the aquarium trade are called allotoca, goodea, characodon, and skiffia.

The family Cyprinodontidae contains oviparous killifishes of several genera from northern South America to the United States, the West Indies, the Mediterranean area, the Nile, and other parts of Africa. Old World genera are *Aphanius,* from various localities around the Mediterranean Sea (to Iran), and *Kosswigichthys,* from Turkey. Possibly related to those genera is *Orestes* of lakes in Bolivia, Chile, and Peru.

There are two species of *Cubanichthys* in Cuba and Jamaica and several species in the southern United States, Mexico, and northern South America. Included are the pupfishes (*Cyprinodon* spp.), the flagfish (*Jordanella*), the goldspotted killifish (*Floridichthys*), the poolfishes (*Empatrichthys* spp.), and springfishes (*Crenichthys* spp.), species of which are found isolated as relicts in southern Nevada.

The Devil's Hole pupfish, *Cyprinodon diabolis,* which is known from one small spring in Nevada, was the center of attention of legal action to save the spring from being dried by pumping water from the aquifer. This served to bring the plight of several potentially threatened species to the attention of the public.

Acanthopterygii: Berycoids Through Snailfishes

Series Percomorpha
Order Stephanoberyciformes
Order Beryciformes
    Suborder Trachichthyoidei
    Suborder Berycoidei
Order Zeiformes
Order Gasterosteiformes
    Suborder Gasterosteoidei
    Suborder Syngnathoidei
Order Synbranchiformes
    Suborder Synbranchoidei
    Suborder Mastacembeloidei
Order Scorpaeniformes
    Suborder Dactylopteroidei
    Suborder Scorpaenoidei
    Suborder Platycephaloidei
    Suborder Anoplopomatoidei
    Suborder Hexagrammoidei
    Suborder Cottoidei

Series Percomorpha

All the remaining orders of fishes constitute the series Percomorpha and include a great array of families that inhabit all seas and many freshwater habitats. This group includes about 230 families, over 2140 genera, and 12,000 species (Nelson, 1994). Percomorphs differ from the series Atherinomorpha in that they have a fourth pharyngobranchial, which the atherinomorphs lack, and usually retain more than two infraorbital bones. The mechanism for lower jaw protrusion differs from that of the atherinomorphs (Lauder and Liem, 1983a). Stiassny (1990) presents the percomorphs as monophyletic because the two halves of the pelvic girdle are fused medially.

There are many problems yet to be solved in the matter of relationships among the percomorphs. Accordingly, there are differing theories and opinions concerning the classification of these fishes. Stiassny (1990) removes the mullets (see Chapter 13) from Percomorpha into the series Mugilomorpha, but she points out that the placement of this group is not certain because the mugiloids have an essentially percomorph pelvic girdle. Johnson and Patterson (1993) present a new view of percomorph relationships and suggest new designations for higher taxa. However, a conservative view of percomorph classification similar to that of Nelson (1994) will be outlined here.

Order Stephanoberyciformes. Ebeling and Weed (1973) presented this group as a separate order (Xenoberyces) because these fishes lack the orbitosphenoid, subocular shelf, and supramaxillaries, which are characteristic of the related Beryciformes (of which this group has usually been considered a suborder). Stephanoberyciformes have a single dorsal fin set well back on the body in most species; dorsal spines, if present, are weak. Pelvic fins are variously placed, from abdominal to jugular.

Moore (1993) does not recognize this order and considers the families assigned to it, and most of those assigned to the next order, to be part of the clade Trachichthyiformes. There seems to be much difference of opinion regarding the relationships of the stephanoberyciforms and beryciforms.

The stephanoberyciforms are marine fishes; some live in the deep sea. Melamphaidae, called melamphids or bigscale fishes, have up to three weak spines in the dorsal fin and three or four spines before the caudal fin, both dorsally and ventrally. The lateral line is not well developed. Scales are cycloid. There are over 30 species in five genera.

The Gibberichthyidae, or gibberfishes, have spines (five to eight) preceding the dorsal and anal fins. A lateral line is present and scales are cycloid. The single genus (*Gibberichthys*) has two species. The larvae and prejuveniles of *Gibberichthys* pass through what is termed a kasoridon stage, during which they have long, strangely branched pelvic fins. (See Keene and Tighe, 1984).

Stephanoberycidae, the pricklefishes, have several short spines preceding the caudal fin above and below. A lateral line is present and scales are cycloid or spiny. There are three genera: *Stephanoberyx* and *Acanthochaenus* of the Atlantic and *Malacosarcus* of the Pacific. Hispidoberycidae contains a single species, *Hispidoberyx ambagiosus,* from waters off China and India. This species has a strong spine on the operculum and spiny scales.

Rondeletiidae, the redmouth whalefishes (Fig. 14–1**A**); Barbourisiidae, the red whalefishes; Cetomimidae, the flabby whalefishes; and some other oceanic fishes

FIGURE 14–1   Representative of order Stephanoberyciformes, **A,** Redmouth whalefish (Cetomimidae). Representative of order Beryciformes, **B,** squirrelfish (*Holocentrus*). Representative of order Zeiformes, **C,** john dory (*Zeus*).

were formerly placed in the order Cetomimiformes by Greenwood et al. (1966) but have been shifted around considerably since that time (see Nelson, 1994). Mirapinnidae (the hairyfish) and Megalomycteridae (the largenose fish), once placed in the Lampridiformes, are now placed with the whalefishes in Stephanoberyciformes (Nelson, 1994).

The two redmouth whalefishes, genus *Rondeletia,* have a large head and mouth and have the pores of the lateralis system of the body disposed in several vertical rows. Length does not exceed 11 cm. None of the whalefishes have fin spines. The single red whalefish, *Barbourisia rufa,* in contrast to the redmouth whalefishes, has spiny skin and reaches over 30 cm long.

There are nine genera and about 35 species of the flabby whalefishes. These have no scales, no pelvic fins, and some species have rudimentary eyes. Some species reach nearly 40 cm long.

The following two families include strange oceanic fishes that some ichthyologists have thought to be closely related to the Beryciformes. They were once placed in an order of their own (Bertelson and Marshall, 1956), and some recent researchers consider them to belong to Lampridiformes (Eschmeyer, 1990; Nelson, 1984; Rosen and Patterson, 1969). Nelson (1994), basing the placement on the findings of Moore (1991, 1993) and others, considers these to be stephanoberyciforms.

The family Mirapinnidae is now considered to have two subfamilies, the hairyfish of Mirapinninae and the tapetails of Eutaeniophorinae (Nelson, 1994). The hairyfish of the Atlantic, *Mirapinna esau,* is scaleless and has skin covered with small villous projections that give the appearance of fur. Its pectoral fins are placed high on the sides well behind the head, and the large pelvic fins are set about where pectorals are usually seen. No mature specimens have been captured; all have been less than 50 mm long. The subfamily Eutaeniophorinae contains three species of elongate fishes that are called tapetails because of a long, ribbon-like extension of the caudal fin in juveniles. These are found in the deep waters of both the Atlantic and Pacific.

The family Megalomycteridae, which was formerly placed in its own suborder, has four genera and five species. These fishes are known as largenose fishes because they show a great development of the olfactory lamellae. They are represented in both the Atlantic and Pacific.

Order Beryciformes (Berycomorphi). This order embraces the roughies, alfonsinos, soldierfishes, and their allies. They are characterized by the presence of spines in the dorsal, anal, and pelvic fins; ctenoid or cycloid scales; thoracic or subthoracic pelvics with one spine and 6 to 13 soft rays; and the orbitosphenoid. However, they lack the mesocoracoid. The gas bladder may be absent or physostomous, but in most species it is physoclistic. The caudal fin has 18 or 19 principal rays with a few procurrent spines both dorsally and ventrally. These marine fishes may be the primitive sister group of the Perciformes (Lauder and Liem, 1983a), and they show trends characteristic of that order; thus they are usually considered as basal acanthopterygians. Their fossil history reaches back to the Cretaceous, at which time they were diverse.

Suborder Trachichthyoidei. Families in this suborder include the deep-dwelling Diretmidae (spinyfins) and the Anoplogastridae (fangtooths), which are usually found deeper than 600 m. The Anomalopidae, flashlightfishes or lantern-eyes, occur in tropical and subtropical waters mainly in the Indo-Pacific region. This is one of the few bioluminescent, shallow water groups. The photophores in these fishes are subocular and are filled with luminous bacteria. The light can be revealed or concealed by everting or retracting the photophore or by advancing or withdrawing a shade. The genera of the family are *Photoblepharon, Anomalops, Kryptophaneron,* and *Kaptotron.*

The Monocentridae, or pineconefishes, of the Indo-Pacific region have large scales firmly united to one another and prominent dorsal fin spines that are alternately oriented to right and left. There are two photophores on the ventral surface of the mandible. There are two genera, *Monocentrus* and *Cleidopus.* The Trachichthyidae (roughies and slimeheads) now includes Korsogasteridae. Some of the roughies contribute to an important mid-water commercial fishery in the South Pacific. The orange roughy, *Hoplostethus atlanticus,* is now common in North American seafood restaurants.

Suborder Berycoidei. Zehren (1979) believed that this group included only the Berycidae (alfonsinos), which appear to be the most primitive members of the order. However, others (Eschmeyer, 1990; Nelson, 1984) have included up to seven families in this suborder. The alfonsinos are usually bright red fishes of moderate depths that have a widespread distribution. The family is composed of two genera, *Beryx* and *Centroberyx.*

A well-known group is the Holocentridae, which contains the big-eyed, brilliantly colored soldierfishes and squirrelfishes of tropical coral reefs (Fig. 14–1**B**). These are active at night and can be found hiding in various shelters during the day.

Order Zeiformes (Zeomorphi). This is the order containing the dories and boarfishes. These fishes are marine and mostly occur in deep water. Fossils are known from the Eocene.

Characteristics include thoracic pelvic fins with a spine and five to nine soft rays, ctenoid scales, and the lack of an orbitosphenoid, subocular shelf, and supramaxilla. The caudal has only 12 or 13 principal rays, and the spinous anal sometimes forms a nearly distinct fin anterior to the soft-rayed section of the anal. The mouth is protrusible, greatly so in some genera. (See Heemstra, 1980, for diagnosis of the order.)

The family Zeidae, the dories, are fishes of moderate depths and are often a part of commercial fisheries. The john dory, *Zeus faber,* has a round dark spot on each side, and some fishermen of various European countries refer to it as "St. Peter's fish," perpetuating a legend that the marks are prints of the saint's thumb and forefinger (Fig. 14–1**C**). The species may reach a maximum length of nearly 1 m. Other families in the order are Grammicolepididae (diamond dories), Parazenidae, Macrurocyttidae, and Oreosomatidae.

The family Caproidae, the boarfishes, which are noted for their greatly protractile mouths, is not closely related to the other fishes placed in the order according to Rosen (1973) and Heemstra (1980), who suggested that these fishes might better be placed with another group, possibly the Perciformes. Later, Rosen (1984) suggested that the entire order should be placed with the Tetraodontiformes, but that has not been widely accepted.

Order Gasterosteiformes (Thoracostei plus Aulostomiformes, Pegasiformes, and Solenichthyes or Syngnathiformes). Many fishes of this order are characterized by rather elongated snouts and spines in the dorsal fin. Pelvic fins are abdominal to subthoracic, and the pelvic bones are not connected to the pectoral girdle. The pelvics are spinous in some species. Most have dermal armor. The gas bladder is physoclistic, and the orbitosphenoid, basisphenoid, and supramaxillaries are absent (Fritzsche, 1984; Pietsch, 1978). There are sufficient differences among the fishes of the order that they have been formerly placed into two or three separate orders. Fishes included are sticklebacks, tube-snouts, pipefishes, seahorses, seamoths, trumpet-fishes, and others, which are mostly marine and prefer warm waters, but a few species reside in fresh waters. Fossils are known from the Eocene. Fritzsche (1984) reviews the difficulties in recognizing the relationships within the group and outlines the classification schemes of Greenwood et al. (1966) and Banister (1967). The classification here is based on those of Pietsch (1978) and Nelson (1994). Infraorders and superfamilies are used in the classification systems of Pietsch (1978), Fritzsche (1984), and Nelson (1994), but for the sake of simplicity they will not be used here.

Suborder Gasterosteoidei. This group includes the sticklebacks and close relatives. In these fishes, the pelvic fin is placed under the pectoral fin in a subthoracic position. Ribs and parietals are present, and the mouth morphology is somewhat different from the other suborders in the order.

The family Aulorhynchidae contains the tube-snouts of the North Pacific. *Aulorhynchus flavidus* has about 18 to 20 separate spines along the back (Fig. 14–2**A**). The hypoptychidae (sand eels) of the western Pacific resemble the sticklebacks but lack free spines (Ida, 1976).

Family Gasterosteidae contains the familiar sticklebacks of northern temperate waters. *Gasterosteus,* the threespine sticklebacks (Fig. 2–15**B**), includes both fresh-water and anadromous species, and the latter have a much greater covering of bony plates along the sides. The members of the genus *Gasterosteus* have been the subject of much study because of their interesting reproductive behavior and nest building. The ninespine sticklebacks of the genus *Pungitius* range from temperate fresh waters into the Arctic, where they are found in both marine and fresh water. Members of this genus and the related *Spinachia,* the fifteenspine sticklebacks, build nests well off the bottom in vegetation. Other genera are *Apeltes* (the fourspine stickleback) and *Culaea* (the brook stickleback). Gasterosteids seldom reach as much as 10 cm long, although *Spinachia spinachia* is reported to reach 18 cm.

Suborder Syngnathoidei. According to Pietsch (1978), the seamoths of the family Pegasidae belong to this suborder. Some investigators, however, place them in their own order, Pegasiformes. The Pegasidae (Fig. 2–7**A**) are among the most peculiar of fishes. The body is covered by sculptured bone, with the tail enclosed in bony rings. The small mouth opens below an odd rostrum composed of modified nasal bones. Pectoral fins are broad, fanlike, and oriented horizontally. The pelvic fins have one spine and one to three soft rays and are subabdominal. The other fins are soft rayed. The gas bladder is lacking. These are small shore fishes of the Indo-Pacific.

Also in this suborder are the pipefishes and seahorses of the family Syngnathidae, peculiar little fishes with no pelvic fins and a body enclosed in bony rings. The pipefishes (*Syngnathus*) are elongate fishes, but the seahorses (*Hippocampus*) have the head flexed ventrally so that it is at a right angle to the body (Fig. 2–7**D**). The tail is flexible and prehensile. Both pipefish and seahorse males are equipped with brood pouches in which the eggs are incubated. These fishes are often well camouflaged

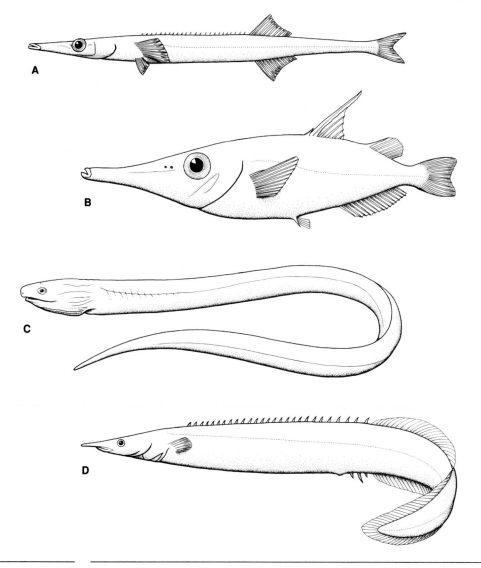

FIGURE 14–2   Representatives of Gasterosteiformes, **A,** Tubesnout (Aulorhynchus); **B,** snipefish (Macro-rhamphosidae). Representatives of order Synbranchiformes, **C,** Synbranchidae; **D,** Masta-cembelidae.

by both color and shape. Perhaps the best example is the seadragon of Australia, *Phyllopteryx foliatus,* which bears many leaflike lobes. The family Solenostomidae, the ghost pipefishes, are less well known than the syngnathids. They have a long tubelike snout, large pectoral and caudal fins, and a high flexible spinous dorsal. The female incubates the eggs between the pelvic fins. The single genus, *Solenostomus,* is found in the Indo-Pacific.

The monotypic family Indostomidae of the fresh waters of Burma is sometimes placed in its own order. Its exact placement may be open to question, but it has features in common with Gasterosteiformes and probably belongs with this order (Greenwood, 1966; Pietsch, 1978). This curious little fish, *Indostomus paradoxus,* has free spines before the dorsal fin, and the slender body is enclosed in bony rings, so it combines some of the characteristics of both sticklebacks and pipefishes.

The trumpetfishes, of the family Aulostomidae, are circumtropical. They have a series of isolated spines along the back and a very long snout. They reach a length of about 60 cm. The family Fistulariidae, the cornetfishes, is also circumtropical, and its members may approach 2 m in length. They are extremely slender and have a long filament formed from the middle rays of the caudal fin. The sides of the long snout are equipped with sharp ridges. Large specimens have been observed to spend up to half an hour moving 60 to 70 cm to get close to a school of small goatfishes in order to prey upon them. During this time, the specimens moved so slowly that they seemed almost motionless.

The shrimpfishes and snipefishes are found in warm seas, often associated with coral reefs. The family Centriscidae, found in the Indo-Pacific region, includes the shrimpfishes (called razorfishes in some areas). They have a thin body with a sharp ventral edge and an armor of plates. Their locomotion is usually by means of undulating fins, and they generally maintain a vertical position, often with the head down. They are found in the Indo-Pacific region.

The snipefishes of Macrorhamphosidae are also Indo-Pacific fishes and, like the shrimpfishes, may swim mainly with the head down (Fig. 14–2**B**). The body is quite deep, the snout is long, and there is a large dorsal fin spine.

Order Synbranchiformes. This order contains the swamp eels of Synbranchoidei and spiny-backed eels of Mastacembeloidei. These are elongate fishes of subtropical and tropical fresh water, although some species are known to enter waters of very low salinity. The phylogenetic relationships of the group have been covered by Travers (1984), who placed the spiny-backed eels of Mastacembeloidei (formerly in Perciformes) into this order.

The synbranchiforms have no pelvic fins or pelvic girdle. The pectoral girdle is set well behind the head and not connected to the cranium because the posttemporal is reduced or lost (Travers, 1984). The caudal fin is small or absent. If present, scales are cycloid.

Suborder Synbranchoidei. The swamp eels have no pectoral fins, and the dorsal and anal fins are reduced to a ridge with no rays. The caudal fin is reduced or absent, and scales are lacking except in the genus *Monopterus.* The gill openings are confluent ventrally. The gills are greatly reduced, and aerial respiration occurs in the pharynx or intestine. Many of the species move overland in or near the swamps where they live. The order contains one family, Synbranchidae, which has about 15 species in tropical areas. Some live in caves in Africa and Yucatan; most are found in freshwater or brackish-water swamps; *Macrotrema caligans* of Malaya is known to enter saline waters. Lauder and Liem (1983a) suggested a close relationship between the Synbranchiformes and the snakeheads of Channoidei.

Synbranchus is the most widespread genus, found in Asia, Africa, South America, and Mexico. The rice eel, *Monopterus albus* (Fig. 14–2**C**), of Southeast Asia and Indonesia, can spend the dry season in holes in the bottom when the swamps dry up. *Amphipnous cuchia,* the cuchia of India, differs from other members of the family in that it has special air sacs that communicate with the pharynx and facilitate aerial respiration. The larvae of *M. albus* can live in waters of low oxygen because they have a small blood–water barrier and skin with many capillaries, in which the main flow of blood is opposite the current of water that the pectoral fins bring from the relatively oxygen-rich surface layer (Liem, 1981).

Suborder Mastacembeloidei. This suborder contains the spiny-backs, which are sometimes placed in their own order or included as a suborder of Perciformes. These are freshwater or brackish-water fishes of Africa and Asia. The best-known representatives are in the family Mastacembelidae, which has a series of sharp spines on the dorsal surface before the dorsal fin (Fig. 14–2**D**). The anterior nostrils form tubes at the tip of the elongate snout and open on each side of a tentacle. Many mastacembelids live in swamps and depend, at least in part, on atmospheric oxygen.

According to Travers (1984), genera of Mastacembelidae include *Mastacembelus,* of Asia and the Middle East, with six species; *Macrognathus,* of Southeast Asia, with 11 species; *Caecomastacembelus,* of West Africa, with 22 species; and *Afromastacembelus,* of East Africa, with about 16 species.

Members of the family Chaudhuriidae lack the isolated spines of Mastacembelidae and have the rostral appendage reduced. They are sometimes placed into a separate order or suborder. They are small freshwater fishes found in China and Southeast Asia. There are two genera: *Rhynchobdella,* with one species in Taiwan, China, and possibly Korea (Travers, 1984); and *Chaudhuria,* with three species in Southeast Asia.

Order Scorpaeniformes. The order Scorpaeniformes (Scleroparei) is a rather large and important group of uncertain relationships. It is generally regarded as a preperciform order. Its 1000 members share the common character of a "suborbital stay," formed by the second infraorbital crossing the cheek, just under the skin, from the orbit to the preopercle. The caudal skeleton is characterized by two platelike hypurals that are sutured to the terminal centrum (Lauder and Liem, 1983a). The members of this order are often characterized by bony plates or spines on the head and body and have large, broad-based pectoral fins. Most members are marine. Many of the species have venomous spines.

There is some lack of agreement among ichthyologists regarding the relationships and classification of this order. Berg (1940) placed the scorpaeniforms into Perciformes as suborder Cottoidei. Gosline (1971) and Lauder and Liem (1983a) regard them as perciform derivatives, but Washington et al. (1983), Nelson (1994), Robins et al. (1991a), and Eschmeyer (1990b) place them before the perciforms.

Suborder Dactylopteroidei. This suborder is composed of the single family Dactylopteridae (= Cephalacanthidae), the flying gurnards. They are characterized by a large, bony head with large spines extending backward from the lower part of the operculum and greatly enlarged pectoral fins, with the first few rays short and separated from the rest of the fin. Those rays and the pelvic fins are used in "walking" over the substrate. In spite of their common name, these tropical marine fishes are usually found on the bottom; and although they have been said to leap from the water and glide, the fins seem too weak to allow gliding, and this activity has not been confirmed. They often swim off the bottom, spread the pectoral fins, and passively "glide" through the water, reaching the bottom some distance from where they started. Wheeler (1975) mentions the display of the pectoral fins by individuals that have been startled. Some members have a bony connection, by means of parapophyses, between the gas bladder and the cranium, an adaptation that might aid in hearing. These fishes have some characteristics in common with various Gasterosteiformes (Pietsch, 1978).

There are two genera, *Dactylopterus* of the Atlantic and *Dactyloptena* of the central and western part of the Pacific and Indian oceans.

Suborder Scorpaenoidei. This suborder contains seven families (Eschmeyer, 1990b; Nelson, 1994). The largest of the seven is Scorpaenidae (Fig. 2–4**A**), the scorpionfishes and relatives, which is divided into 12 subfamilies, with over 300 species in about 60 genera. The species are marine and are found mostly in temperate and tropical waters in the Pacific and Indian oceans. There are about 90 members of this family in North American waters, some reaching nearly 1 m in length. The rockfishes and redfishes (*Sebastes*) of Sebastinae are well-known commercial fishes of both the North Pacific and North Atlantic. They are locally known to sport fishermen on the Pacific coast as "rock cod," "sea bass," and "snapper." Some species are marketed in states where they are sold as "red snapper." *Sebastes fasciatus,* the Acadian redfish, and *S. mentella,* the deep water redfish, both of the Atlantic, are marketed under the name "ocean perch." *Sebastes alutus,* the Pacific ocean perch, is another important species.

Scorpaeninae includes the scorpionfishes of *Scorpaena* and relatives. These live in warmer water than the sebastines, and most species are usually multicolored and well provided with spines, cirri, and fleshy flaps, giving many of them a bizarre appearance. Most scorpaenids have venom-producing tissue along the dorsal fin spines. Wounds caused by these spines are very painful.

The thornyheads, Sebastolobinae, are small, spiny, commercial fishes of the North Pacific. The subfamily Pteroinae (typified by the tropical genus *Pterois*), the lionfishes, and the turkeyfishes are quite venomous and can cause severe illness in persons punctured by the dorsal spines. Stonefishes, of Synanceinae (Indo-Pacific), take their common name from their resemblance to the rocks among which they live. They are reputed to produce the strongest venom of all the fishes in glands at the bases of the dorsal spines. This venom is discharged through grooves in the spines and can be fatal to humans. Other highly venomous fishes with Indo-Pacific distribution are the stingfishes of the genus *Minous* (Minoinae) and the ghouls of the genus *Inimicus* (Chorydactylinae).

Other subfamilies recognized by Eschmeyer (1990b) are the Plectrogeninae, Setarchinae, Neosebastinae, Apistinae, and Tetraroginae (waspfishes). Other families in this order are the Caracanthidae, "coral crouchers"; Congiopodidae, horsefishes; Aploactinidae, velvetfishes; Gnathanacanthidae, red velvetfishes; and Pataecidae, the Australian prowfishes or Indianfishes. Many of the species in these families have the dorsal fin beginning far forward on the head, and the Pataecidae gets its secondary common name "Indianfishes" from the resemblance of the fin to a war bonnet.

The Triglidae, or searobins (Fig. 2–12**E**), are characterized by bony-plated heads and large pectoral fins, with a few of the lower rays detached as separate finger-like tactile and chemoreceptive organs. The pelvic fins are relatively large and strong and aid in "walking" along the bottom. Some species of this widely distributed family are commercial fishes.

Suborder Platycephaloidei. This suborder is composed of three families: Platycephalidae, flatheads; Bembridae, deep water flatheads; and Hoplichthyidae, spiny flatheads. Flatheads are distributed in the Indo-Pacific and the tropical region of the eastern Atlantic. They are food fishes, with some species reaching weights of over 14 kg. They are noted for their long flat heads and extremely protractile mouths. The other two families are found in the Indian Ocean.

Suborder Anoplopomatoidei. This is the suborder containing the sablefish, *Anoplopoma fimbria,* and the skilfish, *Erilepis zonifer,* both of the North Pacific. The sablefish, known also as "black cod," has very oily flesh and is the subject of a commercial fishery.

Suborder Hexagrammoidei. This is a North Pacific marine group that, according to Eschmeyer (1990b), consists of one family, Hexagrammidae, which is divided into five subfamilies. Hexagramminae includes the greenlings, *Hexagrammos* spp., which are characterized by several lateral line canals on each side (Fig. 14–3A).

Pleurogramminae contains the pelagic Atka mackerel. The ling cod, *Ophiodon elongatus,* is placed into the subfamily Ophiodontinae, and the painted greenling and combfishes are in the Oxylebinae and Zaniolepinae, respectively.

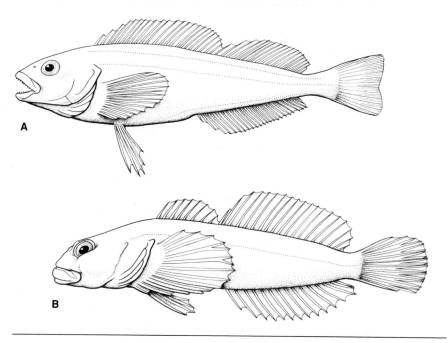

FIGURE 14–3   Representatives of Scorpaeniformes, **A,** Greenling (*Hexagrammos*); **B,** sculpin (*Cottus*).

Suborder Cottoidei. The phylogenetic relationships of this suborder and the families usually ascribed to it are uncertain. Washington et al. (1983) place the greenlings and relatives here. Cottidae, the sculpins and relatives, have been placed into from one to as many as 17 families. Here we present the families recognized by Eschmeyer (1990b), mainly because doing so will convey more information than lumping them into a minimum of families.

Family Normanichthyidae contains only the species *Normanichthys crockeri,* which lives along the coasts of Peru and Chile. This species does not resemble a typical sculpin in that it has a covering of ctenoid scales and no spines or serrations on the head. The suborbital stay does not reach the preopercle, and there are no ribs.

The family Ereuniidae was recognized by Yabe (1981) for the two Japanese genera *Ereunias* and *Marukawichthys.* These have the lower four pectoral fin rays separated from the remainder of the fin and have ctenoid scales on the body.

The Cottidae, sculpins, are usually large headed with large, fanlike pectoral fins, and many are characterized by strong spines on the gill cover. They often have bony plates on the skin and seldom have more than a few rows of scales. They are typically marine fishes of the temperate and cold waters of the northern hemisphere. There are about 70 genera and 300 species, according to Nelson (1983). Freshwater members are mainly in the genus *Cottus,* whose species are common inhabitants of cold streams (Fig. 14–3**B**).

Comephoridae, the Baikal oilfishes, are viviparous pelagic fishes of Lake Baikal in Siberia. The two species of *Comephorus* inhabit deep water but swim to near the surface at night. They are nearly colorless and have large pectoral fins. Unlike most of the fishes in the suborder, the caudal fin is truncate to deeply emarginate. The sex ratio is said to be heavily skewed toward females (Wheeler, 1975).

Cottocomephoridae is a relatively large Lake Baikal group with at least eight genera and 24 species. Most species tend to be heavy-bodied benthic forms, but some exploit open water habitats, so that members of the family make use of most of the habitats in the lake.

The "blob" sculpins or "fatheads" of *Ebinania, Neophrynichthys,* and *Psychrolutes* are retained in Cottidae by Robins et al. (1990a), but Nelson (1982, 1994) recognizes the group as the family Psychrolutidae. These are more or less tadpole-shaped marine sculpins of the Pacific, Indian, and Atlantic oceans. Some species are quite small, but others, such as the flabby, globose *Psychrolutes phrictus* of the eastern North Pacific, may reach 70 cm and nearly 10 kg.

Agonidae contains the poachers and alligatorfishes, curious armored fishes of cold marine waters. They are present in the Arctic, Antarctic, and North Atlantic oceans, along the coasts of southern South America, and are especially numerous in the North Pacific.

The Cyclopteridae contains the lumpfishes and snailfishes, although the latter are often placed in a separate family, Liparidae (= Liparididae) (Kido, 1988). Most of these fishes have the pelvic fins modified into a round suctorial disc with which they cling to rocks and vegetation (Fig. 2–13**G**). They are marine and are found in all cold seas.

Chapter 15

Acanthopterygii: Perches and Relatives

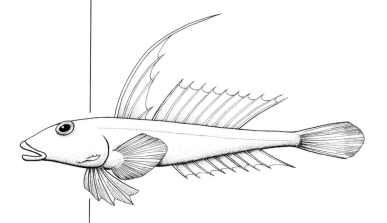

Order Perciformes
    Suborder Percoidei
    Suborder Elassomatoidei
    Suborder Labroidei
    Suborder Zoarcoidei
    Suborder Notothenioidei
    Suborder Trachinoidei
    Suborder Pholidichthyoidei
    Suborder Blennioidei
    Suborder Icosteoidei
    Suborder Gobiesocoidei
    Suborder Callionymoidei
    Suborder Gobioidei
    Suborder Kurtoidei
    Suborder Acanthuroidei
    Suborder Scombrolabracoidei
    Suborder Scombroidei
    Suborder Stromateoidei
    Suborder Anabantoidei
    Suborder Channoidei

Order Perciformes (Percomorphi). This is the largest order of fishes. In fact, Nelson (1994) states that it is the largest order among the vertebrates, with nearly 9300 species in 148 families. The fishes are so diverse that some suborders that are recognized as perciforms by some investigators are considered by other ichthyologists to be separate orders. The diversity includes tiny gobies and tremendous tunas and swordfishes, slender snake mackerels and quillfishes, as well as deep-bodied discus fishes and batfishes. There are fishes of the deep sea and fishes of the surface layers. Although perciforms are the dominant fishes of the oceans, there are thousands of important species from fresh water.

Usually about 20 suborders are included, depending on the arrangements by various ichthyologists. During the past decade, there has been much reevaluation and shifting of families back and forth among suborders, especially those that contain blenny-like species. The suborders presented here are those accepted by Nelson (1994). Lauder and Liem (1983a, p. 167) believe the order to be "clearly polyphyletic."

This order includes most of the spiny-rayed fishes. In addition to fin spines, these fishes have thoracic or jugular pelvic fins, with the pelvic girdle usually connected to the cleithra. Pelvic fins typically have one spine and five (or fewer) soft rays. Pelvic bones are united. The caudal fin typically has 17 principal rays, but some groups within the order have fewer rays. The pectoral fins are usually placed high on the side, with almost vertical fin bases. Branchiostegal rays number five to eight (usually six or seven), with four of these placed on the outer surface of the upper portion of the ceratohyal (anterohyal). There are usually 24 or fewer vertebrae. Perciforms are physoclistic and usually have ctenoid scales. The orbitosphenoid and mesocoracoid are absent, and the maxillary is excluded from the gape by the premaxillaries. Perciforms are known from the Upper Cretaceous.

Suborder Percoidei. This immense group contains over 70 families and about 2900 species. They are mostly marine, with a few very successful freshwater families. Members of some of the families are important as food fishes, others are sport fishes, and many are aquarium favorites. Many of the families are ecologically important in their respective ecosystems. Most percoids have not diverged greatly from the basic, successful morphological groundplan, but they are diverse and have important ecological, behavioral, and reproductive adaptations. Relationships within the group are not well understood.

Among the freshwater representatives of Percoidei are the black basses and sunfishes of the Centrarchidae. These are small- to medium-sized fishes of North America and are especially abundant in the southern half of the United States. The largemouth bass, *Micropterus salmoides* (Fig. 1–1G), is a popular game fish that has been introduced to many areas outside its native range. It may reach a weight of 5 kg or more. The other 30 or so members of the family include the crappies, *Pomoxis* spp., and several members of the genus *Lepomis,* of which *L. macrochirus,* the bluegill, is probably best known. The latter is a good panfish and is often cultured with the largemouth in ponds. The only sunfish found native to the Pacific drainage of the United States is *Archoplites interruptus,* the Sacramento perch, a representative of a genus that has been found in Pliocene deposits of Oregon, Idaho, and Washington.

Although the family Percidae is present in Europe and Asia, it reaches its greatest development in eastern North America, where there are about 115 species, mostly darters of the genus *Etheostoma.* The familiar yellow perch of North America, *Perca flavescens,* has a close relative in the Eurasian *P. fluviatilis.* The sauger, *Stizostedion canadense,* and walleye, *S. vitreum,* have their counterpart in the Old World *S. lucioperca.*

Percoid families that contain important food fishes include the Lutjanidae, snappers; Sciaenidae, the drums and croakers; Carangidae, the jacks and pompanos (Fig. 2–15**C**); and the Bramidae, pomfrets. The numerous sea basses, hamlets, and groupers of most warm seas are in the family Serranidae. The related temperate perches of Percichthyidae are found in both fresh and marine waters of the southern hemisphere. Many ichthyologists place the temperate basses of the family Moronidae with the percichthyids, but Johnson (1984) does not believe there is a close relationship between these two taxa. Some moronids, such as the popular striped bass (*Morone saxatilis*) of North America, are anadromous. The moronid *Dicentrarchus labrax* of Europe and northern Africa is a commercial fish caught in estuaries and along rocky shores.

Other families, some wide ranging but with ranges including North America and usually considered to be in this suborder, are listed in Table 15–1. Space does not allow extended coverage of all the families. Centropomidae (snooks) are found in marine, brackish, and sometimes fresh water in the Pacific, Atlantic, and Indian oceans. Well-known genera are *Centropomus* and *Lates;* the latter includes the giant Nile perch and barramundi perch (also known as Asian seabass) which is well known from Australia. Grammatidae (basslets) of the tropical west Atlantic are small, colorful fishes that are becoming important in the aquarium trade. Priacanthidae (bigeyes) are nocturnal, small- to medium-sized fishes (to about 50 cm) mostly of warm waters. Some species are circumtropical. Many species are red or purplish in color. There are four genera, including *Priacathus* and *Cookeolus.* Apogonidae (cardinalfishes) also are mostly nocturnal, warm-water fishes. They are small, usually not much over 20 cm long, and are noted for their oral incubation of eggs. There are about 300 species, more than a third of these in *Apogon.* Some species are found in fresh waters of Pacific Islands.

Malacanthidae (tilefishes) are marine and of the Pacific, Indian, and Atlantic oceans. *Lopholatilus chamaeleonticeps,* the ocean whitefish of the North Atlantic, once supported a large commercial fishery until a natural incursion of cold water greatly diminished the population. There was no fishery from 1882 until 1915, and only a small fraction of the earlier catch is now realized.

The Bathyclupeidae, which retains a pneumatic duct and has the maxillaries included in the gape, is now placed in the Percoidei but was formerly considered by some scientists to constitute a separate order intermediate between the clupeiforms and galaxiids (Berg, 1940).

Polynemidae contains the threadfins, which are tropical shore fishes. The snout is prominent and reaches well beyond the large mouth; the pectoral fins are composed of an upper part, rather typical of percoid pectoral fins, and a lower section composed of four to eight long filaments that apparently serve as tactile organs (Fig. 2–12**D**). The upper rays attach to the first two actinosts, the third bears no fin rays, and the fourth supports the filamentous rays. The dorsal fins are widely separated. Eschmeyer (1990b) retains this family as a suborder.

TABLE 15–1	**Additional Families of Suborder Percoidei** **(Nelson, 1994)**
Chandidae (Asiatic glassfishes)	Marine, freshwater, Indo-West Pacific
Acropomatidae (temperate ocean basses)	Marine, Indian, Pacific, Atlantic
Ostraberycidae (hornycheek seaperches)	Marine, Indian, West Pacific
Callanthiidae (callanthias)	Marine, East Atlantic, Indian, Pacific
Plesiopidae (roundheads and spiny basslets)	Marine, Indo-West Pacific
Notograptidae	Marine, New Guinea, northern Australia
Opistognathidae (jawfishes)	Marine, Atlantic, Indian, Pacific
Dinopercidae (lampfish)	Marine, Indian, Atlantic off Angola
Banjosidae (banjosid)	Marine, Pacific, Japan, China, Korea coasts
Epigonidae (deep water cardinalfishes)	Marine, Atlantic, Pacific, Indian
Sillaginidae (sillagos)	Coastal marine, Indo-West Pacific
Lactariidae (false trevallies)	Marine, Indo-Pacific
Dinolestidae (long-finned pike)	Marine, southern Australia
Pomatomidae (bluefishes)	Marine, Atlantic, Pacific, Indian
Nematistiidae (roosterfish)	Marine, tropical, eastern Pacific
Echeneidae (remoras) (Fig. 2–9**A**)	Marine, Atlantic, Indian, Pacific
Rachycentridae (cobia)	Marine, Atlantic, Indo-Pacific
Coryphaenidae (dolphinfishes)	Marine, Atlantic, Indian, Pacific
Menidae (moonfish)	Marine, Indo-West Pacific
Leiognathidae (ponyfishes, slipmouths)	Marine, brackish, Indo-Pacific
Caristiidae (manefishes)	Marine
Emmelichthyidae (rovers)	Marine, warm, Indo-Pacific, Caribbean, Atlantic
Lobotidae (tripletails)	Marine to fresh, warm seas
Gerreidae (mojarras)	Marine to brackish, warm seas
Haemulidae (grunts)	Marine to brackish, Atlantic, Indian, Pacific
Inermiidae (bonnetmouths)	Marine, western tropical Atlantic
Sparidae (porgies)	Marine, sometimes to fresh, Atlantic, Indian, Pacific
Centracanthidae ("windtoys")	Marine, eastern Atlantic, Mediterranean, to South Africa
Lethrinidae (emperors)	Coastal marine, West Africa, Indo-West Pacific
Nemipteridae (threadfin breams)	Marine, Indo-West Pacific
Mullidae (goatfishes)	Marine, Atlantic, Indian, Pacific
Pempheridae (sweepers)	Marine, brackish, western Atlantic, Indian, Pacific
Glaucosomatidae (pearl perches)	Marine, Japan to Australia
Leptobramidae (beachsalmon)	Marine, brackish, southern New Guinea, western Australia
Monodactylidae (fingerfishes, moonfish)	Marine, brackish, Africa, Indo-Pacific

TABLE 15–1 *(Continued)*	Additional Families of Suborder Percoidei (Nelson, 1994)	
Toxotidae (archerfishes)	Marine, fresh water, India, Philippines, Australia, Polynesia	
Coracinidae (galjoen fishes)	Marine and brackish, South Africa, Madagascar	
Drepanidae (sicklefish)	Marine, Indo-West Pacific, West Africa	
Chaetodontidae (butterflyfishes)	Marine, tropical, Indo-West Pacific, Atlantic, Indian	
Pomacanthidae (angelfishes)	Marine, tropical Atlantic, Indian, Pacific	
Enoplosidae (oldwife)	Marine, southern Australia	
Pentacerotidae (armorheads)	Marine, southwest Atlantic, Indo-Pacific	
Nandidae (leaffishes)	Fresh water, northeast South America, West Africa, southern Asia	
Kyphosidae (seachubs)	Marine, Atlantic, Indian, Pacific	
Arripidae (Australasian salmon)	Marine, South Pacific	
Teraponidae (grunters)	Marine, Indo-West Pacific, to Japan and Samoa	
Kuhliidae (flagtails)	Marine to fresh, Indo-Pacific	
Oplegnathidae (knifejaws)	Marine, widespread in Indo-Pacific	
Cirrhitidae (hawkfishes)	Marine, Atlantic, Indian, and mostly Indo-Pacific	
Chironemidae (kelpfishes)	Marine, Australia and New Zealand	
Aplodactylidae (marblefishes)	Marine, southern Australia, New Zealand, western South America	
Cheilodactylidae (morwongs)	Marine, southern Atlantic, southern Indian, Pacific to Japan, Hawaii	
Latridae (trumpeters)	Marine, southern Australia, New Zealand, Chile, southern Atlantic	
Cepolidae (bandfishes)	Marine, eastern Atlantic, Mediterranean, Indo-West Pacific to New Zealand	

Threadfins have been observed to swim a spiral course up one piling and down the next with the pectoral filaments fanned out, presumably to detect prospective food items over about a 40-cm-wide area. Examples of genera are *Polynemus*, *Polydactylus*, and *Eleutheronema*, species of which genus may reach 2 m in length. The history of classification of this group is reviewed by de Sylva (1984c). These fishes were considered to constitute a separate order by Berg (1940) and by Lindberg (1974).

Only a few of the families of percoid fishes have been mentioned in the text, but those that have not are listed in Table 15–1, with common names and abbreviated ranges.

Suborder Elassomatoidei. The pygmy sunfishes, *Elassoma* spp., form the family Elassomatidae. These fishes have formerly been placed with the Centrarchidae, although this genus differs from the centrarchids in many characters (Branson and

Moore, 1962; Johnson, 1984). Johnson (1984) believes *Elassoma* should not be placed in the suborder Percoidei, and Johnson and Patterson (1993) place the family in their new taxon Smegmamorpha, which includes synbranchids, mullets, sticklebacks, and atherinomorphs. Nelson (1994) retains the group in Perciformes but in the suborder Elassomatoidei, as is followed here.

These are tiny fishes less than 5 cm long. They have no lateral line of sensory pores and five or fewer dorsal spines. The scales are cycloid. All species occur only in the fresh waters of the southeastern United States.

Suborder Labroidei. This suborder contains fishes with pharyngeal jaws and teeth capable of remarkable grinding activity. According to Kaufman and Liem (1982), the suborder is defined by three characters: the fifth ceratobranchials are fused into a single bone; the bones of the upper pharyngeal jaws are in direct contact with the base of the cranium; and the esophageal sphincter muscle is present as a single continuous sheet with no subdivisions.

Stiassny and Jensen (1987) agree with the diagnosis and add additional characters: there is a lower pharyngeal jaw with a ventral keel for muscle attachment; the lower pharyngeal jaw is suspended directly from the neurocranium by a "muscle sling" made up of several component muscles; and the lower pharyngeal jaw is structurally united into a single unit. Prior to these studies, the suborder was considered to contain three families: the Labridae, wrasses; Scaridae, parrotfishes; and Odacidae, greenbones. Kaufman and Liem (1982) and Stiassny and Jensen (1987) include Scaridae and Odacidae in Labridae and add the following pharyngognathous families to the suborder: Cichlidae, Embiotocidae, and Pomacentridae. These were formerly placed in Percoidei.

The Labridae is one of the largest of marine families, with nearly 60 genera and about 500 species. These fishes are distributed in most tropical to warm-temperate seas and are mostly brightly colored, small species, but there are species that are as long as 3 m (Wheeler, 1975). They have sharp, heavy teeth in the front of the mouth in addition to the crushing pharyngeal jaws. Many of the smaller species are "cleaners" that pick parasites from other species. The life histories of many species involve sex changes (Warner and Robertson, 1978).

Here the Scaridae and Odacidae will be treated as families, as in Nelson (1994), Richards and Leis (1984), and Eschmeyer (1990b). The Scaridae are medium sized to large (approximately 1.2 m and 70 kg) tropical and subtropical fishes usually associated with coral reefs. Their anterior teeth are fused into a parrot-like beak, with which they scrape attached vegetation or bite off chunks of coral. Odacids are fishes of Australia and New Zealand. They have nonprotractile jaws with fused teeth forming a cutting edge. There are about 12 species.

The Cichlidae is an important freshwater family that ranges through warm fresh waters from India, Africa, South America, and Central America north to the Rio Grande. The presence of these fishes on the southern continents and islands, such as Madagascar and Ceylon, suggests that either their ancestors could withstand brackish or even marine waters or that their distribution patterns date from a time (Cretaceous) when those landmasses were close together.

Only one genus, *Etroplus,* occurs in Asia, where *E. suratensis* of India, the pearl-spot, is taken as a food fish. In the Americas, the Rio Grande perch, *Cichla-*

soma cyanoguttatum, is the most northerly representative of the family (Fig. 15–1**A**). Some members of *Cichlasoma* enter the aquarium trade, as do many others in the family. *C. octofasciatum* is known as the Jack Dempsey, *Pterophyllum scalare* is the freshwater angelfish, and *Symphysodon discus*, the discus, is one of the favorites of aquarists.

A well-known food fish is *Tilapia* (= *Oreochromis*) *mossambica*, which has been taken to many parts of the world from its native Africa. Although capable of producing great populations in a short time with a high yield of protein per unit of area, the species often stunts; and due to the small size, those harvested are not readily accepted by some peoples. There are parts of Asia where *T. mossambica* is considered a pest. Other cultured species are *T. nilotica* and *T. zilli*. These species are considered to be in the genus *Oreochromis* by various ichthyologists or in *Sarotherodon* by others (see Trewavas, 1973, 1981; Ivoylov, 1981; and Stiassny, 1991). They are retained in *Tilapia* by Robins et al. (1991b).

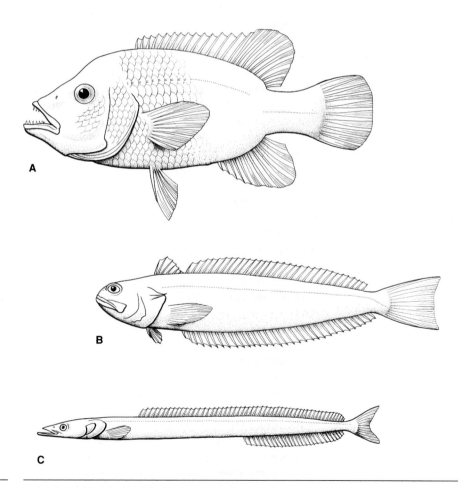

FIGURE 15–1   Representatives of order Perciformes, **A,** Rio Grande perch (*Cichlasoma*); **B,** weever (*Trachinus*); **C,** sandlance (*Ammodytes*).

In some of the great rift lakes of Africa, there are "species flocks" of cichlids, usually of the genus *Haplochromis* and close relatives. There is some evidence that numerous species and some new genera have evolved within certain lakes from a few founding species. Some flocks include well over 100 species and show a remarkable range of trophic adaptations. Many can specialize as sand-plowers, detritivores, herbivores, insectivores, carnivores, eye-biters, fin-biters, scale-eaters, thieves of eggs and larvae, and so on. Even among species with similar feeding habits, subtle differences in reproductive behavior have apparently allowed development of closely related forms that may occupy only slightly different niches in the same lake. Courtship and nesting activities in cichlids follow set patterns of behavior that have been of great interest to ethologists. Many cichlids are oral incubators; others build nests in the substrate.

The Embiotocidae, surfperches, live in the North Pacific, with most of the species on the coast of North America from California northward. Two species occur in Japan. One species, *Hysterocarpus traski* of California, lives in fresh water. These are viviparous fishes that give birth to precocial young. In at least one species, newborn males are mature.

The Pomacentridae are small, tropical marine fishes, usually of shallow waters around reefs. The family includes the damselfishes and anemonefishes. Fishes of this family may be the dominant members of reef communities in some areas. They are often conspicuous because of their numbers and their coloration.

Suborder Zoarcoidei. These are blenny-like marine fishes that were formerly placed with the blennies, then placed with the gadiforms and ophidiiforms (Greenwood et al., 1966), but are now considered to constitute a separate suborder (Eschmeyer, 1990b; Nelson, 1994). However, Anderson (1983) briefly reviewed the history of the classification of the group and considered that the zoarcoids are allied with the Blennioidei. Most of the fishes in the suborder are elongate, with long dorsal and anal fins and more or less compressed bodies, but body form ranges from the filiform quillfish to the rather deep-bodied prowfish. The members of the suborder have a single nostril on each side.

The Zoarcidae (eelpouts) resemble cusk-eels in appearance and somewhat in habit. Many are viviparous. There are about 45 genera and, according to Anderson (1983), probably close to 200 species. Most eelpouts are small, but *Macrozoarces* reaches about 1 m. They are typically deep water species, with some found deeper than 5000 m, although some species live in the intertidal zone. The family is widespread but typical of cold northern waters. Some Antarctic zoarcids have an antifreeze protein in the blood.

The Bathymasteridae, ronquils, are fishes found only in the North Pacific and, according to Matarese (1989), are closely related to Stichaeidae, the pricklebacks, which are found mainly in the North Pacific but also occur in the North Atlantic and Arctic oceans. They typically live in shallow coastal waters. The Cryptacanthodidae, wrymouths, contains a few species in the North Atlantic and North Pacific. The giant wrymouth, *Cryptacanthodes giganteus,* reaches a length of nearly 1.2 m. Pholidae, the gunnels, are shallow water and intertidal fishes of the North Atlantic and the North Pacific, with several species on the Pacific Coast of North America.

The Anarhichadidae, wolffishes, contains at least one species that is taken commercially, *Anarhichas lupus.* The wolf-eel, *Anarrhichthys ocellatus,* of the North

Pacific reaches 2.4 m in length and is noted for its strong dentition of canine teeth in the front of the jaws and wide molariform teeth in back. The diet is primarily shellfish, and the species is often found in the crab pots of Pacific Coast fishermen.

Ptilichthyidae (quillfish), Scytalinidae (graveldiver), and Zaproridae (prowfish) are all monospecific families of the North Pacific.

Suborder Notothenioidei. This suborder contains five families sometimes aligned with the Blennioidei (Gosline, 1968; Stevens et al., 1984) but given subordinal status by Nelson (1994) and Eschmeyer (1990b). About 100 species live in Antarctic or subantarctic seas, with a few entering fresh waters of Australia. The thornfishes, Bovichtidae (= Bovichthyidae; see Eschmeyer and Bailey, 1990, and Robins et al., 1991b) are found in Australia, New Zealand, and southern South America. Nototheniidae, the Antarctic cods or cod icefishes, occur in Australia, Tasmania, and New Zealand as well as the Antarctic. The plunderfishes, Harpagiferidae, spiny fishes that are also referred to as Antarctic sculpins, range north to the Falkland Islands and southern South America. Antarctic dragonfishes, Bathydraconidae, are found in both deep and shallow water of the Antarctic. Icefishes, Channichthyidae, of the seas surrounding Antarctica have no hemoglobin in the blood. Some of the Antarctic groups have glycoprotein antifreeze in their blood as an adaptation against freezing in supercooled waters.

Suborder Trachinoidei. This group includes the weeverfishes, stargazers, and their allies. Trachinoids are generally not as elongate as the blennies but bear structural resemblances to those fishes. An important feature is the jugular pelvic fins. The boundaries of Trachinoidei are viewed differently by various ichthyologists. Some (e.g., Gosline, 1968; Rosenblatt, 1984) place them with the blennies, whereas others consider them a separate suborder but leave some families in Blennioidei and include some usually placed with Percoidei. Stargazers of the family Uranoscopidae are marine fishes with venomous spines at the edge of the opercle. Venom glands at the base of the spines can deliver poison through grooves in the spine. *Uranoscopus* has electric organs, derived from extrinsic eye muscles behind the eyes; they are capable of discharging 50 volts.

The weevers, family Trachinidae, are also venomous and have opercular and dorsal spines equipped with venom-producing tissue (Fig. 15–1**B**). Although the effect of the weever's poison upon humans is severe, it may be less so than that of the stargazers, which are known to cause death.

Most families of trachinoids have the habit of concealing themselves in sand or in other soft bottom materials. Their eyes are generally placed on top of their heads, their mouths are in a superior position, and there are usually fringes or flaps that prevent the intake of sand with the respiratory water. Some of these families are Trichodontidae, the sandfishes of the North Pacific; Trichonotidae, the sand divers; and Leptoscopidae, of Australia and New Zealand. Dactyloscopidae, the sand stargazers of tropical South America, have similar adaptations but are placed with the blennioids by some authorities. Other trachinoid families include Pinguipedidae (formerly Mugiloididae; Rosa and Rosa, 1987), the sandperches; Cheimarrhichthyidae, the torrent fishes, which are placed in the preceding family by some investigators; Percophidae, "flatheads"; Creediidae; and Limnichthyidae. The jawfishes, Opistognathidae; bent-tooths, Champsodontidae; and the swallowers, Chiasmodontidae (deep water fishes that can swallow fishes larger than themselves) are additional trachinoid families that are listed among the percoids in some systems.

The family Ammodytidae includes some small, slender marine fishes with a protruding lower jaw and forked caudal fin (Fig. 15–1C). They lack a gas bladder. Scales are cycloid and the pelvics absent or jugular. The sand lances of Ammodytidae are known to burrow into the bottom sediments. The family is found mainly in temperate to cool seas and may be important as food for predatory species. These fishes are harvested in northern Japan for use as food in fish culture.

Suborder Pholidichthyoidei. This is an Indo-Pacific group that contains only the Pholidichthyidae, or convict blennies. There are probably only two species of these small, dark, elongate fishes. They are reported to live in burrows excavated in soft substrates (Trnski et al., 1989).

Suborder Blennioidei. This group contains a large variety of marine fishes with jugular pelvics and in which each radial of the dorsal and anal fins corresponds with a neural or hemal spine. Most of the blennioids are rather elongate, and many are either taeniform or eel shaped. The group may not be a natural one, and there seems to be a lack of agreement regarding which fishes should be included. The six families included here are represented by nearly 130 genera and close to 700 species.

The Blenniidae (combtooth blennies, scaleless blennies) includes numerous small, shore fishes of tropical and subtropical seas (Fig. 15–2A). They are common in the intertidal zone, and one genus, *Salarius,* contains species that will leap from one tide pool to another when disturbed. The Clinidae contains the familiar kelp-fishes of the Pacific Coast of North America as well as numerous blennies from other shores. These are often perchlike in appearance. The Tripterygiidae, or three-fins (including the cockabullies of Australia), are similar in appearance to the foregoing two families and, like them, are noted for the male's care of the eggs during incubation.

Other families placed in Blennioidei by Nelson (1994) and Eschmeyer (1990b) are Labrisomidae, Chaenopsidae (pikeblennies), and Dactyloscopidae (sand stargazers).

Suborder Icosteoidei. The monotypic ragfish family Icosteidae is found in deep water in the North Pacific. *Icosteus aenigmaticus* is a flabby, limp fish with reduced scales. The young have pelvic fins, but those disappear in adults. The species reaches a length of more than 2 m (Fig. 15–2B). Berg (1940) and, later, Gosline (1971) considered Icosteidae to constitute a separate order.

Suborder Gobiesocoidei. The Gobiesocoidei was formerly considered to be part of Paracanthopterygii but was eliminated from that superorder by Patterson and Rosen (1989). The order contains the single family Gobiesocidae. It is now treated as a perciform suborder (Nelson, 1994), but it has been considered a perciform derivative by some (Allen, 1984) and is placed as a preperciform group by others (Eschmeyer, 1990b; Robins et al., 1991a). The limits and placement of the group are not agreed on by recent authors. The monograph by Briggs (1955) remains an excellent source of information on the group.

The suborder and family, including the formerly recognized family Alabetidae (= Cheilobranchidae; Springer and Fraser, 1976), is characterized by a joint between the interoperculum and the epihyal, a joint between the supracleithrum and the cleithrum, and a greatly modified heart in which the atrium and ventricle lie side by side with the atrium on the left side. The common cardinal veins are expanded as large chambers, and the sinus venosus is reduced (Lauder and Liem, 1983b).

Many gobiesocids typically have a powerful suctorial disc on the ventral surface below the pectorals (Figs. 2–13**E**, 15–2**C**). This consists of the modified pelvic fins and skin folds and can exert enough suction that the specimens are sometimes difficult to remove from a smooth surface.

The gobiesocids are small, soft-rayed, scaleless fishes of tropical and temperate shores; some enter fresh water. Some species of *Chorochismus* (South Africa) reach about 30 cm. At least one South American species of *Tomicodon* is used as food. *Sicyaces sanguineus* of Chile is somewhat amphibious and seeks food intertidally.

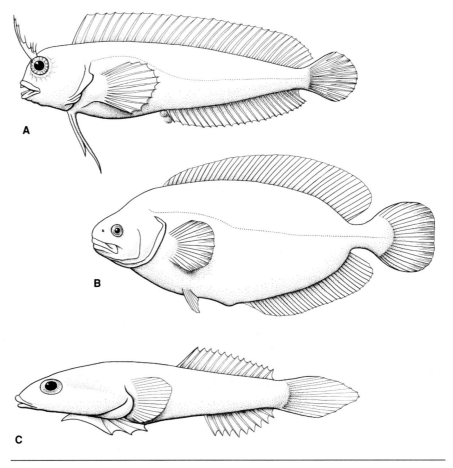

A

B

C

FIGURE 15–2   Representatives of order Perciformes, **A,** combtooth blenny (*Blennius*); **B,** juvenile ragfish (*Icosteus*); **C,** clingfish (*Gobiesocidae*).

The northern clingfish, *Gobiesox maeandricus,* ranges from southern California to Alaska. Other North American genera are *Acyrtops* and *Rimicola.* Genus *Alabes* contains small, eel-like fishes from the seas of Australia. Their sucking disc is greatly reduced, if evident at all, and the gill openings are confluent beneath the head, as in some other members of the family.

Suborder Callionymoidei. The dragonets consist of the Callionymidae (Fig. 15–3A) and Draconettidae, both of warm seas. They are highly colored fishes that are noted for the large and showy fins of the males.

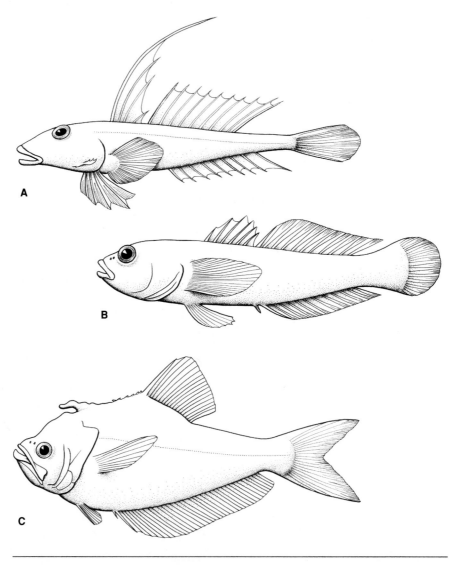

FIGURE 15–3   Representatives of order Perciformes, **A,** Dragonet (*Callionymus*); **B,** goby (*Coryphopterus*); **C,** forehead brooder (*Kurtus*).

Suborder Gobioidei. This suborder consists of eight families with about 270 genera and more than 2100 species, mostly of shallow tropical marine waters but with about 200 species in fresh water (Nelson, 1994). There are many diverse life styles among the gobies. Some can spend considerable time out of the water and obtain food on mudflats; there are freshwater species with larvae that go to sea; many, some of which are blind, burrow in sand or mud. A few gobies establish symbiotic relationships with other animals, and others act as "cleaner" fishes that remove ectoparasites from other fishes. The pelvic fins of some families are united into a sucking disc that can be used to adhere to the substrate.

Springer (1983) lists the following synapomorphies for the Gobioidei: parietals are lacking; a cartilage at the anterior of the pelvic girdle fits between the lower ends of the cleithra; the dorsal end of the symplectic does not meet the dorsal end of the interhyal; and the first basibranchial is cartilaginous. Other synapomorphies listed by Johnson and Brothers (1993) include the following: hypurals 1 and 2 are fused, as are hypurals 3 and 4; the uppermost pectoral ray articulates with the dorsalmost actinost rather than the scapula; supraneurals and the basisphenoid are lacking.

The loach goby family (Rhyacichthyidae) is found in fresh water from China to New Guinea. These fishes have flat heads and an inferior mouth, and the paired fins and lower part of the body are modified into a sucking disc. These are the only gobies with a lateral line. The only genus, *Rhyacichthys,* has two species.

The family Odontobutidae of east Asia contains three genera of freshwater fishes that share some characteristics of the pectoral girdle with Rhyacichthyidae. They apparently are close to the Eleotridae (see Nelson, 1994).

Eleotridae (= Eleotrididae) is sometimes combined with Gobiidae, although these fishes have separate pelvic fins. They are usually bottom fishes, but some live in mid-water and some are pelagic. *Gobiomorus dormitor* of Central America reaches a length of nearly 60 cm. Members of the genera *Oxyeleotris* and *Bunaka* of southeast Asia and the Indo-Australian areas, respectively, may reach lengths of 50 cm or more. *O. marmoratus* is a prized food fish.

Gobiidae (Fig. 15–3**B**) is the largest family in the suborder and includes species from all warm seas, some tropical fresh waters, and a few temperate marine and estuarine localities. According to Nelson (1994), there are about 212 genera and nearly 1900 species in this family. These are mainly small fishes, mostly less than 10 cm long. The smallest (shortest) vertebrate is a goby (*Trimmatom nanus*) of the Chagos Archipelago (Indian Ocean), in which mature females reach no more than about 10 mm. A marine goby of the Marshall Islands, *Eviota zonura,* is nearly as small. Some small gobies of Southeast Asia and the Philippines occur in great enough numbers that they can be harvested, mixed with salt, and fermented to make "bagoong," a sauce that is eaten with vegetables or rice. The bagoong fishery is largely targeted on larval and postlarval fish up to about 25 mm on their migration from marine waters into rivers.

The skipping gobies, or mudskippers, were formerly considered to be in a separate family, Periophthalmidae (Berg, 1940). They are placed in the family Gobiidae, subfamily Oxudercinae, by Nelson (1994). These fishes inhabit tropical shore areas with soft bottoms and are usually seen at the water's edge or on the mud, rocks, or mangrove roots along the shore. They can pull themselves along with their armlike pectoral fins or can flit around with great rapidity by flexing the body.

The eyes of *Periophthalmus* are set high on the head, the gill cavity is expanded, and the skin is very vascular so that it functions in respiratory exchange. Mudskippers can live out of water for over 30 hours if not subjected to excessive heat or desiccation. In high humidity, their ability to stay out of water may be limited by their inability to excrete nitrogenous wastes rather than inability to obtain oxygen. If exposed to the sun, they might live less than an hour. Their upper lethal temperature is close to ambient air temperatures at mid-day in their tropical habitats. One mudskipper, *Boleophthalmus chinensis,* is cultured as a food fish in China and Taiwan. Other genera are *Oxuderces, Apocryptes,* and *Periophthalmodon.*

A group with many freshwater species is recognized as the subfamily Sicydiinae. These are distributed around the world in warm waters and are noted for their ability to ascend waterfalls. Genera include *Awaous, Sycidium,* and *Syciopterus.*

The eel-gobies or eel-like gobies, formerly placed in the family Gobioididae, are nearly circumtropical in fresh, brackish, and marine waters. *Trypauchen* spp., the burrowing gobies, occur in fresh, brackish, and marine waters from Africa to Japan and the Philippines.

On the Pacific Coast of the continental United States, there are 12 native gobies (including *Typhlogobius californiensis,* the blind goby, and *Gillichthys mirabilis,* the longjaw mudsucker, which can remain alive out of water for about a week and which is a favorite bait fish). Two gobies have been introduced to California from Asia, probably in ship ballast water. There are at least 58 gobiids on the Atlantic Coast of the United States, ten of the genus *Coryphopterus* and nine of *Gobionellus.* Two species have been introduced to the Great Lakes from the Black and Caspian seas with ballast water.

Other families of Gobioidei include Kraemeriidae, sand gobies or sandfishes, which range from Hawaii to the Indian Ocean; and Microdesmidae, wormfishes, which are circumtropical. Xenisthmidae, an Indo-Pacific group, is considered a subfamily of Gobiidae by Springer (1983) but is listed as a family by Eschmeyer (1990b) and Nelson (1994).

The family Schindleriidae contains two species. These are small, larvoid (paedomorphic), marine fishes of the tropical Pacific that resemble the Ammodytoidei but also bear a resemblance to the larvae of various other fishes, including microdesmids. They are usually less than 3 cm long. *Schindleria praematurus* is among the most common surface fishes in parts of the tropical Pacific. These fishes were formerly placed in the suborder Schindleroidei, but Johnson and Brothers (1993) demonstrated their relationship to the Gobieoidei.

Suborder Kurtoidei. The family Kurtidae contains the single genus *Kurtus,* with two species of nurseryfishes or forehead brooders. These are tropical fishes in which the egg clusters, held together by a cordlike material, are hung on hooks developed from the supraoccipital of the males. The ribs of these fishes form a tubular ossified structure that encloses the gas bladder. *Kurtus* (Fig. 15–3C) is found in both marine and freshwater habitats in the Indo-Pacific. This family is placed in the suborder Percoidei by Eschmeyer (1990b) but is placed in Kurtoidei here following Lauder and Liem (1983b) and Nelson (1994).

Suborder Acanthuroidei. Except for the pelagic *Luvarus,* this suborder contains fishes typical of reef habitats. They are characterized by 11 synapomorphies, most involving the skeleton (Tyler et al., 1989). Examples of these are as follows: there are four or five branchiostegals; the premaxillaries and maxillaries are bound together so that movement of the upper jaw is limited; the first neural spine is fused to its vertebral centrum; and the lachrymal and second infraorbital join loosely if at all.

The family Siganidae (spinefoots or rabbitfishes) which is often placed in its own suborder, is distinguished by two spines and three soft rays in each pelvic fin. The dorsal and anal spines of rabbitfishes are venomous. Species of *Siganus* are largely herbivorous and show promise in aquaculture.

The Luvaridae was accepted as a member of the Scombroidei prior to the study by Tyler et al. (1989), in which cladistic analysis refuted the relationship to the scombroids and confirmed the relationship to Acanthuroidei. The family has only one species: the louvar, *Luvarus imperialis,* an oceanic fish of warm latitudes noted for its red fins and pink body color. It has been reported to reach a length of about 1.8 m.

The surgeonfishes and tangs, all in the family Acanthuridae, are noted as common and colorful inhabitants of coral reefs. *Acanthurus* (Fig. 15–4A) has an erectile, sharp, forward-pointing blade on each side of the caudal peduncle. Other genera with folding spines are *Zebrasoma, Paracanthurus,* and *Ctenochaetus.* The Indo-Pacific genus *Naso,* the unicornfishes, have a long, spikelike projection extending forward from the forehead. They have one or two fixed spines on each side of the caudal peduncle. The related *Prionurus* has three or more such spines.

The Zanclidae contains the graceful and colorful Moorish idol, *Zanclus cornutus,* which is found around coral reefs in the Pacific and Indian oceans. Its striking color pattern of black bars on a yellow-white background makes it a favorite with aquarists and fish-watching snorkelers. Fishes of this genus have no spines or retractible knives on the caudal peduncle.

Suborder Scombrolabracoidei. The monotypic family Scombrolabracidae contains the widespread deep sea black mackerel, *Scombrolabrax heterolepis,* which has been placed into Scombroidei, Percoidei, or Trichiuroidei by various authors since Roule (1922) first indicated that it should stand in its own suborder. This fish bears a resemblance to Gempylidae but has such percoid characters as protractile premaxillaries and serrations on the opercle and preopercle. The number of vertebrae (30) is lower than in most scombroids, and this fish shows a procurrent spur on the lowermost principal caudal ray (see Johnson, 1975), which allies it to the percoids. Bond and Uyeno (1981) pointed out that *Scombrolabrax* is distinguished from all other fishes by having the fifth through the twelfth vertebral centra hollowed out bilaterally into bulbous bullae that open ventrally and into which evaginations of the gas bladder fit, and they recommended that the family be placed in a separate suborder, Scombrolabracoidei. Johnson (1993a) did not include Scombrolabracidae with the scombroids, and Nelson (1994) places it in Scombrolabracoidei.

Suborder Scombroidei. This suborder, which includes the barracudas, snake mackerels, mackerels, tunas, billfishes, and relatives, is of great interest because of the important commercial and sport fishes it includes. The boundaries of the group have stretched and contracted according to the interpretations made of the structural relationships of the fishes included, which all have in common nonprotractile

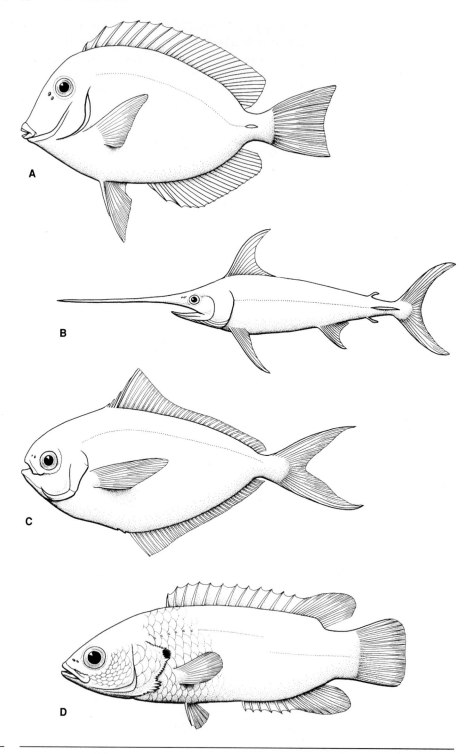

FIGURE 15–4 Representatives of order Perciformes, **A,** Surgeonfish (*Acanthurus*); **B,** swordfish (*Xiphias*); **C,** butterfish (*Peprilus*); **D,** climbing perch (*Anabas*).

premaxillaries with maxillaries firmly attached to them (Lauder and Liem, 1983b; Nelson, 1984, 1994).

Collette et al. (1984) recognized six families in the suborder: Scombrolabracidae, Gempylidae, Trichiuridae, Xiphiidae, Istiophoridae, and Scombridae. Johnson (1986) recognizes Sphyraenidae, Gempylidae, and Scombridae, which he divides into eight tribes. Robins et al. (1991a) consider the Gempylidae as part of the Trichiuridae. Nelson (1994) lists the following families in the suborder: Sphyraenidae, Gempylidae, Trichiuridae, Scombridae, and Xiphiidae.

The barracudas of the family Sphyraenidae are usually considered to be closely related to the mullets (de Sylva, 1984a, 1984b), but Johnson (1986) believes them to be allied to the mackerels of Scombroidei. The barracudas have large mouths and strong, sharp teeth set in sockets. The pectoral fins are set low and the lateral line is well developed. As in the mullets, the dorsal fins are well separated. These are fierce, medium to large predators of warm seas. One species, the great barracuda, *Sphyraena barracuda,* reaches a length in excess of 2 m. Barracudas, although implicated at times in "fish poisoning" (ciguatera) (see Chapter 31), are good food.

Gempylidae, the escolars or snake mackerels, are usually found in deep water, but some are found near the surface. They are usually equipped with long, strong teeth and have a slender, streamlined form. There are usually finlets behind the dorsal and anal fins. The caudal fin is forked. Pelvic fins are reduced to a single spine in some species but have one spine and five rays in others. One species, *Ruvettus pretiosus,* the purgative fish or oilfish, reaches about 2 m in length.

Trichiuridae, the hairtails and cutlassfishes, have exceptionally strong teeth and a compressed body with a long tapering tail (that of *Trichiurus* ends in a fine point). The caudal fin, if present, is small and forked. Pelvic fins are rudimentary or absent. Hairtails are harvested as food in many tropical countries.

Scombridae—the tunas, mackerels, and close relatives—contains a number of commercial and game species (Fig. 2–6**A**). These are mostly swift-moving species of the surface waters, usually in warm seas, noted for their wide-ranging migrations. Many tunas maintain a body temperature several degrees higher than that of the surrounding water because of their rapid metabolic rate and specialized heat exchange mechanisms in the circulatory system.

Mackerels of the genus *Scomber* may reach only a few kilograms in weight but are useful commercial fishes. Examples are *S. scombrus* of the Atlantic and *S. japonicus* of the Pacific. Even smaller species are in the genus *Rastrelliger,* which is of great commercial value in some tropical areas such as the Gulf of Thailand.

The largest tuna is the bluefin, *Thunnus thynnus,* which reaches 4 m in length and may weigh up to about 800 kg. A smaller species of some importance on the Pacific Coast of North America is *T. alalunga,* the albacore. The yellowfin tuna, *T. albacares,* has been historically the most important species in the Pacific tuna fishery but is captured mainly south of the United States. The skipjack tuna, *Euthynus* (or *Katsuwonus*) *pelamis,* although small, is one of the important commercial species, especially in the Pacific. Other genera in the family include *Sarda,* the bonitos, *Auxis,* the frigate mackerels, *Scomberomorus,* the Spanish mackerels, and *Acanthocybium,* the wahoo.

The Xiphiidae contains the swordfish, *Xiphias gladius,* a wide-ranging pelagic predator like the tunas. The swordfish has a flat, bladelike rostrum that makes up about one third the length of the fish (Fig. 15–4**B**). Adults lack scales and teeth.

There is no pelvic fin girdle and no pelvic fin. This is a prized food fish and is sought by harpooners and other fishermen. Swordfishes have been known to attack and pierce small boats. Maximum size is about 540 kg at a length of nearly 5 m.

Billfishes of the family Istiophoridae are among the most popular of the large marine game fishes. These fishes have a rounded bill and pelvic fins and retain scales and teeth as adults. The first dorsal fin is elongate and is very high in the sail-fish, *Istiophorus*. Spearfishes and the white and striped marlins are in the genus *Tetrapturus,* and the blue marlin is in *Maikara*. Blue marlin may reach a weight of about 640 kg, and the black marlin, *Istiompax,* has been recorded at nearly 710 kg.

Billfishes and swordfishes have a remarkable brain heater that elevates the temperature of their brains and eyes. The heater is developed from the superior rectus eye muscle, part of which is adapted to heat generation rather than contraction (Block, 1987).

Suborder Stromateoidei. This suborder is composed of marine fishes that have pa-pillose lateral sacs extending from the pharynx or esophagus behind the gill arches. Many of the species habitually associate with, and perhaps feed on, large jellyfishes. Pelvic fins are subthoracic to jugular and are absent in the adults of some species. The lachrymals are expanded so that the maxillaries are mostly hidden by them, and the scales are cycloid to weakly ctenoid (Haedrich, 1967). There are six families in the suborder (Horn, 1983). Robins et al. (1991a) present all the members of the sub-order covered in that work as belonging to Stromateidae, but Eschmeyer (1990b) recognizes six families.

Amarsipidae is a monospecific family of the tropical Indo-Pacific. The pelagic *Amarsipus carlsbergi* is apparently known only from larvae and juveniles (Nelson, 1994). The medusafishes of Centrolophidae are distributed widely in tropical and temperate seas. The common name is derived from the habit of swimming beneath some of the large jellyfishes. *Icichthys lockingtoni* is found in the North Pacific; *Centrolophus niger* occurs in the northeast Atlantic. Other genera include *Psenop-sis, Tubbia,* and *Seriolella*. The man-of-war fishes, or driftfishes, of the family Nomeidae have teeth in the esophageal sacs and retain the pelvic fins as adults. Some members of *Nomeus* associate with large jellyfishes and swim among the ten-tacles. The family Ariommatidae has one genus, *Ariommus,* with six species in deep waters of warm seas.

Squaretails of the Tetragonuridae are widely distributed in tropical and subtropi-cal areas, with some species ranging into temperate waters. The smalleye squaretail, *Tetragonurus cuvieri,* ranges from British Columbia to Australia and New Zealand. The Stromateidae, the butterfishes, called "white pomfrets" in some areas, are known as good food fishes. They have teeth in the expanded esophagus, and some lack pelvic fins as adults. Species are well distributed in warm-temperate and tropi-cal inshore waters. Examples of genera are *Pampus, Stromateus,* and *Peprilus* (Fig. 15–4C).

Suborder Anabantoidei. This suborder includes the climbing perches and goura-mies. Various ichthyologists have differing views of the composition of the subor-der. Some (Robins et al., 1991b) choose to simplify the arrangement by recognizing only one family, Anabantidae, whereas Eschmeyer (1990b) includes four families, and Lauder and Liem (1983b) recognize five. In this treatment, we follow Nelson (1994) in including five families in this group.

These fishes are distinguished by a labyrinth organ that is developed from the upper part of the first gill arch and occupies much of the gill chamber. The bones of the arch are expanded and folded to present a great surface area in a small space. Oxygen can be extracted from air trapped in this structure, so these fishes are at home in warm waters that may be very low in oxygen. They are freshwater fishes of Asia and Africa. Many species are known for their remarkable territorial courtship and nesting behavior.

The family Anabantidae contains the climbing perches of the genera *Anabas* of Asia and *Ctenopoma* of Africa. *Anabas* is equipped with stout spines on the operculum that aid in pulling the fish along over the ground when it migrates to a suitable habitat during the dry season (Fig. 15–4**D**).

The kissing gourami is in the family Helostomatidae. The giant gourami, *Osphronemus goramy,* is placed in Osphronemidae. It is the largest of the goramies, reaching 60 cm, and is a favorite food fish in many sections of Southeast Asia, where it is the subject of fish culture. Combfishes are in Belontiidae, along with the Siamese fighting fish, *Betta splendens,* the paradise fish, *Macropodus,* and other favorite aquarium fishes, such as *Trichopsis* and *Trichogaster.* The eggs of most anabantoids are incubated at the surface, some float of their own accord, and others are placed in bubble nests that are attended by the male.

The family Luciocephalidae, the pikehead, which Lauder and Liem (1983b) and Nelson (1994) place into this suborder, is considered as constituting the suborder Luciocephaloidei by Eschmeyer (1990b). There is one small species, *Luciocephalus pulcher,* of the Indo-Malayan region that is distinguished by the absence of dorsal and anal fin spines and a gas bladder and the presence of a simple suprabranchial organ and a highly protractile mouth. It captures food by lunging forward and extending the premaxillary forward up to about one third the length of the head. Suction is not necessarily involved (Lauder and Liem, 1981).

Suborder Channoidei. The family Channidae (Ophiocephalidae), the snakeheads, is sometimes placed into its own order because of the unique structure of the air-breathing organ, subabdominal pelvics, and lack of spines. Snakeheads are much more elongate than anabantoids; some species approach 1 m in length. They are voracious predators and, although prized as food, cannot be cultured with food fishes susceptible to predation. A great advantage of snakeheads is that they can be held alive in the markets for days if kept properly moist. There are two genera, *Channa* of Asia (about 18 species) and *Parachanna* of Africa (three species).

Chapter

16 | Acanthopterygii: Flatfishes Through Molas

Order Pleuronectiformes
    Suborder Psettodoidei
    Suborder Pleuronectoidei
Order Tetraodontiformes
    Suborder Triacanthoidei
    Suborder Tetraodontoidei

Order Pleuronectiformes (Heterosomata). The flatfishes are thought to be derivatives of the perciform fishes that have acquired the habit of swimming with the laterally compressed body oriented horizontally instead of vertically. They are mostly marine fishes, common on most coasts. A few enter fresh water, and some are found at great depths in the oceans. They are important food fishes.

Early in their development, after a period of bilateral symmetry, they begin sideswimming, and an eye migrates from what becomes the bottom side to the "upper" side. Policansky (1982) found that in the starry flounder, *Platichthys stellatus,* the direction of eye migration is genetically based and involves one allele. In the starry flounder, metamorphosis begins 27 to 104 days after egg fertilization and eye migration requires about five days.

The side turned toward the bottom is blind and mostly lacks pigmentation. The upper side is generally cryptically colored and capable of color change, allowing some species to become almost invisible to predators or prey. These fishes are usually aided in camouflage by sediments in which they are partially buried or may distribute over themselves.

There is much variation among the nearly 540 species, so classification of the flatfishes presents problems that seem to prevent ichthyologists from reaching a consensus on the relationships within the group. The classification adopted by Nelson (1994) will be followed in this chapter. Workers who have studied the order have disagreed on the number and arrangement of suborders and families to be recognized in the group (Ahlstrom et al., 1984; Berg, 1940; Chapleau, 1993; Eschmeyer, 1990b; Lauder and Liem, 1983b; Lindberg, 1974; Nelson, 1994).

Suborder Psettodoidei. This suborder is composed of one family, Psettodidae, which has spinous rays in the dorsal, pelvic, and anal fins and in which the dorsal fin does not extend onto the head. The pelvic fins are almost symmetrical, and the eyes may be on either the right or left side. The genus *Psettodes* is found in marine waters of tropical Africa, the Red Sea, and the Indo-West Pacific. Because *Psettodes* has several percoid characteristics, some ichthyologists have suggested a relationship with Serranidae, but Chapleau (1993) states that there is little evidence for such a relationship.

Suborder Pleuronectoidei. Fishes of this suborder have no spines in the dorsal or anal fins; there are ten or more abdominal vertebrae; ribs are present; and the edge of the operculum is free.

The fishes of the family Citharidae have pelvic spines. Citharids are found in the Indo-Pacific, the Mediterranean, and West Africa. There are two subfamilies, Brachypleurinae and Citharinae. The former has eyes on the right side (dextral), and the latter has eyes on the left side (sinistral).

The family Bothidae includes the lefteye flounders—species with both eyes on the left side. In this and subsequent families, pelvic spines are absent. This widespread group consists of 21 genera (Hensley, 1986) and nearly 120 species, only a few of which are found in North America. In many species, the eyes are widely separated, and the males of some species show a wider separation than the females. A North American example of that is *Bothus lunatus,* the peacock flounder, which is found from Florida to Brazil. Some slender-bodied tropical species are noted for their large mouths. Examples are *Chascanopsetta lugubris* and *Kamoharaia mega-stoma,* in which the mandibles are longer than the head.

The family Achiropsettidae contains three genera and four species of lefteyed flounders found in Antarctic and subantarctic seas.

Scophthalmidae was recognized as a family by Hensley and Ahlstrom (1984). This sinistral (left-eyed) family contains deep-bodied species from the Black and Mediterranean seas and the North Atlantic. *Psetta maxima* (= *Scophthalmus maximus*), the turbot, which reaches a meter long, is a prized food fish in Europe.

The family Paralichthyidae is included in Bothidae by Robins et al. (1991b), but Nelson (1994), Eschmeyer (1990b), and Hensley and Ahlstrom (1984) retain the group as a family. These sinistral fishes include several food fishes of the coasts of the Atlantic, Indian, and Pacific oceans. Examples are *Paralichthys californicus,* the California halibut; *P. olivaceous,* the olive flounder of the western Pacific; and *P. dentatus,* the summer flounder of the western Atlantic. Other representatives include the widespread *Citharichthys,* the sanddabs and whiffs; *Etropus;* and *Cyclopsetta.*

The family Pleuronectidae, or righteye flounders, has representatives in most seas from the arctic regions to southern Australia. It includes over 90 species that are nearly all dextral. The family is diverse, and four or five subfamilies are generally recognized. Chapleau and Keast (1988) recommend that the subfamilies be elevated to families.

Sakamoto (1984) presented a phenetic revision of the Pleuronectidae in which he recommended several changes within and among genera. These changes were accepted by Robins et al. (1991a, 1991b) but not by others. In the following paragraphs, any mention of fishes for which Sakamoto recommended a change will show his recommended name in parentheses.

The subfamily Pleuronectinae comprises about 60 species, many of which are well-known commercial fishes, such as the halibuts of the genus *Hippoglossus.* These are the largest of the flatfishes; the Atlantic halibut weighs over 300 kg and the Pacific halibut attains about 210 kg. The history of the fishery for the latter is one of early overexploitation followed by careful biological study and, finally, rational management. Some of the members of this family are called "soles" even though they do not show the characteristics of true soles. Examples of pleuronectids called "soles" are the petrale sole, *Eopsetta jordani;* the rex sole, *Glyptocephalus* (= *Errex*) *zachirus;* and the rock sole, *Lepidopsetta bilineata* (= *Pleuronectes bilineatus*). The yellowfin sole, *Limanda aspera* (= *Pleuronectes asper*), has been the object of a successful fishery in the Bering Sea. *Pleuronectes platessa,* the plaice, is an important commercial fish in Europe.

In the starry flounder, *Platichthys stellatus,* a favored sport and food fish of the Pacific Coast of North America, half or more of the specimens have the eyes on the left side (Fig. 16–1**A**).

Other subfamilies of Pleuronectidae are Paralichthodinae of southern Africa, Poecilopsettinae of deep water of the Atlantic and Pacific, and Rhombosoleinae of the southwestern Atlantic, Australia, and New Zealand.

The family Samaridae includes about 20 species of warm marine waters of the Indo-Pacific. These fishes are found mainly in deep waters. Samaridae was formerly considered to be part of Pleuronectidae (see Chapleau, 1993).

The dextral Achiridae (American soles) are fishes of North and South America and are found in rivers as well as in marine habitats, especially in South America.

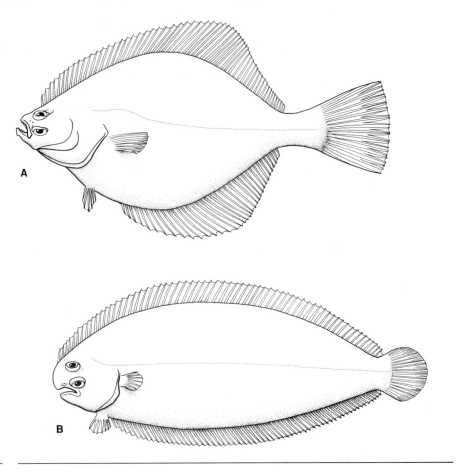

FIGURE 16–1   Representatives of order Pleuronectiformes, **A,** starry flounder (*Platichthys*); **B,** sole (*Solea*).

There are nine genera and about 28 species (Nelson, 1994). They are considered the primitive sister group to Soleidae and Cynoglossidae (Nelson, 1994). The family Soleidae includes the true soles. These have ten or fewer abdominal vertebrae and no ribs, and the edge of the operculum is bound by skin that bridges across the throat region. The edge of the preopercle is covered by skin. Most species lack pectoral fins, but some have the right pectoral fin. The mouth is not terminal but is shifted toward the anatomically ventral position (Fig. 16–1**B**). Eyes are on the right side. The dorsal and anal fins, as in the previous families, are separate from the caudal fin.

The Achiridae and Soleidae were formerly placed into the suborder Soleoidei, which was not recognized by all students of the group. Chapleau and Keast (1988) recommended that the suborder Soleoidei be dropped and that the families be included in the Pleuronectidae. That arrangement is followed by Eschmeyer (1990b). The soles are generally smaller fishes than the flounders and halibuts, but those species large enough to sustain a fishery are highly prized as food.

Members of Soleidae tend to prefer warm waters, but a few are found in temperate seas. Some species are known to enter rivers, especially in Asia to Australia and Africa. The valuable European *Solea solea* is the species that made fillet of sole a popular dish. There are about 20 genera and 90 species.

Tonguefishes of Cynoglossidae, which are considered by some ichthyologists to belong in Soleidae, are found primarily in tropical and subtropical seas. These are lefteyed, slender fishes that have the dorsal and anal fins confluent with the caudal fin. They are used as food wherever large enough species exist. Some species are no more than 50 cm long. Most North American species are in the genus *Symphurus* and are mostly too small to be taken commercially. There are three genera—*Cynoglossus, Paraplagusia,* and *Symphurus*—with a total of about 60 species.

Order Tetraodontiformes (Plectognathi). The molas, boxfishes, puffers, and relatives are distinguished by strong jaws and a small mouth, with strong incisors or a sharp beak composed of modified teeth. The gill openings are restricted, and the branchiostegals are covered by skin that is continuous with that of the body. Members of the order have many modifications of the cephalic osteology (Tyler, 1980). In many species, the maxillae and premaxillae are fused or at least strongly bound together. They lack nasals, parietals, and suborbitals. The operculum and subopersulum are reduced in size, and the preoperculum is elongate in many. The palatines and hyomandibulars are ankylosed to the skull. Scales are usually modified into plates or spines.

The classification of the order has been treated by Winterbottom (1974) on the basis of myology and by Tyler (1980) on the basis of osteology. Most of the differences in phylogenies listed by various authors concern the recognition of some groups as either families or subfamilies and the placement of the Balistidae and Ostraciidae, which Winterbottom (1974) placed into Tetraodontoidei. Winterbottom (1974), followed by Leis (1984), uses the suborder Triacanthodei for the spikefishes because of the placement of Balistidae. The following treatment follows Nelson (1994).

The order has about 100 genera in nine families. There are about 340 species.

Suborder Triacanthoidei. This suborder contains the spikefishes and triplespines. These are fishes of warm waters of the western Atlantic and Indo-Pacific. They have a large pelvic fin spine (that can be locked in place) and one or two pelvic soft rays. The dorsal fin usually has six strong spines. The spikefishes of Triacanthodidae comprise 11 genera and about 17 species. There are four genera and seven species of triplespines in the family Triacanthidae (Nelson, 1994).

Suborder Tetraodontoidei. This group contains diverse families of fishes in which the pelvic spines are lacking or very small. The superfamily Balistoidea includes fishes referred to as leatherjackets because of the appearance of the skin, in which the scales are often hidden. This group comprises over 40 genera and about 135 species. Families in the group are all marine, from the Atlantic, Pacific, and Indian oceans. The families are the triggerfishes, Balistidae (Fig. 16–2A); filefishes, Monacanthidae; and the boxfishes and trunkfishes, Ostraciidae. All members of this superfamily have separate well-developed teeth. With the exception of the Ostraciidae, these are all characterized by large dorsal spines that, in the balistids, can be locked erect. These are mainly shore fishes of the tropics; some have brilliant coloration.

The trunkfishes and boxfishes are enclosed in a bony carapace so that only the eyes, jaws, and fins are mobile. This is also a tropical shore fish group, with few species in temperate waters.

The Monacanthidae is placed as a subfamily of Balistidae by Winterbottom (1974) and Robins et al. (1991a). Tyler (1980) elevates the Aracaninae, a subfamily of Ostraciidae, to family status.

The largest member of the superfamily Balistoidea is the meter-long scrawled filefish, *Aluterus scriptus,* of tropical and subtropical waters.

The superfamily Tetraodontoidea contains fishes that are usually covered by spines and can fill the stomach (or a sac that evaginates from the anterior part of the stomach) with air or water so an individual can inflate itself like a spiny balloon. Inflation is accomplished by coordinated activity of the pyloric and cardiac sphincters and a sphincter controlling the evagination. These fishes have no true teeth, and the sharp edges of the jaw bones serve as dentition. The jaw bones in different groups give the appearance of two, three, or four teeth.

FIGURE 16–2   Representatives of order Tetraodontiformes, **A,** triggerfish (Balistidae); **B,** mola or headfish (*Mola*).

The family Triodontidae contains but one species—the threetooth puffer, *Triodon macropterus,* of seas from Japan to Indonesia.

The Tetraodontidae, or puffers, produce a strong poison, called tetrodotoxin (= tarichatoxin), which can be fatal to humans. The flesh of the puffer is considered a delicacy in Japan, but great care is taken to remove all traces of the viscera that contain the poison. Extracts from puffer skin are reported to be used in Caribbean islands to aid in bringing on "zombie" behavior. Puffers are mostly marine and are found in tropical and subtropical waters of the Indian, Pacific, and Atlantic oceans. Some species live in the fresh waters of southern Asia and Africa.

Diodontidae, the porcupinefishes (considered part of Tetraodontidae by Robins et al., 1991a), have longer spines on the skin than puffers. The inflated and dried skins are familiar curios in many tropical areas.

The family Molidae contains the molas (headfishes or ocean sunfishes). In these oceanic fishes, the caudal portion is restricted during early development so that the fish seems to end abruptly behind the prominent dorsal and anal fins (Fig. 16–2**B**). There are three monotypic genera, *Masturus, Mola,* and *Ranzania.* Although the family is mostly tropical and subtropical, the genus *Mola* ranges into temperate waters. *Mola mola,* which may weigh nearly a metric ton, is called the ocean sunfish, probably because it is often seen at the surface, sometimes on its side as if basking in the sun. Molas range to a depth of over 350 m and feed on a variety of invertebrates and some small fishes. They are usually heavily parasitized, both externally and internally.

Part Three

Organismic Biology of Fishes

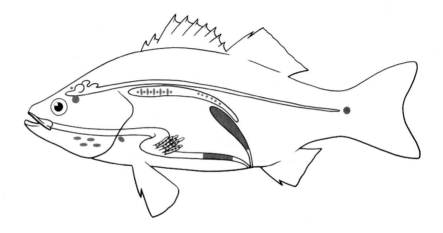

Location of endocrine tissue in bony fishes.

Nervous and Endocrine Systems

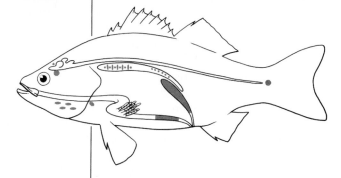

The Central and Peripheral Nervous System

Brain

The brain lies in the lumen of the neurocranium, protected by cartilage and/or bone, the surrounding meninges and cerebrospinal fluid, and a fatty matrix that surrounds the brain and fills much of the cranial cavity. Cranial nerves leave the brain and pass through foramina of the skull to their respective target organs or areas. The spinal cord, with which the brain is continuous, leaves the cranium posteriorly through the foramen magnum. There is a great deal of variation in brain morphology among fishes, even though fish brains all follow a basic vertebrate plan (Fig. 17–1) (Igarashi and Kamiya, 1972; Romer and Parsons, 1978). Some differences in brain morphology are of phylogenetic significance and may be manifested in the complexity and compactness of the organ. Other differences are apparently due to the degree of development of different sensory and motor functions (Tuge et al., 1968; see also Northcutt and Davis, 1983). Differences in the relative sizes of fish brains are also evident. Differences in the relative sizes of parts of the brain among fishes are often due to the relative differences in development of the special senses (sight, hearing, touch, smell, taste, and mechanosensory and electrosensory lateral line).

Brains in some fishes, including the modern coelacanth, weigh considerably less than 0.1 percent of total body weight, while the relative brain weight of many cyprinid minnows may be twice that. Some sharks, with their large olfactory lobes and bulbs, have relatively large brain-to-body ratios (Northcutt, 1989; Smeets et al., 1983). The largest relative size among fish brains is that of the mormyrids (Mormyridae) of Africa, in which the brain may be more than 1 percent of body weight.

The fish brain first develops in three sections (Fig. 17–2)—the forebrain (prosencephalon), mid-brain (mesencephalon), and hindbrain (rhombencephalon)—as in other vertebrates and then differentiates further (see Kent, 1992; Nieuwenhuys, 1962; Romer and Parsons, 1978). The mesencephalon and rhombencephalon together constitute the brain stem or truncus cerebri (Nieuwenhuys and Pouwels, 1983). The forebrain is the site of the olfactory sense. Its anterior part (telencephalon) is characterized by a pair of primary olfactory centers, the olfactory bulbs, from which olfactory nerves extend to the olfactory organ. Caudal to the olfactory bulbs, the telencephalon swells into what are often called "olfactory lobes" (Kent, 1992). These are usually larger than the bulbs and are mainly concerned with nonolfactory functions. In most bony fishes, the bulbs are situated just anterior to the lobes, but in elasmobranchs and in some bony fishes (such as certain catfishes, carp, and cods) the olfactory bulbs are adjacent to the olfactory organ and a long olfactory tract separates them from the lobes (Bernstein, 1970). In such species, and especially in sharks and rays, the bulbs may reach a relatively large size. Part of the telencephalon is developed as the cerebrum in elasmobranchs and bony fishes, although in the latter the cerebral hemispheres are prominent only among the more primitive members. The cerebrum of lungfishes resembles that of amphibians (Romer and Parsons, 1978). The posterior part of the forebrain is the diencephalon, usually set off by a constriction from the telencephalon. The pineal organ, which in many fishes is sensitive to light, arises from the roof of the diencephalon in elasmobranchs and bony fishes (Bernstein, 1970). Lampreys and hagfishes have both a

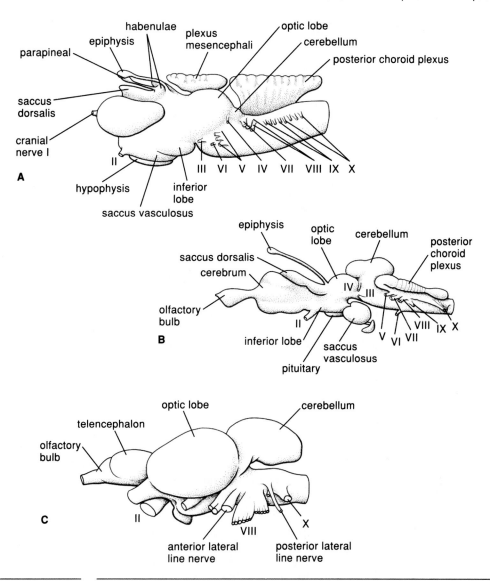

parapineal and pineal organ. The hypothalamus, in the floor of the forebrain, is the site of attachment of the pituitary gland. Also in this area is the saccus vasculosus, which is a vascular evagination with thin walls. The structure is found only in fishes. Its function is not definitely known, but it is lined with hair cells and its sensory fibers go to other brain centers, including the hypothalamus (Kent, 1992).

The optic lobes are the prominent feature of the mid-brain (mesencephalon) and are expecially large in visually oriented fishes. In hagfishes, which have vestigial eyes, the mid-brain is small. The optic lobes of lungfishes are fused into a single median structure. Vision is a major function of the mesencephalon, but it has other

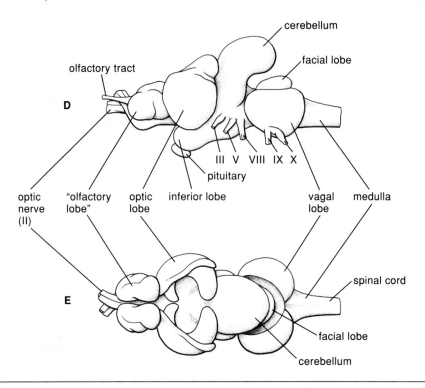

FIGURE 17–1
(Continued)

Brains of **D**, carp (*Cyprinus*), lateral view; **E**, carp, dorsal view.

functions involved in learning and in the relay of sensory messages to motor responses. Afferent nerve fibers from olfactory and gustatory centers appear in the
roof of the mesencephalon or optic tectum. The torus semicircularis is developed in
the floor of the mid-brain and is especially large in fishes that have well-developed
auditory and lateral line senses.

The major feature of the metencephalon (anterior part of the rhombencephalon)
of elasmobranchs and bony fishes is the cerebellum. In bony fishes, the cerebellum
is divided into two major sections: the valvula cerebelli, extending rostrally below
the optic tectum; and the corpus cerebelli, extending anteriodorsally. The valvula are
large in catfishes and minnows and reach a very large size in mormyrids (Nieuwenhuys and Pouwels, 1983). The size of the cerebellum is variable and is especially
well developed in large sharks. In the family Mormyridae, whose members have exceptional auditory and electrosensory systems, the cerebellum is large and complex
and may overlie the forebrain. Catfishes, mackerels, and tunas usually have enlarged
cerebella. In lampreys the cerebellum is very small, and the structure is not recognized at all in hagfishes. Although the cerebellum is derived from the rhombencephalon, it is not considered as part of the brainstem by Nieuwenhuys and Pouwels
(1983, p. 26), even though it is "strongly interconnected." The brainstem consists of
the rhombencephalon and mesencephalon and holds the origins and terminations of
most of the cranial sensory nerves (Nieuwenhuys and Pouwels, 1983).

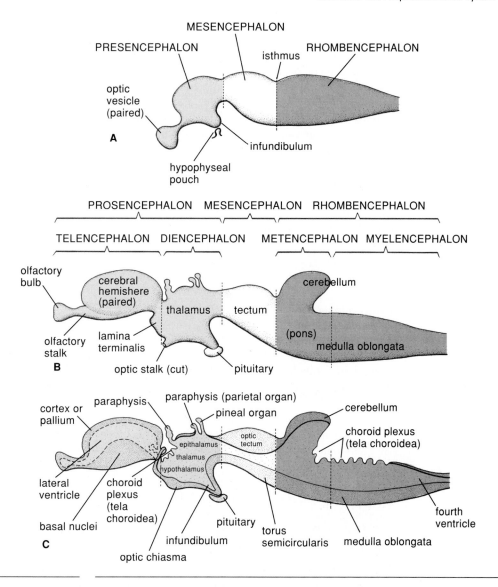

FIGURE 17–2   Stages in development of brain. **A,** Three main divisions developed; **B,** more advanced stage; **C,** median section of **B.** (After Romer, 1970.)

The cerebellum supports coordination of movement, muscle tone, and posture or balance. Many fishes with enlarged cerebella are fast swimmers, but catfishes and mormyrids are slower moving yet have this structure enlarged, so integration of certain sensory information by the cerebellum appears to be an important function in some types of fishes. Mormyrids are noted for their "electrical" sense, and catfishes are equipped with organs for reception of electrical impulses.

At the point where the corpus cerebelli meets the medulla, there are lateral swellings called eminentiae granulares, which may be enlarged in some species (such as in certain catfishes and hairtails [Trichiuridae]). The eminentia granularis appears to be an electroreceptive part of the mormyrid brain (Hopkins, 1983).

The most posterior part of the brain, the myelencephalon, is composed chiefly of the medulla oblongata, which in gnathostomes includes somatic and visceral sensory and motor areas. Several cranial nerves arise from the medulla oblongata, and impulses to and from spinal nerves are also relayed here. Various parts of the medulla are enlarged according to the sensory function and habits of fishes. For instance, certain suckers (Catostomidae) and minnows (Cyprinidae) have characteristic enlargements ("lobes") at the roots of the seventh, ninth, and tenth cranial nerves.

A pair of large neurons, called Mauthner cells, is found in the floor of the medulla of most fishes and aquatic amphibians (Diamond, 1971; Hardisty, 1979; Kuhlenbeck, 1975; Zottoli, 1978). These function in startle and escape reactions (Eaton, 1991). The dendrites are associated with acoustic centers, and the axons travel the length of the spinal cord to caudal musculature (see Zottoli, 1978). Mauthner's cells are lacking in several groups of fishes, including anglerfishes (Lophiiformes), eels (Anguilliformes), pipefishes, seahorses (Syngnathidae), and others (see Bone and Marshall, 1982; Zottoli, 1978).

Cranial Nerves

There is a series of nerves extending from the sensory organs and muscles to the brain. Conventionally, the cranial nerves have been identified by the application of Roman numerals based on their anterior–posterior position in humans. Cranial nerves can be grouped by function. The olfactory (I), optic (II), lateral line (ALLN, MLLN, PLLN), and the acoustic (VIII) are sensory. The terminal nerve (0) is possibly sensory. The oculomotor (III), trochlear (IV), and abducens (VI) are motor, and the trigeminal (V), facial (VII), glossopharyngeal (IX), and vagus (X) have both sensory and motor fibers present. Including the terminal nerve (0), 11 cranial nerves plus the unnumbered lateral line nerves are present in fishes. Actually, the ganglion of the profundus nerve (numbered V_1 in some texts) is fused to the trigeminal ganglion in lampreys, hagfishes, and lungfishes, but the two are not fused in other fishes (Northcutt and Bemis, 1993) and so appear as separate nerves.

Fishes lack distinct accessory and hypoglossal nerves (found in amniotes). Neurons with the homologous function of the accessory nerve of other vertebrates are found in association with the vagus. Hypoglossal neurons are associated with nerves at the anterior end of the spinal cord. The nerves and their major roles are as follows:

0. The terminal nerve is a small nerve which projects to the telencephalon, and is associated with the olfactory nerve. Fibers of the terminal nerve are distributed within the olfactory bulb. This nerve appears to be involved in the control of reproductive functions (see Demski and Schwanzel–Fukuda, 1987; Fujita et al., 1991).

I. The olfactory nerve runs from the olfactory epithelium in the olfactory organ to the olfactory bulb and olfactory lobe of the telencephalon. Its function is chemosensory.

II. The optic nerve projects from the ganglion cell layer of the retina of the eye to the optic tectum. It carries visual information—ultraviolet and visible spectra, color vision in fishes.

III. The oculomotor nerve supplies the inferior oblique muscle and the superior, inferior, and internal rectus muscles. It projects to the mesencephalic brainstem. It is a somatic motor nerve which controls these extrinsic eye muscles.

IV. The trochlear nerve innervates the superior oblique muscle of the eye. This somatic motor nerve projects to the mesencephalic brain stem.

V. The trigeminal nerve is divided into three branches. Two of these, the ophthalmic and maxillary, are somatic sensory nerves. The third, the mandibular, carries somatic sensory fibers from the jaws and motor fibers from muscles that are derived from the first visceral arch. The trigeminal nerve projects to the metencephalon. It carries information from taste buds and tactile receptors and thermal receptors.

VI. The abducens nerve is a somatic motor nerve coursing from the anterior part of the medulla oblongata (myelencephalon) to the external rectus muscle of the eye.

VII. The facial nerve is usually composed of three branches: the superficial ophthalmic, the buccal, and the hyomandibular. These branches supply taste receptors on the head, body touch receptors, and certain head muscles. The facial nerve has components involved with special and general somatic sensory, special and general visceral sensory, and visceral motor functions in the region of the second visceral arch. It projects to the myelencephalon.

VIII. The acoustic (vestibulocochlear) nerve has a special sensory function and serves the inner ear. It connects with the myelencephalon.

IX. The glossopharyngeal nerve is composed of visceral sensory and motor components serving mainly the region of the first gill slit (third visceral arch). A dorsal group of branches serves proprioceptors and, according to Kent (1992), a small segment of the lateral line; branchial branches are involved with taste organs of the pharynx and with branchial muscles. The glossopharyngeal nerve projects to the myelencephalon.

X. The vagus nerve is a large nerve with several branches. Branchial branches travel to the region of the posterior four gill slits (fourth visceral arch and derivatives). The vagus includes both sensory and motor fibers. A dorsal recurrent branch innervates external taste buds, and the visceral branch innervates receptors and muscles associated with internal organs. The site of the root is the myelencephalon.

Lateral Line Nerves. These are unnumbered cranial nerves that are so closely placed to other nerves that they have been considered components of those nerves, but Northcutt (1989) points out that a more plausible hypothesis is that the lateral line nerves are a separate series of cranial nerves.

The anterior lateral line nerve (ALLN) travels with nerves V and VII and has two or three ganglia, depending on species. Branches from these ganglia innervate neuromasts on the anterior part of the head. The middle lateral line nerve (MLLN), if present, innervates a pit line in bony fishes. The posterior lateral line nerve (PLLN) may travel with nerve X and serves mainly the corporal neuromasts, including the lateral line canal (see McCormick, 1983; Northcutt, 1989; Song and Northcutt, 1991).

Spinal Cord and Nerves

The spinal cord, with only a few exceptions, is continuous with the medulla oblongata and extends to the end of the vertebral column. It is essentially a hollow tube, but the central canal is of small diameter compared to the thick walls. Around the central canal, making a pattern in cross section similar to a pair of butterfly wings, is gray matter composed of unmyelinated nerve fibers running longitudinally. In lampreys and hagfishes, all the nerve fibers are unmyelinated and the spinal cord is flattened dorsoventrally. The spinal cord of the lamprey is characterized by 8 to 12 giant "Müller's fibers" on each side (Hardisty, 1979; Kuhlenbeck, 1975; Rovainen, 1978). These are somatic motor axons that run the length of the cord from Müller cell bodies in the brainstem.

The length of the spinal cord can vary significantly. In the ocean sunfishes (Molidae), it does not extend much farther than the hindbrain; and in the goosefish (*Lophius*), the cord is shortened but a long terminal filament extends posteriorly from it.

The paired spinal nerves are arranged segmentally and arise from the gray matter as dorsal and ventral roots that merge and then typically branch into three parts. The dorsal root has a ganglion outside the spinal cord (dorsal root ganglion). Dorsal and ventral branches (or rami) serve the axial muscles and skin, whereas a visceral branch (ramus) supplies the internal organs. Lampreys differ from the other fishes in that they lack the connection between the dorsal and ventral roots. In these forms, the dorsal roots originate opposite the myosepta and the ventral roots opposite the myotomes.

The dorsal roots of spinal nerves in fishes carry somatic and visceral afferent fibers and some visceral efferent fibers. Somatic and visceral efferent fibers enter the spinal cord through the ventral roots. The visceral efferent components of the spinal nerves contribute to the autonomic nervous system, which is involved in the control of smooth muscle and certain glands. Bony fishes have a chain of interconnected, segmentally arranged ganglia. Sympathetic ganglia are found in an irregular series in the trunk region of elasmobranchs. Parasympathetic fibers are largely associated with cranial nerves, almost entirely with the vagus.

Endocrine Glands

The endocrine glands of fishes comprise a system comparable to that of higher vertebrates, but some endocrine tissues do not form discrete glands in fishes and the sites of the tissues may be different from the sites of secretion in higher forms. However, for the most part, the same or very similar hormones are produced. In addition, fishes possess some endocrine tissues, such as the caudal neurosecretory system and the Stannius corpuscles, that do not have homologues in higher vertebrates. Following is a description of the endocrine glands of fishes with the general location of the tissue and a brief mention of the function of the secretions. General locations of the endocrine glands are shown in Figure 17–3.

Pituitary Gland

This gland, also referred to as the hypophysis or hypothalamohypophyseal system, is located beneath the diencephalon (Fig. 17–3) and may be associated with the sacculus vasculosus. Its embryonic origin involves a neural downgrowth from the dien-

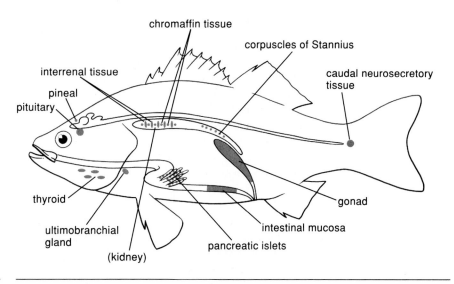

FIGURE 17–3   Diagram showing approximate locations of endocrine glands in a bony fish.

cephalon in conjunction with an ectodermal component growing upward from the dorsal part of the embryonic mouth cavity (stomodeum). The stomodeal component forms an early pouchlike structure (Rathke's pouch) in lower fishes but begins as a solid structure in higher fishes. The neural (diencephalic) component of the adult gland, called the neurohypophysis (pars neuralis), is closely associated with the hypothalamus, which produces neurosecretions that are released into the pituitary. Those secretions have a stimulating or inhibiting effect on the release of certain hormones from the anterior lobe (Kent, 1992).

There are commonly three parts of the neurohypophysis of fishes (and other vertebrates): the infundibular stalk, the median eminence, and the posterior lobe (pars nervosa). The remainder of the complex, formed from the stomodeal component, is called the adenohypophysis (pars buccalis), which consists of histologically distinct sections that are, in general, functionally equivalent to parts of the mammalian pituitary.

The adenohypophysis is divided into an anterior section called the pars distalis, which has a rostral part (pro-adenohypophysis) plus a distal part (meso-adenohypophysis), and a posterior section or pars intermedia (meta-adenohypophysis) (Fig. 17–4). In many fishes, the pars intermedia is intimately related to the neurohypophysis and forms what is referred to as the neurointermediate lobe (see also Pickford and Atz, 1957; Holmes and Ball, 1974; Schreck and Scanlon, 1977; Matty, 1985; and Gorbman et al., 1983).

Agnathans. The pituitary glands of lampreys and hagfishes are considered to be more primitive than those of elasmobranchs and bony fishes (Fig. 17–5A). The gland is flattened and less complex than those of gnathostomes. In hagfishes the neurohypophysis appears as a tubular projection from the brain, whereas in lampreys it is not well developed and is little more than a thin plate of cells in association with

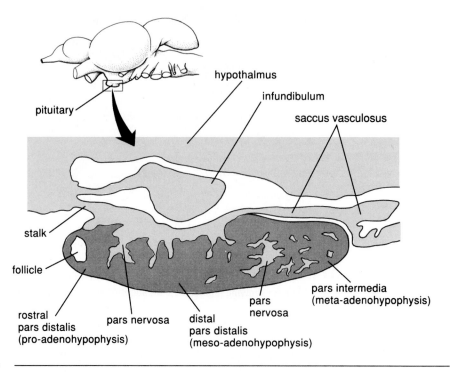

FIGURE 17–4   Diagram of median section of pituitary of rainbow trout (*Oncorhynchus mykiss*).

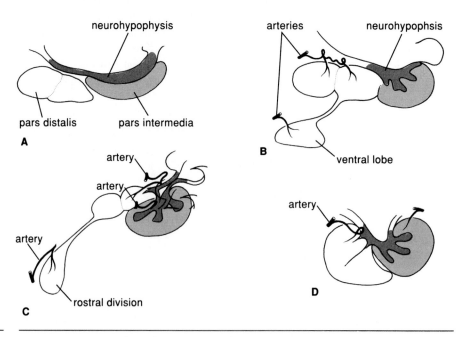

FIGURE 17–5   Diagrams of pituitary glands. **A,** Lamprey; **B,** chondrichthyan; **C,** *Latimeria*; **D,** lungfish. (After Lagios, 1982.)

nerve fibers. The adenohypophysis of lampreys is divided into rostral, distal, and posterior sections, although these may not be homologous with those of gnathostomatous fishes. In hagfishes the adenohypophysis is undivided. Although several types of secretory cells are present, they are not segregated by functional type. The agnathan pituitary probably does not secrete as full a range of hormones as in higher forms, and the divisions are probably not comparable to those of other fishes. The hypothalamo–hypophyseal portal system that constitutes a neurovascular connection between the gland and the brain is not developed in lampreys and hagfishes (Matty, 1985).

Chondrichthyes. In the sharks and rays, the adenohypophysis has a thin forward extension reaching or nearly reaching the optic chiasma (Matty, 1985), and the pars nervosa is mixed with the pars intermedia to form a neurointermediate lobe (Gorbman et al., 1983). The pituitaries of sharks and rays are peculiar in that they have a small ventral lobe attached to the pars distalis by a short stalk and receive a direct arterial supply. A similar structure (Rachendach-hypophyse) in the Holocephali is detached and lies outside the chondrocranium in the roof of the mouth (Fig. 17–5**B**). Sharks and rays have a median eminence with a hypophyseal portal system. There is some evidence of this system in chimaeras.

Bony Fishes. There are considerable differences among the bony fishes in shape and functional organization of the pituitary (Figs. 17–4, 17–5**C** and **D**).

Crossopterygii. The pituitary of *Latimeria* has an elongate ventral or rostral lobe of the pars distalis reminiscent of the ventral lobe of sharks and rays. This is a character that causes speculation about a close relationship between the two groups (Lagios, 1975, 1982) (Figs. 17–5**B** and **C**). A median eminence and hypophyseal portal system are present.

Dipnoi. In the lungfishes, the pituitary complex is more compact than in the aforementioned groups and has a similarity to the pituitary of amphibians (Lagios, 1982; Matty, 1985; Romer and Parsons, 1978). A pars nervosa is formed at the posterior part of the neurohypophysis, and the pars intermedia is distinct. In addition, the median eminence and portal system are developed. There is no sacculus vasculosus in the brain of lungfishes.

Nonteleost Actinopterygians. The bichirs, reedfishes, sturgeons, paddlefishes, gars, and the bowfin all are characterized by a median eminence constituting a portal system between the pars distalis and the floor of the infundibulum (Matty, 1985). As in teleosts, a saccus vasculosus is present. The pituitaries of these fishes are relatively compact, not divided into separate lobes, and the secretory cells are not divided into functional types. In the bichirs, a vestige of Rathke's pouch persists as an orohypophyseal duct, a condition noted in at least one teleost, *Tenualosa* (*Hilsa*) *ilisha,* a herring of the Indian Ocean.

Teleosts. In the teleosts, the median eminence as such is lacking and the adenohypophysis contains branches from the neurohypophysis. There are many differences among the teleosts in the shape and functional organization of the pituitary (Gorbman et al., 1983). The glands are usually somewhat elongate, but they may be roughly globular, as in salmonids and cyprinids (Fig. 17–5C). In the swamp eel, *Monopterus albus,* the adenohypophysis completely surrounds the neurohypophysis except at the connection with the infundibulum (O and Chan, 1974). In most teleosts the pituitary lies close to the hypothalamus, but in some it is on a short stalk. A notable case is the goosefish, *Lophius,* in which the gland is on a long stalk. In the goby *Lepidogobius,* the pituitary is surrounded by the hypothalamus.

Hormones Released by the Adenohypophysis in Fishes

Adrenocorticotropin (ACTH). This hormone stimulates production of steroids in interrenal tissue, which is homologous to the adrenal cortex of higher vertebrates.

Gonadotropic Hormone (GTH). Two distinct hormones may exist, a follicle stimulating hormone (FSH) and a luteinizing hormone (LH). Fishes do not have well-differentiated beta cells (producing FSH) and gamma cells (producing LH) as in higher vertebrates, but some fishes have cells that may approximate these. The action of those hormones is on the gonads and involves production of eggs and sperm and stimulation of production of steroids. GTH is released in response to gonadotropin-releasing hormone (GnRH) secreted by the hypothalamus (Sower, 1990).

Growth Hormone (GH, Somatotropin). This hormone promotes growth in length. An increase in appetite and food conversion appear to result from treatment with GH.

Melanophore Stimulating Hormone (MSH). This hormone acts on melanophores to cause aggregation or dispersal of pigment granules.

Prolactin. This hormone acts on the kidney, gills, gut, and urinary bladder in regulation of osmotic balance. It has been shown to affect permeability of various tissues to water and to modify levels of sodium. Other effects involve xanthophore pigment dispersion and fat metabolism.

Thyrotropin. Stimulation of secretion by the thyroid gland is the action of this hormone.

Neurohypophyseal Hormones. In the various groups of fishes, several peptides are released by the neurohypophysis. Of these, arginine vasotocin (AVT) appears to be present and active in all. Gorbman et al. (1983) list the following neurohypophyseal peptides for the fishes: cyclostomes, AVT only; sharks, AVT plus valitocin and aspertocin; rays, AVT and glumitocin; chimaeras, AVT and oxytocin; lungfishes, AVT and mesotocin; teleosts, AVT and isotocin. AVT appears to be involved in osmoregulation and aids in maintaining proper water balance. It is known to increase blood pressure. The neurohypophyseal peptides act on smooth muscle, and some of them (oxytocin, isotocin, and mesotocin) cause constriction of branchial muscle.

Although the pituitary is usually called the master gland and has been shown to govern many vital activities of the other endocrine glands, a strange cobalt-blue mutant of the rainbow trout (*Oncorhynchus mykiss*) lives to an age of at least five years with almost no pituitary tissue. This form appears occasionally in broods in Japanese hatcheries but is sterile because of abnormal oogenesis and spermatogenesis. It has some metabolic disorders, but the thyroid gland and the interrenal tissue appear normal (Yamazaki, 1974).

Thyroid Gland

The thyroid of fishes consists of epithelial follicles walled by a single layer of cells and usually located in the region of the pharynx or heart. In most groups except the hagfishes, the follicles contain a colloidal material (Matty, 1985). Location and distribution of the thyroid follicles differ among fishes. Those of lampreys and hagfishes are found along the ventral aorta in the branchial region, whereas in elasmobranchs, lungfishes, and *Latimeria,* the follicles form more or less compact glands ventral to the pharynx (Wendelaar Bonga, 1993). The thyroid follicles of many bony fishes are associated with the surface of the heart, ventral aorta, and the lower parts of the branchial arteries. The compact type of thyroid has been described in a variety of bony fishes, such as the mudskipper (*Periophthalmus*), swordfish (*Xiphias*), parrotfishes (Scaridae), and a few others. In some fishes, such as the goldfish (*Carassius auratus*) and the platyfishes (*Xiphophorous* spp.), thyroid follicles may concentrate in the head kidney (anterior hemopoietic section of kidney) as well as in the usual subpharyngeal location.

The iodine-containing thyroid hormones (thyroxin and tri-iodothyrosine) appear to have a variety of physiological effects in fishes. Many of these actions are not yet well understood, but it is certain that they are important in regulating pigmentation of the skin and eye. There is evidence that they affect the rate of oxygen consumption, promote the deposition of guanine in the skin, alter carbohydrate and nitrogen metabolism, and precipitate metamorphosis in flatfishes. In addition, effects on motor activity, skeletal growth, maturation of gonads (Bern and Nishioka, 1985), and function of the central nervous system have been noted. Involvement in osmoregulation is reported (Wendelaar Bonga, 1993), and administration of thyroxin heightens the "preference" of young salmon for salt water. Leatherland (1982) emphasizes that much evidence of effects of thyroid hormones among fishes is contradictory, possibly because of species differences in the complex activities of the gland or in part due to extraneous factors in research.

Interrenal Tissue

In elasmobranchs, interrenal tissue, which is homologous with the adrenal cortex of higher vertebrates, is organized into glands situated between the posterior regions of the kidneys. The interrenal tissue of bony fishes is usually associated with the head kidney (pronephros) and appears as cells or groups of cells scattered there, especially along the cardinal veins. Cells similar to adrenocortical cells are found in the walls of the cardinal veins of lampreys. Small amounts of certain steroids have been found in the blood of lampreys and hagfishes, but little is known of interrenal function in these animals (Matty, 1985).

The secretions of the interrenal tissue are steroids, most notably cortisol, corticosterone, and cortisone (the latter is much more prominent in bony fishes than in elasmobranchs and cyclostomes). The adrenocorticosteroids appear to exert some control over osmoregulatory processes, acting on the kidney, gills, and gastrointestinal tract. Metabolism of proteins and carbohydrates is affected by the corticosteroids, especially in such fishes as the Pacific salmons (*Oncorhynchus*), which make lengthy migrations while fasting and must utilize muscle protein in order to gain sufficient energy to complete their travels and the ensuing spawning process.

Secretions of the interrenal cells are important in the stress response in fishes. Stressful events such as injury, crowding, or abrupt change from one container to another can cause rapid elevation of corticosteroids. These may peak in 1 to 24 hours (Schreck, 1981).

Although regulation of the interrenal is usually by the action of ACTH, it appears that gonadotropins may be involved as well (Schreck et al., 1989).

Chromaffin Tissue

This tissue is homologous with the adrenal medullary tissue of higher vertebrates, but the interrenal and chromaffin cells are organized into a compact gland in only one family of fishes (Cottidae). Usually, chromaffin cells of bony fishes are distributed along the postcardinal veins and may intermingle to some extent with interrenal cells. Chromaffin tissue in elasmobranchs is associated with the sympathetic ganglia and the dorsal aorta anterior to the interrenal tissue. A separation of the two tissues is seen also in the cyclostomes, in which the chromaffin cells appear as strands along the dorsal aorta. Chromaffin cells are derived embryologically from postganglionic cells of the sympathetic nervous system, some cells of which also produce noradrenaline.

Chromaffin cells secrete the catecholamines adrenaline and noradrenaline (epinephrine and norepinephrine), which are important in the "fight or flight" stress response of fishes (Schreck, 1981). Control of heart rate, blood pressure, and blood flow through the gills (Wahlquist, 1981); concentration of melanin in the melanophores; and dilation of the pupils all have been ascribed to catecholamines.

Ultimobranchial Gland

In bony fishes, this gland is located below the esophagus near the sinus venosus, often on or closely associated with the pericardium. In elasmobranchs, the gland is on the left side of the mid-line beneath the pharynx. It secretes the hormone calcitonin, which is involved in the inhibition of bone resorption in mammals and is thought to be involved in calcium metabolism in fishes. Experiments with rainbow trout (Fouchereau–Peron et al., 1986) indicate that calcitonin is involved in the regulation of adaptation to sea water. The gland does not occur in cyclostomes.

Pancreas (Islets of Langerhans)

The pancreatic islets of bony fishes are usually dispersed around the pyloric caeca, small intestine, spleen, and gallbladder. Some teleosts have bodies of pancreatic tissue gathered into "Brockmann bodies," some of which can produce both endocrine and exocrine secretions (Gorbman et al., 1983). A few species have a compact mass of pancreatic tissue on or near the gallbladder. The islet tissue is found in the walls of the intestine in lampreys. In sea lamprey ammocoetes, the islets are located in the gut epithelium at the junction of the bile duct. As the bile duct degenerates during metamorphosis, part of it becomes a caudal endocrine pancreas. An anterior (cranial) pancreas develops from the larval pancreas (Youson and Elliott, 1989). The cranial pancreas is on the dorsal wall of the esophagus, where it joins the intestine (Barrington, 1972). The endocrine pancreas of hagfishes consists of follicles at the juncture of the bile duct and the intestine (Matty, 1985). Elasmobranchs have a discrete pancreas that includes the islets. There are four types of endocrine cells in the fish pancreas, and these produce insulin, glucagon, somatostatin, and pancreaspeptide. A fifth cell of possible endocrine function has been noted in the European bass *Morone* (= *Dicentrarchus*) *labrax* (Carillo et al., 1986).

Secretions of the pancreas are important in governing metabolism. Insulin is involved in the synthesis of protein, conversion of glucose into glycogen, and production of fat from carbohydrate sources. Glucagon is an insulin antagonist and causes glucose to enter the blood (hyperglycemia). It also acts to release glucose from glycogen and is lipolytic (Plisetskaya, 1990). Somatostatin inhibits gastrointestinal motility in the rainbow trout and may slow emptying of the stomach (Chen and Hale, 1992). Somatostatin may also be involved in the release of glucose from the liver in salmonids (Eilertson et al., 1991) and has been shown to inhibit the release of growth hormone in *Anguilla* (Suzuki et al., 1990). Two types of somatostatin cells have been noted in mullets by Lozano and Agulleiro (1986).

Gastrointestinal Mucosa

Several peptides with apparent regulatory function are found in the intestinal mucosal cells as well as in the nerves associated with the gut mucosa. Research on the presence of these substances in fishes is proceeding rapidly and successfully, but the physiological aspects are not as well advanced (Holmgren et al., 1986).

At least 16 types of endocrine intestinal cells are tabulated for fishes by Rawdon and Andrew (1990). The secretions of these appear to be involved mainly in regulating secretions or motility of the digestive tract. Gastrin and related substances, which occur in lampreys, sharks, and bony fishes, cause secretion of pepsin and hydrochloric acid in the stomach. Glucagon, somatostatin, and insulin are produced in the gut as well as in the pancreatic cells. Substance P is found in sharks and bony fishes and has been shown to have an excitatory effect on stomach and gut muscle (Holmgren, 1985; Holmgren et al., 1986). Other peptides that might be involved in stimulation of gut motility are gastrin, bombesin, and serotonin. Some that may be involved in excitatory effects on gastric tissue are enkephalon and neurotensin. Bombesin might stimulate secretion of stomach acid, and vasointestinal peptide may inhibit gut motility (Wendelaar Bonga, 1993).

Gonads

The sex organs of both sexes are involved in the secretion of steroids that are important in the manifestation of courtship, nest building, and other aspects of reproductive behavior as well as in the differentiation of gonads, development and maintenance of secondary sex characteristics, and gametogenesis. Steroids are important in the reproductive cycle of sharks (Tsang and Callard, 1987) and in lampreys (Linville et al., 1987) as well as bony fishes (Redding and Patiño, 1993). The ovary produces estrogens, which have not as yet been well studied in fishes. Investigations have shown positive relationships between ovarian secretions and receptivity to males and development of secondary sex characteristics. Secretions (pheromones) that attract male guppies to female guppies are under the influence of estrogens (Stacey, 1981, 1991). The testes produce androgens, especially testosterone. Other hormones isolated from the testes include dehydro-epiandrosterone and androstenedione. Many studies have shown androgens to be of great importance in the sexual behavior and spawning activity of male fishes.

Caudal Neurosecretory System

Near the termination of the spinal cord in sharks, rays, teleosts and some other bony fishes, such as *Lepisosteus* and *Polypterus,* there are enlarged secretory neurons known as Dahlgren cells. These appear to be of two types (Bhatt and Negi, 1987). The axons of these neuroendocrine cells terminate in a capillary bed that appears to function in the storage and release of secretions (Jaiswal and Belsare, 1973). In teleosts, the capillary network is contained in a well-defined neurohemal structure called the urophysis, which is paired in some species and is on a stalk in some (Matty, 1985). This complex, which includes the terminal filament of the spinal column, is the site of production and release of the peptides urotensin I and II and possibly others that resemble arginine-vasotocin. Although the exact biological activity of the hormones is not well known, experimentation has produced evidence that they appear to influence water balance and sodium regulation in some species (Matty, 1985).

In a year-round study of this gland in an Indian catfish, Sharma and Sharma (1975) noted that stored material disappeared during the breeding season, leading them to surmise that the caudal secretory system is involved in the reproductive cycle. Urotensin II shows an increase in the white sucker, *Catostomus commersoni,* during the spawning season and has been shown to have a contracting effect on smooth muscle in the urogenital systems of some fishes (Matty, 1985).

Corpuscles of Stannius

The corpuscles of Stannius are found in the opisthonephric kidney of holosteans and teleosts. They vary in position among species and are found dorsally, dorsolaterally, or ventrolaterally; they are seldom arranged symmetrically. The featherbacks (*Notopterus*) have only a single corpuscle near the head kidney. The corpuscles may be highly vascularized and lobulated (Belsare, 1973). In eight teleost species studied by Krishnamurty and Bern (1971), the corpuscles were found to have prominent autonomic innervation.

Because removal of the corpuscles brings about changes in plasma composition, their secretion(s) appear to be involved in osmoregulation (Henderson and Jones, 1973). A decline in sodium and a rise in potassium salts produce histological changes in the corpuscle. Pang and Pang (1986) showed that removal of the Stannius corpuscles from the killifish *Fundulus heteroclitus* resulted in hypercalcemia. There appears to be a relationship between the adrenal cortex and the activity of the corpuscles, because injections of corticosteroids bring about nuclear hypertrophy and other evidence of stimulation in cells of the corpuscles. The secretion stanniocalcin probably acts as a calcium-channel blocker to limit entry of calcium into the chloride cells (Wendelaar Bonga, 1993).

Pineal Organ

This body (see Fenwick, 1970), attached to the roof of the diencephalon, is the remnant of the "third eye." It has been demonstrated to have a light sensory function in lampreys and in some elasmobranchs and teleosts. In most bony fishes, a thickening of the pineal epithelium and great vascularization of the structure indicate a glandular function (Belsare, 1975). However, there is great variation among fishes, even those within particular families, in the state of transition from the light-sensitive stage to the glandular situation. A secretion of the pineal is melatonin, which aggregates melanin in amphibians and has been shown to have a similar effect in some fishes. The diurnal pigmentation cycle of lamprey ammocoetes (dark by day, light by night) involves the presence of melatonin as well as nonendocrine entities. The pineal is known to contain and release other materials, such as various peptides and arginine vasotocin (Kavaliers, 1980), and probably is involved in reproduction because of its secretion of melatonin and of serotonin, the concentrations of which change during annual reproductive cycles (Redding and Patiño, 1993; Reiter, 1991). Serotonin is known to stimulate release of gonad releasing hormone in fishes (Redding and Patiño, 1993). The pineal gland can also be involved in gonadal maturation and in control of pigmentation (Tamura and Hanyu, 1980). Removal of the pineal organ from fishes can bring about changes in growth and can result in the stimulation of the pituitary and thyroid glands (Vodcinik, 1978). Pinealectomy also can disrupt circadian activity cycles (Kavaliers, 1980).

Other Glands

Kidney. In teleosts and in some holocephalans, the hormone renin appears to be secreted by certain granular cells (juxtaglomerular cells) associated with the renal blood vessels and apparently derived from arterial cells. Renin in circulation forms angiotensin by acting on a polypeptide. Angiotensin is a pressor and is active in osmoregulation through sodium retention by the kidneys (Brown et al., 1990; Matty, 1986).

Thymus. The thymus gland has its origin in the branchial pouches of fishes and is generally found above the branchial chamber or pockets in lampreys, sharks, and bony fishes. Little is known of its function in fishes, but it is probably not an endocrine gland.

Pseudobranchial Neurosecretory Gland. A pseudobranchial neurosecretory organ has been identified in several Asian airbreathing fishes. It is involved with the anterior aortic arches, but its function is not yet known (Srivastava et al., 1981).

Natriuretic Peptides. These are chemical agents that inhibit reabsorption of sodium and other cations from urine. They have been studied over the past decade in mammals and have been discovered in fishes, including agnathans, chondrichthyans, and bony fishes. They have been found in brains, hearts, and blood plasma (Wendelaar Bonga, 1993). The function of these peptides appears to be regulation of salt content.

Muscles, Locomotion, and Buoyancy

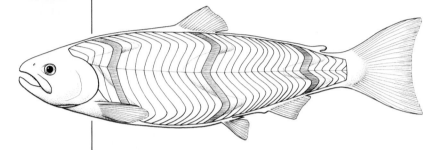

W e usually think of fishes as neutrally buoyant and therefore relieved of the burden of supporting themselves against the downward pull of gravity and consequently not required to expend the amounts of energy that terrestrial animals must in maintaining position. That may be largely true of some fishes, but all fishes are not neutrally buoyant and must maintain position by exerting some force against the water to keep from sinking or rising. Fishes may have to maintain position in currents or move around to a greater or lesser degree while adjusting for positive or negative buoyancy. Any movement in relation to the water demands the expenditure of great amounts of energy, because water is about 800 times as dense as air and significant energy must be expended to push through it, especially if appreciable speed is required (as for escape or pursuit). This chapter will begin with a discussion of skeletal musculature.

Musculature

The skeletal musculature of fishes consists mainly of the large muscles of the trunk and tail. Other smaller muscles are associated with the jaws, the branchial arches, and the fins. The trunk musculature consists of a series of muscle blocks, called myomeres or myotomes, separated by sheets of connective tissue called myosepta or myocommata. These myotomes represent the segmentation seen in all higher animals. Most fishes have both red and white muscle cells (myofibers). The red fibers are oriented more or less parallel with the body axis, whereas the white fibers may deviate as much as 45° from the body axis (Videler, 1993). The myotomes are folded so that, just under the skin, their outer edges resemble the letter W tipped on its side (Fig. 18–1). In lampreys, the angles of flexing of the myotomes are slight, especially anteriorly; but in sharks and bony fishes, the bends are sharper and more evident. The modification and folding of the myotomes is so great that a short and

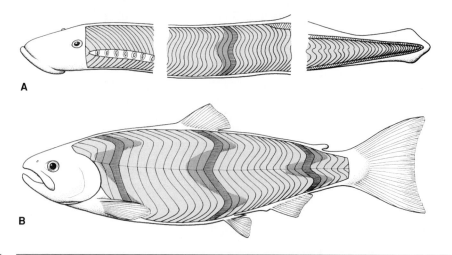

A

B

FIGURE 18–1 Lateral body musculature. **A,** Lamprey, showing myotome patterns in anterior, middle, and posterior sections; **B,** diagram of myotome patterns in salmon (myotome number reduced). Full extent of selected myotomes shown in gray. (**B** based on Greene and Greene, 1914.)

simple description is difficult. Reference to illustrations and to specimens will help in understanding the structure of the trunk musculature.

In lampreys, the myotomes extend forward from their edges on the sides of the body to their origins on the axial skeleton. However, in sharks and bony fishes, the myotomes extend posteriorly toward the axial skeleton from the regions of the backward-pointing upper and lower flexures visible on the surface, each fitting inside another so that a cross section of the trunk or tail cuts through several myotomes on each side, showing myosepta as concentric lines (Figs. 18–2**A** and **B**). There are two myotomes per vertebral centrum in some fishes, but because of the folded pattern a given myotome, depending on species, might have an overall anterior–posterior span of 3 to 12 intervertebral joints (Wainwright, 1983). Each myotome is typically

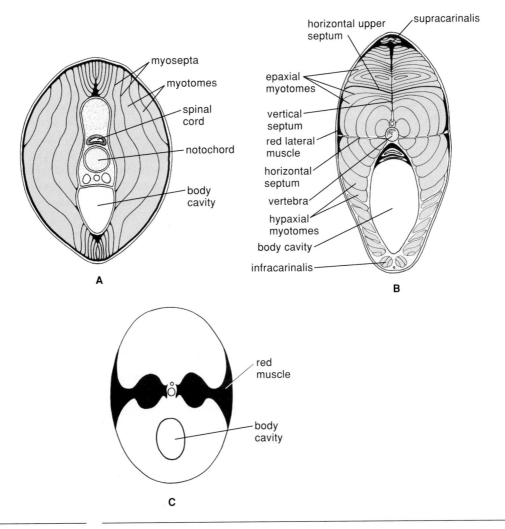

FIGURE 18–2   Diagram of body musculature in cross section. **A,** Lamprey (*Lampetra tridentata*); **B,** chinook salmon (*Oncorhynchus tshawytscha*); **C,** diagram showing approximate extent of red muscle (stippled) in skipjack tuna (*Katsuwonus pelamis*).

divided into four or more portions by myosepta. A vertical septum (mid-dorsal in the abdominal region) separates the muscles into bilateral left and right halves. On each side of the body, a main horizontal septum at the level of the vertebral column divides the myotomes into epaxial (upper) and hypaxial (lower) muscles. Less evident horizontal septa further subdivide the lateral muscles so that posterior to the body cavity there are usually four recognizable muscle bundles on each side (Wainwright, 1983). Lampreys and hagfishes do not have horizontal septa.

The myotomes connect externally to the skin and internally to the median or vertical septum and to the myosepta and horizontal septa (Videler, 1978).

Lamprey myotomes are characterized by flattened white (fast) muscle fibers surrounded by slow fibers and intermediate fibers. Along the sides of most fishes, just under the skin, lie the lateral superficial muscles, which are usually dark in color, well supplied with blood vessels, and with high fat content. These red muscles are used in normal sustained swimming activity and are fatigue-resistant at slow or cruising speed; in tunas and other extremely active fishes, red muscle is more extensive than in sedentary fishes and is highly vascular. The remainder of the lateral musculature is used for sudden bursts of swift or strong swimming, such as during escape or the capture of prey.

Anteriorly the body musculature connects to the pectoral girdle and head; posteriorly the connection is to the caudal fin or to tendons that run to that fin (Lindsey, 1978). The main horizontal septum is made up of two thin layers of tendons. The outer ends of the tendons connect to the superficial red muscle. The inner ends of one sheet attach to the posterior half of the vertebral centra, and the tendons of the other sheet course forward and attach to the front of the centra. Tendons form at the ends of the anterior and posterior cones of the myomeres and are well developed in strong swimmers but are weak or short in most bony fishes. Elasmobranchs have well-developed tendons, and the scombroid fishes have especially strong tendons, especially in the caudal region (Fierstine and Walters, 1968).

The track of fibers in thin superficial red muscles is usually parallel to the body axis, but the orientation of fibers in deeper muscle is usually curved so that they form angles of as much as 40° with the body axis. Alexander (1969) has demonstrated that series of fibers form spiral tracks along the body. This was done by examining the origins and insertions of individual fibers on successive myocommata. A set of helices is formed in each of the arms of the recumbent W that generally describes the shape of myotomes. Apparently, the complicated relationships between the shape of the myotomes, with their cone-in-cone arrangement, and the orientation of the fibers allows the fibers to contract at the same rate regardless of their distance from the vertebral column. The muscles shorten about 5 percent upon contraction regardless of position in the myotome. The force of contraction is thereby transferred to the skeleton efficiently so that maximum power output is gained. About half of a fish's weight is locomotor muscle (Bone, 1978). Red fibers constitute a minor part of the lateral musculature—usually 0.5 to 10 percent—although in tunas and other active pelagic fishes the red fibers may make up nearly 30 percent of the muscle (Greer-Walker and Pull, 1971; Johnston, 1981). In the tunas and close relatives and some of the swift sharks, the red musculature extends from the lateral position to the vertebral column (Fig. 18–2C). Some species, including various salmonids and cyprinids, have intermediate muscle fibers, called "pink" fibers, somewhat concentrated just inside the red musculature (Fig. 18–3D) but also scattered among the white fibers (Gill et al., 1982). Red fibers are found among the white fibers in many fishes. Johnston (1983) reports that five muscle fiber types have been recognized in the spotted dogfish (*Scyliorhinus canicula*) and four in the perch (*Perca*).

The red (slow) fibers and white (fast) fibers differ in several ways related to their respective functions (Bone, 1978, 1989; Johnston, 1981). Red muscle fibers, which are capable of prolonged activity in sustained cruising, are of small diameter, are well supplied with lipids plus some glycogen, have a high mitochondrial volume of about 16 to 35 percent, and are highly vascularized with peripheral capillaries.

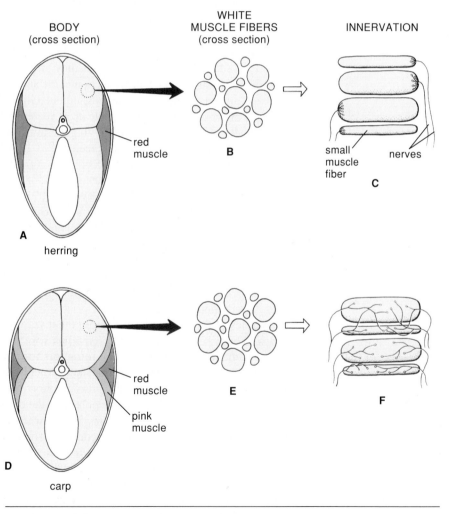

FIGURE 18–3   Diagram showing **A,** cross section of herring (Clupeidae) with red muscle stippled; **B,** disposition of large and small white muscle fibers; **C,** terminal innervation of muscle fibers; **D,** cross section of carp (Cyprinidae) showing position of red muscle (heavy stippling) and "pink" muscle (light stippling); **E,** disposition of muscle fibers; **F,** multiple innervation of muscle fibers. (Redrawn with modification from Bone et al., 1978.)

The red fishes operate aerobically and contain myoglobin, which imparts the reddish color and speeds up oxygen transfer from the muscle to the blood; their enzyme system is aerobic. The white fibers contrast in that they have little lipid, low mitochondrial volume (up to 8 percent), and few peripheral capillaries. White fibers are mostly of larger diameter, have little or no myoglobin, and have an enzyme system for anaerobic glycolysis, although there is a range of aerobic capacity of white muscle among fishes. Tonic muscle fibers that have multiple innervation and do not conduct action potentials in their outer membranes have been described in various teleosts and may be present in the shark genus *Scyliorhinus* (Bone, 1989).

As may be expected, the amount of red muscle in a given species depends on the mode of swimming of the species. Greer-Walker and Pull (1975), who examined the cross sections of several fishes cut about two thirds back toward the tail, found that most species examined had 5 to 15 percent of the cross-sectioned area in red muscle. A few families were found to have more than 15 percent of the cross-sectioned area in red muscle. These included Sparidae, Clupeidae, Carangidae, and Scombridae. Some, including those that mainly utilize the fins for swimming, had 0 to 5 percent red muscle in the cross section at the site. Included were Chimaeridae, Labridae, Caproidae, and Agonidae, plus others. Gill et al. (1989) showed that the percentage of cross-sectional area made up of red muscle in five freshwater fishes sectioned behind the anal fin varied with general habit. The yellow perch, which cruises in schools, had a red muscle percentage of 15, whereas two esocids, which are lurking predators, had only 4 to 5.5 percent. An embiotocid (*Cymatogaster aggregata*), which habitually swims by means of the pectoral fins, has pectoral muscle composed of 90 to 95 percent red fibers (Webb, 1975).

The winglike pectoral fins of skates and most other rays have muscle bundles arranged in two layers, deep and superficial, both above and below the elongate fin rays. The deep bundles next to the fin rays originate on the rays or on the pectoral girdle and attach via tendons to more distant parts of the fin rays. Superficial fibers originate on the pectoral girdle or connective tissue of the deep fiber tendons. Red muscle is located in the outer of the superficial bundles, with red fibers of small diameter extending into the deep bundles.

Muscles of the fins are derived from the myotomes but usually do not correspond with body segments in adults. Carinal muscles on the dorsal and ventral midlines serve mainly as protractors and retractors for the dorsal and anal fins (Fig. 18–4). A thin muscle called the anterior supracarinalis passes from the skull, usually from the supraoccipital, to the first pterygiophore of the dorsal fin. A posterior supracarinalis connects the last pterygiophore of the dorsal with the posterior neural spine or caudal fin supports. Paired anterior infracarinales stretch between the cleithrum and the pelvic bone along the ventral mid-line. Another thin muscle, the infracarinalis medius, passes from the pelvic bone to the first basal pterygiophore of the anal fin, whereas the posterior infracarinalis connects the anal and caudal fins. The caudal fin musculature (see Lauder, 1982, 1989) is dorsoventrally asymmetrical, as is the skeleton of that structure. Important elements of the musculature are the lateralis superficialis muscles, which can move the fin laterally; various flexors, which attach to the fin rays; and the hypochordal longitudinalis, which connects the hypural skeleton to the fin rays of the dorsal portion of the caudal. This complex system can curve the fin rays and change the span of the fin.

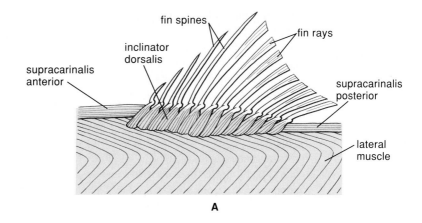

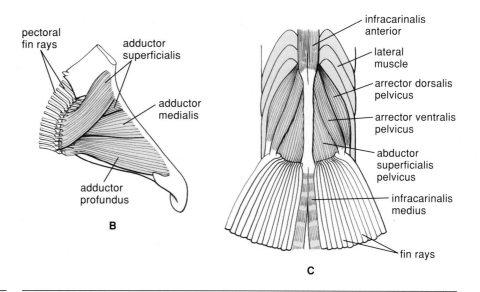

FIGURE 18–4 **A,** Diagram of inclinator muscles of dorsal fin; **B,** medial aspect of pectoral fin muscles; **C,** ventral view of pelvic fin muscles. (**A,** based on Greene and Greene, 1914; **B,** redrawn from Winterbottom, 1974; **C,** based on Greene and Greene, 1914.)

Each median fin ray has a set of erectors and depressors. Soft fins also have inclinators capable of bending the rays. The musculature of the paired fins (Fig. 18–4**B**) is comprised of abductors, adductors, and arrectors, with some fibers attaching on individual fin rays or their basals to give the fin great flexibility. This is especially true of the pectoral fins in some fishes, such as surfperches and wrasses, that use them as a major propulsive unit.

The numerous moving parts of the jaws and branchiohyoid apparatus and the opercular series require complex sets of muscles that may vary significantly in structure among fishes of different phylogenetic lineages and ecological habits. Examples of these muscles are shown in Figure 18–5. The adductor mandibulae, which close the mouth, are the largest of the head muscles and attain a proportionately large size in those fishes that have crushing teeth or those that bite chunks out of prey organisms. For instance, the jaw muscles of the wolf-eel (*Anarrhichthys*) are remarkably large, and the posterior portion of the cranium is greatly compressed and smooth, providing a large place of attachment for the adductor mandibularis.

Opercular musculature is well developed in most fishes, especially those of sedentary habits because of the need for opercular movements to irrigate the gills. Bottom fishes such as sculpins and catfishes generally have better developed branchiostegal muscles than do active species; but swift fishes such as tunas, which depend on continuous swimming movement for gill irrigation (ram ventilation), tend to have small opercular and branchiostegal muscles.

Innervation of muscle fibers varies in the several groups of fishes (Agarkov et al., 1976; Bone, 1988, 1989; Johnston, 1981). Red fibers have multiple innervation in all groups. The various innervation patterns are characterized as focal, multiple, terminal, and dual. Focal innervation is at one site on the muscle fiber. In multiple innervation, two or more axons supply nerve fibers at several places along the muscle fiber(s). Terminal innervation is at one or both ends, and dual is innervation by two axons at adjacent sites at one or both ends of a muscle fiber (Bone, 1989). Dual innervation helps with synchronous nerve firing in sudden, fast muscle activity.

Lampreys have only part of the fast muscle fibers innervated because they are coupled electrically to each other and to the intermediate fibers. The fast fibers have focal innervation, and the slow fibers have multiple innervation in both lampreys and hagfishes. However, in the former, two axons supply each of the innervated fast fibers (Bone, 1989). In elasmobranchs, white muscle fibers are usually innervated terminally or focally. Osteichthyes other than teleosts have terminal innervation of white muscle.

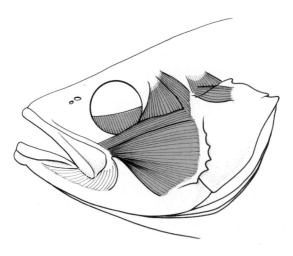

FIGURE 18–5   Examples of jaw and opercular muscles (*Sebastes melanops*).

Among the teleosts, some of the lower groups have terminal innervation. These include Clupeiformes, Anguilliformes, Gonorynchiformes, Hiodontidae, Alepocephalidae, and some catfish families. Acanthopterygians have multiple innervation (Hudson, 1969; Johnston, 1981). Nerve branches from two or more spinal nerves run in myosepta and extend fibers out onto the surface of the myotomes, where sometimes complex nerve endings innervate the muscle fibers (Figs. 18–3C and **F**).

Electromyographic studies indicate that white muscle fibers propagate action potentials but that red fibers show local nonpropagated activity (Bone et al., 1978). In slow, sustained swimming, only red muscle fibers are used by most fishes, although in the species that have pink or intermediate fibers, some of these may be recruited in sustained locomotion. At moderate speeds, there is increased recruitment of pink fibers. In addition, there is some evidence that white fibers may participate to a small extent in slow swimming and to a greater extent at moderate speeds. If burst speed is necessary, the white or fast fibers are used (Bone et al., 1978; Johnston, 1983). Because of their oxidative metabolism and ample capillary network, the red muscles are efficient in slow contraction and can operate for long periods without fatiguing. Pink muscle has high levels of both oxidative and glycolytic enzymes, is fairly fast in contraction speed, and resists tiring more than white muscle, maintaining efficiency at a high rate of work. Burst speed using the white muscle can last only a few minutes in most fishes before fatigue sets in. The highest possible speeds (about 25 lengths per second in fishes about 10 cm long and four lengths per second in fishes about 1 m long) can last only a few seconds. Trout are reported to use 50 percent of stored muscle glycogen in about 15 seconds. Conversion of glycogen to lactate is rapid, but in the white muscle the lactate level decreases very slowly, taking up to 18 hours. Lactate can be oxidized in the gills and apparently can supply part of the energy used in ion exchange. Recovery in red muscle is relatively rapid and takes an hour or less (see Batty and Wardle, 1979; Bone, 1975; Johnston, 1981).

Modes of Locomotion

Swimming

Modes of swimming, whether mainly by body movement or fin action, have been studied at least from the time of Aristotle, and facts and fallacies have built up over the years. Modern instrumentation and cinematography are of great help in winnowing out the fallacies and are enabling scientists to obtain a clearer understanding of the swimming of fishes, although all pieces of the puzzle are not yet in place. Typically, fish swim by undulating the body, sending waves generated by serial contraction of the myotomes from head to tail. The series of muscle contractions alternate from one side to the other, forming the curvature that pushes against the water to generate forward thrust (see Weihs, 1989). Fins are involved in swimming as stabilizers, rudders, and brakes or, in some species, as a means of propulsive locomotion. Fin-powered locomotion is supplemental in some species but the major mode of swimming in others. Although there are many fishes that employ more than one manner of swimming and there is a continuum between those that use full-body undulations and those that use only caudal fin movement, a series of categories describing swimming modes have been useful. These categories are named after fishes that exemplify the particular mode of swimming but have no phylogenetic significance. The terms used here follow Lindsey (1978) and Breder (1926).

Anguilliform Swimming. The type of swimming seen in lampreys, hagfishes, some sharks, eels, larvae of many species, some elongate flatfishes, some blennioids, and other thin-bodied fishes is called anguilliform (Fig. 18–6**A**), after the eels, the best-known practitioners of the mode. In anguilliform swimming, the entire body undulates and more than one wave is present at once (the wavelength is less than the body length). The specific wavelength, which is the wavelength of the waves of the body divided by the body length, is less than 1 (Blake, 1983). Many of the eels, prickle-backs, loaches, and others that swim in the anguilliform fashion are benthic in habit but swim off the bottom for short periods. Eels of the genus *Anguilla* must migrate thousands of kilometers when moving to their spawning grounds. Some pelagic fishes have the slender, elongate shape typical of eel-like swimmers. Sauries and sandlances are examples of surface and shallow water species that swim in the anguilliform fashion. Mesopelagic eelpouts are reported to use that mode, and there is little reason to doubt that other mesopelagic and bathypelagic fishes of slender shape do the same. There are several pelagic fishes of elongate shape among the Stomiiformes, especially in the genera *Chauliodus, Stomias,* and *Idiacanthus* and in the family Melanostomiatidae. The pelagic snipe eels, gulpers, and swallowers are of greatly elongate shape. Other examples of pelagic fishes that seem morphologically suited to anguilliform swimming can be found among the myctophiforms. Anguilliform swimmers with compressed bodies or with dorsal and anal fins that significantly increase the height or span of the fish are more efficient than those of cylindrical bodies or those that are tapered toward the tail. Fishes using this mode are generally slower than those using the subcarangiform mode, which is next in the sequence of swimming types and into which the anguilliform mode grades.

Subcarangiform Swimming. Fishes with a thick forebody have reduced flexibility forward, and undulations are mainly confined to the posterior part of the body (Fig. 18–6**B**) so that usually less than one wavelength is present. The designation of this type of swimming as subcarangiform indicates that it is advanced beyond the anguilliform but not to the level of the speed and efficiency seen in some of the fast swimmers exemplified by the jacks of the family Carangidae. Subcarangiform swimmers are typically of compressed fusiform shape but may vary from fusiform with nearly rounded caudal peduncle to rather compressiform species. Many have a rounded forebody with little flexibility and more compressed mid-body and caudal peduncle.

Caudal fins of subcarangiform swimmers tend to be truncate, rounded, or emarginate, but some stream-living catfishes and cyprinids, for example, have deeply emarginate to forked caudal fins. The caudal fin is of great importance to most fishes that swim in nonanguilliform modes. The caudal has considerable versatility and flexibility and is used for fine as well as coarse changes in direction and power. Loss of the caudal fin in most active subcarangiform swimmers handicaps the individual in maneuvering and in making fast starts and turns, but performance in straight-line swimming in some species is not greatly hindered (Lindsey, 1978). Many species have median fins that effectively increase the body depth and aid in the production of forward thrust. Some of these fins can be depressed for rapid straight swimming and can be erected to accomplish tight turns. This is true espe-

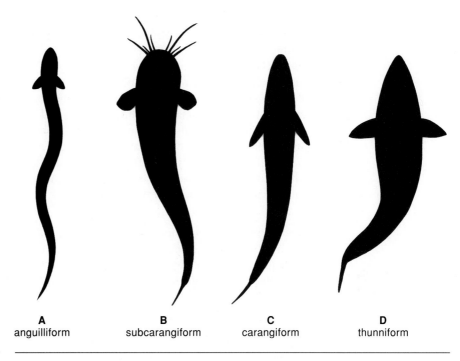

A	**B**	**C**	**D**
anguilliform	subcarangiform	carangiform	thunniform

FIGURE 18–6   Four modes of swimming. **A,** Anguilliform; **B,** subcarangiform; **C,** carangiform; **D,** thunniform.

cially of spinous dorsal fins that are set rather far forward. In some species, the dorsal fins are divided into two or three sections and are not completely depressible. The undulatory nature of the swimming is evident, and usually the arrangement of the vertical fins as stabilizers is not efficient enough to prevent the head from yawing back and forth as the fish progresses. Yawing is less pronounced in the subcarangiform swimmers than in those that employ the anguilliform mode. Subcarangiform swimmers include trouts, cods, goldfishes, basses, and many other familiar fishes.

Carangiform Swimming. In this mode, the anterior one half to two thirds of the body is not very flexible and is bent only slightly during swimming. The flexure that provides forward thrust develops mainly in the back third of the body (Fig. 18–6C) so that usually up to one half a wavelength will be present (Blake, 1983). The posterior part of the body usually tapers sharply to a narrow caudal peduncle and then flares to a strongly forked or even lunate caudal fin. This shape avoids the possibility of strong thrust in advance of the caudal, which transmits the locomotory power to the medium. The length of the wave developed on the body is less than the body length in most of the carangiform swimmers, and usually less than one half of the wave can be present on the body. The caudal fin, which is forked or lunate, has a high vertical span in relation to its area and has what is termed a high aspect ratio, which is the square of the span of the fin divided by the surface area of the fin (Fierstine and Walters, 1968). Aspect ratios among the carangids are typically values of about 3.5. This is somewhat greater than seen in subcarangiform swimmers such as largemouth bass (1.5) or brown trout (about 2). Some of the minnows with forked caudal fins have aspect ratios over 2.0, and adult Pacific salmons have ratios of about 2.7, approaching the carangiform range. Usually the caudal fins with high aspect ratios are less flexible than those with lower ratios and cannot be controlled as well by intrinsic muscles. The functions of guiding and adjusting small direction changes, as seen in the more flexible tail fins of the subcarangiform swimmers, must be taken over by other fins. Although some carangiform swimmers, such as the herrings, show few specializations, others may have the pelvic fins and a spinous dorsal fin moved forward close to a vertical line run through the origin of the pectorals; may have one or more finlets following the dorsal and anal fins; and may have lateral keels developed on the caudal peduncle. The carangiform type of swimming is found in a variety of fish families, from fishes such as herrings and sardines to more advanced fishes including carangids and some of the scombrids.

Thunniform Swimming. This mode of locomotion has received a great amount of attention from scientists because of the speeds achieved and the interesting adaptations of the thunniform swimmers. Relatively few species swim in this fashion, but they include some of the largest teleosts and some of the large sharks. There are remarkable physiological and circulatory adaptations involved with sustained swimming in these fishes, such as the heat-conserving rete, which increases body heat up to as much as 20°C above ambient (Carey and Gibson, 1983), as well as modifications for streamlining.

This mode of swimming may be an improvement in efficiency over the carangiform mode in that the great bulk of the body musculature is used mainly to undulate the extremely narrow caudal peduncle and the high, thin caudal fin to impart tremendous thrust without causing much yawing movement of the body or head (Fig. 18–6**D**). The aspect ratio of the caudal fin ranges from around 4.0 to over 8.5 in tunas and up to 10 or so in marlins and sailfishes. Some control of the span of the caudal fin can be exercised in the tunas, but the stiff caudals of the marlins and sailfishes can be changed little if any by intrinsic muscles. In addition, the lamnoid sharks that swim in this mode have caudal fins with a virtually fixed span. The caudal fins of the thunniform swimmers constitute a powerful propeller that can drive the body forward at great speed or can efficiently move the fish over long distances at moderate speeds. About 90 percent of the locomotor thrust is contributed by the caudal fin. In tunas, the vertebral column is relatively stiff, as is the caudal peduncle. There is, however, a joint at the anterior of the peduncle that allows flexure. Scombrids have posterior oblique tendons that extend posteriorly from myoseptal fibers of epaxial and hypaxial myotomes (over from three to about seven vertebrae, depending on species) before inserting on a vertebra. The first peduncular vertebra is the site of the posterior-most attachment of posterior oblique tendons (Fierstine and Walters, 1968; see also Westneat et al., 1993). By means of this connection, the power of the body musculature can flex the stiff peduncle from side to side (Lindsey, 1978). There is another joint at the end of the peduncle between that structure and the caudal fin. The great lateral tendon extends along both sides of the peduncle and inserts on the caudal fin-ray bases. Various tendons extend past this postpeduncular joint (Lindsey, 1978). The caudal fin is greatly stiffened by overlap of the hypural plate (ural fan) by the strong bases of the fin rays (Weihs, 1989). In species with a bony peduncular keel, the large tendon that has its origin in extensions of the myosepta of the posterior part of the body musculature runs over the keel much as a line runs over a pulley (Lindsey, 1978). This great lateral tendon and others from various groups of posterior myomeres insert on the bases of the caudal fin rays (Fig. 18–7). The placement of the tendons in relation to the joints is such that the angle of attack of the fin—the angle between the fin surface and the direction of movement of the fish—will be close to optimum for generation of thrust.

In many thunniform swimmers, the fins other than the caudal can be depressed into grooves or recesses that make the surface of the fish "clean" from the standpoint of streamlining. Jaw bones fit neatly into fairings, and the bulge of the eye is streamlined by "adipose eyelids." The rather stiff and, in some instances, large pectoral fins can be extended and employed to provide lift and steering. The dorsal fin, which in tunas has a high anterior portion well in advance of the middle of the fish, can be erected and used in quick maneuvering.

Ostraciiform Swimming. This kind of swimming is named for the boxfishes of the family Ostraciidae, which cannot bend the bone-covered body at all. They swim by oscillating a slightly flexible caudal fin on the end of a short caudal peduncle so that practically all the forward thrust is contributed by the caudal fin, although in some species each stroke of the caudal is countered by simultaneous strokes of the dorsal

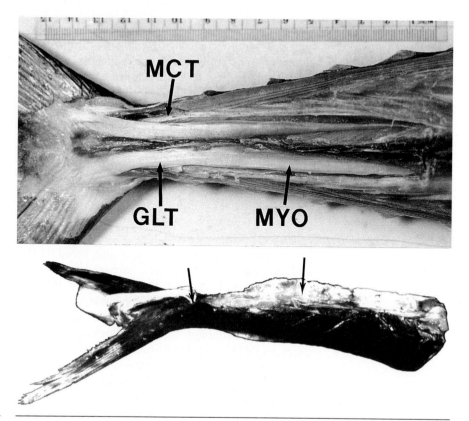

FIGURE 18–7   **A,** *Scomberomorus cavalla* caudal peduncle dissected to show great lateral tendon (GLT), median caudal tendon (MCT), and modified posterior myomeres (MYO); **B,** *Thunnus alalunga,* dissection of caudle peduncle showing keels. Arrows indicate points of flexure. (18–7**A** photograph courtesy of Dr. Mark W. Westneat.)

and anal fins. The caudal fin is swung from the side by alternate contractions of my-omeres on either side. Relatives of the boxfishes that swim in much the same man-ner are the porcupinefishes (Diodontidae) and the puffers (Tetraodontidae). The drag on these species is many times greater than on streamlined fishes, and none of them are fast swimmers. In slow swimming, they commonly use the dorsal and anal fins, but they employ the tail-wagging method when at their highest speed. Few fishes other than those mentioned use this mode of swimming.

Electric rays, because of their laterally inflexible bodies and short caudal por-tion, use a sculling motion in swimming; and those trichiurids with caudal fins, judging from observations on *Aphanopus* (Bone, 1971), might use this method in very slow swimming, such as in stalking prey. Hatchetfishes (Argyropelecidae) have been observed swimming in a manner approximating the ostraciiform mode.

Fin-Swimming Modes. Although most fishes can swim by undulating at least part of the body, the caudal peduncle, and the caudal fin, there are many that use the pec-toral fins or the dorsal and anal fins as their typical means of locomotion at slow or moderate speeds. Those fishes usually hold the body very straight while moving. There are some possible reasons for this, such as minimizing the activity that prey can see when being approached head on or preventing distortion of electric fields in weakly electric fishes that monitor distortions in such fields. Although all fishes do not always conform exactly to the modes that are described, some habitually use a certain fin or combinations of fins so that descriptive terms can be applied. Usually these terms are derived from fishes that typify a given style of fin swimming. For in-stance, many fishes have dorsal fins that extend the length of the back and can swim by sending undulatory waves along that fin to propel the body either forward or backward depending on the direction of the waves. This mode of swimming is called amiiform after the bowfin, *Amia calva*. Other species with long dorsals and small or absent anal fins that swim in this manner are found among the ribbonfishes, the hairtails and scabbard fishes (Trichiuridae), and some sculpins, especially *Nau-tichthys*. The latter has a high spinal dorsal that is slanted forward at about a 45° angle when the soft dorsal is used for amiiform swimming. Other examples include the African *Gymnarchus* and some of its relatives in Mormyridae. These are electric fishes that commonly maintain a straight posture except when swimming rapidly. Other electric fishes that use a single long-based fin for locomotion are the gymno-toids of South America, but these have no dorsal and the anal fin is used. This mode is called gymnotiform (see Lighthill and Blake, 1990). The pipefishes and seahorses have dorsal fins with short bases in relation to the body length, but these flexible fins can undulate with great speed, causing the fishes to move as if powered by an invisi-ble propeller. Although the pectoral fins usually take part in the locomotion, the fishes are said to swim in the amiiform manner.

A great variety of fishes utilize simultaneous undulations of the dorsal and anal fins for locomotion. Because this type of swimming is developed to a high degree in the triggerfishes and filefishes, it is called balistiform (see Lighthill and Blake, 1990). The body is held straight, and undulatory waves traveling down the vertical fins move the fish either forward or backward depending on the direction of the waves. Some variation of this mode is seen in such groups as eels, percoids, flatfishes, and some of the tube-snouted fishes (Aulostomatoidei).

Puffers and boxfishes, along with some of the triggerfishes, can swim by oscillation of the dorsal and anal fins, moving both fins simultaneously toward the same side in what is called the tetraodontiform mode. *Latimeria* is reported by Locket (1980) to use only the second dorsal and anal in slow swimming. The two fins are moved simultaneously in the same direction, and the complete sculling stroke requires twisting of the lobate fins. The pectorals are used for minor adjustments. Observation of *Latimeria* by Fricke et al. (1987) indicated that the animal commonly drifts with currents and uses paired fins for guidance and stabilization. The caudal fin is used for fast starts, and both paired and unpaired fins are used in slow swimming.

Most skates and rays move by undulating the pectoral fins in what is known as rajiform locomotion. The fin movements of the manta rays and eagle rays seem more like the wing movements of birds, but the undulations from front to back, as in the skates, remain as part of the wing flapping. Other fishes also use pectoral undulation as a means of swimming: The chimaeras use the rajiform style, and teleosts such as porcupinefishes, puffers, some of the triggerfishes, and seapoachers employ their broad, vertically based pectorals in the diodontiform mode. Among the perch-like fishes, many species employ narrow-based pectorals as oars or paddles, moving them simultaneously or alternately to provide a propulsive force. In most of these fishes, the base of the pectoral is high on the side and angles forward so the base is not quite vertical. This placement must aid in feathering the fin on its forward stroke. This mode is called labriform after the wrasses. Members of groups other than percoids and their close relatives swim in this manner. Sculpins and snailfishes can use this mode, as can some of the characins and the mid-water pelagic (mesopelagic) *Anoplogaster,* a beryciform. The eelpout, *Melanostigma pammelas,* is reported to swim by means of alternate strokes of the pectoral in its slowest swimming.

Obviously, pectoral fin shape varies considerably among the labriform swimmers. Some of the percoids have long, slightly falcate pectorals with which they can swim rather rapidly. Others have shorter and rounder fins and swim more slowly. The sculpins and snailfishes have broad-based fins.

Nonswimming Locomotion

Although swimming is the primary means of locomotion among the fishes, there are many ways in which fishes move from one place to another. Some of these are more than interesting oddities; they enable ecologically important groups of fishes to exploit niches from which obligatory swimmers are excluded. Some of the modes of travel have been described as follows (see Lindsey, 1978): wriggling (snakelike progression), both in the water and on land; using the pectoral fins as crutches (crutching); flipping; skipping; sucking and hitching; gliding; flying; and drifting passively. Probably all the fishes that wriggle through mud or in and around coral and through the interstices of other hard bottom materials are capable of swimming, but their ability to move in a snakelike manner by forcing bends of their bodies against the bottom allows them to live a much different life from the swimmers in the water above them. The slender and colorful snake eels of the family Ophichthidae, of which there are more than 200 species, can be mentioned as excellent examples. Many other groups have similar abilities, although few have the very slender shape of the snake eels. Lamprey larvae, the eel catfishes (*Channalabes*) of the family

Clariidae, some loaches (Cobitidae), *Erpetoichthys, Electrophorus,* and the members of the Synbranchidae are examples of freshwater fishes that commonly move by wriggling. Marine examples include, in addition to the eels, the following families: Myxinidae, Cepolidae, Ophidiidae, Pholidae, Stichaeidae, Zoarcidae, Scytalinidae, and the wolf-eels of *Anarhichadidae.* Additional families of eels that can be mentioned are the snake eels (Ophichthidae), about 250 species; morays (Muraenidae), about 200 species; congers (Congridae), about 150 species; and the extremely slender spaghetti eels (Moringuidae), about six species.

The freshwater eels (Anguillidae) are good examples of fishes that can move effectively in a snakelike manner on land as well as in the water. Sojourns out of the water are known in other groups as well. Some synbranchids are known to travel overland, and some pholids and stichaeids remain hidden in the intertidal area at low tide and do a fair job of wriggling away when disturbed. Many other fishes are capable of terrestrial locomotion using both body and fins as well as special modifications of various structures. *Anabas,* the climbing perch (Fig. 15–4C), has a series of backwardly directed spines on the operculum. These are used in concert with the paired fins and tail as the fish travels over land. In slow progression, the fish remains upright and extends both opercula and hitches forward; but in a panic situation, such as being placed down in the middle of an expanse of concrete, the fish turns on its side and moves very rapidly by means of the operculum and the caudal fin. Examination of single frames of moving picture film taken of this type of locomotion reveals that the fish has remarkable control of the extension of the opercular spines, especially in moving over a low obstacle.

Periophthalmus, the mudskipper (Fig. 20–7), has highly modified pectoral fins that are armlike with "elbows" separating a rather stiff upper part from a more flexible and expanded lower section. In terrestrial locomotion, the pelvic fins give support as the pectorals are both extended forward. The pectorals are pulled back, and the fish proceeds after the manner of a person using crutches. *Clarias,* the walking catfish, moves on land by extending the pectorals and bending the body to thrust first one pectoral spine and then the other forward. Many fishes can move by flipping, using the tail to propel themselves forward. This mode might seem haphazard, but such fishes as the Gobiidae, Eleotridae, and Clinidae can be very precise in their vigorous jumps, leaving a shallow pool and flipping quickly over mudflats or rocks to dive into some haven. Fishes with ventral suckers (Figs. 2–13 **E–G**) or pads that can be used to adhere to the substrate can usually move over wet surfaces by attaching, hitching the body forward a very small distance, and reattaching. Homalopterids and sisorid catfishes (Fig. 2–12A) are examples of fishes that can overcome artificial barriers, such as low concrete weirs, in this manner. Lampreys, with their powerful oral sucking discs (Fig. 2–3A), can move up wet vertical surfaces at waterfalls or dams by the suck-and-hitch method.

There are a few fishes that walk on the bottom or climb among aquatic vegetation. In some of these, the pectoral fins are modified by having the lower rays separate and stiffened so that they can be used to pull the fish along in a crutching or crawling fashion. Examples are the sea robins (Triglidae) (Fig. 2–12**E**) and certain genera of stonefishes, such as *Inimicus* and *Choridactylus.* The batfishes (Ogcocephalidae) (Figs. 2–12**F** and 12–2**C**) have strong pelvic fins placed in advance of the laterally situated armlike pectorals, which are quite strong and are bent as if they have elbows. With these paired fins plus the short-based, mobile anal fin, these fishes can walk over the bottom as a veritable quintiped (Lindsey, 1978). Some relatives of the batfishes, in the family Antennariidae, use their armlike pectoral fins to climb around in vegetation. The sargassumfish, *Histrio histrio* (Fig. 19–3C), is one of the best-known examples. Various flatfishes can use undulations of the dorsal and anal fins to crawl along the bottom. Tongue soles, Cynoglossinae, have been described as traveling in a millepede-like manner, leaving obvious tracks in the substrate (Clarke and Pearcy, 1968).

Many strong swimmers can leap significant distances. Milkfish (*Chanos*), mullets (*Mugil*), needlefishes (Belonidae), tunas (Scombridae), and various silversides (Atherinidae) are among the fishes known as great jumpers. Eagle rays (Myliobatidae) and mantas (Mobulidae) often leap and return to the water with the flat body cupped to make a loud noise. Leaping activity can be to escape predators, to descend on floating prey, to remove attached parasites, or for reasons unknown to us, including "play" activity. Fishes ascending streams jump over low barriers or falls. The Salmonidae are noted for such ability.

Several kinds of fishes have been reported to skitter over the surface in a kind of richochetal progression by taking short jumps or by applying the force of fins or body to the surface and skipping without full submergence. Individuals that are injured or are in danger from predators sometimes skitter across the water. Lindsey (1978) has reported an interesting set of morphological modifications related to skittering in the genus *Chela*. These Asian cyprinids can bend the head upward to nearly a 90° angle with the dorsal surface. This "neck bending" is accompanied by a downward thrust of the pectoral fins, so that if the fish is at the surface, a rapid succession of the neck bending aids in skittering across the water.

The needlefishes (Belonidae) and the halfbeaks (Hemirhamphidae) have many species that move along the surface with short successive jumps or actually keep the body above the surface while propelling themselves with the caudal fin. Some of these species have elongated lower lobes of the caudal fin. Some have enlarged pectoral fins and are capable of gliding short distances. *Euleptorhamphus longirostris* has been given as an example of a halfbeak that can glide (Myers, 1950). Within the species of the two families (about 60 halfbeaks and 25 needlefishes), there is a progression of adaptations toward gliding flight. Many of these surface-living species that have the gliding adaptation partially developed are successful in their habitat.

Gliding flight is best developed in Exocoetidae, the flying fishes (Fig. 2–12G), members of which can make flights that cover 50 m at speeds of up to 2500 cm s^{-1} (about 90 km h^{-1}) (Rayner, 1981). These flights are generally close to the water's surface, but heights of over 5 m can be reached. Flights can be prolonged by propulsion with the enlarged lower lobe of the caudal fin. Flying fishes accelerate from swimming speed to gliding speed by rapid vibration of the caudal after the body is free of the water. Single flights of up to 400 m have been recorded. Most exocoetids have only the pectoral fin greatly enlarged, with moderate enlargement of the pelvic fins, which are used for maintaining trim or balance. In "four-winged" species, the pelvics are very large and add greatly to the area of the gliding membrane. This decreases wing loading (the weight supported by the unit area of gliding surface) and can result in slower flight because speed of gliding is proportional to the square root of wing loading, according to Rayner (1981).

The Gasteropelecidae of South America, which are called hatchetfishes or "freshwater flying fishes" (Fig. 2–12**C**), not only have enlarged pectorals but have pectoral muscles that make up as much as 25 percent of the weight of the fish. These fishes are known to leap as far as 5 m (Rayner, 1981). Pantodontidae (Fig. 2–12**B**), of African fresh waters, has but one species, *Pantodon buchholzi,* called freshwater butterfly fish or African flying fish. It has enlarged pectoral fins and muscles arranged to move the pectorals forcefully up and down (Greenwood and Thompson, 1960). The species can jump out of the water to escape predators, using "specialized aerodynamic surfaces (pectoral fins) to maximize the distance traveled" (Blake, 1983, pp. 177, 179). Rayner (1981) reports that *Pantodon* can catch flying insects. Whether or not fin flapping is involved in these "flights," which may be as far as 2 m, is unknown.

Other "flying" fishes that should be mentioned are the flying gurnards, Dactylopteridae. These bony-headed, rather clumsy-looking fishes have enlarged pectoral fins and are known to be primarily benthic. The pectorals have been described as being too weak to support the fish in gliding flight, and reports of gliding appear to be mistaken (Blake, 1983). Nonetheless, there are records of dactylopterids being found on decks of boats and reports of them leaping from one tank to another when held in captivity.

Hitch-hikers among the fishes are mainly lampreys (Petromyzontidae) and remoras (Echeneidae). Trichomycterid and other catfishes that parasitize larger fishes should also qualify as hitch-hikers. In some western North American rivers, lampreys attach to salmon on their upstream spawning migration. Remoras attach to large fishes or turtles and are carried along wherever their hosts travel.

Passive drifting is the mode of transportation of the larvae (and floating eggs) of many species. Molas of great size are observed drifting passively at the surface, and the young of tripletails (Lobotidae) are known to emulate drifting mangrove leaves. Adult tripletails are often observed floating on their sides at the surface (Manooch, 1984). Fishes with weak powers of locomotion or those that can gather food with a minimum of swimming may stay in a moving water mass, and although their activity is directed only toward food gathering, they may ride the current across oceans. The sargassum fish lives in drifting seaweed.

Although jet propulsion, caused by forcing water out of the gill chamber through the opercular opening, is a possible means of locomotion, it does not appear to be effective in most instances (Lindsey, 1978). There is no doubt that some forward movement occurs. In observation of fish in aquaria, one can see fish compensate for the force created by opercular jets by movement of the fins. Scientists who have made observations from submersibles at mid-depths in the ocean have noted various species in apparent lethargy oriented with the head up (see Barham, 1971; Clarke and Haedrich, 1968; Clarke and Pearcy, 1968; Clarke and Rosenblatt, 1968). The groups include Bathylagidae, Gonostomatidae, Myctophidae, Paralepididae, Nemichthyidae, Regalecidae, and Trichiuridae. Although the last four families were noted as using activity of the fins or body to maintain position, the myctophids apparently overcome their slight negative buoyancy by strong pumping contractions of the operculum every 2 to 4 seconds (Barham, 1971). There is speculation and some observational evidence that myctophids may migrate to the surface and back by means of opercular movements (Barham, 1971); they have been seen oriented head downward during the time that migration back to depth occurs. In addition, fishes with capacious gill cavities—various sculpins, for example—eject water forcibly when making a fast start. Ejection of water from the opercular openings may have some importance in reduction of drag, as will be discussed later.

Swimming Speed

Because of the dense medium through which they must move, fishes cannot match the sustained speeds that some terrestrial and aerial animals can attain. The highest speeds of fishes can last only a few seconds or minutes before white muscles are fatigued and rest is required. High burst speeds are typically employed in escape responses, pursuit of prey, or overcoming strong currents.

In burst acceleration, fishes often make what investigators have called C starts and S starts. The C starts are "startle" starts involving the Mauthner neurons. In C starts, the fish abruptly bends its caudal region to one side to form a C or L shape and then swings it back in the first of a short series of power strokes (Fig. 18–8**A**). This type of start may give the fish a direction slightly different from its original course, so that compensation is required. More elongate fishes make similar starts after throwing the body into an S curve (Fig. 18–8**B**). There is speculation that fishes such as pikes, with dorsal and anal fins set close to the caudal fin, enjoy a "double tail" effect in increasing the speed of rapid starts (Weihs and Webb, 1983).

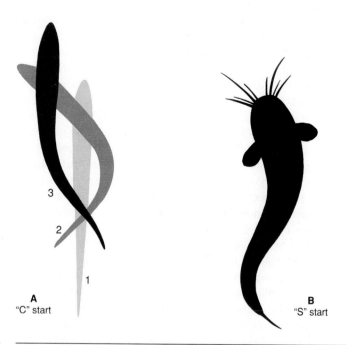

A
"C" start

B
"S" start

FIGURE 18–8   Fast starts; **A,** fish in "C" start—light outline, fish at rest; medium outline, fish in C flexion; dark outline, fish in first power stroke; **B,** fish (*Clarias*) in "S" start.

Increase in swimming speed can be accomplished by increase in frequency or amplitude of body undulation or tail stroke (usually both); although at least one species, the jack mackerel (*Trachurus*), has been reported to maintain a constant amplitude while increasing the frequency (Hunter and Zweifel, 1971).

Burst speeds (Table 18–1) have been measured in some fishes by a variety of methods, ranging from attaching a speedometer to a fishing line to sophisicated methods involving high-speed cinematography or sonar. Swimming performance has been measured by forcing fish to swim in flumes or tunnels through which water is pumped at known speeds, and in circular "doughnut" tanks that are rotated at selected speeds. Some information has been gained from experiments or fortuitous observations involving stream speed or speed of water in culverts. Burst speed in anguilliform swimmers seems to be generally much slower than in subcarangiform, carangiform, or thunniform swimmers, although comparatively few data are available for eel-like swimmers. Whether speed is expressed in a relative term such as lengths per second (Ls^{-1}) or in absolute terms (cm s^{-1}), the anguilliform swimmers are apparently slower than most of the other swimming types. Burst speed in *Anguilla vulgaris* has been reported at about 2 Ls^{-1}. The eelpout *Zoarces viviparous* can swim at about 3 Ls^{-1}, and the flounder *Platichthys flesus* has been reported to swim at nearly 4 Ls^{-1} (Beamish, 1978). A small midwater eelpout *Melanostigma* sp. has been clocked at 4 to 5 Ls^{-1} for a few minutes (Belman and Anderson, 1979). Small fishes can swim at more lengths per second than larger ones, so of the aforementioned fishes, the eel swimming at 2 Ls^{-1} was traveling at about 115 cm s^{-1}; whereas the eelpout and gunnel, at 3 Ls^{-1}, were moving at 30 cm s^{-1} or less. The eel's speed equates to about 4 km h^{-1}.

Much attention has been given the subcarangiform swimmers in the family Salmonidae. Several members of the genera *Salmo* and *Oncorhynchus* have burst speeds of from 6 to 10 Ls^{-1} and actual speeds of 300 to over 500 cm s^{-1}. Rainbow trout have been clocked at over 1000 cm s^{-1} (about 36 km h^{-1}) in an instantaneous burst of speed, but when followed over 10-second bursts the speed was about 200 cm s^{-1}. Other subcarangiform swimmers, such as carp, suckers, and cods, show burst speeds of 100 to 300 cm s^{-1}, swimming at 5 to 8 Ls^{-1} (Beamish, 1978). Among carangiform swimmers, bluefish (*Pomatomus saltatrix*) can swim at about 13 Ls^{-1} and 200 cm s^{-1}, and the jack mackerel (*Trachurus symmetricus*) has shown speeds up to 7.5 Ls^{-1} or about 210 cm s^{-1} in 28-cm fish (Hunter and Zweifel, 1971). This species closely approaches the thunniform mode and can maintain locomotion at near burst velocities for a few hours. Tunas and their close relatives that swim in the thunniform matter are the fastest fishes. The wahoo, *Acanthocybium solandri,* has been clocked to over 20 Ls^{-1} and speeds of 2100 cm s^{-1}, or about 75 km h^{-1}. A sailfish, *Istiophorus,* was once estimated at 84 km h^{-1} from the rate at which a hooked fish stripped line from a reel. Wardle et al. (1989) observed swimming speeds of from 1200 to 3200 cm s^{-1} in caged bluefin tuna (*Thunnus thynnus*). That equates to 0.3 to 1 Ls^{-1} and 3.6 to 10.8 km h^{-1}.

Sustained speeds of fishes are difficult to observe and measure in the field. Methods used to measure or estimate speeds include following schools, following individuals tagged with sonic tags, or tagging and recapturing migrating individuals. Because of the long distances that some species must travel while on migrations, efficient use of stored energy is of prime importance so the speed is geared as closely to the resting metabolism rate as possible. In a study of several species of fish (using

TABLE 18–1	Burst Swimming Speeds of Selected Fishes	
Species	**Ls⁻¹**	**cm s⁻¹**
Anguilliform mode		
Pholis gunnelus	3	30
Anguilla vulgaris	2	115
Zoarces viviparous	ca. 3	18–21
Subcarangiform mode		
Oncorhynchus kisutch	ca. 6–10	287–533
O. mykiss	ca. 3–4	186–226
Carassius auratus	9.4–11	74–200
Gadus merlangus	5–9	70–180
Carangiform mode		
Clupea harengus	ca. 6–10	67–131
Scomber scombrus	5–9	190–300
Pomalobus pseudoharengus	14–16	400–500
Trachurus mediterraneus	16.4	258
Thunniform mode		
Acanthocybium solandri	18.4	2100
Thunnus albacares	6–21	523–2072
T. thynnus	0.3–1	1200–3200[1]

Adapted from Beamish, 1978.
[1] Wardle et al. (1982).

individuals of 700 g or less), Belokopytkin and Shul'man (1989) determined that energy cost was highest at 0.5 Ls^{-1} and at 6 to 7 Ls^{-1} and faster, with the lowest expenditure of energy in most of the species studied at 2 to 3 Ls^{-1}. The migration speeds of a variety of species, from sharks to percoids and flatfishes, have been less than 1 Ls^{-1}. Some of the Pacific salmons travel at close to 2 Ls^{-1}. Quinn (1989), using ultrasonic tracking devices, estimated the average swimming speed of sockeye salmon at 1 Ls^{-1} (2.4 km h⁻¹), which is close to the most efficient speeds of about 0.82 Ls^{-1} (1.8 km h⁻¹) determined in the laboratory. Bluefin tuna tagged in the Gulf of Mexico migrated to Norway in an interval of time that was calculated to equal 760 cm s⁻¹ or about 2.25 km h⁻¹ (Wardle et al., 1989).

In discussing maximum swimming speeds, Wardle and He (1989) and Wardle et al. (1988) report that the "stride length" in fishes is the distance moved with one oscillation of the tail, which is governed by the twitch contraction time of the white lateral muscle. In _Scomber scombrus,_ which has a stride of nearly one body length at a tail beat of 18 Hz and a resulting speed of 550 cm s⁻¹, twitch contraction time was measured at 26 ms and a top speed of 590 cm s⁻¹ was predicted. In a bluefin tuna of 2.26 m, twitch contraction time was 50 ms and tail beat frequency was 10 Hz. At an average stride of 0.65 body length (=1.46 m), a speed of 54 km h⁻¹ could be achieved. Theoretically, if a stride of one body length could be used at high speed, the resulting velocity could be 81 km h⁻¹.

Nonmigrating fishes traveling in schools tend to move faster than migrating fishes. Apparently, migrating fish must maintain speed that is energetically efficient over long distances. Herring swim from less than 1 to nearly 8 Ls^{-1} while schooling. Schooling tuna are known to swim at from 2 to 15 Ls^{-1}, and there is at least one report of 21 Ls^{-1} (Beamish, 1978).

Many laboratory studies have been conducted to determine "critical swimming speed"—that is, the maximum speed at which fish can swim for a selected interval of time (Jones, Kiceniuk, and Bamford, 1974). These tests have most often been carried out by placing the experimental animals in tubes through which water can be pumped at desired speeds and increasing the speed at regular intervals until the fish fail to swim (see Beamish, 1978). Fishes tested have primarily been subcarangiform swimmers of the family Salmonidae, and most of them have been less than 20 cm long. Typical swimming intervals range from 5 to 30 minutes, and velocities range from about 2.5 to 10 Ls^{-1}. This type of study can be useful in assessing the ability of fishes to swim under stress from pollutants, lowered dissolved oxygen, or disease.

Drag Reduction

Investigators studying the hydrodynamics of fish swimming have speculated that a vortex sheet forms behind the trailing edge of the vertical fins and is absorbed by succeeding fins, the last dorsal or the anal passing the vortex sheet on to the caudal fin, from which it is shed from the fish (Blake, 1983; Webb, 1978; Weihs, 1989). Blake (1983) states that this can establish a flow pattern that could be like that caused by a continuous fin, but with less friction drag.

Most fishes that swim at moderate to fast speeds are well streamlined. They have a somewhat conical head that, with the pectoral region, forms a comparatively short "entering" section forward of the point of greatest body diameter. This is followed by an afterbody that tapers to the caudal peduncle. As a body of such shape thrusts against the water, there is typically positive pressure on the entering section and slightly negative pressure along the afterbody. Part of the energy expended by the fish goes into overcoming the pressure drag of any turbulent wake that is formed behind the caudal fin. There is also a frictional component of drag that has to do with the passage of the water over the skin of the fish. There is a thin boundary layer of water next to the skin. Water in this layer can flow smoothly over the skin in a laminar fashion in very small fish or in larger fish as they move slowly. As size and speed are increased, laminar flow in the boundary layer changes to turbulent flow and increases drag. If the turbulent layer separates from the skin, a turbulent wake can result and drag increases. Scientists studying hydrodynamics of swimming fishes predict the nature of the boundary layer by calculating the Reynolds number (Re), which is an expression of inertial forces/viscous forces (see Birkhoff, 1950; Walters, 1963; Blake, 1983). It involves the length of body (L), velocity (V), and the kinematic viscosity[1] (v) of water: Re = LV/v (Videler, 1993; Yates, 1983). This is a dimensionless number, which can be less than 1 but most often is in the thousands or millions for fishes of appreciable size. Blake (1983) states that for streamlined bodies under steady flow, the boundary layer changes from laminar to turbulent as the Reynolds number increases from 5×10^5 to 5×10^6, passing through a transitional or unstable flow in which turbulence increases.

---

[1]Kinematic viscosity is the ratio of dynamic viscosity over density (Videler, 1993).

Drag on the swimming fish is proportional to the square of the velocity, so a fish that must swim rapidly must exert a tremendous amount of energy to maintain speed, and most fast fish have evolved means of reducing drag as well as means to sustain energy output for relatively fast sustained speeds. The drag on a flexible, somewhat compliant fish body can be two to three times the drag on a rigid body of the same size and shape. This drag is increased by the undulations of the swimming fish's body, as flow crosses over from one side to the other and causes separation of the boundary layer. Some fishes minimize that effect when they utilize what is called "kick and glide" or "burst and coast" swimming, in which they accelerate to a given speed or swim up to a given depth and then hold the body rigid and glide for a distance. In fishes heavier than water, the pectoral fins are used as gliding planes to produce lift, but the decreasing speed allows them to sink, so their progression is based on swimming up and gliding down. The saithe (*Gadus virens*) is reported to accelerate up at about 10 Ls^{-1} and then coast down, achieving an average speed of 5 Ls^{-1} (Videler and Weihs, 1982). Those fishes with nearly neutral buoyancy can maintain a nearly constant depth in kick and glide swimming. A gliding fish generally has about one third the drag of a steadily swimming fish at the same speed (Videler, 1981; Weihs, 1974).

The tunas and tuna-like scombroids have excellent morphological characteristics for reducing drag. The maximum body depth is usually at least 60 percent of the body length behind the anterior point (Weihs and Webb, 1983). Their streamlined bodies are shaped so that the generation of adverse pressure gradients is delayed and laminar flow is maintained. They present a clean surface that aids in maintaining a laminar or at least an attached turbulent flow along the skin. The body is held nearly straight during swimming, as most lateral movement is confined to the caudal peduncle and the high-aspect-ratio caudal fin. The caudal peduncle is equipped with flexible finlets that are thought to direct the flow across the peduncle in a manner that minimizes separation of the boundary layer (Lindsey, 1978). The peduncular keels might aid in maneuvering. Ships with such keels are reported to have a much tighter turning radius than those not so equipped (Watts, 1960). Some investigators believe that the corselet of scales that roughen the forebody of some tunas in a characteristic pattern help maintain an attached, if turbulent, boundary layer, for any attached boundary layer produces less drag than if that layer were separated. There seem to be conflicting views as to whether the water flow that results from ram irrigation of the gills helps keep an attached turbulent boundary layer or encourages separation. Some nonscombroids have slots behind the opercular opening that would appear to cause advantageous direction of the branchial flow.

Many fishes present a roughened body surface to the medium, and many of these are rougher than the corselets of tunas. Ctenoid scales with spines that protrude through the mucous coating are encountered on a variety of teleosts. Roughness of any protrusions (Fig. 18–9) that project through the boundary layer can cause small vortices to form and encourage a turbulent boundary layer to remain attached so that drag will be reduced. Some examples of fishes that have special modifications of the scales to provide an especially rough or patterned exterior are the roughscale pomfret, *Taractes asper;* the catalufas, Priacanthidae; the squaretails, *Tetragonurus* spp.; and the oilfish, *Ruvettus pretiosus.* The oilfish has spaces beneath the skin filled with sea water that is injected to the boundary layer as the fish undulates its body (Bone, 1972). This could aid in keeping the boundary layer attached. A different type of injection system has been reported in some ribbonfishes, Trachypteridae, by Walters (1963), who hypothesized that canals in the skin can accept boundary layer water in areas of high pressure and release it farther along the body in areas of low pressure. This system could have the effect of helping maintain a laminar boundary layer.

Possible reduction of frictional drag by mucus of some fishes by reducing the viscosity of the medium has been reported by Rosen and Cornford (1971) and by Hoyt (1975). Most of the experiments involved study of the flow of water mixed with mucus in a tube. There is, however, much variation in the effectiveness of slime from different species. Videler (1993) states that demonstration of the effectiveness

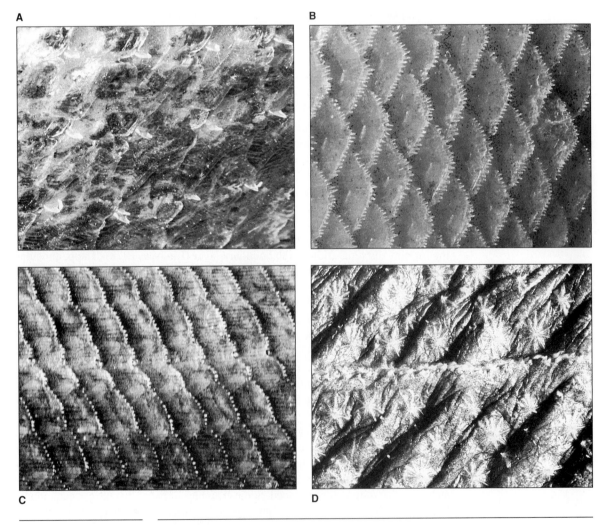

A

B

C

D

FIGURE 18–9   Rough surfaces on fish. **A,** Spiny scales of rough pomfret (*Taractes asper*); **B,** strongly ctenoid scales of popeye catalufa (*Pristigenys serrula*); **C,** ridged and strongly ctenoid scales of small-eye squaretail (*Tetragonurus cuvieri*); **D,** stellate, spinous scales of prickly shark (*Echinorhinus cookei*).

of the mucus in lowering viscosity in a pipe does not prove that the mucus can lower the frictional drag on a swimming fish. However, Videler (1993, p. 84) suggests that the result of experiments by Daniel (1981) with newly killed fish and wax models "leads to the conclusion that mucus plays an important role in overall drag reduction during swimming at low *Re* of at least some species of fish."

Fishes in schools, swimming in a suitable geometric pattern—usually a "diamond" (Weihs, 1973)—can increase swimming efficiency for all but the leaders. Second and subsequent rows in the school have slower tail beat rates than the leaders (Zuyev and Belyayev, 1970).

Buoyancy and the Gas Bladder

In many fishes, the shape of the gas bladder is simple, usually torpedo shaped, but there are many variations. Minnows and carps (Cyprinidae) have anterior and posterior chambers connected by a sphincter. The gas bladder of the featherbacks (Notopteridae) is divided into lateral halves, with the two chambers communicating anteriorly. Many drums and croakers (Sciaenidae) have unusual sexually dimorphic gas bladders, with variously shaped sacs (Fig. 18–10) or branching caeca arranged along each side of the organ.

In the herrings (Clupeidae), the gas bladder has a posterior opening to the exterior near the anus through which gas may be voided. Some fishes have posterior extensions of the gas bladder reaching beyond the body cavity. In viviparous perches (Embiotocidae), for example, the gas bladder extends along the ventral surface of

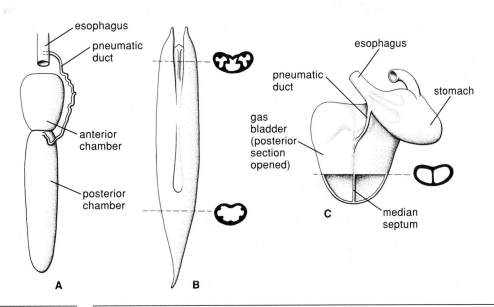

FIGURE 18–10   Examples of gas bladders, ventral views. **A,** Sucker (Catostomidae) showing long and crooked pneumatic duct; **B,** seatrout (*Cynoscion*) showing anterior and posterior chambers in cross section; **C,** channel catfish (Ictaluridae) opened to show median septum (stomach displaced to side).

the vertebrae; whereas in the hairtails (*Trichiurus*), the posterior extension of the gas bladder runs along the concave anterior face of the first interhaemal bone (anal-fin support).

Some bottom-dwelling fishes—such as darters (*Etheostoma*), most flatfishes (Pleuronectiformes), and sculpins (*Cottus*)—that do not require neutral buoyancy lack a gas bladder. Various bathypelagic fishes have lost the gas bladder and achieve buoyancy by other means; the gas bladder is also absent in agnathans and the cartilaginous fishes.

Functions of the gas bladder include hydrostatic balancing, sound production, sound reception, and respiration. Gas bladders for respiration are usually compartmentalized and highly vascular, as in the lungfishes, bowfins, and gars.

Sound production is often accomplished by special vibratory muscles attached to or near the gas bladder. Gas bladders serve in sound reception by acting as a resonator (see Chapter 21).

Buoyancy: Gas Bladder and Retes, Fats, and Oils

Although the gas bladder may have had its origin in the early history of fishes as a lung, its major function among modern teleosts appears to be to help provide buoyancy (lift). Lift is also provided by swimming speed, body shape, and fins, but the gas bladder is the major means of maintaining neutral buoyancy or something close to it. The tissues of a fish generally have densities greater than water. Scales and bone have a specific gravity of about 2.0, and other tissues are in the 1.05 to 1.1 range. Fats and oils have a specific gravity of about 0.9 to 0.93 and tend to decrease the overall density of fishes. The combination of all tissues in a fish without a gas bladder or some other device for maintaining buoyancy causes the specific gravity to be in the range of 1.06 to 1.09. Because fresh water has a specific gravity of 1 and the oceans about 1.026, such fishes are destined to sink or must exert continuous effort to provide enough hydrodynamic lift to prevent sinking.

There are numerous species of demersal fishes that rest on the bottom or even burrow into it. These would be, for the most part, hindered by neutral buoyancy and must have no gas bladder or have it greatly reduced. Sculpins, flatfishes, and clingfishes are examples among the teleosts. Gurnards, which are rather heavily built and live mostly on the bottom, gain lift from large pectorals and a small gas bladder when they swim off the substrate. Sharks and rays have no gas bladders and are negatively buoyant. Most are demersal in habit and must swim constantly to prevent sinking. Large pectoral fins provide dynamic lift in many species of sharks, and hammerhead and bonnethead sharks gain additional lift from the broad head (Bone, 1988). Rays must gain both lift and great physical resistance to sinking from their broad pectoral fins. Many elasmobranchs are aided in maintaining buoyancy by inclusion of much oil and a special hydrocarbon called squalene in their large livers (Bass and Ballard, 1972; Beldridge, 1972). Squalene has a specific gravity of 0.8562. Some sharks have livers that weigh up to 30 percent of the entire weight of the individual, and up to 90 percent of the weight of the liver may be oil (Bone, 1988). Some pelagic fishes in which the gas bladder is small or lacking depend heavily on hydrodynamic lift from pectoral fins, caudal keels, and angle of attack of the bodies.

Although most bony fishes get only negligible buoyancy from fats and oils contained in their bodies, these lipids can be of importance in exceptionally oily species or in those with special inclusions of lipids in the body cavity, liver, or bones. Squalene is present in small amounts in several bony fishes, including *Latimeria,* but most fishes that depend on lipid for buoyancy have wax esters of 0.8578 specific gravity in or around the gas bladder, in the skin, musculature, and even inside the cranium, as in the orange roughy (*Hoplichthys atlanticus*) (Phleger and Grigor, 1990). In *Latimeria,* many myctophids, gempylids, and probably other mesopelagic fishes up to 15 percent of their live weight exists as wax esters. The oilfish, *Ruvettus,* which is charged with oil that has a density of 0.87, is nearly neutrally buoyant. The eulachon, the generic name of which (*Thaleichthys*) means "fat fish," contains about 20 percent lipid, including some squalene. This fish was formerly called "candlefish" because a string wick could be placed in a dried fish to make a crude candle. Lipids may be of relatively greater importance in those species that have reduced bones and watery flesh.

The pelagic *Mola mola,* which sometimes is seen floating at the surface, has no gas bladder, is remarkably watery, and has thin, light bones, a situation seen in numerous deep sea fishes. *Mola* has a rather tough cartilaginous layer beneath the skin. The little lift afforded by body fluids in marine fishes is due to the fact that body fluids have about half the salinity of sea water and are therefore a bit lighter, so that the watery mola is less dense. Marine fishes in which water, on a weight basis, is as much as 86 percent of the tissues gain considerable buoyancy.

Fishes that make their living near the surface, in mid-water, or free swimming close to the bottom gain an advantage from a gas bladder because they are relieved of the necessity of maintaining a chosen depth by muscular effort, which in the absence of a gas bladder amounts to a force of about 5 percent of the body weight in sea water or about 7 percent in fresh water. Gaining and maintaining the advantage of weightlessness is not simple for most species because of the change in pressure with depth, which is about 1 atmosphere per 10 m. A descending fish with a given quantity of gas in the gas bladder becomes less buoyant as the gas compresses and the fish consequently displaces less water. A fish descending from the surface to 10 m has the gas bladder compressed to one half its volume at the surface.

An ascending fish faces a problem that can be very serious: The release of pressure on the gas bladder allows it to expand, and if no relief valve is provided, the other internal organs can be greatly crowded and the stomach can be forced out into the mouth. These drastic results do not generally occur naturally but can be seen in fish brought from deep water by fishermen. Sometimes in a forced ascent, in which the fish is being brought from a sufficient depth, the buoyancy of the expanding bladder can become great enough that the fish cannot overcome it to swim back to the accustomed depth. Furthermore, dragging a fish up from a great depth can result in a ruptured gas bladder. In natural vertical movements, fishes generally do not move rapidly through a great enough depth range to bring about more than a 25 percent change in gas bladder volume. At great depths, the change in pressure in terms of percentage is small for each 10-meter change in depth. For instance, ascending from 100 to 90 m increases the volume of the gas bladder only 10 percent. Those species moving through greater ranges are specially equipped with means of emptying and filling the gas bladder (see Steen, 1970, and Marshall, 1979).

The size, structural modifications, and placement of gas bladders have correlations to the ecology and habits of fishes. Benthic fishes with small or no gas bladders depend on friction with the bottom materials, wedging into tight places, digging in, or holding on with special suction surfaces to maintain position. Certain flatfishes are 5 percent denser than sea water. Those fishes that swim up to capture food above the bottom can derive some benefit from a small gas bladder about 0.3 to 5 percent of their volume.

Most bathypelagic species found below 1000 m lack gas bladders, but benthic fishes found as deep as 7000 m have small gas bladders. At pressures found at the latter depth, the specific gravity of oxygen is 0.7.

Marine fishes that live above the bottom but confine their activities to narrow depth ranges usually have gas bladders of about 5 to 5.6 percent of their volume, compared to about 7 to 10.6 percent in freshwater species. These fishes can hold themselves motionless with comparative ease, using slight fin or body movements to counteract weak current. Many of these species may be slightly heavier than water and sink slowly between adjustments of position. Deep-bodied fishes usually have the gas bladder placed above the center of gravity so that little effort is required to hold the body upright. In some slender fishes, the organ is below the center of gravity and the fish must use fin motion to remain upright. These will go "belly-up" when anesthetized.

Species of fish adapted to flowing water tend to have smaller gas bladders than still-water fishes because the less buoyant condition is favorable to maintaining a given station in the stream. In related stream fishes, those species habitually in the swiftest water have less capacious gas bladders than those that live in slower currents. Furthermore, in laboratory experiments, stream species reared in still water proved to have larger gas bladders than those reared in a current (Gee, 1968, 1972). Within a given species there is a general capability to adjust buoyancy to suit the current encountered. Such adjustment may be made to aid in maintaining a station in swift water (reduction of buoyancy) or to relieve the muscular effort required to maintain position as the current changes to a lower velocity (increase of buoyancy). Over the course of the life history of some species, the relative volume of the gas bladder changes considerably. In the longnose dace (*Rhinichthys cataractae*), the young live in slower-flowing water at stream margins and have a relatively large gas bladder. As the fish become older and move into swift current, the gas bladder grows at a much slower rate than the body of the individual (Gee et al., 1974). In certain anadromous trout and salmon, the young are adapted to a stream life for periods of up to three years or more. During this time, the young are less buoyant than they will be when they begin their downstream migration to the sea. Becoming more buoyant in a current serves as a dispersal mechanism.

Vertical Movement. The problem of vertical movement has been solved by a number of species. Many powerful scombroids lack a gas bladder and govern their depth mainly by rapid swimming, using the pectoral fins for lift when necessary. The larger the pectorals, the less speed that is needed to maintain the lift. Such fishes can make rapid changes in depth. Examples of scombroids without gas bladders are Spanish mackerels and close relatives (*Scomberomorus* spp.), skipjacks (*Euthynnus* spp.), and mackerels (*Scomber* spp.). Larger tunas and their xiphioid relatives (billfishes and swordfish) have gas bladders and consequently can rely on a slower swimming speed than smaller species, although their vertical movements near the surface are restricted unless the gas bladders have extremely tough walls or special adaptation for rapid compensation.

The billfishes, some of which are among the largest teleosts, can make rapid vertical excursions and compensate rather quickly to changing pressure because of the nature of their gas bladders. These are divided into numerous small cells, each of which has its own gas secretion and absorption glands (Robins, 1974) so that pressure can be adjusted rapidly. The bluefish, *Pomatomus,* also has the capability of rapid secretion and absorption of gas (Alexander, 1972).

The yellowfin tuna, *Thunnus albacares,* is interesting in that its gas bladder is not inflated in fish of about 2 kg, and these small fish have a density of about 1.09. Quick growth and inflation of the bladder reduce the density of 10-kg fish to about 1.05.

Many mesopelagic fishes perform daily vertical migrations, so that the greatest concentrations of the species will approach the surface during the night and will be at depths of from 400 to greater than 500 m in the daytime. Others may move from greater depths to a nocturnal depth of about 150 m. The lanternfish family (Myctophidae) has a greater number of species that migrate to the surface at night than the other mesopelagic families. Other families with vertical migrants among the members include Chauliodontidae, viperfishes; Stomiidae, scaly dragonfishes; Astronesthidae, snaggletooths; Melanostomiatidae, scaleless dragonfishes; and Idiacanthidae, stalkeyes. Some families have members that migrate vertically but do not reach the surface. Included are the Melamphaidae, or bigheads, and the Sternoptychidae, hatchetfishes. Certain epipelagic fishes, such as *Clupea harengus* and other herrings, which are usually not neutrally buoyant (Blaxter and Beatty, 1984), are known to go as deep as 150 m and return to the surface in the course of a day, but the herrings can expel gas through a pore behind the anus.

Many of these vertical migrants have gas-filled bladders and face the difficulties inherent in moving captured gas through pressure changes of 15 to 50 atmospheres or more. The problems are greatest for those that move all the way to the surface. Some species have avoided the problems in part by incorporating material of low specific gravity in the body, filling the gas bladder partially or completely with fat, as in the case of the black scabbardfish and certain macrourids (rattails), or by reducing the bladder and investing it with fat, as is found in some of the lanternfishes and anglemouths. Some lipids in or around the gas bladder of lanternfishes have a specific gravity close to that of squalene. For the most part, however, gas is maintained in the bladder by special structures that can secrete or absorb gas (Satchell, 1991; Scholander, 1957). Of course, some physostomous fishes retain the ability to release gas through the pneumatic duct or, in the herrings, through a posterior duct opening behind the anus, but even these must secrete gas into the lumen of the gas bladder if a reasonable hydrostatic balance is to be maintained. There is little evidence indicating that perfect hydrostatic balance is kept throughout vertical migration. Some species of lanternfishes (e.g., *Notoscelopus kroyeri* and *Tarletonbeania crenularis*) are negatively buoyant (Marshall, 1979).

Gas Resorption. Gas bladders are nearly impervious to gas because of the complex structure of the wall. There are usually four layers, the outer one consisting of densely woven but elastic fibers. The next layer is of more loosely organized fibers, and the inner two are smooth muscle and epithelium. In many species, the wall is gas proofed by a layer of guanine crystals just below the outer elastic fiber layer. Tiny overlapping platelets may aid in gas proofing gas bladders of certain poeciliids. Wall thickness is usually in the range of 50 to 300 μm, but exceptions, such as *Aphanopus carbo,* the black scabbardfish, with walls 1.5 mm thick, must be noted (Bone, 1971). This species has the gas bladder tightly enclosed by the ribs (Howe et al., 1980). The problem of removing gas from a gas-tight sac is overcome by special areas in the wall, where a rich bed of capillaries can be exposed to the lumen of the gas bladder. Gas resorption involves diffusion from a high gas tension to a lower tension in the blood, at a rate governed by the tensions, temperature, area of the capillary bed, and rate at which the blood flows through the bed. The capillaries are usually disposed in a subcircular or oval area in the dorsal wall of the gas bladder and can be isolated from the lumen during the nonabsorbent phase of operation by a sphincter. (Because of its shape, the structure is usually called the "oval.") Relaxation of the sphincter and contraction of radial muscles in the oval bring about maximum exposure of the capillaries to the gases in the bladder. There are exceptions to the arrangement described. In some fishes, the capillaries cannot be isolated, but resorption can be prevented by constriction of the capillaries or by thickening of the epithelial lining by muscle contractions.

In the eels of the genus *Anguilla,* the pneumatic duct is modified for resorption of gas (Fig. 18–11**A**). In several percoid fishes, pipefishes, sticklebacks, and others, a diaphragm separates the posterior, or gas-resorbing, part of the gas bladder from the anterior, gas-producing section. Such fishes are called "euphysoclistic," distinguishing them from those "paraphysoclists" in which the gas-secreting and gas-diffusing areas are not well separated (Fig. 18–11**B**). Gas resorption in many euphysoclists is accompanied by contraction of the gas-secreting chamber and may involve anterior displacement of the diaphragm as well as expansion of the resorbent chamber.

The veins leading from the oval or other resorbent structure lead into the cardinal vein in most physoclists. In the eel, blood leaves the resorbent structure via the pneumatic duct vein, which proceeds directly to the heart. Because the blood with its load of gas thus must pass through the gills before being distributed to the systemic circulation, there is an opportunity for excess gas to be diffused into the surrounding water. In considering the rapid ascents made by lanternfishes moving from 500 m to the surface, the resorption process might seem too slow for proper adjustment of gas bladder volume to be made. Certainly the rates of resorption calculated for freshwater fishes would not allow for rapid ascent. However, anatomical studies of the gas resorption apparatus in lanternfishes reveal large areas of capillaries in relation to the volume of the gas bladder as well as capacious arteries and veins serving the oval. Capillaries of the oval are separated from the lumen of the bladder by tissue 1 μm thick or less. Furthermore, most vertical migrants are small, usually less than 100 mm, and have a small gas bladder in relation to their size because buoyancy is often aided by inclusion of lipid.

Gas Secretion. Filling the gas bladder on downward migration cannot be accomplished as easily as the removal of gas. Gas must somehow be secreted from the lower pressure in the blood into the higher pressure in the gas bladder. Oxygen, for instance, found essentially at a pressure of 0.2 atmosphere or less in the blood, must

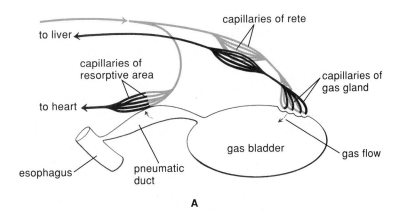

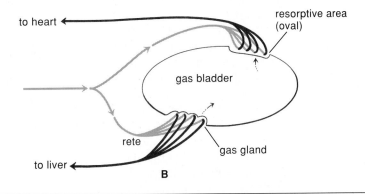

FIGURE 18–11   **A,** Diagram representing gas bladder of *Anguilla,* showing relationships of rete and gas gland and resorptive area on pneumatic duct; **B,** diagram of gas bladder, rete, and resorptive oval of physoclistous fish.

be secreted into the gas bladder to pressures up to hundreds of atmospheres. The apparatus by which the secretion is accomplished consists of bundles of arterial and venous capillaries, closely associated but running counter to each other, so that blood goes to the gas bladder through the arterial capillaries, circulates through the specialized bed of capillaries from which the gases are secreted into the gas bladder, and then flows away through the venous capillaries. The bundles of capillaries comprise the rete mirabile or "wonder net." This name is somewhat inappropriate becase a network is not formed. The parallel alignment of the small vessels forms an efficient countercurrent structure that serves to concentrate gases at the site of the gas gland. Depending on species, there may be from a few hundred to over 200,000 capillaries in the rete.

The process of concentration of gas is aided by the secretion of lactic acid into the blood by the gas gland. This acid increases the partial pressure of carbon dioxide in the blood by releasing CO_2 from bicarbonate, but the greatest effects are on the partial pressure of oxygen. Acid activates the Bohr effect, so higher partial pressures are required to keep oxygen in combination with hemoglobin. The Root effect is operative as the pH is lowered, so the quantity of oxygen that can combine with hemoglobin, regardless of how high the partial pressure, is decreased (see the discussion of Bohr and Root effects, Chapter 23). The Root effect is of great importance in the release of oxygen into the gas bladder. There is, in addition, a "salting out" effect—the acid ions decrease the amount of gas that can be held in solution (Scholander, 1954; Scholander and Van Dorn, 1954).

Because of the aforementioned effects, the blood leaving the gas gland through the venous capillaries of the rete has greater partial pressures of gases, especially oxygen, than the blood in the arterial capillaries. The two sets of capillaries are intimately associated, so that a cross section of the rete would show them in a checkerboard or mosaic pattern. The capillaries of the rete are very long in relation to other capillaries. The length and great number of capillaries allow ample opportunity for the oxygen at high partial pressure in the venous capillaries to diffuse to the blood in the arterial capillaries. Eventually, the pressure in the small vessels is higher than that in the lumen of the gas bladder, so that bubbles of gas are released at the gas gland and serve to inflate the bladder. Despite the considerable amount of research on the subject, the exact mechanisms that allow release of the gas under tension at the gas gland are not fully known (Bone and Marshall, 1982).

The structure and dimensions of retes and gas glands differ among systematic and ecological groups. Some of the upper mesopelagic lanternfishes have retial capillaries 1 to 2 mm long, whereas those living as deep as 1000 m may have retes up to 7 mm long. *Anguilla* has a rete of about 4 mm. *Sebastes norvegicus,* the golden redfish, which lives down to about 600 mm, has retes of 7 to 10 mm in fish 30 to 45 cm long. In benthopelagic grenadiers (Macrouridae), those that are commonly found shallower than about 600 m have retes of 6 mm or less and those found deeper than 2000 m have retes greater than 25 mm long. Relatives living at intermediate depths have retes of intermediate lengths (Marshall, 1979). Many of the Macrouridae have from two to six retes. Multiple retes and gas glands are common among deep sea fishes.

The internal structure of gas glands differs among species. The glandular epithelium is a specialized part of the inner gas bladder lining and may be single celled, multicellular, or folded. Capillaries penetrate between folds or, in some, enter giant specialized cells.

The gases contained in the gas bladders of shallow water fish usually consist mainly of nitrogen and oxygen in the proportions found in the atmosphere—about 80 percent nitrogen and 20 percent oxygen. Members of the Salmonidae tend to have greater proportions of nitrogen, although in some species gas secreted into the gas bladder may be high in oxygen, which is later exchanged for nitrogen by diffusion. Salmonids do not have well-developed retia or gas glands, so gas enters the lumen of the bladder over a wide area of the bladder wall. Gases contained in gas bladders of deep sea fish contain higher proportions of oxygen than nitrogen. A deep sea eel, *Synaphobranchus,* has been noted as having 75.1 percent oxygen, 20.5 percent nitrogen, 3.1 percent carbon dioxide, and 0.4 percent argon in the gas bladder. Various deep water species, including lanternfishes, contain from 76 to 88 percent oxygen in the gas bladder, but lanternfishes captured at the surface show about 43 percent oxygen.

Color, Light, Sound, and Electricity

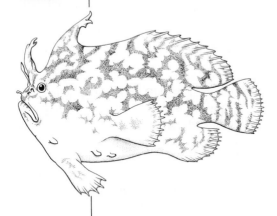

Color

Occurrence of Color in Fishes

Only a few fishes lack or nearly lack skin pigment. Cave-dwelling species, which live in the dark, are notable examples, generally appearing pale to white or slightly colored by blood or other body fluids. Fish larvae are often unpigmented or have only a few chromatophores (color cells) in the head or yolk sac. Larvae of some smelts and gunnels, for example, may remain virtually transparent up to a length of 2 cm or more, and the eel leptocephali may, depending on the species, reach 15 cm or more while still transparent (Figs. 19–1**A** and **B**). Pigment is mostly contained in special cells called chromatophores, but it can be free of these cells in the internal organs, skin, flesh, or bones in some species. For instance, adults of the lamprey, *Geotria australis,* maintain two dorsolateral blue-green stripes caused by deposition of biliverdin (a green bile pigment) during their marine life. Although blue pigment is not common, a blue carotenoprotein is found in the muscles, fin bases, or walls of the alimentary canal of some tropical fish larvae. Chromatophores are named for the color they impart or for the color of the pigment they carry (Fingerman, 1965; Fox, 1957; Fox and Vevers, 1960; Fujii, 1969). Colors imparted by cells are of two kinds, those due to pigments—biochromes—and those due to the reflection of light from a colorless, mirror-like surface and refraction by the tissues—structural colors, or schematochromes (Simon, 1971). These two types of cells are often encountered in combination. Chromatophores classified by color include melanophores, containing a black or brown pigment (melanin); erythrophores, containing reddish pigments (carotenoids and pteridines); xanthophores, containing yellow pigments (carotenoids); leucophores, containing white purines, usually guanine and hypoxanthine, in the form of small crystals that can be moved within the cytoplasm; and iridophores, containing purines, mostly guanine, in large, nonmotile crystals. Iridophores in some fishes have been shown to be motile (Kasakawa et al., 1989). Cells carrying more than one pigment are called compound chromatophores. A type of melanophore (phaeomelanophore) that carries red melanin is found in some cichlids (Avtalion and Reich, 1989).

Chromatophores are located mainly in the dermis but sometimes may be in the epidermis or may even be hypodermal (Hawkers, 1964). Chromatophores are found in the peritoneum (lining of the body cavity) and sometimes around parts of the central nervous system.

Color Change

Color changes involving biochromes and some changes involving structural color depend on the movement of pigments within the chromatophore; these short-term

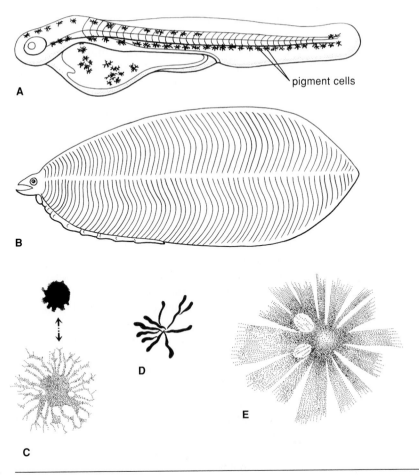

FIGURE 19–1   Larvae of fishes; **A,** showing scattered chromatophores; **B,** unpigmented leptocephalus larva of eel (*Thallassenchelys coheni*); **C,** dendritic chromatophore showing pigment aggregated (top) and dispersed; **D,** chromatophore with pigment in finely radiate pattern; **E,** pigment in coarsely radiate pattern. (After Wunder, 1936.)

and often rapid changes are called physiological or neural color changes. Long-term color changes may be due to increases in the number of chromatophores and a consequent general alteration of pigmentation distribution within cells, referred to as morphological color change.

Pigment occurs in the cells in organelles or, as in the case of the purines, as crystals. Melanin is carried in organelles called melanosomes; erythrophores contain pterinosomes bearing red pteridines. Chromatophores are usually characterized by irregular cell boundaries and dentritically branched processes, although finely radiate and other cell shapes are known (Figs. 19–1C–E). When the pigment-bearing organelles are aggregated in the center of a cell, the skin appears pale; when the organelles are dispersed throughout the cell, the skin takes on the color of the chromatophore. Various hues are made possible by combinations of different chromatophores overlying each other or by the action of compound chromatophores, which can carry separate organelles with pigments of different colors (Fujii, 1969, 1993).

Of special interest are the structural colors caused by the iridophores. Many open-water species have a continuous subdermal sheet of iridophores on the lateral and ventral sides of the body. This stratum argenteum, as it is called, is made up of several layers of cells, so much of the light—up to 80 percent—striking the side of the fish is reflected. In addition, there are usually some iridophores in the dermis. If all these diminutive mirrors are parallel to the surface of the body, the fish generally appears uniformly silvery, but if the reflecting crystals are differentially oriented, the apparent color of an individual may appear dark when viewed from above and light when seen from below (Denton and Nicoll, 1966). This is the mechanism by which countershading is achieved.

In the neon tetra, *Paracheirodon innesi,* changes in color reflectivity can be caused by cytoplasmic movement between reflective plates within iridophores. The plates are like playing cards in a deck (Lythgoe and Shand, 1983). The neon tetra has both physiologically active and physiologically inactive iridophores (Lythgoe and Shand, 1982). The active cells contain broad, thin hexagonal plates, whereas the inactive cells, which participate less in reflective color change, hold thicker, needle-shaped crystals arranged more or less parallel to each other.

Leucophores are dendritic, light-reflecting chromatophores that are present in Cyprinodontiformes and probably in related groups (Fujii, 1993). The chemical composition of the reflecting material, which reflects visible light in all directions, is not known. The reflecting substance is found in organelles termed leucosomes (Fujii, 1993).

Control of Color Change. Aggregation and dispersal of pigments in biochrome chromatophores appear to be under both hormonal and neural control. Although in some fishes one or the other of these is of greatest importance (Naitoh et al., 1985), probably most fishes combine the two types.

The pituitary gland secretes melanocyte-stimulating hormone (MSH) and apparently other substances that cause movement of the melanosomes and thus a change in color. Not all species are affected in the same manner by MSH, which in most causes the pigment to disperse and the skin to darken. Some instances of the same hormone causing aggregation in one species and dispersion in another are known. A melanophore-concentrating hormone (MCH) has been associated with the hypothalamus, and some researchers have suggested that hormones in addition to MSH that affect chromatophores are secreted by various parts of the pituitary (Abbot, 1973).

Adrenalin generally concentrates pigment in melanophores in fish. The same effect can be caused in many species, but not all, by melatonin, a secretion of the pineal body. In addition, there is some evidence that the pseudobranch may be active in secreting or activating a hormone-like substance involved in aggregating pigment in chromatophores.

Nervous control of chromatophores appears to involve the release of chemical transmitters (neurohumors) by the neurons innervating the cells. Some evidence points to the existence of double innervation of chromatophores so that one set of fibers releases a transmitter that causes dispersion of pigment and another set produces an aggregating neurohumor. Leucophores, melanophores, and, to some extent, erythrophores and xanthophores are under nervous control. Iridophores are under the control of the sympathetic nervous system in at least some species (Kasukawa et al., 1987).

Physiological color changes may result from a fish's response to a changed background color in its surroundings or may be due to any one of many responses to social, behavioral, or chemical stimuli (Fujii, 1963). Numerous fishes are known to react to the color of their environment by altering their body color to match it. Shallow water fishes apparently take cues from the light penetrating the surface of the water and light reflected from the bottom or other background (Walls, 1963). The nervous system of the fish seems to take the ratio of light from the two sources into consideration in mediating the response. Light from the surface impinges on the lower part of the retina, whereas that reflected from the bottom strikes the upper part, and the eyes of fishes appear to be specialized to respond to the ratio. Responses may be rapid, taking only a few seconds, or may require hours (Odiorne, 1957). Darkening or paling can be effected by light striking the pineal body in many fishes. There is evidence that other parts of the central nervous system can respond to light, because paling responses have been induced by exposure to light in individuals in which eyes and pineal have been rendered nonfunctional.

Morphological color changes can often be identified in the various life history stages of fish. For instance, young trout and salmon are usually colored to resemble the stream habitat and may have vertical bars (parr marks) on the sides, but as they physiologically prepare to enter the ocean they become silvery on the sides and blue or green on the back (pelagic coloration). Later, as they return to fresh water for breeding, the pelagic coloration is obliterated and a nuptial coloration, often involving very bright colors, appears.

As eels of the genus *Anguilla* mature and appear to enter the ocean migratory phase, they become silvery. Pankhurst and Lythgoe (1982) found that the maturing European eel lost xanthophores and experienced a decrease in the amount of the reflective purines guanine and hypoxanthine, but that reflecting layers were reorganized to produce the silvery camouflage. Change in color can be brought on by changes in diet, especially if carotenoids are involved (Wutiporn, 1988). Diet during the larval stage of flounders has been implicated in the absence of pigment. Juveniles fed rotifers and/or *Artemia* showed a high incidence of albinism (Seikai, 1989). In some instances, parasitism has caused a change in pigmentation (Ward, 1988). Loss of iridophores can be induced by retaining fishes under constant illumination in a light container (Fries, 1958). Keeping fishes on a dark background results in an increase of melanophores (Odiorne, 1957). Melanogenesis seems to be controlled by the pituitary and is brought about by action of the adrenocorticotropic hormone (ACTH) (see Fujii, 1969).

Optical Filters. All fishes do not experience the same view of colors in their environment, because quality and intensity of light striking the retina of the eye can be influenced by color in the lens or cornea. Some cichlids have yellow pigment, which apparently acts as an optical filter, in the lens and cornea. *Argyropelecus* (deep sea hatchetfish) has yellow lenses (McFall-Ngai et al., 1986). Muntz (1983) suggested that yellow lenses in marine fishes might render the camouflage of photophores in other fishes less efficient because such lenses absorb less of the bioluminescence than of the surface light, so the light of the photophores would be brighter than the background.

Greenlings have corneal filters that respond to change in light with change in density of pigmentation (Gnyubkin, 1989). Other fishes with corneal filters include sculpins and wrasses (Gnyubkin and Gamburtseva, 1981).

Significance of Color

Pigment has many significant functions in fishes. For fishes that live in shallow water, pigment may be effective in protecting them from damage from ultraviolet radiation. McArdle and Bullock (1987) report that severe losses of Atlantic salmon (*Salmo salar*) at Irish hatcheries in 1987 were due in part to solar ultraviolet radiation. Exposure of chinook salmon (*Oncorhynchus tshawytscha*) fertilized eggs and fry to various artificial lights indicated that there was damage from ultraviolet light (Dey and Damkaer, 1990).

Color functions also in predator avoidance (by concealment, disguise, mimicry), in advertisement (of toxins, etc.), and in mating signals.

Concealment. A common method of concealment among fishes is countershading (obliterative shading), generally seen in any fishes that are darker on the back than on the sides and belly. Many species of open waters show the type of countershading that is sometimes referred to as pelagic coloration, with dark backs of the general hue of the water as seen from above—usually blue or green—and silver sides and belly. With this arrangement, illumination from above lightens the back, and the light color of the sides lightens the shadows on the flanks so that the entire fish can appear to have a uniform color that blends with the background.

Some smelts and silversides are translucent but have a reflective layer around the body cavity and another around the red lateral muscles. In both instances, the silver does not extend over the top, so the structures are essentially countershaded like the fish itself. Some brightly colored reef fishes are countershaded in that the upper sides and back are black or some other dark tone and lighter, brighter patterns cover the flanks and belly. Several species of butterflyfishes, including *Chaetodon lunula*, have been noted to turn their dark backs toward approaching predators, an act that should render the conspicuous fish less so. Some of the black on the sides of this species serves to conceal an apparent social signal of yellow, which replaces the black during aggressive encounters (Hamilton and Peterman, 1971).

Dark pigment in the peritoneum or gut wall of deep sea fishes might serve to conceal the lights of luminescent prey. The blue pigment in the gut wall of certain tropical fish larvae could mask the presence of the orange or red crustacea upon which they feed (Herring, 1967).

Concealment can be aided by an overall resemblance to the substrate or background. The transparency of larvae may make them less visible to predators (McFall-Ngai, 1990). Some species have hues and patterns similar to those of the bottom (demersal coloration), while some patterns bear a general resemblance to vegetation. Many species can change color or pattern to match their surroundings. Cutthroat trout (*Oncorhynchus clarki*) that live in streams with a heavy overstory of trees tend to be very dark and heavily spotted; those that frequent open riffles are light; and those living in ponds opalescent with colloidal clay can be ghostly pale. Much attention has been given the flounders and other flatfishes, but various blennies, sculpins, scorpionfishes, and others are capable of rapid color change.

Many demersal fishes have flaps and irregular outlines that aid in concealment. Fishes that generally resemble plants include sea snails (*Liparis*), pipefishes and seahorses (Syngnathidae), pricklebacks (Stichaeidae), and gunnels (Pholidae). Many bottom fishes, such as marine sculpins (Cottidae), stonefishes (Synanceinae), and stargazers (Uranoscopidae), show a color resemblance to substrate that consists of stones, shell, and small algae in combination.

Disruptive coloration (Cott, 1940; Muntz, 1990) consists of stripes, bars, ocelli, and other markings that may cause a fish to appear conspicuous out of its accustomed habitat; but in its usual habitat, those markings tend to break up the outline of the individual or to make the eyes or other readily recognizable features less prominent (Fig. 19–2). In both schooling and nonschooling species, disruptive markings can present a confusing pattern. The sight of a large aggregation of horizontally striped fishes such as threadfins in the brightness of a coral lagoon might make you think that your eyes will not focus properly, until an individual fish at the edges becomes evident and the illusion is destroyed. Eye stripes and "hoods" of dark color

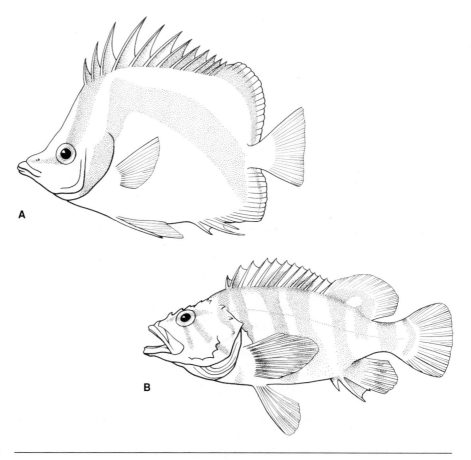

FIGURE 19–2   Examples of fishes with disruptive pigment patterns. **A,** Scythe butterflyfish (*Chaetodon falcifer*); **B,** treefish (*Sebastes serriceps*). (**B,** after Miller and Lea, 1972.)

that cover the top or anterior of the head and the eye are common in fishes. Elongate fishes usually have horizontal eye stripes, and those with deep bodies or blunt heads tend to show vertical stripes or stripes that follow the contour of the head.

Advertisement. Advertisement by bright colors or subtle or conspicuous patterns may serve many purposes for various fishes. Cleanerfishes, which remove ectoparasites from other fishes, usually have distinctive, conspicuous coloration and markings so that they are readily recognized by larger species and are allowed to approach with little danger of being eaten. Differential coloration of the sexes aids recognition of potential mates or attracts one sex to the other. Color may figure significantly in reproductive success, as in the Pecos pupfish, *Cyprinodon pecosensis,* in which Kodric-Brown (1983) found that darker males mated over three times as many females per hour than lighter males. A hypothesis that density of carotenoid pigment is an indicator of male vigor in the guppy (*Poecilia reticulata*) was developed by Nicoletta (1991). Schooling species have markings that probably aid individuals in staying with their own kind. Even countershaded species such as sardines and tunas may have species-specific rows of spots or series of stripes arranged so as not to detract from the obliterative shading to any great degree but still allow species recognition.

Change in color and behavior may coincide. One example in a schooling species is the rudderfish, *Kyphosus elegans,* in which certain individuals change from stripes to spots, round up stragglers from the school, and discourage members of other schools from feeding in their school's area. During their short stints of such activity, the spotted individuals are a marked contrast to the schooled fishes. There are numerous instances wherein color functions to advertise mood or to predict the fighting ability of an individual in intraspecific behavior (Maynard, 1988). (See Chapter 28.)

Often, bright colors and bold patterns advertise unpalatability or dangerous poison or venom (aposematic coloration). An inconspicuous example of this warning coloration is the black area on the dorsal fin of the venomous weevers (*Trachinus* spp.), which are generally cryptically colored. The black dorsal is erected, apparently as a warning, on the approach of animals larger than the weever's prey. Conspicuous examples of fishes with warning coloration are lionfishes (*Pterois* spp.), which are highly colored and highly venomous. One of the conspicuously colored sabertoothed blennies, *Meiacanthus nigrolineatus,* has venom glands in the lower jaw and is rejected by predators, probably after a venomous bite from the large lower teeth. Some surgeonfishes show a distinctive bright spot at the location of their formidable peduncular spines.

Disguise. Although general resemblance and disguise may be indistinguishable in some instances, the concept of disguise is useful because it involves resemblance to specific objects and usually involves not only color but body shape, special appendages, and behavior. An adaptive advantage has apparently accrued to species that bear a close likeness to objects in the environment that elicit neutral reactions in predators or prey. Minicry is a special kind of disguise (see next subsection). Most disguises among fishes cause them to resemble plants (Fig. 19–3). The leaffish of South America, *Monocirrhus polyacanthus,* has a barbel at the chin, simulating a leaf's stem, and has coloration, body shape, and postural behavior that complete the

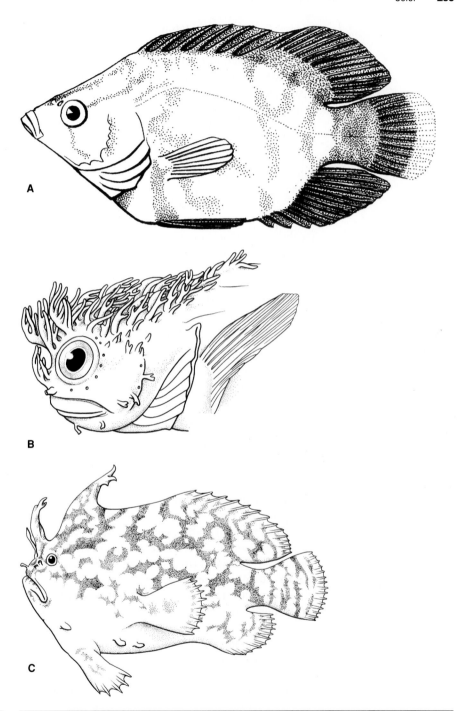

FIGURE 19–3   Examples of fishes with structure and coloration resembling vegetation. **A,** Young of tripletail (*Lobotes* sp.), with general resemblance to mangrove leaf; **B,** mosshead warbonnet (*Chirolophus nugator*), with cirri and flaps resembling marine algae; **C,** sargassumfish (*Histrio histrio*).

illusion. The young of a labrid, *Hemipteronotus pavo,* has a "stem" consisting of the first dorsal fin and posture and coloration resembling a floating or drifting leaf. If color is involved in mimicry, it can be termed chromatic mimesis (Fujii, 1993). Other fishes that have a leaf disguise are the batfishes, *Platax* spp., and the young of a carangid, *Trachinotus falcatus,* both of which combine color and posture in the deceit. Other genera with leaflike young include *Lobotes* and *Oligoplites.* The naked sole, *Gymnachirus melas,* by means of color and serrate dorsal and anal fins looks like a dead, sunken leaf. The filefish, *Aluterus,* holding position with its mouth to the substrate and slender tail upward, resembles eelgrass. Juveniles of many fishes are suitably colored and shaped to resemble floating or drifting plant debris, but one of the most remarkable instances is seen in the genus *Chaetodipterus,* in which the black young resemble the old, sunken seed pods of mangrove with which they occur. An interesting adaptive coloration is the possession of large, realistic eye spots (ocelli), especially in the posterior part of the fish. These spots are believed to misdirect predators or to advertise unpalatability (Neudecker, 1989). When coupled with disruptive bars or stripes that make the fish's eyes inconspicuous, the eye spots effectively disguise the direction in which the fish is heading and may be confusing to a predator. About 91 percent of the species in the butterflyfish genus *Chaetodon* have eye camouflage (often a dark bar across the eye), and nearly half combine it with false eyespots, usually not on the head. Nuedecker (1989) reported that the false eyespots of butterflyfishes are usually situated in a sector of the body that, if injured, would allow a high probability of recovery and survival after an attack. Most eyespots of nonschooling species are in the posterior third of the fish, mostly on the caudal peduncle or dorsal fin. Spots in the anterior third tend to be high on the body just below the formidable dorsal spines. Ocelli may be signals involved in intraspecific behavior or may be camouflage or disruptive coloration. They could be considered instances of mimicry in that they might resemble the eyes of other organisms.

Mimicry. Mimicry may serve many functions among various animals, and several types of mimicry have been recognized, especially among insects. Mimicry is not widespread among fishes, but there are some interesting examples. Some fishes mimic other species (models) that are distasteful or otherwise avoided or at least not preferred by predators (Batesian mimicry). Thus a sole living among weevers lifts a black pectoral fin to mimic the warning signal given by the weever's dorsal fin, or a blenny assumes the coloration of a cleaner wrasse and is able to live near predators with reduced probability of being eaten. Young *Plotosus anguillarus,* themselves venomous, may aggregate to resemble sea anemones. Another type of mimicry, aggressive mimicry, occurs when one species mimics another to gain resources from the model. An example involves a wrasse, *Labroides dimidiatus,* which has access to larger fishes as an ectoparasite cleaner, and one of the saber-toothed blennies, *Aspidontus taeniatus,* which mimics the wrasse and can thereby feed on skin torn from unsuspecting larger fishes. With experience, however, victimized fishes learn to distinguish the cleaner from the outlaw. Another example involves two characoid fishes of South America. *Probolobus heterostomus,* a scale eater, closely resembles, in shape and coloration, some species of *Astyanax.* This mimicry allows the scale eater to school freely with *Astyanax* and prey on their scales (Sazima, 1977). Likewise, some fry-eating (paedophagous) cichlids of Lake Malawi (*Cyrtocara* spp.) are capable of mimicking the color pattern of species they prey on and can make the change from, for instance, a longitudinal stripe to a lateral row of spots within seconds (McKaye and Kocher, 1983). Display of lures to entice prey toward the mouth is a special kind of aggressive mimicry usually involving modified body parts, such as the maxillary barbels of the angler catfish, *Chaca chaca.* The lures of anglerfishes may be luminescent; resemble worms or other edible invertebrates; or, as in a species of *Antennarius,* closely resemble a fish, complete with eye spots and structures resembling fins (Pietsch and Grobecker, 1978).

Another type of mimicry (Mullerian mimicry) is that in which two or more dangerous or unpalatable species assume similar warning coloration. This benefits both species as predators learn to associate the warning coloration with unpalatability. This is a rare phenomenon among fishes; the closest qualifying instance involves some of the saber-toothed blennies (Fig. 19–4). The venomous *Meiacanthus nigrolineatus* is a model not only for the nonaggressive *Ecsenius gravieri* but also for *Plagiotremus townsendi,* a fierce biter that, like *M. nigrolineatus,* is usually rejected by predators (Springer and Smith-Vaniz, 1972).

Some mouth-brooding cichlids display an interesting mimicry that ensures fertilization of the eggs, which are taken into the female's mouth after deposition and before fertilization. The male has a series of egglike spots ("dummy eggs") on his anal fin. He brings these to the attention of the female in the area where she has been depositing eggs, and as she attempts to pick them up he releases sperm, fertilizing the eggs carried in her mouth.

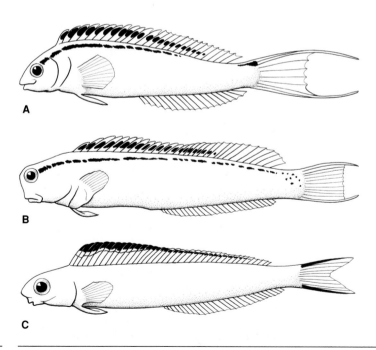

FIGURE 19–4   Mimicry among blennioid fishes. **A,** *Meiacanthus nigrolineatus,* a venomous species; **B,** *Ecsenius gravieri,* a nonaggressive species; **C,** *Plagiotremus townsendi,* an aggressive species. (After Springer and Smith-Vaniz, 1972.)

TABLE 19–1	Families Reported to Contain Bioluminescent Fishes		

		Type of Photophore	
Order	Family	Bacterial	Autoluminous
Squaliformes	Squalidae		x
	Dalatiidae		x
Torpediniformes	Torpedinidae		x
Anguilliformes	Congridae	x	
	Saccopharyngidae		x?
Clupeiformes	Engraulidae		x
Osmeriformes	Opisthoproctidae	x	
	Bathylagidae		x
	Alepocephalidae		x
	Platytroctidae		x
Stomiiformes	Gonostomatidae		x
	Sternoptychidae		x
	Stomiidae[1]		x
	Phosichthyidae		x
Aulopiformes	Chlorophthalmidae	x	
	Scopelarchidae		x
	Paralepididae		x
	Evermannellidae		x
Myctophiformes	Neoscopelidae		x
	Myctophidae		x
Gadiformes	Macrouridae	x	
	Moridae	x	
	Steindachneriidae		x
Batrachoidiformes	Batrachoididae		x
Lophiiformes	Melanocetidae	x	
	Himantolophidae	x	
	Diceratiidae	x	
	Oneirodidae	x	
	Ceratiidae	x	
	Gigantactidae	x	
	Centrophyrnidae	x	
	Linophyrnidae	x	
	Thaumatichthyidae	x	
Beryciformes	Anomalopidae	x	
	Monocentridae	x	
	Trachichthyidae	x	
Perciformes	Acropomatidae	x	
	Apogonidae	x	x
	Leiognathidae	x	
	Sciaenidae		x
	Pempheridae		x
	Chiasmodontidae		?

Based on Herring and Morin, 1978; Morin, 1981; and Pietsch, personal communication.
[1] Includes Astronesthinae, Chaulidontinae, Idiacanthinae, Melanostominae, and Malacosteinae.

Bioluminescence

Occurrence of Bioluminescence

Ability to produce light has been noted in at least 45 families of fish (Table 19–1). Most of these are teleosts, as only two families of elasmobranchs and no lampreys, hagfishes, or nonteleost bony fishes are known to be luminous. Among the teleosts, a wide range of families from soft-rayed osmeriforms to the perciforms and lophiiforms display bioluminescence. The family of lanternfishes (Myctophidae) has more luminescent genera and species than any other, but spookfishes (Opisthoproctidae), grenadiers (Macrouridae), batfishes (Ogcocephalidae), seadevils (Ceratioidei), and stomiatoids (Stomiiformes) have numerous light-bearing representatives, and in some regions the Stomiiformes may outnumber lanternfishes.

There are a few bioluminescent fishes that are permanent residents of shallow water, including slipmouths, midshipmen, and flashlightfishes (Morin, 1981), but most live in or over moderate to great depths, and many move into surface waters as part of a nightly feeding migration. Apparently, most bioluminescent fishes live in the mesopelagic regions of the ocean and are found mostly at depths of from 300 to 1000 m. At some localities, up to 66 percent of the fish species and more than 50 percent of the individuals collected have been luminescent. There are estimates that about 600 to 700 species of mesopelagic fishes are luminescent (Marshall, 1979). The light organs and modes of light production are more diverse in fishes than in any other marine animals (Herring, 1982).

Production of Light

Light production usually takes place in special organs called photophores (Fig. 19–5). Light is produced chemically, without heat, through a chemical reaction of the enzyme luciferase with the substance luciferin (a heterocyclic phenol) in the presence of an oxygen source and adenosine triphosphate (ATP).

Although most luminous fishes produce their own light (self-luminous), many depend on symbiotic luminous bacteria nurtured in special glandlike structures. At least one species of fish depends on luciferin obtained through diet as a basis for luminescence. The plainfin midshipman, *Porichthys notatus* of the Pacific Coast of North America, is one of the few shallow water species with self-luminous photophores, of which it has about 700. In the southern part of its range it is luminescent, but north of San Francisco, where its occurrence is discontinuous, it is not. However, luminescence can be induced in individuals from Puget Sound after dietary intake of luciferin from the ostracod *Vargula hilgendorfi* (Thompson et al., 1988; Thompson and Tsugi, 1989). Luminescence of this dietary nature occurs also in *Parapriacanthus* (Pempheridae) and in some genera of Apogonidae (Haneda, 1986).

Light from Symbiotic Bacteria. Most luminous fishes of shallow water have only bacterial photophores, usually in the region of the eye or gut (Morin, 1981), but fish from deep water may have bacterial light organs in other parts of the body. Herring (1977) reports that bacterial bioluminescence is found in relatively few fishes as compared with the great numbers of species that are self-luminous and that there are

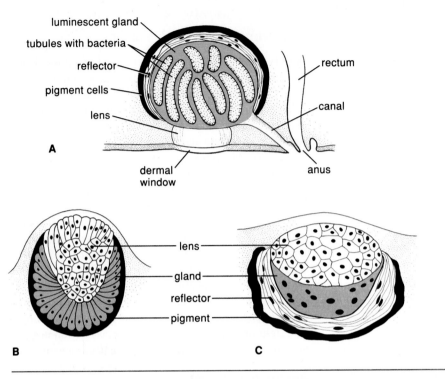

FIGURE 19–5 Diagrams of examples of photophores. **A,** bacterial photophore in which luminous bacteria are nurtured in tubelike structures; **B,** self-luminous organ with lens and pigment sheath; **C,** self-luminous organ with lens, reflector, and pigment sheath.

usually no more than four bacterial luminous organs per individual. There are about 65 genera with bacterial photophores and 130 genera that are self-luminous (Hastings and Morin, 1991). Many of the self-luminous genera contain numerous species.

The bacterial luminous organs usually open to the exterior. Certain bacteria appear to be restricted to given groups of fishes. Some are said to be species specific, although *Photobacterium leiognathi* is reported to live with members of Leiognathidae, Percichthyidae, Apogonidae, and squids (Fukasawa et al., 1988). The pinecone fish *Monocentrus japonicus* harbors the luminous bacterium *Vibrio fisheri*.

Luminous structures can be associated with a variety of body parts. Grenadiers, some cods (Moridae), steindachneriid hakes (Steindachneriidae), and roughies (Trachichthyidae) are examples of groups with luminescent structures at or around the anus along the ventral surface close to the anus. In some of these, gelatinous tissue under the skin allows the light to diffuse over a relatively wide area. Some of the light glands release luminous materials that can spread along the exterior of the fish. Slipmouths (Leiognathidae) have a bacterial light gland around the esophagus, a lens that concentrates the light and directs it into the gas bladder, and a pigmented shutter that can conceal the light. A species of the genus *Garra* shows luminescence from the mouth as well as from the ventral surface and can project luminescence as a discrete beam (McFall-Ngai and Dunlap, 1982). Other fishes with bacte-

rial luminescence associated with the digestive system, anus, or ventral musculature are the spookfishes (Opisthoproctidae), lanternbellies (Acropomatidae), and
genus *Siphamia* of the cardinalfishes (Apogonidae). Most of these can light up
their thoracic or ventral areas because of internal reflectors and translucent tissue
under the skin. The lanterneyes (Anomalopidae) bear bacterial light organs under
the eyes. The two best-known species are *Photoblepharon palpebratus* and *Anomalops kaptotron,* both of which are shallow water fishes of the Indo-West Pacific.
Anomalops has been observed to flash light on and off in a rather regular manner
by rotating its light organ inward and down into a black-pigmented pocket under
the eye. *Photoblepharon* shows a more steady light but can conceal it by drawing a
fold of black tissue over it. Most members of the suborder of deep water anglerfishes (Ceratioidei) are luminous, with bacterial photophores on the barbels, the
esca, and elsewhere on the body.

Nonbacterial Luminescence. Self-luminous fishes commonly possess a series of
light organs along the ventral aspect of the body that direct the light downward.
However, a few have some dorsal lights, and many of the stomiiforms have an abundance of small, simple organs along the dorsal surface that enable the entire fish to
be silhouetted in light. Some have light organs within the mouth as well as on the
jaws. There may be numerous (1000 or more) photophores per individual and more
than one type of light organ on an individual. Photophores are not always open to
the exterior and are not usually associated with the gut (Herring, 1977).

Light organs vary from very small unpigmented structures, such as those on the
back and fins of the stomiiforms, to very complex ones, with the glandular portion
surrounded by an efficient reflector that directs the light through a lens, which can
concentrate the beam as it is emitted. Some have iris-like structures that control the
amount of light emitted (Herring and Morin, 1978).

Members of the family Platytroctidae (Searsidae) have a light gland above the
pectoral fin with an opening to the surface. Apparently, luminous material can be released voluntarily from the gland. Similar glands are known in the ceratioids. This
type of luminescence is called extracellular and contrasts with the intracellular type,
in which the luminous material is confined to the cells (photocytes) that produce the
light.

Control of Luminescence. In the anomalopids, some stomiiforms, and pinecone
fishes (Monocentridae), control of the display of light is indirect, by concealing or
screening the luminous tissue. In most self-luminous bony fishes, the photophores
are innervated and appear to be under direct nervous control. Injections of adrenaline usually cause activity of luminous tissue in most species tested, but the exact relationships of the nervous control and possible hormonal controls have not yet been
discovered (see Hastings and Morin, 1991).

Significance of Bioluminescence

Considering that about two thirds of deep oceanic fishes produce light, this ability
must confer some adaptive advantage on these fishes. Identification of the selective
advantages is difficult for most species because of their habitats. Experimentation
and observation are difficult. The functions of light are thought to parallel in part the
functions of color and, of course, color is involved in bioluminescence. Many

species have color filters built into the light organs so that emitted lights may have different wavelengths. According to Morin (1981), functions of luminescence are obtaining prey (by luminescent predators), evading prey (by luminescent prey), and communicating between conspecifics. These can involve concealment, advertisement, and disguise.

Concealment (Crypsis). Placement of photophores and other luminous tissue in a ventral position in most luminous species has led to the theory that a mid-water luminous fish can match the background of light coming from above, so that predators hunting from below will have less chance of seeing the silhouette. There is evidence that various lanternfishes and other species with ventral photophores regulate the intensity of light emission according to the light coming from above (Young and Ropper, 1977). Many luminescent species, including representatives of Myctophidae, Opisthoproctidae, and Stomiiformes, apparently adjust luminescence by comparing light intensity from above with light emanating from photophores that shine into the eyes (Lawry, 1974). Some species are known to switch on ventral photophores when illuminated from above. The color filters in the ventral light organs of hatchetfishes have a transmission band close to 480 nm that matches blue-green light, which has great penetration into sea water (Denton, 1970). McFall-Ngai and Morin (1991) found that leiognathids, which have bacterial photophores around the esophagus, increased the intensity of luminescence as illumination from above was increased, although not in direct proportion to the increase in down-dwelling light.

McFall-Ngai (1990) reported that pelagic fishes can gain concealment by three major types of crypsis: transparency, reflection of light, and counterilluminating photophores, or combinations of these.

Advertisement. Communication involving luminescence is active in the reproduction of some species. Midshipmen, *Porichthys,* are known to display the light from photophores during courtship (Wheeler, 1975), and other fishes might do the same. Some lanternfishes have sexually dimorphic patterns of luminous organs; such different patterns might serve in recognition of mates (Lagler et al., 1977).

There are several other advantages that might accrue to fishes that advertise themselves with lights. Species recognition could aid in keeping schools together or, on the other hand, could aid individuals of nonschooling species in maintaining territories. Bioluminescent communication can involve more than one species of fishes or even other taxa (Herring, 1990).

Sudden displays of light by a single fish or by an aggregation of small fishes might serve to startle or confuse predators. Defense by confusing visually oriented predators is considered by Hastings and Morin (1991) as probably the most important function of bioluminescence in fishes. The Atlantic midshipman, *Porichthys plectrodon,* possessor of a venomous spine, is known to flash its lights upon approach of predators, thus warning them away.

McAllister (1967) has suggested that the dense pigment of the peritoneums or stomachs of deep sea predators serve mainly to guard against the lights of recently swallowed prey shining through the body wall, thus preventing the advertisement of the predator.

Several predators, including stomiiforms and anglerfishes, have luminous organs or tissue close to the mouth, in the mouth, or on barbels or illicia. In some or all of these cases, prey could be attracted to the mouth by a show of photophores on fins and the dorsal surface of stomiiforms such as *Echiostoma,* which is also equipped with ventral camouflage photophores. The dorsal photophores would seem to silhouette the body, but such advertisement is of unknown significance (Somiya, 1979).

Disguise. Some photophores tend to disguise fishes by creating a resemblance to other objects in the environment. If the aforementioned baits closely resemble some organism that is habitually eaten by the anglerfishes' prospective prey, this might be considered disguise. Some investigators have suggested that the luminous tissue of certain fishes mimics various luminous invertebrates. Perhaps a school of small luminous fish might take on the appearance of a single organism large enough to be intimidating.

Special Considerations. Placement of light organs on the head so that objects in the most effective visual field of the fish can be illuminated should make possible the floodlighting of prey, and the observation of feeding stomiiforms appears to confirm this. Some melanostomiatids combine a red-sensitive retina (see Chapter 20) with a large red postorbital photophore that allows them to see nearby prey, especially those with red pigment (Denton et al., 1970). Because few deep sea animals have red-sensitive retinas, the red spotlight of the stomiiform is not conspicuous (Marshall, 1979). One of the loosejaws (Malacosteinae) is known to have red and green photophores and visual pigments sensitive to both those colors (O'Day and Fernandez, 1974). If an individual of a prey species illuminates a nearby predator, others of the species could avoid it. Alternatively, certain types of signals from individuals of prey species might alert conspecifics to the presence of a predator. The term *burglar alarm* has been applied to these hypothetical cases. In some species, it is probable that there are multiple uses of the photophores in predation, intraspecific communication, and avoidance of predators.

Production of Sound

That fishes and other animals living in the sea produce sounds has been recognized from ancient times, and in some areas fishermen have located certain species by listening for their characteristic sounds.

Water is a good medium for the transmission of sound. The speed of sound in water is about 1500 m/sec and varies with temperature, density, and salinity, so that the speed in sea water can approach 1540 m/sec, nearly five times its speed in air (Tavolga, 1971). Sounds can be carried long distances in water, being reflected off the bottom, the surface, and density boundaries caused by temperature or salinity differences.

How Sound Is Produced

Nature of Sounds. Sound-producing structures among fishes include teeth, skeletal elements, muscles, and the gas bladder, and sounds made by fishes have been described by a great variety of terms. Sounds of schooling fish swimming have been called rustles or roars. Stridulation produces sounds reminiscent of clicks, rasps, scratches, and so on when not aided by the gas bladder, and croaks, grunts, and knocks when the bladder acts as a resonator. Frequency of stridulatory sounds in the freshwater drum (*Aplodinotus grunniens*) can range from 150 to 2000 cycles per second (Hz) (Schneider and Hasler, 1960), although the gas bladder–aided sounds are generally below 1000 Hz and unaided stridulation produces frequencies usually in the 1000 to 4000 Hz range.

Sounds made by vibrating the gas bladder have been described as hoots, boops, grunts, yelps, knocks, and croaks, with the toadfishes, *Opsanus* spp., being known for their boat whistle sounds. Gas bladder sounds are harmonic and usually of low frequency, from 40 to 250 Hz, with the great majority in the 75 to 100 Hz range; but some scaenids can produce frequencies up to 2000 Hz.

Stridulation. Grinding, snapping, or rubbing teeth together is the most common type of stridulation among fishes (Tavolga, 1971). Pharyngeal teeth appear to be important sound producers, especially because they are close enough to the gas bladder that the sounds can be amplified (Takemura, 1984). In the cichlid, *Tilapia mossambica,* in which special muscles run from the occipital region and the first vertebra to the upper pharyngeals, sounds made in the absence of the gas bladder were essentially of the same amplitude as those produced by the intact animal (Lanzing, 1974). Jaw teeth are often used in sound production; some filefishes (Monacanthidae) have ridges, apparently effective in stridulation, on the backs of the front teeth, and various perciform fishes can snap their teeth together, making a sharp sound. Other mechanisms of stridulation include movement of fin spines against their sockets in catfishes (Fig. 19–6A), triggerfishes, filefishes, sticklebacks, surgeonfishes, and others; contact between the first dorsal interspinous bone and modified neural spines in some sisorid catfishes; and friction between other skeletal parts in triggerfishes, seahorses, and clownfishes.

The Gas Bladder and Sound Production. Incidental sounds may be made by release of air from the lungs or other cavities used in air breathing. Some fishes, such as the Atlantic eel and some catfishes, cause sounds by release of gas from the gas bladder. Such sounds are common, but their importance is unknown.

Most gas bladder sounds are due to vibrations set up by some special means. Several species are equipped with muscles that vibrate the gas bladder directly or indirectly. Some triggerfishes vibrate the gas bladder by rubbing or drumming the pectoral fin against an area where the gas bladder is close to the body wall. Sound production is common among the Sciaenidae; the freshwater drum, *Aplodinotus,* produces sound by vibrating special muscles of the body walls that attach to broad tendons stretching over the gas bladder, whereas another sciaenid genus, *Micropogon,* has tendons attached directly to the gas bladder (Schneider and Hasler, 1960). Several species have muscles that originate on the skull or vertebral column and insert on the gas bladder itself or on ribs or other structures associated with it. Certain fishes of the families Macrouridae, Priacanthidae, Brotulidae, Serranidae, Scorpaenidae, Theraponidae, and Holocentridae have been noted as having such arrangements. One of the best-developed structures is the "elastic spring mechanism" seen in certain catfishes (*Galeichthys, Bagre*). This consists of the plates formed from the first few vertebrae and placed in contact with the gas bladder wall dorsally. Muscles

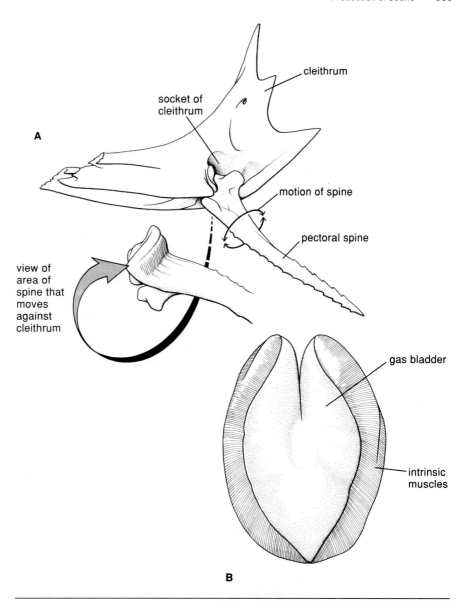

FIGURE 19–6   **A,** right cleithrum and pectoral spine of sea catfish (Ariidae) showing roughened flange with which stridulatory sounds are made; **B,** gas bladder of toadfish (*Opsanus*) with musculature along outer edges.

stretching between the skull and this springy apparatus can vibrate it rapidly, setting up audible vibrations of the gas bladder (Tavolga, 1971).

Intrinsic sonic muscles are incorporated into the walls of the gas bladder in several species of toadfishes, Batrachoididae; in searobins, Triglidae; in flying gurnards, Dactylopteridae; and in *Zeus faber*. In most of these fishes, the bladder tends to be divided into left and right chambers, and it is heart shaped in the toadfishes (Fig. 19–6**B**) and nearly completely divided in gurnards. Usually an internal diaphragm divides the bladder into anterior and posterior sections. The diaphragm is typically perforated by an opening surrounded by a sphincter muscle. The relationship of the diaphragm to the sounds produced is not understood, and experimentation has produced no clear picture of its function.

The sonic muscles of fishes are red (except in Theraponidae). Very vascular, well supplied with myoglobin, and resistant to fatigue, they are among the most fast-acting of vertebrate muscles and can be artificially stimulated to act at more than 100 contractions per second (Tavolga, 1971).

Miscellaneous Sources of Sound. The act of swimming can cause production of sound because of the movement of parts of the skeleton on each other and the turbulence caused by movement through the water. Large schools of fish have been noted to cause sounds as they veer and turn (Moulton, 1960). Feeding activities are often noisy. Parrotfishes (Scaridae), wrasses (Labridae), surfperches (Embiotocidae), and many others that crush molluscs or crustaceans with their strong teeth in the jaws or throat make much incidental sound. Fishes that break the surface of the water, whether in feeding or for other purposes, cause sounds. Many fishermen are attracted to the exact spot where the quarry is feeding by hearing the smacking sound of the bluegill or the splash of the trout. Many know the double splash of the leaping carp. Eagle rays are known to make prodigious leaps that result in loud noises as they fall back to the water. Opening and closing of oral and opercular valves in large fishes can be audible, as can the release of air from the mouth or anus of air-breathers. *Misgurnus fossilis,* which utilizes the alimentary canal as a respiratory surface, is known as "squeaker" because of the sounds made as respiratory gas is expelled. Release of bubbles because of reduced pressure during rapid ascents probably causes sounds.

Significance of Sound Production

Although incidental sounds are of significance to fish in that such noises may be detected by potential prey or predators, most sounds produced volitionally by fishes appear to play a role in communication and are generated with the purpose of gaining some adaptive advantage for the sender of the sound (Myrberg, 1981). Six classes of receivers of sounds from sound-producing fishes are suggested by Myrberg (1981). These are predators, prey, mates, companions, competitors for mates, and competitors for other resources.

Reproductive behavior is accompanied by sounds in several marine and freshwater species. In most instances, the males seem to generate most of the sounds and are usually better equipped to do so than the females. In fresh water, sounds are used in courtship—or at least in the presence of prospective spawning partners—in such families as the sunfishes (Centrarchidae), minnows (Cyprinidae), cichlids, characids, and anabantids. Several marine fishes, including cods, seahorses, gobies, blennies, damselfishes, hamlets, and parrotfishes, are also known to use sounds as part of courtship (Lobel, 1991, 1992). Some of the most remarkable courtship sounds are made by toadfishes (*Opsanus*), in which the male produces growls, grunts, and boat whistles; strangely, the latter sound appears to be at a higher frequency than the toadfish's best auditory capability (Fine, 1981). The plainfin midshipman, *Porichthys notatus,* shows definite sexual dimorphism in the size of the gas bladder and, more surprisingly, in the properties of the sonic musculature (Bass and Marchaterre, 1989).

The male of the freshwater drum does not develop sound-producing capabilities until the third year of life, and it produces sounds only during the spawning season. The naked goby (*Gobiosoma bosci*) produces sounds under various stimuli, but only during the reproductive season (Mok, 1981). Females of *Pomacentrus partitus* can locate nest sites by the courtship sounds of the males and seem to be able to distinguish among, and evaluate, the males by their sounds (Myrberg et al., 1986).

A variety of species emit sounds when defending or maintaining territories. Several marine percoid families, including soldierfishes, gobies, wrasses, damselfishes, and tigerperches, are included. Other marine fishes included are sea catfishes, toadfishes, triggerfishes, and cods. Among the freshwater fishes, the cichlids are notable for making sounds of this nature.

"Alarm" or "startle" sounds made when disturbed are attributed to nearly 40 families by Myrberg (1981). The sounds are mostly produced internally, but some are made as a consequence of swimming. There is some evidence and much speculation on the function of these sounds in startling predators, and there is further speculation that some of the sounds might act as an alarm signal to conspecifics or other fishes receiving them. Sound signals are thought to be of significance to schooling fishes and to aid them in maintaining contact. Gudgeon (*Gobio gobio*), which commonly live in small groups and emit creaking sounds, became quiet when held alone in aquaria but produced sounds when in groups (Ladich, 1988).

Production of Electricity

The Electric Fishes

Three kinds of fishes that generate strong electrical shocks have long been known, although this power was a mystery until the nature of electricity was discovered. The electric rays of the family Torpedinidae are said to have been used by ancient Roman physicians in an early form of electrotherapy. The electric catfish, *Malapterurus electricus,* was featured in Egyptian hieroglyphics as early as 2750 B.C. and has been noted as having an Arabic name translating as "father of thunder." In South America the electric eel, *Electrophorus electricus,* the most powerful producer of electricity among fishes, was apparently well known to the indigenous people before its discovery by European explorers. Other groups with weak electrical powers were confirmed as being electric only after instrumental means of studying electricity were developed, and the strong electrical capability of the electric stargazers was not recognized until the twentieth century.

Electric fishes, although members of diverse groups, share certain convergent attributes. They are generally slow moving or sedentary, are active at night or in murky waters of low visibility, and have thickened skin that serves as a good insulator. Most have reduced eyes, and some electric rays are blind. Generally the cerebellum is enlarged, greatly so in the mormyrids (Nelson, 1994).

Electric organs are present in six orders of fishes, suggesting that the ability to produce electricity evolved independently at least six times (and possibly more, depending on the phylogeny accepted; see Feng, 1991). Electric organs consist of modified striated muscle fibers, except for those of Sternarchidae, which arise from nervous tissue. They are usually flat cells with innervation on one side. They are arranged to allow a summation of potentials from depolarization of membranes and generate a current. Strongly electric fishes are the electric rays (Torpedinidae, ten genera), electric catfish (Malapteruridae, *Malapterurus* spp.), electric eel (Electrophoridae, *Electrophorus electricus*), and the electric stargazers (Uranoscopidae, *Astroscopus* spp.). The electric rays, or torpedoes, are widespread in marine waters, some living at considerable depths. They are benthic and slow, and some of the larger species, such as *Torpedo nobiliana,* are capable of delivering a shock of 220 volts. Electric catfish live in the murky water of African rivers. They are known to reach a length of about 1 m and to produce shocks of 350 volts. The electric eel, an Amazonian gymnotiform species, is a sluggish fish like the latter two and lives in water of low visibility. This relatively large fish (at least one specimen was measured at nearly 3 m long) can generate pulses of up to 650 volts, but 350 volts is a more usual maximum. Electric stargazers, marine fishes of the western Atlantic, have a habit of burrowing in sand and can deliver 50 volts.

Weak electric fishes are in families mostly found in tropical fresh waters, but one is marine. All are either benthic or semibenthic and rather sluggish. The skates (Rajidae), well known and nearly cosmopolitan in the oceans, include numerous species of small to moderate size. Mormyrids, the elephantfishes and relatives, are freshwater fishes of Africa, as are their close relatives, the gymnarchids. The mochokid (synodontid) catfishes are also from Africa. Many of these fishes are nocturnal in habit, and most live in waters of relatively low visibility. In the fresh waters of South America, the knifefishes of the order Gymnotiformes (which also includes the electric eel) form a group of weakly electric fishes that are strikingly convergent with the African mormyriforms. They are the families Gymnotidae, Apteronotidae, Sternopygidae, Hypopomidae, and Rhamphichthyidae. Some taxonomists consider the order to comprise but one broad family, the Gymnotidae (Robins et al., 1991b).

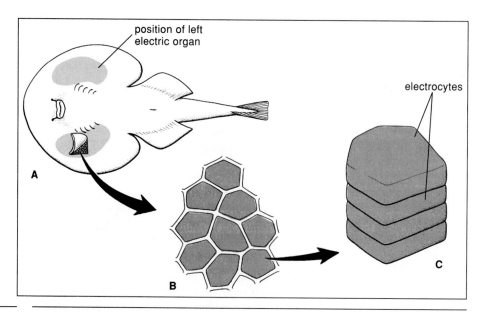

FIGURE 19–7   **A,** ventral view of electric ray (*Torpedo*), showing position of electric organs; **B,** shape of electrocytes as viewed ventrally; **C,** diagram of arrangement of one "stack" of electrocytes.

Electric Organs

Electric organs are made up of specialized cells, called electrocytes, that have evolved from muscle cells with the exception of those of the South American Apteronotidae, which evolved from nervous tissue (Bass, 1986; Bennett, 1970, 1971a). Electrocytes are typically thin (about 10 to 30 μm in torpedoes) and wafer-like and are arranged in bundles or stacks (Fig. 19–7). One surface of a typical electrocyte is heavily innervated, and the opposite face is irregular with numerous papilla-like projections. Gelatinous material surrounds the bundles or columns of electric cells, and the electric organs are rich in blood vessels, nerves, and connective tissue. Electrocytes of skates are of two types, flat and cup shaped. They are not packed tightly, and in some species they retain the striations of muscle cells. One family, Apteronotidae (= Sternarchidae), differs from the other electric fishes in that the electrocytes are modified spinal neurons, and the electric cells of muscle origin have been lost. These enlarged neurons pass forward after entering the electric organ and then loop back; they can reach more than 100 μ in diameter in both the forward- and backward-running sections (Bennett, 1971a).

Among the strong electric fishes (Figs. 19–8A–**D**), the electric eel has three separate electric organs forming a large part of its bulk. The hypaxial section of the long caudal region is made up mostly of the main electric organ with a smaller one,

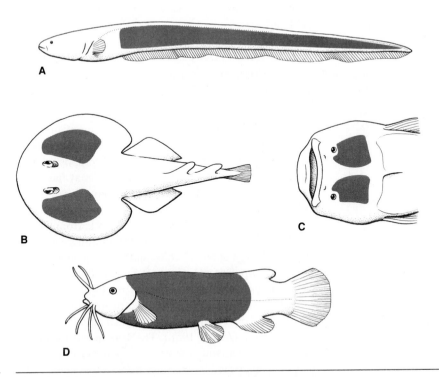

FIGURE 19–8   Diagrams illustrating positions of electrical organs in electric fishes. **A**, electric eel (*Electrophorus*); **B**, electric ray (*Torpedo*); **C**, stargazer (*Uranoscopus*); **D**, electric catfish (*Malapterurus*).

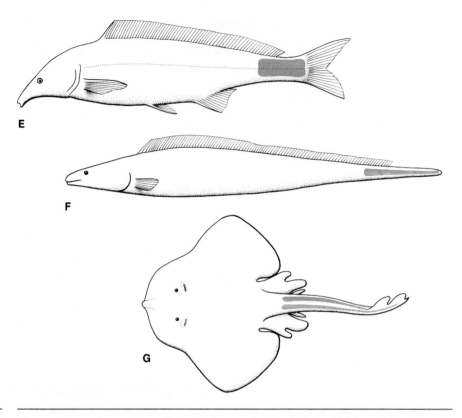

FIGURE 19–8
(Continued)
Diagrams illustrating positions of electrical organs in electric fishes. **E,** Mormyridae; **F,** Gymnarchidae; **G,** Rajidae. (**A** through **D** strongly electric, **E** through **G** weakly electric.)

the organ of Hunter, running along its ventral surface. The organ of Sachs is posterior to the main organ. These organs have formed from axial musculature, and the electrocytes are ribbon-like, flattened anteroposteriorly, and extend from the medial septum out toward the skin. A large adult may have over 100,000 electrocytes in the main organ on each side, as there can be up to 6000 vertical arrays of up to 25 of the ribbon-like cells. The posterior surfaces of the electrocytes are innervated from spinal nerves. Current flow in the organ is from back to front (Bone and Marshall, 1982), with reverse flow in the surrounding water. The organ of Sachs produces weak pulses of about 10 volts, and the frequency depends on the activity of the fish, from a few pulses per minute at rest to 30 per second while active. Hunter's organ apparently is capable of generating both strong and weak pulses (Kramer, 1990).

In the electric catfish, the electric organ is only a few millimeters thick and lies in the skin between the epidermis and a fatty layer that covers most of the body musculature (see Grassé, 1958). The several million electrocytes are disclike, about 1 mm in diameter, with a short stalk on the posterior, innervated face. Each side of the organ is innervated by branches from a large neuron in the anterior part of the spinal cord. The electric organ is derived from pectoral musculature. Current flow is from anterior to posterior.

Electric rays have a large, kidney-shaped electric organ in each side of their disc, adjacent (lateral) to the head and branchial region. The columns of hexagonal to roughly circular electric cells (electrocytes) that make up the organ are oriented vertically and reach from the dorsal to the ventral surface. The electrocytes, which reach 7 mm in diameter in some species, are derived from branchial musculature. Current flow is from ventral to dorsal. The genus *Narcine* has an accessory organ at the posterior part of each main organ.

Like the electric rays, the stargazers (*Astroscopus*) have dorsoventrally flattened electric organs in the head region. Their relatively small organs are situated posterior to the eye and are derived from extrinsic eye muscles. The flattened electrocytes may be as much as 5 mm in diameter. Current direction is dorsal to ventral.

Weakly electric fishes tend to have one or more elongate electric organs along each side in the caudal peduncle (Figs. 19–8**E–G**). Mormyrids have two columns of cells in each side of the caudal peduncle. *Gymnarchus* has four thin columns per side in the posterior half of the caudal region, and skates are equipped with one pair of organs running most of the length to the tail. In *Gymnotus,* the electric organ extends from below the head to the tip of the tail, coursing along the ventral aspect of the body. In addition to elongate, ventral main organs, some gymnotoids (*Steatogenys, Gymnorhamphichthys*) have small accessory organs under the skin of the chin region. *Steatogenys* has another small organ in the pectoral region. Apteronotids have nerve-derived lateral electric organs reaching from above the pectoral fin to the base of the caudal fin. Current flow is forward in *Gymnarchus* and backward in *Raja,* and both directions are possible in the gymnotoids and in some mormyrids (Bennett, 1971). Mochokid (synodontid) catfishes have an electric organ apparently derived from a sonic muscle above the gas bladder (Hagedorn et al., 1990).

Some weakly electric fishes (gymnotiforms and mormyrids) have larval electric organs (Bass, 1986). These are similar in structure to the adult organs but do not develop into the organs found in the adults. The larval organs reach from near the head to the caudal peduncle, whereas the adult organs are generally more posterior in location.

Functions of Electric Organs

The function of strong electric organs appears to be in the stunning of prey and discouragement of intruders or predators. Use of electricity to obtain prey has been observed in the electric rays, and its use for this purpose in others seems probable, considering the circumstances under which the electric species live. All are secretive, living near the bottom in situations that probably allow prey to approach closely without alarm (sit-and-wait predators). Although the stargazer can be especially well concealed, allowing small crustaceans or fish to move onto the sand under which the predator is buried, there is doubt that the electric discharge is of sufficient duration to stun prey (Pickens and McFarland, 1964). Electric catfishes (*Malapterurus* spp.) are known to feed on schooling clupeids and schilbeids, which would seem to be agile enough to escape from the sluggish predators. Use of the electric organs to stun several prey fish at a time was inferred in a study of the food habits of the catfish (Sagua, 1979). Rankin and Moller (1986) report that *M. electricus* uses electrical discharges against intruders in defense of shelter sites, especially against members of other species.

Electric discharges in the weakly electric fishes function both in electrolocation of nearby objects and in communication with other fishes (see Chapter 21).

Intraspecific communication seems to be the main function of electric discharge in the skates (Bratton and Ayres, 1987). The mormyrids, gymnarchids, and gymnotoids live mostly in turbid waters and some are nocturnal in habit, so vision might often be of limited use. Other senses—hearing, olfaction, and mechanoreception—must aid their orientation to their surroundings, as in other nocturnal species, but the possession of a system to generate and detect electrical fields sets these animals apart. For the most part, these electric fishes hold their bodies rigid and straight, depending on undulations of fins for propulsion. The straight posture ensures symmetry in the electrical field that they generate (Fig. 19–9). Interference with, or distortion of, the field by nearby objects can be detected by electroreceptors (see Chapter 21). In addition, electric fields can be detected by conspecifics and fishes in other species and may indicate species identity, sex, age, and reproductive state. Electrolocation forms the basis of a sophisticated communication system.

Electric organs discharge (EOD) is characteristic of individual species, with frequencies and other features of the pulses or waves differing from one species to another. Although many electric fishes have the ability to vary their EOD, oscillograph tracings show characteristic species-specific shapes and amplitudes. The EODs of mormyrids and most gymnotoids are pulsed. *Gymnarchus niloticus* and some gymnotoids, such as *Apteronotus, Eigenmannia,* and *Sternopygus,* have an EOD in the form of a wave. These are sometimes called pulses and hummers. Mochokids can produce discharges continuously or in bursts. Various mormyrids discharge from one to six pulses per second as a rule but can accelerate their discharge up to about 130 pulses per second. Voltage output in mormyrids generally ranges from about 9 to about 16. *Gymnarchus* is reported to operate at about 4 to 7 volts, discharging at about 300 pulses per second. Investigators have demonstrated a wide range of pulse frequencies in gymnotoids—from 2 up to approximately 1000 per second.

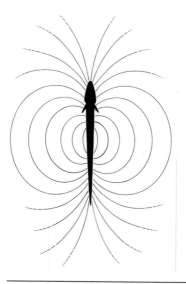

FIGURE 19–9   Diagram illustrating electric field around a gymnotoid fish.

range of pulse frequencies in gymnotoids—from 2 up to approximately 1000 per second.

Control of electrical discharges in weakly electric fishes appears to be both neural and hormonal and involves a "pacemaker nucleus" located in the brainstem (Dye and Meyer, 1986; Keller et al., 1991). Frequency of discharge can be altered by administration of steroids. Androgens decrease frequency of discharge in *Sternopygus* and *Apteronotus* when given in large doses. On administration of low doses by daily injections or by implantation, testosterone and dihydrotestosterone (DHT) can bring about large decreases of frequency in *Sternopygus,* whereas a slight increase is caused by estradiol. Long-term effects in *Apteronotus* are reversed from those in *Sternopygus* in that a decrease in frequency is caused by estrogen and an increase by DHT. This is apparently the basis for specificity of EOD depending on age and reproductive state (Dye and Meyer, 1986).

Vision

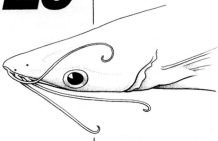

General Morphology of Fish Eyes

Although there are many modifications of eye shape and structure among fishes, the general plan is similar throughout. As with other vertebrates, the major features of the eye are an anterior chamber, an iris, a lens, and a vitreous chamber containing the vitreous humor and lined by the retina (Fig. 20–1). The entire structure is covered by the sclerotic coat, a tough covering of connective tissue that is transparent in the region where the eye is in contact with the water. The transparent section is the cornea. The sclera of elasmobranchs and teleosts may be stiffened by cartilaginous structures or, in the case of the latter, by scleral ossicles.

The eye is generally nearly spherical, but usually the corneal surface is flattened so that the spherical lens is nearly in contact with the cornea. Some fishes have nonspherical lenses. Stingray lenses are flattened to the extent that the equatorial diameter exceeds the axial diameter by 18 percent. In many sharks, the equatorial diameter exceeds the axial diameter by 12 to 16 percent. Some deep water teleosts have slightly flattened lenses, and *Trachipterus* (Trachipteridae) has the equatorial diameter greater than the axial diameter (Sivak, 1990). Some highly compressed species, such as butterflyfishes, have extremely flattened eyes—to the point that they are described as disclike (Bauchot et al., 1989). A pigmented, vascular, choroid layer separates the retina and the sclera. The choroid layer is continuous with the iris and prevents blurring of images caused by internal reflection of the retinal images

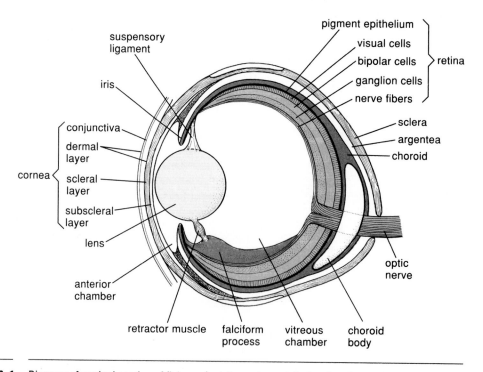

FIGURE 20–1   Diagram of vertical section of fish eye (not drawn to scale), showing the relationships of its parts.

back to the retinal photoreceptors (Ali and Klyne, 1985). A choroid body or "gland," actually a rete mirabile, is prominent in the choroid of many teleosts.

Lampreys differ from other fishes in that there is no circumorbital sulcus, the invagination of skin around the outward portion of the eyeball. Instead, the skin is continuous over the eye, forming a dermal cornea or "spectacle" separated from the scleral cornea. The lamprey eye is rather solidly attached to the rim of the eye socket; the eyeball lacks any cartilaginous or bony reinforcement, although the eye socket has some cartilage medially and ventrally (Nicol, 1989).

Suspension of the slightly flattened lens of the elasmobranch eye is by a band of gelatinous tissue that attaches around the equator of the lens. This suspensory material has a relatively robust dorsal portion. Some species have a ventral papilla called the pseudocampanule that aids in suspension of the lens and may contain smooth muscle fibers that can protract the lens (Nicol, 1989).

The lens in bony fishes is suspended mainly by a dorsal ligament but is attached by soft material all around its circumference. The retractor muscle (*campanula Halleri*) of the teleost lens, which is ventral to the lens and is attached to the anterior end of the falciform process, is an outgrowth of the choroid that extends through a fissure in the retina (Fig. 20–1). The retina consists of a pigment epithelium; the visual receptor cells (rods and cones); bipolar cells that connect the visual cells to the ganglion cells, which are closest to the vitreous humor; and the nerve fibers leading to the optic nerve. Horizontal cells with large cell bodies connect between visual cells; amacrine cells, which lack axons, form horizontal connections between ganglion cells (Hawryshyn, 1992; Wagner, 1990).

Cone photoreceptors differ from rods in that they have short outer segments that are conical instead of long, cylindrical ones. The inner segments of the cones approach the outer segments in size, whereas the inner segments of the rods are smaller than the outer segments (Fernald, 1993). Cones in teleosts differ in length, with the longest sensitive to long light waves and the shortest to short light waves (Fernald, 1993).

Function

The Eye and the Light Environment in Water

The eyes of fishes show many structural adaptations to their visual environment. Fishes live in a medium that has optical properties much different from those of the atmosphere. Depending on the angle of incidence of light, a calm water surface can reflect up to 80 percent or more of light striking it. If the water is rough, there is great variation in the transmission of light regardless of the angle of incidence. The bending of light rays entering water is such (approximately 48.625°) that a fish in water with a perfectly smooth surface views objects above the water through a circle ("Snell's window") subtended by a 97.2° cone above each eye (Fig. 20–2). Nearly all objects from horizon to horizon appear in the circle, which is surrounded by the reflective undersurface seen beyond the limits of the cone. In rough water, the circular window in the surface is broken up and light is transmitted through ever-changing patterns (Walls, 1963).

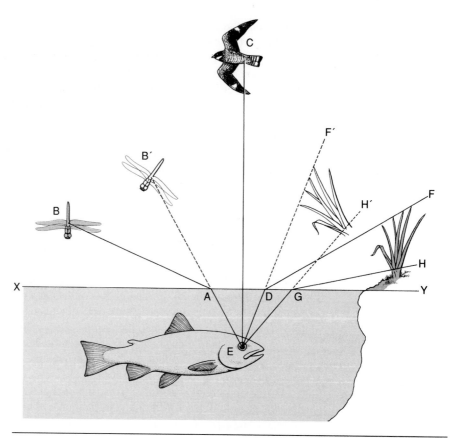

FIGURE 20-2 Diagram showing refraction of light entering with a perfectly flat surface (*XY*). Because of the bending of the light rays, the fish's eye (*E*) does not receive light striking the surface above the shaded area. The bird at *C*, directly above *E,* is seen in its actual position. The insect at *B* is perceived as if at *B´*, and the angles *EDF* and *EGH* cause the plant to be seen as if the top were at *F´* and the bottom at *H´*.

Light that enters water is absorbed rapidly and differentially, with red, for instance, attenuating rapidly and blue penetrating to greater depths. Absorption is defined by Loew and McFarland (1990, p. 3) as "the change of electromagnetic energy into some other form," and light that is redirected by reflection, refraction, and diffraction is said to be scattered. The various species of fishes must cope with life in a variety of habitats, including the bright surface, brilliant coral reefs, dimly lit caves, sheltered forest streams, dark bogs, and ocean depths where sunlight cannot penetrate and the only light is that of bioluminescent creatures. Some even invade the land or commonly peer above the water's surface and encounter problems of vision in the very different medium of air. Considering the many habitats in which fishes exist, it is evident why they have evolved visual systems that can be of use in so many ways. In many fishes, sight is not the primary sense, and other senses (such as olfaction and the mechanosensory lateral line or electrosensory system) are of greater importance.

The eyes of most fishes are placed on the sides of the head so that they have wide lateral fields of vision. The lateral placement does not preclude binocular vision in certain segments of the visual field, for eye placement or adaptations of various parts of fish eyes allow for binocular vision. Some species (Fig. 20–3) specialize in the inspection of more restricted parts of their surroundings and, for example, have eyes set forward (*Gigantura* and others) or upward for binocular vision (*Argyropelecus* and others). Some species have the eyes positioned for a wider field of vision below (*Hypophthalmichthys*) or above, as in many top-minnows that feed from the surface or many bottom fishes that have no need of vision below. The four-eyed fish of the ocean depths, *Bathylychnops* (Fig. 20–3**D**), has a small accessory retina that monitors the environment below the fish while the main section of the eye looks obliquely upward. Some bottom-living flatfishes and stargazers have eyes on short stalks (Fig. 20–3**E**).

Eyelids, except for those of some elasmobranchs, are not well developed in fishes. Eyelids of most sharks move very little, but in some genera (such as *Ginglymostoma, Galeorhinus,* and *Cephaloscyllium*) the lids can partially or completely cover the eye (Nicol, 1989). Some sharks have a third eyelid, the nictitating membrane, which is attached at the anteroventral aspect of the eye and fits beneath the

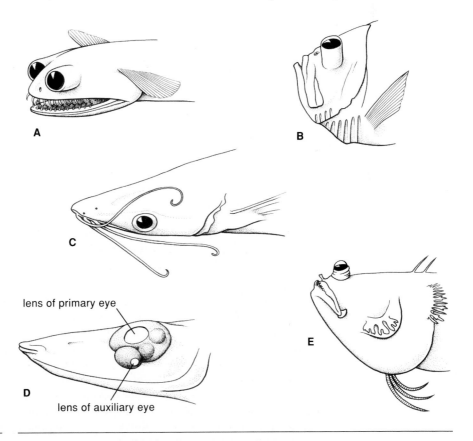

lens of primary eye

lens of auxiliary eye

FIGURE 20–3   Examples of fishes with unusual eyes. **A,** *Gigantura;* **B,** *Argyropelecus;* **C,** *Hypophthalmus* (eyes directed obliquely downward); **D,** *Bathylychnops;* **E,** *Dactyloscopus.*

lower lid, from which it is developed. It can be moved upward and obliquely backward to cover and protect the surface of the eye. Coverage is complete in some carcharinids. The so-called adipose eyelids of some sharks and some teleosts (herrings and tunas, for example) are immobile and serve to streamline the slight bulge of the eye beyond the surface of the head.

In most fishes, appreciation of a large field of view is achieved by placement of the spherical lens so that it bulges through the opening of the pupil and nearly touches the cornea. The lens can thus gather light from nearly all of each lateral field.

The light-gathering ability of fish eyes can be appreciated by considering the relationships of some of their physical dimensions. One of these is a relationship called Matthiesen's ratio, which is the focal length divided by the lens radius. This ratio commonly ranges in various species from about 2.17 to 2.56, with the ratio for a given species being fairly constant. The *f*-number of the eyes of fishes (Matthiesen's ratio divided by 2) is about 1.1 to 1.3, which is comparable to a fast camera lens. The short focal lengths of fish lenses allow for a sufficient depth of field.

The cornea of teleosts is usually made up of four layers (Lythgoe, 1975a). Externally there is a multicellular epithelial layer, and inside that there is a collagenous stroma that is separated from the inner layer (endothelium) by the thin Descemet's membrane. The cornea has a refractive index approximating that of water, whereas the lens is constructed to have an effective refractive index of about 1.67 (Munz, 1971). Fish lenses studied by Fernald and Wright (1983) proved to have a gradient of refractive index, with the less compact outer cells showing an index of 1.38 and the cells near the center having an index of 1.56. These authors pointed out that because light curves in a spherical lens, the gradient of refractive index makes sharp focus possible.

Visual Cells and Pigments

Visual cells of fishes include rods, cones, and double cones (including "identical twin" cones), with subtypes in various species (some have both long and short single cones). Furthermore, the cones within some species can respond to blue, green, or red. This variety indicates that certain cells are specialized to respond to specific wavelengths and light intensities. Discrimination of wavelengths has been demonstrated in the goldfish by behavioral training (Naumeyer, 1986), reaction to polarized light by recording physiological response. Hawryshyn and McFarland (1987) used conditioned response to demonstrate that cones sensitive to ultraviolet, green, and red wavelengths showed evidence of sensitivity to polarized light, whereas blue-sensitive cones showed no such sensitivity. The presence in the Japanese dace (*Tribolodon*) of pigment that absorbs ultraviolet light has been shown by microspectrophotometric analysis (Harosi and Hashimoto, 1983).

Although rods dominate in the retinas of deep-dwelling fishes, cones are known in deep water members of the following families: Chlorophthalmidae, Diretmidae, Agonidae, Omosudidae, Notosudidae, Scorpaenidae, and Zoarcidae. However, only double cones are present in these families. In the splitnose rockfish (*Sebastes diploproa*), a species that has shallow-living larvae and small juveniles that stay near the surface for about a year prior to moving to deep water, Boehlert (1978, 1979) has shown ontogenetic changes in number and type of cones. As the *S. diploproa* grows and as the individuals migrate to deep water and vision adapts to a dimmer environment, cones decrease in density and rods increase. Single cones disappear, apparently by fusion into double cones.

Cones are set in regular patterns in many species and are often arranged in a mosaic of squares (Wagner, 1990). Other patterns may approximate rows, circles, crosses, or triangles. This regular arrangement ensures that the most important parts of the retina are supplied with each type of cell specialized for the various wavelengths of importance to the vision of the species (Lythgoe, 1979).

Rods function in dim light and outnumber the cones in fishes with duplex retinas (retinas with both rods and cones). Many elasmobranchs and deep sea teleosts have either pure rod retinas or retinas with only a few cones. Several elasmobranchs (representing about ten families) that have duplex retinas show rod–cone ratios of 5:1 to 100:1, with most between 7:1 and 13:1. In species that commonly feed in dim light, numerous rods may connect with only a few bipolar cells, which in turn connect to a single ganglion cell. This convergent adaptation aids in summing the visual input and increases sensitivity in subdued light. Several groups of fishes, including lampreys, cusk-eels, congers, and various deep sea families, have two to several layers of visual cells (multiple-bank retinas) that could increase the efficiency of light interception. In some species, there are more layers of cells at the fovea than in other parts of the retina. In *Chauliodus,* the number of retinal layers increases with size of the fish, and this is correlated with shifts to a deeper habitat (Locket, 1980). The numbers of rods in nocturnal or deep sea species can reach up to 20 million per mm^2. Some of these fishes have retinas that absorb 90 percent or more of the light striking them, so the thresholds at which they can detect light are probably lower than that of humans, in which the retina absorbs about 30 percent of the blue-green light striking it.

The photosensitive pigments in visual cells of vertebrates are rhodopsins and porphyropsins. When exposed to light, changes in these pigments cause excitation of nerve cells and result in the sensation of vision. These compounds are rhodopsins, based on retinal from vitamin A_1, and porphyropsins, based on an aldehyde of vitamin A_2 (3-dehydroretinal), plus a protein (an opsin) (see Bowmaker, 1990; Munz, 1971).

Freshwater fishes generally have mainly porphyropsin in their retinas and marine fishes have mainly rhodopsins, but many freshwater species and a few marine species have both of the two pigments. Elasmobranchs have rhodopsin almost exclusively, as do many cyprinids and a few other teleosts. Changes in pigments and pigment ratios have been documented in several species in response to quality of seasonal light, to temperature, hormone level, life history stage, and diet (Beatty, 1984; Muntz and Mouat, 1984). For instance, some species of anadromous fishes—for example, *Petromyzon marinus, Morone americana* (Ali and Klyne, 1985), and *Oncorhynchus* spp. (Muntz, 1971)—may have a preponderance of one pigment over the other at certain life history stages. Pacific salmon are known to change from rhodopsin to porphyropsin on the spawning migration. Among freshwater fishes there is much variation in the proportions of the two pigments, and these proportions vary in given species with age, season, and temperature (Beatty, 1984).

As young pollack move from shallow to deep water at a length of about 50 mm, a progressive change in the maximum absorbance of light (λ_{max}) from the violet range to blue takes place (Shand et al., 1988).

Most fishes are thought to have color vision, and Ali and Klyne (1985, p. 171) state that no fish that has been "properly investigated has been shown to be colour blind." This may apply even to species with pure-rod retinas, and Bowmaker et al. (1988) state that *Malacosteus* and other deep sea genera with paired visual pigments (see Bowmaker, 1990) have potential for color vision although they have rod-only retinas. Pigment pairs arise because the same opsin can form two different pigments depending on whether it binds with a retinal from vitamin A_1 or A_2.

That some species of fish do respond to color has been demonstrated by means of behavioral experiments, but some early attempts may not be acceptable because of failure to match brightness of light in the trials (Douglas and Hawryshyn, 1990).

Visual pigments of fishes respond to light of a considerable range of wavelengths. Maximum absorbance ranges from about 360 nm (UV) (Harosi and Hashimoto, 1983) to 625 nm (red) (Levine and MacNichol, 1979). However, the actual spectral sensitivity may be broad; for instance, Bowmaker (1990) mentions that the cod *Gadus morhua* has double cones with λ_{max} of 517 nm and single cones with λ_{max} of 446 nm but has a range of spectral sensitivity from about 400 nm to over 650 nm. From the standpoint of absorbance of wavelengths, Bowmaker (1990) considers marine fishes to fall into three broad ecological categories: (1) deep sea species with most visual pigments adjusted to the dim blue light of the environment but some capable of absorbing red light from bioluminescent photophores; (2) coastal species living in blue–green light with visual pigments absorbing maximally at 440 to 460 nm (violet to blue) and 520 to 540 nm (green); and (3) tropical reef species in an environment of strong blue light and having maximum absorbance at about 495 nm and about 513–530 nm (green). Many reef fishes have yellow corneas or lenses that reduce the strong blue light.

Most freshwater species have pigments with maximum wavelength absorption (λ_{max}) of 498 nm to 535 nm, but the portion of that range differs with species, habitat, food habits, and age. Levine and MacNichol (1979) suggested four overlapping categories of freshwater fishes according to behavior and ecology. Group I comprises diurnal species from shallow water. These typically have eyes with some single cones that absorb maximally at wavelengths down to about 410 nm in addition to double cones and twin cones that have λ_{max} of up to about 580 nm (yellow). *Anableps anableps* has single cones with λ_{max} at about 409 nm but has two types of double cones. One type of double cone is not identical and has one member with about λ_{max} 463 nm and the other at 576 nm. The other type of double cone is identical, with both members at 576 nm.

Group II consists of the many diurnal mid-water species with generalized habits, which generally have eyes sensitive to blue and green. Single cones have λ_{max} at about 460 nm. A pigment in double cones absorbs maximally at 540 nm (green), and other double cones are sensitive to red (λ_{max} = 580–629 nm). Some species have small single cones that absorb in the ultraviolet range (355–390 nm).

Group III comprises mostly crepuscular mid-water fishes of predatory habits. These have single cones sensitive to greens in the range of about 500–540 nm and double cones absorbing at wavelengths in the red range. *Amia* has λ_{max} of 554 and 613 nm in double cones. Pigments sensitive to blue are absent in this group. This may be advantageous in improving visual acuity (Bowmaker, 1990).

Group IV is made up of nocturnal and crepuscular benthic species with but two types of visual cells—single cones and rods. One type of cone has λ_{max} at about 530–540 nm, another at over 600 nm. Rods absorb maximally at 540 nm.

Hawryshyn (1992) points out that many fishes, including carp, have four types of cones. Short cones have peak sensitivity at about 460 nm, middle cones at about 530 nm, long cones at about 600 nm, and ultraviolet cones at around 380 nm. There is considerable overlap in sensitivity, with long cones absorbing relatively well in the UV range. Cones sensitive to ultraviolet showed a greater response to vertically polarized light than the middle and long cones, which were better at sensing horizontal polarization (Hawryshyn, 1992).

The usual range of maximum absorption shown by visual pigments of marine species is from 477 nm to 522 nm (Munz, 1971). Deep sea species tend to have pigments that maximize vision in a dimly lighted blue environment. Some of these apparently have some porphyropsin along with rhodopsin. Most of the deep-dwelling species studied by Partridge et al. (1988) had pigments ranging in maximum absorption from 475 nm to 488 nm. Two species had two visual pigments each, one with λ_{max} of 466 nm and 500 nm, the other with 478 and 485 nm. Levine and MacNichol (1979) point out that fishes living in the same general environment will have differing spectral absorbancies depending on microhabitat, niche, and other factors. Species from shallow waters, however, tend toward maximum absorbance values in shorter wavelengths than those of benthic or crepuscular species.

Focus, Accommodation, and Regulation of Light

Focus and Accommodation. Several species of fishes cannot focus images sharply because of spherical aberration of the lens (Kreuzer and Sivak, 1984). Light transmitted through the periphery of the lens tends to be focused at a slightly different point from the light transmitted through the center of the lens, thereby creating a blurred image. Visually oriented predators (trout and pike) have lenses optically superior to those of fishes more dependent on chemical senses (bullhead, carp, goldfish).

Chromatic aberration resulting from light of different wavelengths focusing at different distances from the lens is small in fishes because of the gradient of refractive index in the lens, ranging, among fishes studied, from about 2 percent of focal length in a cichlid to 5.3 percent in goldfish, with most fishes studied at 4 to 5 percent. This aberration is probably not large enough to cause serious problems in vision (Fernald, 1990; Sivak, 1990).

Although the lens of the typical teleost eye is spherical, or nearly so, the vitreous chamber is not, giving the retina an ellipsoid shape (Fig. 20–1). As a result, relatively distant objects lateral to the fish are in focus and objects close to the fish are not. Objects that are close and right in front of the fish in the binocular field are in better focus than more distant objects. Accommodation by teleosts to distant vision in the anterior field is accomplished by moving the lens posteriorly by means of the retractor lentis muscle. The accommodative movements of teleosts are generally nasal-temporal (see Fernald and Wright, 1985), but some move the lens slightly toward the back of the eye (*fundus oculi*).

Fishes differ in their ability to accommodate. Most teleosts (Somiya and Tamura, 1973) were found to have well-developed, triangular lens muscles and to accommodate very well. A few freshwater species (largemouth bass, bluegill, and snakehead) shared the well-developed muscle and good accommodation, but most freshwater fishes had less well-developed muscles and moderate powers of accommodation. The lens muscle appears to be of small diameter in many freshwater species, and in some with poor accommodation (*Anguilla,* some catfishes) the muscle is slender and weak. Among marine species, *Mugil cephalus* was noted as having a nonfunctional lens muscle, and a few other species showed little or no accommodation.

Lampreys differ from other fishes in the mechanism for accommodation to near and distant vision. The lens is not suspended from the interior of the eyeball, as in other fishes, but rather is held in place by the pressure of the fluid in the vitreous cavity. In accommodation for distant vision, the lens is forced back by contraction of a muscle that flattens the cornea.

Although elasmobranchs are said to accommodate by moving the lens forward, Somiya and Tamura (1973) noted no lens movement or deformation of the eyeball in the four elasmobranchs in their study. Sivak (1991) cites work showing accommodative ability in two sharks and a stingray. In some fishes, including skate, some rays, mudskippers, and some flatfishes, there is a type of inactive accommodation (Ali and Klyne, 1985). The eyeball diverges considerably from the spherical, placing the lower part of the retina close to the lens so that distant objects above the level of the eye are in focus and close objects at the level of the eye or below it are also in focus. This "ramp retina" allows the fish to accommodate, if necessary, by changing the position of the head. Many elasmobranchs can reduce the pupil to a very small aperture (or two or more separated small apertures) when in bright light. This probably creates an effect similar to a pinhole camera, giving reasonably good focus to both close and far objects. In the barrel-shaped eyes of certain deep sea species, the immobile lens allows for no accommodation, but the retina is specialized by division into two parts so that objects at two different distances can be seen in focus.

Generally, elasmobranchs have been thought to be hypermetropic (farsighted) and teleosts myopic (nearsighted) because of their mechanisms for accommodation. Retinoscopy (study of the refraction of the eye) by several investigators has demonstrated that many teleosts are apparently emmetropic or are slightly farsighted, having relatively small refractive errors. However, because some researchers have measured to the nearest surface of the retina and some have measured through to the back, there is still doubt about the degree of significance of the error. Judging from the distribution and architecture of visual cells in the retina, the most acute vision is aimed at anterior objects. In fishes with a duplex retina, cones are most numerous in a retinal area in the posterior (temporal) part of the eye, where images from the anterior field of view are focused. In deep sea fishes with rod-only retinas, the rods of the posterior area of the retina have elongated, light-sensitive sections.

Several species of teleosts, including shallow and deep water marine species and some freshwater fishes, have been shown to possess a fovea, which is a small depression in the posterior part of the retina at the site of the greatest concentration of visual cells. The fovea, depending on species, may be shallow and saucer shaped or deeper with something of the shape of a trumpet bell. The shallow shape is said to aid in maintaining a visual fix, and the deeper type can refract the light through the convex sides to a great number of visual cells and may, because of its optical properties, aid the fish in judging distance in monocular fields (Harkness and Bennet-Clark, 1978).

In many teleosts, the iris is shaped to allow the greatest possible oblique view forward (Fig. 20–4) even to the point of providing an opening large enough to let light coming from the lateral field strike the anterior part of the retina without pass-

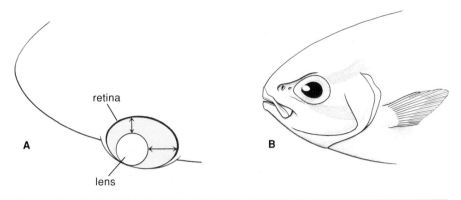

FIGURE 20–4 **A,** Diagram of eyeball shape and placement of lens in relation to retina in many teleosts. Close objects in the anterior field can be in sharp focus, whereas the lateral field is adapted to more distant vision. **B,** Diagram of elliptical eye shape (Girellidae) that allows for a large anterior field of vision by a laterally placed eye. Directly in front of the eye is a groove that facilitates forward vision.

ing through the lens (Muntz, 1971; Walls, 1942). Fernald (1993) states that eyes of fishes function at such low *f*-stops that the iris cannot keep aberrant light from reaching the retina, and some species have irises with diameters larger than the diameter of the lens. The effect of these aphakic apertures on vision is still under investigation. However, in deep sea fishes in which aphakic apertures occur around the lens, the apertures are thought to increase the illumination of the central part of the retina and to aid in perception of lateral objects under conditions near the threshold of vision. If the lensless aperture is forward of the lens, the posterior part of the retina receives the extra illumination.

Regulation of Light. Regulation of light as it enters or after it enters the eye and adaptation to light or dark is accomplished by several means: (1) Retinomotor mechanisms can move pigment or visual cells to cover or expose those cells; (2) contractile irises can reduce the amount of light admitted; (3) pigment in the cornea, iris, or lens can filter light; (4) a reflective tapetum lucidum in the choroid or retina can reflect light back through the visual cells; and (5) from a behavioral standpoint, the fish can swim to or away from the source of light.

Light and dark adaptation in many teleosts is accomplished by retinomotor movements of pigment and visual cells (Fig. 20–5). Pigment cells in the outer layer of the retina contain processes through which melanin can move to or from the outer parts of the visual cells. Under bright illumination, the eye adapts as a result of the movement of melanin toward the visual cells and movement of the outer segments of the rods into the pigmented area, where they are shielded from the light. In dim light, the pigment is drawn back and the contractile or myoid part of the rods pulls the rods away, allowing the receptors to be exposed to light. Movement of the cones is opposite that of the rods, but the cones are not usually hidden by the pigment. In wrasses (Labridae), a red pigment is prominent in the pigment cells, which also contain melanin. Cylinders of pigment are extended to screen rods and long cones, and

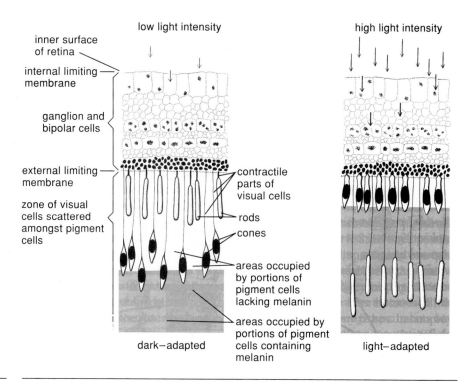

low light intensity

high light intensity

inner surface of retina

internal limiting membrane

ganglion and bipolar cells

external limiting membrane

zone of visual cells scattered amongst pigment cells

contractile parts of visual cells

rods

cones

areas occupied by portions of pigment cells lacking melanin

areas occupied by portions of pigment cells containing melanin

dark—adapted

light—adapted

FIGURE 20–5   Diagram illustrating movement of rods, cones, and pigment in the retina of teleosts. In light adaptation (right), the rods are moved away from the light and are protected by forward movement of pigment.

in light adaptation (photopic vision) the red pigment, which mainly absorbs rays under 560 nm and when extended does not allow any other light but red from reaching the long single cones and rods (Fineran and Nicol, 1974). In dark adaptation (scotopic vision), the rods shorten and are not screened by the red cylinders. The time required for pigment movement in the tapetum of sharks or the shifting of pigment and visual cells in teleosts is considerable—about 2 hours for advancing or receding in the shark tapetum, and about 30 minutes for light adaptation and 1 hour or more for dark adaptation in teleosts (*Oncorhynchus*).

Most elasmobranchs and a few teleosts have contractile irises that can control the amount of light entering the eye. Contraction of the iris is usually more rapid than dilation. Some teleost species have mobile pupils under nervous control, but the eels of the genus *Anguilla*, like elasmobranchs, appear to have irises that contract in direct response to light (Seliger, 1962; Young, 1933). Most rays, many flatfishes (especially bothids), stargazers, and the armored catfishes (*Plecostomus*) have a specialized iris that forms a pupillary operculum that can expand to leave a thin U-shaped aperture and cut off most of the light reaching the pupil. Similarly, the pupillary operculum of rays may have a fringe with up to a dozen or so projections that form several small apertures in a crescent or U-shaped configuration around the lateral and ventral aspects of the pupil, effectively diminishing entering light but maintaining a greater visual field. (See Murphy and Howland, 1991.)

The eyeshine of fishes, including sharks and *Latimeria,* can be caused by various layers of reflective material associated with the choroid or retina or, in some fishes, the cornea (Cohen, 1990). A choroidal tapetum lucidum is present in most elasmobranchs, and a retinal tapetum is present in many freshwater teleosts (see Best and Nicol, 1980). These tapeta are layers that usually involve guanine crystals as reflectors, which are as efficient as a good mirror. A different type of choroidal tapetum consisting of fibers that are reflective like a tendon is seen in some marine fishes (Muntz, 1971). The tapetum acts to reflect light that has passed through the visual cells back into the system and is an effective adaptation for sight in dim light. In sharks, the cells in which guanine provides a reflective surface are parallel to the plane of the retina throughout the fundus of the eye but are oblique peripherally, so they are always perpendicular to the light entering the eye, thus increasing their efficiency as reflectors. Melanin is associated with the guanine reflectors and can migrate along the reflectors and mask the reflectivity. This has two possible functions: One is to adapt the eye for vision in bright light, while the other may be to reduce eyeshine and make the fish less conspicuous. Extraneous light is prevented from entering the eyeball through its walls, by the stratum argenteum, a reflective layer outside the choroid. This is of importance in fishes with translucent tissue around the orbits, such as the catfish genera *Ompok* and *Kryptopterus.* Irises usually have reflective or opaque layers.

Some fishes have pigment in the cornea or lens that acts as a filter and eliminates certain light, especially short-wavelength light. Some benthic species in several families have specially constructed iridescent corneas that protect the eyes from strong down-dwelling light without greatly affecting light entering the eye from the lateral field of view (Lythgoe, 1975a). Iridescence is produced when light passes "from one medium through a thin layer of another which has a different refractive index" (Lythgoe, 1975b, p. 263). There are about six types of iridescent corneas, all with lamellae less than a lightwave thick, which, because of their regular arrangement and thinness, refract and reflect light to cause iridescence when illuminated from above; this reduces the light that enters the eyes (Lythgoe, 1975a, 1975b).

The semiterrestrial blenny *Dialommus fuscus* has a photosensitive pigment in its nonconstricting iris and in its aqueous humor, which, when the fish is out of water in bright light, turns dark. This change in optical density apparently protects the retina from overillumination in air. Gnyubkin (1989) showed that the yellow cornea of *Hexagrammos stelleri* reacts to the level of illumination by dispersing or aggregating pigment, but the reaction is slow, requiring about 100 minutes.

Visual Adaptation and the Life of Fishes

Dependence on sight varies greatly among fishes, as does the relative size of the eyes. Sight-feeding diurnal fishes such as trout, bass, and sunfishes have prominent eyes with a diameter equal to about one fifth or one sixth the length of the head. The pike is a daytime predator with eyes that seem small (about one tenth the head length) but has an extremely long head, not an especially small eye. Crepuscular or nocturnal fishes that hunt by sight and at least partly sight-dependent deep sea fishes tend to have eyes that are one third to one half the head length. Wide pupils and the tremendous numbers of cones (up to 20 million per mm^2) in the retina aid in the increase of sensitivity in such eyes (Marshall, 1979). Those nocturnal or deep water fishes that depend largely on olfaction, taste, the mechanosensory lateral line, or electroreception tend to have reduced eyes. These small-eyed fishes include such freshwater fishes as mormyrids, gymnarchids, and various catfishes that are nocturnal or found in turbid waters. In the marine environment, bathypelagic fishes, such as gulper eels, snipe eels, and ceratioid anglerfishes, as well as many demersal species have relatively small eyes (Marshall, 1966).

The quality of eyes of the same size can differ markedly from species to species. The number, disposition, and types of visual cells, connections of the cells to the optic neurons, mechanisms for accommodation, effectiveness of the tapetum lucidum, and so on determine the efficiency of the eye. In a study of 31 shallow water species off New Zealand, Pankhurst (1989) found that, based on photoreceptors and eye morphology, fishes of differing ecological function or distribution had differing visual structure and abilities. Carnivores had larger eyes than herbivores, relative to size of the body. Diurnal species appeared to have better visual acuity than nocturnal species, but, based on retinal features, the nocturnal species had better sensitivity to light. Relatively large eyes were present in small nocturnal species and in planktivores.

Some of the most remarkable adaptations of eyes are seen in marine fishes living in moderate to great depths. In some, the parts of the eye retain approximately the same proportions as those of shallow water species but are enlarged and occupy a greater portion of the head. In others, the eye is tubular with a greatly enlarged lens and a large pupillary opening admitting light to the large lens, which focuses it on a relatively small retina. In these greatly modified eyes, as in virtually all fish eyes, the distance from the center of the lens to the retina (focal length) is about 2.55 (in some as low as 2.10) times the lens radius, conforming to Matthiesen's ratio (Walls, 1965), so optical properties may be similar to more "normal" fish eyes. These eyes are fixed, lack mechanisms for accommodation, and mostly receive clear images from one direction only. Some have two effective parts of the retina, or an accessory retina, and can form images coming from two different directions. Usually a kind of accessory lens (lens pad) aids in directing light to the accessory retina. The genus *Ipnops* lacks eyes but has visual cells located on top of the flattened head (Marshall, 1979).

Some deep water fishes, including hatchetfishes, *Argyropelecus;* scaleless dragonfishes, *Echiostoma;* spinyfins, *Diretmus;* greeneyes, *Chlorophthalmus;* and pearleyes, *Scopelarchus,* have yellow lenses. Although a yellow filter reduces the amount of light reaching the retina, it may function to make photophores that emit the light that matches the dim daylight seem brighter so that functions such as intraspecific communication or predation can be better accomplished (Muntz, 1976). Somiya (1979) suggests that the yellow filter functions mainly in cutting down short-wavelength light in species that have some special strategy for use of the dim long-wave light. For instance, *Echiostoma barbatum* has yellow lenses, red photophores, and red-sensitive visual pigment. Objects such as potential prey or predators illuminated by red light from *Echiostoma*'s photophores, or other fish generating red light, would be visible to *Echiostoma.* Other genera that have red photophores and red-sensitive visual pigments are *Pachystomias* and *Malacosteus.* Because most deep sea fishes are not sensitive to red light and cannot see the red photophores, these three genera have an advantage in illuminating prey predators without being seen clearly (Munk, 1982).

An eelpout from the Bering Sea, *Opaeophacus,* which lives near the limit of light penetration, has a strange modification of the eye lens. There is a vertical trench in the lens, filled with a jelly-like material that is suspected to have a refractive index less than that of the lens, so that light striking the trench would be distributed over a wide area of the retina and visual acuity would be diminished. This might be an aid in detecting movement of objects silhouetted against the dim light (Bond and Stein, 1984). (Unfortunately, the few specimens available were all preserved, so no live or fresh material could be studied.)

Fishes that apparently cannot focus a sharp image on the retina because of a flattened lens are reported by Munk (1984). These include five species of anglerfishes and the gulper eel. Some mormyrids are known to have flattened lenses. Munk believes that these small, optically nonadjusted eyes would be of best use in detecting movement.

A few fishes are adapted to aerial vision, and some genera have eyes greatly modified so that both aerial and aquatic vision is good (Fig. 20–6). The four-eyed fish of Central America and northern South America (*Anableps*) swims at the surface with the upper half of the eye exposed to the air. The iris is modified so that two flaps divide the pupil of the eye at the level of the water surface. The lens is egg shaped, so light entering from above the water's surface passes through a short axis, which compensates for the added refraction of light by the aerial cornea (Schwassman and Kruger, 1965), and light entering from the water passes through a long axis. The upper part of the cornea is thicker and is curved more than the lower. Sivak (1976) reports the refractive index of the *Anableps* cornea to be 1.51. That is greater than the refractive index of the corneas of other fishes; Charman and Tucker (1973) state that the cornea of the goldfish has a refractive index of about 1.33. The lower part of the retina has more receptor cells than the upper. *Anableps* must submerge its eyes frequently to prevent drying. A blenny (*Dialommus*) of the Galapagos Islands has eyes that are divided by vertical septa. The two corneal surfaces are flat and are at a 110° angle to each other. When this fish is out of the water, it can focus clearly on objects because of the 1.0 index of refraction of its eye fluids, but the two angled corneal surfaces cause two images to be formed. Under water, there is a single image (Stevens and Parsons, 1980). A similar eye is seen in a clinid, *Mnierpes macrocephalus* (Graham and Rosenblatt, 1970). Some flyingfishes have three flattened surfaces on the cornea, but Ali and Klyne (1985) believe that there is little difference between aerial and aquatic images. Mudskippers (*Periophthalmus*), which spend much of their time on mudflats or among mangrove roots, have prominent eyes set high on the head. The lens of the eye is flattened more than in most fishes, so that aerial vision is good. A pocket below the eye carries moisture, and the eye can be retracted into this pocket to prevent drying. An Indian mullet, *Rhinomugil corsula,* is convergent with *Anableps* in the body and head shape and the placement of the eyes (Hora, 1938); Walls, 1963). *R. corsula* can hold its eyes above water as it swims at the water surface. Aerial vision appears to be good, for the slightest movement of an observer is sufficient to send this nimble fish below the surface to pop up again at a safe distance. The archerfish *Toxotes jaculator,* although it keeps its eyes

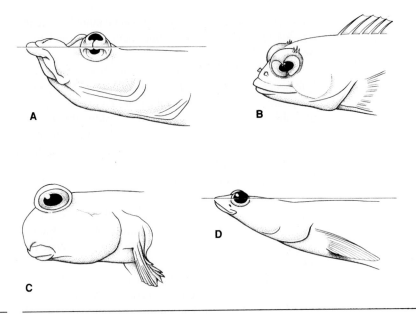

FIGURE 20–6   Fishes with eyes adapted to aerial vision. **A,** *Anableps,* in which the eye is modified to have an aerial (above) and an aquatic (below) aperture; **B,** *Dialommus,* in which the angled cornea with two flat surfaces causes double images on the retina in aerial vision; **C,** *Periophthalmus,* which spends much time completely out of water; **D,** *Rhinomugil corsula,* which often swims with the eyes above the water surface.

below the surface, has excellent aerial vision and can squirt water at insects and other small prey with great accuracy. The eyes of the archerfish are little modified. It has a narrow (14°) field of binocular vision and no fovea. Walls (1963) attributes its accuracy to a type of range-finder in the brain.

The ability to sense polarized light (see Cameron and Pugh, 1991; Dill, 1971; Hawryshyn, 1992) occurs in many fishes, although the functional advantage to the fish is not definitely known. There is speculation that navigation of migratory species is in part facilitated by that ability and that the fact that many objects, including fish scales, polarize light under water might mean that certain objects of biological importance can be recognized readily.

Many fishes in several taxonomic groups are functionally blind. A few are completely eyeless, such as some of the whalefishes (Cetomimidae) and *Phreatichthys andruzzii* from Somalia. The latter lacks optic nerves and has markedly reduced optic lobes. Blindness does not necessarily mean a complete lack of sensitivity to light, for light sensors may exist in the skin of some blind species and are associated with the central nervous system of many. Photoreceptors in the caudal region of larval lampreys (ammocoetes) appear to be part of a system that triggers burrowing movements when the animal is exposed to light. Light-sensitive areas in hagfishes are the skin of the tail, and in *Eptatretus burgeri* the white line that extends down the back. The degenerate eyes of hagfishes are not especially light sensitive (Patzner, 1978). Blind cavefishes possess photoreceptors.

The most important site of extraocular photoreception is the pineal organ and associated structures. Histological studies have demonstrated secretory cells in pineal bodies of various species and in the parapineal of *Latimeria* as well (Locket, 1980; McNulty, 1981). Electrophysiological studies have shown receptor potentials in the pineal nerve following stimulation by light. The role of the pineal complex in regulating chromatophores has been demonstrated in many experiments. Cave-dwelling species initiate swimming movements away from a light source stimulating the pineal region. Most fishes fall into three categories regarding illumination of the pineal region (Breder and Rasquin, 1950). Fishes of category 1 have transparent or translucent tissue covering the pineal complex and usually react positively to light. Category 2 fishes have an opaque covering over the pineal and usually are light negative. Category 3 fishes can control entry of light to some extent by the action of chromatophores above the pineal complex; these vary in their reaction to light.

The pineal window in some sharks can allow the transmission of up to seven times as much light as adjacent parts of the head. In some species, the threshold for detection of light at this site is below the energy level of moonlight. Tunas have, in addition to a window, a tubelike translucent structure that directs light to the dorsal part of the brain. About 25 percent of incident light can be transmitted to the vicinity of the apparently photosensitive pineal. *Heteropneustes fossilis,* a catfish of India, has a lenslike structure in the pineal window. *H. fossilis* is a nocturnal species, but it is thought that the combination of the "lens," the pineal window, and a pineal fossa allows a sufficient concentration of light on the pineal to stimulate photoendocrine function (Srivastava and Srivastava, 1991).

Experiments with the eyeless fish, *Phreatichthys,* show that its central nervous system is sensitive to light. Sensitivity of the pineal to light is involved in thermoregulatory behavior, according to experiments by Kavaliers (1980), in which white suckers (*Catostomus commersoni*) with the pineal shielded spent significantly more time in lighter and warmer parts of the experimental environment than did individuals that were unshielded. The suckers appeared to show both a fast (neural) and a slower (hormonal) response.

Auditory, Mechanosensory, and Electrosensory Systems

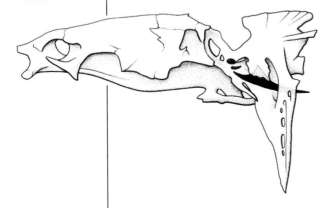

Structure

The auditory, mechanosensory, and electrosensory systems, also called the octavo-lateralis system, are usually considered to include the inner ear (fish have no middle or outer ear), the neuromasts and canals that make up the lateral line system, and the ampullary and tuberous organs of the electrosensory lateral line. The functions of the ear appear to be primarily balance and sound reception. The organs of the lateral line respond mainly to displacement of water and to pressure. The ampullae and tuberous organs sense electrical fields and biologically generated electric signals.

Membranous Labyrinth

The ear of bony fishes is composed of the osseous labyrinth, a cavity including ducts within the bones of the otic capsule, and the membranous labyrinth within the osseus structure (Lewis et al., 1985). The membranous labyrinth typically consists of two or three more or less distinct chambers—the utriculus, sacculus, and lagena—and three semicircular canals (Fig. 21–1). The canals and the utriculus constitute the pars superior of the organ and are mainly involved in balance and detection of angular acceleration, and the sacculus and the lagena are the pars inferior and are mainly involved in hearing (Fay and Popper, 1980; Popper, 1983). These two parts of the inner ear are nearly separated in some minnows (Cyprinidae) and are completely separate in featherbacks (Notopteridae) and some gobies (Gobiidae). The size of various parts of the ear varies among groups or species; in the bowfin (*Amia*) and ostariophysines, the lagena is larger than the sacculus, although the converse is true for most groups. The semicircular canal system is reported to be larger than the pars inferior with its otoliths (ear stones) in the flyingfishes (Exocoetidae) and goosefish (*Lophius*) (Platt and Popper, 1981).

Otoliths differ greatly in size and shape among species and are often so distinctive that they can be used for identification. The accretion of material to otoliths as

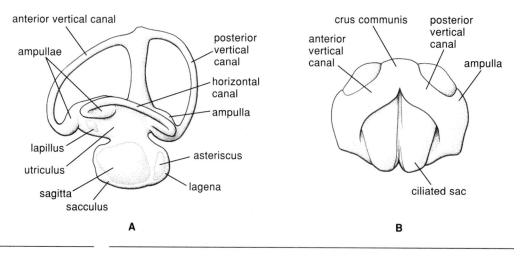

FIGURE 21–1   Membranous labyrinth of **A**, cutthroat trout (*Oncorhynchus clarki*), and **B**, Pacific lamprey (*Lampetra tridentata*), both showing the left labyrinth from left.

they grow is usually in a regular pattern of layers, so that the ages of individuals of many species can be assessed by study of the otoliths using appropriate methods (usually by microscopic examination of a cut and polished cross section).

In sharks, rays, and chimaeras, an endolymphatic duct communicates with the exterior. In bony fishes, this duct is shortened or lacking. The utriculus is the place of attachment of the semicircular canals, each of which is oriented in a different plane in relation to the others. One is horizontal, situated on the lateral aspect of the utriculus. The other two are vertical, one posterior and one anterior, at right angles to each other, placed so that each is at an approximately 45° angle to the axis of the body as viewed from above. The vertical component of the labyrinth to which the semicircular canals are attached dorsally, termed the *crus communis,* is continuous with the utriculus. Lewis et al. (1985) report that in some elasmobranchs the semicircular canals do not connect to the utriculus.

Each semicircular canal has an ampulla at its junction with the utriculus (the ampulla of the horizontal canal is at the anterior connection). Rising from the floor of each ampulla is an eminence called a crista, upon which is a gelatinous ridge or cupula covering hairlike, ciliary extensions of hair cells, which are the receptor cells of the inner ear. The sensory organ (macula) consists of a group of sensory hair cells and supporting cells (Montgomery, 1988).

The crista and cupula form what is regarded as a flexible diaphragm in the ampulla (Platt, 1983). Movement of the head in any direction will cause the endolymph of the canals to deform one or more of the cupulae, bending and thus stimulating the hair cells of the maculae.

The sacculus is attached ventrally to the utriculus; the lagena, which attaches to the posterior part of the sacculus, may be well delineated but is not distinct in many species. In all three of these organs, the hair cells are located on epithelial maculae, upon which the otoliths lie. The maculae consist of a basement membrane that supports numerous hair cells, each with its bundle of filaments. Each bundle has a long, true cilium (kinocilium) with several shorter filaments (stereocilia), graded in length, alongside it (Platt, 1983; Popper and Coombs, 1980).

Otoliths of elasmobranchs consist of calcareous granules (otoconia) in a soft matrix and have been reported to include mineral particles, such as sand grains, that enter through the endolymphatic duct (Montgomery, 1988). The exogenous particles are called otarena. In most bony fishes the otoliths (Fig. 21–2) are hard structures, and those of most species have a characteristic shape and size. These are generally held in place by a gelatinous membrane. The otoliths of the utriculus, sacculus, and lagena are called, respectively, the lapillus, sagitta, and asteriscus. In the labyrinth of lampreys, elasmobranchs, and many bony fishes, there is a macula (macula neglecta) without an otolith (Lewis et al., 1985).

Hair cells on the maculae are disposed in groups with similar orientation so that the kinocilium is on the same side of each of the cells. Because of the contours, shapes, and placement of the various maculae and the pattern of polarity, there are hair cells oriented in different planes in relation to the axis of the body (Fay and Popper, 1980; Lewis et al., 1985; Platt and Popper, 1981). The saccular macula of teleosts can have four or more regions of different hair cell orientation; the lagenar macula generally has two regions of opposite orientation (Platt et al., 1989; Popper, 1983). The cilia bundles of the hair cells differ in length and arrangement in various sensory areas. Those in the semicircular canals have kinocilia much longer than

FIGURE 21–2   Photograph of medial aspect of otolith of rainbow trout (*Oncorhyncus mykiss*), showing growth rings. (Photograph courtesy of John McKern.)

those in the maculae, and the stereocilia range up to about half the length of the kinocilium. The bundles of cilia on otolithic organs of several species of bony fishes studied by Platt and Popper (1983) generally varied regarding length and relative lengths of the kinocilia and stereocilia.

The membranous labyrinth of hagfishes consists of a lower chamber that has a single macula associated with a membrane that carries mineral inclusions (mostly calcium phosphate, or apatite, with some calcium carbonate), which serve the function of an otolith (Lewis et al., 1985). Above the lower section, a continuous canal forms an arch. This has been called a single semicircular canal but has also been considered to represent the anterior and posterior vertical canals, because ampullar swellings with annular cristae appear near the junctions with the pars inferior. Ross (1963) notes a tiny endolymphatic duct. In lampreys the labyrinth is more complicated. The common macula is functionally divided into anterior horizontal, vertical, and posterior horizontal parts that appear to correspond to the sacculus, utriculus, and lagena (Lowenstein, 1971), but it is still covered by an "otolith" made up of calcareous crystals, mostly apatite. There are two vertical semicircular canals with divided cavities. Two ciliated sacs that comprise a large portion of the labyrinth are not seen in any other vertebrate. The function of these structures is poorly known.

The otolithic maculae and ampullar maculae of the inner ear are innervated by cranial nerve VIII, the auditory, acoustic, or octavus nerve. Each hair cell is in contact at its base with an afferent and an efferent fiber ending. The ratio of hair cells to neurons entering the maculae in the burbot (*Lota lota*) and in the bowfin (*Amia calva*) is about 10 to 1 (Popper, 1983).

The Lateral Line System

Mechanosensory Organs. Lateral line organs may be free neuromasts on the skin or in pits or may be located on the canals or grooves on the head and body (Fig. 21–3). The lateral line canals typically open to the surface through pores penetrating bones or scales (Figs. 21–3 and 21–4). The cephalic lateral line canal courses through a distinct subset of skull bones. Usually these canals through the bones are of small diameter, but in some groups, such as the drum family (Sciaenidae), they are wide and cavernous. Different sizes of canals in some fishes may allow for tuning to different frequencies (Denton and Gray, 1988, 1989; Platt et al., 1989).

Herrings and their close relatives (Clupeidae) are peculiar in that they possess a tubelike intercranial space called the lateral recess (recessus lateralis), which connects the head canals of the lateral line system with the inner ear and anterior bullae of the gas bladder. It is situated between the pterotic and frontal bones above and the prootic below. A membrane separates the recess from the labyrinth of the auditory organ. Extremely small changes in position of the membrane due to changing pressure can cause a flow within the perilymph around the labyrinth of the inner ear (see Blaxter et al., 1989).

Electrosensory Organs. The ability to respond to weak electric fields is a primitive characteristic in fishes. Presumed electroreceptors have been noted in fossils of agnaths, acanthodians, osteostracans, lungfishes, and rhipidistians (see Hopkins, 1983). Among extant fishes, electroreception is known in lampreys, elasmobranchs, some primitive nonteleosts, and a few teleosts.

There are two general types of electroreceptors, ampullary and tuberous. (Lampreys have electroreceptors that differ from those of other fishes.) The elasmobranchs have electroreceptors called ampullae of Lorenzini (Fig. 21–5). Similar ampullary organs are found in other nonteleost fishes, including sturgeons, paddlefishes,

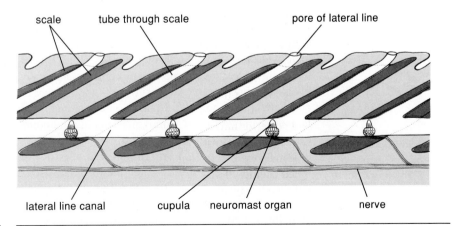

FIGURE 21–3   Diagram showing relationship of lateral line canal to scales in a typical teleost. The drawing represents a horizontal section with the thickness of the scales and the size of the lateral line exaggerated.

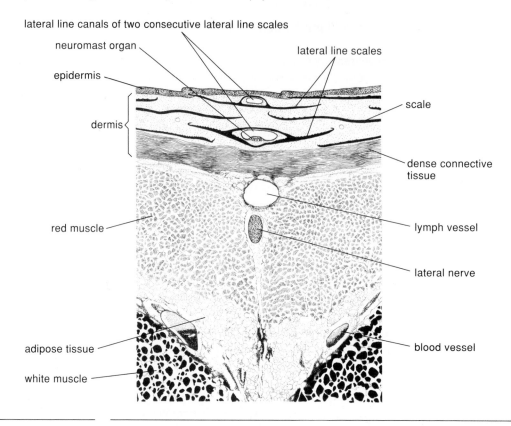

lateral line canals of two consecutive lateral line scales

neuromast organ

lateral line scales

epidermis

scale

dermis

dense connective tissue

lymph vessel

red muscle

lateral nerve

blood vessel

adipose tissue

white muscle

FIGURE 21–4   Drawing showing relationship of lateral line to scales, skin, and muscles.

lungfishes, *Latimeria,* bichirs, and reedfishes, but are lacking in hagfishes, gars, and bowfin. Among teleosts, ampullary organs are known in catfishes, gymnotiform knifefishes, the xenomystine knifefishes (of Notopteridae), and the mormyriforms (Jørgensen, 1989; Zakon, 1986). In elasmobranchs, the ampullae are usually distributed in clusters on the head and on the pectoral fins of skates. These clusters take their names from their respective placements and innervation—supraorbital, buccal, hyoid, and mandibular. At the surface of the skin are pores that open into the canal or tubule of each ampulla. In the Rajidae and many other rays, the canals are both large in diameter (about 1 mm) and long in relation to those of sharks, and canals spread widely from the clusters. In freshwater stingrays, such as Potamotrygonidae of South America or *Dasyatis garouaensis* of Africa, the ampullae are smaller and less complicated than those of their marine relatives. They are not arranged in clusters but are distributed over the body. The canals are short so that the microampullae are nearly at the surface (Raschi and Makanos, 1989; Zakon, 1986). Canal length in ampullae is apparently an evolutionary response to the differing resistivity of the fish body in relation to the fresh or salt medium. In salt water, the body fluids are more resistant than the medium and a long duct aids in increasing the potential along the receptor organ. In fresh water, the canals are shorter.

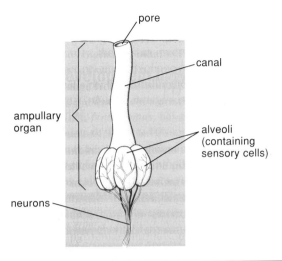

FIGURE 21–5   Diagram of ampulla of Lorenzini.

A conductive, gelatinous material fills the canal of ampullary organs, but the walls are electrically resistive (Montgomery, 1988) or at least passive (Zakon, 1986). The sensory cells are set into the epithelium of the walls of the ampullae, with a kinocilium (in elasmobranchs) or apical microvilli (in teleosts) exposed to the lumen of the ampulla (Bullock, 1981). The receptor cells are separated by supporting cells, which are arranged so that current will not flow between adjacent cells but through them. There are hundreds of sensory cells per ampulla (Zakon, 1986).

Lampreys have electroreceptors that are not of the ampullary type. Instead, they are end buds that formerly were thought to be taste or lateralis receptors. The receptor cells are set in the epidermis with the apical ends (of which there are up to 25) extending to the surface. Electroreceptors are somewhat concentrated on the head of lampreys but are distributed sparsely on the body as well (Ronan, 1986). Electroreceptors are well distributed over the head and body in electroreceptive bony fishes except in sturgeons, paddlefishes, bichirs, and the South American and African lungfishes, in which they are on the head only. Ampullary organs have not been demonstrated in *Latimeria,* but there is a large rostral organ that is suspected to be electrosensory (Northcutt, 1986; Northcutt and Bemis, 1993).

Cutaneous electroreceptors are found in mormyroids, gymnotoids, and various other freshwater fishes (see Bullock, 1981, for a list of types of electroreceptors). One type of electroreceptor is ampullary in nature, each ampulla with a canal about 100 to 200 µm long leading to the surface. Another type of electroreceptor is the tuberous organ, which typically has no opening to the exterior. The "canals," if present, are filled with loose epithelial cells (Zakon, 1986). This type appears to be more sensitive to higher electrical frequencies than the ampullary type. There are differences in the histology of tuberous organs found in various gymnotiforms and mormyrids, so that kinds called knollenorgans, gymnotomasts, and mormyromasts are recognized (Kramer, 1990; Szabo, 1974; Zakon, 1986).

Other sensory structures, the organs of Fahrenholz (found in lungfishes and bichirs), have been shown to be electroreceptors and are homologous to electrosensory ampullae of Lorenzini (Northcutt, 1986).

Other Sensory Receptors. Additional cutaneous sensory receptors found in some elasmobranchs are the vesicles of Savi, the function of which is still under investigation. These are found on the snout region and along the anterior edge of the pectoral fins of electric rays, mostly on the ventral surface. Similar vesicles are found in some other families of rays (Barry and Bennett, 1989). Usually there are about 200 vesicles, with only 30 to 40 on the dorsal surface of the snout. According to Barry and Bennett (1989), they are mechanoreceptors sensitive to vibrations up to 350 Hz (cycles per second).

Additional organs of probable proprioceptive function are the spiracular organs of sharks, rays, and such primitive bony fishes as sturgeons and bichirs (Barry and Bennett, 1989). Proprioceptors respond to changes in position, balance, and use of muscles. These organs are similar to the typical lateral line sensory organ in structure.

The physical stimuli of touch and temperature are primarily received by fishes through free nerve endings in the skin. Although the ampullae of Lorenzini have been shown to react markedly to changes in temperature, that has nothing to do with their primary function. In addition, some sharks, searobins, and eels have corpuscle-like structures that may be specialized touch receptors.

Function

The inner ear and the lateral line system differ in the type of stimuli to which they respond. The function of the inner ear is balance and the reception of pressure waves (sound). The inner ear is active in the detection of changes in position due to acceleration in any plane and of position changes with respect to gravity (equilibration); it is also active in detection of sound. The lateral line responds mainly to movements of water over the receptors. Water movements may arise from many sources, such as currents, activity of other animals, or movement of the fish itself.

Membranous Labyrinth

Sound Reception. Because water is so much denser than air (about 800 times as dense), a greater amount of energy is required to cause sound in water; but when sound is propagated it travels at about 4.5 times the speed of sound in air (approximately 1500 m s^{-1} versus 330 m s^{-1}) and is not rapidly attenuated. Sound energy in water consists of compression waves (particle displacement is sensed by the lateral line). Low-frequency sound propagation is relatively easier in water than propagation of high frequencies.

Many natural physical processes cause sound in water. Earthquakes, volcanism, winds and the water movements caused by them, rain, movement of bottom materials by currents or waves, and action of ice all provide background noise. Anthropogenic sound may be evident in certain places and may consist of noise from ships and submarines and from industrial operations both on shore and on the sea. Biological sound may result from sound production, incidental or deliberate, by animals.

Marine mammals, molluscs, crustaceans, and fishes are all sound producers. Sounds from all biological sources may have significance for fishes (in finding mates or prey or in escaping from predators). There is a possibility that dolphins can debilitate fishes with sound (Norris and Mohr, 1983).

At one time, there was a prevalent belief that fishes were deaf. However, careful study over the years has shown that fishes are sensitive to sounds and that in most species the energy is received by the sacculus and lagena (Jenkins, 1989). Compression waves cause movement of the maculae in relation to the otoliths that rest on them. Because fishes have a density very close to that of water, pressure waves cause them to vibrate along with the water. The heavier otoliths do not vibrate at the same time or rate as the rest of the fish, so there is movement in relation to the beds of hair cells upon which the otoliths lie and the sensory cilia are bent and stimulated. Response to a wide range of frequencies has been measured in fishes, but the response is related to the auditory equipment of the various species. Generally, species with a functional connection between the swimbladder and the ear (otophysic connection) have greater sensitivity than those without such a connection and respond to a greater range of frequencies. With some exceptions, such as the hardhead catfish (*Arius felis*), otophysan members of the Ostariophysi appear to have exceptional powers of hearing both in terms of sensitivity and range of frequency received (Popper and Coombs, 1980). In these fishes the Weberian apparatus, which consists of bones modified from the first few vertebrae and their processes, forms a connection between the anterior chamber of the air bladder and the labyrinth (Fig. 21–6). The ossicles through which the vibrations are conducted are the tripus (a crescent-shaped structure in contact with the anterior part of the air bladder) and the smaller intercalarium, scaphium, and claustrum. The claustrum is in contact with the walls of the membranous labyrinth, which in these fishes is modified so that the left and right organs coalesce posteriorly to form the sinus impar (unpaired sinus). In otophysans, the lagena and lagenar otolith (asteriscus) are larger than the sacculus and the saccular otolith (sagitta), but the sensory epithelium of the sacculus may be as extensive as that of the lagena (Platt and Popper, 1981).

Examples of otophysan fishes are the goldfish, *Carassius auratus,* and the loach, *Nemacheilus barbatula,* which are known to respond to sounds up to about 3500 Hz, and the minnow, *Phoxinus,* which in various experiments has responded to a range of frequencies from 20 to 7000 Hz. Another otophysan, the brown bullhead, *Ameiurus nebulosus,* has been reported to have absolute frequency limits of 60 to 10,000 Hz (Poggendorf, 1976). Sensitivity is usually greatest in much lower frequencies than the upper limits. For instance, the goldfish and brown bullheads have hearing thresholds in the 200 to 1000 Hz range (Hawkins, 1981; Tavolga, 1971).

In a variety of fishes, there are direct connections between the swimbladder and the ear. The codlings or deep sea cods (Moridae) have large branches of the swimbladder in contact with the skull. In the mormyrid fishes, small portions of the swimbladder become separated during early development and are enclosed within the skull in contact with the ear. Herrings and anchovies have gas-filled bullae, attached to diverticula of the swimbladder, in close connection with the utricular area of the ear. A membrane within each bulla separates the gas from a section of the bulla filled with perilymph, which is in intimate contact with the utriculus. An elastic thread extends from the membrane to the ear. The recessus lateralis is also in

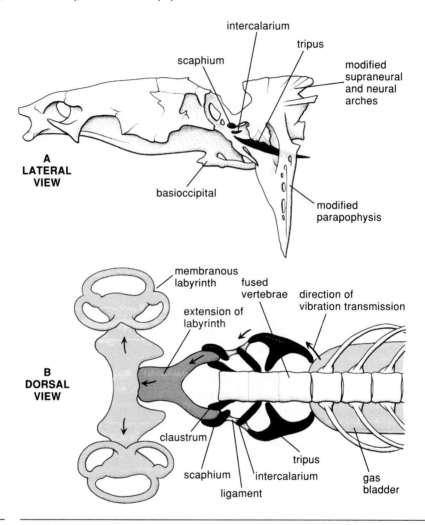

FIGURE 21–6 Weberian apparatus. **A**, Skull of *Catostomus* showing the modified first three vertebrae and the relationship of the left Weberian ossicles (in black, claustrum not shown) to them; **B**, diagram of the relationship of the left Weberian ossicles (in black) to gas bladder, vertebrae (shown here unmodified and unfused), and the extension of the sinus impar of the membranous labyrinth.

close contact with the utriculus. Herrings respond to frequencies from 0.01 to 1000 Hz, a range that extends from depth-related pressure changes to high-frequency sound (Blaxter et al., 1981). Some soldierfishes (Holocentridae) of the genus *Myripristis* have a close connection of the swimbladder with the ear and exceed many Otophysi in response to high frequencies. They are able to detect frequencies up to about 3000 Hz (Coombs, 1981).

Featherbacks (Notopteridae) have an otophysic connection but do not have especially acute hearing. Other families with otophysic connections are tarpons (Megalopidae), porgies (Sparidae), and bigeyes (Priacanthidae). Drums (Sciaenidae) and triggerfishes (Balistidae) have the swimbladder in close relationship to the skull. A strange adaptation for contact between the swimbladder and the skeleton is seen in the Scombrolabracidae, in which several vertebrae develop hollows into which diverticula of the swimbladder extend. Whether that has a relationship to hearing is unknown (Bond and Uyeno, 1981).

Species with airbreathing chambers in the head, such as climbing perch and snakeheads, maintain a bubble of air in the suprabranchial cavity close to the auditory region. This could act as a resonator. Liem (1967b) notes a tympanum on the exoccipital foramen of the saccular bulla of *Luciocephalus*. This is near the suprabranchial cavity. In addition, there are cranial lobes of the swimbladder that are in close proximity to the suprabranchial cavity. Gas bladders can increase the sensitivity of fishes to sounds and increase the range of frequency detected (Platt et al., 1989). A gas-filled suprabranchial cavity might have a similar effect.

Fishes without a close connection between the swimbladder and the ear usually respond to a smaller range of frequencies than those with this otophysic linkage. Responses to sound are elicited either by conditioning the fishes to show a specific behavior upon receiving a sound or by inserting an electrode around the auditory nerve and measuring microphonic potentials resulting from stimulation of the ear by sound. The highest frequencies causing response in these fishes are mostly below 1200 Hz, often in the 300 to 600 range (Tavolga, 1971).

Species without swimbladders appear to be less responsive to high-frequency sound than those that have this organ. In one study that involved implanting electrodes in the ears of a tilapia and a channel catfish, both species had similar sensitivities to underwater vibrations up to 600 Hz. The response of the tilapia diminished above that frequency, but it responded to 900 Hz. The response of the catfish went to 4000 Hz before diminishing. Deflation of the swimbladder made little difference in the tilapia, but the catfish lost some sensitivity at frequencies over 100 Hz. Removal of gas from the swimbladder is known to reduce the sensitivity to sound at a given frequency, so sounds must be of higher amplitude to be heard. Placing a small gas-filled balloon in contact with the head of a fish that naturally lacks a swimbladder will increase both sensitivity to sound and the frequency response (see Blaxter, 1981).

Although early experimentation with directional hearing seemed to show that orientation to near-field sounds received by the lateral line was possible, that ability is doubted (Schuijf and Buwalda, 1980). Kalmijn (1989) considers the lateral line to be a hydrodynamics detector. More recent research on the ability to localize sound in the far field has shown that directional hearing involves particle displacement or velocity as well as pressure. Because of the differential orientation of the maculae and the orientation of hair cells in directional groups, researchers theorize that stimulation of the two ears can result in output that, when compared by the central nervous system, allows the fish to determine the direction from which the sound originated (Popper, 1983; Popper et al., 1988). Sharks (Corwin, 1981; Myrberg and Nelson, 1991) and some bony fishes have been observed to swim to a sound source on many occasions.

Fishes of various types have been shown to distinguish tones, with ostariophysines responding to frequency differences of from 3 to 5 percent over the range of 200 to 500 Hz. Fishes other than ostariophysines are less able to discriminate frequencies, and some, such as sharks and eels, respond only to frequency differences of 50 percent or more. Fishes can also distinguish intensity of sound.

Equilibrium

Angular Acceleration. This is sometimes referred to as dynamic equilibrium and is detected by the ampullar macula of the semicircular canals. The hair cells in that sensory epithelium, like the hair cells in other parts of the inner ear and mechanical lateral line system, are pear shaped to cylindrical, and each has a single kinocilium (a true cilium) that is longer, larger, and more complex than the numerous shorter stereocilia that form a sloping series, with the longest next to the kinocilium (Fig. 21–7). Bending the ciliary bundle toward the kinocilium is inhibitory, and bending away from the kinocilium is excitatory.

The vertical canals respond to rotation in any direction, but there appears to be the possibility of greater stimulation by movement on horizontal axes (Lowenstein, 1971).

Gravistatic Function. Detection of gravity, or static equilibrium, involves the utriculus in most fishes, although there is evidence that the utriculus is active in sound perception in some groups (such as the herrings) and that in others the lagena or part of the sacculus is involved in maintenance of equilibrium. Impulses from the utriculus control a series of reflexes that govern posture. Some that are easiest to observe are the curling of fins and rolling of eyes as a live specimen is held at an angle. Change of position in relation to gravity is apparently detected by means of deformation of sensory hairs as the utricular otolith (lapillus) shifts. If a fish is rotated, there will be initial stimulation of semicircular canal cristae as well as of the utricular maculae, but when the endolymph of the canals has stabilized the otoliths will continue to deform the sensory hairs. Platt and Popper (1981) have reported that such a great diversity

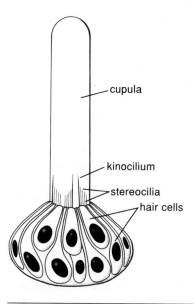

— cupula

— kinocilium

— stereocilia

— hair cells

FIGURE 21–7   Lateral line organ (neuromast).

exists among fishes that generalizations about division of function among parts of the inner ear and about a "typical" teleost ear may be untenable.

Mechanosensory Lateral Line System

Typical neuromast receptor organs of the lateral line system are found not only in canals (canal neuromasts) (Fig. 21–4) but also on the skin (superficial neuromasts, pit organs). These organs may differ in size and in the number of hair cells and number and morphology of stereocilia in each hair cell. They respond to deformation by mechanical stimuli, particularly water movement in relation to the fish. Thus, currents from whatever source (e.g., the fish swimming), water displacements caused by other organisms, and the small displacements caused by sources of sound at close range can be detected. The responses of canal and superficial neuromasts have been studied in cichlids and the rainbow trout (Munz, 1989). The canal neuromasts extend their sensitivity to higher frequencies than the superficial neuromasts. In the trout, canal neuromasts respond to water acceleration and superficial neuromasts respond to water velocity. In cichlids, superficial neuromasts respond to water velocity, but both types respond to water acceleration (Munz, 1989). The frequency range of 50 to 100 Hz seems to be optimal for reception by the lateral line, although it can respond to frequencies from below 10 Hz to about 500 Hz (Coombs and Janssen, 1989; Kalmijn, 1989).

Superficial neuromasts are of several types, some sitting flush with the epidermis with hair cells extending into the water, some in pits either in the dermis or epidermis, and some on papillae (see Coombs et al., 1988, for terminology). Superficial neuromasts are usually arranged in series either as accessory lines oriented with the lateral line canals or as replacement lines (or groups) that take the place of evolutionarily lost canals (Coombs et al., 1988).

The hair cells of a neuromast are covered by a single cupula. Adjacent hair cells within a neuromast usually show opposing orientation (e.g., with the kinocilia oriented headward in some and tailward in the others). Water movement striking the side of the fish sets up an impulse in the fluid of the lateral canal. This causes slight deformation of the cupulae and thus bending of the hair cells, which elicits a response transmitted into the nerve innervating the neuromast. Kalmijn (1989) shows that lateral line canals are hydrodynamic detectors that respond to acceleration (not sound) close to the source of vibration.

The lateral line sense is important because it allows fishes to detect the presence of predators or prey, to locate sources of surface waves, to orient properly to the current, to maintain position in a school, and to avoid obstacles. These abilities would seem to be especially important among species of nocturnal habit or those living in caves or the deep ocean. The latter two groups certainly show great modification of the lateral line canals and organs.

Electrosensory System

The weak electric organs of mormyroids and gymnotoids were once referred to as "pseudoelectric," probably due to a lack of sensitive instruments to measure the discharges, although as early as 1841 they were assumed to have an electric function. Once the discharges were confirmed, about 1880, the search began for the function of the organs. Matching the production of electrical fields with the presence of receptors responsive to the fields required clever experimentation and solid scientific reasoning. Because some electroreceptors were found to be sensitive to changes in temperature, they were first considered to be thermoreceptors.

Some apparently nonelectric fishes (sharks, rays, catfishes, and others) possess the electrosensory ampullae of Lorenzini or similar electrosensitive organs. These fishes are said to have passive electrosensory systems because they react to externally generated electric stimuli only. In active systems, the electric fishes provide stimuli to which their own tuberous electroreceptors are sensitive. The receptors in active systems can also react to external stimuli.

Electroreceptors are located in the skin, in patterns that vary with species. In general, ampullary organs, equipped with a canal opening to the exterior, are responsive to low-frequency stimuli of 0.1 to 50 Hz over long periods. These are called tonic receptors in contrast to the tuberous organs, which do not open to the exterior and are sensitive to higher frequencies up to 2000 Hz. The tuberous organs are phasic receptors, are not sensitive to direct current, and become insensitive to prolonged stimuli. Electroreceptors are innervated by the lateral line nerves.

The tuberous organs, found only in Gymnotiformes and Mormyriformes, show structural differences that apparently reflect differences in function. Some fishes have the ability to sense differences in wave form and frequency, and some have receptors programmed for low or high thresholds.

Sources of electrical stimuli are both biological and environmental, caused not only by activity of electric organs or by other processes (such as secretory or muscular activity) of aquatic animals but also by movements of water masses, atmospheric processes, and various geological and electrochemical processes.

Functions of electroreception include, but may not be restricted to, location of objects, communication, and navigation (Hopkins, 1983; Kramer, 1990). Fishes with active electric systems hold their bodies straight and swim by undulating the dorsal, anal, or pectoral fins. Some species move backward equally as well as forward. The tail-first approach, seen often in gymnarchids, is probably advantageous in that the posterior sections of the electric organs are somewhat electrically isolated from the rest of the body so that maximum current density is set up around the tip of the tail. Distortions of the field by objects are thus maximized. In addition, only the tail is exposed to the possible dangers of a new situation. The straight posture allows fishes to establish a symmetrical electric field around themselves, and the extent of the field is governed by the resistance of the water and the nature of the electric discharge. The field approximates a dipole field in some species and in the young of others. In apteronotids and rhamphichthyids, the anterior three quarters of the fish act as a distributed source of current. The tail, where the ends of the electric organs are close to the surface and are not surrounded by body fluids, acts more like a point source (Fig. 19–19). Objects encountered in the field, whether good or poor conductors, distort the field. The high-frequency receptors sense the change in impedance, and the fish reacts appropriately to the information received. Although the range of this system is not great and probably operates best within only a few centimeters of the fish (Kramer, 1990), it is apparently of significance to nocturnal fishes and to those fishes that live in turbid waters.

Communication between electric fishes may have significance in such aspects of life as reproduction and spacing of individuals. Range of communication may be from less than 50 cm to nearly 7 m, depending on water conditions and the species involved. Some species can quickly change the frequency of their pulses to avoid "jamming" by interfering frequencies. That change is termed the jamming avoidance response (JAR) (Heiligenberg and Rose, 1985; Viete and Heiligenberg, 1991).

Passive electrolocation allows fishes to react to the fields that emanate from living organisms in the water and inanimate sources. Activity of muscles establishes very small alternating current (AC) fields, and electric organs set up larger fields. Direct current (DC) fields from organisms originate from potentials involving the body fluids and the surrounding medium. Wounds are reported to strengthen the DC fields, so that a wounded organism can be detected electrically at a greater distance than can an entire organism. Certain sharks have been shown to locate flatfish concealed by covers that allowed passage of an electrical field but prevented the passage of odors (Kalmijn, 1971). Many electrosensory species react to magnets or to other inanimate sources of electrical fields, and sharks will readily seek out a dipole field from electrodes set out to imitate the field around a flatfish.

Although there is no firm evidence, use of their sensitivity to weak electrical and magnetic fields for orientation is suspected in salmon and eels and is supported by some experimental evidence. Kalmijn (1978) was successful in training stingrays to respond to magnetic fields. Some skates have been shown to be sensitive to the earth's magnetic field. The elasmobranchs apparently can detect electrical gradients as low as 5×10^{-4} millivolt/cm (Feng, 1991), which is low enough that they can use the gradient caused by their swimming in the magnetic field of the earth for orientation.

Walker et al. (1984, p. 755) trained yellowfin tuna to discriminate between two "Earth-strength" magnetic fields, ambient and altered. The gradients were equal but opposite. Although this species has no ampullary organs, it has a concentration of magnetite in the ethmoid region that could respond to magnetic fields. Magnetite is known also from otoliths of certain elasmobranchs (Vilches-Troya et al., 1984).

Olfaction, Gustation, and Other Chemical Senses

Introduction

The chemical senses of fishes respond to molecules in solution. The dissolved substances may originate far from the fishes and be carried by currents or, more slowly, by diffusion to the receptor organs. The stimuli travel more slowly than gases in air and may persist around the fish for a longer time. Water soluble, nonvolatile compounds of small molecular weights may be detected and may be of importance in the behavior of fishes (Hara, 1992a). The excellent solvent properties of water allow myriads of organic and inorganic substances to be carried in solution, and even though many are only sparingly soluble, they may still be detected by the acute olfactory and gustatory organs of fishes.

Fishes are well equipped for sensing chemicals in solution. Many have taste buds well distributed over the body, and some have especially sensitive olfactory organs. A common chemical sense that aids in the detection of certain ions is reported to be located in the skin. Taste is important at close range, while olfaction may be of greater importance in locating sources of more distant stimuli. In the absence of current, a gradient of dissolved material will exist that may allow orientation to the source of the odor or taste. Currents or turbulence may alter the state of the gradient and change the possibilities of orientation (Kleerekoper, 1969).

Olfaction interfaces with behavior in fishes in four main areas, according to Hara and Zielinski (1989): feeding, homing migration, reproduction, and fright reaction. Taste is of primary importance in food selection and swallowing and may be involved in other aspects of behavior. Fishes are known to detect and recognize the smell of prey (Little, 1983), predators (Rehnberg and Schreck, 1986), individuals of their own species (Todd et al., 1967), and even specific small streams (Hasler and Wisby, 1951). Reaction of the fish to the odors depends not only on the species involved but can involve the previous learning experiences of the individual.

Reproductive behavior is often triggered by the detection, usually by olfaction, of pheromones. The definition of pheromone is still subject to argument (Sorenson, 1992), but pheromones are generally considered to be chemicals released by organisms and have specific effects on the behavior or physiology of members of the same species. Detection of pheromones may be important in homing behavior.

Olfaction

Structure of Receptor Organs

The organs of olfaction in fishes are located in sacs or pits on the anterior part of the head, usually directly in front of the eye in bony fishes. In elasmobranchs, the nostrils are usually in front of the mouth. The olfactory receptor cells are in the epithelium lining the sacs. The epithelium is columnar and is underlain by a basement membrane. In addition to the sensory epithelial cells, the olfactory epithelium includes the ciliated nonsensory cells, supporting (sustentacular) cells, and mucous cells. Olfactory receptor cells differ from other sensory cells in that the receptor cells themselves are neurons that extend an axon directly to the brain. The receptor cells are located on series of lamellae (folds) that are disposed in the epithelial lining of the nasal sac. The epithelium on the lamellae contains two types of receptor cells: ciliated receptor cells and microvillous receptor cells (Caprio, 1988). Water must flow through the sac so that dissolved material can come into contact with the sensory cells.

Rod cells that apparently do not develop into receptor cells are present (Zeiske et al., 1992). The ciliated cells are somewhat elongate, and a small swelling at their distal ends reaches the lumen of the olfactory sac into which the cilia extend (Hara, 1971). The sensory cells taper at their bases to form very thin (approximately 0.2 μm) axons. These are grouped into fascicles that extend to the olfactory bulb, where they form synapses with dendrites of mitral cells. Fish species differ in the size and number of cilia per olfactory receptor cell. There are about 20 per cell in the burbot (*Lota lota*), and these appear to be 20 to 30 μm long (Gemne and Döving, 1989). Ciliar movements observed in vitro have shown no regular pattern; bending and straightening occur in unorganized sequences. The relationship of the cilia and their movements to the olfactory sense has not yet been described fully, but they seem to be the site where chemical substances come into contact with the appropriate nervous tissue so that the smell impulse can be generated (see Cagan, 1984).

The number of receptor cells per mm^2 of olfactory epithelium is usually in the range of 50,000 to 95,000, but densities up to 500,000 cells/mm^2 have been reported (Caprio, 1984; Yamamoto, 1982).

In cyclostomes, elasmobranchs, and a variety of bony fishes (including some catfishes, cyprinoids, cods, and mormyroids), the olfactory bulb, a part of the olfactory lobe of the forebrain, is adjacent to the olfactory organ. In many of these, the olfactory bulb is close to the rest of the brain; but in elasmobranchs, *Latimeria,* and some teleosts, the bulb is distant from the forebrain. In *Squalus,* the olfactory bulb is so closely associated with the olfactory epithelium that there is no discrete olfactory nerve (see Kent, 1992). The axons of the large cells of the mitral cell layer of the olfactory bulb contribute to the olfactory tract, which extends to the olfactory lobe. Each tract has two bundles, lateral and medial. In most bony fishes the olfactory bulb is close to the olfactory lobe so that a long olfactory nerve connects the olfactory organ to the bulb (Hara and Zielinski, 1989; Kleerekoper, 1969; Meisami, 1991).

Kleerekoper (1969) recognizes three arrangements of olfactory chamber, olfactory nerve, olfactory bulb, olfactory tract, and olfactory lobe. First, found in lampreys, hagfishes, chondrichthyans, Australian lungfish, mormyrids, cyprinoids, many catfishes, and gadids, is a type with the bulb close to the olfactory epithelium so that the two are connected by short filamentous nerves instead of a single olfactory nerve. In lampreys, hagfishes, and many teleosts, the bulbs are close to the olfactory lobes. In chondrichthyans, the coelacanth, and some teleosts, the bulbs attach to the forebrain by a long olfactory tract. In another type, found in chondrosteans, gars, and most teleosts, the bulbs are close to the olfactory lobes, so that a long olfactory nerve connects them to the olfactory epithelium. In the third type, found in some gadiforms and characins, the bulbs are in an intermediate position between the olfactory chambers and the lobes.

The thin olfactory nerve fibers conduct impulses more slowly (0.2 m/s) than most vertebrate axons (Meisami, 1991). The identity of the olfactory cells as ganglionic cells and the direct connection with the brain has led some morphologists to conclude that the olfactory receptor is essentially a primitive structure that has remained relatively unchanged (Kleerekoper, 1969).

The olfactory epithelium, which is ectodermal in origin and developed from the nasal placodes, is typically arranged on a series of olfactory lamellae on a longitudinal ridge called the raphe, forming what is usually termed the olfactory rosette in most bony fishes (Zeiske et al., 1992). In elasmobranchs, the raphe are transverse and support olfactory lamellae on each side of each raphe. There is great variation in the arrangement of olfactory lamellae (Fig. 22–1), and the number of olfactory lamellae varies greatly among fishes (Yamamoto, 1982). Some, such as clingfishes and seahorses, have none. Others have but one or two, and familiar fishes (including salmonids, minnows, and pikes) have fewer than 20. Various eels have numerous olfactory lamellae; *Anguilla,* for instance, has around 90, but a greater number has been reported for the porgy (*Haplopagrus guentheri*), which has about 230 (see

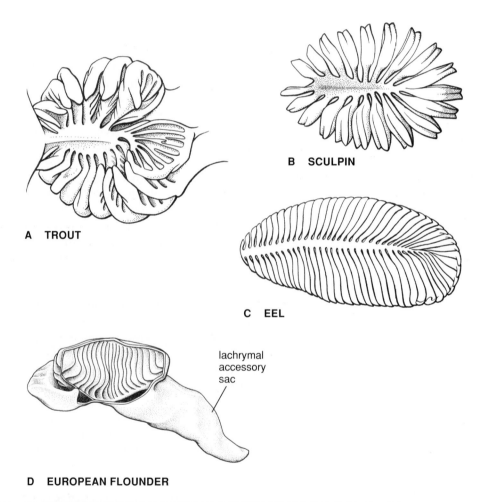

A   TROUT

B   SCULPIN

C   EEL

lachrymal accessory sac

D   EUROPEAN FLOUNDER

FIGURE 22–1   Dorsal views of olfactory rosettes: **A,** trout (*Oncorhynchus mykiss*), showing concave rosette with finger-like processes; **B,** staghorn sculpin (*Leptocottus armatus*), convex rosette; **C,** American eel (*Anguilla rostrata*), flat rosette; **D,** European flounder (*Pleuronectes flesus*), with accessory sacs. In eels, incurrent water moves from left to right of figure; in others, water impinges near middle of rosette. (**D** after Kleerekoper, 1969).

Kleerekoper, 1969). In some species, the number increases as the individual grows. Fishes with fewer lamellae (and presumably fewer receptor cells) generally have weaker powers of olfaction (microsmatic). However, although the relationship seems somewhat obscure (see Hara, 1971), olfactory powers are probably related more to the area of the olfactory epithelium in relation to the surface area of the body and the disposition of receptor cells in the olfactory epithelium.

Some general observations have been made that fishes with oval rosettes that have an intermediate number of lamellae have intermediate powers of olfaction and that those with long rosettes have great olfactory acuity (macrosmatic condition). Macrosmatic fishes tend to have continuously distributed olfactory epithelium on the lamellae, with a dense packing of sensory and ciliated nonsensory cells (Yamamoto, 1982). However, Zeiske et al. (1992) find no correlation between the number of receptor cells and olfactory acuity.

The majority of fishes have external nares only, which do not communicate with the pharynx. Typically there is an anterior and posterior naris on each side (see Fig. 2–5). Fishes that have communication between the external sacs and the oral cavity or pharynx include lungfishes, hagfishes, some eels, and stargazers (Uranoscopidae) (Atz, 1952a, 1952b). The internal nares are not homologous in these diverse forms because they develop differently. The lungfishes are considered to have true internal nares (choanae). Unusual anatomy is seen in some tetraodontiform fishes, *Tetraodon* and *Diodon,* in which the olfactory epithelium is situated in perforated "nasal lobes" or on bifid "nasal tentacula" protruding from the surface of the head at the usual position of olfactory organs (see Kleerekoper, 1969). Some angler fishes have tentacular nostrils (Marshall, 1967).

For chemical substances to come into contact with the olfactory epithelium in the olfactory sacs, the sacs must be irrigated with water. Respiratory movements can facilitate the irrigation of the olfactory sacs of those fishes with internal nares, but maintaining a flow of water in most fishes requires some special hydraulic engineering. Circulation of water within the nasal sac is accomplished in one of three ways: by forward movement of the fish in relation to the water (ram ventilation); by the action of cilia in the sac or extensions of it; and by pumping effected by direct or indirect constriction of the sensory nasal sac and accessory sacs.

Fishes that depend on their forward motion or on water currents for movement of water through the olfactory organ usually have flaps or ridges behind the anterior nares to guide the water over the olfactory epithelium. These cutaneous structures are easily seen in common freshwater fishes such as trout, minnows, and suckers. Sharks and rays are notable for their occasionally elaborate complement of flaps and grooves associated with the nares (and mouth) (Fig. 2–3). These flaps are arranged so that they accept a current of water over the olfactory epithelium as the animal swims or, in some, as it pumps a respiratory current (see Bell, 1993).

In eels, in some catfishes, and probably in other fishes that have long nasal sacs and widely separated anterior and posterior nostrils, cilia are important in moving water through the system, even though contraction of facial muscles that would facilitate movement of water into olfactory sacs may be involved. Many such fishes have tubular extensions of the nares. In nettostomatid eels, the anterior nostrils are tubular but the posterior nostrils are pores or slits, usually opening in front of the eye; but in the genus *Nettenchelys* they open behind the head at a distance of up to two times the length of the head, depending on the species (Smith and Castle, 1982).

Latimeria has tubular anterior nostrils and slitlike posterior nostrils in front of the eyes. There is a five-sectioned, complicated rosette bearing the lamellae. Cilia apparently move water through the nasal cavity. Bichirs and reedfishes have remarkably complex nasal organs with an anterior tube and a valvular posterior nostril. Both nasal cavities are divided into large and small chambers, which house multiple rows of lamellae. Kinocilia apparently move water through the organ (Zeiske et al., 1992).

Modifications for pumping water in and out of the olfactory chamber in a cyclic sequence are common among bony fishes but can also be seen in the lampreys. Lampreys have a single olfactory opening leading to an olfactory rosette that could represent a fusion of two organs of smell (or be a consequence of the lack of the organ splitting into two parts embryologically). The olfactory nerve is double in cyclostomes (see Kleerekoper, 1969). However, the nasal portion of the canal is continuous with a blind tube called the nasohypophyseal canal or pouch, which runs posteriorly under the brain and ends between the brain and the pharynx. Contractions of the branchial muscles apply pressure to the pouch and cause rhythmic emptying and refilling. As water flows into the pouch, some is shunted into the olfactory organ by a valve at its entrance (Kleerekoper, 1969).

In bony fishes, accessory sacs continuous with the nasal sac are commonly located under the skin in the region lateral to the nostrils, and some species have such sacs under the skin between the nasal organs. Respiratory and other movements involving the jaws and dermal bones of the face affect the volume of the nasal and accessory sacs and cause water to flow in and out over the olfactory epithelium as the sacs are compressed and relaxed by those movements. The anterior nostril is usually incurrent and the back excurrent (Fig. 22–2). Some bony fish groups have a single nostril on each side. This can result from the loss of the posterior naris, as in eelpouts (Zoarcidae) and snailfishes (Liparididae), or from coalescence of anterior and posterior nares, as in sticklebacks (Gasterosteidae). Other groups that have single nostrils are the cichlids (Cichlidae) and damselfishes (Pomacentridae).

The terminal nerve system is associated with the olfactory apparatus in vertebrates. Although there is no direct evidence that this system has a chemosensory

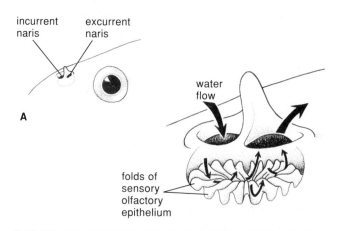

FIGURE 22–2   **A,** Position of left nares of Cyprinidae; **B,** water flow over olfactory rosette of Cyprinidae.

function (Meredith and White, 1987), experiments with various animals (including goldfish) seem to show a relationship to reproductive function, especially in the detection of pheromones (Meisami, 1991; Yamamoto, 1982). However, Fujita et al. (1991) refer to this nerve as having no known function and demonstrate that the olfactory system is responsible for chemosensory responses to pheromones in the male goldfish.

Function and Significance

Olfactory Thresholds. Early research on thresholds involved behavioral experiments in which individuals trained by reward or punishment were conditioned to select or avoid water that held the chosen chemical and then were tested on more and more diluted concentrations. More modern methods may involve conditioned heart rate, electroencephalograms, recordings from nerve tracts, and other electrophysiological methods. Thresholds from various sources for a number of fish species and chemical substances are tabulated by Kleerekoper (1969), Little (1983), Caprio (1984), Meisami (1991), and Hara (1992b). Examples of these thresholds are presented in Table 22–1. Although there are lower thresholds reported, Hara and Zielinski (1989) state that the lowest olfactory threshold identified in fish by modern electrophysiological means is 10^{-13} M for a preovulatory pheromone (17-α, 20-β-dihydroxy-4-pregnen-3-one) in goldfish. Fishes show generally low thresholds for amino acids, many of which contribute to food odors. Ictalurid catfishes show low thresholds (10^{-9} to 10^{-6} M) for amino acids (see Hara, 1992b; Little, 1983).

Eels of the genus *Anguilla* have especially acute olfaction and have been used successfully in experiments involving conditioned responses to extremely diluted solutions of food extracts or other substances. For example, only one molecule of β-phenylethyl alcohol (10^{-18} M) in the olfactory sac has been reported to cause a conditioned response in a trained eel (see Kleerekoper, 1969). The olfactory organ of the eel is relatively large and is equipped with many folds in the epithelium, which, unlike most fishes, is pigmented.

TABLE 22–1   **Olfactory Thresholds of Fishes for Various Substances**

Substance	Fish	Threshold	Source
Amino acids	hagfish	10^{-6} to 10^{-5} M	Döving and Holmberg (1974)
L-methionine	lemon shark	10^{-8} to 10^{-7} M	Zeiske et al. (1986)
Amino acids	catfishes	10^{-9} to 10^{-6} M	Caprio (1982)
Methionine	grayling	1.3×10^{-6} M	Döving and Selset (1980)
Bile acids	grayling	6.3×10^{-9} M	Döving and Selset (1980)
Bile acids	goldfish	10^{-9} M	In Meisami, 1991
Steroids	goldfish	10^{-13} to 10^{-12} M	In Meisami, 1991
Phenylethyl alcohol	A. anguilla	2.9×10^{-20} M	In Little, 1983
Phenylethyl alcohol	rainbow trout	10^{-9} M	In Little, 1983
Sucrose	Phoxinus	1.2×10^{-3} M	In Little, 1983

Döving et al. (1980) found that arctic char and grayling (*Thymallus arcticus*) responded to the odors of bile acids at thresholds much lower than for amino acids. They suggested that two different olfactory receptors might be involved in sensing the two different kinds of acids because response could be measured only from the medial part of the olfactory bulb when bile acids were presented and only from the lateral part when amino acids were presented. Fish generally have low olfactory thresholds for steroid hormones (see Hara, 1992b).

Detection of Food. Detection of food is a major function of olfaction in most fishes and may be of special importance in species that feed in dim light or search through bottom materials or vegetation for edible objects. Among bony fishes, most active open water predators are primarily sight-oriented hunters, but even among these (as in sharks) olfaction is of great importance, along with other chemical senses. Attraction of sharks to baits or to wounded fish upstream from the sharks even in gentle currents has been observed by both scientists and laypeople, and knowledge about this phenomenon has been refined by experimentation. Some sharks lost the ability to find food placed into a tank with them when both nostrils were plugged with cotton, but they sought and found it when only one nostril was occluded, even though the pattern of search differed from that followed when both nostrils were clear. In many of these experiments, the possibility of sighting the food was removed by concealing the food in cheesecloth or some other substance through which odor could penetrate. In other experiments, sharks were blinded but still would perform the typical figure-eight search pattern in locating food (Tester, 1963).

Some of the most interesting studies have involved the introduction of extracts, dilution, or washes of various substances into experimental tanks with sharks. Extracts of fish flesh, especially those with a considerable oil content, caused search and feeding activity even when presented in very diluted concentrations. Flesh rotted to foulness appeared to repel sharks to some extent. In studying the response of sharks to living fish, Tester (1963) found that water flowing over intact but distressed or excited fish caused a greater response in the test animals than did water flowing over quiescent fish. Starved sharks were more responsive than well-fed specimens, and some detected the odor of food at a concentration of 1×10^{-4} ppm, a dilution that may be stronger than the undetermined threshold level. Amino acid thresholds as low as 1×10^{-14} M are reported by Montgomery (1988) for elasmobranchs. Hueter and Gilbert (1991) mention studies that indicated that blacktip and reef sharks could sense extract of grouper flesh at 1 part per 10 billion.

Because of the occasional shark attacks of humans, there has been interest in the attractiveness of human odors to sharks. Human urine appears to be detected by some sharks but causes no particular activity, whereas human blood attracts some of the species tested. In one series of experiments, human sweat caused what Tester (1963) termed "aversion" in sharks at about 1 ppm. Repulsion of sharks by odors has been studied for many years in the hope that a suitable repellent can be found. Certain dyestuffs and copper acetate have shown limited promise, but research on the shark-repelling poison (pardaxin) found in a tropical flatfish (*Pardachirus marmoratus*) has led to the discovery that certain surfactants (especially sodium dodecyl sulfate) appear to be effective in repelling sharks if discharged into the shark's mouth (Nelson, 1991).

Fishes other than sharks and rays that are known to orient to food by olfaction include lampreys, hagfishes, African lungfishes, and many teleosts. Among the latter are numerous minnows, catfishes, eels, perches, wrasse, and cods. Field observations and examination of the relative sizes and degree of development of olfactory organs and olfactory centers have led investigators to believe that certain plankton feeders, of both shallow and deep waters, and some swift predators orient toward food at least partially by olfaction.

Orientation. Orientation in the environment is probably achieved partly through the sense of smell in many fishes, possibly in more than are presently recognized. Some minnows can be taught to distinguish between the odors of streams that differ in geology or in organic components, such as species of aquatic plants, indicating that they might be able to recognize localities by very diluted olfactory cues (Hasler and Wisby, 1951).

A variety of fishes (including several salmonid species and some centrarchids—including sunfishes, *Lepomis*) apparently can locate their home areas by means of the olfactory sense. The ability to do so has been demonstrated for Pacific salmon (*Oncorhynchus*), which can return to the stream locality from which they migrated months or years earlier (Hasler and Scholtz, 1983).

Olfaction plays a prominent part in the salmon's homing behavior during the spawning migration. Field experiments have shown that blinded salmon taken from a spawning stream and displaced downstream can make the correct choices of tributaries and return if allowed full use of the olfactory organs but distribute in a random manner if the nostrils are plugged (Wisby and Hasler, 1954).

Electroencephalographic studies have shown that strong impulses can be recorded from the olfactory bulb of salmon stimulated by home stream waters, although some fish reacted to nonhome natural waters as well (Oshima et al., 1969). Imprinting salmon smolts with the organic compounds morpholine or phenylethyl alcohol, which do not occur naturally, and then, at the time of spawning migration, placing the appropriate attractant (morpholine or phenylethyl alcohol) in a stream other than the one that the smolts descended has resulted in attracting the returning adults to the nonhome stream (Cooper and Hirsch, 1982). Salmon apparently depend on olfactory cues in nonreproductive homing as well as reproductive homing (Stabell, 1992).

Studies by Selset and Döving (1980) showed that mature arctic char (*Salvelinus alpinus*) showed a preference for water that contained the odor of smolts from their home stream population. The ability of several species of fishes, including salmonids, to recognize the odor of their own kin is reviewed by Olsén (1992). There are two hypotheses involved in the relationship of olfaction to homing. One is that the fish respond to an imprint[1] of odors of soils, vegetation, and other such materials present in the stream from which they migrated. The second is that there is an innate response to pheromones released by fish of their own particular strain in the stream from which they came. Pheromones important to homing are thought to be produced in the liver and released with the feces (Stabell, 1992).

---

[1]Imprinting is defined by Cooper and Hirsch (1982, p. 344) as "a process of rapid, irreversible learning of a particular visual, auditory or olfactory stimulus that occurs at a 'critical' or 'sensitive' period during development that influences the behavior of the animal."

Other Behavior. Olfaction has been implicated or suspected to be important in many aspects of fish behavior (Little, 1983). In reproductive behavior, in addition to homing to the spawning stream, the sense of smell is important in locating mates, triggering certain phases of the spawning act, recognizing young, and defending territory. Pheromones related to reproductive behavior can be detected at very low thresholds. Olfaction is involved in the social behavior of fishes in many ways, including recognition of members of the same species, collectively in schools or individually.

Detection and avoidance of predators is of great importance. In some fishes, especially the ostariophysines, the presence of predators is noted indirectly by the sensing of alarm substances released from injured individuals of conspecifics and closely related fishes. Minute amounts of the fright substance (*Schreckstoff* in German) issuing from damaged mucous cells of a wounded fish cause almost immediate fright reaction and retreat in members of the same species. A dilution of skin extract of about 0.02 part per trillion can be sensed (see Little, 1983). Fright reaction is caused in Pacific salmon by rinses or extracts of mammalian skin containing L-serine (Brett and MacKinnon, 1954). This may represent recognition of the odors of potential predators. There is evidence that minnows surviving attacks by predators show fright reactions when exposed to the odor of the predator species (Little, 1983).

Taste

Taste Receptors

The receptors of the gustatory sense are called taste buds, which are groups of specialized epithelial cells. Taste buds are somewhat oblate or pear-shaped structures made up of about 100 to 150 elongate cells, including basal cells and supporting cells as well as the receptor cells (Fig. 22–3). The role of the basal cells is not clear. There may be chemical synapses between the sensory cells and the basal cells, which in turn may synapse with afferent nerve fibers. There is a possibility that the basal cells may be mechanoreceptors. Taste buds, which are about 30 to 80 μm long and 20 to 50 μm wide, are set in the epithelium (above the basement membrane) with the apical end at the surface of the epithelium (see Caprio, 1984). The apical ends of the receptor cells have microvilli that protrude from the surface of the receptor cells. Histologically, there are two types of cells involved, light sensory cells and dark supporting cells (Meisami, 1991; Reutter, 1982, 1992).

Taste buds are generally concentrated in the mouth, pharyngeal region, and gill arches of bony fishes. Some species, notably carps, have a great concentration of taste buds on a specialized palatal organ, which may be of principal importance in gustation (Hara, 1971). Many species, including catfishes, have external taste buds on specialized structures such as barbels, elongate fin rays, and certain areas of the body surface. Some catfishes have taste buds over much of the body. Distribution of external taste buds differs among species. In cyprinids, taste buds are more numerous toward the head than toward the tail and increase from dorsal to ventral (Gomahr et al., 1992). Atema (1971) reports concentrations of taste buds of up to 50 per mm^2 on surfaces of gill arches in *Amieurus natalis* and a total of around 175,000 on the body of a 25-mm specimen. Taste buds commonly occur on the lips and head of bottom-feeding fishes, but that is by no means universal (Livingston, 1987). Inner-

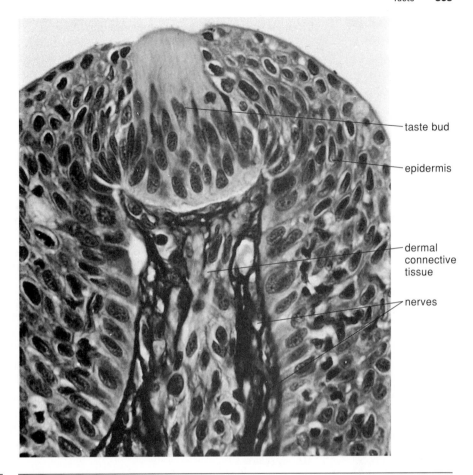

taste bud

epidermis

dermal
connective
tissue

nerves

FIGURE 22–3   Taste bud of juvenile rainbow trout (*Oncorhynchus mykiss*). (Photomicrograph courtesy of Professor. Joseph H. Wales.)

vation of the taste buds of the oropharyngeal cavity is usually by the glossopharyngeal (IX) and the vagus (X) nerves. The facial (VII) nerve innervates the external taste buds and, in some fishes, some buds at the anterior part of the mouth (Caprio, 1984; Hara and Zielinski, 1989).

Function and Significance

Sapid Substances.   Some sharks that have a keen sense of taste, have been observed to prefer certain food fishes in taste tests (Tester, 1963), and are known to react to substances that are described by humans as bitter and sour. Some react to salt, seemingly through the gustatory sense. Although there has been progress in research on sensory physiology of sharks in the last three decades (Heuter and Gilbert, 1991), comparatively little research has been carried out on their sense of taste.

Much more is known of the gustatory capabilities of bony fishes. Various species are known to react to the four categories of taste known to humans (bitter, sweet, salt, and sour) as well as to many other tastes of greater biological importance.

Through electrophysiological studies and training for conditioned responses, various teleosts have been shown to respond to numerous substances placed in contact with the taste receptors (see Marui and Caprio, 1992). Gustatory response is elicited by numerous substances, including amino acids, aliphatic acids, nucleotides, many salts, quinine and related bitter materials, saliva, milk, extracts of earthworms and silkworm pupae, and other food items. The palatal organ of the carp responds strongly to carbon dioxide (Hara, 1971). Experimental evidence has shown that taste receptors on various parts of the body have different sensitivities and thresholds and that there are notable specific differences among fishes. Even within the same species, the responses can differ between strains. For example, Japanese carp reacted more strongly than did the Swedish strain to quinine and extracts of worms and silkworm pupae, whereas the Swedish strain responded only weakly to quinine but much more strongly to sucrose than did the Oriental strain (Konishi and Zotterman, 1963).

In testing responses of single nerve fibers, investigators have noted some specialization among the receptors; some react to many tastes but others only to specific tastes or combinations (Kiyohara et al., 1975; Tucker, 1983). Carp appear to have seven, and puffers at least three, kinds of receptors.

Taste Thresholds. The thresholds at which fish can respond to tastes vary from species to species, but generally speaking fishes react to smaller quantities of sapid substances than humans (Table 22–1). In the minnow (*Phoxinus*), for instance, the threshold reported for sucrose is 1.2×10^{-5} M and that for fructose is 1.6×10^{-5} M (see Little, 1983). The Mexican blind cavefish, *Anoptichthys,* is reported to have a much greater taste sensitivity than minnows—up to thousands of times better for the four basic taste qualities (see Hara, 1971). *Phoxinus* can detect sodium chloride at 4×10^{-5} M, and *Ameiurus* can taste quinine at about 1×10^{-4} M. Nurse and lemon sharks are known to react to the chemical betaine at concentrations as low as 1×10^{-9} M (Carr, 1982).

Amino acids are detected by fishes at low concentration and often elicit feeding responses. Johnson et al. (1990) obtained electrophysiological responses to several amino acids by *Tilapia zillii,* an herbivore, at a test concentration of 1×10^{-6} M, but no thresholds were sought. Caprio (1984) and Marui and Caprio (1992) tabulate thresholds for amino acids of several fishes. For instance, channel catfish can detect L-alanine and L-arginine at 1×10^{-9} to 1×10^{-11} M. *Pseudorasbora* can detect L-alanine and proline at about those same levels.

Feeding and Other Behavior. Experiments with chemical stimulation of feeding behavior in a variety of fishes, including the percoids *Lagodon* and *Orthopristis* (Carr, 1982), have disclosed that amino acids and the compound betaine are responsible for much of the feeding stimulus. Species vary in their responses to extracts of marine organisms and to artificial mixtures.

Olfaction and gustation are both important in sensing more or less distant sources of stimuli, which may be of great importance in food location and various reproductive activities. Sharks are well known for sensing food odors from miles away (Hueter and Gilbert, 1990), but taste can be involved in this as well as in the eventual selection and ingestion of food. Bullheads (*Ameiurus*) are known to rely heavily on the external gustatory sense in finding food at a distance (Bardach et al., 1976a), and the same appears to be at least partially true in other fishes with numerous taste buds over the body and fins. In most fishes, the taste receptors on the lips, in the mouth, and on the branchial arches are instrumental in the final detection of food items, in initiation of the reflexes involved in seizing and swallowing, and in the rejection of unwanted items (Atema, 1971). Some function of taste is suspected in the courtship of certain fishes—cichlids and gouramies, for instance—because of the mouth and fin contact during mate selection and other processes. Recognition of young may be in part dependent on taste, although olfaction may be of greatest importance.

Common Chemical Sense

Other Chemosensory Receptors

Although olfaction and taste are probably the chemical senses of greatest importance to fishes, there are others that might be of basic importance. A common chemical sense, attributed to free nerve endings in the skin, has been suggested as a sensor of solutions of salts, acids, and alkaline materials (Whitear, 1992). This sense has been studied very little and its true significance is not well known.

Solitary chemosensory cells, which are structurally similar to the receptor cells in taste buds, have been described from the epidermis of teleost fishes. Comparable cells are known from lungfishes, selachians, lampreys, hagfishes, and larvae of frogs (Whitear, 1992). The cytology of these bipolar cells is described by Whitear (1992). The cells are often found at concentrations of over 1000 per mm^{-2}. Kotrschal (1991) found densities of 2000 to 4000 per mm^{-2} in some European cyprinids. These cells are especially numerous on the dorsal fin rays of the rockling, *Gaidropsarus mediterraneus,* where up to 6 million of the cells may be found (Whitear, 1992).

The searobins (Triglidae), which show a well-developed chemical sense located in the modified lower pectoral rays, have no taste buds in the skin of those rays but apparently have solitary chemosensory cells, which are neither olfactory nor taste cells (Silver and Finger, 1984). These are innervated by spinal nerves (Bardach et al., 1967b; Finger, 1982). Fishes with modified fin rays that bear taste buds innervated by cranial nerves are the hakes of the genus *Urophycis,* the rocklings (*Ciliata*), and the gouramies (Belontiidae). Rocklings' fins have input from sensory spinal nerves as well as from the facial nerve (Whitear and Kotrschal, 1988).

The apparent response of pit organs and free neuromasts of the lateral line system to certain ions has been reported by Katsuki and Yanigisawa (1982), but Whitear (1992) believes that those studies are flawed.

Circulation, Respiration, and the Gas Bladder

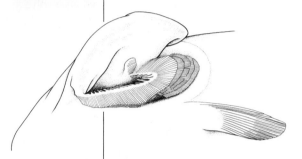

Because fishes live in an environment that is oxygen poor compared to the atmosphere we breathe and because their simpler hearts must force blood past a capillary bed in the gills before its distribution to the tissues of the body, their adaptations to problems of circulation and respiration are of great interest. Special adaptations involve the composition of the blood, the morphology of the circulatory apparatus, behavioral responses to oxygen levels, and the structure and function of the gills and other respiratory surfaces. Some of the more interesting adaptations involve the direct utilization of atmospheric oxygen. The lungfishes (Dipnoi), because of their special pulmonary circulation, differ markedly from the other bony fishes in this aspect.

Circulation

Vascular System

Heart. A typical fish heart has an atrium, a thick-walled ventricle, a lightly muscled sinus venosus, and, depending on the species, either a contractile conus arteriosus or an elastic bulbus arteriosus (Fig. 23–1A). The heart is contained within a pericardial

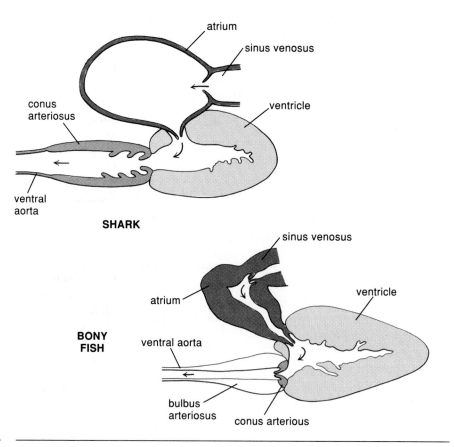

FIGURE 23–1 Diagrams of heart in shark and bony fish.

cavity located below the gills. Although the various groups of fishes have many features of the heart in common, there are some characteristic features that warrant considering the groups separately.

The hagfish (Myxini) heart is the most primitive among the fishlike vertebrates. The well-developed sinus venosus is partially divided into anterior and posterior portions by a fold of tissue. It receives the left anterior cardinal vein, the caudal vein, and the portal vein. The left anterior cardinal vein and the posterior cardinal empty into the portal heart. The atrium empties into the ventricle through a narrow passage. The ventricle pumps blood into a noncontractile but elastic portion of the ventral aorta, the bulbus arteriosus (Fig. 23–1**B**). A conus arteriosus appears to be lacking in hagfish; the semilunar valves that prevent backflow from the bulbus arteriosus are set into the walls of the ventricle itself. The elastic bulbus arteriosus helps to maintain a constant flow of blood to the gills. In addition to this branchial heart, hagfishes have three additional hearts of differing design that aid in circulating the blood. A cardinal heart in the head pumps venous blood toward the branchial heart, as does a caudal heart, located near the end of the tail. A portal heart pumps blood through the two lobes of the liver (Satchell, 1991).

The heart in the lampreys (Petromyzontidae) is relatively large in comparison to that of most fishes. The sinus venosus is a small, vertical, tubular structure. The atrium overlies the ventricle, and a bulbus arteriosus is present at the base of the ventral aorta. As in the hagfish, a conus arteriosus is lacking. Only the right common cardinal vein (duct of Cuvier) is present and empties into the atrium.

In elasmobranchs the sinus venosus, atrium, ventricle, and conus arteriosus are well developed, although the sinus venosus has very little cardiac muscle and may not be as important in filling the atrium as in bony fishes. The muscle of the ventricle has a compact outer layer (compacta) and an inner spongiosa. The compact layer brings about higher efficiency in the function of the ventricle (Tota, 1989). In the "warm-bodied" sharks, the compacta may comprise over 40 percent of the mass of the ventricle, whereas in poikilothermic sharks it may make up as little as 15 percent. The larger compact layer in the endothermic sharks may be functionally related to the need for greater vascular output (see Tota, 1989). Unlike lampreys and hagfishes, both right and left common cardinal veins are developed. The bulbus arteriosus is absent.

The hearts of bony fishes are more variable in structure than the other groups because of the great evolutionary diversity in living forms. The typical bony fish heart has a thin-walled sinus venosus that receives blood from the ducts of Cuvier and the hepatic veins. Blood empties into the atrium from the sinus venosus and is pumped to the muscular ventricle. Very few bony fishes have a compact layer in the ventricle. From the ventricle the blood is pumped into the bulbus arteriosus (or in some primitive species, the conus arteriosus) through which it will pass into the ventral aorta. Sturgeons and paddlefishes (Acipenseriformes), bichirs and reedfishes (Polypteriformes), gars (Lepisosteidae), bowfins (Amiidae), and some lower teleosts (tarpons, etc.) retain a contractile conus arteriosus with two or more rows of internal valves. Lungfishes have a conus arteriosus that retains some proximal cardiac muscle and valves, but in all three genera of lungfishes the conus bends sharply and twists 270° (Burggren and Johansen, 1987).

In most teleosts, however, the conus arteriosus is reduced, nonmuscular, and bears only one set of valves between the ventricle and the nonmuscular, elastic, bulbus arteriosus. The bulbus arteriosus may appear as a small, white dilation in a dissected fish, but when the heart is pumping the bulbus expands to the size of the ventricle. The bulbus arteriosus is reported to be more than 30 times more expansible than the aorta of humans (Licht and Harris, 1973) and can expand 700 percent in carp (Satchell, 1991). Blood is prevented from flowing back into the ventricle by valves located within the reduced conus arteriosus. Thus, the elastic walls of the bulbus maintain pressure on the blood flowing to the gills, maintaining an almost continuous flow, in contrast to the pulsatile flow from a conus arteriosus (Johansen and Gesser, 1986; Satchell, 1971).

The atrium of the lungfish heart is essentially divided into two parts by an incomplete septum (the pulmonalis fold), and thus the lungfishes are described as having a functional three-chambered heart similar to amphibians. The right chamber is generally larger than the left. In the South American lungfish (*Lepidosiren*), the two sides are nearly completely separated by an atrioventricular cushion, which also serves as a valve between the atrium and ventricle (Burggren and Johansen, 1987). The right division of the atrium receives deoxygenated blood via the sinus venosus, and the left side receives oxygenated blood via the pulmonary vein. Virtually complete separation of these two blood supplies is thought to be maintained through the atrium, and mixing in the ventricle is minimized by another incomplete partition. *Lepidosiren* has the best-developed separation of the right and left halves of the heart, and the Australian lungfish (*Neoceratodus*) has the least developed. The conus arteriosus of lungfishes is provided with a peculiar spiral fold that divides the separate oxygenated and deoxygenated blood supplies coming from the ventricle. In both *Lepidosiren* and the African lungfishes (*Protopterus* spp.), the fold starts in the ventral proximal wall of the conus arteriosus and continues along its length. It meets a second fold arising from the opposite wall farther along in the organ so that the oxygenated and deoxygenated blood are further separated. In *Neoceratodus,* the spiral fold is little developed proximally but sufficiently developed distally so the blood flow is divided (Burggren and Johansen, 1987). Oxygenated blood is guided mainly to the first and second gill arches (which lack gill tissue) and thence to the dorsal aorta.

Branchial Arteries. The ventral aorta of fishes extends forward beneath the pharynx. In lampreys, the aorta remains single to the fourth gill pouch, dividing into right and left branches at the septum between this and the third pouch. Eight pairs of afferent branchial arteries branch from the aorta and enter the walls of the gill pouches (Fig. 23–2**A**).

Afferent branchial arteries in elasmobranchs arise from the single ventral aorta and enter each of the branchial arches to supply blood to the holobranchs (complete gills) borne by these arches as well as to the hyoid hemibranch (half gill) (Fig. 23–2**B**). There are five pairs of branchial arteries in most rays and sharks, more in those having six or seven gill slits. Teleosts are similar in this respect to the elasmobranchs, except that the afferent branchial artery leading to the hyoid hemibranch is absent. Sturgeons more closely resemble the sharks, and some nonteleost actinopterygians are intermediate between the sharks and bony fishes in their patterns of branchial vessels. In lungfishes, the ventral aorta is short, so the afferent branchial arteries branch from the conus arteriosus close to the heart.

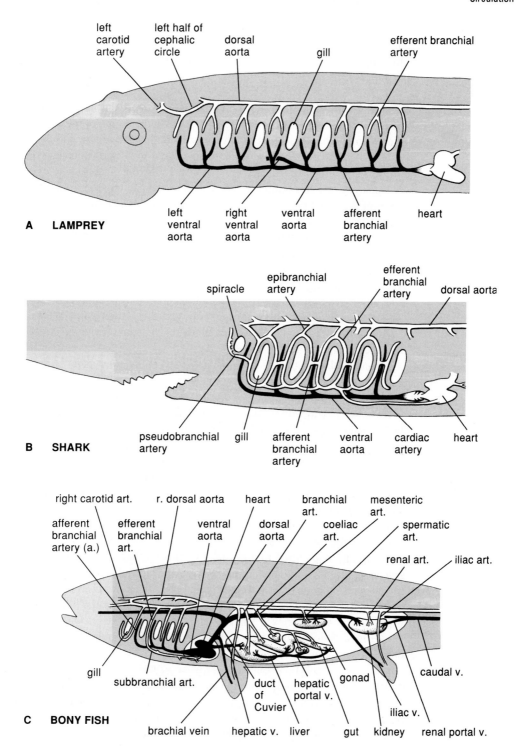

FIGURE 23–2   Diagrams of blood circulation. **A,** Lamprey, branchial arteries; **B,** shark, branchial arteries; **C,** bony fish, showing major blood vessels.

In the gill arch, the afferent branchial arteries give rise to arterioles that terminate in capillaries or in open spaces (lacunae) in the gill lamellae. After passing through the gills, the oxygenated blood flows through efferent branchial arteries to the dorsal aorta.

General Circulation. Lampreys and hagfishes have a single, median dorsal aorta and a peculiar section called the cephalic circle in the region of the first gill pouch or anterior to it, from which major arteries supply the head region with blood. In most jawed fishes, the dorsal aorta is paired anteriorly, with extensions continuing to the head as the internal carotid arteries. Posteriorly, the dorsal aorta is unpaired through the trunk region and continues into the tail as the caudal artery, through the hemal arches of the vertebrae. Arteries supplying the viscera and musculature are branches from the aorta and caudal artery.

The caudal vein runs through the hemal arches ventral to the caudal artery. In lampreys and hagfishes, the caudal vein splits to form the paired posterior cardinal veins. In jawed fishes, the caudal vein enters the kidneys through the renal portal system. Postcardinal veins receive blood from the kidneys and gonads and from the musculature as they run forward to join the common cardinal veins (ducts of Cuvier). Precardinal veins, subclavian veins, and the jugular veins enter the ducts of Cuvier. Blood from the ducts enters the sinus venosus, which also receives blood from the hepatic portal system (Fig. 23–2C).

Venous blood return in elasmobranchs and many bony fishes (except the Acanthopterygii) is facilitated in part by caudal pumps or caudal hearts (Satchell, 1991). Caudal pumps, as in some sharks, consist of left and right caudal sinuses and a series of vessels equipped with valves; so as the fish swims, the bending of the caudal fin to the left and right will alternately compress the sinuses and force blood from the tail toward the caudal vein. The caudal heart of sharks is arranged in such a way that the caudal sinus can be compressed by serial contraction of muscle in the tail so that blood is pumped forward even when the fish is not swimming. The teleost caudal heart is made up of two chambers, one on each side of the caudal fin skeleton, connected by a foramen (Satchell, 1991). The origins of the skeletal muscles that operate these hearts are on the vertebrae, and the insertion is on the hypural plate. In the eels (*Anguilla* spp.), blood from the caudal fin and cutaneous veins is received into the right side of the caudal heart and is pumped first to the left side and then to the caudal vein. In some species, blood flows into both chambers from the tail and is pumped from both into the caudal vein, although there is a connection between the two chambers.

In elasmobranchs, venous blood return is aided by the "hemal arch pump," which consists of a series of valves at the entry of each segmental vein to the caudal vein. Blood flow through the valves is aided by the contractions of the myotomes during locomotion (Satchell, 1991).

Satchell (1991) presents a description of the secondary circulation in fishes. This system receives blood through anastomosing short vessels arising from the dorsal aorta: the efferent branchial arteries and segmental arteries. In some species, the mouths of these interarterial vessels have microvilli that can extend and screen out most of the erythrocytes. This system was regarded as lymphatic until it was determined that its source of blood is primary arteries and that blood cells are sometimes passed through its vessels (see Satchell, 1991).

Most of the vessels of the secondary system are involved with the skin, although they are also found beneath the epithelium of the gut and in the mouth cavity. In the skin, the system is found in a network of capillaries under the epidermis of the exposed portions of the scales. These capillaries form from arteries flowing posteriorly beneath each scale, and they carry the almost-clear blood forward over the scale and join veins under the preceding and overlapping scale. Blood of this system is pumped toward the heart by the caudal heart in species that have this organ (Vogel, 1985).

Although the secondary system may contain more than one half the blood plasma in some species, its function and importance are not well known. There is speculation that it might be involved with osmoregulation in addition to the transport of nutrients (Satchell, 1991).

The Blood of Fishes

The blood transports a variety of materials, including inorganic ions and a number of organic constituents, such as hormones, vitamins, and several plasma proteins that may make up from 2 to 6 g/100 ml. These proteins, involved in immune responses, may include two forms of alpha globulin, two forms of beta globulin, and gamma globulin, as well as albumin, transferrin, and others. These proteins buffer against pH changes and aid in maintenance of osmotic pressure in body fluids.

The osmotic concentration of blood in fishes varies according to habitat and osmoregulatory capabilities. The osmolarity of bony fish blood ranges from somewhat below 200 milliosmoles (Mosm) in freshwater fish to more than 400 Mosm in marine species. Sodium and chloride ions are the main contributors to the total ion concentration, with lesser contribution from potassium, calcium, and magnesium, urea, and free amino acids. The freezing point depression (Δ_{fp}) of bony fish blood ranges from about –0.6 in freshwater fishes to about –0.75 in marine species without special antifreeze protection.

Some Arctic fishes live in waters of –1.7°C; Antarctic waters are as cold as –1.86°C. Fishes that live in such habitats are protected from freezing by blood glycoproteins or glycopeptides, which may account for one half or more of the osmolality of the blood. In Arctic fishes, the Δ_{fp} ranges down to –1.0°C, so those living in the coldest water have supercooled body fluids and, with few exceptions, will freeze if brought into contact with ice. A similar situation occurs in the Antarctic, where most species have freezing points of –0.9°C to –1.54°C. Certain nototheniids freeze only at temperatures lower than –2.2°C because of elevated levels of glycopeptides in the blood (Eastman, 1993). Nototheniid antifreeze consists of eight separate glycopeptides that have an extensive range of molecular weights (2600 to 33,700 daltons) in one species. The activity correlates positively with molecular weight (Eastman, 1993). Many temperate and polar fishes are known to adjust osmolality of their blood seasonally (Duman and DeVries, 1974). Such a fish is the winter flounder, *Pseudopleuronectes americanus,* which can withstand temperatures as low as –1.6°C (Fletcher, 1977). Some fishes increase the concentration of NaCl when adapting to cold; others increase the concentration of organic compounds (DeVries, 1971; Feeney and Hofman, 1973).

Cellular constituents of the blood are the red blood cells, or erythrocytes, and the white cells, or leukocytes. Erythrocytes obtain their characteristic color from hemoglobin, made up of the colorless protein globin and the iron-containing red–yellow pigment heme. Hemoglobin molecules of elasmobranchs and bony fishes are tetrameric—with four peptide chains—and have molecular weights of about 66,000 to 68,000 (Fänge, 1992; Satchell, 1991). The type of hemoglobin of lampreys and hagfishes is much like myoglobin, a form of hemoglobin found in muscle tissue, in that it has only one chain (monomeric). Hagfish hemoglobin has a molecular weight of 16,500 to 17,000 (Satchell, 1991). Although the heme units of fish hemoglobins appear to be the same in different species, the proteins may differ among species and within species, so more than one type of hemoglobin can be found in some species (Satchell, 1991). Differences in hemoglobins apparently aid in physiological adaptation to different environments (Fänge, 1992). Hemoglobins in a species may differ in many features, such as the composition of amino acids, affinity for oxygen, electrophoretic mobility, and the extent of the Bohr effect. Some salmonids may have up to 18 different hemoglobins during their life cycle (Satchell, 1991).

Hemoglobin transports oxygen in combination with the ferrous iron of the heme, to which it is loosely bound. The affinity of hemoglobin is controlled in part by nucleoside triphosphates, such as adenosine triphosphate and quanosine triphosphate (Fänge, 1992). The combination of oxygen and hemoglobin is reversible, depending on the partial pressure of the oxygen.

Only a few fishes lack hemoglobin; for instance, some channichthyids of the Antarctic and leptocephalus larvae of eels have colorless blood, in which oxygen is transported in solution. Channichthyids compensate for lack of hemoglobin by living at low temperatures, reducing muscular activity, and maintaining very vascularized skin and fins for cutaneous gas exchange. They have increased circulatory volume and a rapid movement of blood through the respiratory system (Hemmingson, 1991).

Most fishes have only nucleated erythrocytes that are usually oval; relatively few species have nearly round cells. Lampreys have round red cells about 9 μm in diameter. Elasmobranchs have large erythrocytes, the length ranging from 20 to 27 μm and width from around 14 to 20 μm. Erythrocytes of bony fishes generally range from 12 to 14 μm in length and 8.5 to 9.5 μm in width, but lungfishes have large red cells, about 36 μm long. Some deep sea teleosts with blood vessels of exceptionally small diameter have nonnucleated red cells about 5.5 μm long and 2.5 μm wide (Fänge, 1992). Examples are the sternoptychid genera *Maurolicus* and *Valencienellus* and the phosichthyid genus *Vinciguerria*.

With notable exceptions, fishes have a smaller blood volume than other vertebrates; the volume usually ranges between about 2 and 4 ml/100 g in bony fishes, compared with mammals, which have volumes of 6 ml/100 g or more (Lagler et al., 1977). Lampreys appear to have greater volumes than these (about 8.5 ml/100 g) and hagfishes even more (17 ml/100 g), and elasmobranchs are reported to have blood volumes of 6 to 8 ml/100 g (Satchell, 1971). Salmonids approach the blood volumes of elasmobranchs, with from 5 to more than 7 ml/100 g (Smith, 1966). Tunas (Scombridae) have a high blood volume, ranging from about 8 ml/100 g in 9-kg fish to 13 ml/100 g in 4.5-kg fish; smaller individuals are reported to have even greater volumes. Some investigators have indicated that there may be a phylogenetic trend toward a decrease in blood volume throughout the fishes (Satchell, 1971). The higher bony fishes possess a more efficient circulatory system and thus need less blood for transport of oxygen and other materials.

Generally, there is an inverse relationship between size of red blood cells and their number per unit of volume of blood, with sharks and rays having fewer than half a million cells per cubic millimeter. However, some gobies may have similar counts—for instance, *Gobius exanthonemus* is reported to have a count of $0.425 \times 10^6/mm^3$. An Antarctic fish, *Trematomus,* has from 0.66 to $0.80 \times 10^6/mm^3$. Most bony fishes have red cell counts of 1 to $3 \times 10^6/mm^3$, with a majority under $2 \times 10^6/mm^3$, but some active marine fishes have higher numbers, ranging from 4 to $6 \times 10^6/mm^3$.

The percentage of the blood that consists of red cells (the percentage of packed red cells) is called the hematocrit and is correlated with the red cell count. Humans have hematocrits of about 47 percent. Hagfishes have about 13 percent, elasmobranchs usually have hematocrits under 25 percent, and the spiny dogfish (*Squalus acanthias*) has about 13 percent. Most teleosts are in the 20 to 30 percent range; some marine species that apparently require large oxygen-carrying capacity—for instance, Atlantic mackerel (*Scomber scombrus*), bluefin tuna (*Thunnus thynnus*), and Atlantic herring (*Clupea harengus*)—have hematocrits of 51 to 52.5 percent (Satchell, 1991).

Hemoglobin concentration in fish blood, expressed as g/100 ml, is usually 7 to 10. Red blood cells, and consequently the hematocrit and hemoglobin concentration, can vary with season, temperature, and nutritional state and health of the fish. Circadian changes have been noted in some species (Riggs, 1970).

White blood cells (leukocytes) are not as numerous as red cells and usually number fewer than 150,000 per cubic mm ($0.15 \times 10^6/mm^3$) in most fishes (Mulcahy, 1970). The range within a single species may be great; counts for the common carp (*Cyprinus carpio*), for example, have been reported as ranging from about $0.032 \times 10^6/mm^3$ to $0.146 \times 10^6/mm^3$. There are four kinds of white cells: granulocytes, thrombocytes, lymphocytes, and monocytes.

Thrombocytes are involved in blood clotting; they carry a compound that promotes the conversion of prothrombin to thrombin and they are more numerous than the other white cells in many marine fishes, constituting about half the total. Clotting time in fishes is extremely rapid.

Granulocytes include four types of cells, named for their staining properties: neutrophils and three types of eosinophils (Satchell, 1991). Neutrophils are common in most species, but eosinophils are not always present. Granulocytes are phagocytic, involved in combatting disease, and may increase in number when the fish is infected by bacteria.

Lymphocytes include the phagocytic macrophages, plasma cells, and small lymphocytes, which may be active in protein production. More than 90 percent of the white blood cells in carp and trout can be lymphocytes. Monocytes are mononucleated macrophages.

Blood cell formation (hemopoiesis) occurs at several sites in fishes. The spleen is usually the most important site of erythrocyte formation. The kidney produces both red blood cells and various leukocytes (Satchell, 1991). In elasmobranchs, the organ of Leydig, most often associated with the wall of the alimentary canal (commonly along the esophagus), is a site of leukocyte formation. In some elasmobranchs, there is similar tissue (epigonal organ) associated with the gonads. Similar tissue may occur in various places in fishes—in the gut wall, orbit, meninges, base of the cranium, and in the cranium above the hindbrain in teleosts. The thymus is the site of lymphocyte production as well. In lampreys, larval hemopoiesis occurs mainly in the typhlosole, with some occurring in the nephric fold. In adult lampreys, the "fat column" that extends along the dorsal surface of the spinal cord is the site of hemopoiesis (Potter et al., 1982).

The oxygen capacity of fish blood includes oxygen carried in solution as well as that carried in combination with hemoglobin in the erythrocytes. *Chaenocephalus aceratus,* the blackfin icefish of the Antarctic, which has no hemoglobin, has a reported blood oxygen capacity of 0.45 to 1.08 ml/100 ml (Hemmingson and Douglas, 1972), whereas most teleosts have capacities in the 8 ml/100 ml to 12 ml/100 ml range. Very active fishes, such as the pelagic scombroids, and species adjusted to oxygen-poor waters have blood oxygen capacities up to 20 ml/100 ml. The oxygen capacity of the blood of sharks and rays is typically less than that of teleosts, usually ranging from 3.5 to 6 ml/100 ml.

The actual oxygen content of the blood depends on many factors, including the partial pressure of oxygen in the water, the partial pressure of carbon dioxide, pH, temperature, and the activity of the fish (Fry, 1947). Normally, blood from the dorsal aorta is at 85 to 95 percent saturation, while venous blood usually carries oxygen at 30 to 60 percent saturation. Trout undergoing strenuous exercise have been reported to carry no oxygen in the venous blood as it returns to the heart.

Bohr and Root Effects. The relationships among CO_2, pH, and the oxygen affinity of the blood is of special interest. One of these relationships, the Bohr effect or Bohr shift (see Burggren et al., 1991), involves a decreased affinity of hemoglobin for oxygen at low pH due to altered configuration of the hemoglobin molecule by binding hydrogen ions. (The oxygen dissociation curve is shifted to the right; Fig. 23–3.) This augments offloading of O_2 at respiring tissues, where CO_2 is high. Because CO_2 is rapidly lost at the gill, H^+ dissociates from hemoglobin, allowing effective loading of O_2. This effect is prominent in fishes that are adapted to habitats with high oxygen content and low CO_2. A distinct advantage exists for those species in that at the gills, in the presence of a low partial pressure of CO_2, the blood can easily load oxygen even at low partial pressure; then, in the tissues, at higher partial pressure of CO_2, oxygen can be released independently of its partial pressure. A disadvantage of the Bohr effect in fishes adapted to low CO_2 and high oxygen is that if the CO_2 content of the medium rises, an increasingly greater dissolved oxygen content becomes necessary to facilitate loading of the hemoglobin. Fishes such as bullhead catfishes (*Ameiurus* spp.), which exhibit low activity and are adapted to slow-water habitats where low pH and low oxygen content is normal, have blood that has a small Bohr effect (Moyle and Cech, 1982).

An extension of the Bohr effect is the Root effect, which is mostly present in fishes with gas bladders. This effect involves a decrease in the oxygen capacity of the blood with rising partial pressure of CO_2. A decrease in pH will render fish blood incapable of becoming 100 percent saturated with oxygen regardless of the pressure of oxygen. This has been demonstrated in experiments in which oxygen pressures up to 140 atmospheres were used.

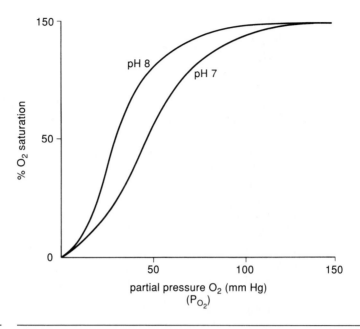

FIGURE 23-3   Effect of blood pH on oxygen dissociation curve. At a selected partial pressure of oxygen, the percent saturation will be much lower at pH 7 than at pH 8 (Bohr effect).

Gills

Agnatha

The gill tissue of lampreys and hagfishes is arranged as a series of radiating ridges inside expanded pouches that are internal to the branchial skeleton. This skeleton consists of an elaborate, basket-like arrangement that has considerable elasticity. The pouches are flattened anteroposteriorly and are somewhat separated from each other. Lampreys have seven pouches per side, each opening separately to the exterior via a short tube (Fig. 23-4**A**). Internally they open into a special respiratory tube or "pharynx" beneath the esophagus. In larval lampreys (ammocoetes), the pharynx is continuous with the oral cavity anteriorly and with the esophagus posteriorly. It is similar to the pharynx of hagfishes, but during metamorphosis of the ammocoetes the pharynx disconnects from the esophagus posteriorly, with an ensuing separation of the two tubes. Anteriorly, the entrance to the respiratory tube is guarded by a valvular velum. This special respiratory tube is unique among vertebrates.

Depending on the species, hagfishes may have 5 to 15 pairs of gill pouches that open internally into an elongate pharynx. External openings are separate in some genera (Fig. 23-4**B**), but in others some or all of the excurrent tubes from the gills may connect with a collecting tube, which conveys excurrent respiratory water to an external pore on each side. Hagfishes usually have the respiratory apparatus set well back behind the head. They are unique in possessing a pharyngocutaneous duct that connects the pharynx with the exterior, opening on the left side behind the last branchial opening or, in *Myxine,* into the single gill opening.

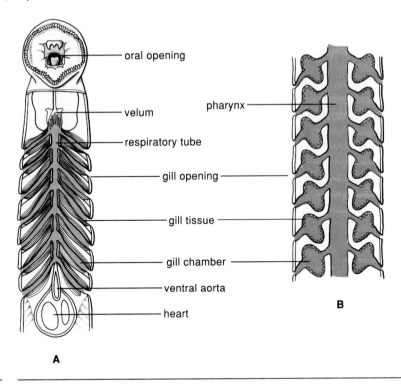

oral opening

velum

pharynx

respiratory tube

gill opening

gill tissue

gill chamber

ventral aorta

heart

A

B

FIGURE 23–4 **A–D,** Diagrams showing arrangement of gills in frontal section. **A,** Lamprey (*Lampetra*); **B,** hagfish (*Eptatretus*).

Gill irrigation in the cyclostomes (lampreys and hagfishes) is accomplished by contracting muscles around the branchial area that force water out of the several gill pouches. Elastic recoil of the cartilaginous branchial basket aids in filling the pouches. Water can enter and leave the individual pouches through the separate external openings of lampreys and of those hagfishes that possess separate openings. It can also enter through the mouth or through the pharyngocutaneous duct, which opens just behind the last gill pouch of hagfishes. Gill irrigation in hagfishes is accomplished in part by a velar pump in the "velar chamber" just posterior to the mouth. This scroll-like structure creates a current by alternately rolling tightly and unrolling (Hardisty, 1979; Johansen and Strahan, 1963).

Gnathostomata

In gnathostomatous fishes, the gill tissue occurs in the form of filaments or ridges on interbranchial septa borne on the gill arches (Figs. 23–4C and **D**). Both the septa and the gills are external to the branchial skeleton, contrasting with the agnaths. In gnathostomes, the branchial apparatus is concentrated into a smaller proportion of the body than in the lampreys and hagfishes.

Cartilaginous Fishes. The interbranchial septa of sharks and rays extend to the body wall, so each branchial chamber is entirely separated from the others and each has its own opening (gill slit) to the exterior (Fig. 23–4C). Sharks typically have five

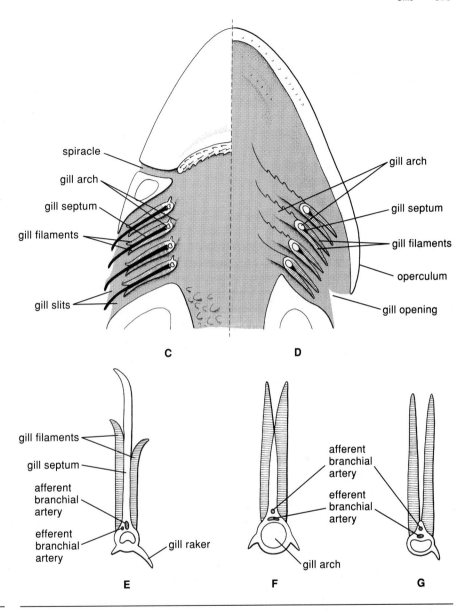

C, Shark; D, bony fish. E–G, Relationship of branchial septum and gill tissue. E, Shark; F, sturgeon; G, teleost. (C based on Weichert, 1951.)

FIGURE 23–4
(Continued)

openings, but six or seven openings are characteristic of some sharks. Cartilaginous branchial rays extend from the gill arch outward within the septum. Each arch and septum bears a series of gill lamellae on both the anterior and posterior faces. The gill lamellae on one side of a septum constitute a hemibranch or "half-gill." The two hemibranchs on a gill arch are called a holobranch. In most sharks, the posterior hemibranch of the hyoid arch is present in the first gill pocket, so that an odd number of hemibranchs occurs on each side—9, 11, or 13, depending on whether the fish has five, six, or seven gill pouches. A remnant of the mandibular gill, called the mandibular (or spiracular) pseudobranch, is associated with the spiracle, anterior to the functional gills in most species (Goodrich, 1930; Laurent and Dunel-Erb, 1984).

Chimaeras (Chimaeriformes) have four branchial arches and four gill pouches covered by a fleshy operculum. Thus, the gill septa do not extend to the body wall and are only slightly longer than the gill filaments. Adult chimaeras have no spiracle, and the pseudobranch is absent. There are even numbers of hemibranchs on each side. A hyoid hemibranch is followed posteriorly by holobranchs on the first, second, and third branchial arches and an anterior hemibranch on the fourth branchial arch. In general, the branchial apparatus of chimaeras is not as long relative to that of the sharks.

Bony Fishes. In bony fishes, gill septa are progressively reduced (Figs. 23–4**E, F,** and **G**). Some of the more primitive ones, such as sturgeons (Acipenseridae) and gars (Lepisosteidae), have slightly reduced septa, with the tips of the gill lamellae extending beyond as free filaments (Fig. 23–4**F**). In teleosts, the septa become greatly reduced to no more than small ridges along each gill arch, from which the gill tissue extends as long filaments (Fig. 23–4**G**). The gill apparatus is thus more compact than in the sharks, rays, and chimaeras. The gill arches and filaments are closely apposed, and the entire chamber is covered on each side by the bony operculum. Loss of the septa results in greater respiratory efficiency because the flow of water through the secondary lamellae is facilitated.

Among lungfishes, the greatest modification of gill apparatus is seen in the African lungfishes. These fishes retain a hyoidean hemibranch, have no gills on the first or second branchial arches, and have holobranchs on the third and fourth branchial arches and an anterior hemibranch on the fifth. The Australian lungfish has a hyoid hemibranch and four holobranchs. None of the lungfishes retains a pseudobranch or spiracle.

Bony fishes typically retain a pseudobranch at the site of the hyoid arch and holobranchs on each of the four branchial arches. In primitive forms such as sturgeons, gars, and coelacanths (Latimeriidae), a hyoidean hemibranch as well as a pseudobranch is retained. The deep sea eel *Eurypharynx* has five complete gills. Reduction of gills has occurred in some air-breathing teleosts and others. Some members of the family Synbranchidae (swamp eels, etc.) have only one well-developed holobranch.

The pseudobranch is evidently the remnant of the primitive gill of the mandible (or first visceral) arch (see Kent, 1992; Romer, 1970). The mandibular pouch and slit may be retained as the spiracle in many sharks, all rays, and some primitive bony fishes, such as sturgeons, paddlefishes, and bichirs. The pseudobranch is associated with the spiracle in the sharks, rays, and sturgeons, but in the gars the hyoidean hemibranch and the pseudobranch are adjacent on the inner side of the opercular base at the anterior portion of the branchial chamber. In the teleosts, that is the usual location of the pseudobranch (Fig. 23–5).

Although the pseudobranch closely resembles a functional gill in some species, it receives only oxygenated blood and therefore cannot function in respiratory gas exchange, but it may have that function in the leptocephalus larvae of eels (Laurent and Dunel-Erb, 1984). Moreover, in most bony fishes the pseudobranch may be reduced to the appearance of a glandular organ situated beneath the skin. The pseudobranch

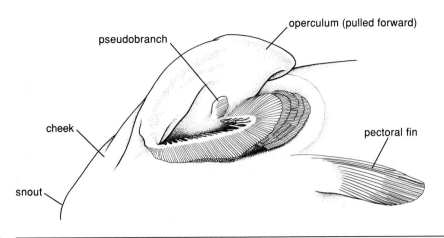

FIGURE 23–5   Position of pseudobranch in trout (Salmonidae).

may be involved in the function of the choroid rete, which secretes oxygen to the retina of the eye (Beatty, 1975). Blood flows directly from the pseudobranch, which is rich in carbonic anhydrase, to the ophthalmic artery. Although the contribution of the pseudobranch to the concentration of oxygen by the choroid is not known, removal of the pseudobranch lowers the concentration of oxygen in the eye (Satchell, 1991). In some species, it has been shown that secretion of gas into the gas bladder may be facilitated by secretions of the pseudobranch (Wittenberg and Haedrich, 1974). However, according to Laurent and Dunel-Erb (1984), the functions of the pseudobranch are uncertain, but the organ may function in pressure detection.

Branchial Circulation and Function

The Branchial Sieve

Effective exchange of gases in a gill-breathing fish depends on bringing the blood and respiratory water into close apposition on either side of a membrane through which the gases can diffuse. Such a system works best if the blood and water flow opposite to each other, and fishes have developed this countercurrent system. The system requires a force to move the water and one to move the blood and finely divided channels through which these fluids can flow. The channels are provided by the structure of the gills.

Each gill arch bears numerous filaments, so the total in the entire apparatus is several hundred; the actual number varies with factors such as size and surface area of the fish and the general habits of the species. Active fishes generally have more filaments than sluggish species (see Hughes, 1984). Small, bottom-living darters (Percidae) have about 300 to 500 filaments in total (Branson and Ulrickson, 1967). Mackerel weighing 800 g have about 2400, and perch (Percidae) of 30 g have about 1500. Each gill filament carries an abundance of secondary lamellae at right angles to the long axis. The lamellae, distributed on both the upper and lower surfaces of the filaments, are fragile ridges with thin walls. These walls constitute the barrier between the blood and the surrounding water and have three layers of cells: a relatively thick epithelial layer; a basement membrane; and pillar cells (Satchell, 1991). The thickness of these walls (the respiratory membrane through which gases must diffuse) varies with mode of life of the different species. Tunas and their close relatives have very thin lamellar walls of about 0.53 to 1.0 µm. Most bony fishes have walls of 2 to 4 µm, and some demersal species have respiratory membranes 5 to 6 µm thick. Most elasmobranchs for which data are available have lamellar thicknesses of 5 to 11 µm (Hughes, 1984).

The number of lamellae in marine fishes studied by Hughes (1966) ranged from 52,000 to 689,000 and depends on the number of filaments and the count of lamellae per unit length of filament as well as the size of the fish. Slow-moving fishes usually have from 10 to 20 lamellae per millimeter of length, whereas active fishes have 30 to 40 per millimeter. Most bony fishes are in the 15 to 30/mm range. Some air-breathing species have fewer lamellae, with a water–blood barrier 10 µm thick in *Anabas testudineus;* whereas the air–blood barrier of the aerial respiratory surface is 0.21 µm. In the mudskippers (*Periophthalmus* spp.), the blood–water barrier of the gill lamellae is thin because the gills, moistened by water held in the branchial chamber, are the site of gas exchange (Hughes, 1984).

A great surface area for exchange of respiratory gases is the result of having numerous filaments bearing the small but numerous lamellae. As with other respiratory features, active and slower moving fishes differ with respect to gill area. Some examples of estimates of gill area are as follows: (1) Less active fishes—such as whitefish (*Coregonus*), weighing 1000 g—had a gill area of 290 mm^2/g; bullheads (*Ameiurus*), weighing 50 g, had an area of 158 mm^2/g; and (2) more active fishes—such as herring, weighing 11 g—had a gill area of 636 mm^2/g; mackerel, weighing 80 g, had a gill area of 533 mm^2/g. S. de Jager and Dekkers (1975), to facilitate comparison among fishes, converted published gill measurements to that expected in 200-g fish and showed that, for fish of that size, scombrids generally have gill areas of more than 1000 mm^2/g, with various tunas having 1500 to 3500 mm^2/g. Other active pelagic fishes have 500 to 1000 mm^2/g, and most bony fishes for which measurements were available were found to be in the range of 150 to 350 mm^2/g.

The effectiveness of the gill area in exchange of respiratory gases depends on the contact made with water being pumped through the system. In each hemibranch, lamellae of adjacent filaments meet to form tiny channels through which water is forced (Fig. 23–6). The filaments are equipped with muscles that hold the tips of the filaments of posterior hemibranchs of each arch against the tips of the filaments of the anterior hemibranch on the following arch, so that all water must pass through the lamellar channels (Fig. 23–7). Langille et al. (1983) tabulate dimensions of lamellar channels (pores) and velocities of water in the gills in relation to several sizes of largemouth bass and a skipjack tuna of 1667 g. The number of pores in the skipjack tuna was 7,186,000, and the number in a bass of 837 g was 302,700. Flow of water over the gill of the bass is shown as 17.6 cm^3/sec during activity and 2.15 cm^3/sec at rest. For the skipjack tuna, the corresponding figures were 143 cm^3/sec and 24.6 cm^3/sec. As the space in the oral and branchial cavities changes with respiratory movements, there is compensatory change in the space occupied by the gill mass due to the action of gill arch and gill filament musculature. The tips of the filaments from adjacent arches remain in contact most of the time and are separated

A

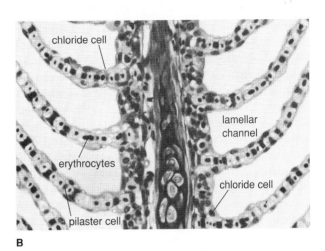

chloride cell

lamellar channel

erythrocytes

chloride cell

pilaster cell

B

FIGURE 23–6   Photomicrographs of sections along gill filaments of rainbow trout showing **A,** arrangement of secondary lamella to form numerous channels with great surface area; **B,** pilaster cells, chloride cells, and erythrocytes. (Photos courtesy of Professor Joseph H. Wales.)

briefly, at least in some species, during part of each opercular cycle. The tips are separated during coughing and during bypassing of excessive water flow.

In agnathans and elasmobranchs, the filaments are bound to the gill septa but the secondary lamellae stop short of the septa (Fig. 23–8), so passages or canals are formed next to them. Water passes through the interlamellar channels to these canals along the septa and then toward the gill slit. This arrangement allows a countercurrent flow of blood and water. This system may not be as effective as the countercurrent flow in teleosts (see Butler and Metcalfe, 1988).

Internally, the secondary lamellae of the gill filaments are divided into numerous capillary-sized channels by pillar or pilaster cells that are disposed in more or less regular rows. A marginal channel rims each lamella (Fig. 23–9). These small spaces receive blood from capillaries branching from the afferent arterioles of the filaments and pass the blood, counter to the flow of water outside the lamellae, to the efferent arterioles. There is a central sinus in the filament through which some investigators believe the blood can be bypassed without passing through the lamellae. Shunting of blood through the central cavity or around the tips of the filaments under conditions of abundant dissolved oxygen has been suggested by certain investigators, but others believe that no shunting occurs and that the only control of blood flow through the filament is by the muscles of the arterioles. Shunting of blood around the gill tissue is necessary in air-breathing fishes while in oxygen-poor

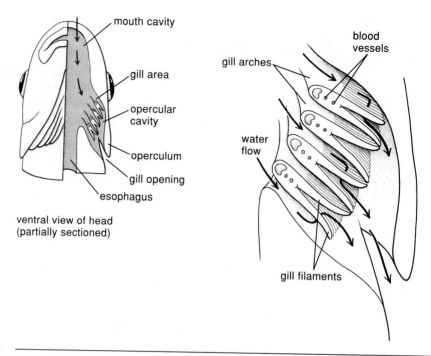

mouth cavity

gill area

opercular cavity

operculum

gill opening

esophagus

ventral view of head
(partially sectioned)

blood vessels

gill arches

water flow

gill filaments

FIGURE 23–7   Diagram of frontal section through gill area of teleost, showing tips of filaments from adjacent arches held together so water flow must cross filaments through lamellar channels. (Lamellae, which are at right angles to the axes of the filaments, are not shown.)

water, or else they would lose oxygen from the blood to the water. In these cases, only enough blood to take care of ammonia excretion and ion exchange must go through the relatively thick-walled lamellae (see Bone and Marshall, 1982).

Blood Flow in Gills

The heart provides the major force to move the blood through the ventral aorta, up the afferent branchial arteries, and into the gill filaments. There are two pathways that blood can follow in the lamellae: the arterioarterial and the arteriovenous pathways (Nilsson, 1986). In the former, blood passes through the lamellae, is oxygenated, and proceeds to the efferent branchial artery and then to the suprabranchial artery, which in teleosts develops into the carotid artery anteriorly and the dorsal aorta posteriorly. In the arteriovenous pathway, blood can pass from the filamental arteries to arteriovenous anastomoses and thence to the central filamental venous system and to possibly nutritive vascular beds in the gill tissue.

Contractions of the ventricle generate a considerable pressure that is attenuated as the blood is forced through the intralamellar channels and into the dorsal aorta via the efferent branchial arteries. Farrell (1991) tabulates blood pressures of various fishes from several sources. Examples of systolic pressures in the ventral aorta are given in Table 23–1. The bulbus arteriosus, because of its elasticity, maintains a positive pressure on the blood even though the diastolic ventricular pressure may drop to zero.

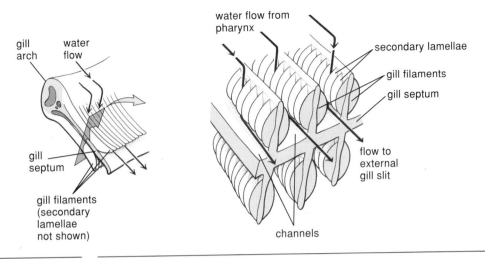

FIGURE 23–8   Diagram of gill filaments and secondary lamellae of dogfish (*Squalus acanthias*), showing path of respiratory water between lamellae, then through channels next to gill septum.

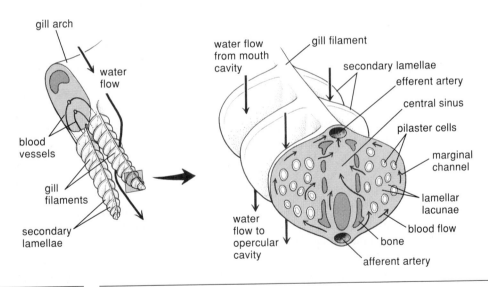

FIGURE 23–9   Diagram of section through gill filament at level of secondary lamella. Arrows depict flow of blood through spaces (lacunae) among pilaster cells.

The pressure drops markedly as the blood passes through the gills. Pressure in the dorsal aorta is usually 40 percent to 50 percent of that in the ventral aorta in bony fishes, but some, such as the carp, maintain from 50 to 70 percent of the ventral aortic pressure in the dorsal aorta (Ngan, 1971). The recorded dorsal aortic pressures of the carp, 22 to 32 mm Hg, contrast with the low figures of 8 to 12 for the Antarctic icefish. Systolic pressure in the dorsal aorta of various elasmobranchs has been measured at 25 to 40 percent of that of the ventral aorta. Pressure in the dorsal aorta of

TABLE 23–1	Systolic Pressures in Ventral Aorta for Selected Fishes	
Species	**Pressure (mm Hg)**	**Source**
Myxine sp.	6.8	Satchell (1991)
Myxine sp.	17	Farrell (1991)
Eptatretus cirrhatus	10.8	Satchell (1991)
Squalus sp.	30	Farrell (1991)
Raja sp.	16	Farrell (1991)
Oncorhynchus mykiss	45	Farrell (1991)
O. mykiss under exercise	72	Farrell (1991)
Ictalurus sp.	40	Farrell (1991)
Gadus morhua	38	Farrell (1991)
Ophiodon elongatus	39	Farrell (1991)

the hagfish, *Myxine,* ranges from about 2 to 14 mm Hg; the higher pressures apparently are due to the contraction of the gill pouches. The gill contractions superimpose pressure on the pulse caused by the heart.

Volume of blood flow through the gills is variable, depending in part on gill resistance. Control of blood flow through the gills involves both neuronal and humoral factors (Nilsson, 1986). Position of filaments is controlled by striated abductor and adductor muscles, which can change the angles of the structures rapidly, as well as by smooth adductor musculature. There are vascular sphincters at the bases of the filamental efferent arteries that could govern blood flow to some extent (Nilsson, 1986).

Another factor in gill blood flow, of course, is the rate at which the heart pumps blood. Typical bony fishes have cardiac outputs in the range of 15 to 20 ml/kg/min, but outputs from 5 to 100 ml/kg/min have been reported. Elasmobranchs have cardiac outputs in the range of 20 to 25 ml/kg/min.

Although the heart supplies the major force that maintains blood flow in the dorsal aorta, systemic arteries, and veins, other structures may assist in the flow, such as the gill contractions mentioned earlier in regard to *Myxine.* Serial contractions of body musculature aid the blood flow, and one of the consequences of heart contraction inside a nonelastic pericardium is a sucking action that hastens venous blood flow. Among the special structures that help maintain flow of body fluids are lymph hearts and caudal hearts of hagfishes, some sharks, and eels. These hearts are muscular and function only in pumping lymph or blood (see Satchell, 1991). Hagfishes also have portal (hepatic) hearts. Swimming activity aids in moving venous blood toward the heart, as in the caudal and hemal arch pumps in some bony fishes and sharks. A branchial pump that forces blood into the duct of Cuvier as water is pumped over the gills is present in many fishes (Satchell, 1991).

Branchial Ventilation

The respiratory pump that forces the water across the gills consists of the buccopharyngeal cavity plus all the mechanisms for opening, enlarging, and constricting it, and the parabranchial cavity (between the gills and the operculum), which can be enlarged or constricted by action of the operculum and the branchiostegals (see Saunders, 1961). Coordinated action of these two cavities can produce a continuous flow of water across the gills. As the mouth is opened and water is sucked in by the enlarging of the buccal cavity, the parabranchial cavity is rapidly enlarged, but with its opening closed. This causes a negative pressure that draws water across the gills. When the mouth is closed, the oral valves prevent escape of the water past the lips, and the water is forced across the gills, through the parabranchial cavity (which is now constricting), and through the opercular opening.

In hagfishes, respiratory water is pumped to the gills by a velar pump situated in a chamber posterior to the buccal cavity. Water enters the chamber through the nostril and is forced toward the gill pouches by the action of paired, scroll-like structures that unroll into the velar chamber from the top and then force the water out of the chamber as they rapidly roll up again, with the edge of the roll in contact with the lateral walls of the chamber. This action also sucks water into the chamber from the nostril (Hardisty, 1979).

Adult lampreys apparently can draw a respiratory current through the mouth when free swimming but usually depend on a tidal action to move water in and out of the external branchial apertures. This mode is necessary when attached to prey or other objects. The branchial region can be constricted to force water out of the gills, and the elasticity of the branchial skeleton then greatly aids in expanding the gill chambers to draw water in. A valvular arrangement carries incoming water to the mesial side of the gill pouch so that it can pass through the gill lamellae counter to the flow of blood. Larval lampreys utilize a velar pump in moving respiratory water, in addition to contractions of the branchial basket.

The relationships between the habits of fishes and their respiratory apparatus have long been recognized, and a classification of branchial pumps of teleosts was proposed by Baglioni (1908). Most teleosts fit into the system, which points out the increasing use of the branchiostegal apparatus in species with more and more sedentary habits. Pelagic species tend to depend mainly on opercular movements in pumping respiratory water. (Some swift pelagic species irrigate gills by holding the mouth open as they swim.) Less active species that spend some time resting on the bottom tend to have the branchiostegal part of the branchial pump better developed and combine opercular and branchiostegal movements in gill irrigation. Demersal species depend greatly on the branchiostegal apparatus and may have the opercular elements reduced. Some fishes, such as eels, pipefishes, and others with unusual branchiostegal configuration, were placed into a miscellaneous group by Baglioni.

Swift species such as tunas and billfishes can depend on their swimming speed to force water over the gills (ram irrigation), and some have nearly lost the ability to irrigate the gills by pumping. The branchial apparatus is highly modified in some species by having the tips of filaments of adjacent arches connected and by having connecting tissue between the branchial arches. Ram irrigation usually takes place at speeds over 1 km h^{-1} and has been calculated to take less than 1 percent of the total energy expended by a swimming skipjack tuna (*Katsuwonus pelamis*). Several marine fishes, including some demersal forms, can ram irrigate if their speeds reach 1.5 to 2 km/h (Roberts, 1975).

Some stream fishes are known to hold the mouth and opercula open to irrigate the gills passively while maintaining position in swift water. Some stream fishes, such as algae eaters (*Gyrinocheilus*), cling to the substrate while allowing the swift current to force water over the gills.

The volume of water pumped over the gills (respiratory volume) varies with factors such as morphology, size, temperature, carbon dioxide content of the water, oxygen content, and activity. Fishes respond to stresses of activity, lowered oxygen, etc., by increasing the number and amplitude of respiratory movements, so that the respiratory volume can be increased greatly. Experiments performed with the sucker (*Catostomus catostomus*), which showed a respiratory volume of about 50 ml/min/kg under certain conditions, obtained volume of up to 6000 ml/min/kg under exercise and 12,900 ml/min/kg when the dissolved oxygen in the medium was decreased. Relationships of breathing rate, amplitude of movement, and volume can be seen for the rainbow trout, in which the normal breathing rate is about 80/min. With exercise, the rate was found to increase to about 100/min, but the respiratory volume increased from 594 to 3042 ml/min/kg. Efficiency of oxygen removal from the water is usually impaired at high irrigation rates because not all the water pumped through the gills comes in close enough contact with the lamellae for a sufficient time. However, the blood passing through the gills is usually 85 to 95 percent saturated with oxygen (Saunders, 1961, 1962; Shelton, 1970).

Oxygen Uptake

Resting oxygen consumption for some familiar freshwater fishes is shown in Table 23–2. Under favorable conditions, fishes can remove about 85 to 90 percent of the dissolved oxygen from water passing over the gills counter to the flow of blood in the gills. Such efficient removal occurs when dissolved oxygen is high and respiratory volume low. More typically, the range is from 50 to 60 percent; and under conditions of low dissolved oxygen, high temperatures, and increased respiratory volume, the

TABLE 23–2	Representative Resting Oxygen Consumption in Selected Freshwater Fishes		
Species	**Temperature (°C)**	**O^2 Consumption (mg/kg/hr)**	**Source**
Carassius auratus	10	15.7	Beamish and Mookherjii (1964)
	20–22	30–160	Beamish and Mookherjii (1964)
	32–35	127–262	Beamish and Mookherjii (1964)
Cyprinus carpio	10	17	Beamish (1964)
	20	48	Beamish (1964)
	30	104	Beamish (1964)
Salmo trutta	10	81	Beamish (1964)
	20	128	Beamish (1964)
	20	282	Beamish (1964)
Cottus spp.	15	92–157	Original
	25	150–264	Original

utilization may fall to 10 to 20 percent or even lower. Actual oxygen consumption in fishes depends on many factors, including size, temperature, activity, the standard metabolism of the species, oxygen pressure, carbon dioxide pressure, pH of the medium, and salinity. The history of the individual in regard to acclimation or acclimatization to the aforementioned factors is important to oxygen consumption, as are circadian and seasonal cycles.

Some fishes can resort to anaerobic catabolism when using stored carbohydrates in burst or prolonged swimming beyond their aerobic capacities (Beamish, 1980), and some appear to be in the anaerobic mode in moderate swimming (Duthie, 1982). Anaerobic metabolism incurs an oxygen debt, and the fish must rid the blood of excess lactic acid. Maintaining optimum water conditions for valuable fish life is important because of the possible combined effects of poor conditions. When the dissolved oxygen content of the water is lowered, for instance, the fish responds by increasing the rate of gill ventilation. This added activity requires a greater consumption of oxygen, so the individual can be placed in the position of trying to extract a greater amount of oxygen from a smaller supply. A rise in temperature or increased carbon dioxide content could combine with the lowered oxygen and increased activity to make the situation intolerable.

Extrabranchial and Aerial Oxygen Uptake

Cutaneous Respiration

Most fishes are capable of absorbing oxygen from the water through the skin, although in sharks and many bony fishes the uptake might not be enough to satisfy the local cutaneous oxygen requirement. Many species can obtain from 5 to 30 percent of their required oxygen through the skin (Feder and Burggren, 1985a). In most fishes, cutaneous respiration is of importance mainly during periods of low activity or of relatively low temperatures. Many amphibious fishes depend on cutaneous absorption of oxygen for significant proportions of their respiratory requirements. Eels of the genus *Anguilla* obtain about 12 percent of their O_2 requirements through the skin; gobies of the genera *Boleophthalmus* and *Periophthalmus* obtain 36 and 48 percent, respectively; and *Neochanna* (mudfish of New Zealand) obtains 43 percent (Feder and Burggren, 1985b). The reedfish (*Erpetoichthys*), even though covered by heavy ganoid scales, obtains about 32 percent of its oxygen via the skin (Sacca and Burggren, 1982). Larval fish obtain oxygen via diffusion through the skin, and some, such as the larvae of *Monopterus albus,* have a countercurrent arrangement in which the blood in the skin flows forward counter to the water propelled posteriorly by the pectoral fins (Bone et al., 1995). Some larvae have gill filaments that extend from the gill opening, and special external gills are found in the larvae of polypterids and lepidosireniform lungfishes.

Air Breathing

The ability to extract oxygen from the atmosphere is a primitive characteristic of bony fishes and occurs as a convergent specialization in many teleost groups. The habit is encountered from the tropical swamps and beaches to the freezing Arctic bogs where Alaskan blackfish live. Overall, there are many more airbreathers among warm-water fishes than temperate or cold-water types, with the habit most common among the tropical swamp dwellers.

Air breathing by means of lungs is an ancient characteristic of fishes and probably originated in oxygen-poor environments during the late Silurian and early Devonian periods. Although lungs may have arisen in hypersaline Silurian seas, they were common among freshwater fishes of the Devonian swamps, and the heritage has been passed on to our present-day lungfishes and other nonteleosts, such as *Lepisosteus* spp., *Amia,* and *Polypterus* spp. In contemporary fishes, not only the lung or gas bladder is used by airbreathers, but several other structures have been modified for the purpose.

Obviously, waters permanently low or lacking in oxygen can be inhabited by fishes only if those fishes can derive their oxygen from an alternate source. Throughout the tropics, swamps of high organic content and heavy vegetative cover support year-round populations of fishes, some of which are obligate airbreathers. Swamps and streams that provide good dissolved oxygen supplies during part of the year, but that stagnate and even dry up at other times, may maintain a complement of specialized fishes that are facultative or obligate airbreathers. These can cope with the drying of the water either by burrowing and aestivating or by moving overland to more permanent bodies of water. Some tropical mountain streams support species that can withstand the torrents of the rainy season and the oxygen-poor pools of the dry season.

There are some fishes that expose themselves to air even though the surrounding water contains sufficient dissolved oxygen. Certain blennioids habitually remain in place under rocky cover as the tide recedes and returns, living in a dewatered but damp habitat for 2 or 3 hours every tidal cycle. Eels (*Anguilla* spp.) increase the living space available to them by moving overland through wet vegetation to isolated ponds and lakes. Species of walking catfishes owe part of their geographical distribution to overland forays that place them in new bodies of water. The spread of the exotic walking catfish (*Clarias batrachus*) in Florida is an example. Several species of airbreathers, including gobies, the blennioid *Dialommus,* swamp eels, and mudskippers, actively seek food while out of the water. The mudskippers are at home on sunny mudflats, where they move about freely, engaging in aggressive displays and other social behavior.

Adaptation to the aerial mode of respiration demands that air be brought into contact with highly vascular tissue of considerable surface area. Usually, some cavity in the head or body is modified for the purpose of oxygen uptake. In some instances, an existing cavity is modified; in others a new cavity is formed; and in others existing structures are modified to provide the requisite surface area. In addition, air-breathing species must be equipped to carry on the functions of osmoregulation, release of carbon dioxide, and excretion of ammonia. The pumping of blood through very thin gill tissue to facilitate one or more of these functions could lead to loss of oxygen from relatively oxygen-rich blood to the oxygen-poor water. Development of thicker gill lamellae, reduced branchial irrigation, and other modifications have aided in overcoming these problems. Some fishes have pathways in the gills that shunt oxygenated blood through without coming into contact with hypoxic water (Burggren and Roberts, 1991; Randall, 1985; Steen and Kruysse, 1964).

Many bony fishes that commonly spend time out of the water absorb oxygen through the skin. The eels, *Anguilla* spp., while in air, absorb 30 to 66 percent of required O_2 via the skin (Berg and Steen, 1966; Feder and Burggren, 1985b). Amphibious species that absorb much of their oxygen from the air through the skin are *Boleophthalmus* spp. (43 percent); *Erpetoichthys* (41 percent); and *Periophthalmus* spp. (76 percent) (Feder and Burggren, 1985b). Other species listed by Johansen (1970) as obtaining significant amounts of oxygen from the air through the skin are the longjaw mudsucker, *Gillichthys mirabilis,* and the bluntnose knifefish, *Hypopomus brevirostris,* although the latter also has gills modified for air breathing. An eliotrid (*Dormitator latifrons*) has a specialized vascularization on the top of the head that can absorb oxygen from the air when held at the water surface.

Most air-breathing species rhythmically or periodically empty and fill a specialized cavity with air, so the atmospheric oxygen can come in contact with highly vascular tissue in that cavity (see Liem, 1980a, for mechanisms involved). In the case of the electric eel (*Electrophorus electricus*), the cavity is the mouth and pharynx, where the lining is folded and otherwise modified to provide a large surface rich in blood vessels (Johansen, 1970). The mouth lining of some of the swamp eels, including *Synbranchus,* plays a significant part in respiration, although it is not much modified for an increase of surface area. The common carp brings a bubble of air into contact with a specialized part of the palate when in oxygen-poor water. Gee (1983) has observed similar behavior in gobies. The pharyngeal walls of mudskippers are reported to be a respiratory surface in air as well as the gills and skin (Gordon et al., 1969; Teal and Lacy 1967).

The walls of the pharynx are enlarged by diverticula in several species, including all of the snakeheads, Channidae, and the synbranchoid, *Amphipnous cuchia* (Hughes et al., 1974). The branchial chamber is the site of respiratory epithelium in many airbreathers. Some, such as the walking catfishes (*Clarias*) and the labyrinth fishes of the suborder Anabantoidei, have parts of gill arches modified into firm structures of large surface area bearing a respiratory epithelium. The labyrinth fishes include the climbing perch (*Anabas* spp.), the pikehead (*Luciocephalus pulcher*), and the gouramies of three families. In the labyrinth fishes, the respiratory organ is in an enlarged cavity above the gills and consists of a number of folded and crenelated plates. The corresponding organ of the catfish develops from the second and fourth gill arches and is arborescent in nature, taking the name "gill tree." *Saccobranchus fossilis,* an Asian catfish, has a pair of saclike diverticula leading from the branchial chamber into the lateral musculature.

One of the most remarkable adaptations for the use of atmospheric oxygen is the modification of parts of the alimentary canal for respiratory purposes. The swamp eel, *Monopterus albus,* has a modified esophagus that is used in aerial respiration (Liem, 1967a). Some South American armored catfishes—*Plecostomus* spp. and *Ancistrus* spp., for example—use the stomach as a respiratory organ. Many loaches, Cobitidae, and several of the armored catfishes are intestinal breathers, using a large section of the gut exclusively as a "lung."

Bony fishes, unlike the cyclostomes and elasmobranchs, apparently developed the lung and gas bladder early in their history. The lungs of lungfishes, like the lungs of other vertebrates, connect with ducts from the ventral wall of the alimentary tract, whereas the gas bladders of other fishes have a dorsal pneumatic duct. Lung tissue of lungfishes is quite like that of the lungs of higher vertebrates. Many physostomes use the gas bladder for aerial respiration. Some, such as the reedfish, obtain up to 40 percent of their oxygen through the lungs while in the water (Sacca and Burggren, 1982). Other fishes that breathe air include the bowfin, the arapaima (*Arapaima gigas* and some other osteoglossoids), mudminnows (*Umbra* spp.), the tarpons (*Megalops*), and the aimara (*Hoplerythrinus unitaeniatus*). The latter is a characoid of South American swamps and, along with the arapaima, is notable for having secondarily acquired a gas bladder structure similar to a lung.

The best development of the gas bladder as a lung is in the lungfishes, especially in the African and South American species, which have the lung divided into right and left sections. The Australian lungfish has an undivided lung. All have a special pulmonary circulation (Burggren and Johansen, 1987).

Several airbreathers have lost the capability to keep themselves supplied with oxygen from the water, even when in well-aerated situations, so if restrained from reaching the surface they soon die. Such obligate airbreathers include South American and African lungfishes, arapaima, electric eel, the snakeheads, and *Hoplosternum* spp., South American armored catfishes. Magid and Babiker (1975) report that large walking catfishes of the species *Clarias lazera* cannot survive even in oxygen-saturated water if prevented from breathing air. Many others must rely on air breathing during periods of low dissolved oxygen or during excursions out of the water, but their gills can maintain their respiratory needs only if the dissolved oxygen content of the water is high enough (Johansen, 1970).

Those species that aestivate, spending the dry season buried in the mud, in burrows, or even in mucous cocoons as in the case of the lungfishes, must be able to maintain themselves by breathing air over periods lasting for months. Other than the African and South American lungfishes, fishes known to aestivate during summers or dry seasons include *Amia,* the bowfin; *Umbra limi,* the central mudminnow; *Clarias* spp., the walking catfishes; *Synbranchus* spp. and *Amphipnous* spp., the swamp eels; Channidae, the snakeheads; *Anabas,* the climbing perch; *Lepidogalaxias salamandroides,* the salamanderfish; and species of *Neochanna,* the mudfishes of New Zealand.

There are few marine fishes capable of aerial respiration other than certain species typically found at the ocean's edge, where they may expose themselves during the changes of the tide. The amphibious clingfish (*Sicyases sanguinus*) of Chile is an example. Tarpon, the young of which are commonly found in brackish or fresh water, have a lunglike gas bladder. Some sharks of the genus *Chiloscyllium* have been observed to gulp air.

Muscle and Choroid Retes

Retes concerned with secretion of gas at the gas bladder are discussed in Chapter 18, but there are other retial systems in fishes, involving aspects of circulation. Retial systems involved with lateral muscles are found in the scombroids and the lamnid sharks, groups in which most members must swim constantly and vigorously to stay at a given depth. These retes are associated with the lateral red muscle, the site of nearly constant activity that is favored by a temperature higher than the medium in which the fish swims. The red muscle, which is surrounded by white muscle (see Chapter 18), is supplied with blood by large cutaneous arteries and drained by lateral cutaneous veins. A rete is imposed between those blood vessels and the red muscle, forming a countercurrent system that functions in the exchange of heat (Fig. 23–10).

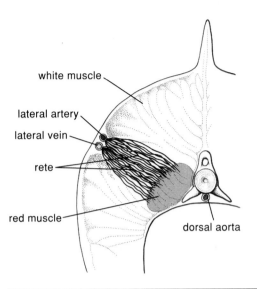

white muscle

lateral artery

lateral vein

rete

red muscle

dorsal aorta

FIGURE 23–10   Diagram of partial cross section of porbeagle shark, showing relationships of lateral blood vessels and rete to deep-seated red musculature.

The heat generated by the activity of the red muscle is conserved deep within the fish (Carey et al., 1971). In some primitive tunas, such as the skipjacks and the frigate mackerels, the rete is located in enlarged hemal arches (Schafer, 1975).

Within the viscera of some lamnoid sharks and certain tunas, there is a retial heat exchanger that maintains warmth in the body cavity. In the bluefin and bigeye tunas, the rete is formed from branches of the coeliomesenteric arteries. In sharks, the visceral rete forms from pericardial arteries (Carey et al., 1971).

In the superior rectus muscle of the eyes of swordfishes and billfishes, there are a glandlike structure and a countercurrent heat exchanger that maintain a constant temperature of about 28°C in the eyes and brain. The brain heater is a modification of the muscle to produce heat (as muscles do) without contractions. The blood supply for the rete comes from the carotid (Block, 1987).

Another retial system is found in the choroid of the eyes of teleosts and the bowfin. The choroid rete forms a horseshoe-shaped body around the optic nerve and serves to maintain a high interocular partial pressure of oxygen. Predators that depend heavily on sight have large choroid retes and interocular partial pressures up to more than 1300 mm Hg. More sedentary fish may have partial pressures only 25 percent of that figure. Elasmobranchs do not have choroid retes and may have an interocular partial pressure of oxygen as low as 30 mm Hg.

Excretion and Osmoregulation

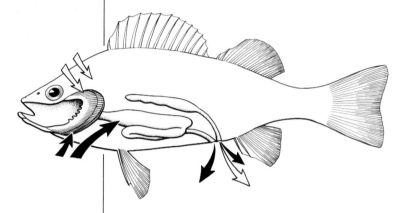

Urinary System

Kidneys

Kidney structure and function in fishes is an extensive and complex subject due to the wide evolutionary span of the animals involved and their myriad adaptations of form and physiology. Various modes of life in fresh and salt water have required structural adjustment of the kidneys to accommodate changing function. In some fishes, there is a close association of the kidneys and the genital system, but the systems are virtually separate in most bony fishes. The degree of utilization of the degenerate anterior portion of the kidney as a hemopoietic area varies from group to group. These and other considerations make it obvious that only a portion of this highly interesting and important subject can be treated here. Following a general orientation, a few examples of gross kidney structure and relationships will be given.

Although the coelacanth *Latimeria* is a notable exception in that it has a ventral unpaired kidney, the kidneys of most fishes are slender, elongate, dark red organs extending along the dorsal aspect of the body wall just ventral to the vertebrae (see Figs. 25–6 and 25–7). When viscera are removed from the body cavity, the kidneys can be seen through the peritoneum. Kidneys are paired but usually placed close together in most bony fishes; fusion along the mid-line is not uncommon. Excretory function is concentrated in the posterior section. The anterior part of the kidney is subject to modification both in structure and function. In male elasmobranchs, chimaeras, and nonteleost bony fishes such as sturgeons, gars, and bowfin, the anterior part is involved in the reproductive system. In most teleosts, the anterior part of the kidney has a concentration of lymphoid and hemopoietic tissue, with chromaffin (suprarenal) and interrenal tissue distributed along the postcardinal veins. (See Daniel, 1934; Goodrich, 1930; Romer and Parsons, 1978; Wake, 1979.)

The structural unit of the kidney is the nephron, which consists of a renal corpuscle and a convoluted tubule; the latter leads to ducts that convey urine to the exterior (Hickman and Trump, 1969). The renal corpuscle is made up of the double-walled Bowman's capsule and a glomerulus, a mass of capillaries coiled within the capsule. The lumen between the walls of Bowman's capsule is continuous with the remainder of the tubule, which consists of several segments (Romer and Parsons, 1978). Examples of tubule structure will be given later.

The excretory kidney of most fishes is an opisthonephros, which resembles the mesonephros that appears in the embryonic stages of amniotes (Hildenberg, 1988) (Fig. 24–1). In an opisthonephros, the tubules are not arranged segmentally and may be concentrated posteriorly. Hagfishes, in which the essential features of the mesonephros are retained (Fänge, 1963), are an exception. The renal corpuscle is present and the nephrostomes are absent. In general, the functional kidney of an adult anamniote is termed an opisthonephros (Hildenberg, 1988).

A more primitive pronephros, with segmentally arranged tubules opening by means of ciliated funnels (nephrostomes) to the abdominal and pericardial cavities, is present during embryonic development of fishes and persists in modified functional form in hagfishes and in larval lampreys. In both of these groups, the nephrostomes empty into the pericardial coelom. A few bony fishes, such as certain lanternfishes (Myctophidae), eelpouts (Zoarcidae), and clingfishes (Gobiescocidae), retain a pronephros (Weichert, 1965), which is always anterior to the opisthonephros, sepa-

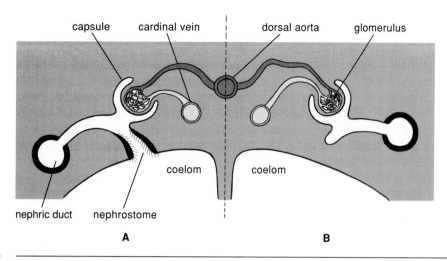

capsule   cardinal vein   dorsal aorta   glomerulus

coelom   coelom

nephric duct   nephrostome

A   **B**

FIGURE 24–1   Diagrams of kidney types. **A,** Pronephros; **B,** opisthonephros. (Based on Goodrich, in Lankester, 1909.)

rated from it in hagfishes but intergrading in most others. A kidney retaining both pronephric and opisthonephric elements is sometimes called a holonephros, but others reserve this term for a kidney in which each trunk segment has a tubule.

Each kidney is provided with a duct that drains posteriorly to a juncture with its fellow to form a median duct, or to a bladder or sinus. The major duct draining each kidney is called the archinephric (or nephric) duct in most fishes, although a different urinary duct draining the kidney has been developed by the elasmobranchs and chimaeras.

Hagfishes. The hagfishes retain a pronephric kidney (Fänge, 1963). These are small, paired, and situated some distance anteriorly to the opisthonephros and dorsally to the heart on the pericardial wall. Usually one large renal corpuscle (glomus) consisting of three glomeruli occurs in each pronephros posteriorly. There appears to be no connection of the corpuscles and tubules to a functional archinephric duct or to any other excretory duct. The nephrostomes enter the pericardial coelom, which is continuous with the body cavity, and the ciliated tubules connect with the pronephric vein. The function of the pronephros in the adult hagfish is problematical. It may have a hemopoietic (blood-forming) function and a lymphatic function. The functional kidney of hagfishes is long and slender, extending nearly the length of the body cavity. The 30 or more renal corpuscles are arranged irregularly along the length of the kidney. Tubules are unciliated and simple in structure, consisting of a short neck segment and a short proximal segment that empties into the archinephric duct or "ureter," which leads to the urinary sinus (Fänge, 1963). Urine received by the urinary sinus from the archinephric ducts is conducted to the exterior through a urogenital papilla.

Lampreys. Lamprey larvae begin life with paired pronephric kidneys that function through the prolarval stage. When that stage ends and the burrowing ammocoete stage begins, a pair of larval opisthonephroi develop posteriorly and eventually become the functional kidney tissue as the pronephric kidneys lose their function. The larval kidneys apparently do not contribute to the formation of the adult kidney. They degenerate, and an entirely new structure forms during transformation (Youson, 1981).

The long, straplike opisthonephric kidneys of adult lampreys are suspended from the dorsal body wall. They are somewhat comma shaped in cross section, with a dorsal portion tapering to a thin lower edge (Youson, 1981), along which the archinephric duct courses. The bulk of the organ in most species consists of a single, elongate renal corpuscle that has numerous tubules. This compound glomerulus or glomus has no typical Bowman's capsule; however, the capsule is present in the larvae, which have several renal corpuscles, with the posterior ones combining into glomeruli (Youson, 1981; Youson and McMillan, 1970). *Lampetra fluviatilis* is estimated to have 2300 to 2800 functional glomeruli (see Rankin et al., 1983).

The tubule of the lamprey kidney consists of a ciliated neck segment, a proximal segment divided into convoluted and straight portions, an intermediate segment, a distal segment with straight and convoluted parts, and a collecting segment. There are two types of intermediate segments: a short type and a longer type that may be involved in urine formation. The collecting tubules communicate with the archinephric duct, which has muscle tissue in the walls, possibly for the expulsion of urine. The archinephric ducts from each side merge and enter a urogenital sinus, which opens to the exterior via a urogenital papilla.

Elasmobranchs. The opisthonephric kidneys of sharks and rays are usually flattened band-shaped or strap-shaped structures that are wider posteriorly (Fig. 24–2). Kidney shapes vary, with some species having lobules along the lateral edges. This lobate form is especially prominent in rays, in which the kidney may be confined to the posterior part of the body cavity in females. The anterior part of the female kidney is usually reduced in sharks as well. The anterior part of the male kidney in elasmobranchs is enlarged and functions as part of the genital system (Callard, 1988; Goodrich, 1909).

Nephrons of the anterior part of the male kidney may become modified to secrete seminal fluid, which is received by the archinephric duct. The archinephric duct is usually converted into a sperm duct for the transport of sperm and seminal fluid, but in some species it receives urine from separate urinary ducts. In other species, collecting tubules convey urine to urinary ducts that have no connection with the archinephric duct. In male elasmobranchs, the archinephric duct (sperm duct) usually enlarges into a seminal vesicle that, along with the urinary duct, empties into the urogenital sinus. In females, the archinephric ducts typically drain into a urinary sinus.

The kidneys of chimaeras are somewhat similar to those of sharks and rays, but the males have multiple archinephric ducts that usually reach the urogenital sinus separately (Hickman and Trump, 1969).

Kidney tubules of sharks and rays consist of a long neck segment, an initial proximal segment composed of two cytologically distinct regions, a second more homogeneous proximal segment, and a distal segment followed by a collecting tubule. In chimaeras, the neck tubule is shorter and the glomeruli not as vascular, but overall the structure bears a resemblance to the kidney of *Squalus* (Hickman and Trump, 1969).

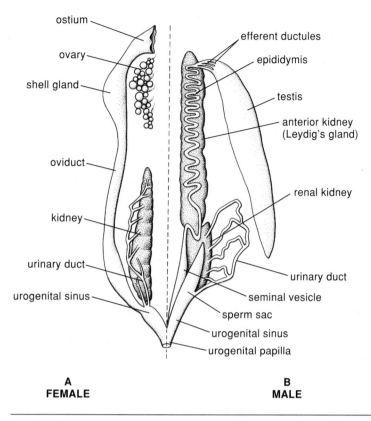

ostium

ovary

shell gland

oviduct

kidney

urinary duct

urogenital sinus

efferent ductules

epididymis

testis

anterior kidney
(Leydig's gland)

renal kidney

urinary duct

seminal vesicle

sperm sac

urogenital sinus

urogenital papilla

A
FEMALE

B
MALE

FIGURE 24–2   Diagram of urogenital organs of shark. **A,** Female; **B,** male. (Based on Goodrich, in Lankester, 1909.)

Bony Fishes. Sturgeons and paddlefishes have elongate kidneys extending the length of the body cavity (Goodrich, 1930; Romer, 1970). The pronephric portion, or head kidney, is included. The kidneys are fused posteriorly but taper forward separately. Anteriorly, the kidney of the male is associated with the genital system, receiving many efferent ducts from the testes. Many of the tubules are modified for sperm transport. The nephric ducts from each side meet posteriorly in an expanded, bladder-like section. In *Polypterus,* the bichir, it is the posterior section of the kidney that is most intimately associated with the testis, but there is no connection of the sperm duct with the archinephric duct short of the urogenital sinus.

In the bowfin (*Amia*), gars (*Lepisosteus* spp.), and lungfishes (*Neoceratodus, Protopterus, Lepidosiren*), the relationship between the testes and the kidneys resembles that of sturgeons and paddlefishes. In the lungfishes, the nephric ducts from the left and right kidneys run separately into the cloaca except in males of *Protopterus* spp., in which the nephric ducts unite before entering the cloaca. A urinary caecum or bladder is present in lungfishes (Romer and Parsons, 1978; Wake, 1987).

Teleosts, which are more diverse than other fish groups and have wider geographic and ecological distribution, have a variety of kidney morphologies. The nephrons of teleosts (Fig. 24–3) are generally composed of the following: a glomerulus, a neck segment, a two-part proximal segment, an intermediate segment, a distal segment, and the collecting tubule. Freshwater fishes typically have all of these components; marine fishes usually lack some of these, especially the distal segment. Actually, the only nephron structures that all teleosts have in common are the second proximal segment and the collecting tubule. Marine fishes, which lose water by osmosis across many body surfaces besides the kidneys, usually have smaller and fewer glomeruli than freshwater species (Hickman and Trump, 1969). This aids in preventing excessive extrarenal loss of water.

Although there is probably insufficient information on the thousands of species for a definitive classification of teleost kidneys, Ogawa (1961) proposed a classification based on the configuration or shape of marine teleost kidneys and suggested general relationships to phylogeny and habitat, as some of the types are encountered in freshwater species.

Ogawa's type I is exemplified by salmons and trouts (Salmonidae), herrings (Clupeidae), and the ayu (*Plecoglossus*). These have kidneys fused throughout their lengths, and the opisthonephros is continuous with the head kidney, which is the nonrenal remnant of the embryonic pronephros (Romer, 1970). The head kidney may be somewhat expanded laterally, especially in the salmonids. It consists of lymphoid tissue, with some suprarenal tissue included, and has lost the typical renal function. The nephron is typical of marine fishes, although many of the clupeoids

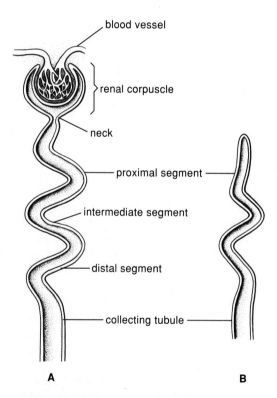

blood vessel

renal corpuscle

neck

proximal segment

intermediate segment

distal segment

collecting tubule

A   B

FIGURE 24–3 Schematic diagrams showing components of two types of nephrons found in bony fishes. **A,** Glomerular; **B,** aglomerular (found in some marine fishes).

and salmonoids live in fresh water. The glomerular capsule is large (ca. 85–105 μm), as in freshwater species, but the neck segment and distal segment are lacking.

Many minnows (Cyprinidae), at least one loach (*Misgurnus*), some catfishes, and most eels have kidneys fused to each other anteriorly and posteriorly, but separate through the mid-section (Ogawa's type II). The head kidney is usually expanded laterally. These fishes have large glomerular capsules (ca. 60–95 μm), and all segments are present in the freshwater species and in the marine catfish genus *Plotosus*. Morays (*Gymnothorax* spp.) lack the distal segment.

Ogawa recognized his type III in perchlike fishes (Percoidei), gobies (Gobioidei), barracuda (*Sphyraena* spp.), blennies (Blennioidei), the medaka (*Oryzias*), the snakeheads (*Channa* spp.), lanternfishes (Myctophidae), mackerels (Scombridae), sculpins (Cottidae), flounders (Pleuronectidae), and others. In these, the kidneys are found only posteriorly, and the head kidneys are well differentiated from the opisthonephric kidney. The glomerular capsule is relatively small (ca. 40–70 μm), and the distal segment is absent in the marine forms. *Channa* and *Oryzias,* both of which are freshwater forms, retain the distal segment.

Pipefishes and seahorses have narrow kidneys connected only at the most posterior end, and the head kidney is not developed. This represents Ogawa's type IV structure. The nephron is greatly reduced, with only the second portion of the proximal segment and the collecting segment present.

Type V kidneys are completely separated from one another. The head kidney is developed and may even retain pronephric glomeruli in some anglerfishes (Lophiiformes). This type also occurs in puffers (*Fugu, Canthigaster*) and boxfishes (*Ostracion*) and other closely related fishes. Some species are reported to lack glomeruli: the toadfish (*Opsanus*), the sargassumfish (*Histrio histrio*), and the porcupine fishes (*Diodon* spp.) are examples. These retain only the second proximal segment and the initial collecting tubule. Species that have glomeruli lack the distal segment.

Function of Kidneys

Fish kidneys play their major role in osmoregulation, maintaining body fluids at the proper osmotic concentration. The urine of freshwater fishes is much less concentrated than the blood plasma, whereas the urine of marine species is much more concentrated, approaching the osmolality of the blood plasma. Freshwater fishes produce a dilute urine at rates up to about 14 ml per hour per kilogram (Hickman and Trump, 1969). Production of urine by marine fishes is about 10 percent of that or less. In freshwater fishes, the osmolality of urine is typically in the range of about 20 to 37 milliosmoles per liter (mOsm l^{-1}), and in marine fishes the urine has osmolality in the range of 300 to 400 mOsm l^{-1} (see Hickman and Trump, 1969).

Freshwater fishes have large glomeruli that can filter much water from the blood. Some of this water is reabsorbed from the proximal tubule, but most is passed as urine. Monovalent ions are reabsorbed at both the proximal and distal tubule segments. In marine fishes, glomeruli are typically small, nonfunctional, or lacking. The proximal tubule segment is long. The first section, next to the glomerulus, if present, functions much like the proximal segment of the freshwater fish. The more distal part of the proximal segment secretes divalent ions (mainly Mg^{2+} and SO_4^-) into the tubule.

The kidneys of chondrosteans are specialized to retain urea and trimethylamine oxide (TMAO) to the extent that, like the osmoconforming hagfishes, they have blood plasma nearly isosmotic to sea water. The coelacanth (*Latimeria*) also retains urea and TMAO.

The major ions and solutes in the urine of the spiny dogfish (*Squalus acanthias*), in millimoles per kilogram, are sodium, 240; chloride, 240; urea, 350; and sulfate, 70. TMAO is present at 10 mmol kg^1. The osmolality of the urine is 800 mOsm kg^1 (Hickman and Trump, 1969).

Osmoregulation

The Osmotic Problem

Some of the most interesting adjustments that fishes of all kinds must make in their particular environments concern the maintenance of proper water and salt balance in their tissues. Failure to maintain the proper balance would result in dehydration of salt-water fishes and lethal dilution of body fluids of freshwater fishes. Few fishes have internal salt concentrations that closely match the water in which they swim, so they must prevent excessive gain or loss of water physiologically. The body fluids of freshwater fishes have a higher osmotic concentration than their medium, and marine species have more dilute fluids than the sea water surrounding them. Both ecological types, then, are at an apparent osmotic disadvantage that must be overcome for fishes to occupy the earth's waters. Regulation of osmotic concentration is accomplished by the kidney, the gills, some special organs, and, to some extent, by the integument in its role as a barrier.

A simplified review of the expressions used in discussing salt and water balance is in order at this point. One gram mole of a substance in 1 liter of solution is called a molar solution, whereas 1 gram mole per liter of solvent is referred to as a molal solution. In expressing osmotic activity, which depends on the number of undissociated molecules and ions per unit volume or weight of solvent, the term osmol is used. One gram mole per liter (kg) of water of a substance that does not dissociate can be said to have an osmolality of 1 osmol kg^{-1} and exerts an osmotic pressure of 22.4 atmospheres. Compounds that dissociate have a higher osmolality, corresponding to the degree of dissociation. One mole of sodium chloride kg^{-1} has an osmolality close to 2 Osm kg^{-1} because the compound dissociates nearly completely in solution. In dealing with the body fluids of animals, it is convenient to use smaller units than osmols, so the milliosmol (mOsm) is used.

Osmotic concentration can also be expressed in terms of freezing point depression (Δ_{fp}) of aqueous solutions. A molal solution (1000 mOsm) freezes at $-1.86°C$, which approximates the freezing point of surface water in temperate seas. Table 24–1 shows comparisons between freezing point depression and salinity, expressed as parts per thousand (ppt).

A solution with a smaller amount of salt per unit of volume than a solution to which it is being compared is said to be hypo-osmotic to the more concentrated solution, which, of course, is hyperosmotic to the less concentrated solution. If the solutions have the same osmotic pressure, they are said to be isosmotic.

Although diffusion will be regarded in simple terms for the sake of this treatment, it is a complex process involving such things as the nature of the membranes

TABLE 24–1	Freezing Point Depression of Water at Selected Salinities		
ppt		Δ_{fp} (°C)	mOsm/kg
5		−0.29	155
10		−0.58	312
15		−0.87	444
20		−1.13	608
25		−1.45	780
30		−1.72	925
32 (sea water)		−1.86	1000
35		−2.03	1091
40		−2.35	1263

being penetrated, concentration of solutes, and electrical charges of particles. This discussion will focus on the net results of the process.

Osmoregulation in Freshwater Fishes

Because the osmotic concentration of typical freshwater fish blood as expressed in mOsm kg^{-1} is in the range of about 265 to 325 (Δ_{fp} = −0.5° to −0.61°C), freshwater fishes are hyperosmotic to their medium and tend to gain water by diffusion through any semipermeable surface. If unchecked or uncompensated, the inward diffusion would dilute the body fluids to the point that the necessary physiological functions could no longer be accomplished, a state referred to by some as "internal drowning." Waterproofing the body would appear to be a means of preventing the diffusion but can be only partially successful. A thick scale covering (or bony armor) or even large amounts of connective tissue in the skin might afford some protection, but any site that maintains circulation of blood near the surface of the skin will be a chink in the armor. The gills, obviously, cannot be waterproofed and provide a great surface for diffusion of water as well as gases so, overall, water cannot be kept out.

If water in excess of the needs of the organism is driven in by inexorable osmosis, a balance must be maintained by driving the water out through some other means. The task of removing water is accomplished by the kidney. Blood from the dorsal aorta is led to the kidney by the renal artery, where it passes through the capillaries of the glomeruli and then through capillaries surrounding the kidney tubules before leaving via the renal vein. Blood from the renal portal vein joins a network of capillaries around the tubules.

The glomerulus is a filter that allows blood plasma containing dissolved materials to pass into the space between the walls of Bowman's capsule and thence into the kidney tubule. Blood pressure provides the force for glomerular filtration. Blood cells and large molecules, such as proteins, cannot pass the filter. There would be no osmoregulatory advantage gained by freshwater fishes if the filtrate removed from the blood at the glomerulus were to be excreted with its normal complement of salts. The advantage is gained by excreting urine more dilute than the plasma. (Further physiological advantage is gained in some fishes by the excretion of small amounts of nitrogenous waste in the urine.) As the fluid passes down the tubule, substances are resorbed at specific locations. Glucose is resorbed in the proximal tubule, and salts are resorbed in the distal tubule, the walls of which are impermeable to water in many fishes.

The urinary bladder appears to function in osmoregulation in freshwater and euryhaline teleosts (Agarwal and John, 1975; Evans, 1980). Water permeability of its walls is low, but Na^+ and Cl^- are actively reabsorbed through the walls. The resultant urine is dilute, with osmotic concentrations in various species and conditions from about 16 to 55 mOsm kg^{-1} ($\Delta_{fp} = -0.03°$ to $-0.09°C$). The urine contains small amounts of nitrogenous compounds such as uric acid, creatine, and ammonia.

Although the concentration of salt in urine is low, the copious flow causes a significant amount of salt to be lost. Salts are also lost by diffusion from the body. These losses are balanced by salt intake in food and by active absorption through the gills. Uptake of Na^+ and Cl^- at the gills is involved in an exchange of $NH4^+$ and $HCO3^-$. The actual site and mechanisms of the uptake are not understood, although evidence is accumulating. Some freshwater teleosts drink water, even though this places an added burden on the kidney: removing excess water (Evans, 1980).

Lampreys in Fresh Water. Although lampreys have not been the subject of as much physiological study as elasmobranchs and bony fishes, their capabilities as osmoregulators have been explored. Osmotic concentration of the body fluids of lampreys in fresh water is about 230 (*Lampetra fluviatilis*) to over 500 mOsm kg^{-1} (*Petromyzon marinus*) (Holmes and Donaldson, 1969). About 70 percent of the body fluid is intracellular, about 7 percent is in the plasma, and the remainder is interstitial. The skin and gills of lampreys are permeable to water, but research findings on the extent of permeability have been variable. *Lampetra fluviatilis,* the river lamprey, is reported to absorb water at a rate equivalent to about one third the body weight per day. *L. planeri,* the brook lamprey, apparently absorbs water at about twice the rate of the river lamprey (see Morris, 1972).

Normal urine flow in freshwater lampreys is apparently not well known because of difficulties with laboratory methodology. Many values shown in the literature, ranging up to 360 ml/kg/day for the river lamprey, are considered high by some authorities, but all agree that flow of the very dilute urine is copious. Osmolarity of *L. fluviatilis* urine in fresh water has been reported as about 25 to 38 mOsm l^{-1} (see Youson, 1981) and as 20 to 30 mOsm l^{-1} (Evans, 1993). The amount of sodium excreted in the urine is about 117 mole/h; chloride is excreted at about 5 mole/h. Only small amounts of nitrogenous substances are excreted by the kidney. Most nitrogenous excretion is at the gills.

Salts lost by lampreys in the urine and through the body surface are balanced by salts in food and by direct absorption from the water. The gills actively remove chlorides from the water and release them into the blood. Sodium, potassium, and calcium enter through the gills when made available as chlorides.

Freshwater Bony Fishes. Permeability of bony fishes to water (see Fig. 24–4) is generally less than that of lampreys, although some of the former have been noted as absorbing up to one third of their body weight per day. Scales and other armor in the bony fishes aid in retarding water uptake, as armor must have done in the extinct relatives of the lampreys. The eel, *Anguilla,* is often singled out as having a nearly impervious skin. Eel skin is reported to be so thick that it equals about 10 percent of the body weight. For bony fishes, there is evidence that most of the water absorbed comes through the gills (see Black, 1957; Conte, 1969). The body water of freshwater teleosts makes up about 70 to 75 percent of body weight. Sturgeons, paddlefishes, gars, and the bowfin generally have a similar proportion of water. In freshwater teleosts, according to Thorson (1961), intracellular water accounts for about 60 percent (55 to 63 percent) of the total body water; about 12 to 16 percent is extracellular water; and plasma accounts for about 2 percent of body water. Figures given by Parry (1966) are somewhat higher: 74 to 80 percent as intracellular water, and 2.5 to 3 percent of the body fluids are plasma.

Osmotic concentration of freshwater fish blood is usually slightly less than 300 mOsm l^{-kg}, but at least one species—the reedfish, *Erpetoichthys calabaricus*—has a dilute plasma of 199 mOsm l^{-kg} (Lutz, 1975a). Many anadromous species in fresh water may maintain concentrations of around 325 mOsm l^{-1}. Bony fishes move relatively large amounts of water through the kidneys. Urine flow varies with species, temperature, etc., but many determinations have shown between 50 and 150 ml/kg/day. Urine volumes as low as 16 ml/kg/day have been measured for the pike, and volumes as high as 330 ml/kg/day have been reported for the goldfish (Hickman and Trump, 1969). Osmotic concentration of bony fish urine is usually between 30 and 40 mOsm l^{-1}. A range of 20 to 80 is given by Bone and Marshall (1982).

Composition of the urine of freshwater fish varies greatly among species and conditions, but commonly the following solutes appear in the amounts given (mmol l^{-1}): sodium, 5 to 17; potassium, 1 to 5; calcium, 0.5 to 1; chloride, 2 to 10. Small amounts of nitrogenous compounds are excreted in the urine of freshwater fishes. In addition to urea and ammonia, which are excreted mainly through the gills, the urine may contain creatine, uric acid, amino acids, and creatinine (Hickman and Trump,

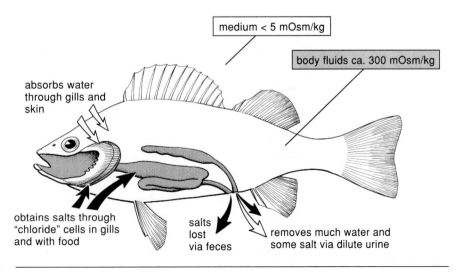

FIGURE 24–4 Diagram summarizing osmoregulation of freshwater teleost.

1969). In addition, the African lungfishes, *Protopterus* spp., produce urea during aestivation and store it in the blood, producing no urine for several months (Forster and Goldstein, 1969; Smith, 1930).

As in lampreys, loss of salts in the urine and by diffusion is compensated by uptake from ingested food and by absorption through the gills. Uptake of sodium is in part related to the excretion of ammonia at the gills. Ion exchange mechanisms operating in branchial cells facilitate exchange of ammonia for sodium and bicarbonate for chloride. Under suitable conditions, Na^+ is exchanged for H^+ (Maetz, 1974). The exchange rate for sodium in freshwater species is small compared with the exchange that occurs in saltwater species. Only about 1 percent of the exchangeable sodium in the body of the threespine stickleback (*Gasterosteus aculeatus*) is exchanged per hour in fresh water. The rate is 20 times that for the same fish in sea water. The total exchange of sodium for goldfishes has been reported to be 27 millequivalents/100 g/h.

Osmoregulation in Marine Fishes

The osmotic concentration due to salts in marine fishes (except hagfishes) is less than that of sea water. Bony fishes adapted to marine life maintain blood at the osmolality of about 380 to 470 mOsm kg^{-1} ($\Delta_{fp} = -0.7°$ to $-0.87°C$). The salt content of elasmobranch blood is slightly higher than that but forms only a part of the blood's osmolality; the remainder is due to retention of nitrogenous substances. Thus, the maintenance of proper water and salt balance in salt water requires different mechanisms from those in fresh water. There are at least three methods evident among marine fishes: Hagfishes are essentially osmoconformers, although they maintain a plasma concentration of sodium higher than that of sea water (Evans, 1993); elasmobranchs and *Latimeria* regulate by retaining urea; and the remaining bony fishes and lampreys osmoregulate by special mechanisms.

Lampreys in Salt Water. While in salt water equivalent to one half to full-strength sea water, river lampreys (*Lampetra fluviatilis*) and sea lampreys (*Petromyzon marinus*) can maintain body fluids at about 285 to 330 mOsm l^{-1}. Because the body fluids are hypo-osmotic to the medium, diffusion tends to remove water from the body through the gills and skin. To recoup the loss, lampreys swallow sea water (50 to 220 ml/kg/day), which is absorbed from the gut.

Also absorbed from the gut are monovalent ions, mainly Na^+ and Cl^-, which are in excess of the lamprey's needs. Divalent ions from the sea water are not absorbed by the gut and are eliminated with feces. Divalent ions are excreted in the meager but concentrated urine. The excess chloride is excreted through the gills, apparently through special excretory cells. These cells will be mentioned later in the treatment of bony fishes.

Hagfishes. These are characterized by a permeable skin and virtual lack of the ability to regulate sodium chloride, so that exposure to water of higher or lower salinity than sea water results in rapid changes in osmotic concentration of body fluids. The body fluids of hagfishes are practically isosmotic to their medium, although some investigators have shown a slight hypertonicity (see Robertson, 1963). The ionic content of hagfish plasma differs from that of sea water in that Ca^{++}, Mg^{++}, and

SO4⁻⁻ are in lower concentration and Na⁺ is in higher concentration. The concentration of Cl⁻ in hagfish plasma has been reported as lower than that of sea water by some and as higher by others. The concentration of K⁺ is reported to be nearly the same as in sea water. The freezing point depression of hagfishes has been determined to range from 1.74 to 1.98°C in normal sea water (Robertson, 1963). Urea appears to be present in very small amounts. Divalent ions and K⁺ are excreted in the scant urine.

Elasmobranchs. As mentioned earlier, the osmoregulation of elasmobranchs (Fig. 24–5) involves bolstering the osmotic concentration due to salts in the blood by retention of urea and smaller amounts of other nitrogenous compounds, especially trimethylamine oxide. Urea, an end product of nitrogen metabolism, produced in the liver, is excreted only in relatively small amounts via the urine of sharks and rays. As the glomerular filtrate passes along the kidney tubule, special segments resorb much (70 to 90 percent) of the urea so that the blood contains about 350 mmole l⁻¹ urea in a typical marine elasmobranch. The urea content of the blood is usually given as 2 to 2.5 percent. The urine of marine elasmobranchs usually contains about 100 mmole l⁻¹ urea. Trimethylamine oxide (TMAO), another nitrogenous waste product, appears in the blood at about 70 to 100 mmole l⁻¹ and is therefore of secondary importance to the osmolality of the blood. It is reabsorbed in the kidney tubule, and its concentration in the urine is about 10 mmole l⁻¹. The gills of elasmobranchs are relatively impermeable to urea and TMAO. The importance of TMAO and other methylamines in the body and intracellular fluids may be mainly the counteracting actions of these to the destabilizing action of urea to proteins (Shuttleworth, 1988).

The concentrations of the various solutes—mostly sodium, chloride, urea, and TMAO—in the blood of elasmobranchs combine for an osmolality of about 1000 to

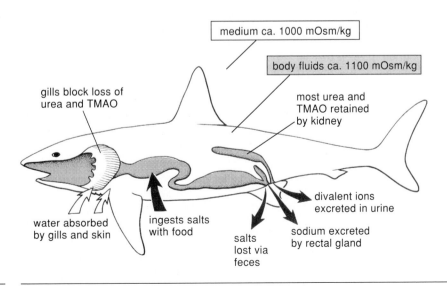

FIGURE 24–5   Diagram summarizing osmoregulation of salt-water elasmobranch.

1100 mOsm kg^{-1} in animals living in sea water of about 930 to 1030 mOsm kg^{-1}. Sharks and rays typically have an osmotic concentration in the blood from 50 to 100 mOsm kg^{-1} higher than the concentration of the medium. Being hyperosmotic to sea water, elasmobranchs tend to gain water by diffusion. Influx is mainly through the gills, because the skin is nearly impervious. This excess water is excreted as urine, the flow of which is typically from 1 to 1.5 ml/kg/h. Salts enter the marine elasmobranchs by diffusion and via ingested food because their body fluids have a more dilute concentration of salts than the medium. In addition, drinking of sea water has been confirmed in some elasmobranchs, but at a rate one tenth to one fifth that of teleosts.

Salts are excreted via two main pathways. The urine is the most important medium of excretion of divalent ions. The concentration of magnesium and phosphate in urine is more than 30 times the concentration in plasma, and the concentration of sulfate is nearly 140 times that of plasma. Sodium and chloride account for most of the osmotic concentration of the urine but appear in nearly the same concentrations found in plasma (Table 24–2).

Sodium is excreted by the rectal gland, an organ that was of unknown function until about 1960 (Burger, 1962; Burger and Hess, 1960). The normal concentration of NaCl in rectal gland fluid is nearly twice that of plasma, although the osmolality of the two fluids is the same. Small amounts of potassium and even smaller amounts of calcium and magnesium are excreted by this gland. The urea concentration of rectal gland fluid is about 4 percent of the concentration in the plasma. Volume of flow from the rectal gland is variable, usually 0.5 to 0.8 ml/kg/h. When the rectal gland is experimentally rendered nonfunctional, the kidneys appear to compensate by releasing copious amounts of urine (Shuttleworth, 1988).

Elasmobranchs possess chloride cells in the gills, which would suggest an active ion exchange at this site (Bone and Marshall, 1982; Kirschner, 1991), but although ions apparently can be lost and gained by the gills, Shuttleworth (1988)

TABLE 24–2

Concentrations of Selected Solutes in Plasma and Urine of the Spiny Dogfish (*Squalus acanthias*)*

Solute	Concentrations (mmole/kg)		
	Sea Water	Plasma	Urine
Sodium	440	250	240
Potassium	9	4	2
Calcium	10	3.5	3
Magnesium	50	1.2	40
Chloride	490	240	240
Sulfate	25	0.5	70
Phosphate	0	0.97	33
Urea	0	350	100
TMAO	0	70	10
Osmolality	930	1000	800

*Adapted from Hickman and Trump, 1969.

believes that any net elimination is insignificant compared to the amounts released by the kidneys and rectal gland.

A number of elasmobranchs enter water of low salinity or even fresh water, and the freshwater stingrays (*Potamotrygon* spp.) of the Amazon and some African and Asian dasyatids are found mainly in fresh water. Obviously, excess water would enter the body if they maintained the same osmotic concentration as that maintained by their marine counterparts. Some elasmobranch species can be acclimated to about half-strength sea water. They adjust by reducing the amounts of chloride, urea, and TMAO in the blood until the osmotic concentration is reduced to about 600 mOsm kg^{-1}. Even in the face of this reduced osmotic concentration, their hypoosmotic state requires increased urine flow (Evans, 1993).

Elasmobranchs that often move into large tropical rivers from the sea include the smalltooth sawfish (*Pristis pectinata*); the bull shark (*Carcharhinus leucas*), sometimes called the freshwater or Nicaragua shark; and others. These species can reduce the osmotic concentration of the blood to from 485 to 550 mOsm l^{-1}, retaining only about 100 to 180 mmole l^{-1} urea in the blood. Urine flow in the smalltooth sawfish has been measured at an average of 10.4 ml/kg/h (Thorson, 1967). The rectal gland of the bullshark regresses by decreasing the number of glandular tubules while in fresh water. The freshwater stingray (*Potamotrygon* spp.) has nearly broken the urea habit after its long history as a freshwater genus; the blood contains slightly more than 1 mmole l^{-1} of urea. Osmoregulation is essentially like that of freshwater teleosts. The rectal gland is nonfunctional.

Holocephali. Like sharks and rays, chimaeras retain a high content of urea in the blood and maintain their internal osmotic pressure at, or slightly above, that of the surrounding sea water. They differ from sharks and rays in the relative proportions of solutes in the blood: more salt and less urea and TMAO are retained (Table 24–3) (Read, 1971).

A discrete rectal gland has not been recognized in holocephalans but is considered by various investigators to exist in primitive form (Fänge and Fugelli, 1962). Secretory cells similar to those of rectal glands are present along with ducts entering the rectum.

Latimeria. This ancient marine fish osmoregulates in much the same manner as the two preceding groups. It is tolerant of urea and TMAO and retains these in the blood so that the nitrogenous solutes are relatively more important at keeping a high

TABLE 24–3

Constituent (mmole/l)	Serum	Urine	Sea Water
	Concentration of Selected Solutes in Serum and Urine of *Hydrolagus*		
Na$^+$	300	162	400
Cl$^-$	306	268	476
Urea	245	51.6	—
TMAO	5.5	—	—
Osmolality	897	844	892

*Adapted from Read, 1971.

osmotic concentration than they are in sharks, rays, and chimaeras. Sodium chloride concentration in the blood of *Latimeria* is less than that in the cartilaginous fishes. Examples of concentrations in serum of *Latimeria* are Na$^+$, 197 mmole l^{-1}; Cl$^-$, 87 mmole/l; urea, 377 mmole l^{-1}; and TMAO, 122 mmole l^{-1} (Griffith et al., 1974). The figures for Na$^+$ and Cl$^-$ are only slightly higher than for some teleosts but much lower than for chondrichthyans. Urea and TMAO are higher in *Latimeria* than in the cartilaginous fishes. Although the osmolality of *Latimeria* serum has been reported to range from 923 (lower than sea water) to 1181 mOsm l^{-1} (higher than sea water), the usual level is hypo-osmotic and *Latimeria* must drink sea water to maintain its fluid level. Water loss is probably retarded by the relatively small gill area, which in *Latimeria* is less than other fishes. Salt is probably secreted by the rectal gland, although the evidence for this is circumstantial (Griffith and Pang, 1979). The kidney does not provide a high concentration of salt in the urine, and chloride cells in the gills are not numerous.

The similarity in osmoregulation between elasmobranchs and *Latimeria* is not indicative of a close phylogenetic relationship, according to Griffith and Pang (1979). *Latimeria* differs from chondrichthyans in that it is unable to resorb urea in the kidney, a character shared with the crab-eating frog, another animal that retains urea as an osmoregulatory aid.

Marine Teleosts. These have slightly less total body water than freshwater teleosts (Fig. 24–6) and have a blood osmolality of 380 to 450 mOsm kg^{-1} in an external environment of 800 to 1200 mOsm kg^{-1}, so that water constantly diffuses from them to the medium. Some species have been shown to lose from 30 to 60 percent of their intake by osmosis. This loss is compensated for largely by drinking sea water. The rate of drinking varies with species, and within species it varies with salinity. The higher the salinity, the greater the rate of drinking. Marine species commonly swallow sea water reported to amount to from 7 to over 35 percent of their body weight per day (Conte, 1969; Johnson, 1973; Kirschner, 1991). From 60 to 80 percent of the ingested water is absorbed through the alimentary canal, beginning in the esophagus, where *Anguilla* spp., for example, absorbs a significant proportion of the swallowed water (Conte, 1969; Kirschner, 1991). Absorbed with the water are the monovalent ions Na$^+$, K$^-$, and Cl$^-$. Divalent ions remain mostly in the gut; usually less than 20 percent of those swallowed are absorbed.

The excess monovalent ions imbibed with the water are excreted mainly through the gills via "chloride cells," which resemble the salt-secreting cells of other animals (Kirschner, 1991; Vickers, 1961). These cells are concentrated in the gills, but in many species they appear in the mouth lining and the skin of the inner surface of the operculum as well as in the skin of the head and anterior body. Guppies (*Poecilia reticulata*) adapted to salt water showed an increase of chloride cells not only on gill tissue but in the mouth, inner lining of the operculum, and even in pockets under scales (Vickers, 1961). The chloride cells are extremely rich in mitochondria and have a remarkable network of tubular structures throughout the cytoplasm. These tubules are continuous with the basal surface of the cell; are rich in Na$^+$, K$^+$, and adenosinetriphosphatase (ATPase); and are likely the site of N$^+$ extrusion. Each ion-secreting cell has a cavity at its apex, opening to the exterior. Chloride has been found to be secreted from these cavities (Bone et al., 1995). Sodium chloride likely

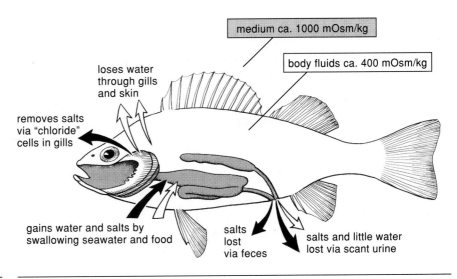

medium ca. 1000 mOsm/kg

body fluids ca. 400 mOsm/kg

loses water
through gills
and skin

removes salts
via "chloride"
cells in gills

gains water and salts by
swallowing seawater and food

salts
lost
via feces

salts and little water
lost via scant urine

FIGURE 24–6 Diagram summarizing osmoregulation of marine teleost.

enters via connections with adjacent cells, but Na$^+$ is kept at a low level in the chloride cell by an Na$^+$/K$^+$-activated ATPase pump. As Na$^+$ is removed from the cell at its base, it moves paracellularly around the chloride cell and into the medium. The Cl$^-$ is extruded from the apex of the cell, perhaps as a vescicle, into the water.

Sodium excretion balances the intake through the gut, and the sodium–potassium exchange at the chloride cell appears to be important to the loss of sodium. Studies on the flounder *Platichthys flesus* showed that no sodium was excreted through the gills in potassium-free sea water (Maetz, 1969).

Much remains to be learned about the mechanisms of ion exchange in fishes. Ranges of amounts of materials excreted or absorbed can be measured and verified, but processes are not yet well understood (Kirschner, 1991). The use of the electron microscope has done much to bring about an appreciation of the structure of the secretory cells, and biochemical studies have led to the knowledge that certain enzyme systems, especially ATPase, seem to play a strong role in exchange. The function of the chloride cell involves activity of Na$^+$/K$^+$ ATPase located on the plasma membrane at the base of the cell (Maetz, 1974).

The cobbler (*Cnidoglanis macrocephalus*), a plotosid catfish, apparently depends on a special salt gland (dendritic organ) for maintenance of salt balance. The organ is external, is attached just posterior to the urogenital opening, and has an ultrastructure similar to the rectal gland of sharks. Ligature of the organ results in severe ion imbalance (Kowarsky, 1973).

The kidneys of marine teleosts, as in freshwater fishes, are the major site for excretion of divalent ions, but their role as a water pump is diminished because of the extrarenal losses of water. Consequently, the glomeruli are generally smaller and fewer than in freshwater fishes and elasmobranchs (Table 24–4). Several species of marine teleosts have lost the glomeruli and have minimized renal loss of water. Among these are species of gulpers (Saccopharyngidae), seahorses and pipefishes

TABLE 24–4

Number and Size of Glomeruli in Selected Fishes*				
Species	Weight (g)	Number of Glomeruli in One Kidney	Average Diameter of Glomeruli (μm)	Relative Volume (mm³) of Glomeruli per m² of Body Surface
Freshwater Teleosts				
Ameiurus nebulosus	89	18,160	100	126.5
Cyprinus carpio	221	24,310	82	50.8
Perca flavescens	116	4,870	102	30
Marine Teleosts				
Gadus morhua callarias	670	16,250	37	1.49
Lutjanus griseus	544	31,860	55	10.99
Pseudopleuronectes americanus	160	5,300	50	3.14
Elasmobranchs				
Mustelus canis	485	4,400	185	60.93
Raja erinacea	1060	1,200	190	10.38

*Data adapted from Nash, 1931.

(Syngnathidae), dragonets (Callionymidae), scorpionfishes (Scorpaenidae), poachers (Agonidae), sculpins (Cottidae), puffers (Tetraodontidae), toadfishes (Batrachoididae), clingfishes (Gobiesocidae), anglerfishes (Lophiidae), frogfishes (Antennariidae), and batfishes (Ogcocephalidae). In addition to diminution, degeneration, or loss of glomeruli, marine fishes lack the distal segment of the kidney tubule (see Hickman and Trump, 1969).

Urine flow in marine teleosts is scant, usually amounting to 1 to 2 percent of the body weight per day. In some species, most of the water in the urine enters through the kidney tubule. In others, water is filtered out by the glomerulus, and most is subsequently reabsorbed by the tubule. The osmotic concentration of urine is slightly less than that of the blood. The urinary bladder is involved in adjusting the concentration of the urine. It has greater water permeability than that of freshwater teleosts and does not absorb salts.

Principal divalent ions in the urine of marine fishes are, in decreasing order of concentration, magnesium, at 50 to more than 100 times the plasma concentration; sulfate, at concentrations up to more than 300 times that of the plasma; and calcium, at 4 to 10 times the plasma concentration. Phosphate occurs in concentrations less than that of calcium. There apparently is active secretion of magnesium, sulfate, and, sometimes, phosphate into the tubule. In some marine species, calcium and magnesium are precipitated as salts in the tubular lumen and thus no longer influence the osmotic gradient. Chloride may be virtually absent from the urine in some species but can appear in concentrations up to 85 percent of the plasma in others, especially when an individual is in a diuretic state (see Black, 1957; Hickman and Trump, 1969).

Diadromous and Other Euryhaline Fishes. Many species are capable of living in both fresh and salt water. Anadromous fish are hatched in fresh water, subsequently move to the sea to feed and grow, and then return to fresh water to spawn. Catadromous species reverse the two media. Many other species have a wide tolerance for salinity and can move freely between fresh and salt water. All of these must be able to adjust osmoregulatory mechanisms more or less rapidly depending on the speed with which they change habitat. Diadromous species generally undergo progressive changes that may alter appearance as well as physiology. At a given time, depending on the stage of life history, these fishes are suited mainly to one medium or the other and do not usually change back rapidly. Some euryhaline species, however, can make excursions rapidly through a range of salinities.

The young of several anadromous species of salmonids, upon attaining a characteristic minimum size (often 10 to 15 cm in length), undergo a transformation from the parr stage to the salt-tolerant smolt stage. The critical size range (which in nature might not be reached for one, two, or more years, depending on the species) can be reached in only a few months of rearing in a hatchery. The fishes change to a silvery color, become slimmer in form, and have a tendency to migrate downstream into the ocean. If they are restrained from migrating, the salt tolerance of some of the species—such as rainbow trout (*Oncorhynchus mykiss*)—regresses after several weeks. Chum salmon (*O. keta*) and pink salmon (*O. gorbuscha*) migrate seaward at a small size and are salt tolerant soon after the yolk sac is absorbed. This salt tolerance does not normally regress.

When the diadromous species are adapted to salt water, the osmotic concentration of blood is higher than in the same species adapted to fresh water. Examples of a few species follow. Concentrations are given in mOsm/kg, first that of the animal adapted to fresh water and then a dash and the corresponding figure for salt-water adaptations: (1) *Petromyzon marinus,* 280–317; (2) *Oncorhynchus tshawytscha,* 304–350; (3) *Salmo salar,* 328–344; (4) *Anguilla* spp., 350–430.

Adaptation of euryhaline fishes to salt water generally requires drinking of the medium. Fishes such as rainbow trout and eels, which swallow little or no water while in fresh water, may drink about 4 to 15 percent of their body weight per day in salt water. *Tilapia mossambica* (= *Oreochromis mossambicus*), the Mozambique tilapia, may swallow nearly 30 percent of body weight per day (Johnson, 1973). Drinking rate is in part dependent on temperature, and the rate increases with the temperature (Maetz, 1974). Many species respond to the saline medium by changes in kidney function. The glomerular filtration rate may diminish dramatically, and the tubular reabsorption of water usually increases. Urine flow decreases to 10 percent or less of the flow in fresh water. Usually the blood of a fish physiologically ready for hypo-osmoregulation stabilizes within 1.5 to about 5 days.

The osmoregulatory capacity of some species is truly remarkable. The inanga, *Galaxias maculatus,* lives in salinities from less than 1 to 49 ppt and can acclimate to salinities of 62 ppt. The Mozambique tilapia has been acclimated to 69 ppt (Chessman and Williams, 1975). Many members of the Poeciliidae and Cyprinodontidae can tolerate high salinities. The sheepshead minnow, *Cyprinodon variegatus,* of the southern United States has been found alive at 142.4 ppt, which is about four and one half times as salty as sea water.

In the tropics especially, many members of typically marine families move into and out of fresh water, some at certain life history stages, but some nearly randomly.

These must be able to switch abruptly from conserving water to filtering out large volumes through the kidney, and they must turn from excretion of excess salts to conservation. Several marine families have given rise to vicarious freshwater forms. A few of these forms seem to be unlikely candidates for a freshwater life inasmuch as they have few or no glomeruli. For example, there are freshwater species of pipefishes and toadfishes. These species apparently are impermeable and can increase urine flow through a mechanism that is not well known. A marine toadfish, *Opsanus tau,* which can adapt to low salinity, has been noted as having plasma osmolarity of 392 mOsm kg^{-1} in sea water and 250 mOsm kg^{-1} in fresh water (Lahlou et al., 1969).

Eggs and Larvae. Eggs of fishes, at the time of deposition, are essentially isosmotic to the body fluids of the female. In elasmobranchs, urea and TMAO are present in sufficient quantities that there is no quick change of osmotic balance encountered by the eggs of oviparous species when they are deposited. These embryos, as well as those of ovoviviparous species, maintain their levels of urea and TMAO through the sometimes lengthy incubation and are apparently not at an osmotic disadvantage (Shuttleworth, 1988). Hagfish eggs are nearly isosmotic to the environment. On the other hand, eggs and sperm of teleosts are generally placed in environments with either higher or lower osmotic concentrations. Under conditions normal for the species involved, exposure of the gametes to a medium differing in osmolality exerts no ill effects. Furthermore, results of experimentation with gametes of salmon, herring, and flounder have shown that fertilization in some species can occur over a wide range of salinity. The percentage of successful fertilization decreases at low (< 15 ppt) salinity in the marine spawners and at high (> 24 ppt) salinity in the salmon. Most freshwater species have low rates of fertilization and hatching in saline water (see Holliday, 1969).

Fertilized eggs, although they imbibe water through the chorion, are somehow able to regulate their salt concentration to a great extent. There appears to be some effect of the salinity at which fertilization takes place on the subsequent ability of the developing egg to develop optimally. This seems to be due to some initial influence on the physical properties of the perivitelline fluid by the salinity of the medium.

There are some examples of incubation and hatching at unusual salinities. The sheepshead minnow has been known to hatch at 110 ppt, the desert pupfish (*Cyprindon*) and the fourbeard rockling (*Enchelyopus cimbrius*) at 70 ppt, and the herring (*Clupea harengus*) at 60 ppt (Holliday, 1969). Chum salmon (*Oncorhynchus keta*) eggs, when transferred at the eyed stage to full-strength sea water, showed 50 percent survival through hatching, although the alevins did not survive (Kashiwagi and Sato, 1969). The eggs of a few primary freshwater teleosts are known to incubate normally and hatch well at salinities up to 5 ppt.

Knowledge of the relationships between osmoregulation and growth in salmonids and various other fishes used in aquaculture is of great practical importance. Culture of salmon and trout in net cages set in bays and fjords has certain advantages over rearing in fresh water, so the time or growth stage at which the young fish can be transferred to a saline environment without undue mortality or altered metabolism must be known. Numerous studies have addressed parr–smolt transformation and freshwater–sea-water transfer (see Hansen et al., 1989).

Morgan and Iwama (1990) incubated rainbow trout eggs and reared fry at several salinities and found that a hypertonic medium caused decreased hatch and increased mortality of alevins. Fry grew slower and showed elevated metabolic rates at hypertonic salinities. Fry reared in isotonic water showed no increase in growth rate.

Research on Atlantic salmon smolts has shown that a loss of appetite and a slowing of growth ensues after transfer to sea water, and those effects last beyond the adjustment of smolts to osmoregulation in salt water. Metabolism changed from a tendency to deposit lipids while in fresh water to a tendency to deposit proteins after transfer and adaptation to salt water (Usher et al., 1991).

Newly hatched larvae of some marine species (herring, flounder) tolerate wide ranges of salinity but reduce this tolerance with age. Alevins of chum salmon can survive in quarter-strength sea water one day after hatching but do not adapt well to full-strength sea water until the yolk sac is absorbed at 60 or more days after hatching. Cells rich in microtubular structures, resembling the chloride-secreting cells of mature fishes, appear in the epidermis of larval herring. Possibly osmoregulation is begun in such cells prior to the full development of gills, gut, and kidneys (see Holliday, 1969).

Endocrine Secretions and Osmoregulation

Although there is a probability that hormones are of great importance to the osmoregulatory powers of lampreys and hagfishes, details and proof are largely lacking. That state of affairs is not due entirely to lack of attention by physiologists. Results of research with those animals have not been clearcut. Rankin et al. (1983) note that adrenaline and arginine vasotocin are diuretic in lampreys.

Many questions regarding hormonal control of osmoregulation in elasmobranchs are still to be answered. A peptide called rectin appears to be a factor in the control of secretion of the rectal glands of *Squalus, Scyliorhinus,* and *Raja.* Adenosine stimulates secretion in rectal glands, and somatostatin may be inhibitory of secretion. There is also evidence that steroid hormones are important in regulation of the rectal gland and of urea content and permeability to water. Thyroid hormones and prolactin are involved as well (Shuttleworth, 1988). Much knowledge is accumulating concerning the relationship of the endocrine secretions to osmoregulation in bony fishes, especially teleosts. As expected, there have been differences in the responses of various species to such conditions as experimental administration of hormones and the blockage of endocrine secretion, but there appears to be firm evidence that certain hormones affect osmoregulation.

Secretions of the pituitary are probably both directly and indirectly involved in control of salt and water balance. Prolactin has been shown to decrease permeability of membranes and to have the effect of sodium retention in freshwater fishes (Avella et al., 1990; Lahlou, 1980; Young et al., 1989). Coho salmon smolts adapted to salt water and then transferred to fresh water increase plasma prolactin concentrations up to ten times that maintained in salt water. At the same time, plasma osmotic pressure changes from about 360 mOsm kg^{-1} to about 325 mOsm kg^{-1}. Coho smolts adapted to fresh water and then transferred to salt water reduce plasma prolactin levels rapidly and change osmolality to about 360 mOsm kg^{-1} (Avella et al., 1990).

Atlantic salmon (*Salmo salar*) undergo a decrease in plasma prolactin during the parr–smolt transformation (Prunet and Boeuf, 1989). Prunet et al. (1985) showed that the plasma prolactin in the rainbow trout decreased from 10–15 ng/ml to 3–5 ng/ml after one day following transfer to sea water. Reciprocal transfer resulted in restoration of the higher levels. In the euryhaline Mozambique tilapia, blood and pituitary prolactin levels were shown to be lower after transfer to salt water (Nicoll et al., 1981).

Arginine vasotocin influences kidney function and sodium permeability in both marine and freshwater species (Hickman and Trump, 1969). Growth hormone is thought to influence the onset of salt tolerance in the young of anadromous salmonids. It shows effects on salinity preference as well as tolerance, but the exact role is not known. (See Butler, 1966; Holmes and Donaldson, 1969; Johnson, 1973.) Bolton et al. (1987) noted that growth hormone influenced osmoregulation in the rainbow trout by reducing plasma levels of calcium, sodium, and magnesium.

Adrenocorticotropin and thyrotropin, secreted by the pituitary, may influence osmoregulation by stimulating secretions of the respective target glands. The thyroid is active in migrating diadromous species, but the specific effect of thyroxin on water and ion balance is poorly known. The interrenal tissue produces adrenocortical steroids that act on renal and extrarenal systems to aid in regulating the body fluids in both marine and freshwater species, but again details of activity are largely lacking. Cortisol may be the most important steroid involved. Redding et al. (1991) implicate cortisol in osmoregulatory processes of juvenile coho salmon during adaptation to salt water. Plasma concentrations of sodium were lower in both fresh and salt water after injection with cortisol, and concentrations of potassium were higher. Cortisol had direct effects on the number and biochemical characteristics of chloride cells in Mozambique tilapia opercular membranes when these membranes were removed and held in vitro (McCormick, 1990). The effect on restoring numbers of chloride cells was dose dependent. Cortisol has a part in control of absorption of water by the alimentary canal in the salt-water phase of eels. It also acts on the permeability of the urinary bladder in salt-water–adapted fish. In his review of corticosteroids in osmoregulation, Dharmamba (1979) refers to cortisol as "the dominant hormonal factor" involved in the osmoregulation of marine-adapted fishes. However, Wendelaar Bonga (1993) notes the importance of cortisol in the osmoregulation of fishes transferred from sea water to fresh water as well. The corpuscles of Stannius have been implicated in ion and water balance by experiments on *Anguilla anguilla*. When sexually maturing "silver" eels were transferred to sea water, they increased the osmotic concentration of the blood serum, mainly by increases of Na^{2+} and Cl^-. On removal of the corpuscles of Stannius, there was a decrease in the concentration of Na^+ and increases of Ca^{2+} and K^+. When eels with corpuscles removed were injected with a preparation of corpuscles, serum concentrations of Ca^{2+} and K^+ were reduced and Na^+ was increased (see Holmes and Donaldson, 1969).

Feeding, Nutrition, and Growth

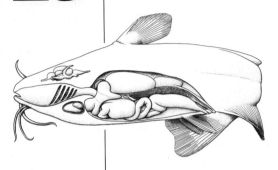

The Alimentary Canal: Functional Morphology

Mouth, Teeth, and Pharynx

Adaptations for diverse manners of feeding are seen in many groups of fishes, and these adaptations naturally involve the size and placement of the mouth and the size and kind of teeth in the mouth or pharynx. Most fishes have the mouth at, or very near, the front end of the head; but numerous bottom feeders, such as suckers or sturgeons, have subterminal or inferior mouths. Superior mouths are possessed by relatively few fishes—those that capture food from the surface or those that wait at the bottom to catch prey passing overhead. The mouth can provide a clue to feeding habits, especially when considered along with size and placement of teeth.

Teeth. Pikes (Esocidae), handsawfishes (Alepisauridae), and many sharks are equipped with the large mouths and big, sharp teeth that identify them as predators on rather large prey that can be swallowed whole. Some sharks have dental arrangements that enable them to bite large chunks out of animals too big to swallow. Piranhas (*Serrasalmus* spp.), some other characins, and other fishes, such as barracudas (*Sphyraena* spp.), may do the same. A variety of deep sea predators have dagger-like teeth that help them grasp relatively large prey and hold it until it can be swallowed.

Many largemouthed fishes that have small teeth or none at all in the mouth may be equipped with other structures that can hold prey or strain plankton out of the water. Pads of small conical or cardiform teeth on the jaws or several bones (such as the vomer and the palatines) are seen in many species that are opportunistic in capturing a variety of prey. The largemouth bass (*Micropterus salmoides*) and many species of catfishes are examples of successful predators with small teeth.

Fishes with specialized feeding habits may depart from the usual in remarkable fashion (Fig. 25–1). Wolf-eels (*Anarrhichthys*), which habitually feed on shelled invertebrates, have strong canine teeth in the front of the jaws for grasping their prey and have blunt molars for crushing the shells. Lungfishes have tooth plates that are used for holding or crushing prey. Parrotfishes (Scaridae) can bite off chunks of coral with a beaklike structure formed by the fusion of the front teeth.

Many butterflyfishes (Chaetodontidae) have small mouths at the end of a thin snout, an arrangement that is useful in removing food items from crevices. The mouth and teeth of butterflyfishes have been studied by Motta (1982, 1984a, b, 1985), who found that protrusion of the mouth in this genus (*Chaetodon*) involves three different complex couplings of bones. The teeth of butterflyfishes contain iron. Those species that feed on the hardest prey have more iron in the teeth than those that feed on somewhat softer materials (Motta, 1987).

Slipmouths (Leiognathidae), dories (Zeidae), mojarras (Gerreidae), and several other families are capable of protruding the mouth an exceptional distance to siphon in prey.

Teeth are borne on several of the head and face bones. Those in the upper jaw include the premaxillary and the maxillary in most of the soft-rayed fishes, but the maxillary does not bear teeth in most of the higher teleosts. Additional teeth are commonly the vomer and palatines. Many species bear teeth on the pterygoids and

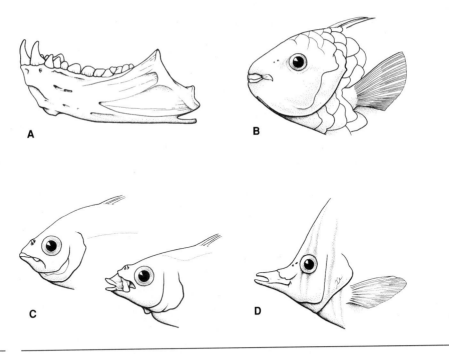

FIGURE 25–1   Examples of teeth and mouths. **A.** Dentary bone of wolf-eel (*Anarrichthys*); **B,** beaklike teeth of parrotfish (Scaridae); **C,** protrusible mouth of slipmouth (Leiognathidae); **D,** small, specialized mouth of butterflyfish (Chaetodontidae).

parasphenoid. In the lower jaw, the dentaries are usually the main toothed bones, but teeth may be present on the tongue (glossohyal) and the basibranchials (Fig. 25–2).

Attachment of teeth to the supporting bone falls into four general types (Fink, 1981). Type 1 is the most primitive and features a strong mineralized connection between tooth and jaw or pharyngeal bone. This pattern is seen in bichirs, paddlefishes, gars, bowfin, lower teleosts, and a few higher fishes. In type 2 attachment, common to many teleosts, mineralization is incomplete and the tooth is connected to the jaw by collagen. Type 3 teeth, found mainly in the Stomiiformes, are hinged and depressible so that captured prey can be moved toward the esophagus but is prevented from escaping when the teeth are erected. The base is not fully mineralized anteriorly in the area of the hinge. In type 4 attachment, collagen attaches the tooth to the posterior part of the base and acts as a hinge. When the tooth depresses, the anterior edge lifts off the base, exposing the pulp cavity. This type is found in pikes, some stomiiforms, and several groups of higher teleosts.

Pharynx. The gills lie just behind the oral cavity in the pharynx. There are typically four pairs of gills in bony fishes, but sharks and rays may have gills on five to seven arches (see Chapter 23). The gill arches may be equipped with inwardly directed projections called gill rakers that aid in filtrating food from the water (Fig. 25–3). Fishes that consume large prey may have gill rakers that are few in number and

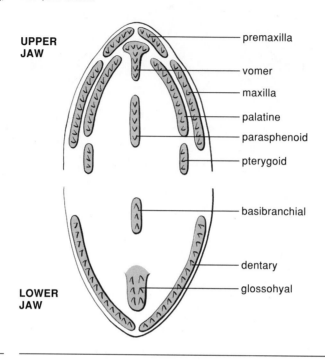

UPPER JAW
premaxilla
vomer
maxilla
palatine
parasphenoid
pterygoid

basibranchial

dentary
glossohyal

LOWER JAW

FIGURE 25–2   Diagram of positions of bones that can bear teeth in bony fishes.

small, but many carry rough prominences or denticles that aid in holding and swallowing prey. Plankton feeders usually have an extensive straining sieve formed of long, slender gill rakers. Some examples of salmonids with the two types of gill rakers are the chinook salmon, which feeds often on herring and other such fishes, and the sockeye salmon, which feeds mainly by filtering plankton.

An epibranchial or crumenal organ that may be for collection and concentration of small food particles is present in certain groups of fishes that feed on plankton and similar materials. This organ is epibranchial or postbranchial in position. It is found, for instance, in plankton-eating herrings (*Clupea* spp.), the herbivorous milkfish (*Chanos*), the suborder Stromateidoidei, and a number of deep sea fishes, including *Bathylychnops*, the four-eyed fish (Stein and Bond, 1985).

The fifth gill arches of bony fishes are usually reduced to a single lower element (the fifth ceratobranchial) on each side. This bone bears teeth that bite against opposing teeth borne on the upper elements (pharyngobranchials) of one or all of the four branchial arches. In minnows (Cyprinidae) and suckers (Catostomidae), the lower pharyngeal bones bite against a pad borne on an extension of the basioccipital bone. The monophyletic group Labroidei includes damselfishes (Pomacentridae), cichlids (Cichlidae), surfperches (Embiotocidae), wrasses (Labridae), parrotfishes (Scaridae), and Odacidae. These are examples of some fishes with pharyngeal jaws (pharyngognathous teleosts). The fifth ceratobranchials are fused and bear teeth in those families (Lauder and Liem, 1983a). In the cichlids, the lower pharyngeals bite against the second to fourth pharyngobranchials (Liem, 1991). Pharyngeal teeth are varied in size and shape, ranging from small conical points to grinding plates (Fig. 25–3).

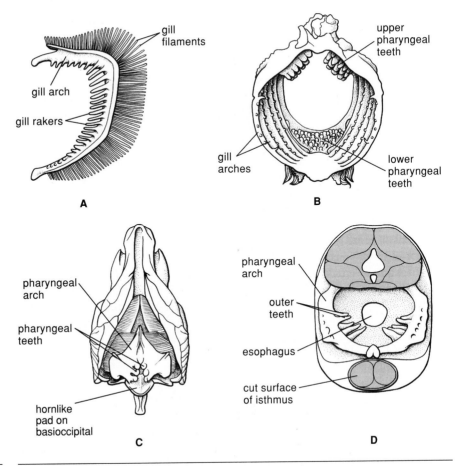

FIGURE 25–3 Examples of gill rakers and pharyngeal teeth. **A,** Diagram of gill arch with rakers and gills; **B,** anterior view of gill arches and pharyngeal teeth of surfperch (Embiotocidae); **C,** ventral view of the pharyngeal region of a carp (*Cyprinus*), with pharyngeal arch displaced anteriorly to expose the basioccipital pad; **D,** anterior aspect of pharyngeal teeth of the squawfish (*Ptychocheilus*), cross section behind the last gill arch with musculature and other soft tissue removed.

Esophagus, Stomach, and Intestine

Esophagus. In general, the esophagus in fishes is short and distensible so that relatively large objects can be swallowed, but microphagous fishes have less distensible tubes than those of predatory fishes. Esophageal walls are generally equipped with both circular and longitudinal muscles, which function in swallowing. The lining of the esophagus consists of stratified epithelium and columnar epithelium with numerous mucous cells or glands. Taste buds are probably present in all species. Gastric glands appear in the posterior part of the esophagus in some mullets (Mugilidae) and sculpins (Cottidae). In more primitive bony fishes, the esophagus is the site of the connection of the swimbladder with the alimentary canal via the pneumatic duct.

Several modifications of the esophagus are known. The butterfishes (Stromateidae) and their close relatives have muscular sacs connected to the esophagus. In some stromateid genera (*Pampus, Nomeus*), these esophageal sacs are lined with teeth, which are attached to thin bones in the walls of the sacs. The sacs of stromateids serve various functions in different species, such as mucus production, food storage, or preparation of food by trituration (grinding or crushing). In some other fishes, the esophagus is modified for respiration. The rice eel (*Monopterus alba*) and the Alaskan blackfish (*Dallia pectoralis*) are examples of the latter.

Stomach. In most fishes a stomach is present, varying in shape and structure according to the diet of the various species. Usually the stomach is a bent, more or less muscular tube shaped like a U or V (Fig. 25–4**A**). Another common form is a bag-shaped stomach with anterior openings to the esophagus and gut. Heavy-walled, gizzard-like stomachs are found in mullets (*Mugil*), gizzard shad (*Dorosoma*), and a few others (Fig 25–4**B**).

The stomach is lacking in lampreys, hagfishes, chimaeras, and some bony fishes, including, for example, minnows (Cyprinidae), pipefishes (Syngnathidae), sauries (Scomberesocidae), and parrotfishes (Scaridae). In these fishes, gastric glands are absent and the esophagus empties directly into the intestine. Hagfishes have a sphincter between the branchial region and the digestive gut.

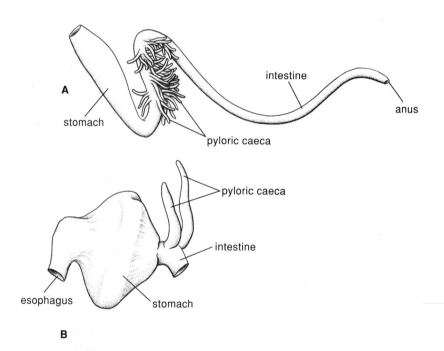

FIGURE 25–4   Examples of stomachs and pyloric caeca (anterior to left). **A,** stomach, caeca, and intestine of trout (Salmonidae); **B,** stomach and pyloric caeca of mullet (Mugilidae).

Intestine. The length of the intestine in fishes is generally correlated to the amount of indigestible material ingested (L. Smith, 1989). Carnivores generally have short guts (Table 25–1, Fig. 25–4). Herbivorous fishes and those that eat detritus and mud may have guts several times the body length. These long intestines are usually folded into distinct patterns in the body cavity (Fig. 25–5). Omnivorous fishes have guts of intermediate lengths (Fig. 25–6). Sturgeons, lungfishes, polypterids, *Latimeria*, the bowfin (*Amia*), and gars (*Lepisosteus*) as well as the sharks and related cartilaginous fishes possess a spiral intestine (Fig. 25–7), in which the absorptive surface is increased by the corkscrew course of a fold of tissue down the length of the organ. A modification is the scroll-like rolled valve present in some sharks.

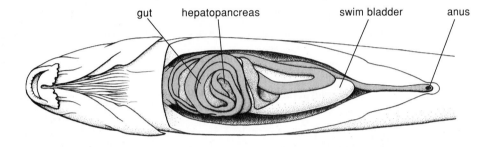

FIGURE 25–5   Elongate gut of sucker (*Catostomus macrocheilus*), a microphagous fish.

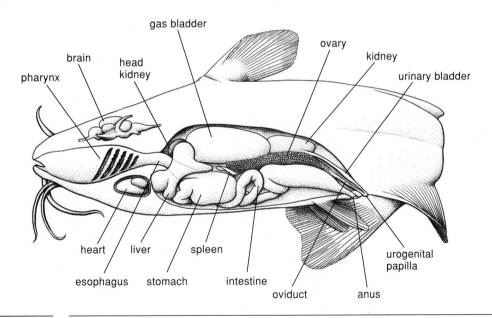

FIGURE 25–6   Bullhead (*Ameiurus*), head sectioned slightly to left of midline, body cavities opened to show relative positions of internal organs. (Note that the head kidney is separated from the renal kidney by the gas bladder.)

TABLE 25-1	Ratio of Intestine Length (*I*) to Body Length (*B*) in Selected Fishes		
Species	**I/B**	**Remarks**	**Source**
kanyu, *Elopichthys bambusa*	0.63	Piscivorous	Kapoor et al. (1975)
black sea bass, *Centropristis striata*	0.71	Piscivorous	Blake (1930)
northern squawfish, *Ptychocheilus oregonensis*	0.78	Piscivorous	Weisel (1962)
Pyrrhulina filamentosa	1.0	Insectivorous	Kapoor et al. (1975)
northern sea robin, *Prionotus carolinus*	0.89	Benthivorous	Blake (1936)
goatfish, *Mulloides auriflamma*	1.03	Benthivorous	Al-Hussaini (1946)
longnose sucker, *Catostomus catostomus*	2.29	Benthivorous	Weisel (1962)
Xenocharax spilurus	2.0	Eats plants and invertebrates	Kapoor et al. (1975)
goldfish, *Carassius auratus*	2.3	Eats phytoplankton, other plants, detritus	McVay and Kaan (1940)
Citharinus congicus	4.0	Microphagous	Kapoor et al. (1975)
Citharinus citharus	6.0–7.5	Microphagous	Kapoor et al. (1975)
nase, *Chondrostoma nasus*	2.0	Herbivorous	Junger et al. (1989)
grass carp, *Ctenopharyngodon idella*	2.5	Herbivorous	Kapoor et al. (1975)
silver carp, *Hypophthalmichthys molotrix*	13.0	Phytoplankton	Kapoor et al. (1975)
Labeo horie	15–20	Algae, detritus	Kapoor et al. (1975)

An interesting contrast is seen in the guts of some jawless fishes. The intestine of parasitic lampreys, which feed on the blood and juices or finely divided flesh of their prey, is extremely thin walled and can be greatly extended, but when empty it appears as a thin cord. The hagfishes ingest larger pieces of their prey, and the hagfish intestine has a thick wall and an extensively folded lining.

In most fishes, the vent or anus represents just the posterior opening of the gut. Other than in a few rare exceptions (such as the female of *Nerophis*, a pipefish), only the sharks, rays, and lungfishes have a cloaca, which receives the end opening of the gut as well as those of the urinary and genital systems. A few families, such as the herrings (Clupeidae), have an opening to the exterior from the swimbladder posterior to the anus.

The pyloric caeca, liver, pancreas, and swimbladder are attached to or associated with the intestine.

Pyloric Caeca. On the intestine of most bony fishes, just beyond the pyloric end of the stomach, there may be from one to many blind sacs, called pyloric caeca. A few groups, such as topminnows (Cyprinodontidae), pikes (Esocidae), and some catfishes (Ictaluridae), lack these structures. The bichir (*Polypterus*) has only one, the yellow perch (*Perca flavescens*) has three, and in other groups, such as flatfishes (Pleuronectiformes), the pyloric caeca are also few in number, with usually no more than five. In others, such as mackerels (Scombridae), salmons (Salmonidae), and snailfishes (Liparidae), the number of these caeca may be 200 or more. Caeca of different species vary considerably in size, state of branching, and the connection with the gut. In the sturgeons (Acipenseridae), the many caeca form a large mass, but only a single duct leads to the intestine. In salmon, each caecum communicates directly with the gut (Fig 25–4). The functions of pyloric caeca probably involve both digestion and absorption. Digestive enzymes have been isolated from the caeca of many species. Trypsin is secreted in the pyloric caeca in some species.

Liver. The liver is a large gland in all fishes, but sharks and rays may have extremely large livers that comprise about 20 to 30 percent of the total body mass, especially in some pelagic sharks (Fig. 25–7). The liver usually lies over or partially surrounds the stomach. It is typically bilobed but may have only one lobe, as in salmon, or three, as in mackerels and *Squalus*. In hagfishes, the liver is in two distinct parts, with separate ducts leading to the gallbladder. Adult lampreys have no bile ducts or gallbladder, but in most other fishes the bladder is present and functions to store liver secretions. Ordinarily, one hepatic duct originates from each lobe of the liver and joins the cystic duct from the gallbladder to form the bile duct. Liver function includes bile secretion and glycogen storage, in addition to several other biochemical processes.

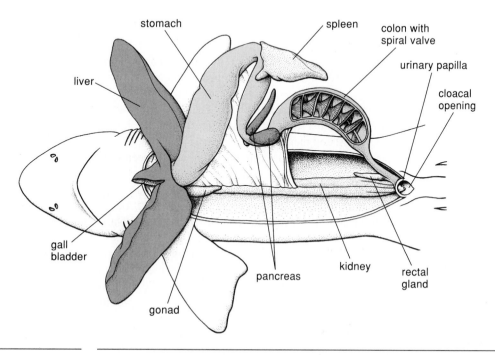

FIGURE 25–7 Diagram of viscera of shark (spiral valve opened to show internal structure). (Based on Daniel, 1934.)

Pancreas. Hagfishes have a small pancreas with several ducts that empty into the bile duct. Lampreys have pancreatic tissue (endocrine only?) located throughout the liver and intestinal wall. Among the bony fishes, the pancreatic tissue is usually diffuse and located in or around the liver. This is especially true of the pancreas of the advanced fishes, in which the pancreas and liver are combined into a "hepatopancreas." The lungfish (*Protopterus*) and many of the soft-rayed bony fishes have a discrete pancreas. In sharks and rays, the pancreas is a compact organ, usually consisting of two lobes. The pancreatic duct may reach the small intestine separately from the bile duct, as in the sharks, or may discharge into the bile duct, as in the gar (*Lepisosteus*) and lungfish (*Protopterus*).

The pancreas secretes several enzymes that are active in digestion. In addition, the pancreatic islets have the endocrine function of producing insulin. (See also Chapter 17.)

Spleen. The spleen is usually recognized as a dark red, often pyramidal structure lying on or behind the stomach, to which it attaches by a bandlike ligament. Although it is associated with the digestive organs, it has no digestive function, but rather is instrumental in blood-cell formation. The function of red blood-cell destruction has also been ascribed to the spleen of advanced bony fishes. In lampreys and hagfishes, which do not have a compact spleen, spleenlike tissue is diffused along the intestine. Lungfishes lack the spleen.

Feeding and Nutrition

Feeding

Figure 25–8 shows a simplified diagram of some generalized trophic relationships of fishes. Additional information on the subject is provided in Chapters 28 and 29.

Detection and Selection of Food. Feeding is carried out daily by most fishes and may be the most frequent of voluntary activities. Flight from enemies, reproduction, migrations, and many other activities might be occasional or periodic, but feeding is usually part of the daily routine and in some species may require extended periods of time. Most fishes are especially adapted for specific food gathering but are not necessarily narrow in this selection. Competition can be reduced or avoided by switching to other foods even though the new foods are not customary. As Liem (1980b) remarks, even specialists can be jacks-of-all-trades if necessary to find food. In diet switching, individuals of a given species with two choices of food will feed on the more abundant of the two until a point is reached at which the rate of food intake is lessened to a low threshold; then the fish switches to the other food supply. This type of diet switching is part of the optimum foraging theory, but there is also diet switching of an unpredictable nature not included in optimal foraging (Gerking, 1994).

Although many structures serve several functions in the life of fishes, feeding may be an important, if not the principal, use for these structures. Some may be of greater importance in certain fishes than in others, and all the senses can be instrumental in the detection and selection of food. The locomotory powers and structures

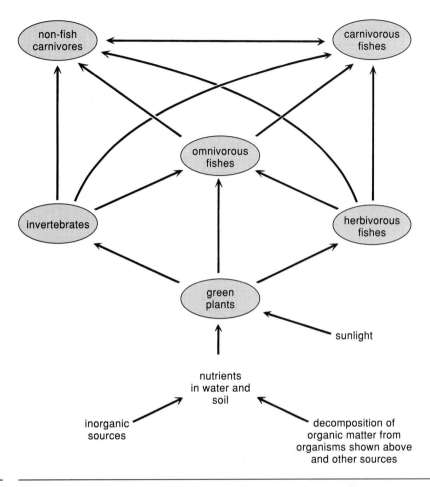

FIGURE 25–8   Simplified diagram of hypothetical trophic relationships of fishes.

of fishes are used in great measure for finding and gathering food. In earlier chapters, there were many references to specialized structures involved in feeding, mostly concerning the head and mouth but including lures for attracting prey.

Detection of possible food items at a distance can be through chemical senses, the eyes, the auditory organs, the lateral line system, or the electrosensory system. Most species are visual feeders and find food mainly based on visual cues; examples are many and range from most fish larvae to large-eyed pelagic predators such as scombroids. In many species of fishes, the position of the eyes in the head serves the particular feeding habits.

Olfaction and taste both play major roles in feeding. Many species that appear to feed mainly by sight are known also to have acute olfaction. Fishes such as lungfishes, eels, the spiny eels of the Mastacembelidae, and others have well-developed olfactory systems, as do sharks, rays, and many deep sea fishes. Because olfaction is a general alert system, many species get the first signal that food is in the vicinity via the sense of smell and are able to follow a chemical gradient to the source.

Taste, along with the tactile sense, appears to be significant in the final selection of food and its retention for preparation and swallowing. Although taste buds are typically in the mouth, distribution of external taste buds over the skin of a wide variety of species is well-known. Many fishes with barbels, including catfishes and cyprinids, are examples. Fishes appear to separate food from unwanted detritus at the level of the pharynx. Taste buds are abundant on the gill arches, gill rakers, epibranchial organ, and the tissue surrounding the pharyngeal teeth. In many instances, material is ejected through the gill openings after being subjected to a final test by sensory facilities in the pharynx. Large particles are usually ejected through the mouth with a "coughing" action. Gill rakers, pharyngeal teeth and bristles, epibranchial organs, and special pharyngeal musculature all serve mechanical functions in retaining or rejecting ingested material. The black surfperch (*Embiotoca jacksoni*) has pharyngeal muscles that enable it to winnow (sort) small food organisms from other small particles (Schmitt and Holbrook, 1984).

The amounts of food ingested per day and the amount of time spent feeding depend on many factors. Active predators, with high metabolic rates, require more food energy than do sluggish fishes. If a predator feeds habitually on small organisms, a relatively great amount of time must be spent gathering prey in order to acquire this energy. However, if the predator can catch and swallow large organisms, it might satisfy itself with one or two captures per day. Filter feeders might take in large amounts of biomass in relatively short times. Because metabolic rate varies with temperature (See Jobling, 1994), cold water predators that endure winter conditions should require less food than warm water predators, such as tunas, that avoid cold water. Deep sea predators (such as anglerfishes and gulpers) in constant cold water probably require infrequent large meals (Marshall, 1966).

Daily and seasonal temperature fluctuations affect food intake in most species. The predator, feeding on the bodies of other animals, such as insects, crustaceans, squids, and other fishes, is ingesting high-protein, high-calorie food that provides nutritional requirements without much bulk. On the other hand, the herbivore or detritivore must ingest large quantities of less concentrated foodstuffs, sometimes including a large proportion of undigestible material. Consequently, feeding activity in these fishes must require a greater period of time per day.

Some species feed mainly by sight and are active by day, although peaks of feeding activity occur in morning and evening. Other fishes that depend more on chemical senses can feed effectively in twilight or at night, so they may be most active in early morning and late evening. Other differences in feeding activity are tied to seasons, cycles of migratory or reproductive activity, or age and size.

In relative terms, small individuals consume more per day in relation to their body weight than large individuals (Jobling, 1994). The daily food intake of juvenile brown trout weighing 5 g at 15°C is about 11 percent of body weight per day. The same species, at 50 g at the same temperature, eat 6.6 percent of body weight per day; and those at 500 g, under the same conditions, have food intake of 4.3 percent body weight per day (Jobling, 1994). Many fishes show a narrow average daily food intake despite fluctuations when feeding is measured over an extended period. In addition, food intake appears to be adjusted to the nutritive value of available food, so greater amounts of less concentrated foods are eaten.

Variety of Foods and Foraging Activity. Most species of fishes are predatory, feeding on live animals or parts thereof. But in some habitats, especially in the tropics, 10 to 20 percent of the species present and nearly half of the individuals may depend primarily on plant material for food; and in ocean areas and lakes where soft bottom materials accumulate, there may be detritus feeders (more in the tropics than in temperate regions). Other than specialists that are restricted by their feeding apparatus to a narrow range of foods, most species tend toward opportunism and will take advantage of a wide variety of foods. Although we may tend to consider some species as mainly carnivores or herbivores for the purposes of some discussions, they may often tend toward omnivory.

Fishes are often categorized by their foods and feeding strategies. Informative treatments of the subject are found in Keenleyside (1979), Hyatt (1979), and Gerking (1994). The following is a brief outline of feeding types in fishes.

Carnivores. The benthic invertebrate fauna provides a significant portion of the food for carnivorous fishes. For the most part, organisms such as aquatic insects (including larvae and pupae, small crustaceans, molluscs, and worms) constitute the main food source of most of the fishes; whereas larger organisms are safe from all but large or specialized fishes. Some benthic predators live on the bottom and hunt by searching and capturing individual organisms as they are encountered. Others simply remain still and capture invertebrate prey that comes close. Many bottom feeders swim above the substrate and search by sight for prey, and a few others supplement sight feeding by other senses. Nocturnal foragers may use vision only minimally and depend on taste, olfaction, the lateral line system, tactile sense, and electroreceptors (if present) to locate edible bottom organisms. Barbels or elongate fins carrying organs of touch and taste aid in the search.

Predaceous fishes may seek prey individually, in schools, or in the company of other species. Some pelagic predators, such as various sharks, salmons, tunas, jacks, and dolphinfishes, hunt by sight and pursue and capture prey. In some instances, groups of predators have "herded" prey against the shore or into tightly packed schools or "balls," thus preventing effective escape. Wetherbee (1990) mentions that the blackfin reef shark (*Carcharhinus melanopterus*) wriggles ashore to eat the fishes that have been herded to the shallows and onto the shore. This may be evidence of cooperative pack hunting by the predators.

Many predators, such as sculpins, stonefishes, groupers, morays, and flatfishes, lie and wait in a hiding place or on a substrate, where they are inconspicuous, and then ambush prey that comes near. Benthic lophiiforms, such as goosefishes, batfishes, and frogfishes, employ lures (illicia) to draw prey close to their mouths. The ceratioid angler fishes and the viperfishes, *Chauliodus*, have elaborate lures (some of which are luminescent), developed from the dorsal fin, for the attraction of prey. The squarehead or angler catfishes, genus *Chaca*, use wormlike barbels at the corners of their broad mouths as lures.

Some predators are aided in food capture by strong electric organs that stun or immobilize prey. Electric rays and stargazers are examples. The electric catfish (*Malapterurus electricus*) feeds on schooling clupeids and schilbeid catfishes, apparently by stunning them with some aid of the electric organs (Sagua, 1979).

Fishes that stalk prey may be benthic (e.g., flatfishes, which can crawl with some stealth along the bottom before attacking). They may also be neutrally buoyant species that can hang nearly motionless in the water and approach prey gradually. Gars, *Lepisosteus*; pikes, *Esox*; trumpetfishes, *Aulostomus*; and some piscivorous cichlids are all slender-bodied or compressifom predators that disclose little view of their actual size as they slowly approach the prey head on. Trumpetfishes are known to swim with a larger fish or a school of smaller fishes and then dash out to capture prey.

A special kind of dependence on other fishes as a source of food is seen in cleaning symbiosis. In this relationship, host fishes are cleaned of parasites and, in some cases, damaged tissue by cleaner fish. In some tropical areas, there are species that mimic the cleaners and make their living by biting off parts of fins or skin. The ichthyoborids of Africa are known as fin biters because they pursue a parasitic existence by feeding on the fins of other species. Scale-eating fishes are known from both South America and Africa. Certain characoids of South America and several cichlids of the African lakes have dentition adapted for removing scales from other fishes. The slender piranha (*Serrasalmus elongatus*) is reported to eat both scales and parts of fins, as well as some flesh.

Certain paedophagous cichlids of African lakes eat the young of others by dislodging broods of fry from the mouths of females either by ramming the hyoid region, as in the species of *Cyrtocara* (McKaye and Kocher, 1983), or by engulfing the snout of the female and sucking out the young, as in an unnamed species reported by Wilhelm (1980).

Lampreys are generally termed parasitic feeders but are really predatoparasitic, for although a lamprey that is small in comparison to the host can suck out fluids without killing the host, those that are close to the host in size or those (*Lampetra ayresi*, for example) that rasp out and ingest viscera must kill their victims rather quickly (Beamish, 1980). Several years ago, the sea lamprey (*Petromyzon marinus*) in the Great Lakes was calculated to cause up to 56 percent of the mortality in lake trout older than nine years (Swanson and Swedberg, 1980) and up to 75 percent of the mortality in whitefish (Spangler et al., 1980). Lamprey control methods have lessened the problem.

Although lampreys usually discard a lifeless victim, Beamish (1980) presents evidence of continued feeding on dead prey. One small species (*Lampetra minima*), now extinct, commonly remained with prey until all soft tissue was consumed (Kan and Bond, 1981). Hagfishes are mostly saprophagous, feeding as scavengers. Certain catfishes of the families Cetopsidae and Trichomycteridae, from the Amazon, apparently parasitize the gills of pimelodid catfishes and other large species, causing a flow of blood on which they feed.

Many species feed on zooplankton, some by straining the water through gill rakers and others by selecting individual organisms. Of course, given the wide range of size in zooplankters and in fishes, plankters are visually detected and selected individually by tiny fishes, or fish larvae might be strained from the water by larger fishes and not retained by the straining apparatus by larger fishes. Fish of a given size might filter out plankton up to a certain size but select somewhat larger organisms individually. Janssen (1978) noted four different ways that alewifes capture plankton. They capture individual organisms in two ways—sucking in rather passive prey in simple particulate feeding and darting after more vigorous organisms. Two ways employed in multiple capture are filter feeding and gulping in the presence of a high density of prey. Such switches in feeding modes may be related to optimal foraging so that a reasonable return of energy gained for energy expended can be realized (Crowder, 1985). Crowder noted that in studies involving several species, shifting from particle feeding to filter feeding occurred when density of small prey, such as *Artemia* nauplii, reached predictable levels. Gibson and Ezzi (1985) found that Atlantic herring filter fed on *Artemia* nauplii at high density.

Young Sacramento perch (*Archoplites interruptus*) feeding on individual plankters used a leisurely approach in capturing passive prey but used a fast start (from a slightly S-shaped body flexion) and pursuit to capture evasive prey. The energy cost of capturing evasive prey was eight times that of capture of passive prey (Vinyard, 1982).

Herbivores. There are few or no fishes that are life-long herbivores, because larvae and juveniles usually begin life feeding on zooplankters or extremely small benthic and pelagic animals. In addition, species that feed on filamentous algae or vascular plants also ingest whatever animals—snails, crustaceans, insect larvae, etc.—are attached to the vegetation. Nonetheless, there are many species that depend, throughout adult life, on the intake of plant material and show appropriate structural modifications for gathering, processing, and digesting it.

Those species that feed on phytoplankton have, for the most part, long, fine, and closely spaced gill rakers and engage in filter feeding. Some cichlids, herrings (*Sardinella, Dorosoma*), menhaden, and cyprinids (*Hypophthalmichthys*) are feeders on phytoplankton.

There is a number of species that browse on large filamentous or thallose algae, or on vascular plants. The grass carp *Ctenopharyngodon*, is famous for its ability to ingest huge quantities of vegetation, which it breaks loose with its rather hard, toothless mouth and shreds with its pharyngeal teeth. *Puntius javanicus* is another browser. Others include *Myleus* spp. of South America, certain *Tilapia* species, and the characoid *Distichodus* of Africa. In marine waters, the milkfish, *Chanos*, is herbivorous, apparently feeding on phytoplankton as well as filamentous algae. Adults of the globehead sculpin, *Clinocottus globiceps*, and some of the associated stichaeids feed mainly on algae. Rabbitfishes (Siganidae) are browsers on filamentous algae, and various reef fishes, such as triggerfishes, sea chubs, damselfishes, and surgeonfishes, utilize algae. Some surgeonfishes—for example, the orangespot surgeonfish (*Acanthurus olivaceus*)—have thick-walled, gizzard-like stomachs (Wheeler, 1975).

Numerous species in both marine and freshwater habitats are adapted to scrape films of diatoms and other vegetation from substrates. This grazing is accomplished by means of modified jaw edges in some and by teeth in others. Some have only the lower jaw modified and scrape by a unidirectional chiseling action, whereas others employ both jaws in scraping. In some of these the teeth are bristle-like, whereas they are chisel-like in others. Examples of grazers and scrapers are the chiselmouth (*Acrocheilus*), the ayu (*Plecoglossus*), various suckers (Catostomidae), the stoneroller (*Campostoma* spp.), various cichlids (*Pseudotropheus* spp., *Gephyrochromis* spp.), parrotfishes (Scaridae), and suckermouth catfishes of Loricariidae.

Some special feeding habits have developed in the Amazon basin. Many characiform fishes there commonly feed on fruit, nuts, and flowers of plants growing over the waterways or in the seasonally flooded forests.

Omnivory. Although they may be specialized for a narrow (stenophagous) diet, some species turn to a wider choice of food when the opportunity arises. A variety of herbivores can be caught with animal bait. The chiselmouth, marvelously adapted for scraping the substrate, can be captured on a hook baited with an earthworm, or on a small spinner, sometimes a meter or more above the bottom, or even on a floating artificial fly. Ayu, with maxillary and mandibulary teeth set on the outer surface of the bones and aligned to form long scraping edges, adapt quickly to feeding on floating pellets in hatcheries. Tui chub (*Gila bicolor*) are ordinarily pickers of small prey and algae among vascular plants or along the bottom but will move to the surface by the thousands to feed on unusually large hatches of insects.

The prickly sculpin (*Cottus asper*) is well adapted to a benthic life, but small individuals of 5 or 6 cm will swim well off the bottom to capture individual water fleas.

Many of the cichlids of the African Great Lakes have teeth, jaws, and pharyngeal teeth that are morphologically specialized for feeding on a narrow range of foods but nonetheless are opportunistic and will take advantage of abundant and easily obtained foods. Greenwood (1974) and Liem (1980b) have noted such versatility. McKaye and Marsh (1983) studied the feeding behavior of two algae scrapers, *Petrotilapia tridentiger* and *Pseudotropheus zebra*, in Lake Malawi. Although the territorial males of *P. tridentiger* engaged almost exclusively in algae scraping, females, juveniles, and nonterritorial males switched frequently to eating zooplankton, as did both sexes of *P. zebra*. Because the males do not leave their territories, their ability to specialize on the algae available in the territories is of great importance (McKaye and Marsh, 1983). There are numerous examples, from chance observation and from food habit studies, that indicate that dietary opportunism is common among fishes.

Some species appear to be omnivorous, commonly taking a variety of foods in their daily fare. The common carp (*Cyprinus carpio*) is perhaps the best example of a fish that will eat various kinds of animals, including fishes, but eats plant material as well. The rainbow trout is primarily a carnivore, but most studies of its food habits disclose filamentous algae in stomachs. The channel catfish is among many other species with a wide taste in foods.

Feeding: Structure and Function. Larval lampreys have jawless mouths with a dorsal hood that does not seem capable of closing the oral opening. The oral cavity is filled with a fine, many-branched filter that retains fine particles. Adult lampreys have an oral disc set with horny teeth and a "tongue" that can act to cut and rasp because of its longitudinal and transverse tooth laminae. At the same time as it rasps, it must act as a pump plunger, which, as it retracts in the rasping motion, reduces pressure and serves to suck out fluids and tissues from the prey.

Hagfishes lack the oral disc of the lampreys and have lingual teeth arranged bilaterally so that, when everted, they and the teeth in the roof of the mouth can "bite" the prey and tear a hole in skin or pull off pieces of flesh. Hagfishes can use their "knot-tying" behavior to aid in feeding (Hardisty, 1979). A simple overhand knot is tied, beginning at the tail, and shifted forward over the head, which is thereby forced away from the prey, and the portion of the prey that is engaged by the teeth is pulled free and swallowed.

Jawed fishes show great variety in structure and function of the feeding apparatus. There are remarkable evolutionary advances seen from the early gnathostomes to the teleosts, and even within the Actinopterygii there is a progression toward greater complexity and efficiency in the feeding mechanisms (Lauder, 1982a). The jaws appear to have had few moving parts in primitive gnathostomes. Probably some could elevate the cranium by the function of the intracranial joint at the same time that the mandible was lowered, with little or no lateral movement possible in the lower jaw and no forward movement possible in the upper jaw. Some of the living sharks with amphistylic or orbitostylic jaw suspension (*Chlamydoselachus, Heptranchus*) may approach this arrangement. Most modern sharks have a type of hyostylic jaw suspension that allows the upper jaw to swing forward and down.

Among bony fishes, there has been a progression from the comparatively simple long jaws, which mainly caught and held prey, to suction feeding aided by protrusible jaws. (Lungfishes have developed effective suction feeding without protrusible jaws [Bemis, 1987].) The evolutionary sequence leading to protrusible jaws involved development of movable maxillaries and premaxillaries, enlargement of the premaxillaries, exclusion of maxillaries from the gape, and development of a long ascending process of the premaxillaries. Lauder (1983) remarked that the feeding apparatus of fishes can consist of more than 30 movable bony parts and over 50 muscles.

Generally, in lower bony fishes, there is little movement in the premaxillary and the head of the maxillary is attached to the neurocranium just posterior to the premaxillary. In opening the mouth, contraction of the epaxial body muscles elevates the neurocranium and contraction of the muscles along the ventral aspect of the head (mainly the sternohyoideus but including, in some fishes, the geniohyoideus and hypaxial body muscles) lowers the mandible. In *Amia* and the teleosts, there is an additional coupling that serves to open the mouth. This is powered by the levator operculi muscles that raise the opercular apparatus, which is coupled to the mandible through the interoperculomandibular ligament. Another coupling is seen in higher teleosts, which have an interoperculohyoid ligament (Lauder, 1982a). The ligamentous coupling between the maxillary and the mandible causes the maxillary to pivot on its connection to the neurocranium and swing downward and forward, thus forming the lateral edges of the open mouth.

Usually, the mouth-opening sequence is accompanied by the lowering of the floor of the mouth by depression of the hyoid apparatus. During this time, the opercular opening is closed. This can lower the pressure in the orobranchial cavity and aid in capture of food. Opercular expansion contributes to the suction but usually occurs as the mouth begins to close. Closure of the mouth is accomplished mainly by the adductor mandibulae. In *Amia*, hyoid depression occurs after mouth closure (Lauder, 1979). In *Latimeria*, neurocranial elevation is facilitated by the intracranial joint (Lauder, 1980a). Discussions of mouth function in lower fishes can be found in Lauder (1980b), Lauder and Liem (1980), and Rand and Lauder (1981).

In cypriniforms and in most higher teleosts, the mobile premaxilla makes jaw protrusion possible, and suction feeding is well developed. Protrusion of the premaxilla, depending on species, can (1) add to the velocity of approach to the prey, (2) allow the fish to "reach" for food, (3) aid in forming a more efficient tunnel for suction, and, (4) if left protruded while the mandible is adducted, allow the mouth to close faster.

The sequence of events during the suction feeding of a percoid, *Gymnocephalus cernua*, as reported by Elshoud-Oldenhave and Osse (1976), consists of a preparatory phase, in which the volume of both oral and branchial cavities is decreased; phase I, in which abduction pressure is exerted with the mouth closed; phase II, in which the mouth is opened and the oral and branchial chambers are enlarged; and phase III, in which the mouth is closed and compressed to force water and ingested material posteriorly, where water is forced out by compression of the posterior chamber and food is retained on gill rakers. Lauder (1983) presents suction feeding as consisting of a preparatory phase; an expansive phase, which involves opening the mouth and expanding it to full gape; a compressible phase, which lasts from full gape to the closure of the mouth; and a recovery phase, as the mouth parts reach the initial position. During the suction phase, the mouth cavity is said to be like a truncated cone with the base at the posterior, with pressure becoming more negative as the wider base is approached (Liem, 1990).

Different mechanisms have evolved for protrusion of the premaxilla in different groups of fishes (see review by Motta, 1984). A common type involves a ligamentous coupling of the maxilla to the mandible so that as the mandible is depressed, the maxilla is partly rotated and a process on its head engages the premaxilla and forces it forward. In some advanced forms, the premaxilla is attached to the mandible so there is a direct pull on it as the mandible is depressed.

The speed at which small prey can be engulfed by various fishes using suction feeding usually is very rapid. Most species can complete oral expansion in less than 100 msec, and many require less than 50. Small prey can be engulfed in as little as 4 msec by frogfish (*Antennarius* spp.) (Grobecker and Pietsch, 1979) or about 5 msec by the swamp-eel *Monopterus* (Liem, 1980a). *Antennarius* can complete the entire feeding sequence in as little as 25 msec. Grobecker and Pietsch suggest that such ultrafast feeding is common to "sit-and-wait" predators with large mouths.

Gill rakers are of great variety, usually reflecting the diet of the species involved (Fig. 25–9). The branchial sieve is probably not a passive filter in fishes that eat plankton or other small organisms. Space between adjacent gill arches can be controlled, and gill rakers and the grooves between them are controlled by musculature, which can govern the ability to retain particles of differing sizes. In certain cyprinids, the palatal organ works in concert with the gill rakers to govern the dimensions of the "branchial slit," which allows water to pass from the mouth to the branchial basket (Hoogenboezem et al., 1991). Apparently, not all fishes require gill rakers for retention of very small particles. *Tilapia galilaea* could retain and ingest particles as small as 0.07 mm after gill rakers and microbranchiospines were surgically removed (Drenner et al., 1987). Obviously, the understanding of filter feeding will require much additional study with specialized equipment and methods.

Crumenal organs are apparently used in consolidating small particles prior to swallowing. Pharyngeal teeth are adapted in many species simply to aid the oral teeth and the gill rakers in holding prey and perhaps in forcing prey into the esophagus. In other species, pharyngeal teeth are modified and specialized for crushing, grinding, and shearing and otherwise are related to the diet (Fig. 25–10).

Preparation of Food for Digestion. Many predatory fishes, such as the largemouth bass (*Micropterus salmoides*), have cardiform teeth in relatively broad pads both in

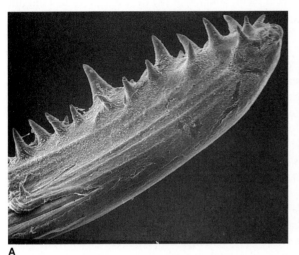
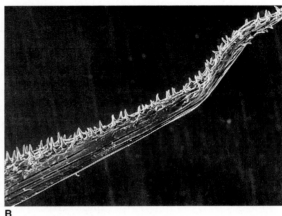

A

B

FIGURE 25–9   **A,** Distal half of gill raker of bocaccio (*Sebastes paucispinis*), a piscivore, showing coarse teeth; **B,** tip of gill raker of pygmy rockfish (*S. wilsonsi*), a crustacean eater, showing numerous fine teeth. (Scanning Electron Micrographs by A. H. Soeldner and G. Pequeno.)

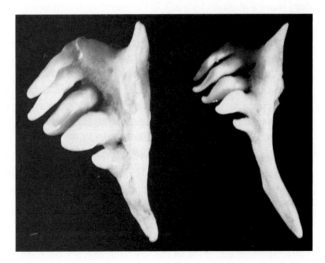

FIGURE 25–10   Pharyngeal teeth of tui chub (*Gila bicolor*), *left,* spring-living race that feeds on gastropods and large insects; *right,* lacustrine race that feeds on zooplankton and small insects. The specimens were of equal length.

the mouth and on the pharyngeal bones. These teeth act to catch and hold prey and to aid in swallowing it, but they do not prepare the prey for the digestive process. Other predators, such as salmon or trout, have rows of sharp teeth that commonly tear or break the skin of the prey at the time of capture or during the swallowing process. The squawfishes (*Ptychocheilus* spp.) have no teeth in the mouth but are equipped with strong pharyngeal teeth, which tear the prey as it is forced into the esophagus. A few predators, such as sharks and piranhas, bite pieces out of the prey.

Although a number of fishes (lungfishes, chimaeras, wolffishes, croakers, and others) have tooth plates or molariform teeth in the mouth and can grind or mash food with them, pharyngeal teeth appear to be the major apparatus used for mastication among the teleosts. Some of the greatest development of pharyngeal teeth can be seen in species that habitually feed on various shellfishes or coral. Many of these, such as the surfperches (Embiotocidae), have pharyngeal mills that can triturate the food organisms past recognition. Some herbivorous fishes that feed on macroscopic substances must tear or grind plants extensively so that digestive enzymes can act on the cell contents. The grass carp (*Ctenopharyngodon idella*) has relatively long, rough-edged pharyngeal teeth that intermesh while tearing the soft plants on which the species feeds.

Mucous cells or glands are present in the mouth and pharynx of most fishes, but the greatest mucus production is usually in the esophagus. Mucus aids in retention of small particles at the gill rakers, facilitates the process of swallowing by lubricating large particles, and may aid in holding finely divided particles together for swallowing. Butterfishes (Stromateidae) and their close relatives have expanded esophageal sacs that store food. They have teeth in the lining of the sac, to triturate the food. The sacs are well equipped with mucous glands. The digestive process appears to begin in the esophagus of certain mullets (Mugilidae) and sculpins (Cottidae) that have gastric glands in the posterior section.

Digestion

In most fishes, the chemical digestive process begins in the stomach, which differs from the esophagus in the composition of the walls and in the type of glands in the mucosal lining. In addition to secreting a protective mucus and pepsin, a protease, the glands of the stomach secrete hydrochloric acid, which maintains the pH of the stomach contents in a range suitable for the action of pepsin. Pepsin shows a peak of activity at a pH of about 2, and pH values of about 1.5 to 4 are common in the stomachs of predatory or insectivorous species. Some herbivorous fishes have acid stomachs with a range suitable for digestion of algae (Horn, 1989). Pepsin is not found in cyclostomes, chimaeras, lungfishes, and many teleosts that lack a stomach (Stevens, 1988).

Fishes secrete gastric enzymes in addition to pepsin (Fänge and Grove, 1979). In some fishes, there are apparently proteases that are optimally active at pH 3 to 5. There are also enzymes in the stomach that act on foods other than proteins. Chitinase has been obtained from insectivores and from species that commonly feed on crustacea. Amylase has been identified from some clupeids, and lipase from clupeids and the Mozambique tilapia.

Flow of gastric juices is initiated by the act of feeding and especially by distension of the stomach wall. Secretion is, to some extent, under the control of the vagus.

The stomach generally acts to store food and initiate digestion by mixing the ingesta with the gastric juices. Depending on the food habits of the species, the organ can be large and distensible or small and capable of passing small food items along during an extended feeding period each day. The stomachs of some mullets (Mugilidae), the milkfish (*Chanos chanos*), some herrings (*Clupeidae*), and characins (Characinidae), all microphagous fishes, are modified into gizzards. In many of these, the gizzard involves only the pyloric part of the stomach, and the secretory function of the stomach is lessened. In the gizzard shad (*Dorosoma cepedianum*), the gizzard is divided into cardiac and pyloric sections. Many microphagous fishes have unspecialized stomachs or none at all. There are great differences in the shape and development of the stomach, even among closely related species or those with similar food habits. Motility of the stomach is, in many instances, related to the degree of fullness, so that ingesta is removed more rapidly from a full stomach than from one that is partially full.

As noted earlier, a variety of fishes lack a true stomach. Stomachs are not recognized in lampreys, hagfishes, chimaeras (Holocephali), lungfishes (Dipnoi), some gobies (Gobiidae), minnows (Cyprinidae), suckers (Catostromidae), sauries (Scomberesocidae), pipefishes (Syngnathidae), wrasses (Labridae), and in at least some of the members of the following families: Cobitidae, Cyprinodontidae, Poeciliidae, Atherinidae, Belonidae, Mugilidae, Cichlidae, Scaridae, Bleniidae, Callionymyidae, and Gobiesocidae. Some of these may maintain some gastric function because gastric glands have been noted in pipefishes and in the needlefish *Xenentodon* (Fange and Grove, 1979). There are several hypotheses explaining the loss of the stomach in various kinds of fishes, one being that the condition is neotenic, inasmuch as larval fishes generally lack a stomach. At the low concentrations of chlorine available in fresh water, the ability to digest food in a completely alkaline system would seem to be advantageous because the fishes would be freed of the burden of producing hydrochloric acid. In the absence of acid in the digestive tract of the cunner (*Tautogolabrus*), whole molluscs ingested by the fish are known to have survived the half-day journey from ingestion to defecation. Stomachless fishes of predatory habits usually have an expanded portion of the intestine in which large morsels can be stored while undergoing digestion. These expansions are often mistaken for stomachs, inasmuch as one of the important functions is being served.

The pyloric caeca, if present, branch from the intestine near the pyloric end of the stomach. There are conflicting reports on whether digestive enzymes are produced by the caeca. Although some histological studies of fishes indicate that enzyme-secreting cells are absent from the caeca (Fänge and Grove, 1979), the caeca are considered by Prosser and DeVillez (1991) to be sources of both carbohydrases and proteinases. However, the paper by Glass et al. (1987), cited by Prosser and DeVillez (1991), reported high levels of digestive enzymes (trypsin, amylase) in the caeca but attributed the secretion of these enzymes to the diffuse pancreas. No secretory function is ascribed to pyloric caeca by L. Smith (1989). The caeca may have a function in absorption as well as increasing the area of digestive membrane. Pyloric caeca have been shown to be active in converting fatty alcohols into fatty acids (Cowey and Sargent, 1979).

As mentioned earlier, among the teleosts the intestine varies in length and conformation with food habits and, to some extent, with individual diet. Fishes that habitually ingest a large proportion of indigestible material with their food appear to have the longest relative gut length, and in these fishes the gut is usually flexed or coiled in an elaborate manner. The added length increases the retention time of the food and allows for more efficient digestion of materials that are hard to digest. Table 25–1 shows a comparison of the relative gut lengths of several species. In many species, the length (and thus the volume and absorptive area) of the gut increases by a factor greater than the increase in body length, so that large adult individuals have a markedly greater gut length to body length ratio than do small individuals.

Digestion proceeds in the intestine in a neutral to alkaline medium. Enzymes involved are secreted by the pancreas, intestinal mucosa, and possibly the pyloric caeca. The types and amounts of enzymes present in the digestive system of a given species are related to the general food habits of that species (see Fänge and Grove, 1979). The pancreatic tissue is probably the source of many of the digestive enzymes, but the diffuse nature of the pancreas in most fishes presents physiologists with great difficulty in locating the exact source of the enzymes. One of the most important proteases, trypsin, is secreted by the pancreas. Trypsin is an endopeptidase, which breaks up polypeptides at the bonds adjacent to specific amino acids (phenylalanine and tyrosine, in this case). Other endopeptidases, chymotrypsin and elastase, are produced by fish pancreatic tissue. Exopeptidases, which break up large peptides and remove some terminal amino acids of the chains, are secreted by the pancreas of fishes (Stevens, 1988). Amylase is secreted by the pancreas of fishes of all kinds, as are lipases. Chimaeras, and possibly other fishes, produce chitinase in the pancreas, although this enzyme is more commonly secreted by the gastric mucosa.

Surface epithelial cells of the intestine in vertebrates secrete a variety of enzymes that act on carbohydrates, but knowledge of the intestinal enzymes is by no means complete. Maltase is known from the intestines of ayu, various salmonids, the common carp, and *Pagrus*, a sparid. In addition, the carp appears to secrete sucrase and cellobiase (Stevens, 1988).

The liver secretes emulsifiers, carried to the intestine in bile, that aid in fat digestion. Lipases and esterases cannot hydrolyze fats that are not in solution or nearly so.

Rates of digestion are variable, depending on type of foodstuff, species of fish, temperature, and amount of food ingested. There is some indication that small fishes of a given species digest food more rapidly than larger individuals. Studies of the rate of gastric plus intestinal digestion have presented some problems, but satisfactory results have been obtained in a number of experiments. These have usually involved timing the passage of a meal from ingestion to defecation of the waste resulting from the meal; this has been accomplished by feeding and subsequent observation following a fasting period, or by feeding materials colored by inert dyes. Gastric digestion has been measured more directly, usually by feeding a measured amount of food and studying the rate of disappearance from the stomach. Fishes can be sacrificed and dissected at intervals, the stomachs can be pumped, or X-ray techniques can be used in these studies.

Because various investigators have used different test species, different methods, and different test foods, there is some difficulty in comparing results of digestion rate studies. Temperatures have usually been reported for the experiments, but

TABLE 25–2 | **Time Necessary to Empty Digestive Tract for Selected Fishes**

Species	Temperature (°C)	Time to 100% Empty (hr)	Reference
Salmonidae			
Oncorhynchus mykiss	15	40	Grove et al. (1978)
Esocidae			
Esox lucius	12	72	Lane and Jackson (1969)
Cyprinidae			
Pimephales promelas	20	12–24	Lane and Jackson (1969)
Carassius auratus	20	60–72	Lane and Jackson (1969)
Cyprinus carpio	23	48	Lane and Jackson (1969)
Gibelio catla	28–30	18–54	Renade and Kewalramani (1967)
Cirrhina mrigala	28–30	18–60	Renade and Kewalramani (1967)
Ameiuridae			
Ameiurus nebulosus	20	60	Lane and Jackson (1969)
Embiotocidae			
Phanerodon furcatus	23–26	10–12	Bray and Ebeling (1975)
Centrarchidae			
Micropterus salmoides	20	60	Lane and Jackson (1969)
Cottidae			
Cottus gobio	10	100	Western (1971)
Scombridae			
Katsuwonus pelamis	23–36	14	Magnuson (1969)
Cichlidae			
Tilapia nilotica	25	7–15	Moriarty (1973)

these have not always been discussed in relation to the ecology of the test species. Temperature influences such phenomena as the rate of secretion and the activity of digestive enzymes, the absorption rate of the digested food, and the muscular activity of the digestive tract. The amount of food fed has an effect on the rate of digestion; usually a large meal is digested at a more rapid rate than a small one. Some results of digestion rate studies are shown in Table 25–2.

L. Smith (1989) indicates that carnivores have a relatively slow food passage in comparison to herbivorous fishes. Examples include virtually complete digestion of salmon fry in 24 hours (at 15°C) by Dolly Varden (*Salvelinus malma*) and a similar time for total digestion of food by rainbow trout (9–10°C). Tuna empty the gut in about 14 hours. Sharks generally are slow in digesting. Wetherbee (1990) states that the lemon shark (*Negaprion brevirostris*) will feed for about 11 hours and then not eat for about 32 hours.

The herbivorous grass carp passes food in about 8 hours. Rabbitfishes (*Siganus*), following a meal of algae, empty the gut in 3 hours or less. Some herbivorous fishes, such as some sea chubs (*Kyphosus*), have microbial flora and their food undergoes fermentation. The gut takes 20 hours or more to empty (Rimmer and Wiebe, 1987).

Digestion efficiency in carnivores ranges between 70 and 90 percent; in herbivorous species the range is lower, usually 40 to 50 percent. However, L. Smith (1989) cites studies that indicate a higher digestibility of fats and protein in the diets of herbivores. These digestibilities range from about 65 to 70 percent for lipids and about 60 to 71 percent for proteins. Cai and Curtis (1989) reported a digestion efficiency of about 50 percent for grass carp fed elodea, and a digestive efficiency of 67 to 68 percent for the same species fed commercial catfish food.

Nutrition

Fishes, like other animals, require the common components of food—proteins, carbohydrates, fats, minerals, vitamins, and water. Specific requirements and optimum levels of these in diets have been studied in only a few species, and research on these is not complete. Most information available on fish nutrition pertains to species that are reared in captivity. Included are several members of the Salmonidae, the Japanese eel (*Anguilla japonica*), some members of Cyprinidae, the madai (*Pagrus major*), and the channel catfish (*Ictalurus punctatus*). Because fish culturists are continually seeking means of producing more and better fishes at lower costs, there is a practical value in studying fish nutrition. Even though species may be specific in their requirements, knowledge gained from studying one species can provide a start for the nutritional needs of others.

Fishes differ from warm-blooded animals in that their metabolism is directly influenced by temperature and in that some species are adapted to cold waters and others to warm waters, each showing characteristic changes in metabolic rate over their tolerance range. Furthermore, fishes of various species, especially carnivores, can utilize proteins and fats for energy sources more efficiently than mammals.

Protein Requirements. Most species of fishes subjected to intensive aquaculture are known to require high dietary protein levels. Protein requirements of salmonid fishes have been studied to some extent. These differ with temperature; less protein is required at low temperatures than at higher ones. Wilson (1989) tabulates figures that show protein requirements in prepared diets for chinook and coho salmon at 40 percent of the diet, for sockeye salmon at 45 percent, and for rainbow trout at 40 to 45 percent. Natural foods of trout have as low as 12 percent protein and appear to be quite efficient (Brown, 1957).

Channel catfish production diets contain 32 to 36 percent protein. Common carp diets contain similar amounts. Diets for tilapias have from 28 to 32 percent protein, although young tilapia may grow best when fed 50 to 56 percent protein (see Wilson, 1989; Lovell, 1989). The juveniles of the madai (*Pagrus major*) and the buri or yellowtail (*Seriola quinquiradiata*) are estimated to need 55 percent protein in the diet. Protein requirements of fishes as compared to other vertebrates are treated by Bowen (1987).

In general, age and temperature are considered to have effects on the total dietary protein required by fishes. Fry of channel catfish need about 40 percent protein and fingerlings about 30 to 35 percent. Salmonid fry require 45 to 50 percent protein and yearlings about 35 percent. Considerable differences have been observed in the change in protein requirements with age in species of tilapia. Fry need 35 to 50 per-

TABLE 25–3

Ten Essential Amino Acids, with Requirements in Grams per 100 g of Diet for Chinook Salmon (Mertz, 1962)	
Amino Acid	*Requirements per 100 g of Diet for Chinook Salmon*
Arginine	2.4
Histidine	0.7
Isoleucine	0.9
Leucine	3.9
Lysine	2.0
Methionine (cystine present)[1]	1.6
Phenylalanine	2.1
Threonine	0.9
Tryptophan	0.2
Valine	1.3

[1]If cystine is present, the requirement for methionine is lower because of the sparing action of cystine on methionine.

cent, whereas fishes exceeding 25 g need 20 to 25 percent protein. (See Wilson, 1989.) Some species, including chinook salmon and striped bass, increase protein requirements with increasing water temperature (Wilson, 1989).

Quality of protein is of great importance. Fishes need ten essential amino acids for proper growth, as do most animals. Table 25–3 lists these with the requirement, in grams per 100 g of diet, for chinook salmon (Mertz, 1969).

The amino acid requirements are generally greater than those for terrestrial mammals. Cowey and Sargent (1979), reporting the work of S. Arai and T. Nose, showed that the Japanese eel and common carp require less arginine than the chinook salmon; the eel's requirements for isoleucine, leucine, threonine, and tryptophan are higher than those of the chinook. In general, the carp's requirements are similar to the chinook's.

Wilson (1985, 1989), in reviewing amino acid requirements, lists 12 fish species definitely known to require all ten of the essential amino acids. Quantitative requirements for all ten are known for only a few species.

Carbohydrates. Inclusion of carbohydrates in the diet can spare some protein for use in growth rather than for energy expended in activity. Carbohydrates are usually much more inexpensive than proteins, so there is an economic advantage if they can be fed to cultured fishes.

Omnivorous species, such as carp, can digest carbohydrates better than trout or other carnivores and can have a higher percentage in the diet. Carp can adapt to diets up to 50 percent carbohydrates, whereas the carnivorous buri cannot adapt to 40 percent carbohydrate in the diet. Carnivorous species may show nutritional disorders if fed an excess of digestible carbohydrates.

Digestibility of carbohydrates by trout ranges from 100 percent for some simple sugars to about 30 percent for raw starch and 0 percent for fiber. Salmonids usually respond to high levels of dietary carbohydrate by depositing an excess of glycogen in the liver and excess fat, including infiltration of fat, in the liver and kidneys. Trout are naturally diabetic and will retain high levels of glucose in the blood when fed excess carbohydrates (Comm. on Anim. Nutr., 1993).

Hatchery diets for trout contain from 12 to 20 percent digestible carbohydrates and less than 20 percent fiber. Channel catfish have a better ability than trout to digest carbohydrates, and they show no problems with excess glycogen or fat in the liver when fed diets high in starch or sugar (Comm. on Anim. Nutr., 1993). A similar condition has been noted in the madai, which tolerates only low amounts of carbohydrate in formulated diets (Yone, 1976). Diets containing as much as 20 percent dextrin were found to retard the growth of the madai (Yone, 1976). Channel catfish have responded to test diets containing 30 percent of either starch or dextrin, showing good growth.

Lipids. Fats provide an energy source for fishes but can be used only in limited amounts. If fed in excess, fat will infiltrate the liver and possibly cause death. Fats differ greatly in digestibility (those with high melting points are difficult for fishes to digest). When digestible fats are used in balanced diets, some of the dietary protein is spared for growth or other purposes in addition to energy. Rancid fats are harmful in fish diets; fishes fed oxidized fats have been noted to develop fatty degeneration of the liver (Halver, 1989). Some naturally occurring oils are known to contain substances toxic to some fishes. For instance, cottonseed oil contains cyclopropene fatty acids, which are harmful to trout if fed above a certain level. As in higher animals, there is a requirement for essential fatty acids in fish diets.

Fishes are not capable of synthesizing all the fatty acids necessary for growth, but a relatively few have been studied either qualitatively or quantitatively. Several studies, most utilized in aquaculture, are known to require essential fatty acids, and requirements have been established for some. A linolenic acid with 18 carbons, three double bonds, and the first double bond at the third position from the methyl end of the acid (18:3n-3) is essential to the diets of some salmonids, the ayu, the common carp, and the Japanese eel. The ayu apparently can substitute 20:5n-3, and that acid is essential to the madai. The eel, chum salmon, and common carp require an acid of the linoleic series (18:2n-6) in addition to the 18:3n-3. Some tilapias require 18:2n-6 or 20:4n-6 acids (Kanazawa, 1985; Sargent et al., 1989). Requirements are usually from 0.5 to 1 percent of the diet (Kanazawa, 1985).

Fats can constitute as much as 20 percent of properly compounded diets for salmon and trout; this level approximates the highest level of lipids usually found in natural diets of trout. Most trout foods contain 15 percent or less of fats.

Vitamins. Vitamins are needed in the diets of all the fish species for which this aspect of nutrition has been studied. The most thorough studies have been made of salmonids, especially trout (*Oncorhynchus mykiss, Salmo trutta, Salvelinus fontinalis*) and the chinook and coho salmons (*O. tshawytscha, O. kisutch*). Essential vitamins for trout include ascorbic acid (C), thiamine (B_1), riboflavin (B_2), pyridoxine (B_6), vitamin B_{12}, biotin (H), choline, folic acid (folacin), inositol, niacin, pantothenic acid, tocopherol (E), and vitamins K and A. The requirements of chinook salmon are similar. Generally, studies with the aforementioned species and with other cultured species, such as channel catfish, carp, Japanese eel, madai, and buri (*Seriola quinquiradiata*), have shown similar qualitative needs for vitamins, but deficiency symptoms differ from species to species and quantitative requirements may differ (Halver, 1989).

Vitamin requirements are usually determined by feeding experimental lots of fish diets, from which specific vitamins have been withheld, and noting deficiency symptoms that may occur in the experimental fish but not in the control groups. Experiments may involve restoring the missing vitamin to the diet and noting any possible recovery of the deprived fish. Deficiency of ascorbic acid in trout can cause abnormal spinal curvature (scoliosis, lordosis) (Fig 25–11) and internal bleeding. Poor or no growth is a consequence of withholding several vitamins, and a high mortality rate accompanies a deficiency of tocopherol, biotin, thiamine, and especially pyridoxine. Excesses of some vitamins are detrimental. For instance, excess of vitamin A (hypervitaminosis A) causes pathological changes in trout, including enlarged liver and abnormal growth of bone.

Some examples of daily dietary requirements of water soluble vitamins for trout, expressed in mg/kg of dry diet, are ascorbate, 100–150; thiamine, 10–12; riboflavin, 20–30; niacin, 120–150; and pyridoxine, 10–12. Requirements of the fat soluble vitamins A and D are 2000–2400 and 2400 IU/kg dry diet, respectively. Requirements of warm water fishes for the vitamins listed are usually less than the requirements for trout, about one half for several (Halver, 1989).

Minerals. There has been some difficulty in assessing the dietary requirements of minerals and other trace elements in fishes because of their ability to absorb elements directly from the water. Calcium, chloride, cobalt, phosphorus, strontium, and sulfate can all be taken out of the water by trout, and probably by other species as well. In addition, many elements are required in such small amounts that quantitative assessment is difficult (Lall, 1989; Piper et al., 1982).

Minerals and related substances required in large quantities by most animals are calcium, phosphorus, magnesium, potassium, sulfur, and chloride. Many others

FIGURE 25–11 Rainbow trout (*Oncorhynchus mykiss*), showing severe scoliosis resulting from diet lacking ascorbic acid.

are known to affect the health and growth of animals, even though they are present in trace amounts. Some that are considered essential in fish are iodine, iron, copper, manganese, molybdenum, cobalt, zinc, selenium, and fluorine (Comm. on Anim. Nutr., 1993). Others that are less known but may be essential in the diet are nickel, vanadium, silicon, arsenic, lead, cadmium, bromine, and tin (Lall, 1989).

Calcium and phosphorus are important to bone growth, and deficiency of either can result in abnormal skeletons. Both have other important roles in metabolism. Most of the required calcium can be absorbed directly from the water (Lall, 1989) at the gills, mouth lining, and fins, but a low percentage (0.34 percent per kilogram feed) is required in diets. Phosphorus is absorbed by fishes, but the low concentrations in water cannot supply the requirements, so the food must supply the major part. The phosphorus requirement in the diet is about 0.4 percent per kilogram feed (Lall, 1989). Magnesium is essential to the proper development of bone and to normal growth and appetite. Freshwater fish need about 0.05 percent magnesium in their diet.

Iron is essential for cellular respiration and is usually sufficiently supplied to fish in diets both natural and artificial. Iodine is necessary for production of hormones from the thyroid, and deficiency of this element in the diet can cause goiter. (See Lall, 1989, for review of minerals in fish nutrition.)

Diets for Fish Culture

An economic advantage can be gained by the fish culturist who produces healthy, market-sized fishes in a short time, and much of the ability to maximize growth depends on food. The culturist must provide proper space, sufficient water of high quality and correct temperature, and protection from infectious diseases so that the fishes can utilize the food to best advantage. The food must supply all the materials that the fishes are unable to absorb from the water in sufficient quantity. Vitamins, minerals, and other food elements concerned with metabolism must be present, but the culturist is especially concerned that the diets provide energy for basal metabolism, activity, and growth. The energy, expressed in kilocalories (kcal), that can be used for growth depends not only on the amount but also on the kind of food given to the fish. As noted, most commercially reared fish species do not tolerate large amounts of fats and carbohydrates in the diet. The type and balance of protein with other ingredients can make a considerable difference in the amount of food that will produce good growth. For instance, in one study with brook trout (*Salvelinus fontinalis*), natural food that contained about 640 kcal/kg proved more than twice as efficient as a compounded dry diet containing 1540 kcal/kg; with the dry diet, 4600 kcal were required to produce 1 kg of trout whereas only 2000 kcal were required with the natural food.

Only a portion of the energy ingested is available for growth, for several reasons. First, not all of the energy may be in materials digestible by the fish, or some materials may otherwise escape assimilation. This portion of the food energy is passed out of the alimentary canal as feces. The materials absorbed through the intestinal wall contain some energy in nitrogenous compounds that cannot be metabolized and are excreted at the gills or kidneys. Estimates of energy in fecal and other wastes are usually in the range of 15 to 20 percent of energy ingested. The remaining metabolizable materials are available for necessary metabolism and growth.

Needs of the organism that must be met before energy in appreciable amounts can be used for growth include standard metabolism of the resting animal, any swimming or other activity over the resting condition, and the energy of what is called specific dynamic action (SDA) (Roberts and Bullock, 1989). SDA includes energy used in deamination of amino acids that are not used in growth, as well as energy utilized in the digestion and assimilation of food. This is called the "heat of nutrient metabolism" (R. Smith, 1989). SDA has been studied in a few fishes. The amount of energy attributed to it in the bioenergetics of the cutthroat trout has been estimated to range from about 14 to about 47 percent of energy consumed by individual fish. Usually, SDA in fishes ranges from 10 to 18 percent of energy consumed.

A general energy budget that takes the fate of all ingested energy into consideration can be expressed as (Warren, 1971)

$$Q_c - Q_w = Q_g + Q_s + Q_d + Q_a$$

where Q_c = energy in food consumed; Q_w = energy in waste (feces, urine, etc.); Q_g = energy in materials added to body (growth); Q_s = energy of standard metabolism; Q_d = energy of SDA; and Q_a = energy used in activity over that of standard metabolism. If the three latter qualities are combined and termed respiration, Q_r, the equation can be simplified: $Q_c - Q_w = Q_g + Q_r$.

Another consideration of energy and growth in fishes is presented by R. Smith (1989):

$$DE = IE - FE$$
$$ME = IE - (FE + UE + ZE)$$
$$RE = ME - HE$$

where DE is apparently digestible energy, IE is intake of energy in food, ME is metabolizable energy, FE is waste energy in feces, UE is waste energy in urine, ZE is energy excreted through gills, RE is energy retained as growth of useful products such as gametes, and HE is total heat production, including basal metabolism, activity SDA, etc. (see also National Research Council, 1981).

In trying to maximize growth while minimizing the other quantities, the fish culturist must plan diets with great care, providing the right balance of all the required vitamins, minerals, fats, carbohydrates, and high-quality protein for growth—all at the least possible cost. At one time, various animal foods from slaughterhouses were available, and many trout hatcheries fed fish on liver, spleen, lungs, and other viscera, as well as day-old calves. These products could be refrigerated or frozen and the daily ration could be ground, mixed with wheat middlings or other grain products, and fed fresh. These diets were cheap and often effective, although nutritional problems were common. As mink ranches and the pet food industry began to compete for these meat products and other, more profitable uses were found for some of them, the supply available to fish hatcheries at low prices dwindled. Hatchery owners, nutritionists, and fishery biologists had long been experimenting with fish diets, trying to eliminate nutritional disorders and formulate inexpensive dry diets that would provide complete nutrition and require a minimum of storage cost. During the 1950s and 1960s, dry pelletized fish foods largely supplanted fresh diets and have continued to improve. These foods are formulated from a variety of fish meals and other meals of animal origin, such as blood meal, plus vegetable products, such as soybean meal, corn gluten meal, cottonseed meal, and alfalfa meal. Dried milk products, byproducts from the brewing and distilling industries, vitamin mixes, and mineral mixes are other ingredients (see Lietritz and Lewis, 1976, and Piper et al., 1982).

Moist pellets, made up of a mixture of typical dry ingredients plus marine fish, and byproducts (such as pasteurized salmon viscera and tuna viscera), in the ratio of about 70 percent dry ingredients to 30 percent wet ingredients, have been very successful in salmon culture. Moist pellets are quick frozen immediately after manufacture and can be held frozen for a reasonable length of time without loss of quality. Oregon moist pellets (OMP), with less moisture and with preservative added, can be held under refrigeration without freezing (Lovell, 1989).

Diets for trout usually contain from 38 to 50 percent protein, up to 12 percent carbohydrate, and from 5 to 8 percent fat. Diets for channel catfish generally contain about 30 percent protein, up to 25 percent carbohydrate, and about 5 to 8 percent fat. Some of the best catfish food contains about 1870 kcal/kg. Trout of about 15 cm held in 14°C water are fed dry diets at about 2 percent of body weight per day. Smaller fish are fed more and larger fish less; more food is given in warmer water and less in colder water. Channel catfish held in 25°C water are fed at the rate of 2.5 to 3 percent of body weight per day. Conversion of food into fish is usually from 1.6 to 2.5 kg of dry food to 1 kg of fish produced.

Growth

Although most fishes have indeterminate growth, individuals of a given fish species have a genetic potential for reaching a characteristic maximum size under the most favorable circumstances. Some gobies never reach more than a few centimeters in length, but sharks and tunas, for instance, reach many meters. The characteristic upper size for a species is reached in a relatively short time in short-lived species, but it may be attained only after decades in long-lived species. Under less favorable environmental conditions, fishes reach a size smaller than physiologically attainable for the species. The sigmoid pattern of increase in size with age generally exhibited by fishes is illustrated in the theoretical curve in Figure 25–12. The curve represents growth from hatching to the maximum possible in a given environment. Actually, fish growth is more irregular than shown in the idealized curve. Growth is usually faster in warm weather than in cold and may decrease during migrations or spawning, sometimes even being negative when metabolic demands exceed the food energy intake (Fig. 25–13). In addition to annual fluctuations, fish growth may occur in "stanzas" during the normal life history, with each stanza being defined by a sigmoid curve showing a slowing down of growth before resumption of more rapid growth with entry into the next stanza. Growth stanzas generally result from physical, physiological, or ecological changes. A migratory fish might end its first growth stanza, begin its second by moving from a stream into a lake or the ocean, and begin a third when it is large enough to feed on other fishes rather than small invertebrates.

Factors involved in irregular growth or in the limitation of growth may be environmental or may be concomitants of the fish's physiology. Environmental factors include water quality, which embraces such entities as dissolved oxygen, carbon dioxide, salinity, ammonia, pH, and chemical pollutants; plus other physical factors, such as temperature and light, both of which change with the season and can exert ef-

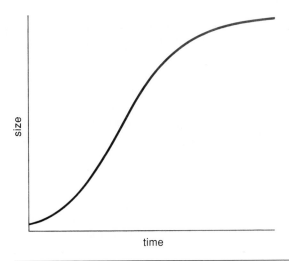

FIGURE 25–12   Idealized sigmoid growth curve.

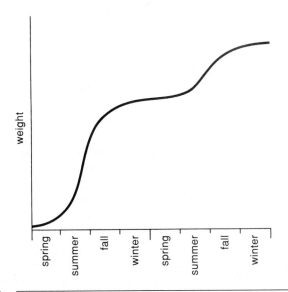

FIGURE 25–13   Theoretical growth curve of fish in temperate climate, showing cessation of growth in winter.

fects directly on the fish or on the abundance, composition, and availability of food. Physiological factors that can influence growth include age, sexual maturity, and state of health. Behavioral factors involved might include spawning activity, migrations, and defense of territory. Production of gametes and reproductive activity can require significant proportions of ingested energy. Waiwood and Majkowski (1984) estimate that cod (*Gadus morhua*) of age group 6 to 10 utilize about 19 percent of total food biomass in reproduction and age classes 11 to 15 use about 38 percent.

    Brett (1979) covers the effects of environmental factors on growth of fish, discussing these in terms of the factors presented by Fry (1971): (1) controlling factors, such as pH and temperature; (2) limiting factors, such as oxygen; (3) masking factors, such as salinity; and (4) directing factors, which may, for instance, be entities such as light or temperature but may bring on genetically set responses of the fishes.

Generally, food and temperature favorable for growth occur during spring, summer, and fall in temperate areas, but most species do not grow at a constant rate during these periods. Summer-spawning species such as the bluegill may put on most of the year's growth in the spring and then grow slowly through the spawning period and the warmest part of the summer. Spring spawners grow best following spawning, but growth may slow during the hot summer. Some, such as the large-mouth bass, feed heavily in late summer and early fall.

In addition to food source and temperature, growth rate is regulated by hormones, with the pituitary growth hormone being of foremost importance. Thyroid hormones are thought to bring about increase in fish growth by increasing appetite and by increasing the efficiency of food conversion. There is evidence that a combination of 17-methyltestosterone and thyroid hormone might enhance growth rates of hatchery salmonids (Donaldson et al., 1979). Physiological changes influenced by hormones affect growth rate during migrations, spawning, and wintering, all natural segments of the life cycle. Hormonal influence can be seen in the differential growth of the sexes. In some species, males are of much smaller maximum size than females. This occurs in many species that have internal fertilization, but the most extreme examples of sexually dimorphic size are the ceratioid anglerfishes, in some of which the males parasitize the females. Fertilization is external in the ceratioids. In those species in which the males are larger than females, size appears to be related to function, and the males are involved in building or guarding nests or in other activities that require size and stamina.

Hormones are involved in the response of fishes to stressful situations. Most fishes grow at a slow rate under stresses such as overcrowding. Even when sufficient food is made available, fishes held in overcrowded ponds grow poorly. That is probably caused by tertiary effects of stress (Jobling, 1994).

The study of growth is not only of great scientific interest but is also of considerable practical importance in the management of fisheries. Growth may be studied in terms of nutritional bioenergetic considerations, with the expectation that the knowledge gained will aid in growing fishes faster on less expensive rations than currently used. Or growth may be considered in relation to age for various species, or stocks within species, in order, for instance, to gain information on the influence of population size on growth so that catch regulations making the best use of the resource may be set. Comparing size at first maturity among genetically similar stocks can provide insight into possible limitations on growth in various bodies of water. Various relationships of length to weight are used by fishery managers to compare the general conditions of fishes from separate stocks, of fishes of the same stocks at different times of the year, or of fishes from different bodies of water.

Growth and age of fishes are studied by several methods. Length frequencies plotted for large samples of individuals generally show a multimodal distribution, with the modes representing prominent lengths of each year class. This method can be useful for short-lived species or for the first few years of life of long-lived species, but it has limited application for slow-growing fishes that live a decade or longer.

Many marine species live to advanced ages. For instance, the shortraker rockfish, *Sebastes borealis*, is known to live 120 years, and the rougheye rockfish, *S. aleutianus*, reaches 140 years (Beamish and McFarlane, 1987).

In many species, interpretation of recognizable marks left periodically in scales or other hard structures can result in determination of age. In those species that lay down compressed circuli in the scales during winter or other periods of seasonal slow growth and form annuli (see Fig. 2–15), age can be determined. Then, assuming that there is a relationship between growth of the scale and growth of the fish, measurement of a selected scale dimension, such as the radius to each annulus and to the edge of the scale, can allow a back calculation that results in an estimated length of the fish at each annulus. For long-lived species, age determination and back calculation of length are better accomplished by using structures such as cross sections of otoliths, fin rays and spines, opercular bones, or vertebrae (see Beamish and McFarlane, 1987; Carlander, 1987; Lagler, 1956; Royce, 1972).

Validation of the results of such studies can be accomplished by research on individuals that are captured, measured, marked, released, and later recaptured. Comparison of scales or fin ray sections taken at initial capture and recapture of the same individual can be useful to validate or invalidate the use of these methods for estimation of age and growth for the species involved. Marks can consist of numbered metal or plastic tags if individual fishes are identified, or of some mutilation such as clipped fins. Other marks can involve injection or feeding of fluorescent chemicals, such as oxytetracycline, which leave deposits in bone or other calcified structures. These marks can be identified by the use of ultraviolet light after the structure is removed from the fish and sectioned.

Age and growth in fishes are being studied by such relationships and methods and RNA–DNA ratios (Bulow, 1987), uptake of radioactive amino acids (Adelman, 1987), and aminoacetic acid (Busacker and Adelman, 1987). Caillet and Radtke (1987) and Caillet et al. (1986) report the use of electron microprobe analysis of calcium and phosphorus in vertebrae of sharks to verify seasonal deposition of growth bands. Some sharks are known to live up to 100 years (Pike, 1990).

Reproduction

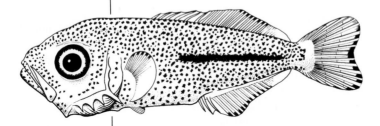

Anatomy of the Reproductive System

The gonads of fishes are usually elongate structures suspended by mesenteries from the dorsal aspect of the abdominal cavity. Their relationship to the kidneys and their associated ducts differs widely among groups. Examples will be given for several groups, and in each notable exceptions from the general plan will be mentioned.

Hagfishes

A single elongate gonad is present in both sexes; this is located to the right of the gut (Brodal and Fänge, 1963). In *Myxine* there are few true males, and most retain both testicular and ovarian tissue in the gonad. Some have sterile gonads (Hardisty, 1979). In *Eptatretus,* the sex ratio is close to normal, although a tendency toward hermaphroditism is seen in some females (Hardisty, 1979).

The testis is irregular and lobate, without a sperm duct, and sperm are released into the body cavity. Unlike the lampreys, which deposit small eggs, hagfishes produce large eggs with tough shells and hooks for attachment to each other or to the substrate. There are no oviducts or sperm ducts. Both eggs and sperm reach the exterior by passing through one of two abdominal pores that open into a sinus, which communicates with a single genital pore just behind the anus. In hagfishes, the gonads are suspended by a mesentery from the gut.

Lampreys

The gonads are single, suspended by a peritoneal fold (mesorchium in the male, mesovarium in the female), and extend along most of the length of the body cavity. They tend to be to the right of the intestine and have no connection with the kidneys. In immature specimens, they appear as thin lobulate structures, but at maturity they may virtually fill the body cavity, crowding the other viscera. The gut, which may be greatly distended during the feeding stage, becomes small and nonfunctional in mature fishes and may be almost completely hidden by the ripe ovary or testis. The single ovary of lampreys is the result of fusion of the primordia (Hoar, 1969). No sperm ducts or oviducts are present; at spawning, both the eggs and sperm are shed into the body cavity, from which they exit through paired abdominal pores (that open before spawning occurs) to the urogenital sinus (Hardisty, 1979). A prominent urogenital papilla is developed in mature specimens of males.

Eggs are somewhat over a millimeter in diameter at extrusion. Fecundity depends largely on body size in lampreys. Small nonparasitic species may produce from 400 to 9000 eggs, but large anadromous species deposit hundreds of thousands—124,000 to 260,000 in *Petromyzon marinus* (Hardisty, 1971).

Sharks

The testes are paired and usually placed anteriorly in the body cavity, suspended dorsally by means of a mesorchium (see Daniel, 1934; Goodrich, 1909, 1930; Romer, 1970). Often the right testis is larger than the left. Sperm discharges into a central canal network that communicates with the anterior part of the kidney through efferent ducts traversing the mesorchium. The front part of the kidney is modified into a glandular epididymis, where the archinephric duct receives the ductuli efferentes and runs posteriorly via a coiled and tortuous path. The time spent in traversing the epididymis may be involved in the maturation of sperm. Just posterior to the efferent ductules from the testis, the kidney is modified into Leydig's gland (Fig. 24–2), in which the tubules secrete a seminal fluid into the archinephric duct.

As the archinephric duct runs posteriorly, it courses along the functional kidney as the vas deferens and then enlarges into a seminal vesicle from which a sperm sac opens dorsally. The vesicles and sperm sacs open into the urogenital sinus, which in turn empties into the cloaca. From the cloaca, sperm enter the grooves of the claspers, through which they can be transferred to the female. Associated with the claspers, under the skin of the pelvic fin and abdomen, are glandular sacs called siphons that secrete a lubricating fluid.

Ovaries are paired, but the left one may be greatly reduced in size in some species. Like the testes, they are placed well anteriorly in the body cavity; each is suspended by a mesovarium. The oviducts accept the eggs anterior to the ovaries, usually through a common mouth or funnel. Eggs are released into the coelom, proceed into this funnel, and then travel down the oviduct to the region of the shell gland (nidamental gland), where fertilization occurs and a horny shell or membrane is secreted (Romer, 1970). In oviparous (egg-laying) species, the shell is tough and protects the developing embryo. In viviparous (live-bearing) species, the shell is slight or vestigial, and the young develop in the posterior, uterine portion of the oviduct.

Chimaeras

In the male, the testes are compact structures placed forward in the body cavity (see Dean, 1906). *Hydrolagus colliei* in early maturity has testes about 4.5 cm long that weigh about 13 g (Stanley, 1961). The system differs from that of sharks in the complexity of the network of efferent ducts traversing the mesorchium. These tubules join so that a smaller number (four to seven) then enter the testis canal, which is in the mesorchium at the base of the testis (Stanley, 1961). In addition, there is an expansion of the posterior part of the archinephric duct into a structure called an ampulla (or vesicula seminalis). This is glandular in nature, is partially compartmentalized in some species, and has been described as a receptacle for maturation of the sperm in the spermatophores coming from the epididymis. The ampulla receives a number of ducts from the posterior part of the kidney.

In males, the urogenital pore opens medially behind the anus; adjacent to the opening are pelvic claspers that convey the seminal fluid to the genital openings of the female. Anterior to the anus is a pair of abdominal claspers that apparently function during copulation.

Females of chimaeras are unique in that the two oviducts open separately to the exterior, not connecting to each other or to the urinary system. A shell gland and a uterine portion are present in the oviducts.

Bony Fishes

In most bony fishes, the testes are whitish, lobate organs lying along the gas bladder, although in some groups (such as salmonids) the organs appear smooth and entire. In most forms, there is no connection between the reproductive and urinary systems, and there are separate openings to the exterior for the two systems, with the urinary pore posterior to the genital pore. In some, the sperm ducts connect with the urinary system in a urogenital sinus located at the posterior end of the body cavity. Primitive bony fishes may differ from this plan (Fig. 26–1) (see Goodrich, 1909, 1930; Hoar, 1969).

In *Polypterus,* the testes are closely bound to the kidneys, and although the sperm ducts enter that organ, they do not join the urinary system until the urogenital sinus is reached.

In lungfishes, the testes are elongate and may stretch the length of the body cavity. A longitudinal collecting duct or network of ducts lies along the medial edge of the testis, and in the posterior section efferent ducts connect with the kidney. In *Protopterus,* the central ducts from each testis merge to form a median structure posterior to the testes, and ducts from this tube enter the kidneys.

The testes of *Amia* are in the anterior part of the body cavity and are closely associated with the kidneys. A longitudinal duct is situated along the medial edge of the testes, and numerous efferent ducts extend from this to the anterior part of the kidney. *Acipenser* and *Lepisosteus* are similar, but the testes are more elongate and communicate with a greater length of the kidney.

Testes of teleosts are of two types: Most have a structure in which spermatogenesis occurs along the length of a lobule, with a central lumen into which the sperm are shed; in atheriniforms, the spermatogenesis is confined to the distal end of the tubule (Billard et al., 1982; Grier, 1981; Jamieson, 1991).

Fish spermatozoa are varied in shape and structure (see Jamieson, 1991). Most sperm of bony fishes have roundish heads usually about 2 to 5 μm, made up mainly of nucleus and a lesser amount of mitochondria. Usually there is a single flagellum for propulsion, but a few species have two. Agnaths, chondrichthyans, lungfishes,

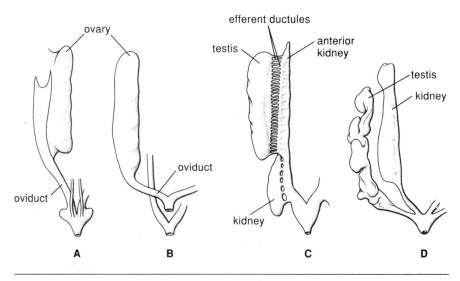

FIGURE 26–1 Diagrams of gonads and ducts in bony fishes. **A,** Bowfin female (Amiidae); **B,** representative teleost female; **C,** bowfin male (Amiidae); **D,** representative teleost male. (Based on Goodrich, in Lankester, 1909.)

Latimeria, bichirs, and sturgeons tend to have long, slender heads with a structure on the front called the acrosome.

In contrast with most vertebrates, most teleosts have oviducts that are continuous with the covering of the ovaries so that the ova are not shed into the body cavity. This is called the *cystovarian* condition (Hoar, 1969). This saccular ovary–oviduct system arises in one of two ways during development. Some species show a condition in which two folds along the edge of the genital ridge meet and merge to enclose a hollow, which becomes the central cavity of the ovary and extends behind this as the oviduct. In others, there is lateral growth of the edge of the genital fold, so that it curls upward to fuse with the body wall. This process also captures a bit of the coelom, which becomes the cavity of the ovary and oviduct. Oviducts reach the exterior via a pore between the anus and the urinary pore.

Most nonteleosts show the *gymnovarian* condition, in which the ovaries open into the body cavity and the ova are conveyed through an open funnel to the oviduct. A few teleosts have gymnovarian ovaries. These include osteoglossiforms, Anguillidae, the loach (*Misgurnus*), Salmonidae, and Galaxiidae. The nonteleost gars (Lepisosteidae) do not release eggs into the body cavity (Hoar, 1969).

Ova of fishes are usually a few millimeters in diameter, but some may be exceptionally large, especially in chondrichthyans and in *Latimeria.* Numbers of eggs deposited usually vary with the size of the female, but ovoviviparous or viviparous species may have few as compared with oviparous species, with exceptions.

The ovaries of fishes are usually well separated, but partial or complete fusion of the right and left organs can be seen in some percoids. In the largemouth bass and some darters, the ovaries join posteriorly to produce a V-shaped structure. Ovaries of the yellow perch, *Perca flavescens,* are so completely fused that they give the appearance of a single organ. This ovary is fused to the body wall just posterior to the anus, and eggs are extruded when this area ruptures, so that oviducts are not functional. The rupture of the body wall heals soon after oviposition (Parker, 1942). A few fishes, such as needlefishes (*Strongylura* spp.), have single ovaries.

Function and Reproductive Patterns

The investment of energy in reproduction is referred to as the reproductive effort (see Kamler, 1992). There are several ways in which reproductive effort can be measured, including number of eggs per female or biomass of eggs per female, but one of the most useful is the gonadosomatic index, or the weight of the gonads expressed as a percentage of the body weight. There have been numerous studies on reproductive effort based on the energy budget, which is expressed as follows (see Ricker, 1968):

$$C = P_g + P_r + R + U + F$$

where C is energy consumed, P_g is that used in growth of the body, P_r is that used in reproduction, R is metabolism, U is energy lost as urine and through other nonfecal excretory processes, and F is energy in feces.

Reproductive Strategy

Reproductive strategy in fishes consists of the reproductive traits that fish will try in order to leave some offspring (Wootton, 1984). Reproductive traits are variations in the strategy made to respond to environmental fluctuations. Important traits are fecundity according to size and age, reproductive age, size of gametes, reproductive behavior, seasonal timing of reproduction, sex change, and the number of times spawning occurs in the life of the female (parity).

Fishes are well known for their high potential fecundity, and most species utilize the primitive condition of releasing thousands to millions of eggs annually (Hoar, 1969). The environment takes its toll of eggs and hatched young. Thus, the minimum requirement of reproduction, if a species is to maintain itself in stable numbers, is eventual replacement of each spawning pair by an equally successful pair. Stability of population numbers is seldom actually achieved, and numbers fluctuate depending on the pressures of environmental factors. Fluctuations may be episodic or cyclic, depending on these factors, and not all species in an area will be equally affected by the same environmental changes.

Fish species have evolved reproductive methods and attendant physiology that allow them to be successful under a great variety of conditions. The entire combination of habits, physiology, and behavior—the overall approach to reproduction—is termed the reproductive strategy. Strategies may include great numbers of eggs, as mentioned, or fewer eggs with greater opportunity for survival. Strategies must ensure survival of a portion of the eggs—through force of numbers, concealment, protection of nests, or retention in the body; strategies must place the earliest feeding stage of the young in the proximity of ample and suitable food and must ensure that the juvenile fish have eventual access to the living space of the adults. Time and location are generally of great importance.

Within a species there may be annual, latitudinal, or altitudinal differences in the timing of spawning, and these differences may be tied to temperature, light, or occurrences such as rainy seasons and freshets. There are species that live in environments that offer favorable conditions for eggs and young over much of the year. These may have an extended spawning season in which females release eggs in small numbers over an extended period. Extended spawning activity can occur near the equator, where seasonal change is slight; among deep sea species, in which young do not depend on seasonal abundance of food in upper layers; or in some thermal environments. Stein and Pearcy (1982) have reported evidence that some North Pacific macrourids are reproductively active throughout the year. The Borax Lake chub (*Gila boraxobius*) of southeastern Oregon lives in a thermal lake that undergoes only moderate seasonal changes in temperature. The species shows a peak spawning period in the spring but is reproductively active year round (Williams and Bond, 1983).

Placement of eggs or young in the optimal place at the right time is due to the response of the endocrine system to environmental cues such as temperature and light, so that gametes are matured and spawning migrations are undertaken (Lam, 1983). In some species, reproductive readiness may be achieved by the influence of long photoperiod and warm temperatures, and spawning may be brought on by shortened photoperiod and decreasing temperatures. In others, the converse is true. According to Billard (1981), most teleosts fall into three groups: Group I, in which gametogenesis is completed in summer and fall on decreasing temperature and photoperiod and in which spawning is performed in the cold season; group II, in which gametogenesis begins in the fall but is arrested during winter so that maturation and spawning are completed in spring or summer; and Group III, in which gametogenesis takes place and is completed on increasing temperature and photoperiod and spawning takes place in spring or summer.

In some species, light may be more important than temperature, whereas others may be more responsive to temperature. Water flow and flooding, availability of food, salinity, and the lunar and tidal cycles may have reproduction-related impacts on the endocrine system.

The hypothalamus appears to be the site of release of a gonadotropin-releasing hormone (GnRH) (Peter, 1981; Sundararaj, 1981). Gonadotropic hormones (GtH), which are secreted by the proximal pars distalis of the pituitary, promote the development of eggs and sperm and stimulate the production of androgenic and estrogenic steroids, which control sexual behavior and the development of secondary sex characteristics. In temperate and other fishes that spawn once per year, there is an annual cycle of endocrine activity in what Crim (1981) terms the hypothalamo–pituitary–gonad axis, resulting in preparation for spawning, spawning, and a quiescent postspawning period. Crim points out that a daily cycle of endocrine activity is present in some fishes. The pineal body (and eye) has a role in the cycle of reproductive activity, as shown by Vodicnik et al. (1978, 1979). Possible pathways and control mechanisms are shown in Figure 26–2.

Fish culturists have learned to influence or control the maturation of brood stock in hatcheries or fish farms. By controlling photoperiods, temperature, and, for some species, water flow, fishes can be brought into spawning condition earlier or later than the normal reproductive season. More direct control of reproductive readiness can be obtained by injection of gonadotropic materials that promote maturation and spawning (Donaldson and Hunter, 1983). A variety of materials has been used (see Lam, 1982), including carp pituitary extract, partially purified salmon gonadotropin, and various mammalian gonadotropins. An effective substance for many species is human chorionic gonadotropin. The development of synthetic analogs of luteinizing hormone-releasing hormone (LHRH) has been a great aid to aquaculturists working with a variety of species in many countries. Analogs of LHRH by itself are not effective in bringing about ovulation in salmonids (Lin and Peter, 1986), so salmon gonadotropin is used in conjunction with LHRH in inducing spawning in salmon (Sower et al., 1982). Analogs of LHRH are effective in inducing ovulation in several cyprinids (Lin and Peter, 1986) as well as some percoids (see Matsuyama et al., 1993).

Semelparity and Iteroparity

Individuals of semelparous (monocyclic) species make a one-time investment in the future of the species in that they spawn only once before they die (see Cole, 1954). Obviously, development of semelparity would be possible mainly under conditions sufficiently stable that reasonable success would be assured, or under conditions that would assure that some compensatory mechanism would make up for failures brought on by a fluctuating environment. The risk of failure would seem to be greatest in annual fishes, in which there is no overlap of generations (e.g., the ayu, *Plecoglossus altivelis*). In that diadromous species, straying of the young as they return from the ocean could compensate for failure of reproduction in any given stream. Pacific salmon of the genus *Oncorhynchus* are semelparous, but most have

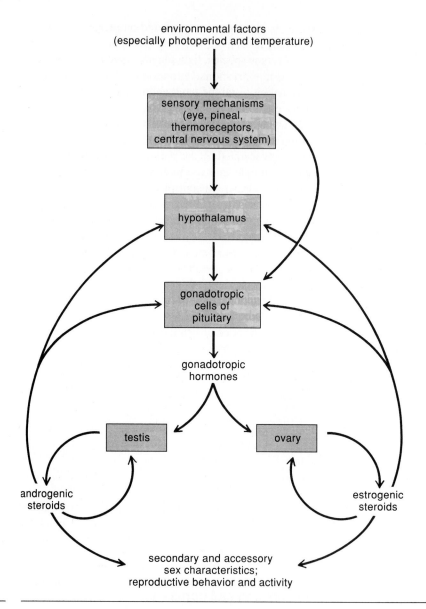

environmental factors
(especially photoperiod and temperature)

sensory mechanisms
(eye, pineal,
thermoreceptors,
central nervous system)

hypothalamus

gonadotropic
cells of
pituitary

gonadotropic
hormones

testis

ovary

androgenic
steroids

estrogenic
steroids

secondary and accessory
sex characteristics;
reproductive behavior and activity

FIGURE 26–2   Possible relationships among environmental factors, receptors, endocrine organs, and reproductive activity.

overlapping generations caused by variation in the number of years spent in both the freshwater and marine environment. In addition, there is a certain amount of straying among returning adults.

Iteroparous (polycyclic) species, in which individuals might spawn two to several times during a lifetime, have a better chance of succeeding in an unpredictable environment in that failure in one spawning season has a chance of being compensated for in some succeeding spawning season. Long-lived species would be expected to have a greater chance at eventual success than short-lived species. Up to the onset of senescence, the older, larger females tend to increase the number of eggs released, so that a decline in numbers of individuals is at least partially offset by an increase in eggs. An example is the Borax Lake chub, mentioned previously. Few females of this dwarf species live into the fourth year of life, but those that do grow to about three times the length of the usual females and can carry up to 80 times the eggs developed by fishes in the second year of life (these constitute almost all the spawners).

An interesting instance of adjustment of reproduction strategy to environment was reported for *Cottus gobio* by Fox (1978), who contrasted a northern population living in acid moorland waters with a southern chalk-stream population. Females of the unproductive northern water had a long life span, matured at age two (some reached nine years), and spawned once a year. Fewer than 5 percent of the females in the southern stream reached two years of age. They matured at age one and were capable of spawning at least four times during that year. A similar adjustment to environment has been noted for American shad, *Alosa sapidissima,* by Carscadden and Leggett (1975).

Finding Mates

Central to reproduction in fishes is the matter of males and females coming together on spawning or mating grounds. Schooling fishes or species in which males and females migrate upon environmental cues to the same area may have no problems in getting the sexes together. Mate selection can depend on visual cues involving color, pattern, size, or placement of photophores; upon sonic cues; or upon chemical signals (pheromones). Many demonstrations leave no doubt that pheromones produced by the ovary or kidney of females attract males to them (Liley, 1982). Attempts to characterize pheromones have not been entirely successful. Some evidence points to estrogen-like materials, but studies have given conflicting results. Pheromones may have important roles in courtship, as suspected in the glandulocaudine characid *Corynopoma riisei* (Atkins and Fink, 1979). Glands on the caudal peduncle of males of that species apparently secrete a mucopolysaccharide that acts as a chemical signal during courtship.

There is circumstantial evidence that pheromones are of importance in some deep sea fishes. Males of ceratioid anglerfishes, the gonostomatid genus *Cyclothone,* and the benthopelagic halosaurs have enlarged olfactory organs. Investigators infer that such males make use of the oversize organs in locating females (Marshall, 1979).

Oviparous Fishes

Eggs of oviparous (egg-laying) bony fishes are usually small, typically between 1.5 and 3 mm in diameter, but there is a tremendous range in size. Kamler (1992), in reviewing data on egg size in fishes, notes that the eggs of *Chlamydoselachus,* which range from 90 to 97 mm in diameter, have a wet weight 34 million times that of the eggs of *Cymatogaster,* which are only 0.3 mm diameter. Some fairly large eggs are seen in some species of trout and salmon, which have eggs exceeding 5 mm, and ariid catfishes, which commonly have eggs of from 15 to 25 mm. The eggs of a given species or even a given female may vary in size with such factors as race, population, age, or nutrition of an individual (see Kamler, 1992). The elongate eggs of hagfishes are up to 30 mm long, and oviparous sharks, rays, and chimaeras deposit eggs of 60 to 70 mm in cases up to 300 mm long. Some of the larger egg cases contain two eggs. Elasmobranch egg cases are of various shapes, from spindle-like to purselike, as shown in Figures 26–3**A** and **B**. Eggs of bony fishes are typically round, but elongate eggs are known in anchovies, some gobies, and clownfishes.

Most fishes lay eggs that are heavier than water (demersal eggs), but many produce buoyant eggs that may be hydrostatically adjusted by oil inclusions, by imbibed water in a large perivitelline space, or by a high ratio of surface to volume to float at the surface or at some intermediate depth, depending on the species involved. Pelagic eggs of some species drift freely; those of some others attach to each other or to vegetation by means of tendrils (Fig. 26–4). Tendrils, hooks, or other attachment devices also occur on demersal eggs and are present on elasmobranch and hagfish eggs. Demersal eggs may be adhesive and deposited in clumps, such as eggs of sculpins and darters, which stick together through the incubation period, or they may be attached singly to some substrate. Some are temporarily adhesive, like the eggs of trout and salmon. Such eggs adhere to the substrate and to each other for a short period and then separate. The practical advantage of temporary adhesion for a species that constructs gravel nests in running water and takes several minutes to cover the eggs is obvious.

Some fishes engage in mass spawning, with no pairing. Numerous males and females release gametes together in a suitable environment. Numbers of eggs are high, and they may be left to drift in the open waters or to be carried and buoyed by the turbulence of a stream, allowed to settle on the substrate, or released so that they may adhere to vegetation. Spawning often ensues after migration to a suitable site, often against a current that will carry eggs and larvae back to a nursery area. Some

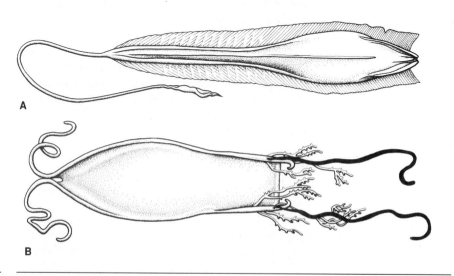

A

B

FIGURE 26–3   **A,** Egg of chimaera (*Hydrolagus*); **B,** egg of shark (Scyliorhinidae).

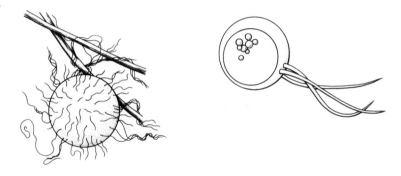

FIGURE 26–4   Teleost eggs with filaments and tendrils.

herrings are exemplary of open sea mass spawners. Others migrate to shore areas, and some, like the shad, are anadromous.

Polyandrous spawning is exhibited by many species in both fresh and marine situations. Males position themselves on the spawning ground and surround females as the latter swim into their midst. Eggs and sperm are released simultaneously, sometimes with violent activity on the part of the spawners. In some suckers (Catostomidae), the activity consists of vigorous vibration strong enough to dislodge stones of the substrate and allow eggs to sift into crevices. The eggs of yellow perch are enclosed in a gelatinous rope that often festoons vegetation or debris in the spawning area. Many polyandrous spawners of the open ocean leave the eggs to drift.

Pairing is common in many species, such as tunas and carangids, that spawn in open waters or over unprepared sites on the bottom. Other species pair for spawning after one or both members of a pair prepare a site for reception of the eggs. Preparation may range from merely fanning silt from stones to the construction of simple to elaborate nests. Some, such as salmon and trout, bury their eggs in gravel and then abandon the nest.

Blumel (1979) reviewed parental care among fishes, noting more than 80 families of bony fishes in which some kind of parental care was confirmed. Guarding of eggs was the type of parental care in most families, followed by nest building or cleaning of spawning substrate and fanning eggs. Other behavior listed among the 15 types of parental care includes internal gestation, oral brooding, burying of eggs, and splashing water on eggs deposited above the water line.

Nest building is known in lungfishes, the bowfin, osteoglossiforms, minnows, catfishes, sunfishes, snakeheads, and many others. In numerous species, from primitive to derived, the nest is guarded by one or both parents.

A few species leave their eggs in the care of other animals. Some snailfishes (*Careproctus* spp.) deposit eggs in the gill chambers of crabs of the genera *Paralithodes* and *Lopholithodes*. The tubesnout *Aulichthys* places eggs into the peribranchial cavity of ascidians, and the bitterlings, *Rhodeus* spp. and close relatives, introduce eggs into the siphon of freshwater mussels so that the eggs are incubated in the bivalve's chamber. *Percilia gillissi*, of Chilean lakes, places its eggs in the outlet canals of a freshwater sponge.

In a number of pairing species, the eggs are carried and protected by one of the parents after external fertilization. Eggs are carried in the mouth or branchial cavity in about ten families, but this is most prevalent in the Cichlidae, in which numerous species are mouthbrooders. The breeding habits and behavior of cichlids are of great interest to scientists and aquarists alike and have been subject to considerable study (Barlow, 1991; Keenleyside, 1991).

At least 17 species of cardinalfishes (Apogonidae) and 13 of sea catfishes (Ariidae) are mouthbrooders (Breder and Rosen, 1966). Other families are Osteoglossidae, Liparidae, Opisthognathidae, Anabantidae, Belontiidae, Luciocephalidae, and Malapteruridae. The female amblyopsids carry their eggs in the branchial cavity. The catfish *Tachysurus* (Ariid) is reported to incubate eggs intestinally (Blaxter, 1969). Some banjo catfishes (Aspredinidae) and some loricariid catfishes carry the eggs embedded in the skin of the lower surfaces of the body and fins. Males of some pipefish and seahorse species bear eggs on the skin, but most species carry eggs in brood pouches. In their close relatives, the Solenostomidae, the female is equipped with a brood pouch formed by the pelvic fins (Orr and Fritzsche, 1993). The males of *Kurtus* have a hooklike structure projecting forward from the forehead, from which the clustered eggs hang during incubation (see Breder and Rosen, 1966).

Some oviparous fishes fertilize the eggs internally. This is encountered, for example, among skates, chimaeras, many sharks, some characins, catfishes, and rockfishes. Egg retention for a short period, after fertilization takes place in the follicle or ovarian lumen, naturally leads into the conditions of ovoviviparity (internal incubation) and viviparity.

Egg Retention, Internal Incubation, and Viviparity

There is a continuum of conditions leading from the deposition of internally fertilized eggs in the cleavage stage to the release of large, well-nourished juveniles or young adults. Several possible advantages are conferred by these patterns. The first of these is protection. The eggs and embryos are safe from predators except those large enough to overwhelm the female. They are protected from adverse water conditions; desiccation, anoxia, and injurious temperatures are not dangerous unless the female is unable to escape these conditions. Young are protected against loss by drifting, as eggs or larvae, away from suitable rearing areas. Another possible advantage might be conservation of energy, although careful study would be necessary to confirm energy saving in specific instances. Certainly nest building is unnecessary, and the size of the male can be reduced. An emphasis on large numbers of eggs is generally not required, and usually there is no need for long migrations to a specific breeding site. Usually fertilization is ensured, so that few eggs are wasted. Viviparity (in its broad sense) may be of some advantage in species dispersal, as a single pregnant female might be able to accomplish an extension of range. Similarly, survival of one pregnant female after a catastrophe might allow continued survival of the species.

Because there is some difficulty in fitting uncomplicated definitions to complicated natural processes that grade into each other, there have been several versions of definitions for ovoviviparity and viviparity. A definition that viviparity, in a broad sense, includes all conditions in which hatched young are liberated from the female might be most suitable, but the term *ovoviviparous* is useful for emphasizing those species that incubate eggs and liberate live young without providing any maternal source of nourishment other than that in the egg. Some recent treatments of the matter of definitions emphasize the trophic relationships of the embryos. For instance, some (see Wourms, 1981) favor the term *lecithotrophy*—a condition in which eggs are retained until hatching and all nourishment is derived from the reserves in the yolk—over the term *ovoviviparity*. Dependence on the mother for nourishment is called matrotrophy (Wourms, 1981).

Viviparity in the broad sense is widespread among elasmobranchs (Wourms, 1977, 1981). Wourms (1981) recognizes viviparity in 40 of the 98 families accepted in Compagno's classification (1990). In bony fishes, there are about 15 viviparous families out of about 480. These include Latimeriidae, Zoarcidae, Bythitidae, Ophidiidae, Aphyonidae, Hemirhamphidae, Goodeidae, Jenynsiidae, Anablepidae, Poeciliidae, Comephoridae, Embiotocidae, Labriosomidae, and Clinidae.

As mentioned earlier, some oviparous species set the stage for viviparity in fertilizing eggs internally and releasing them in an early stage of development. Internal fertilization requires modification of behavior and the development of the means of introduction of spermatozoa into the genital orifice of the female. Spermatophores have developed in Poeciliidae and Horaichthyidae; those of the latter are barbed for sure attachment. Some of the most remarkable intromittent organs among fishes are seen in oviparous species (Fig. 26–5). Chimaeras, skates, and oviparous sharks have pelvic "claspers," while the anal fin is modified into a gonopodium in the oviparous poeciliid *Tomeurus* and in Horaichthyidae. Phallostethidae and Neostethidae have elaborate structures developed for use in clasping and inseminating the female. Some sculpins have fleshy genital papillae that reach a large size. *Pantodon* has a hollow tube on the leading edge of the anal fin.

From the short-term retention of eggs to incubating and hatching them internally is only a short step. According to Wourms (1981), there are 14 families of bony fishes that are viviparous to some degree. These are, according to the classification used in this book, Latimeriidae, Ophidiidae, Bythitidae, Aphyonidae, Hemirhamphidae, Goodeidae, Anablepidae, Poeciliidae, Embiotocidae, Zoarcidae, Clinidae, Labrisomidae, Scorpaenidae, and Comephoridae.

Many species of rockfishes (Scorpaenidae) are ovoviviparous, releasing newly hatched larvae that weigh less than the eggs from which they developed. The rockfishes show little sexual dimorphism and have no particular specialization of the ovary, where the eggs remain during incubation. In an ophidiid genus, *Dinematichthys,* eggs hatch from the ovarian follicles and develop into advanced larvae in the lumen of the ovary. The term *larviparous* has sometimes been applied to the condition of giving birth to larvae.

Another type of mainly lecithotrophic viviparity is seen in most Poeciliidae, in which the young are retained until the juvenile stage is reached (Thibault and Schultz, 1978). Young remain in the follicles of the ovary, with sufficient yolk for development. In some species, including those of *Gambusia* and *Xiphophorus,* the pericardial sac expands into a hoodlike or straplike structure that folds over the embryo's head; this structure is thought to have an absorptive function.

Many of the poeciliid species are capable of retaining more than one brood at a time. This is called superfetation and is a result of continued maturation of eggs during the development of previously fertilized eggs. Some species can store live sperm

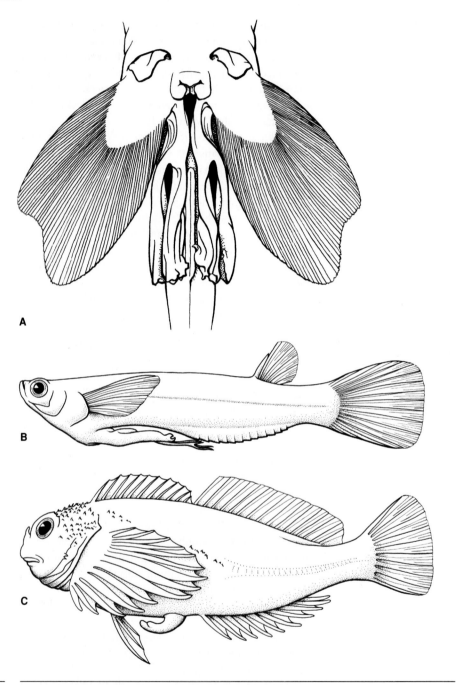

FIGURE 26–5 Examples of male oviparous fishes with large intromittent organs. **A,** pelvic claspers of a chimaera (*Hydrolagus*); **B,** *Horaichthys;* **C,** *Clinocottus.*

for several months so that successive batches of oocytes can be fertilized as they are ready. There may be as many as five broods at once in *Poeciliopsis prolifica* (Thibault and Schultz, 1978). A member of the Clinidae, *Clinus superciliosus,* called super klipfish in South Africa, may harbor up to 12 broods in what might be the extreme in superfetation (Veith, 1980).

The coelacanth, *Latimeria chalumnae,* has the largest eggs among the bony fishes—8.5 to 9.0 cm in diameter. In the early days of study of *Latimeria,* because of the egg size and because of evidence of live-bearing reproduction in fossil coelacanths, some scientists believed that the species would be shown to be viviparous. This was proven in 1975 by the discovery of five advanced young, up to 33 cm long, in the oviduct of a specimen at the American Museum of Natural History (Thomson, 1991).

Viviparous fishes nourish developing embryos through a variety of adaptations, most involving the secretion of nutritive materials by the female but including formation of pseudoplacentae or placentae. In some of the aplacental species, the developing young subsist exclusively on the yolk for a time, and then the yolk is supplemented or supplanted by secretions ("uterine milk") of the female and, in a few species, material from dead eggs and embryos. Any secretions, cells, and cell debris other than blood that comes from the female and is available to the embryo is called histotroph. In several species of sharks, the nutritive secretions are taken in through the mouth or the spiracles and swallowed. In some, the villi (trophonemata) that secrete the nutrient fluid are elongate and extend through the spiracles of the embryo to the gut.

In placental sharks, there are three phases of nutrition according to Hamlett (1989), who applies the following terms: vitellogenesis, the phase in which yolk is the energy source; histotroph secretion, the phase involving nutrition secreted by the female; and hematotrophic placentation. Placental sharks are found in the families Carcharhinidae and Sphyrnidae. Yolk sac placentae, formed by the close apposition and eventual interdigitation of thin tissues of the yolk sac and the uterine oviduct, are present in such carcharhinid sharks as *Mustelis canis, Carcharhinus falciformis,* some species of *Scoliodon, Prionace, Hemigaleus,* and in some hammerheads of the genus *Sphyrna* (Wourms, 1977). Placental species vary in dependence on the placentae (Dodd, 1983). In some, the early development is supported mainly by yolk, and a placenta is not developed for several months. In some sharks, the umbilical cord is provided with delicate vascular structures called appendiculae, which may function in the absorption of nutrient materials as well as perform other functions (Hamlett, 1989). In others, placentation occurs early in gestation and is responsible for most of the embryonic nutrient. An example is *Scoliodon laticaudatus,* which is believed to have the smallest eggs of any elasmobranch (1 mm diameter) (Hamlett, 1989).

A grisly type of matrotrophy is known in some families of shark, mainly lamnoids. Advanced embryos devour newly ovulated eggs (oophagy) or their weaker siblings (adelphophagy) (Wourms, 1981). Considering the life style of most lamnoids, this interuterine cannibalism is good training for the future. Adelphophagy has been reported in the ophidioid genus *Cataetyx.* The young of *Latimeria* are believed to reach their large size at birth not only because of the large yolk but also because they might ingest egg debris and histotroph secretions (Balon, 1991) or possibly smaller embryos (Balon, 1984).

There is a great variety of adaptations for nourishment of embryos among the viviparous teleosts (Hoar, 1969; Wourms, 1981). Many of these involve supplemental structures of the female or the embryo. In the eelpout, *Zoarces viviparous,* the only special absorptive structure other than the skin is the hypertrophied hind section of the intestine. Some other examples are the expanded pericardial sac in some Poeciliidae and follicular pseudoplacentae in others; expansion and specialization of the fin membranes in Embiotocidae; structures called trophotaeniae that radiate from the anal region of embryos of most Goodeidae and some ophidioids; and the great expansions of the gut in Anablepidae. Some embiotocid species have enlarged hindguts that protrude from the anus of developing young. Some of these specializations allow the retention of young until an advanced stage is reached. In an extreme case, the males of the embiotocid *Cymatogaster aggregata* are born sexually mature (Weibe, 1968), even though this species has eggs that are among the smallest known in fishes—about 0.3 mm (Eigenmann, 1894).

Placenta-like connections are formed in some teleosts and myliobatoids. A "branchial placenta," in which trophonemata or similar structures invade the branchial chamber of the embryo and form close association with the gill tissue, has been described for *Jenynsia* (Jenynsiidae) and *Gymnura* (Dasyatidae) (see Wourms, 1981).

Reproductive Guilds

The varied reproductive methods of fishes have inspired a few attempts to classify them from ecological or behavioral standpoints. Balon (1984) has proposed a classification that includes 34 reproductive guilds (see also Balon, 1975b, 1981). The guilds are arranged in three sections. Section A, *Nonguarders,* has two subsections: open substratum spawners, which leave their eggs exposed, and brood hiders, which conceal the eggs. The first group varies from pelagic spawners to those that spawn on rock and gravel, plants, sand, and on land. The second group hides eggs in sand, rock and gravel, cavities, and in invertebrates. Section B, *Guarders,* includes substratum choosers and nest spawners. Substrata chosen include open water, rocks, plants, and above-water materials. Nests range from froth to rock and gravel, plants, sand holes, and even anemones. Section C, *Bearers,* includes those that bear eggs (and larvae in some) externally in pouches, the mouth, branchial cavity, etc., or internally, from lecithotrophic to placental situations. Although the full reproductive habits of only a small fraction of the fishes are known, and although Balon's classification may be incomplete or may not provide for some exceptional or versatile species, the guilds form a framework useful in presenting extensive information on breeding habits.

Differences Between the Sexes

With many important exceptions (to be covered later), the individuals of most species of fish function as either male or female throughout their adult life. That is, most species are bisexual, or gonochoristic, as opposed to hermaphroditic, a condition in which an individual produces both eggs and sperm at some stage of its development. Establishment of sex depends on the sex chromosomes, designated X and Y for most fishes (see coverage of sex determination in Chapter 27, on page 503).

Although the sexes have very similar appearances in many species, sexual dimorphism or dichromatism is common in fishes and may be especially well marked in those species with internal fertilization or elaborate reproductive behavior. The differences between the sexes may involve secondary sex characters necessary for the accomplishment of copulation, oviposition, or incubation (requisite characters) or may be so-called accessory characters, which may not be directly involved in the mechanics of reproduction but are important to recognition, courtship, or other reproductive behavior. Requisite secondary sex characters include the claspers of elasmobranchs and the various gonopodia of the males of phallostethiforms and cyprinodontiforms, such as hemirhamphids and embiotocids. Even some oviparous species with external fertilization, such as sculpins, have large genital papillae (Fig. 26–5). Brood pouches and specialized ovipositors are requisite characters.

Accessory secondary sex characters are many and varied and are usually sexually dimorphic. Many structures and colors change with the reproductive state of the individual, while others are more or less permanent throughout the year. Males of many species are more brightly colored than females and may have larger fins and bolder markings. Sexual dichromatism is seen in salmon and trout, in which the males are more colorful; in the bowfin, in which the male has a caudal ocellus; and in many minnows, characins, cichlids, and others. Sexual dimorphism is often seen in fishes. Longer fins are characteristic of the males of many fishes. Suckers, gobies, dragonets, and climbing perches are examples. The color and larger fins of males can be significant in courtship or aggressive displays toward rival males, or the fins can, in certain species, aid in holding the spawners together. This is especially true in species having nuptial tubercles or contact organs on the fins or body. These structures are prominent on the fins and scales of suckers and reach large sizes on the heads of certain minnows (Fig. 26–6). They are most common in species that spawn in flowing water. Nuptial tubercles are formed from both keratinized and nonkeratinized epidermal cells. The horny caps of the former type are often pointed. Whitefish, grayling, smelts, ayu, retropinnids, kneriids, phractolaemids, most cyprinoid families, a few characoids, the mochokid catfishes, and percids have nuptial tubercles. Contact organs are small, bony, spinelike structures usually associated with scales or fin rays; they are present in needlefishes, certain cyprinodontoids, characins, and sculpins (Wiley and Collette, 1970).

Size differences between the sexes are evident in many species. The reproductive pattern of species determines which sex is the larger, but commonly the female, carrying the bulky eggs, is larger than the male. A notable exception is the lungfish, *Protopterus aethiopicus,* in which the male is reported to be twice the length of the female (Greenwood, 1987). The greatest disparity in size occurs in the ceratioid anglerfishes, in which the female can be many times larger than the male.

In certain species, the male ceratioid grasps the female's skin with his teeth, literally grows to her, and becomes a testis-filled parasite, available for service at spawning time. Pietsch (1976) reviewed the reproductive strategies of ceratioids and showed that males of Ceratiidae, Linophrynidae, and probably Neoceratiidae are obligate sexual parasites. The males of Caulophyrnidae and the oneirodid genus *Leptacanthichthys* are thought to be facultative parasites. Sexual parasitism is not known in the remainder of the suborder.

Hermaphroditic fishes have attracted attention as a rich resource of physiological and genetic information as well as for their general interest (Atz, 1964). Occasional hermaphrodites are found in many gonochoristic species as an abnormality, but there are numerous species that are normally hermaphroditic, some even capable of self-fertilization. Synchronous (or simultaneous) hermaphrodites have ripe ovaries and testes at the same time but usually spawn with one or more other individuals, alternately taking the role of male and female. *Serranus subligarius,* a marine species

FIGURE 26–6   Nuptial tubercles on **A**, scales, anal fin, and caudal fin of sucker (*Catostomus*); **B**, scales, head, and pectoral fin of Oregon chub (*Mylocheilus*).

of Florida, has fertilized its own eggs in captivity (Yamamoto, 1969). *Rivulus marmoratus* can fertilize its eggs internally prior to oviposition. Fishes in which the same individual can be of both sexes at the same time (synchronous hermaphrodites) are known from the following families: Chlorophthalmidae, Alepisauridae, Paralepididae, Evermannellidae, Cyprinodontidae, Serranidae, Maenidae, and Labridae. Individuals of other families may occasionally be hermaphrodites. A familiar example is the striped bass, *Morone saxatilis,* in which occasional hermaphrodites are seen.

Sequential hermaphrodites are either first male (protandrous) or first female (protogynous) (Yamamoto, 1969b). Many of the species begin life with undifferentiated gonads that contain both male and female elements. The protandrous condition is known in members of the Gonostomatidae, Serranidae, Sparidae, Maenidae, Labridae, Platycephalidae, Pomacentridae, and Centropomidae. Protogynous hermaphrodites are known in Scaridae, Sparidae, Emmelichthyidae, Synbranchidae, Serranidae, Maenidae, and Labridae. In the latter family, there are species in which some males do not pass through the female stage (primary males) as well as those secondary males that change to male from female. The secondary males are brightly colored, but the primary male may be dull, or at least colored like the females. In one species, the two types of males exhibit different spawning behavior, the primary males spawning in groups with a single female and secondary males pair-spawning with females in turn.

Sex reversal and hermaphroditism are controlled by the endocrine system, which is genetically "programmed" in normally hermaphroditic species to act on the gonads in response to the proper stimuli: internal, external, or both. Experimentally, administration of androgens to genetic female fish has changed them into functional males, and administration of estrogens has changed genetic males into functional females (Guerrero, 1979). The breeding of sex-reversed, functional males (with no Y chromosome) to normal females can result in all-female progeny. This can be significant to fishery management, especially when no reproduction of early maturing pond fishes is desired or when a useful but potentially troublesome species, such as the grass carp, *Ctenopharyngodon idella,* is to be stocked.

Naturally occurring all-female species are known among the Poeciliidae; the best known example is probably the Amazon molly, *Poecilia formosa.* This fish originated through hybridization with other species of *Poecilia* and is a permanent diploid species that produces diploid eggs (Schultz, 1973). The species is perpetuated through matings with males of other species that contribute sperm, which trigger development of the eggs but do not contribute any genetic material (a condition known as gynogenesis plus hybridogenesis). Gynogenesis is the production of all-female offspring. Other gynogenetic species are known, and some have been produced in the laboratory.

Experimental production of all-female broods has been accomplished with several species for various purposes. Some laboratories "clone" females of a species by activating eggs with sperm made impotent by irradiation or some other means and then restoring the diploid chromosome set contributed by the female by treating the activated eggs with pressure or temperature alteration to inhibit the first or, if desired, the second meiotic division (Streisinger et al., 1981). This induces retention of the second polar body (Chevassus, 1983).

Hybridization

Fish hybridization in nature has been recognized for many years, and hybrids have long been produced in laboratories and hatcheries. (See also the section on hybrids in Chapter 27, on page 501.) Schwartz (1972, 1981) listed 3759 articles on the subject. Natural hybrids can occur by two or more species spawning in proximity at the same time. Fish spermatozoa remain active for many seconds to a few minutes, and sperm from a spawning pair or group closely upstream from another group can invade the nest or spawning area of those downstream. If of different species, hybrids may result. Hybridization involving pairing of different species is known, but usually there are reproductive barriers—behavioral, physical, or temporal—that will preserve the integrity of species, especially those living synmpatrically. Occasionally, these are not effective.

Selective Breeding

Aquaculturists, whether involved in pond culture, hatchery operations for stock enhancement, "salmon ranching," breeding of ornamental species, or other activities, are constantly seeking to maintain high quality in the product or to improve the quality. Maintenance of genetic variation in captive "wild" animals requires attention to known genetic principles and requires use of an adequate number of breeders and minimization of certain types of inbreeding (Frankel and Soulé, 1981), so that some selection of breeders must be practiced. Maintenance of high quality and variation in semiwild stocks, such as those used in salmon ranching (in which brood stock is selected from survivors of marine life), carries some difficulty in that, although large numbers of individuals are involved, they have been subjected to the artificial selection of hatchery life, and without some elaborate methods of identification, there is great difficulty in identifying family lines.

Most selective breeding with fishes is accomplished in situations that allow the retention of a captive broodstock from which selected matings can be made. Purposes of selective breeding in domesticated stocks are varied within the general goal of increasing productivity (Gjedrem, 1983; Gjerde, 1993; Kirpichnikov, 1981). Fish are selected for resistance to adverse environmental factors encountered in artificial surroundings, including diseases; extremes of temperature, dissolved oxygen, pH, and other physical and chemical factors; stress brought on by crowding; and build-up of metabolic wastes in rearing facilities. Selection can be made for rapid growth on either manufactured or natural food, for efficiency in weight gain per unit of food consumed, for increased fecundity, for delayed or accelerated sexual maturity, and for color, squamation, body conformation, and many other attributes that could make culture easier, increase production, and enhance acceptability of the product in the market.

Kirpichnikov (1981) suggests that selection can be used to improve anadromous or marine fishes. He mentions faster growth of young, increased fertility, disease resistance during freshwater life, better growth and survival in the marine habitat, and shorter duration of marine life in anadromous species.

Selection can be accomplished by inspecting prospective breeders for desirable phenotypic attributes, with little or no knowledge of the genotype, and choosing those that appear to have the best balance of the characteristics wanted. Selective breeding requires that the stocks from which the prospective breeders are to be selected are subjected to the same environmental factors, so that differences among individuals will have less chance of being environmentally induced.

Although inbreeding is a part of breeding programs, it generally leads to reduced performance and is avoided (Gall, 1983; Kincaid, 1983). Crossing of unrelated strains of a species is often used to increase heterozygosity and to produce heterosis. This is especially important in breeding programs with carp, *Cyprinus carpio* (Kirpichnikov, 1981).

Embryological and Early Development in Fishes

Embryology

In normal fertilization, only one sperm enters the egg through a tiny passage in the chorion known as the micropyle. In the pink salmon, the external funnel is about 15 μm but narrows to about 1 μm as the tube passes through the chorion (Depêche and Billard, 1944). Even if more than one sperm may enter an egg (a condition known as polyspermy), as is common in elasmobranchs and a few bony fishes, only one sperm is involved in fertilization (see Ginzburg, 1972). The great majority of fishes release sperm into the water in the vicinity of the eggs. Because the sperm of freshwater species live only a short time following release into the water, it is important that the fishes bathe the eggs in a heavy concentration of sperm as the eggs are being extruded. Fish sperm can be held alive for a short time in physiological (0.9 percent) saline solution. Sperm of marine fishes have a longer active life than those of the freshwater species. Fish culturists are interested in methods of long-term storage, and techniques are constantly being improved. Sperm of several species have been successfully stored for several days at temperatures near freezing in suitable conditions (see Stoss, 1983). Cryopreservation of sperm with liquid nitrogen has been accomplished with a number of species, including carp (Lubzens et al., 1993), wels (*Silurus glanis*) (Linhart et al., 1993), yellow perch (Ciereszko et al., 1993), rainbow trout (Wheeler and Thorgaard, 1991), and others (see Harvey, 1993). Sperm frozen in appropriate supporting media at the temperature of liquid nitrogen have yielded some success in fertilization of fish eggs (Stoss, 1983).

The development of fertilized fish eggs follows much the same pattern as in other vertebrates. Most fish eggs have a relatively good supply of yolk (are telolethical), but there are wide differences among the fishes regarding the actual amount of yolk. A few species have only a small egg and a small enough supply of yolk so that the entire egg divides during cleavage (holoblastic cleavage). This is the case in lampreys, some sturgeons, and lungfishes. The cleavage is unequal, with larger cells at the vegetal pole, where the yolk is concentrated. Some fishes have eggs with the cytoplasm thinly distributed around the relatively large yolk, while others have cytoplasm concentrated at the animal pole. The cytoplasm forms a polar cap at the site of the nucleus following fertilization, and this begins to divide, forming the embryo on the surface of the yolk (meroblastic cleavage). Some stages in the development of a bony fish are shown in Figure 26–7.

Teleosts and lampreys share a peculiarity of embryology in that the central nervous system forms by the hollowing of a medullary keel instead of by the formation of medullary folds. Another interesting peculiarity in the development of some bony fishes is the delay (diapause) undergone by eggs of certain annual cyprinodontids, which deposit fertilized eggs in the bottoms of drying ponds. These eggs do not complete development until the ponds hold water again. *Nothobranchius* and *Aphyosemion* are genera with annual species in which eggs might not hatch for several months.

Usually the incubation period is governed by temperature. Within the optimum range for normal development, the period shortens as the temperature increases. For

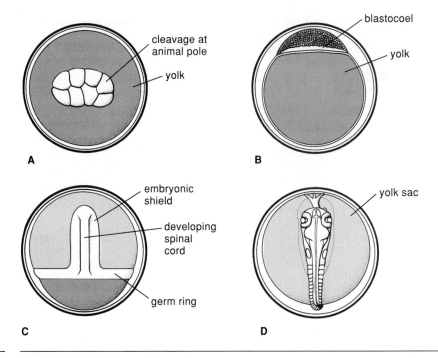

FIGURE 26–7   Examples of embryonic stages in teleosts. **A**, Cleavage; **B**, blastula (sectioned); **C**, embryonic shield; **D**, organogeny.

instance, trout and salmon eggs will hatch in about 50 days at 10°C, but at 2°C, incubation requires about six months. The eggs of the common carp incubate normally at temperatures of from 15° to 30°C, and hatching occurs in about a week at the lower temperature and one day or a few hours less at the higher temperature. Within the genetic capability of the species, incubation temperature influences the meristic features of the individual. These are the features primitively tied to segmentation of the body, especially vertebrae, scales, and fin rays. As a general rule, those individuals of a selected batch of eggs incubated at low temperature will have more vertebrae, scale rows, and fin rays than their siblings incubated at higher temperatures. There are a few instances in which the opposite has been shown to be true.

Hatching in many bony fishes is aided by secretions of special glands on the head or inside the mouth. The secretions are generally enzymatic in nature and weaken or even liquefy the chorion.

Parthenogenesis

Although development of the fish egg normally begins upon fertilization by a spermatozoon, it has been experimentally induced without sperm, especially by various chemical and physical techniques. The artificially activated eggs do not usually develop normally (Blaxter, 1969). Parthenogenesis (development of unfertilized eggs) does occur naturally in some fishes (see Echelle and Mosier, 1981; Solar et al.,

1991). There are two types of parthenogenesis: androgenesis, in which the egg is activated by sperm but develops without any contribution from the egg nucleus; and gynogenesis, in which sperm activates the egg but contributes no male genetic material to the developing embryo (see Chapter 27).

Early Life History

Largely because the subject is of such great importance to the science of fisheries, the published literature on early life histories of fishes is growing tremendously. The following references can provide an entrée to the literature: Russell, 1976; Lasker, 1981a; Moser et al., 1984; Balon, 1985a; Hoar and Randall, 1988a and b; and Kamler, 1992.

Newly hatched fishes of oviparous species may be tiny, unformed creatures destined to undergo considerable additional development, as in the lampreys, or may be essentially small replicas of the adults, as in oviparous sharks, rays, hagfishes, and some bony fishes. Discussions of early life history stages of fishes have suffered somewhat because of a lack of standard terminology, and the diverse nature of very young fishes seems to prevent the ready acceptance of any set of terms. Generally, early development is divided among the egg, larval, and juvenile developmental stages (Fig. 26–8) (Kendall et al., 1984). According to Kendall et al. (1984), the egg

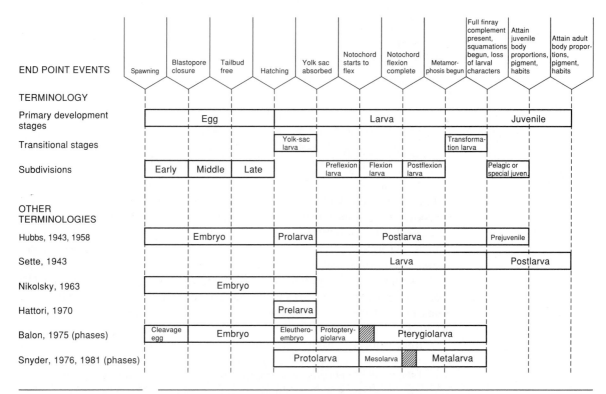

FIGURE 26–8   Terminology of life history stages of fishes. (From Kendall et al., 1984, Early life history stages of fishes and their characters, p. 13, Fig. 6, in *Ontogeny and Systematics of Fishes.* Used with permission.)

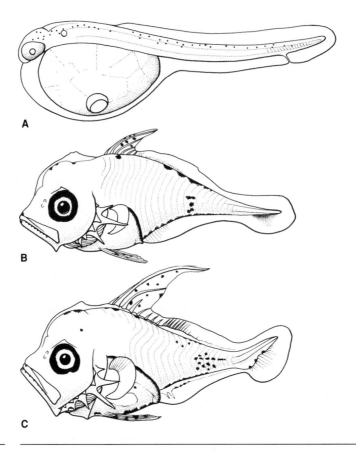

FIGURE 26–9 Examples of fish larvae. **A,** Yolk sac larva of *Brevoortia patronus* (2.6 mm); **B,** preflexion larva of *Selene vomer* (3.2 mm); **C,** postflexion larva of *Selene vomer* (3.9 mm). (**A** from Hettler, 1984, *Fish. Bull. U.S.,* 82, p. 87; **B** and **C** from Aprieto, 1974, *Fish. Bull. U.S.,* 72, p. 433.)

stage is divided into early, middle, and late subdivisions; these end with blastopore closure, freeing of the tailbud from the yolk, and hatching, respectively. Following hatching, there is a transitional stage, the yolk sac larva (Fig. 26–9**A**), which ends with the absorption of the yolk sac. The next developmental subdivision is the preflexion larva, which ends when the notochord begins to flex upward (Fig. 26–9**B**). That is followed by the flexion and postflexion larval subdivisions, following which the larva begins metamorphosis in the transformation transitional stage, which ends when the fish has its full complement of fin rays and loses larval characters. It then acquires juvenile characteristics.

Some workers consider the embryonic period to last until hatching or parturition, and others consider it to last until the young begin to feed for themselves. In one of the least complicated sets of terms, proposed by Hubbs (1943), larval stages are those that are beyond the embryo but well differentiated from the juvenile, which is essentially like the adult. Hubbs recognized two larval stages: the prolarva, which retains a yolk sac, and the postlarva, which has absorbed the yolk sac but is still unlike the juvenile stage. The upright-swimming larvae of flatfishes or the leptocephali of eels are examples of postlarva. If yolk-bearing larvae transform directly into a juvenile, as is the case in many salmonids and certain sculpins, these larvae are called alevins. They are the "sac-fry" of salmonid culturists.

A terminology suggested by Balon (1975b) divides the embryonic period into three phases—cleavage egg, embryo, and eleutheroembryo (free embryo), which is free from the egg. The term *free embryo* is not universally accepted. A newly hatched fish is usually called a larva (see Kamler, 1992). The early larval phase, with undifferentiated fin folds, is called apterolarva by Balon. The later larval phase, with fin rays forming, is called pterolarva (Balon, 1985b).

Balon (1985b) discusses the theory of saltatorial ontogeny in fishes, which questions whether a typical fish life history comprises a succession of barely noticeable small changes or a stepwise series of longer steady states interspersed with fast changes in function and form. He recognizes three life history models—indirect, transitory, and direct—and divides the saltatory ontogeny into up to five periods, and these into phases and steps. The embryo period has three phases in all three models: cleavage, with three steps; embryo phase, with four steps; free embryo, with three phases. The indirect model has a larval period that consists of the finfold phase and the finformed phase. The transitory model has what Balon terms the alevin period and phase, whereas the direct model has no larval or alevin period. The juvenile period is common to all three models, but in the transitory model it is divided into parr, smolt, and juvenile phases. Other periods in the three models are adult and senescence.

The Life of Larval Fishes. A marvelous array of adaptations to the environment is seen among fish larvae. Some of these involve structures and shapes entirely unlike those in the juvenile stage and require an extensive metamorphosis. Lampreys, for example, undergo great internal changes as well as some obvious external modifications. They lose the functional gallbladder and bile ducts, grow a new esophagus as the respiratory tube disconnects from the alimentary canal, gain functional eyes, lose the oral hood and filtering sieve (replacing them with an oral disc set with horny teeth), and acquire larger fins. The larval mesonephric kidney is replaced by the adult mesonephros (Hardisty, 1979).

Eels change from the toothy, leaflike, transparent leptocephalus (Fig. 26–10), to transparent "glass eels," to the juvenile or elver (Castle, 1984; Tesch, 1977). Both lampreys and eels shrink considerably in length from larva to juvenile.

Examples of remarkable metamorphosis are numerous, including the following: the flatfishes, which change the entire architecture of the head to get both eyes

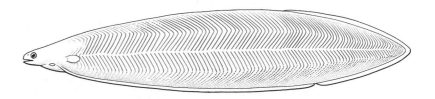

FIGURE 26–10   Leptocephalus of *Anguilla*.

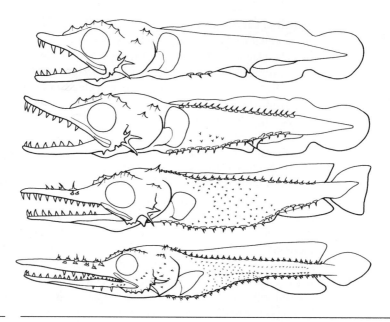

FIGURE 26–11   Spinous larvae of swordfish, *Xiphius gladius*. (From Potthoff and Kelley, 1982, *Fish. Bull. U.S.*, 80, p. 180.)

on the same side (Ahlstrom et al., 1984); the molas, in which most of the caudal region of the body is lost during larval development so that a normal caudal fin does not form; and the swordfishes, which as larvae have prolonged toothed jaws. As larvae, many fishes—including the molas, swordfishes, and many other perciform and scorpaenid genera—have very large spines in relation to their size (Fig. 26–11). Spines are thought to represent a protection against predation (Moser, 1981).

Many larvae show special adaptations for respiration, ranging from highly vascular fin folds, pectoral fins, or yolk sacs to the feather-like true external gills of *Polypterus, Protopterus,* and *Lepidosiren* (Goodrich, 1930). Gill filaments project from the gill openings of several species in the embryonic or larval stages of fishes, including mormyroids, loaches, and some elasmobranchs. Larvae of *Protopterus annectans* begin to breathe air at 23 to 27 mm (Greenwood, 1987).

Attachment or adhesive organs are present on larvae of *Protopterus, Lepidosiren, Amia,* and *Lepisosteus.*

Many pelagic larvae are specially modified for flotation so that they maintain a specific depth or range of depth. Oil globules are effective in conferring hydrostatic balance and are common in drifting larvae. Inclusion of a large proportion of water in the flesh is another common flotation device, found among leptocephali. In the larvae of many fishes, sinking is retarded by a high ratio of surface to volume. In these, the fins may be of exceptionally large size, the fin rays or guts may extend into long trailing filaments (Fig. 26–12), or the body may be covered with spines, although those may have more importance in deterring predators.

Some larvae have elliptical eyes that are believed to have better rotation around the long vertical axis than round eyes so that a larger field of vision can be attained

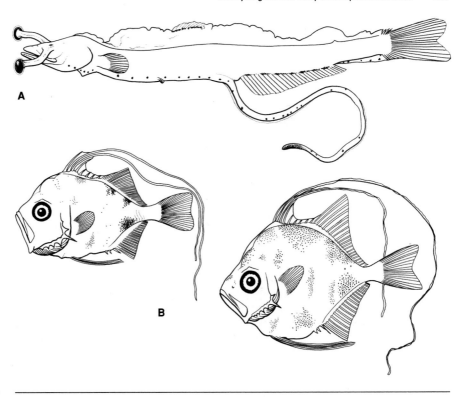

FIGURE 26–12   Examples of larvae with filamentous structures. **A,** *Myctophum aurolateratum* with stalked eyes and filamentous gut (26 mm); **B,** *Selene vomer* with filamentous fin rays (7.7 mm and 9.0 mm). (**A** from Moser and Ahlstrom, 1974, *Fish. Bull. U.S.,* 72, p. 399; **B** from Aprieto, 1974, *Fish. Bull. U.S.,* 72, p. 433.)

(Moser, 1981). The stalked eyes of lanternfish and stomiatioid larvae may aid in seeing prey over a wide area (Fig. 26–13).

Interesting melanophore patterns occur in the larvae of several groups of fishes. Mostly, those with heavy pigmentation, either black or yellow, live close to the surface and may require protection from ultraviolet light (Moser, 1981). Some patterns of pigmentation may prevent refraction of light from gut contents or from gas bladders. Others may aid in breaking up body outlines or otherwise making the larvae less noticeable to predators (Fig. 26–14).

In exceptional instances, fishes with larvoid characteristics mature sexually and reproduce. This phenomenon, known as neoteny, is common in the icefishes (Salangidae) of the coasts of China and Korea. Certain sauries and needlefishes are neotenic, and a tendency toward this condition is seen in some brook lampreys, in which the ammocoetes go a considerable way toward developing gonads prior to transformation. Some of the best examples of neotenic fishes are the members of the gobioid genus *Schindleria* of the central Pacific Ocean. These tiny transparent fishes reach about 20 mm in length and retain several larvoid characters, including a functional pronephros, opercular gills, and a nonfolded heart. The atrium is behind the ventricle instead of being folded over it (Bruun, 1940; Gosline, 1959; Watson et al., 1984).

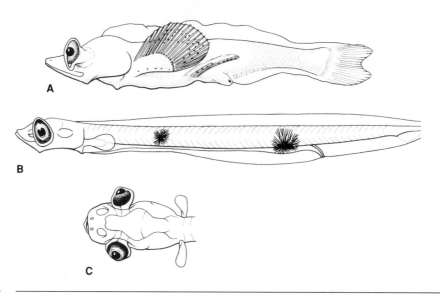

FIGURE 26–13   Examples of larvae with modified eyes. **A,** *Symbolophorus californiense,* with elliptical eyes on short stalks (9.6 mm); **B,** *Leuroglossus schmidti,* with elliptical eyes (11.7 mm); **C,** dorsal view of head of *L. schmidti* (SL 14.6 mm). (**A** from Moser and Ahlstrom, 1974, *Fish. Bull. U.S.,* 72, p. 399; **B** and **C** from Dunn, 1983, *Fish. Bull. U.S.,* 91, p. 27.)

Ecology of Early Life History Stages. In this section, there can be only a general treatment of the complex ecology of the early life history stages of fishes. For marine fishes, a stable ocean environment appears to be necessary for optimum survival (Lasker, 1981b). Unusual or unseasonable storms, disruption of upwelling patterns, and changes in currents can change water quality or the abundance of suitable food organisms. The young of freshwater species can be impacted by many weather-related circumstances, such as scouring floods and fluctuating water temperature or water level.

Although yolk sac larvae and early feeding larvae may seem capable of only minor directed movements, experiments have shown that larvae of walleye pollock (*Theragra chalcogramma*) are able to move in relation to environmental gradients (Olla and Davis, 1990). Larvae showed positive phototaxis by moving horizontally from a darkened area of a tank to an area of low light intensity, but they demonstrated negative phototaxis under high light intensity. They showed a daily periodicity in response to light, moving toward the surface in the dark and toward the bottom in the light. They avoided surface turbulence and moved upward in the water column when chilled water was introduced on the bottom.

The transition from dependence on yolk to exogenous feeding (mixed feeding period) has been called one of the "critical periods" in the life of a fish (Kamler, 1992), although Miller et al. (1988, p. 1666) suggest that there may be mortality at this time because "small larvae are simply more vulnerable to starvation than larger larvae." Given that the larva has survived early predation and any transport to waters of insufficient oxygen, unsuitable temperatures, or other unfavorable water quality, it must now ingest small digestible particles that will supplement and then replace the yolk. Food must be available and abundant.

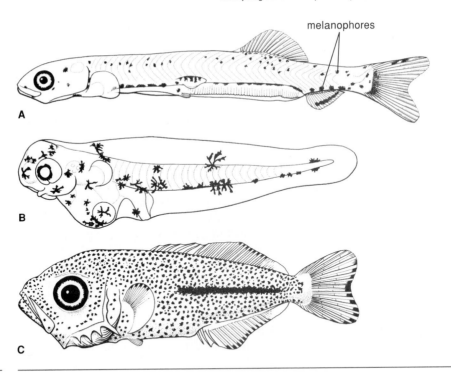

melanophores

FIGURE 26–14   Examples of melanophore patterns in larvae. **A,** Larva of gulf menhaden, *Brevoortia patronus* (16.5 mm), showing melanophores over intestine and gas bladder as well as body and fins. Note postanal series above anal fin. **B,** Larva of white croaker, *Genyonemus lineatus* (2.38 mm), showing melanophore pattern that may tend to break up body outline; **C,** heavily pigmented larva of leatherjacket, *Oligoplites saurus* (SL 5.7 mm). (**A** from Hettler, 1984, *Fish. Bull. U.S.,* 82, p. 89; **B** from Watson, 1982, *Fish. Bull. U.S.,* 80, p. 406; **C** from Aprieto, 1974, *Fish. Bull. U.S.,* 72, p. 436.)

Food for fish larvae at first feeding must be small because many larvae do not exceed 3 mm long at that stage. Initial food of larvae may include phytoplankton, copepods, ciliates, and mollusc larvae. Phytoplankton is usually ingested for only a short time, but the northern anchovy, *Engraulis mordax,* can feed on dinoflagellates for up to 20 days, although the growth rate is depressed (Hunter, 1981). Zooplankton, especially copepods, make up the most important segment of the food of larvae.

The width of prey taken at the beginning of feeding for most clupeoid larvae is 50 to 100 μm, but the upper limit increases to about 200 μm as the larvae grow from about 3–4 mm to 7–10 mm. Larvae of piscivorous fishes such as mackerels tend to feed on larger prey than that eaten by the clupeoids and may become cannibalistic at 5 to 10 mm (Hunter, 1981).

Density of prey is of great importance to the survival of fish larvae at the onset of feeding. Locomotor powers of larvae are not great, and prey is detected at one body length or less during early feeding. Laboratory studies cited by Hunter (1981) indicate that from 220 to 4000 copepod nauplii per liter are necessary for 50 percent survival of larvae stocked at rates of from 0.2 to 50 per liter. Open sea concentrations of nauplii are usually less than 40 l^{-1}. In enclosed waters such as estuaries and lagoons, average copepod concentrations can be 200 l^{-1} or more. Distribution of plankton is by no means uniform, so that fish larvae that are found in dense patches of plankton are generally well fed (Hunter, 1981).

Predation on eggs, larvae, and juveniles is of great importance to the survival of fish to maturity (Hunter, 1981; Nellen, 1986). Many animal groups are capable of ingesting drifting eggs and yolk sac larvae, and some can pursue and catch swimming larvae. Predators include the following: tunicates such as salps, doliolids, and pyrosomes; jellyfishes (including siphonophores and chondrophores); comb jellies (Ctenophora); arrowworms (Chaetognatha); squids; pelagic hyperiidian and calanoid copepods; euphausids and fishes, which are probably the most important predators (Nellen, 1986).

Vulnerability to predation depends on many factors. Where and when the eggs are deposited is of great importance. Many fishes are nocturnal spawners and spawn in areas of drift or current, which gives pelagic eggs a chance to disperse before being exposed to diurnal predators. Diurnal spawners tend to release eggs into currents or in parts of the ocean with few large planktonic predators (Hunter, 1981). Fishes that deposit demersal eggs that hatch into pelagic larvae usually spawn where currents will aid in the dispersal of the larvae.

As mentioned earlier, many species, both marine and freshwater, guard demersal eggs at least until hatching; and some guard demersal young, so that predation is minimized. In some of these, such as the freshwater sculpins (in which the young are not guarded for any length of time), the yolk sac larvae find refuge under stones downstream from the nest stone. The alevins of trout and salmon remain buried in the gravel nests until the yolk is nearly used up and they are able to swim and seek food.

Size of eggs and larvae is important in that small invertebrate predators may not be able to ingest larger eggs and larvae. Rapid growth of larvae will place them beyond the abilities of smaller predators to ingest them. Maturation of sensory organ systems aids in detection of predators. Escape from predators is facilitated by development of the locomotor apparatus so that the larvae can swim faster and longer.

Nellen (1986) emphasizes the importance of predation by fishes on eggs and larvae and advances the hypothesis that high fecundity and cannibalism are of considerable importance in the ecology and recruitment of fishes.

Chapter 27

The Genetics of Fishes

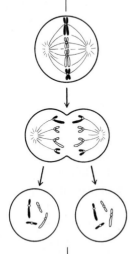

Introduction

The mechanisms and concepts of genetics are the bases for every aspect of biology; moreover, they tie all biological fields into a coherent picture. The particular aspects of biological disciplines for which genetic principles provide explanations are inheritance, gene expression, and evolutionary processes. Inheritance is the process by which genetic information is transmitted between parents and offspring and explains why offspring generally resemble their parents and why individuals within a species share many characteristics. Gene expression is the means by which genetic information carried by an individual is revealed and provides the mechanisms for ontogenetic progression and physiological response. Evolutionary processes provide insight into the causes of genetic divergence that may result in speciation. Systematic descriptions of species ordinarily attempt to follow the evolutionary relationships among species, **phylogeny,** which are based on the descent from a common ancestor; in other words, individuals in a taxon share a genetic history. Applications of genetic methods to understanding the biology of and unraveling the relationships among fishes are developing rapidly. This chapter focuses on the principles of genetics as they pertain to fish and on the genetic methods that are currently used or are being developed to study the biology of fishes.

Inheritance is accomplished by the transmission of deoxyribonucleic acid (DNA) sequences from parent to offspring. DNA is carried in each cell of an individual and encodes the potential structures, functions, and even behaviors for each individual. The resultant individual is the outcome of genetic expression in a particular environmental context. Genetic variation arises from changes in the DNA sequence, **mutations.** Mutations provide genetic alternatives that may be acted on by natural selection. Most gene changes are rejected because they are not as effective in their contribution to the species' overall ability to meet and solve environmental challenges. An occasional mutation, however, provides an even better solution that is manifested by an increased ability to contribute offspring to subsequent generations; that is, it increases **fitness.** Because the environment experienced by a population continually changes, from both year-to-year weather fluctuations and longer term climate cycles, there is usually no single best genetic solution for a population. As a result of this environmental uncertainty, the populations usually display genetic diversity, which reflects the previous environmental history of the population and may ensure that at least a portion of the population will be able to withstand future environmental extremes. Genetic diversity is essential for persistence of species because it provides raw materials for natural selection and may allow the species to evolve (Stearns, 1992).

Over time and space, species are not fixed entities; rather, they are dynamic and reflect results of response to evolutionary pressures that have acted on populations of the species. To comprehend why there are so many kinds of fish (more than 24,618; Nelson, 1994) and to understand the origins—and perhaps destinies—of a particular fish species, one must have an appreciation for the genetic basis and the related evolutionary processes that underlie fish diversity. Every aspect of the genetics of fish, from the molecular structure of genes and proteins to the fish's interaction with the biosphere, plays a role in understanding ichthyology.

Fundamental Concepts

Gene Expression and Genome Organization

Genes are expressed when the information encoded in the DNA sequences, the **genotype,** is manifested in the organism. Gene expression may be observed at many levels ranging from synthesis of a particular biomolecule to exhibition of complex behaviors. The result, which is often influenced by the environment, is the **phenotype.** Expression of the information encoded in the genetic material of an organism results in the phenotype we recognize as characteristic of a particular species. The most fundamental level of organization of a **genome** (an organism's entire set of genetic information) is the sequence of adenine (**A**), thymine (**T**), guanine (**G**), and cytosine (**C**) nitrogenous bases in the DNA molecule. DNA is a very long, double-stranded helical structure held together by hydrogen bonds between **A-T** and **G-C** pairing between the two strands. The amount of DNA carried in each molecule is enormous, approaching 1 billion nucleotide pairs. Base pairing stabilizes the molecule and provides a mechanism for accurately **replicating** the information carried on the molecule; each strand can direct the construction of new double-stranded molecules identical to the original. The result is two precise duplicates of the genome of an individual as needed for cell division and reproduction (Fig. 27–1).

The information encoded in the DNA base sequence includes both structural genes and regulatory genes. The DNA sequence in structural genes specifies the order of amino acids in a protein or the sequence of ribonucleic acid (RNA) nucleotides in housekeeping molecules such as ribosomal RNAs (rRNA) or transfer RNAs (tRNA), which are involved in protein synthesis. Protein synthesis occurs at **ribosomes** in the cytoplasm and follows directions carried by messenger RNA (mRNA) copies of structural gene base sequences coded in the nuclear DNA. RNA synthesis, **transcription,** is the construction of ribonucleotide polymers complementary to the DNA sequences or **gene** that specifies the structure (Fig. 27–2).

Regulatory genes determine when and how long a particular structural gene will be expressed. During development, entire batteries of genes must be turned on or off at different stages; differentiation of cells during development results in different tissues. Although erythrocytes (nucleated in fish) and muscle cells carry the same information, differential expression of the information results in cells possessing very different functions and capabilities. Adaptation of individuals may require changes in gene expression later in life to optimize chances for survival and reproduction. Examples of such adaptation are changes in osmocompetence, which allow diadromous fish to move between salt and fresh water, and changes in sex of nonsimultaneous hermaphrodites. The organization of genes in the genome is complicated. In fact, genome reorganization, rather than accumulated changes in single

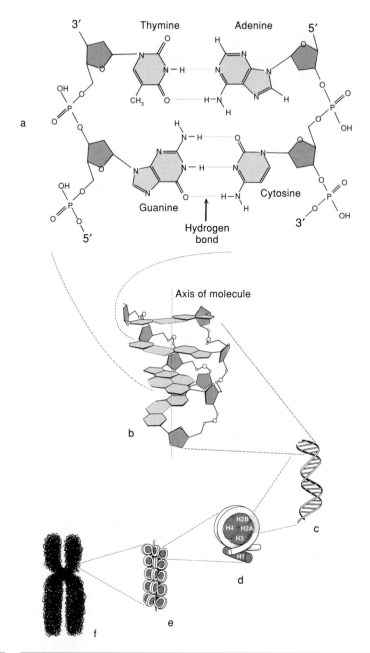

FIGURE 27–1 Deoxyribonucleic acid (DNA) is double-stranded and constructed from the deoxyribonu-
cleotide triphosphate subunits possessing the nitrogenous bases adenine (**A**), guanine (**G**),
cytosine (**C**), and thymine (**T**). Hydrogen bonding between **A-T** and **G-C** base pairs (a) holds the
two strands together. The stack of base pairs is located at the center of the molecule (b) and
alternating deoxyribose sugars and phosphates connect the base pairs and wind helically
around the outside (c). The DNA is packed into the nucleus of a eukaryote by histone proteins
(H1, H2A, H2B, H3, and H4) around which it winds (d). The histones and other chromosomal
proteins organize (e) the DNA into chromosomes (f).

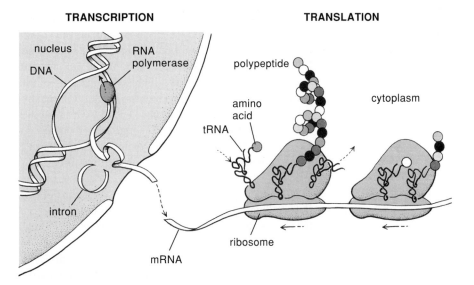

TRANSCRIPTION	TRANSLATION

FIGURE 27–2 Gene expression as protein synthesis results from transcription of DNA to RNA in the nucleus and translation of the RNA transcript in the cytoplasm into a polypeptide. RNA polymerase catalyzes the synthesis of RNA using one strand of the DNA as a template. Following transcription, the RNA is processed and introns are excised. The resulting messenger RNA (mRNA) is transported across the nuclear membrane to the cytoplasm where the instructions it carries in its nucleotide sequence are translated into an amino acid sequence by ribosomes using transfer RNA's (tRNA) to mediate the positioning of the amino acids specified by the instructions in the sequence.

genes, probably accounts for many of the differences between species (King and Wilson, 1975). The organization of the regulatory and structural genes and many other interspersed sequences in the DNA is a second level of genomic organization.

Gene expression can be detected at a number of levels. These levels trace the process of expression starting with the mRNA transcribed from a particular gene; to the protein translated from that message; and then to a simple phenotype such as coloration, which may be a result of expression of the gene, or to more complex phenotypes, like size or fecundity, which result from the combined expression of a number of genes and are referred to as **polygenic traits.**

Mitosis and Meiosis

Packing the enormous amount of DNA found in each very small cell creates some problems in cell division. For example, the armored catfish (*Corydoras elegans*) has about 6 picograms (10^{-12} grams) of DNA in each cell (Hinegardner and Rosen, 1972), which is about 5.5 billion nucleotide pairs or 1.8 meters (6 feet) of DNA. A typical cell is about 10 micrometers in diameter; and the DNA is localized within the cell in a much smaller organelle, the nucleus. Just think of the difficulty you would have unraveling 43 kilometers (27 miles) of very thin spaghetti packed into a 25-centimeter-diameter (10-inch) bowl. Worse yet, while unraveling it you must split it into two pieces lengthwise. This happens to the DNA at each cell division, during which the DNA is accurately replicated and apportioned to daughter cells. This is possible because DNA is organized into separate chromosomes, another level of genome organization beyond the nucleotide sequence and gene order. The DNA of the armored catfish is subdivided among 50 chromosomes (Hinegardner and Rosen, 1972), each of which has about 3.6 centimeters (1.4 inches) of DNA. Chromosomal proteins organize the DNA into highly compacted structures during cell division (Fig. 27–3). Morphologically, chromosomes have **arms** joined at a **centromere,** the location to which spindle fibers attach during cell division. Chromosomes are described by their size and the position of the centromere (Fig. 27–4). Genes are arranged linearly along chromosomes, always in the same order in a species. As a result of this colinearity, a DNA sequence or gene for a particular trait is often conceptualized as its location on a chromosome, its **locus** (plural is loci). Physical maps of the positions of genes on their chromosomes have been developed for well-studied species like mouse, fruit fly, and tomato; and they are being developed for a number of fish species.

Most fish, like most other vertebrates, are diploid (2N), which means that they possess two sets of chromosomes and, therefore, two copies of each gene (locus). During **mitosis,** normal proliferative or equational cell division, DNA replicates so each chromosome, now a pair of DNA molecules, separates and the two new chromosomes are distributed to each daughter cell. Each daughter cell thereby receives a full diploid complement of chromosomes (Fig. 27–5A).

In vertebrates, gamete production (sex cell formation) results from **meiosis,** which is a modified mitotic process. One of the modifications is that meiosis involves two cell divisions. As in mitosis, DNA replication occurs prior to the first division. However, the products of chromosomal (and DNA) replication, called sister chromatids, remain joined at a single centromere during the first division. Unlike in mitosis, the homologous chromosomes (those carrying genes for the same traits) join together at metaphase to form **tetrads** or **bivalents,** which are the four copies of a particular chromosome. During this first division (Meiosis I), the homologous chromosomes (pairs of sister chromatids) are drawn to opposite poles so that homologous chromosomes are segregated. The second division (Meiosis II) is a reductional division (2N to N). During Meiosis II, the centromere of each pair of sister chromatids replicates and each member of the resultant chromosome pair moves toward the opposite pole. The products are haploid (N) gametes (Fig. 27–5B), each of which bears a single complement of genes. When two gametes unite at fertilization to form a zygote, the diploid (2N) complement is restored.

Mendelian Inheritance

The rules for inheritance, or Mendel's Laws, describe the expression of a simple trait specified by a single locus among progeny of diploid parents. The rules reflect the segregation of chromosomes (and their genes) as a result of meiosis. An example of a simple Mendelian trait is albinism in medaka (*Oryzias latipes*) (Yamamoto, 1969). Mendel's First Law describes the behavior of two **alleles** at a single locus. Alleles at a locus are alternative sets of information for a particular trait. Normally pigmented medaka possess at least one allele for normal pigmentation, whereas

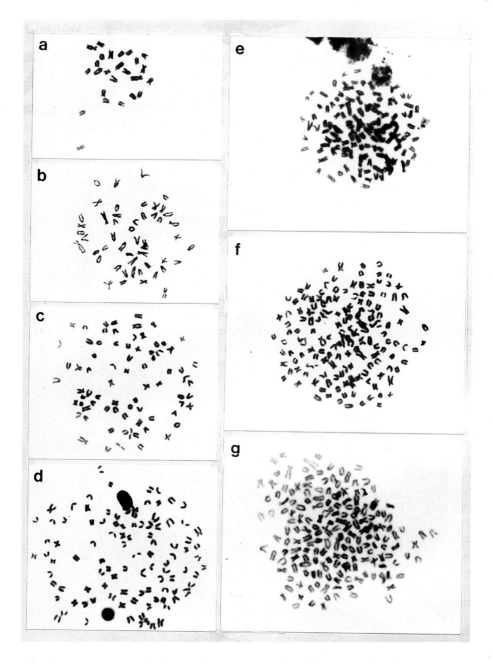

FIGURE 27–3 Karyotypes of metaphase haploid (a), diploid (b), triploid (c), tetraploid (d), pentaploid (e), hexaploid (f), and heptaploid (g) loach (*Misgurnus anguillacaudatus*). The haploid spread is from an embryo, the heptaploid spread from fry cells, and the others from 5-7 month old fish (from Matsubara et al., 1995).

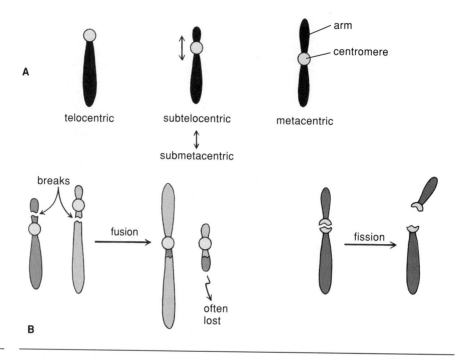

FIGURE 27–4 Chromosomes are categorized according to their configuration (top). They are *metacentric* if the ratio of the longer arm to the shorter arm is between 1.0 and 1.7, *submetacentric* if the ratio is between 1.7 and 3.0, *subtelocentric* for ratios between 3.0 and 7.0, and *telocentric* if the ratio is greater than 7.0. Changes in chromosome number can result from fusion resulting from reciprocal translocations of chromosomes (bottom left) or fission of a chromosome (bottom right). Telocentric chromosomes are counted as a single arm; fission and fusion change the chromosome number without changing the arm number.

albino medaka have alleles that specify no pigmentation. Each individual carries two alleles at a locus, one on each chromosome of a homologous pair. If the alleles in an individual are indistinguishable[1] (e.g., **AA**, **aa**, or **A′A′**), the organism is referred to as **homozygous**. If the two alleles it carries are distinguishable (e.g., **Aa**, **AA′**, or **A′a**), an individual is **heterozygous**. The actual allelic composition of an individual is its **genotype**. The result of the allele combination actually observed in the individual is the **phenotype**. Mendel's First Law predicts that the two copies of a trait in each diploid (2N) parent are distributed equally among its haploid (N) gametes; an offspring has a 50:50 chance of receiving one particular copy.

Often the effect of one allele overrides the effect of the second allele, or is **dominant**; pigmentation in the medaka is dominant over albinism. If the **A** allele is dominant, the phenotype of heterozygous individuals (e.g., **Aa**) will resemble that of

---

[1]"Indistinguishable" may depend on the level of resolution used; that is, how the trait itself is detected. For example, at a color-determining locus, two alleles may be indistinguishable based on the resultant color but may differ in amino acid or DNA sequence. Therefore, an individual may be homozygous for the color allele but heterozygous for the amino acid or DNA sequence alleles.

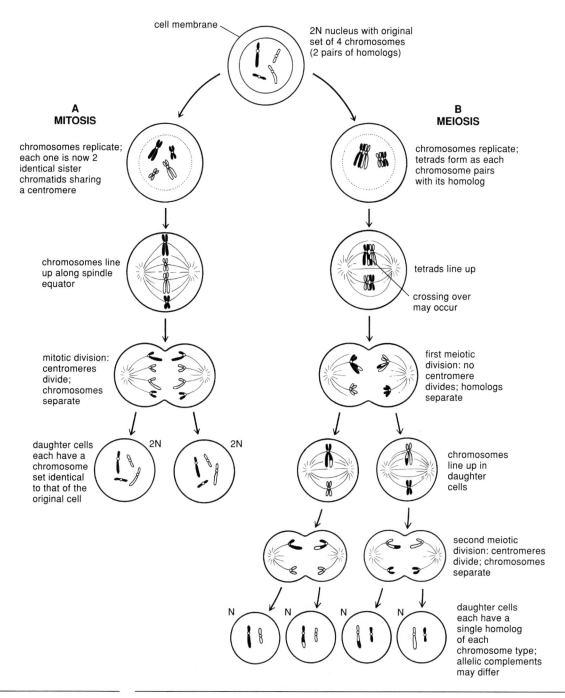

cell membrane

2N nucleus with original
set of 4 chromosomes
(2 pairs of homologs)

A
MITOSIS

B
MEIOSIS

chromosomes replicate;
each one is now 2
identical sister
chromatids sharing
a centromere

chromosomes replicate;
tetrads form as each
chromosome pairs
with its homolog

chromosomes line
up along spindle
equator

tetrads line up

crossing over
may occur

mitotic division:
centromeres
divide;
chromosomes
separate

first meiotic
division: no
centromere
divides; homologs
separate

daughter cells
each have a
chromosome
set identical
to that of the
original cell

2N   2N

chromosomes
line up in
daughter
cells

second meiotic
division: centromeres
divide; chromosomes
separate

N   N   N   N

daughter cells
each have a
single homolog
of each
chromosome type;
allelic complements
may differ

FIGURE 27–5   **A.** Mitosis in eukaryotes is the mechanism that ensures that each product of cell division has a complete complement of information (chromosomes) identical to that of the parental cell. **B.** Meiosis in eukaryotes is the process that reduces the diploid complement of chromosomes and ensures that each gamete has a complete haploid complement.

the homozygous dominant (**AA**) ones. Alleles like **a** are referred to as **recessive** and their phenotype appears only in individuals homozygous for that allele (**aa**). Additional examples of traits exhibiting dominance or recessive expression include coloration in channel catfish (*Ictalurus punctatus*—albinism is recessive), common carp (*Cyprinus carpio*—recessive phenotypes are blue, gold, and gray), guppies (*Poecilia reticulata*—recessive phenotypes are blonde, gold, and albino), and rainbow trout (*Oncorhynchus mykiss*—albinism is recessive); spotting in jewel tetras (*Hyphessobrycon callistus*), green swordtails (*Xiphophorus helleri*), and platyfish (*Xiphophorus maculatus* has at least nine alleles); and spinal abnormalities in guppies and Japanese medaka (*Oryzias latipes*) (summarized in Tave, 1986).

Progeny of matings between an albino (**aa**) medaka (*O. latipes*) and homozygous normal (**AA** orange-red) phenotype carry one allele from each parent and so are genotypically heterozygous (**Aa**) but phenotypically resemble the pigmented parent. Among the progeny of matings between heterozygous fish, all three possible genotypes will occur (**AA** in one quarter, **Aa** or **aA** in one half, and **aa** in one quarter) and both normal and albino phenotypes will be apparent, but in a 3:1 ratio of dominant (normal—**AA, Aa,** and **aA**) to recessive (albino—**aa**) types (Fig. 27–6).

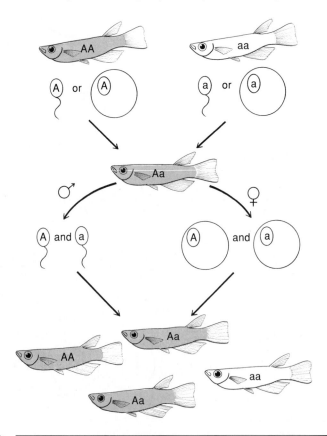

FIGURE 27–6   Inheritance of albinism in medaka follows Mendel's Law of segregation. The phenotypic frequency in the second generation of a cross of a homozygous pigmented fish and an albino is 3:1.

In other modes of inheritance involving a single locus, heterozygotes are distinguishable either because they are intermediate between the two homozygous types (**partially dominant**) or because they exhibit the characteristics of each homozygous type (**codominant**). One example of partial dominance is coloration in Siamese fighting fish (*Betta splendens*); one homozygote is steel blue, the heterozygote is blue, and the other homozygote is green (Wallbrunn, 1958). Normal (**SS**) brown trout (*Salmo trutta*) mated with a fine-spotted trout (**S´S´**) produce offspring (**SS´**) with an intermediate phenotype. Crosses between heterozygous individuals result in one quarter normal (**SS**), one half intermediate (**SS´** and **S´S**), and one quarter fine-spotted (**S´S´**) (Skaala and Jorstad, 1988) (Fig. 27–7).

Variants of proteins that can be detected electrophoretically are an especially useful and widely used group of codominant traits. These allelic variants result from amino acid substitutions (mutations) that create small electrical charge differences between their protein products. The differently charged proteins can be separated in an electric field; the technique is termed **electrophoresis.** Electrophoretically detectable alleles at a locus are referred to as **allozymes.** Allozyme alleles differing by a single amino acid (frequently the result of a single DNA nucleotide change) can often be resolved. Protein electrophoresis ordinarily uses a starch or polyacrylamide gel as a support medium. Histochemical stains specific for the activity of a particular enzyme are used to detect the location of that enzyme on the gel (Murphy et al., 1990). Individuals homozygous for an allele at a locus produce a single electrophoretic band, but heterozygous individuals produce multibanded patterns reflecting the different enzyme products (Fig. 27–8). Whereas no more than a handful of simple Mendelian traits may be available for morphological characteristics in a species, it is relatively easy to identify electrophoretically dozens of protein coding

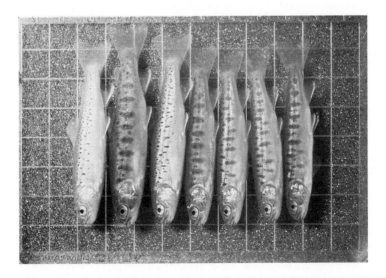

FIGURE 27–7   Inheritance of spotting pattern in *Salmo trutta* exhibits partial dominance; the phenotypes of both homozygous types are partially expressed in the heterozygote. From the left, 1 and 3 are homozygous for fine spotted, 2 and 4 are normal, and 5, 6, and 7 are heterozygous (from Jorstad et al. 1991).

loci. More than 30 loci have been used in electrophoretic studies of a wide variety of fish species, including poeciliids, salmonids, cyprinids, mugilids, clinids, and gasterosteids (Buth, 1984a; Buth and Haglund, 1994; Campton and Mahmoudi, 1991; Haglund et al., 1993; May and Johnson, 1990; Mayden and Matson, 1992; Morizot, 1990; Morizot et al., 1991b; Stepien and Rosenblatt, 1991; Wood and Mayden, 1992).

Mendel's Second Law extends the behavior of alleles at a locus to the inheritance of more than one trait. Mendel observed that what takes place at one locus does

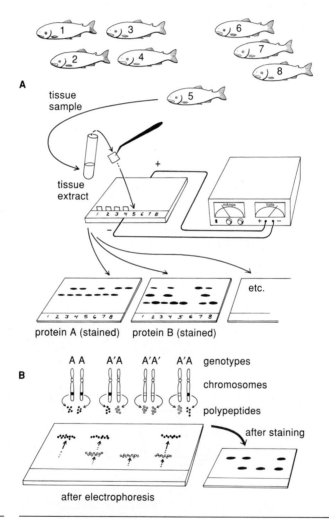

FIGURE 27–8 **A.** In protein electrophoresis genetically determined differences in the expression of proteins are examined. Tissues are sampled from individual fish, cells of the tissues are broken, and samples of the cytoplasm are introduced by filter paper wicks into a supporting matrix of starch or a similar substance. **B.** An electrical field causes the proteins (charged particles) to move through the matrix. Differently charged (genetically different) polypeptides (solid and unfilled circles in the figure represent differently charged gene products) migrate at different rates. The positions of the proteins in the matrix are determined by histochemical stains (from Gharrett and Utter, 1982).

not depend on what happens at the other locus—that is, the loci behave independently. Mendel's First Law predicts that three quarters of the progeny of a cross between individuals heterozygous for two traits (**AaBb**) will have the dominant phenotype (**A-** is **AA** or **Aa**) for the first trait; three quarters of those (3/4 × 3/4 or 9/16 of the total) will also have the phenotype that is dominant for the second trait and be **A-B-.** The other one quarter of the **A-** offspring (3/4 × 1/4 or 3/16 of the total) will be recessive (**A-bb**) for the second trait. The remaining one quarter of the progeny are recessive for the first trait (**aa**). Of the latter, three quarters are dominant for the second trait (1/4 × 3/4 or 3/16 are **aaB-**) and the remainder (1/4 × 1/4 or 1/16) are double homozygous recessives (**aabb**). The phenotypic ratio expected as a result of independent assortment of alleles at two loci (9:3:3:1) applies to traits expressed in a dominant–recessive relationship and would be reflected by the cross-hatched portions of Figure 27–9 if **A′** and

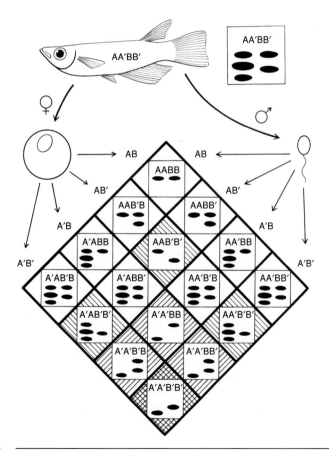

FIGURE 27–9   The expected phenotypic and genotypic frequencies which result from crossing two individuals, each of which is heterozygous for two electrophoretically detectable traits. The genotype and electrophoretic pattern of the parents and each progeny type is shown. One of the electrophoretic patterns results from a monomer and shows one (homozygote) or two (heterozygote) bands. The other results from a dimer and shows one (homozygote) or three (heterozygote) bands. Electrophoretically detectable traits are often codominant. Diagonal lines indicate which progeny would be undetectable if the traits were not codominant and show the 9:3:3:1 ratio predicted by Mendel's Law of independent assortment.

B′ were recessive alleles. Much of the data available from fish, however, involve codominant alleles (for instance, allozymes) that permit us to infer genotypes for individual fish. The genotypic ratios for a cross between double heterozygotes is 1:2:1:2:4:2:1:2:1, which is far more informative than the 9:3:3:1 ratio (Fig. 27–9).

During the past decade, methods have been developed to detect allelic differences in DNA nucleotide sequences. This variation can be detected directly by sequencing target DNA fragments isolated by cloning or amplified by polymerase chain reaction techniques (reviewed in Wright, 1993) or detected indirectly by restriction fragment length polymorphism (RFLP) analysis. RFLP analysis is done by digesting isolated DNA sequences with restriction endonucleases that recognize short, but specific, sequences in the DNA and cleave the DNA. Endonuclease digestion repeatedly produces a set of fragments whose length is determined by the nucleotide sequence. An alteration (mutation) in one nucleotide of an endonuclease recognition site will alter the restriction fragment set. After digestion, the restriction fragments can be separated by size electrophoretically. The positions of the DNA fragments in an agarose or polyacrylamide gel can be detected using DNA-specific dyes, radioisotopes, immunochemicals, or the ability of the fragments to base pair with labeled DNA that has complementary nucleotide sequences. Single-locus DNA variants have been reported for Atlantic salmon (*Salmo salar*), Nile tilapia (*Oreochromis niloticus*), *Xiphophorus* spp., Atlantic cod (*Gadus morhua*), and chum salmon (*Oncorhynchus keta*) (Bentzen et al., 1991; Franck et al., 1992; Harless et al, 1990, 1991; Harris et al., 1991; Taggert and Ferguson, 1990; Taylor et al., 1994; Wright, 1993). Because methodology for examining DNA sequence variation is relatively new, there are not yet many examples of its use in fishes. Access to the entire complement of genetic information and rapid improvement in the methods, however, will make these DNA techniques increasingly valuable tools for studying fish.

If Mendel's Laws were all there were to genetics, we would find them cited as facts (important ones nevertheless) buried in biology texts. In fact, Mendel's Laws describe the normal situation, and the rest of the genetics of inheritance elaborates on these rules, describes the underlying mechanisms, or considers the exceptions.

Gene Mapping

One important exception to Mendel's Laws results because during meiosis the distribution of genes to daughter cells is mediated by chromosomes. As a result, traits that are encoded by genes located close together on the same chromosome are **linked.** Alleles for linked traits generally are inherited as a pair, violating Mendel's Second Law (independent assortment). Linkage can be detected from deviations in the phenotypic or genotypic ratios predicted by Mendel's Second Law.

Recall that although the order of genes on homologous chromosomes is identical, the chromosomes often carry different alleles for those genes. Consequently, for a species to adapt to a dynamic environment, a mechanism that can produce different combinations of alleles on a chromosome is advantageous. In fact, such a mechanism does exist; it is called **recombination,** and it takes place during Meiosis I, when homologs are assembled as bivalents. The likelihood of recombination between two loci depends largely on their physical separation on a chromosome; the further they are apart, the more often recombination will occur between them. The frequency of recombination observed in offspring is used to infer the distances separating the loci. Linkage maps have been made for a variety of fishes, including some salmoniform, atheriniform, and perciform species (May and Johnson, 1990; Morizot, 1990). A recent map for *Xiphophorus* spp. assigns 56 loci, mostly protein coding loci, to 17 linkage groups (Fig. 27–10). Although these loci reflect only a minute number of the genes in fish (estimated to be between 10,000 and 100,000), there are several linkage groups that are conserved not just among piscine species but among vertebrates in general (Morizot, 1990), even after 400 million years of divergence. Recently, the gene maps of *Xiphophorus* spp. have been extended to include loci detected by DNA sequence differences (Harless et al., 1990, 1991).

Non-Mendelian Inheritance

Most traits result from expression of diploid nuclear genes; however, another set of genes—mitochondrial genes—is inherited independently of nuclear genes. The

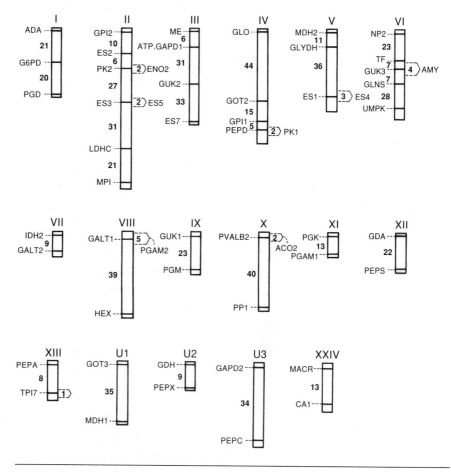

FIGURE 27–10   The linkage map of *Xiphophorus* spp. Abbreviations for protein loci are represented by locus abbreviations. Roman numerals designate linkage groups and Arabic numerals indicate recombination percentages. (Morizot et al. 1991a).

mitochondrion is a subcellular organelle that carries the electron transport apparatus responsible for aerobic metabolism in eukaryotes. Mitochondria in fish possess a DNA molecule, a circle of about 16,500 base pairs or so (reviewed in Meyer, 1993) that specifies several mitochondrial proteins (the rest of the mitochondrial proteins are specified by nuclear genes and transported in from the cytoplasm) as well as some of its own protein synthesis machinery, including rRNAs and tRNAs (Fig. 27–11). Unlike nuclear DNA sequences, mtDNA is haploid (N). Although the number varies considerably, each cell carries about 1000 mitochondria. During reproduction in animals, most or all of the mitochondria received by an offspring come from the yolk of the egg; that is, mtDNA is derived almost entirely from the female parent as a clone.

Mitochondrial DNA comparisons have become an important tool both in phylogenetic and population genetic studies in fish. One advantage is that the rate of change in nucleotide sequences of the mitochondrion appears to be higher than the rate of change in many nuclear sequences (Brown et al., 1979). Other advantages are that mtDNA is a relatively simple molecule (small and haploid) and can be easily isolated. Because there is no known mechanism for recombination between mtDNA molecules, each different mtDNA sequence (clone) is treated as a complex allele rather than an assemblage of alleles.

The methods used to study mtDNA are similar to those used for nuclear sequences, except that a particular nuclear sequence must be resolved from an enor-

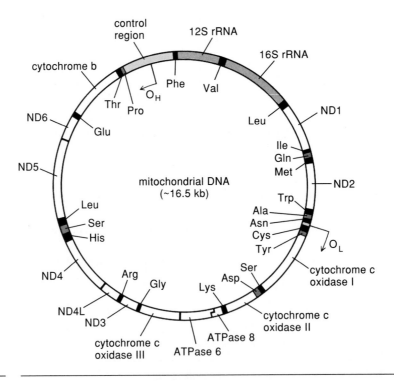

FIGURE 27–11   The mitochondrial DNA genome includes genes for structural proteins, transfer RNA's, and ribosomal RNA's.

mous background of other sequences. The relatively simple techniques used in mtDNA analyses have made the method available to studies of a broad range of fish species including, but not limited to, northern pike (*Esox lucius*) and chain pickerel (*Esox niger*), Atlantic cod (*Gadus morhua*), common carp (*Cyprinus carpio*), American eels (*Anguilla rostrata*), skipjack tuna (*Katsuwonus pelamis*), salmonids (Salmonidae), bowfin (*Amia calva*), American shad (*Alosa sapidissima*), the South American lungfish (*Lepidosiren paradoxa*), and the coelacanth (*Latimeria chalumnae*) (Avise et al, 1986; Bentzen et al., 1988; Bermingham et al., 1986; Carr and Marshall, 1991; Chang et al., 1994; Graves et al., 1984; Herke et al., 1990; Johansen et al., 1990; Meyer, 1993; Meyer and Wilson, 1990; Shedlock et al, 1992; Thomas and Beckenbach, 1989).

Chromosomes

Chromosome and Arm Number in a Species

A constant number of chromosomes is usually characteristic of a species. The chromosome numbers reported for different fish species range from fewer than 20 (16 for the chocolate gourami, *Sphaerichthys osphromenoides*) to well over 400 (446 estimated for the schizothorazine cyprinid *Diptychus dipogon,* which is polyploid) (Table 27–1) (Calton and Denton, 1974; Chiarelli and Capanna, 1973; Gold et al., 1980; Kirpichnikov, 1981; Yu and Yu, 1990). Very small **microchromosomes** are found in the more ancient lineages Agnatha, Elasmobranchii, and Chondrostei. Because of the difficulty in counting them, it is likely that even larger chromosome complements exist.

A characteristic chromosome number may vary when chromosome arms rearrange without changing the amount of genetic material. For example, a metacentric chromosome may result from fusion (translocation) of two telocentric chromosomes, or two telocentric chromosomes may result from fission of a metacentric chromosome (Fig. 27–4). As a result of rearrangement, the chromosome numbers observed in some species vary over a small range although the chromosome arm number remains constant. For example, the diploid chromosome number in rainbow trout varies from 58 to 64, but the chromosome arm number in each case is 104 (Thorgaard, 1976, 1983).

Among closely related species, the chromosome number is usually similar, but an even greater similarity is often observed for chromosome arm number. In Salmonidae the chromosome number varies between 52 and 84, but the range in arm number is much more restricted, ranging between 100 and 108. Other similarities in arm number can be seen, for example, among centrarchids and percids (Table 27–1). Scrutiny of the number of chromosome arms characteristic of a species suggests that many osteichthyan species have about 50 arms, whereas others have about 100. Although the chromosome number or arm number may be similar among species of cyprinids, closer scrutiny reveals centromere rearrangements, which indicate chromosomal differentiation.

Species that have a larger number of chromosome arms also tend to have about twice as much DNA (Hinegardner and Rosen, 1972), which led Ohno (1970) to propose that some lineages arose through the complete duplication of a chromosome set (e.g., 2N to 4N). An event in which the entire chromosome complement is

TABLE 27–1

	Chromosome and Arm Numbers (when Available) for a Variety of Piscine Species			
Taxon	*Order*	*Species*	*2N*	*Arms*
Agnatha				
	Myxiniformes	*Epatretus stouti*	48*	48
		Myxine glutinosa	42*	42
	Petromyzontiformes	*Petromyzon marinus*	164–168*	—
Gnathostomata				
Chondrichthyes				
Holocephali				
	Chimaeriformes	*Hydrolagus colliei*	58*	—
Elasmobranchii				
	Squaliformes	*Squalus acanthias*	62*	—
	Rajiformes	*Narcine brasiliensis*	28	50
		Raja radiata	98*	104
Osteichthyes				
	Acipenseriformes	*Scaphirhynchus platorynchus*	112*	—
		Polyodon spathula	120*	—
		Acipenser naccarii	239±7*	—
	Lepisosteiformes	*Lepisosteus platostomus*	68*	—
	Osteoglossiformes	*Hiodon alosoides*	50	90
	Anguilliformes	*Anguilla rostrata*	38	—
	Clupeiformes	*Alosa pseudoharengus*	48	48
		Clupea pallasi	52	60
	Cypriniformes	*Astyanax mexicanus*	50	90
		Cyprinus carpio	104	168
		Diptychus dipogon	est. 446	—
		N. A.[a] Cyprinidae	48–52	80–100
		N. A. Catostomidae	96–100	—
	Siluriformes	N. A. Ictaluridae	40–72	62–94
	Esociformes	*Esox* sp.	50	50
		Umbra limi	22	44
	Salmoniformes	*Dallia pectoralis*	78	118
		Coregonus spp.	80	92–108
		Oncorhynchus spp.	52–74	104–106
		Prosopium spp.	64–82	100
		Salvelinus spp.	80–84	100–102
	Myctophiformes	*Synodus lucioceps*	46	76
		Myctophidae	48	48
	Gadiformes	*Pollachius virens*	40	50
	Mugiliformes	*Mugil cephalus*	48	48
		Mugil curema	28	48
	Cyprinodontiformes	N. A. Cyprinodontidae	32–48	48–52
		N. A. Poeciliidae	46–69	46–69
		Nothobranchius rachowi	16 or 18	—
	Gasterosteiformes	N. A. Gasterosteidae	42–46	54–78
	Scorpaeniformes	*Hexagrammos octogrammus*	48	48
		Cottus pygmaeus	48	48

TABLE 27–1
(Continued)

Chromosome and Arm Numbers (when Available) for a Variety of Piscine Species				
Taxon	Order	Species	2N	Arms
	Perciformes	N. A. Centrarchidae	46–48	46–48
		N. A. Percidae	48	48–94
		Sphaerichthyes osphromenoides	16	30
	Pleuronectiformes	*Citharichthys spilopterus*	28	48
		many others	48	48

Asterisks indicate the presence of microchromosomes that may be incompletely enumerated.
[a]N. A. is North American.

duplicated is called **polyploidy.** Such an event might result from failure in one cycle of cell division in an early embryo. If the embryo came from parents of the same species, the event is called **autopolyploidy;** if the parents are different species, it is called **allopolyploidy.**

Cell DNA content, chromosomal complements, and duplicated loci that are electrophoretically and immunologically similar but map to different linkage groups (called **isoloci;** Allendorf and Thorgaard, 1984) indicate that the species in the family Salmonidae probably descend from an autotetraploid ancestor (Allendorf and Thorgaard, 1984; Buth, 1983). In contrast, members of the family Catostomidae appear to be descended from an allotetraploid (Ferris, 1984; Uyeno and Smith, 1972). Based on chromosome complements, DNA content, and the presence of isoloci or multiple loci for particular enzymes, other taxa that include polyploid species are some Old World cyprinids, probably the silurids, ascipenseriforms, and possibly a single perciform (reviewed in Buth, 1983). In the oriental loach (*Misgurnus anguillicaudatus*), naturally occurring diploid (50 chromosomes), triploid (75 chromosomes), and tetraploid (100 chromosomes) individuals have been observed (Arai et al, 1991). Haploid (25 chromosomes), pentaploid (125 chromosomes), hexaploid (150 chromosomes), and heptaploid (175 chromosomes) loach have been produced by ploidy manipulation methods (Matsubara et al., 1995) (Fig. 27–3).

Induced Polyploidy

Intentional manipulation of the number of sets of chromosomes characteristic of an organism, **ploidy manipulation,** began early in this century (Ihssen et al., 1990). Ploidy manipulation is accomplished by interfering with normal fertilization or cell division. If cell division is interrupted at an early stage of embryogenesis when cell division is synchronous, it is possible to double the ploidy of the embryo, say from 2N to 4N. Processes that disrupt cell division to increase ploidy include heat or cold shock, pressure shock, and some chemical or antibiotic treatments (Thorgaard and Allen, 1987). All these treatments interfere with spindle formation, preventing distribution of chromosomes to the poles of the spindle apparatus, subsequent cytokinesis, or both. A ploidy increase can also be induced by polar body retention. Treatments that interfere with cell division can also cause a zygote to retain its second polar body. Knowing this, it is possible to manipulate the ploidy in fish much as has been done with plants in agricultural applications.

Common products of ploidy manipulation are all-female stocks, triploid (3N) stocks, and tetraploid (4N) stocks. All-female stocks are produced by stimulating

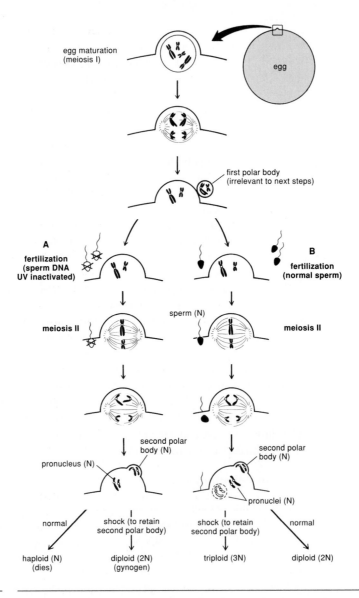

FIGURE 27–12 Chromosome manipulation is accomplished by disrupting cell division at fertilization or very early in development. **A.** Production of all female individuals (gynogenesis) is done by using sperm whose DNA has been inactivated with ultraviolet light or ionizing radiation and shocking the fertilized egg to induce retention of the second polar body produced during oogenesis. The resultant individual has only maternal chromosomes. **B.** Triploidy can be induced by using sperm whose DNA has not been activated.

fertilization with sperm whose DNA has been rendered useless (e.g., by ultraviolet light treatment) without inactivating sperm motility or its ability to recognize a micropyle. After fertilization, the zygote is heat or pressure shocked to induce polar body retention and produce a diploid complement (the sperm's complement was inactivated) (Fig. 27–12**A**). In many fish species, the female has two X chromosomes (sex-determining chromosomes); all the resultant embryos of those species are female **gynogens,** which means that all the chromosomes are maternally derived (Fig. 27–12**A**). An all-female stock can have several advantages in an aquacultural context. For developing a brood stock, the number of eggs is usually limiting. For some species like pink salmon (*Oncorhynchus gorbuscha*), the roe may at times be more valuable than the flesh, so an all-female stock is considerably more valuable. Finally, females often do not exhibit the pronounced secondary sex characteristics observed in males and are of greater economic value. An all-female stock can be maintained by using steroids to reverse the sex of some of the females. In species in which males are heterogametic (see the section on sex determination later in this chapter), these phenotypic males have only X-producing sperm and produce only female offspring.

Triploids are produced in a similar manner, but using normal sperm (Fig. 27–12**B**). Triploid products have one paternal chromosome complement and two maternal complements (Fig. 27–12**B**). Tetraploids are produced using heat or pressure shock after fertilization to block one of the early cleavage divisions. Triploids can also be produced by crossing diploid (2N) and tetraploid (4N) individuals; haploid (N) and diploid (2N) gametes combine to produce a triploid zygote.

Polyploid fish have applications in aquaculture and stocking because fish that possess an odd complement of chromosomes (3N, 5N, etc.) often are sterile or nearly so. An odd ploidy ordinarily disrupts meiotic pairing and may arrest gametogenesis, particularly oogenesis. If oogenesis is partially or completely arrested at an early stage, the individual may not mature on schedule (or at all) and may convert energy ordinarily destined for gamete production into growth (Thorgaard and Allen, 1987). Delayed maturation can lead to a prolonged life in **semelparous** species (which spawn once and die) like the Pacific salmon. Early maturation results in revenue losses from precocious males and deteriorating flesh quality in maturing fish, especially males. To avoid this problem, all-female triploid fish are used in some Atlantic salmon pen-rearing operations. In addition to sterility, triploids may outperform diploids after the age of maturation, when energy ordinarily devoted to gamete production in diploid fish may be directed toward additional growth in triploid individuals. Stocking sterile triploids into barren systems may produce large fish without the concern of their establishing a self-perpetuating population. Herbivorous sterile triploid grass carp (*Ctenopharyngodon idella*) are expected to reduce the abundance of the aquatic plant hydrilla (*Hydrilla verticillata*) in waters infested by this weed in the southern United States without establishing self-perpetuating populations. The hope of producing nonmaturing, trophy-sized fish has stimulated experimental releases of triploid chinook salmon (*Oncorhynchus tshawytscha*) in the Great Lakes. Another important reason for using sterile fish in some applications is that genes of sterile fish cannot introgress into and disrupt wild gene pools.

Hybrids

The nature of the chromosomal complement in a species can be responsible for reproductive isolation. If alignment of chromosomes in a hybrid fails during meiosis because the chromosomal organization of the two parental species differs, the hybrid may be unable to produce functional gametes and will be sterile. The results of hybridization vary from cross to cross (Chevassus, 1983); hybrids between some species of *Xiphophorus* are often fertile (many share the same chromosome and arm number) (Morizot et al., 1991a) whereas other interspecies hybrids are inviable.

Natural hybridization occurs between some sympatric species. Dowling and Moore (1984, 1985) observed extensive hybridization between the cyprinid sibling species *Luxilus cornutus* (formerly *Notropis*) and *L. chrysocephalus* in the Midwest, but because the hybrids are less fit, the species are able to maintain separate gene pools. Hybridization can also occur between endemic and introduced species. In Texas, smallmouth bass (*Micropterus dolomieu*), northern largemouth bass (*M. salmoides salmoides*), and Florida largemouth bass (*M. s. floridanus*) have been stocked in regions endemic to the Guadalupe bass (*M. treculi*). Genetic surveys indicate substantial introgression from the introduced bass species into the endemic Guadalupe bass populations (Morizot et al., 1991b). Rainbow trout have been stocked extensively in western North America. Where native cutthroat trout (*O. clarki*) are endemic, hybridization often takes place. Hybridization that occurs as a result of introductions can jeopardize the genetic integrity of endemic species. If habitat loss or degradation (e.g., logging, dams, urbanization) also affects the species, the combined effect may threaten the existence of the endemic species in that area (Campton, 1987).

Hybridization resulting from ill-advised transfers of fish may jeopardize some native stocks, but some purposeful hybridizations may be useful. As with sterile triploid fish, sterile hybrids have advantages for recreational "put and take" fisheries and aquaculture. Because the increase in abundance of a sterile population of fish is under the control of the manager or aquaculturist, competition for limited food resources can be avoided and fish can realize their growth potential. Use of sterile hybrid sunfish in farm ponds (Childers, 1967) and hybrid tilapia in intensive culture are examples of such an application.

The concepts of hybridization and polyploidization are brought together in triploid hybrids, produced by interspecific fertilization followed by polar body retention. Triploid hybrids, like many interspecies hybrids and triploids, would be expected to be sterile, especially because both sterility-producing processes are involved. In addition, there may be opportunity for combining the advantageous traits of two species. For example, triploid hybrids between pink (*O. gorbuscha*) and chinook (*O. tshawytscha*) salmon result in a fish that has the early sea-water tolerance of the pink salmon, but the size of the fish and its flesh quality are nearer that of the chinook. An added bonus is that the fish has a faster growth rate than either species (Joyce et al., 1994).

Although ploidy manipulation provides a useful aquacultural tool, caution must be exercised before embarking on a program that uses hybrid or polyploid sterile fish. First, it must be determined that surviving fish are indeed hybrids (Chevassus, 1983); reports have been made of triploid, gynogenic, and androgenic (the genome is paternally derived, possibly by multiple insemination) among presumed hybrids. Second, the fish must be unequivocally sterile, incapable of interbreeding with other hybrids or backcrossing with indigenous species. Third, although the hybrids and polyploids may not be able to produce viable offspring, they may still participate in spawning activities and reduce the productivity of the wild population, as has been purposely done to control the Mediterranean fruit fly in California. Finally, the potential ecological effects of releasing a new organism must be considered carefully.

Sex Determination

Most fish species have separate sexes, although there are a number of **unisexual** and **hermaphroditic** species. Unisexual fish are all female and produce clonally (asexually), whereas hermaphroditic individuals produce both eggs and sperm either simultaneously or sequentially. A variety of sex-determining mechanisms have been discovered in fishes.

Chromosomal Determination

The most common form of sex determination results from sex-determining genes carried on particular chromosomes. In humans, for example, females carry two **X** chromosomes and males carry an **X** and a **Y** chromosome. These chromosomes are homologous pairs during meiosis. The **Y** chromosome carries male-determining genes, but the **X** chromosome carries many other essential genes not carried by the **Y** chromosome. Many species of fish (e.g., salmonids) share this means of sex determination, although the degree of physical divergence observed between human **X** and **Y** chromosomes is not observed.

Some fish species share with many avian species an alternative chromosomally determined sex-determining mechanism. In this mechanism, the female has two different chromosomes, referred to as **W** and **Z,** whereas the male has two **Z** chromosomes. This mode of sex determination is observed in some poecilids, some *Tilapia* sp., and the fourspine stickleback (*Apeltes quadracus*).

The sex that produces gametes that carry both sex chromosomes (**X** and **Y** chromosomes or **W** and **Z** chromosomes) is referred to as the **heterogametic** sex. The heterogametic sex can be determined cytologically if the sex chromosomes have differentiated sufficiently (e.g., sockeye salmon, *Oncorhynchus nerka,* and rainbow trout, *O. mykiss;* Thorgaard, 1977, 1978). Another method for determining the heterogametic sex is to make gynogenetic females, reverse the sex of some of them with steroids to produce phenotypic males, and cross the two sexes. If normal males of the species are heterogametic (**XY**), then all the offspring will be females (**XX**$_{♀♀}$ by **XX**$_{\text{phenotypic}♂♂}$ → **XX**$_{♀♀}$ offspring). However, if the females are heterogametic (**WZ**), then the progeny will be a mixture of males and females (**WZ**$_{♀♀}$ by **WZ**$_{\text{pheno-typic}♂♂}$ → **WW**$_{♂♂}$ + **WZ**$_{♀♀}$ + **ZZ**$_{??}$ offspring). The heterogametic sex can also be determined from crosses between normal and sex-reversed fish.

Some species have more complicated chromosome-based sex-determination mechanisms. The freshwater goby (*Gobionellus shufeldti*), for example, has multiple **X** chromosomes (Pezold, 1984), the characiform *Parodon affinis* has multiple **W** chromosomes (Filho et al., 1980), and southern platyfish (*Xiphophorus maculatus*) have a combination of **X, Y,** and **Z** chromosomes (Gordon, 1946).

Environmental Determination

Our ability to reverse sex using steroids suggests that in some species of fish similar changes may occur naturally and may be regulated physiologically (hormonally). In most species of fish, sex is chromosomally determined and the sex ratio is 1:1 because the heterogametic sex produces nearly equal numbers of **X-** and **Y-** (or **W-** and **Z-**) carrying gametes. However, in addition to sex-determining genes on **X, Y,** or **W** chromosomes, some species have autosomal chromosomes that influence sex determination (Kosswig, 1964). The sex of the atherinid Atlantic silverside (*Menidia menidia*) is governed by both the genotype of the fish and the environment (Conover and Kynard, 1981). During the summer, the sex ratio varies. In spring and early summer, 70 to 80 percent of the fish are female; but later in the summer, the proportion is nearer to 50 percent or less. Differential mortality has been ruled out, and experiments in which developing embryos were subjected to different temperature treatments indicate that in silversides, temperature influences sex determination. There also appears to be a genetic component because progeny of different females react differently to the temperature treatments. Another species in which sex determination is influenced by temperature is *Poeciliopsis lucida* (Schultz, 1993; Sullivan and Schultz, 1986).

Hermaphroditism

A number of coral reef fishes are sequential hermaphrodites. These fishes spend part of their lives as one sex and part as the other. The change is usually triggered behaviorally. The most common form is **protogyny,** in which females turn into males (Warner, 1984). At least 14 families have protogynous species, and such species are especially common in the families Labridae, Scaridae, and Serranidae. Some harem-living species like the cleaner wrasse (*Labroides dimidiatus*) have a rigid social hierarchy. This fish feeds by cleaning the skin, mouth, and gills of larger fishes at "cleaning stations." Each male has about five or six females in his harem. When the male is lost, the dominant female rapidly changes sex and takes over the harem. Within 10 days, the new male is actively producing sperm. In species that have a less rigid social structure, like the saddleback wrasse (*Thalassoma duperreyi*), change from female to male is determined by the relative numbers and size of conspecifics in the vicinity (Ross et al., 1983, 1990). Protogyny is often found when being a large male is advantageous. Large males can better defend breeding sites or harems and, as a result, have the opportunity to inseminate many more eggs than they could produce as females.

A less common form of sequential hermaphroditism is **protandry,** a change from male to female. The Sparidae, Pomacentridae, and Muraenidae families have protandrous species. The clownfishes or anemonefishes (genus *Amphiprion*) live near or within stinging anemones, which serve as their protector and provider of food. These fishes are generally found in pairs, a larger female and smaller male. The loss of the female may result in the male becoming a female and joining with a smaller male (Warner, 1984). In social structures in which mates are paired, larger females can produce more eggs, so protandry would be advantageous.

Some simultaneous hermaphrodites have been reported. The hamlets (genus *Hypoplectrus*) found in the Caribbean alternate male and female roles during their mating. Many of the deep sea fishes that rarely have high densities are simultaneous hermaphrodites. Because encountering another member of the species may be rare for such dispersed fishes, hermaphroditism enables these fishes to spawn with any mature conspecific they meet.

All-Female Species

Several species of fishes are exclusively female. Most of these species, such as the Amazon molly (*Poecilia formosa*), belong to the family Poeciliidae, which consists of live-bearers. Unisexual species often arise as interspecies hybrids that have developed mechanisms to avoid genetic recombination (expected in Meiosis I) and keep their ancestral genome intact. Development of ova is triggered by sperm from males of another species (often one of the progenitor species) that are enticed into inseminating the ova. Of course, the male genome does not become part of the zygote. The unisexual Texas silverside (*Menidia clarkhubbsi*) is an oviparous species that analysis by protein electrophoresis suggests probably originated from a hybridization between the inland silverside (*M. beryllina*) and tidewater silverside (*M. peninsulae*) (Echelle et al., 1983).

Quantitative Genetics

Many traits, including life history characteristics such as fecundity, size, and developmental and maturation timing, are subject to natural selection but are determined by the combined action of many loci and influenced by the environment. These traits would be expected to reflect adaptive differences among populations inhabiting different environments. Natural selection on polygenic traits is probably the most important evolutionary force that natural populations experience. Understanding how polygenic traits respond to selection is essential to understanding evolutionary processes. Moreover, these are the kinds of traits ordinarily of interest to aquaculturists. Since early times and long before Darwin and Mendel, humans have been empirical animal breeders who first domesticated animals and then selected for desirable attributes like size, shape, and temperament. Sometimes these efforts were successful and sometimes they were not, but today the legacy of some of their efforts is reflected by the large variety of canine and bovine breeds. Comparable results can also be achieved with fish with the advantage that we now understand the process and have some predictive tools that can help determine the feasibility of selective breeding and the amount of effort that would be required to achieve a particular goal.

Polygenic Traits

Many of the agricultural advances that have been made since the turn of the century resulted from artificial selection. The potential for similar improvements exists for aquatic species, but relatively little has been done to tap that potential. Many of the economically important traits in fishes result from the combined expression of alleles at many different loci and are referred to as **polygenic** or **metric** traits.

Although each locus contributing to a polygenic trait obeys Mendel's Laws, it is not possible to follow its individual effects from parent to offspring because the effects are obscured by actions of the other contributing loci and contributing environmental effects. Polygenic traits can be measured and have statistical properties; the mean and variance of the trait can be estimated for a sample of individuals. As a result, demonstration of the genetic basis for inheritance of a polygenic trait requires statistical analysis (hence the term *quantitative genetics*). The environment also influences the expression of polygenic traits; even monozygotic (identical) twins are not perfect replicates. The ultimate expression (phenotype), therefore, results from the interpretation of the available genetic information (genotype) in the particular environment experienced by the individual:

$$\text{Genotype} \xrightarrow{\text{Environment}} \text{Phenotype}$$

Heritability

The objective of quantitative genetics is to determine the relative importance of the genotypic and environmental influences on the phenotype. For a group of individuals (e.g., a population), the variation observed in the magnitude of a particular trait (V_P) will in part be due to genetic determinants (V_G) and in part to environmental influences (V_E), so we can refine our conceptual relationship into an equation:

$$V_P = V_G + V_E$$

The **heritability** (in the broad sense, H^2) of a particular trait is the proportion of the total phenotypic variability (V_P) that is genetic in nature (Fig. 27–13**A**):

$$H^2 = \frac{V_G}{V_P} = \frac{V_G}{V_G + V_E}$$

Alleles at a locus can have dominant, partially dominant, or recessive effects on a phenotype. In addition, alleles at different loci can interact to produce **epistatic** effects. The overall genetic variation (V_G) accounts for all of these effects, but it can be partitioned into the genetic variation that results from considering each allele as independently contributing an increment toward the phenotype (remember, it is metric) and the variation attributable to dominance effects at a locus or interaction between loci. These are called the **additive** variation (V_A) and the **nonadditive** variation (V_N), respectively. The additive variation represents the portion of the genetic variation that can be predictably used for selective breeding, and the portion of the total phenotypic variation (V_P) that is additive genetic variation (V_A) is referred to as heritability in the narrow sense (h^2):

$$h^2 = \frac{V_A}{V_P} = \frac{V_A}{V_G + V_E} = \frac{V_A}{V_A + V_N + V_E}$$

Two basic approaches are used to estimate heritabilities: **correlation between relatives** and **realized heritability.**

The idea underlying the correlation between relatives is that related individuals possess some alleles that are identical replicates of the DNA sequence, either because one relative provided the genetic information to the other (e.g., parent to offspring) or the information was inherited from a common ancestor (e.g., siblings). Similarities in phenotypes of related individuals (statistical correlations) that are not observed among nonrelated individuals reflect genetic influence. For example, an individual receives one half of its genes from each parent. Therefore, parent and offspring share one half of the same information, and phenotypic similarities between parent and offspring not observed between the offspring and the population in gen-

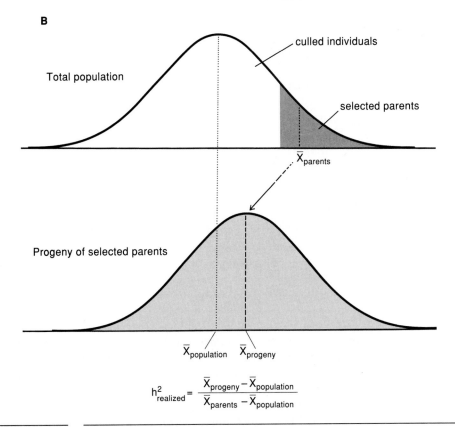

$V_G \{$

V_A

V_N

V_P

V_E

$H^2 = \dfrac{V_G}{V_P}$

$h^2 = \dfrac{V_A}{V_P}$

B

Total population

culled individuals

selected parents

$\bar{X}_{parents}$

Progeny of selected parents

$\bar{X}_{population}$ $\bar{X}_{progeny}$

$$h^2_{realized} = \frac{\bar{X}_{progeny} - \bar{X}_{population}}{\bar{X}_{parents} - \bar{X}_{population}}$$

FIGURE 27–13   **A.** The total phenotypic variability observed in a population can be partitioned into genetic $(V_G = V_A + V_N)$ and environmental (V_E) components. From these components it is possible to estimate the extent to which the traits are determined genetically, that is, are heritable. **B.** Response to selection can be estimated from the change in population mean of progeny produced by selected parents.

eral reflect heritability. Care must be exercised in selecting a particular experimental design because factors, such as egg quality or rearing conditions that produce environments that are common to members of a family but differ between families, can bias results. Heritabilities estimated from these breeding experiments can be used to predict the extent to which breeding individuals at a phenotypic extreme (**artificial selection**) can alter the phenotype of a population.

Realized heritability is an empirical way to estimate heritability. In this method, the mean phenotype of progeny of parents selected from one tail of a phenotypic distribution is compared to the mean of the population in general and the mean of the parents. The relationship of the mean of the progeny to the mean of the population as a whole and the mean of the selected parents determines the realized heritability:

$$h^2_{\text{realized}} = \frac{\overline{x}_{\text{progeny}} - \overline{x}_{\text{population}}}{\overline{x}_{\text{parents}} - \overline{x}_{\text{population}}}$$

If the progeny mean is the same as the mean of the population, it is clear that the variation observed is entirely environmental. If the mean of the progeny equals that of the parents, all the variation observed is genetic (Fig. 27–13**B**). Prediction of change of mean of a population in response to selective breeding can be made by rearranging the equation for realized heritability using heritability estimates from previous experiments.

Heritability estimates have been made in cultured species, including channel catfish (*Ictalurus punctatus*), several tilapia species, mosquitofish (*Gambusia affinis*), common carp (*Cyprinus carpio*), several salmonid species, and the velvetbelly shark (*Etmopterus spinax*), for traits as diverse as size at age, vertebrae number, DDT tolerance, pyloric caecae number, and disease resistance (summarized in Tave, 1986).

Although relatively little directed selective breeding has been done with aquacultural species, most cultured strains have experienced genetic selection resulting from culture. The successful Norwegian Atlantic salmon (*Salmo salar*) broodstock was domesticated from an initial mixture of several Norwegian stocks. Many domestic rainbow trout strains derive from the McCloud River in California, and many farmed channel catfish (*Ictalurus punctatus*) strains originated from the Red River in Oklahoma. Often the domestication process is accompanied by considerable divergence among domestic strains that result from unintentional selection by culturists. One indication of the involvement of genetic selection in the domestication process is that both domesticated rainbow trout and catfish have faster growth rates and higher survivals in the hatchery environment than wild strains do. Domesticated stocks also appear to be more susceptible to anglers than wild fish (Dunham, 1986).

Directed broodstock improvement is possible in intensive aquaculture, in which the environment can be closely controlled. A wide variety of traits hold potential for genetic improvement in an aquacultural setting. Siitonen and Gall (1989) advanced the spawning time of a captive rainbow trout stock an average of nearly seven days per generation for six generations. From breeding experiments, Iwamoto and his colleagues estimated that the heritability for weight during freshwater rearing of intensively cultured coho salmon (*O. kisutch*) was between 0.2 and 0.6. This means that between 20 and 60 percent of the size variation observed in those fish is attributable to additive genetic variation that can be selected for. If we assume an intermediate value, 0.4, and use the largest 20 percent of the fish as breeders, we can expect to increase the average size about 30 percent per generation (Iwamoto et al., 1982).

When one considers the scope of application of selective breeding of agricultural species, it is clear that the potential for genetic improvement of aquacultural species has only begun to be tapped. In contrast to many of the organisms used in agriculture, many wild populations of potentially culturable aquatic species still exist. Each population possesses inherent genetic variability that is essential to the long-term success of the species; and in the near future, that genetic variation also may be invaluable in improving and maintaining aquacultural strains.

Evolution and Systematics

Evolution acts at the population level. Individuals either survive or perish, but the population evolves in response to changes in the environment that alter natural selection pressures. Adaptation of populations to environmental differences as well as random changes in the genetic complement of a population result in divergence among them. The amount and nature of the genetic variation within a population, and by extension within a species, contribute to the resilience of a population or species to environmental changes. Most of the adaptively important characteristics are polygenic traits. The response to selection that can be achieved with artificial selection on a genetically determined trait in a cultured population also takes place in wild populations as a result of environmental pressures. Individuals possessing phenotypes that increase their ability to contribute progeny under prevailing environmental conditions contribute more genes. If the phenotype has a genetic component (is heritable), the population will respond to natural selection by increasing the incidence of the favored phenotype in the population.

Population Genetics

Although most of the traits that are acted on by natural selection are polygenic traits that express complex phenotypes such as life history characteristics, polygenic traits are difficult to study in wild populations. Consequently, the majority of studies of fish population genetics focus on single-gene characters such as those that can be resolved by protein electrophoresis or, more recently, from mitochondrial or nuclear DNA sequence variation.

Although alleles for some of these traits produce phenotypes that are acted on by natural selection or are closely linked to loci that are, many of them (although there is some controversy about the extent) probably have little or no influence on fitness and are referred to as **neutral.** Neutral or nearly neutral traits can reveal information about the movements of fish between populations (**migration** or **gene flow**) and about the extent of genetic diversity existing within and among populations. Reproductively isolated populations tend to diverge genetically over time both as a result of natural selection and random processes (**random genetic drift**), reflecting the error in sampling the alleles that are transmitted between generations. The smaller the population, the greater the chance of missampling. Flipping an unbiased coin provides a good analogy to random genetic drift. Although the chance of observing tails on any one toss is 50:50, in a single toss you only see one or the other because a coin landing on edge is not allowed. If the coin is tossed ten times, you will see exactly five tails only 24.6 percent of the time. At larger sample sizes (more tosses), the chances of observing tails exactly one half of the time decreases because there are many more possible outcomes; but, on the average, the proportion of tails observed gets closer and closer to 50 percent as the number of tosses increases.

Population genetics theory predicts that the divergence among populations will reflect the balance between gene flow (homogenization of populations) and random genetic drift (divergence of populations) for selectively neutral loci. One of the most frequent uses of allozyme data by fish geneticists has been to examine the genetic structure of populations and use the extent of genetic divergence to quantify dispersal and infer colonization patterns. In a study of ten marine shore species of southern California and Baja California, Waples (1987) observed that genetic divergence was inversely correlated with presumed dispersal ability. For example, estimates of gene flow among black perch (*Embiotoca jacksoni*) populations are low, consistent with black perch being a live-bearer and not having marine larvae. In contrast, for the halfmoon (*Medialuna californiensis*), which has larvae that live offshore for a few months and can be widely dispersed by oceanographic influences, estimates of gene flow were high. Estimates of gene flow for species whose larvae are found mostly inshore, like the wooly sculpin (*Clinocottus analis*) and the island kelpfish (*Alloclinus holderi*), were intermediate.

Genetic divergence among populations is also used to infer biogeographic relationships. Included among a wealth of studies are the northern studfish (*Fundulus catenatus*), the threespine stickleback (*Gasterosteus aculeatus*), Pacific herring (*Clupea pallasi*), Pacific salmon (*Oncorhynchus* spp.), bluegill sunfish subspecies (*Lepomis machrochirus machirochirus* and *Lepomis machrochirus purpurescens*), and Tennessee shiner (*Notropis leuciodus*) (Avise and Smith, 1974; Beacham et al., 1988; Buth and Haglund, 1994; Gharrett et al., 1987; Grady et al., 1990; Grant and Utter, 1984; Mayden and Matson, 1992; Utter et al., 1989). Genetically, Pacific herring fall into two obvious races: Bering Sea and northern Pacific Ocean. Grant and Utter (1984) speculate that those differences reflect repeated Pleistocene glaciation along the Alaskan coast, particularly when the Aleutian chain separates the two bodies of water. Divergence is also observed among the widely separated populations of threespine stickleback (*Gasterosteus aculeatus*) found in the Atlantic and Pacific basins (Buth and Haglund, 1994). Another more divergent group includes most of the samples collected from the Japan Archipelago.

Genetics and Fisheries Management

The differences among populations in frequencies of allozymes provide markers that can be used for stock identification. Species of Pacific salmon (*Oncorhynchus* spp.) are intercepted by fisheries in marine waters as they migrate to their natal streams to spawn but before they have segregated into discrete stocks. Management of other species, like the witch flounder (*Glyptocephalus cynoglossus*), is also complicated by mixed-stock harvests (Fairbairn, 1981). One of the goals of fisheries managers is continued productivity of the harvested species, which in turn depends on the productivity of each contributing population. Mixed-stock harvesting can result in different exploitation rates for different component populations. In the extreme, some populations run the risk of severe overharvest or extirpation. For some species, naturally occurring genetic markers exist that can be used to estimate the stock composition of a catch and provide managers with information on stock abundances (Pella and Milner, 1987; Shaklee et al., 1990). When naturally occurring differences do not exist, it may be possible to "genetically mark" a population by altering the allozyme frequencies—for example, by selective breeding (Gharrett and Seeb, 1990; Skaala et al., 1990). Allozyme markers have also been used to assess the success of stocking of walleye (*Stizostedion vitreum vitreum*) (Schweigert et al., 1977).

Although allozymes can be used to obtain information about the genetic structure of populations in a species, local adaptation primarily involves response to selection of quantitative traits like life history characteristics. An example of local adaptation is orientation in the water current of newly emerged sockeye salmon (*O. nerka*) fry. Sockeye usually rear in freshwater lakes for one or more years before their seaward migration. Fry emerging from redds in tributary or outlet streams must navigate to the rearing lake. Experiments show that newly emerged fry from populations that spawn in outlet streams move against the current, whereas fry from populations that spawn in tributaries move with the current (Raleigh, 1967). There also appears to be a relationship between stream temperature and spawning timing in sockeye salmon. Timing of fry emergence is critical to the survival of salmon because the window of food availability is limited; however, the rate of embryonic development, which determines when the fry can emerge, depends largely on the incubation temperature; and a longer incubation period is required in colder water. The obvious solution to the problem is to coordinate spawning timing with the incubation temperature and emergence timing. Natural selection favors phenotypes that adopt such a strategy, and sockeye salmon in the Fraser River system in British Columbia exhibit a strong relationship between spawning time and average incubation temperature (Brannon, 1987).

Conservation geneticists are concerned with maintaining both intra- and interpopulational variation, loss of which poses a threat to a species (Nelson and Soulé, 1987). Of course, genetic changes occur naturally over time as a result of environmental changes; but humans have the ability to cause large changes in the genetic structure of fish populations in a relatively short time as a result of harvest practices, urbanization and population growth, and hatchery practices (Goodman, 1990; Hindar et al., 1991; Riddell, 1993; Skaala et al, 1990; Thorpe, 1993). Erosion of genetic diversity within a species compromises that species' ability to adapt to normal environmental changes and perhaps its existence.

Detection and Study of Species

Speciation takes place when populations have diverged to the extent that reproductive barriers form. In some cases, the barriers are incomplete and natural interspecific hybrids can occur. Even when barriers to reproduction are complete, it may be difficult to distinguish between species. Often allozyme or DNA markers can be found that have fixed frequency differences and can be used to determine unequivocally the species of a specimen. Such markers can be used as a systematic tool (Buth, 1984b) to resolve new species, detect naturally occurring hybrids, or identify larval forms that have not been adequately described for visual identification. The all-female atherinid *Menidia clarkhubbsi* studied by Echelle et al. (1983) is a cryptic species that is found in brackish waters with *M. beryllina* and *M. peninsulae*. The species was recognized and subsequently described because its allozyme profile differs from both of the other species with which it was observed. The observation of fixed differences at allozyme loci in Hawaiian bonefishes (*Albula* sp.) led to a thorough morphological examination in which two species possessing different morphotypes clearly corresponding to the two different genetic profiles were identified (Shaklee and Tamaru, 1981).

Allozymes have contributed valuable information in the study of cyprinids by providing another source of information. The large number of species, the similarity and sympatry of many of them, and occasional hybridization often confuse their taxonomic status. Allozyme data have corroborated separation of the Rio Grande silvery minnow (*Hybognathus amarus*) from the Mississippi River (*H. argyritus*) and Atlantic slope (*H. regius*) forms of silvery minnow (Cook et al., 1992). The level of reproductive isolation between the common shiner (*Luxilus cornutus*) and striped shiner (*L. crysocephalus*) has been estimated from characteristic electrophoretic differences between species (Dowling and Moore, 1984, 1985). Analyses of mtDNA variation provide yet another character set for use in inferring phylogenetic relationships. In some instances, mtDNA variants provide discriminating characters for some cyprinid taxa (Meagher and Dowling, 1991) as well as some of the numerous cichlid taxa that have emerged recently in the East African Great Lakes (Meyer et al., 1990; Schliewen et al, 1994); but in other cases they are inadequate (Dowling et al., 1992; Moran and Kornfield, 1993).

Gene duplication and polyploidy provide species with evolutionary flexibility. For example, because not all loci expressing a particular protein are essential, some of the "extra" loci may reflect species differences resulting from genetic divergence and altered or lost activities. Differences in regulation of genes that cause alterations in tissue specificity for particular loci also can be useful (Buth, 1984b). An example of differential expression between two allotetraploid catostomids, the Tahoe sucker (*Catostomus tahoensis*) and cui-ui (*Chasmistes cujus*), includes fixed allelic differences, the number of genes specifying the enzyme fructose–biphosphate aldolase, and the tissue specificity of genes for nicotinamide adenine dinucleotide phosphate (NADP)-dependent malate dehydrogenase (Buth et al., 1992).

The Big Questions

The focal point of phylogeny (and evolution) is that existing species descended from a common ancestor. The nucleotide sequences in the DNA carry the "operating instructions" for an organism, and descended from the common ancestor, modified by DNA nucleotide changes, insertions or deletions of sequences, and chromosomal rearrangements. Sequence changes accumulate over time, but the "instructions" still reflect the ancestral sequence. Some genes tolerate much less change in sequence than others; that is, they are much more conservative. Because genes carry historic information of a taxon, the nucleotide sequences can serve as a tool for systematics. Most previous phylogenetic work has used structural features (presence or absence of jaws, number of gill slits, location of paired fins, etc.) as characteristics that indicate common ancestry. Ancestral forms preserved in the fossil record provide much of the historic record.

Because the fossil record is often sketchy or nonexistent, many questions about the higher levels of systematic relationships remain unanswered. Careful comparisons of DNA sequences that ultimately provide instructions for the structures may provide a molecular paleontological tool to address some of the questions that cannot be answered with existing fossil and comparative structural data and permit us to examine some of the big questions, such as the origin of vertebrates, the origin of tetrapods, and the relationships between the major groups of fish.

It is not yet feasible to compare the entire DNA sequence of several billion base pairs of two organisms, but it is possible to focus on a limited number of sequences. There is an enormous variety of sequences in a genome. Some genes code for structures that tolerate little modification, and their sequences are strongly conserved. For example, recognizable homology exists between the rRNA sequences of eukaryotes and prokaryotes. A comparison of rRNA sequences of lampreys and hagfishes suggests that they form a monophyletic group (Stock and Whitt, 1992), although morphological characters do not support that conclusion (Forey and Janvier, 1994). Conserved sequences can be used to examine questions involving relationships between distantly related taxa. One such questions is, From what group of organisms did vertebrates evolve? Two recent pieces to that puzzle confirm the position of the amphioxus (*Branchiostoma floridae*) as a protovertebrate. Clusters of genes (*Hox* genes) involved in embryological development are found in both invertebrate and vertebrate taxa. The amphioxus has only a single *Hox* gene cluster, in contrast to four clusters in mammals; but the complexity and organization of that single cluster are similar to that of mammals, and ten of the genes in the *Hox* cluster are homologous to mammalian genes (Garcia-Fernandez and Holland, 1994). Another piece to the puzzle comes from analysis of amphioxus mtDNA. The gene order in mtDNA is nearly the same in all vertebrates that have been examined, but that gene order differs considerably from a variety of arrangements observed in invertebrates (Meyer, 1993). The gene order in mtDNA of amphioxus is very similar to the "consensus" order in vertebrates, except for the positioning of three tRNA's (Szura, Brown, and Gharrett, unpublished data). In addition, codon analysis suggests that the amphioxus uses an invertebrate mtDNA genetic code. Nucleotide sequence studies of other genes and taxa may provide additional pieces to the puzzle.

The origin of the tetrapods is another question obscured by an inadequate fossil record. A molecular paleontological approach examined the relationships among the lungfish, coelocanth, and tetrapods to determine if the South American lungfish (*Lepidosiren paradoxa*) or coelacanth (*Latimeria chalumnae*) was more closely related to early tetrapods (Meyer and Wilson, 1990). This study compared the sequences of the small ribosomal RNA subunit from mitochondria and concluded that the coelacanth sequence was more divergent from the African frog (*Xenopus laevis*) than the lungfish; therefore, it is likely that the lungfish line is more closely related to the line that gave rise to tetrapods than is the coelacanth.

Other Applications

Some genes are less conservative than the rRNA genes. At the other extreme are sequences carried by most eukaryotes, which appear to have little or no function and which can diverge freely as a result of mutation or recombination. The variety of genes and the degree to which their sequences are conserved makes it necessary to choose carefully genes for examining the divergence between two taxa. More conserved sequences would be appropriate for ancient taxa and less conserved sequences for more recent taxa. Some sequences may vary radically among individuals of the same species and provide the basis for DNA fingerprinting, which is used forensically and can be applied in paternity determinations in ichthyological applications as well as for humans. We saw earlier that comparisons of allozyme mobilities may be useful in distinguishing subspecies, such as in bluegills (Avise and Smith, 1974), or detecting closely related species, such as some of the atherinids (Echelle et al., 1983). The latter example is special and may provide an example of speciation that conflicts with our concept of a single common ancestor because this all-female species appears to have emerged in more than one location as a result of hybridization between two similar species. It is important to keep in mind that nucleotide sequence data are just one of several tools that can be brought to bear on a systematic question, and that the results of some sequence comparisons may be inconclusive and may even conflict with results from other sequence comparisons.

The recent revision of the classification of the family Salmonidae (Stearley and Smith, 1993) is a good example of an application of available morphological, paleontological, and molecular data for determining the coancestry of a group of species. In addition to the 119 morphological characters used to examine 33 extant and four fossil taxa, the study included data available for mitochondrial DNA and chromosome complements. One important result of this review was the reclassification of Pacific drainage trouts (formerly *Salmo*) as *Oncorhynchus*.

Some of the gaps in the fossil record may be patched using the living record encoded in the DNA sequences. However, only by using all the tools available will we be able to formulate the most accurate phylogeny of fishes; and even with all of the tools possible, it is unlikely that we will be able to answer unequivocally all the current questions. Moreover, as we address some questions we will undoubtedly expose new ones.

Recent Directions

Fishes as Model Systems

Fishes are the oldest, most numerous, and most diverse vertebrates, but they share many aspects of structure, function, and gene expression with higher vertebrates. Both the similarities and differences between fishes and other vertebrates are sources of insight into the biology of vertebrates. Comparative studies of vertebrate genes and gene complexes improve our knowledge of their expression and function (reviewed by Powers, 1991). Information from fishes has been important for studies of the immune system, of hormones such as vasotocin and insulin, and of the precursor of egg yolk protein, vitellogenin (Chan et al., 1993; Heierhorst et al., 1993; Lazier and MacKay, 1993; Litman et al., 1990; Marchalonis and Schluter, 1990; Pohajdak et al., 1993). From comparative studies, researchers can extract information about the evolution of a gene or gene system, including the variation among genes of different species that may be necessary for appropriate function in different biological or ecological contexts. In addition, since regulation of gene expression depends on the organization of genes in complex expression systems, differences in organization can provide clues to the modes of expression. The comparison of the immune system of vertebrates provides an example in which data from fishes reveal an alternative gene organization. Vertebrates share many features of their immune system that do not occur in invertebrates. Of particular interest are circulating antibodies characteristic of all vertebrates, including the agnaths. The organization of gene complexes that produce circulating antibodies, however, varies. Studies of the horn shark (*Heterodontus franscisci*) reveal a different, and presumably primitive, arrangement of the battery of genes that produce immunoglobulins (Litman et al., 1990). The gene organization of more advanced fishes like the ladyfish (*Elops saurus*), rainbow trout (*Oncorhynchus mykiss*), and channel catfish (*Ictallurus punctatus*) resembles that of higher vertebrates (Pohajdak et al., 1993).

Combined with the similar gene complement and relative ease of culture in some species, fishes serve as important models for some kinds of research. Both genetic and physiological experiments that cannot be done with humans can be performed on fishes. For example, the fishes of the *Xiphophorus* species are particularly important in cancer research. In some crosses, melanomas develop from genes that control pigment pattern (Vielkind et al., 1989). These crosses provide a model for understanding the genetics of tumor production. Other cancer-inducing genes are being studied in fish species, including rainbow trout, Atlantic tomcod (*Microgadus tomcod*), winter flounder (*Pseudopleuronectes americanus*), and goldfish (*Cyprinus auratus*) (reviewed in Van Beneden, 1993).

The genome structures of vertebrates vary enormously from polyploid species, which have multiple copies of genes, to species that are streamlined, with few redundant genes or highly repeated sequences. The pufferfish (*Fugu rubripes rubripes*) has the smallest known genome of any vertebrate, less than 15 percent the size of the human genome (Brenner et al., 1993). Keeping in mind that most of the genes required for their existence and function are shared by all vertebrates, the streamlined genome of the pufferfish offers an excellent model for learning about the organization and expression of vertebrate genomes. Sequencing and analyzing this compact genome would be an effective way for learning about organization in more complicated genomes (for instance, human) that are cluttered with sequences that appear to be superfluous.

Genetic Engineering

The development of molecular techniques makes genetic engineering possible and provides exciting possibilities for developing aquacultural products. An entire industry has emerged based on genetically modified bacteria that produce biochemicals. For example, the human insulin gene has been cloned into bacteria that now produce insulin identical to that produced by humans and that is used to treat diabetics. The process involves introducing a functional gene from an exogenous source and inducing the recipient (or **transgenic**) organism to express that gene. A number of experiments have been conducted to construct transgenic fishes. The results of those experiments indicate that transgenic modification of fishes is possible, but not necessarily easy. Moreover, it is not always possible to predict the outcome of such an experiment.

In a successful and permanent gene transfer, a gene (DNA) must be introduced into the nucleus of a recipient cell and be integrated into its genome, transmitted to its progeny, and expressed in the progeny. Although the process is simple conceptually, there are many challenges to overcome. Introduction is ordinarily accomplished by microinjection or by electroporation, analogous to electrophoresis through the cell membrane (Cloud, 1990; Inoue et al., 1990). Ova or embryos are ordinarily used because there are only one or a few nuclei, the nuclei are accessible, and the tissue has not differentiated irreversibly into cell lines other than the germ line. Unfortunately, there is no control over where the gene integrates into the genome; but expression of the introduced gene depends on where it is inserted in the recipient's genome. Further, integration of a gene into the recipient genome is no guarantee that it will be expressed because gene expression is a result of regulation that ordinarily involves regulatory sequences. Therefore, most genes that are introduced include a **promoter** sequence that, as its name indicates, promotes the expression of the gene. If the gene is successfully integrated, it may disrupt an essential portion of the genome or may not end up in the germ line and be inherited.

The promise of genetic manipulation is the ability to design organisms that have characteristics not found in nature. This promise of genetic engineering also raises serious concerns. Engineered organisms, whether planned or unplanned, may have a competitive advantage over native species and displace them or even disrupt the ecosystem. Because of the unpredictability and the potential adverse impacts, caution must be exercised in developing and testing transgenic fishes (Kapuscinski and Hallerman, 1990).

Growth hormones from various species have been successfully introduced into fishes, and expression of those genes has been documented. In many cases, faster growth has been observed, but much more work is required to ensure that there are no additional unexpected deleterious results, such as sterility, lethargy, and weakness, which accompanied growth hormone gene transfer in swine (Marx, 1988). In recent experiments, growth hormone and a promoter derived from sockeye salmon (*Oncorhynchus nerka*) were spliced together and transferred into coho salmon (*O. kisutch*). The transgenic coho averaged more than 11 times as heavy as non-transgenic fish after a year (Devlin et al., 1994); whether or not the gene is inherited by and expressed in the second filial generation (F_2) has not yet been determined. A gene for protein antifreeze that is synthesized by some species of Arctic fishes, like the winter flounder (*Pseudopleuronectes americanus*), has been successfully introduced into Atlantic salmon (*Salmo salar*) with the hope that it might increase the survival of pen-reared salmon in very cold waters.

The economic potentials of transgenic organisms are boundless; only our imagination limits the possibilities. If we apply our knowledge wisely, during the next few decades we should be able to benefit from natural production and enjoy exciting new aquacultural products.

Behavior: Doing What Fishes Do Best

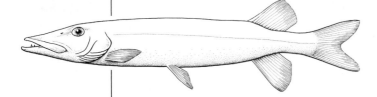

Perhaps an acceptable, working definition of *behavior* would be that it is the sum of all the motor responses of the organism to all of the external and internal stimuli acting on it. These would include locomotion, changes in color or appearance, or secretion of compounds for the purposes of defense or attraction of others. Probably, for many people, the notion of behavior includes those things that we may assume have been acquired in the evolution of ourselves; and when fishes behave, if that is something that they actually do, their actions are rather simplistic and lacking in the subtlety that we mammals possess. If we believe the preceding statement, the next time we are in the bathtub we should try to discern the nuances of the aquatic realm with the senses we possess. We certainly do not have at our disposal a plexus of canals on our surface imparting to our brains an array of electromagnetic or acoustic stimuli. Our ears, although they are partitioned into outer, middle, and inner compartments, do not receive stimuli from our bowels, as they do in ostariophysians. One of the key evolutionary developments in this amazingly diverse group of primarily freshwater fishes is the presence of the Weberian apparatus connecting the swim bladder with the inner ear, thus enabling them to resonate with their surroundings. Obviously, a fish's behavior is the result of the unique way in which it is attuned to its environment, in ways that we could never hope to perceive.

Early humans no doubt were practitioners of behavioral biology, for their lives depended on the ability to optimize their foraging activities like any other organism. To be able to tell when the salmon or suckers were due on spawning grounds would become the focus of significant ceremonial activity among Northwest Indian tribes. Mass migrations of mammals, birds, fishes, and insects are conspicuous examples of behaviors that the ancients held as important knowledge. As our abilities to discern the mysteries of fish behavior developed, we began to understand the importance of such intrinsic actions in the lives of fishes.

The behavior of fishes has been the subject of much classical ethological work (ethology is the study of behavior). Descriptive studies of the complex ritualistic actions of sticklebacks have given us an insight into the biological basis of behavior. In the past twenty years, fish behavior has become an integral component of studies on fish ecology and evolution as we seek to explain the total adaptation of fishes to their environment.

As a means of coping with such a diverse array of intrinsic actions in fishes, we can first group them into several areas of consideration. Our consideration of the topic of responses of fishes to stimuli will focus on locomotor responses to various kinds of stimuli. Such behavioral studies give us insight into the psychology of fishes. Intrinsic behaviors associated with self-preservation in environments as fluctuating as estuaries or headwaters of streams or as consistent and unvarying as the ocean deeps suggest another connection, the role of behavior in the structuring of ecological communities. Special areas of consideration, with direct application to the understanding of evolutionary and ecological adaptations, are those behaviors associated with feeding and reproduction.

Locomotor Responses to Stimuli

We will begin our investigation of the manifold aspects of fish behavior by first considering the various kinds of stimuli that prompt responses (chiefly locomotor) in fishes. Fishes are attracted or repelled by a variety of external environmental cues as well as by internal signals. These result in changes in the motivational state of the animal, thus resulting in some net response observed as motion. The simplest and most predictable locomotor response is termed a kinesis. This involves a simple enhancement of locomotor activity in response to some stimulus. Although most frequently associated with invertebrate phyla, there are instances of kineses recorded among vertebrates. For example, the ammocoetes of lampreys, though blind, have light-sensitive cells in the skin of the tail. The larvae normally live buried in sediment. Removal prompts burrowing or swimming movements that will eventually return the larvae to their proper location away from illumination. The larvae of pelagic bony fishes also exhibit kinetic responses.

Directed, nonrandom movements are termed taxes (singular taxis). Taxes are differentiated from trophisms because the latter term is used to denote slow, growth-related orientations in response to stimuli as witnessed in plants or sessile invertebrates (Arnold, 1974; Lyon, 1904). Taxes can be positive or negative in relation to the stimulus initiating them. Specific taxes to be considered include phototaxis (orientation with respect to light), rheotaxis (current), chemotaxis (chemicals), geotaxis (gravity), and thigmotaxis (contact with substratum or some other object). Given the diversity of habitats occupied by fishes, it would be natural to assume that the habitats differ widely in their influence on the lives of fishes. For example, thigmotaxis would be of little relevance to a solitary, pelagic predator, such as a shark, but would be essential to burrowing forms, such as garden eels (Fig. 28–1).

Phototaxis

Fishes, for the most part, are highly visual creatures. It is therefore not surprising that phototaxis plays such an important role in their lives (see Guthrie in Pitcher, 1986, for a comprehensive review of this topic). Positive phototaxis is apparent in many species of diurnal pelagic fishes in that they will swim toward lights at night.

FIGURE 28–1   A colony of garden eels. Large numbers of these conger eels live together in areas with a fine sand bottom. They face into the current and pick out small organisms from the water flowing by. (Courtesy of Paul Humann; *Reef Fish Identification — Florida, Caribbean, Bahamas,* New World Publications, Jacksonville, FL.)

Some fishery techniques exploit this behavior by displaying lights near nets or pumps to gather fishes as they make the final phototactic response of their lives. Fish harvesters have learned that the response of the target species may vary with the intensity of illumination or that some fishes are more attracted to blue lights than to white. Other pelagic fishes may exhibit negative phototaxis, as in the case of mesopelagic species such as lanternfishes, hatchetfishes, and bristlemouths, which undertake extensive diurnal vertical migrations. This behavior brings large aggregations of these species closer to the surface at night, where their planktonic food sources are more abundant. As ambient light levels increase at dawn, the fishes return to the depths. Negative phototaxis is also associated with the cryptic behavior seen in many species, such as those found in small streams. Here, such species as madtom catfishes (*Noturus* spp.), sandroller (*Percopsis*), sculpins (*Cottus* spp.), and rock basses (*Ambloplites* spp.) retreat from light and spend their days beneath cutbanks, rocks, vegetation, or other suitable cover. Stream pollution in the form of discarded cans or other refuse can often have a positive effect on species abundance by providing a greater amount of cover for such negatively phototactic species (Burr and Mayden, 1982; Kottcamp and Moyle, 1972).

Light also operates as a *zeitgeber* (or "time giver") in the sense that it will be responsible for the entrainment and maintenance of endogenous rhythmicity in fishes, although there also appear to be circadian rhythms that function independently of light (Müller, 1978a, 1978b; Palmer, 1974). Seasonal and daily aspects of reproductive physiology are governed, at least in part, by light in many fishes, especially those in temperate latitudes (Billard and Breton, 1978). North American and freshwater eels (*Anguilla* spp.) on their reproductive journey swim at depths greater than 400 m in daytime but ascend to 50 to 215 m at night (Tesch, 1978). In general, light sensitivity plays a major role in governing the time of day (or night) in which fishes will be active (cf. Glass, Wardle, and Mojsiewicz, 1986). Squirrelfishes (*Holocentrus* spp.) and morays (*Muraena* spp.) are members of the coral reef "night shift" that comes out and forages when most of the reef fish community have retired for the evening. It is interesting to note that the most conspicuous and brightly colored members of the coral reef community tend to be diurnal and that grazing herbivores, such as surgeonfishes (Acanthuridae) and parrotfishes (Scaridae), and cleaners, such as wrasses, also restrict their activities to the daytime hours (Goldman and Talbot, 1976). Many species are classified as crepuscular in that their peak activity occurs at dusk or dawn. This is especially true of piscivorous species (Helfman, in Pitcher, 1986). In a few cases, fishes may seasonally shift from day-active to night-active life styles. This is especially true for those species living at high latitudes, such as burbot (*Lota lota*) and sculpins of the genus *Cottus*, which are nocturnal during the summer months but shift to a diurnal mode in mid-winter (Erikson, 1978; Müller, 1978a). Finally, a brief mention should be made of those species living in the mesopelagic and bathypelagic realms, where ambient light level is very low to nonexistent. Bioluminescence plays a significant role in the lives of these fishes—mate recognition often is a function of perception of species-specific patterns of photophore emissions. The 1989 symposium "Light and Life in the Sea" included several noteworthy contributions on the subjects of bioluminescence, visual systems, and behavior of marine fishes (Herring et al., 1990).

Geotaxis

The posture of most fishes is governed by reaction to the aforementioned light stimuli acting in combination with a sensitivity to gravity. Gravity awareness is mediated through the membranous labyrinth. Fishes will tend to maintain their body position with the dorsal side up, a positive phototaxis of a sort termed the dorsal light reaction. If the light source is shifted to the side, the fish will adjust its body position accordingly (Fig. 28–2). Geotaxis will prevent the fish from achieving extreme angles of inclination, but this can be overridden if the labyrinth is removed. In some cases, fishes in which the inner ears have been removed will orient their dorsal surface to the bottom if the light source is positioned there. Upside-down catfishes of the family Mochokidae contradict these general observations because their normal mode of behavior is to maintain body position with the ventral side up. The unusual orientation of mochokid catfishes has resulted in a reversal of the pattern of obliterative countershading normally seen in fishes. In this case, the ventral surface is darker than the dorsal surface.

Electrotaxis and Magnetotaxis

Among the most mysterious sensory modalities that fishes possess, one that we are unable to appreciate truly (because it is generally lacking in tetrapods) is electroreception. Largely because of the presence of the lateral line, fishes are capable of detecting emitted electrical signals. As such, they display electrotaxis, also known as galvanotaxis. The knowledge that most fishes will swim toward the positive electrode of a direct current field has been used in a variety of ways, particularly in

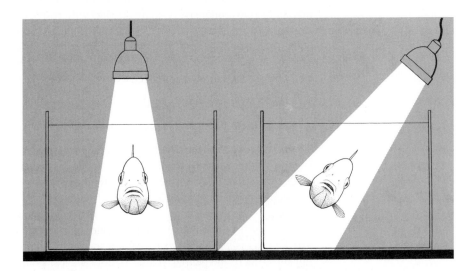

FIGURE 28–2   Representation of dorsal light reaction. When light is directly overhead, the fish assumes a normal upright position, but as the light is moved to the side the fish maintains the same relative position in regard to the light by assuming an oblique posture.

fishery techniques that employ direct current systems, often in conjunction with lights, to attract fishes to nets or pumps.

Although there are only a few species of fishes—such as torpedo rays (Torpedinidae), electric catfishes (Malapteruridae), and the notorious electric eels (Electrophoridae)—that generate an electrical current powerful enough to stun prey, the capacity to emit and detect weak electrical signals is much more widespread, perhaps ubiquitous, among fishes. In some families—the elephantfishes (Mormyridae), gymnarchid "eels" (Gymnarchidae), and knifefishes (Gymnotidae), for example—emission and detection of weak electrical fields is well known as a valuable means of orientation and communication among individuals that often live in muddy or turbid waters, where visual contact is minimal. In the mormyrids, such impulse-generating capacity, found in modified muscle tissues, is associated with an especially large cerebellum that is instrumental in coordination of the system. Remarkably, the electroreceptive capabilities of mormyrids enable discrimination between the ohmic and capacitive properties of an object—detection of capacitive properties is especially useful in exploration of properties of living tissue (Heiligenberg, 1993; von der Emde, 1990).

The capacity of fishes to respond to magnetic fields is a more recent discovery and one that has provoked much debate and discussion. Training experiments have demonstrated the ability of elasmobranch fishes to orient to magnetic fields (Kalmijn, 1977, 1985). Magnetic particles have been found in the snout of yellowfin tuna (Walker et al., 1984) and in the lateral line of Atlantic salmon (Moore et al., 1990). The possibilities are intriguing. Are these fishes, characteristically long-distance migrators, using magnetotaxis to orient their migration in response to cues from the earth's magnetic field? There are many who claim that fishes, as well as other vertebrates (such as birds), are doing exactly that. Elasmobranchs remain among the most intensively studied fishes with respect to magnetotaxis. Although magnetic particles have been detected in the inner ear of some sharks (Hanson et al., 1990), sharks do not appear to utilize specific magnetodetectors but rely on their highly tuned electroreceptors. These provide a means of magnetoreception by detection of electrical currents generated as the fish swims through magnetic fields (Heiligenberg, in Evans, 1993).

Thigmotaxis

The way that most fishes orient themselves spatially in the environment seems to suggest that they usually keep a certain amount of distance between themselves and other individuals. As we shall see in our consideration of schooling behavior, this orientation is often remarkable in its precision and coordination. Spawning behavior will bring schooling species into contact with each other or with the substratum, if only for a short time. A few pelagic species, such as the man-of-war fish (*Nomeus*) and the sargassum fish (*Histrio*), show a special affinity for the different kinds of living substrata of the pelagic realm. Benthic fishes, as might be expected, display a high affinity for particular features of a given substratum, and this is consistent with their cryptic mode of existence. Sculpins, pricklebacks, worm eels, darters, and loaches will lurk among rocks and vegetation; whereas some species, such as skates, weeverfishes, lizardfishes, or flounders, will bury themselves in loose sediments (Fig. 28–3). Sculpins are a remarkable case study in substratum affinity. When

FIGURE 28–3   Sand diver buried in the sand. The combination of cryptic coloration and the habit of burying itself makes this predator almost invisible. From this lair it will dart out and snatch unsuspecting prey. (Courtesy of Paul Humann; *Reef Fish Identification — Florida, Caribbean, Bahamas,* New World Publications, Jacksonville, FL.)

placed in a rectangular glass aquarium lacking in available cover, individuals will squeeze themselves into a corner to achieve contact with as many surfaces as possible. When several individuals are placed in these conditions, they will pile up in an attempt to maximize contact with some other object. Benthic fishes are able to use their lateral-line systems to detect minute vibrations in the substratum that would betray the presence of prey. Mottled sculpins (*Cottus bairdi*) that were experimentally blinded were shown to be able to home in on vibrations elicited by prey. They accomplished this by placing their mandibles on the substratum and moving toward the source of vibrations in a series of short hops accompanied by biting actions directed toward the substratum (Janssen, 1990). Such behavior is obviously of benefit in nocturnal foraging species. Some species, such as madtom catfishes, are observed clustering under a single shelter even when multiple shelters are available. This suggests that thigmotactile behavior may be preferentially directed toward conspecifics as well as the substratum.

Rheotaxis

Reaction to a current of water is commonly observed among stream fishes and is even recognized in some still-water forms (Arnold, 1974; Lyon, 1904). Some species will cruise as a school around a tank without a current, proceeding in seemingly random directions, but will break up and orient as individuals into a current when water is made to flow into a tank. Rheotaxis is an important component of the life history of anadromous species such as salmon. In fact, the orientation either toward or away from the current may be genetically determined. Rainbow and cutthroat trout, obtained either from streams draining into a lake or from those draining out of a lake, were shown to possess different orientations to current that corresponded with their natural habitat. Trout from inlet streams have to migrate downstream to reach feeding areas in the lake, whereas those in outlet streams have to travel upstream to reach the lake (Raleigh, 1971). In the catadromous *Anguilla*, a positive rheotaxis is evident in the elver stage, when the newly transformed individuals first enter fresh water. This facilitates their eventual arrival at suitable upstream habitats. Near the end of the eel's life, a negative rheotaxis helps guide the animal back downstream to the ocean. Ocean currents as powerful as the Gulf Stream assuredly influence the migration of coastal species by providing rheotactic cues. Various species of herrings may under- take contranatant migrations to reach spawning grounds from which the eggs and lar- vae can drift with the current back to the rearing areas.

Knowledge of rheotactic responses in fishes is useful for various harvesting techniques. Carp can be induced to move from one holding pond to another by low- ering the water level in one pond, thus causing a flow between it and another. Rheo- taxis is exploited in salmonid hatcheries to separate breeding stock that tend to exhibit a greater rheotactic response than individuals that are not in breeding condi- tion. Reaction to water currents has also been used to guide fish away from intakes at dams, thus increasing survival of downstream migrants.

Optomotor Response

A phenomenon called optomotor response is, in a sense, a combination of phototac- tic and rheotactic behaviors. Fishes exhibiting optomotor response may hold a posi- tion in a current by getting a visual fix on some object on the stream bottom or side. Thigmotactic responses may also be used because some species may brush their fins against the bottom. Optomotor response is best seen in experimental situations, in which the response is often used to induce fish to swim at a given velocity. A typical test device consists of a circular tank with a transparent outside wall and a circular partition inside to form a narrow swimway in which the fishes can move adjacent to the side. If a circular curtain marked with vertical light and dark stripes is rotated around the tank, test fishes of many species will predictably move with a given stripe or will move at a slightly different speed than the stripes. They will always move in the direction of the stripes, however.

As might be expected, the optomotor response is especially well developed in stream fishes, but it is evident in pike, perch, and other species that inhabit still wa- ters. Smelts and several species of marine fishes, including certain herrings, cods, and jacks, show the response as well (Shaw and Tucker, 1965). Optomotor response is not as apparent in benthic species, and young fish show a stronger response than adults (Harden-Jones, 1963). Fishery scientists are keenly aware of the potential for optomotor response as exhibited in commercially desirable species. Fishes demon- strating this behavior can outpace and eventually avoid trawl nets; this suggests that the most efficient trawling nets might be those that do not contrast with the back- ground, thus eliminating a visual reference point for optomotor response.

Chemotaxis

Given what is known about the sensitivity of the olfactory and gustatory senses, as described in Chapter 22, it is not surprising that chemical responses figure signifi- cantly in the lives of fishes. They are important for the location of food, for breeding and parental care, and for avoidance of danger. The odor of a pike is known to elicit violent swimming, darting to the surface, or tonic immobility in minnows and poeciliids. This reaction appears to be innate and can be elicited in individuals with no previous experience with predatory fishes. At least in certain parts of their range, Pacific salmons show a strong escape response in the presence of the odor of bears, seals, and humans (Brett and MacKinnon, 1954). When extracts of mammalian skin or solutions of an amine that appears to be the stimulus are placed in fishways or natural streams containing migrating adult salmon, many of the fish will abandon their positions and retreat downstream.

Chemical cues are also essential components of the environment that enable both long- and short-term homing behavior in many species. Characteristic of the ostariophysine fishes is the release of a fright substance, or "Schreckstoff," when they are injured or severely stressed (Pfeiffer, 1962). This substance is released from special cells in the skin when it is damaged. It can be sensed in minute quantities and usually elicits flight from the area, tightening of schooling orientation, or concealment (Heczko and Seghers, 1981). The phenomenon is widespread in the Ostariophysi, with some species producing secretions that are highly species-specific in elicitation of response and other species producing substances more generally recognized among members of the superorder (Verheijen and Reuter, 1969). Just as Schreckstoff acts as a stimulus to elicit avoidance behavior, pheromones emitted by fishes may serve to attract potential mates. Their application has been discussed in detail in Chapters 22 and 26.

Homing Behavior

Any or all of the aforementioned taxes may figure significantly in the ability of fishes to seek out and maintain home ranges. This often involves, as in the well-known case of salmon, migrations of several thousand miles. Homing behavior is one of the most extensively researched areas of fish ethology. It is an integral component of the spectrum of migratory behaviors exhibited by fishes (Wootton, 1992). Homing ability, specifically the capacity to return to recognizable habitats, figures significantly in the successful transit from one habitat type to another (Fig. 28–4).

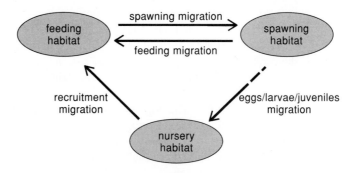

FIGURE 28–4   The role of migrations in the transitions between feeding, spawning, and nursery habitats in fishes. (Adapted from Wootton, 1992.)

Fishes of widely divergent phylogenetic affinities, occurring across a broad spectrum of habitats, have been shown to possess a home range where they spend most of their time. In some instances, this range may be as small as a few meters of stream or a tide pool or as large as several kilometers of coastline or lakeshore. That fishes return to their home territories after being displaced has been demonstrated for intertidal rocky shore species (Dooley and Collura, 1988; Gibson, 1969, 1982; Horn and Gibson, 1988), bass in stream pools (Gerber and Haynes, 1988; Gerking, 1958; Hasler and Wisby, 1958), and rockfishes on offshore banks (Carlson and Haight, 1972). The cues by which fishes orient and navigate must include a combination of physical and chemical entities as well as simple random search patterns. Searching along the shore of a small lake would eventually take a fish home, but this does not account for movement toward the home range across a large lake or from one rock reef across deep ocean water to the home reef. Whereas displacement experiments using marine shore fishes typically demonstrate homing capacities over fairly short distances, some species, such as the cunner (*Tautogolabrus adspersus*) and the yellowtail rockfish (*Sebastes flavidus*), will return to home sites from distances of 4 and 22.5 km, respectively; these return journeys often involve transits across deep water (Carlson and Haight, 1972; Green, 1975). Inhabitants of rocky intertidal shores are typically benthic species with very limited home ranges (Ralston and Horn, 1986), yet they exhibit the ability to return to home areas even when displaced considerable distances. Tidepool sculpins show an ability to return to home areas when displaced up to 100 m (Green, 1971; Horn and Gibson, 1988). Shallow water blennioids of the Canary Islands were able to locate home areas when displaced up to 200 yards (Dooley and Collura, 1988). Cunner, yellowtail rockfish, and possibly tidepool sculpins (*Oligocottus maculosus*; Green, 1971) demonstrate homing capabilities even after being held in artificial surroundings for several months.

A Case Study in Homing: The Salmon. Studies on the seasonal movements of animals should consider two different kinds of causative agents: proximate causes and ultimate causes (Lack, 1954; Honore and Klopfer, 1990). Proximate factors are those that trigger and guide the migratory response. We shall consider these in the context of perhaps the best-known examples of homing as a correlate of migratory behavior in fishes—the salmons of the genera *Oncorhynchus* and *Salmo*. What morphological and physiological mechanisms do fishes have at their disposal to interpret the environment and make appropriate responses such that return to home streams is possible after spending several years in the open ocean? By ultimate factors we mean those historical or selective factors of the environment that result in the establishment and maintenance of homing and/or migratory behavior. What caused the development of such remarkable homing capabilities in salmon?

Proximate Factors. The migratory behavior of both anadromous and catadromous fishes takes them over the breadth and depth of both the freshwater and marine realms. As expected, a variety of environmental cues can be perceived by species such as salmon. These cues enable navigation over great distances of the ocean by salmon until they find themselves in the vicinity of their natal streams. They then ascend these streams, making the right choices at each tributary until they eventually return, often to the same stretch of gravel bed that they emerged from as alevins. It is apparent that young salmon acquire an "imprint" of their natal stream sometime during their migration downstream to the ocean. Fishes probably use their powerful olfactory capabilities here such that the most recognizable feature of the natal environment is its smell. Researchers have demonstrated that salmon react to minute quantities of a chemical, morpholine, dissolved in the water. When fishes are artificially imprinted with this compound, it can be used to guide their movements to particular streams (Hasler and Scholz, 1978). Olfactory imprinting was first advanced by A. D. Hasler and W. J. Wisby (1951) as a means of orientation. Since then, it remains the most likely explanation for how fishes are able to discriminate different tributaries during their homeward migrations. When Atlantic salmon were released directly into the ocean and thus deprived of olfactory cues that they would normally acquire during their downstream migration as smolts, they failed to return to a particular river (Mills, 1989). The mechanism that is most supported by the data on homing in streams is one in which imprinting is a sequential process by which the smolt receives a continuous set of cues as it makes its way to the ocean. These then provide a continuum of reinforcing signals when the fish makes its return several years later as an adult. Other cues may prove essential in triggering upstream migration. For example, as early as the middle of the last century, it was known that Atlantic salmon do not initiate upstream movements until the stream has been freshened by a heavy fall of rain (Williamson, 1843, in Mills, 1989).

In addition to olfactory cues, salmon on the high seas may employ a host of other cues, including water currents, temperature and salinity gradients, position of the sun during the day or the stars at night, the pattern of polarized light in the daytime sky, and variations in the earth's magnetic field (Wootton, 1992). Researchers have demonstrated that fishes can perceive polarized light and that specialized cone cells in the retina that are sensitive to ultraviolet light appear to be involved in this process (Hawryshyn, 1992; Land, 1991). Some researchers have suggested that salmon do not require sophisticated orientation mechanisms on the open ocean; by simply engaging in a more or less random search pattern, they will eventually locate the appropriate coastline that will take them to the outlet of their home stream.

Ultimate Factors. Although we may never be able to discover just how migrations and homing behavior developed as they did in salmon, we can speculate on why they evolved as they did. Fishes evolved migratory behavior and the consequent homing capacities for a variety of reasons (McKeown, 1984). Fishes may migrate as a means of availing themselves of highly productive feeding areas located at some distance from natal areas. By such means, fishes were able to respond rapidly to food resources that may have been seasonal in their availability. Fishes, especially those in temperate to polar regions, may migrate as a means of avoiding unfavorable conditions. In addition to unsuitable features of the physical environment, such conditions may also include the arrival of predators. Bluegill sunfish undertake seasonal migrations offshore in order to exploit available vegetation as cover from predators (Goodyear and Bennett, 1979). Migrations may serve as a way to optimize reproductive success. Much of our concern over degradation of coastal estuarine environments stems from the dependence of many coastal species on these environments as nursery areas. In such places, egg and larval stages may be ensured a greater measure of security at critical times in the life history of the species. We must also consider the geologic history of the earth's surface in our speculations on the origin of migrations, especially in the case of salmon. Habitat quality may change over time such that fishes may be compelled to seek alternative habitats. It is generally agreed that Pleistocene glaciation in the northern hemisphere was a significant factor in influencing the development of migration patterns of salmonids and may even be the impetus behind colonization of the marine environment. Subsequent adaptation to the feeding opportunities of the marine realm may have set salmon on their current evolutionary "course"—one in which they developed their typically anadromous life style even after deglaciation meant that their ancestral homes in fresh water were again available for habitation. In the evolution of migratory behavior, salmon (as well as other fish species) made use of their inherent sensory capacities—especially visual, olfactory, and acoustic senses—in order to orient in the continuum of environmental change experienced along their migratory routes.

Social Behavior in Fishes: The Individual, Small Groups, Shoals, and Schools

Behavior of Individual Fishes

The fundamental social unit in fishes, as it is in all organisms, is the individual (Keenleyside, 1979). Individual behavior is best investigated in the context of specific activities, such as feeding and reproduction. In the pursuit of these and other activities, fishes will, as a matter of course, be brought into contact with other organisms. In recent years we have witnessed the increasing significance of two disciplinary approaches to the study of behavior: behavioral ecology and sociobiology. We will investigate the behavior of individuals in the context of the former—that is, how does the behavior of an individual fish contribute to its adaptive capacity, its ability to exist in a given environment? One of the principal proponents of sociobiology, Edward O. Wilson, defines it as "the systematic study of the biological basis of all social behavior" (Wilson, 1975, p. 4). It is in this context that we will attempt to evaluate the nature of those behaviors that fishes engage in when interacting with members of their own species or other species.

Feeding Behavior. Compared to endotherms, the energy requirements of fishes are much lower and the onset of hunger a more gradual process. Feeding motivation appears to arise from a combination of intrinsic contributions, including gut fullness and metabolic balance (Colgan, in Pitcher, 1986). The significance of food as a motivational force is obvious in the amount of behavioral research in which food is used as a reinforcement for learning. Given the diversity of form and function among fishes, it is obvious that feeding behaviors are as diverse as their practitioners.

The behavioral "strategies" associated with individual feeding patterns naturally represent an application of the specific morphology associated with a particular feeding type. Although the range of feeding types has been thoroughly explored in Chapter 25, a few of the more remarkable behaviors associated with feeding deserve mention here.

As mentioned in Chapter 25, macrocarnivory can be expressed in a diversity of fishes that encompass a range of body forms, including fusiform pursuit predators, such as the pelagic sharks and scombroids, as well as ambush predators, such as the cryptic and sessile ceratioids (Fig. 12–2**B**) or the swift, dart-shaped gars and esocoids (Fig. 28–5).

What may seem, at first glance, to be a simple act of grabbing a large chunk of food is for macrophagous forms such as sharks, actually a complex sequence of behaviors associated with detecting, grasping, and dismembering prey (Bres, 1993; Frazzetta, 1994). White sharks (*Carcharodon carcharias*) have been observed to practice a "bite and spit" behavior in which the prey is released after an initial bite. This apparently promotes death by blood loss, after which the shark returns to finish its meal (Klimley, 1994; McCosker, 1985). Engana and McCosker (1984) suggest that this is the reason why North American divers have a higher rate of survival from shark attack than do their Chilean counterparts. Divers in North America are more conscientious in their practice of the "buddy system"—a companion can effect a rescue during the 5- to 10-minute interval before the shark returns. The lesser fat content of humans as compared to seals and sea lions also apparently makes them a less palatable prey item. White sharks also appear to be somewhat sensitive to prey freshness and will refuse decomposed carcasses (Klimley, 1994). Other large predators, such as tunas and dolphins, often forage below the deep scattering layer of the ocean—a concentrated band of vertically migrating marine organisms. These large predators were occasionally found with wounds of a curious origin—almost perfectly cylindrical holes about 1 cm in depth, as if made by a cookie cutter. The culprit was determined to be a miniature shark, *Isistius brasiliensis*, which has since become known as the cookie cutter shark. The dreaded piranha feeds in much the

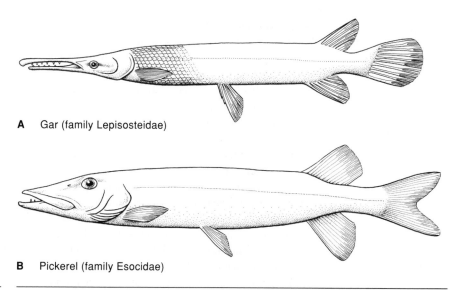

A   Gar (family Lepisosteidae)

B   Pickerel (family Esocidae)

FIGURE 28–5   Diversity of body form in ambush-type predators. The gar and pickerel are mid-water forms that rely on a fusiform shape with medial fins placed well to the posterior. This enables quick forward darts often coupled with equally rapid lateral swipes at their prey. Compare the body form of these types of ambush predators with that shown by anglerfishes that possess devices to lure prey into their vicinity (Fig. 12–2**B**).

same fashion, taking bite-sized chunks from larger prey. Even herbivores may employ such grasping and tearing behaviors. The monkeyface prickleback, a large blennioid, feeds by grasping blades of seaweed and, with corkscrew-like motions, twisting them off rocks to which they were attached. Some species have large, molariform teeth, which enable them to crush otherwise indigestible food items. Another blennioid family, the wolf eels (Anarhichadidae), is able to feed on hard-shelled invertebrates such as bivalves, crustaceans, and sea urchins because of this kind of dentition; whereas parrotfishes are able to bite off chunks of coral with sharp, beaklike incisors and crush the ingested material with powerful pharyngeal teeth. Parrotfish grazing is easily detected from the characteristic scars that they leave on the coral reef.

For many species, feeding actions are facilitated by the generation of a vacuum in the immediate vicinity of the oral cavity. By such means, food can be brought into the mouth by a sudden enlarging of the oral and buccopharyngeal cavity. Many microcarnivorous or planktivorous species feed by simply opening their sometimes enormous mouths as they pass through dense concentrations of food. In these, the gill rakers serve as filters to trap small prey items. Such feeding behavior is observed in species as small as the anchovies and herring or as large as the paddlefish. The largest of all fishes are the microcarnivorous basking sharks (Cetorhinidae) and the whale sharks (Rhincodontidae). In 1976, a most remarkable shark was discovered in deep waters off Hawaii when it became entangled in a parachute deployed as a sea anchor on a research vessel. The animal was rediscovered under similar circumstances off the coast of California in 1984. This extraordinary fish (*Megachasma pelagios*), dubbed "megamouth," is distinguished by a large (4.5 m), relatively flabby body and an enormous head terminating in a wide mouth lined with reflective crystals, which may be associated with bioluminescent capabilities. It is surmised that this creature feeds by opening this cavernous mouth and drawing in hordes of tiny, deep-ocean crustaceans attracted by its blue-green glow (Taylor et al., 1983).

The marvelous sensory arsenal of fishes generally allows them a large degree of flexibility in their feeding behavior. Certain environments and habitats favor the use of some senses over others, but usually most senses are functional to some degree and available for use. A fish might be alerted to the presence of potential food through the acoustico-lateralis system, move in the direction of the food by following a scent, and finally capture the food by perceiving visual, gustatory, touch, or electrotactic stimuli.

Visual detection of prey is of great importance in the feeding of most fishes and depends on a number of factors. An important consideration is the nature of the optical mechanisms in fishes, particularly the visual field. This involves the placement and mobility of the eye; visual acuity, which is a function of the size of the eye (larger eyes usually mean sharper vision); and brightness and color discrimination, which is associated with contrast and color perception. In the case of color perception, a correlation has been observed between development of cone cells in the retina of young salmon and trout and the onset of feeding behavior (Noakes, in Burghardt and Bekoff, 1978). Environmental factors are significant in determining the effectiveness of visual prey detection. Feeding may be suppressed when waters become turbid as a result of rainfall or wave action. Most fishes feed during daylight hours, and their activity diminishes as light intensities approach 10^{-1} lux. Some species are crepuscular, feeding during twilight and even at intensities of 10^{-3} to 10^{-5} lux (Blaxter, 1980). J. R. Brett, in his classic work on the energetics of sockeye salmon, demonstrated that growth is optimized by a daily cycle of vertical migration, which brings young sockeye to the surface of the lake to feed at dusk and dawn. When not feeding, the salmon conserve energy and thus maximize growth potential by remaining at depths below the thermocline (Brett, 1971). The ecological application of certain behaviors is seen in another example, that of longnose dace (*Rhinichthys cataractae*). Whereas most species of minnows are diurnal or crepuscular feeders, dace have, as a means of optimizing prey consumption and avoiding salmonid predators or cyprinid competitors, shifted to nocturnal foraging (Culp, 1989).

Olfaction enables fishes to detect food beyond the limits of vision, and it is of primary importance to bottom-feeding species such as catfishes, loaches, and eels. The olfactory capacities of sharks and their relation to feeding have been discussed earlier. Even such visual feeders as tunas will orient positively to the direction of food odors (Atema, 1980). Fishes often react to food odors by moving upstream toward the source of the odor and then initiating a search pattern until the source is approached. Taste receptors located externally, such as those on the barbels of catfishes, or vision can aid in the final location and capture. The lateral-line system may also assist in near orientation to food by detection of electrical fields or acoustical disturbances emitted by the prey.

Quests for food may involve active searching over a large area, as in pelagic predators, or may be confined to a restricted home range. Some species will occupy stations in the vicinity of food sources and await their chosen prey. Some species, such as anglerfishes, facilitate this by presentation of lures that cause prey to approach (Fig. 12–2**B**). Feeding is usually not a continuous activity, however. There may be definite periods during the day (or night, in the case of nocturnal feeders) in which feeding activity is concentrated. As mentioned earlier, light levels may be important cues signaling the onset of feeding in diurnal or crepuscular feeders. Many species of fishes living in shallow estuaries will engage in feeding activities that correspond to the rise and fall of the tides. Staghorn sculpins (*Leptocottus armatus*) can be observed foraging immediately behind the advancing tide line in water that is just a few centimeters in depth. Diurnal or tidal rhythmicity in feeding behavior is entrained in some species such that individuals will display rhythms in activity and physiology even when held in a constant environment. There are cycles of feeding activity attuned to the seasons and life histories of fishes. Fishes of temperate waters usually maximize their food intake during temperatures that are optimum for growth, reducing their feeding during the winter and during the hottest part of the summer. For headwater species, summer may coincide with intermittent flow conditions, when fishes become concentrated in the remaining pool environments. Cessation of feeding at this time may be an accommodation to the restricted feeding opportunities imposed by the environment.

Fishes with Other Fishes: Social Interactions

Two assumptions are fundamental to the sociobiological approach to studying the behavior of organisms: first, that behavior is included among the heritable traits, and second, that individual fitness is measured, not in terms of the survival of the individual, but in the survival of its genes. The adaptive fitness of a certain gene pool will then be a function of the extent of successful interaction among the individuals contributing to that gene pool. This is what sociobiology seeks to explain: To what extent does the social behavior of the organism contribute to its adaptive capacity? For this purpose, it is important that we take up the subject of the social behavior of fishes in more detail.

We can begin by continuing our discussion of feeding behavior and focusing on its social aspects. Indeed, feeding may be facilitated by group foraging behaviors. We will have much more to say about the nature of schooling behavior later, but for now we can recognize the benefits of such behavior in foraging. Schooling by predators may make the rounding up of prey an easier task. Some bottom-feeding species may cooperate in dislodging and removing objects that individuals alone could not tackle successfully. There is some evidence that suggests that fishes are able to identify which among them are the best at locating food. Studies on foraging behavior of bluegill sunfish (*Lepomis macrochirus*) reveal that individuals prefer to associate with those with which they have had the greatest success in feeding in the past (Dugatkin and Wilson, 1992). Game theory models of animal behavior are based on the assumption that animals are capable of strategic behavior as a means of optimizing adaptation to the environment. Fishes, in being able to recognize individuals and distinguish their "value," are certainly capable of such behavioral flexibility.

Group interactions may facilitate feeding in schooling species to the extent that some species may be able to communicate food resources by visual signals, such as color changes, or by other actions. In other species, feeding is such a solitary activity that conspecifics in the immediate vicinity are always regarded as intruders. Herbivorous damselfishes (Pomacentridae) will cultivate gardens of algae among coral reefs and zealously guard them against intruders. In this case, we can see how the distribution of the primary production in a given environment can be the consequence of such territorial behaviors.

Communication: Signals and Social Behavior. Implicit in our discussion of feeding behavior is the acknowledgment of the importance of communication. All behaviors of fishes that bring them into contact with other species will result in the elicitation of specific modes of communication. Just what is the "language" of fishes? To what extent are we able to translate it? Fishes, being mainly visual creatures, depend primarily on the sense of vision for the cues, signs, and signals that trigger and maintain social behavior, but other senses can be significant or even more important than sight. Visual signals can range from those as simple as perception of the characteristic form and color of a given species to complex ones that involve specific actions, such as postures, dances, approaches, or flight. Some of these movements are designed to display specially colored parts of the body, including fins, bellies, mouth linings, and opercular or branchiostegal membranes. In deep ocean fishes, the presentation of bioluminescent flashes may occur as indicators of the presence of a given species. In some species, these colors or patterns are present only at breeding season or at other significant times of the year. A good example is that of the stoneroller minnow (*Campostoma anomalum*), one of the most common cyprinids in eastern central North America. Breeding males of this species undergo a dramatic modification of body morphology, including enlargement of the head, acquisition of breeding tubercles, and development of bold coloration (Fig. 28-6). In others, the color may be apparent but may not be displayed until required by breeding or other behavior. Reef dwellers, such as the wrasses (Labridae), and damselfishes are among the most conspicuous of fishes because of their brilliant coloration. They frequently show dramatic changes in these coloration patterns at different stages of their life histories.

Although significant colors and patterns are a seasonal phenomenon in some species, other species change signals to suit the activity or "mood" of the moment: paling in fright; darkening in anger, with sexual motivation, or extreme physiological stress; or changing from stripes to spots according to the stage of feeding activity. Some cichlids can display a fright pattern within a second of receiving the stimulus. A variety of patterns, including stripes, ocelli, and bars, may be used. The Asian fish *Badis badis* has about a dozen patterns that are used to signal behavior, but sometimes activity changes faster than the patterns such that there might be a lag between actions and the color signals associated with them (Barlow, 1963). A sand tilefish from the Philippines was observed to exhibit 24 changes in color in 15 seconds (Klauswitz et al., 1978). Ocelli are among the most intriguing of color patterns.

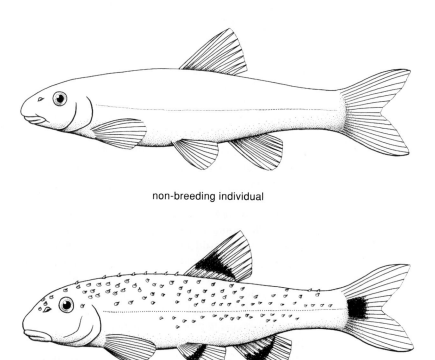

non-breeding individual

breeding male

FIGURE 28–6   Sexual dimorphism in stoneroller minnows of the genus *Campostoma*. During the breeding season, males develop much more pronounced coloration, a more pronounced head profile, and tubercles scattered over the body surface.

These "false eyes" have long been held to be directive marks serving to direct predator attacks away from more critical areas of the body. Reversing the perceived profile of the fish may be beneficial in other ways. Young piranha are known to engage in fin nipping on other species. Kirk Winemiller has demonstrated the predator-deceiving capabilities of ocelli in tests in which cichlids with ocelli near the base of the caudal fin experienced fewer attacks on their fins by young piranha that those species lacking ocelli (Winemiller, 1990; Winemiller and Kelso-Winemiller, 1993).

Mention has been made (Chapter 19) of the significance of sounds in the life of fishes. Threats, warnings, and signals to prospective mates can be produced by vibrations of the gas bladder, movement of one hard part against another, or by other mechanisms. Electrical signals during certain behaviors are known in fishes with active electrical systems. Gymnotoids include bursts of electricity in their repertoires of aggressive activity, along with serpentine motions, head butts, and bites. Dominance hierarchies have been shown to be correlated with characteristic electric organ emission spectra in the central American gymnotoid genus *Hypopomus* (Hypopomidae) (Hagedorn and Zelick, 1989). Certain types of schooling behavior disappear in mormyrids when nerves to the electric organs are cut (Moller, 1976).

Chemical signals may be incorporated into the behavioral repertoire of most fishes, but they have not been studied to the extent that has been accorded other means of communication. The bulk of the studies have focused on the remarkable secretions of ostariophysans, such as the Schreckstoff mentioned earlier. Bullhead catfishes (family Ictaluridae) have developed a complex hierarchy based on characteristic odors of individuals and their ability to release pheromones that communicate social position to neighbors. Such chemical signals communicate peaceful coexistence among these normally gregarious fishes. When water from a tank in which a peaceful group coexisted was added to another tank in which agonistic behavior was present, this behavior was terminated and the apparent perception of chemical signals from a peaceful aggregate was sufficient to instill similar behavior in an unrelated group (Todd, 1971; Todd et al., 1968).

Reproductive and Parental Behavior. The subject of reproduction is one of great complexity, embracing aspects of morphology, physiology, as well as behavior of the individuals considered. With respect to behavioral aspects, a chronology of activities can be considered. First, we might investigate behavioral phenomena associated with the onset of sexual maturation and ripening of the gonads, which, we assume, occurs simultaneously in males and females and at a time propitious to the survival of the eggs and larvae. Behaviors will then be directed toward the selection of suitable breeding grounds, preparation of the site for breeding (this can be a very complex series of actions for some nest-building species), courtship and spawning activities, and finally behaviors associated with the care and maintenance of the fertilized eggs and larvae. Recognizing the tremendous diversity of fish species, we can expect considerable variation in the extent to which each of these behavioral components is manifested. Ethologists, ecologists, fishery biologists, and aquarists have all made valuable contributions to the study of reproductive behavior, with Breder and Rosen's (1966) massive treatise perhaps being the definitive work on the subject. Many species of fishes will take on a decidedly different appearance when reproductive activities are imminent. Male stoneroller minnows acquire their blunt profile, breeding tubercles, and brilliant colors in the early spring as breeding time approaches (Fig. 28-6; the name *stoneroller* comes from their nest-building activities, which involve moving large numbers of small stones about). Breeding tubercles play an important role in reproductive behavior. They are keratinized epidermal structures used for maintenance of body contact between the sexes, tactile stimulation of the female during spawning, and defense of nests or spawning territories (Wiley and Collette, 1970). At least 15 families of bony fishes develop breeding tubercles during the spawning season.

The gathering and movement of numbers of fishes at a given time of the year are among the most conspicuous behavioral phenomena associated with reproduction. In general, reproductive migrations are synchronized with the onset of gonadal maturation such that the fishes arrive on the spawning grounds ready to breed. Seasonal changes in the environment are detected and physiological and behavioral transformations are mediated through a series of endocrine pathways. Typically, fishes will migrate and gather on the spawning grounds and complete the reproductive process in response to a prescribed set of environmental stimuli. For instance, rainbow trout, over much of their range, will migrate and spawn in the spring under the influence of increasing photoperiod and rising water temperature. On the other hand, the related brown trout migrates and spawns in the fall under the influence of decreasing light and temperature. In another salmonid species, the chinook salmon, the situation is complicated by the apparent separation of the environmental cues and behaviors associated with migration from those associated with the actual events of spawning. This species consists of relatively distinct races that enter fresh water at different times. The stream-type race apparently responds to increasing photoperiod and warming temperatures during the spring and summer months by migrating from the ocean to natal streams and is thus resident in fresh water months before spawning. Once in the streams, these fish will take up residence in deep pools for the duration of the summer, during which gonadal development will take place. Spawning occurs in late August and September during a time of decreasing day length and cooling temperatures. The ocean-type race moves into streams in late summer and fall and spawns soon after entering fresh water. Migration into natal streams and gonadal maturation are thus apparently stimulated simultaneously by the changes in photoperiod and temperature regime taking place in the fall. Age-specific sensitivity to environmental cues may also maintain the observed differences in stream residence time of juveniles of the two races. Stream-type ("spring chinook") juveniles spend comparatively more time in fresh water than do the ocean type ("fall chinook") (Healey, in Groot and Margolis, 1991).

The onset of migration means that fishes turn from their usual pursuits and take heed of the environmental cues that move them in the direction of their spawning grounds. The world of fishes is full of potential cues that can be sensed, and any or all may be significant in orientation and migration. Water movement (especially current and tidal fluctuation), solutes in the water, temperature, solar or other celestial light, the earth's magnetic field, sounds, and shoreline or bottom configurations all may be involved in stimulating and guiding migrating fish. Behavioral reactions to these stimuli may be as simple as movement upstream in opposition to current flow or as complex as the utilization by salmon of a variety of environmental cues as they move through several hundred miles of oceans and rivers to reach their natal streams. As mentioned earlier, these species are imprinted with features of natal environments and possibly migratory routes early in their lives. Information on sights, smells, tastes, degrees of azimuth, magnetic fields, and other cues were encoded in the nervous system, where they await the proper endocrine cues that arise and focus the lives of the fish on completion of the life cycle. They may then be sequentially "replayed" in a manner that guides the fish back to where it all began. How this entire process is carried out remains one of the consummate mysteries of the aquatic realm.

At the end of the migration or other movement toward a reproductive area, fishes select the actual arena in which reproductive activities will be carried out. The complexity of the chosen site varies with the species and reproductive requirements of the species involved. For pelagic species, this may amount to nothing more than the patch of open sea to which their migrations have brought them. Here, eggs are broadcast for external fertilization. Fertilized eggs and larvae will then drift with the currents as members of the planktonic community for a period of time. In catadromous species, such as eels (*Anguilla* spp.), the young will eventually be drifted back to the vicinity of their "home" streams, at which time they will undergo the process of transformation into elvers and ascend these streams. For most species, reproduction means selection of some suitable patch of the bottom for the purposes of defining a territory or engaging in nest building. Selection of a suitable site may entail evaluation based on a number of environmental variables, such as depth, streamflow, substrate texture, availability of nest building materials, or availability of cover. Fishes will test the environment in several ways, exploring nooks and crannies, mouthing potential nesting materials, or settling on the bottom to feel it and make test excavations. Sometimes a partially cleaned or excavated nest site will be abandoned if deemed unsuitable for any reason.

Mate selection or courtship activity may accompany nest preparation or may occur after a suitable site has been prepared. In many species, breeding is communal or polygamous, with no distinct pairing or nest preparation. In others, males may wait on the spawning site and approach en masse any female of their species that approaches. In general, courtship is a highly visual process, with colors, contrasting patterns, and configuration of body and fins providing cues. For deep ocean species existing in virtual darkness, vision may still be an essential sensory modality and photophore placement may give cues about species and sex. In those fishes in which pairing takes place, or in those cases in which a male entices and spawns with a few to several females in succession, visual stimuli may be coupled with ritualized actions or postures and olfactory, auditory, or other modalities.

Sex pheromones have been demonstrated to promote gonadal maturation in several species of fishes. These compounds have all been identified to be either sex steroids or prostaglandins, and electrophysiological studies have demonstrated specific actions on the olfactory system (Van Weerd and Richter, 1991). Chemical cues are important in triggering nest building in certain members of the Anabantidae (bettas or Siamese fighting fish), a family containing many bubble nest builders. Although the olfactory and gustatory senses have been demonstrated to be significant in the courting ritual of the stickleback (Segaar et al., 1983), these fishes of the family Gasterosteidae are noteworthy in the simplicity of their olfactory structure. Nor does olfaction appear to play a key role in their feeding behavior (Hara, 1975, 1993; Kleerekoper, 1969). Cods, croakers, toadfishes, and others employ sound in their reproductive activity, and the midshipman (*Porichthys*) adds the light of photophores to its nuptial noises. Courtship posturing and other rituals serve as sign stimuli in many pairing fishes and have become so ritualized that they are recognized as important mechanisms that ensure reproductive isolation of closely related species.

Courting behavior usually begins with some kind of stylized approach, the male sidling toward the female or otherwise displaying his form, color, or posture. If the female is receptive and is ready to spawn or copulate, she responds by approaching or allowing the male to approach. Nuzzling, butting, lateral contact, "dancing," or other requisite acts may ensue before mating is consummated. The aforementioned breeding tubercles are important in facilitating these activities. If a nest or other prepared site is involved, the male must cause the female to follow him to it. The well-documented zigzag dance of the threespine stickleback is a classic example. Many fishes pair off before the nesting site is prepared, and both sexes may work at building or cleaning the spawning site, as in the Cichlidae. In the salmonids, the females usually dig the nest, called a redd, in suitable gravel of the stream bed while the males spend their time warding off intruders.

Territorial behavior is usually enhanced during the breeding season, with many normally nonterritorial species acquiring and defending territories only during breeding season. Such behavior is usually reinforced by the presence of bright colors and threatening postures, which might include displays of the mouth lining, flaring of the opercula, or spreading of the fins. Although actual contact rarely occurs, intruders may be nipped, butted, or grabbed by the defender of the territory. Male-to-male agonistic encounters may sometimes involve locking of jaws with attendant pushing, twisting, or tugging in an apparent test of strength (cichlids may even employ mouth-to-mouth contact and jaw locking as a means of mate selection). Such contests are of usually short duration, with the loser quickly fleeing the scene. In the case of the bettas, males will fight to the death. In some cultures, the fighting ability of individual fishes is highly esteemed and wagers are placed on the outcome of such contests, much in the same way as they are in cock fights or dog fights. Some species of fighting fish are so belligerent that they will attack and kill the female after all the eggs have been laid in the bubble nest if she does not immediately vacate the premises. Pacific salmon engage in aggressive encounters on the spawning grounds in defense of territories. The kype of the male, with its arching curve and large, recurved teeth, can inflict potentially serious injury on an intruder. A favorite ploy is to grasp the opponent's caudal peduncle and twist it violently. Agonistic encounters may be terminated by the elicitation of so-called displacement behavior on the part of the loser. Such activity takes the form of some action unrelated to fighting, such as nest building, which appears either to allow some relief to the frustrated combatant or may actually be a signal of submission.

The act of spawning can be triggered by a variety of releasing mechanisms acting alone or in concert. Tactile stimulation of the female by the male is probably the most common mechanism. Head butts, nips, lateral strokes and quivers, and passes beneath with the male's dorsal fins in contact with the vent region of the female have all been noted in one or more species. The sight of the male or of the nest may be enough to induce the spawning sequence. An interesting form of deception is practiced among some genera of cichlid fishes—that of egg mimicry. Female haplochromine cichlids brood the eggs in their mouths. Males possess spots on their anal fin that are, to the human eye, virtually indistinguishable from newly laid eggs. When the female approaches the anal fin presented by the male, he will eject sperm, which will then fertilize the eggs held in the female's mouth (Goldschmidt, 1991). Not only do chemical secretions induce gonadal maturation, as previously mentioned, they may also serve as cues to spawning behavior; secretions from either sex can have a stimulatory effect on others of the same or opposite sex.

Parental care in fishes ranges from virtual indifference to eggs once they are spawned to intense, complex behaviors that have evolved to optimize survival of the young. Pelagic species, lacking a substratum with which to orient, build nests, or bury eggs, typically exhibit the least amount of parental care. Parental care in salmonids extends to the building of redds in which eggs are buried. Other species (cichlids, for example) engage in a variety of egg- and larval-maintaining behaviors. In addition to vigilant guarding of the eggs and larvae against predators, cichlids will fan the eggs to ensure adequate oxygenation, remove dead eggs or those infected with fungus, and continue to clean and maintain the nesting area. Although it is generally held that visual cues are the most important in eliciting egg care, recent studies on nocturnal egg care in cichlids suggest that olfactory or gustatory cues can trigger and maintain egg care in the absence of visual cues (Reebs and Colgan, 1992).

Many species of fishes brood the eggs on some part of their anatomy. Male pipefishes and seahorses carry eggs in brood pouches; cardinalfishes, some cichlids, sea catfishes, and other species incubate the eggs in the buccal cavity. Male humpheads (family Kurtidae) carry the eggs about on a peculiar hook that projects forward above the head. Once hatched, the larvae may continue to swarm about the parents, who continue to guard them vigilantly. Because many species of cichlids are easily reared and maintained in captivity (a feature that contributes to their popularity with home aquarists), the parenting activities of this family are probably the most well documented of all species. They are generally aggressive by nature, especially in defense of nests and young. In mouthbrooding species, the young can find a safe haven in the buccal cavity of the parents. The parents are apparently capable of distinguishing their own young by odor. Some species will even provide nourishment in the form of mucal secretions from the skin surface.

The obvious benefit of reproductive behaviors is that they optimize the number of young that survive. Although much of the behavior is tied to launching a great number of young of the right size at the right time, other behaviors help to prevent the flooding of an area with young in numbers exceeding the carrying capacity of the habitat. Some types of behavior ensure that the most suitable males, presumably with the genome most conducive to survival of the species, have the best chance of spawning. Other behaviors protect the genetic integrity of the species, preventing hybridization, or preventing the waste of gametes in infertile alliances. Noakes (in Pitcher, 1986) provides a good introduction to this rapidly developing area of study. Some perciform fishes have been observed consuming eggs from nests that they are tending. Such forms of "filial cannibalism" might seem counteradaptive but, in fact, are believed to promote survival of the species (Elgar and Crespi, 1992). Males of a species of pomacentrid, the garibaldi (*Hypsypops rubicundus*), will consume eggs in later stages of development from the nest in order to promote continued visitations and spawning by females. In this way, utilization of the spawning substratum, which in this case consists of a mat of carefully tended red algae, is maximized (Sikkel,

1989, 1994). Some forms of post-spawning behavior serve to lure adults away from the vicinity of hatching young, thus preventing intraspecific predation and competition. Intraspecific competition may also be avoided by distinctly different feeding behavior in adults and their young.

A behavior that is probably adaptive but in a somewhat indirect manner is the defense of the young of others. Certain Central American cichlids are known to adopt young of other parents and even kidnap young from nearby broods. This tends to decrease the possibility that attacks on the young would harm offspring of the guarding pair (McKaye and McKaye, 1977). The cichlid, *Cichlasoma nicaraguense*, an herbivore, has been noted guarding the young of a predator, *C. dovii*. The advantage to *C. nicaraguense* appears to be that *C. dovii* is a predator on the major competitors of the herbivorous species (McKaye, 1977). An even more unusual relationship is the guarding of the young of cichlids by bagrid catfishes, which, however, keep the cichlids on the periphery of the school of their own young, where they presumably are more vulnerable to attack (McKaye and Oliver, 1980).

Interspecies Relationships

There are numerous examples of social relationships involving two or more species of fishes. These relationships are usually commensalistic in nature in that only one individual benefits. These benefits usually take the form of enhanced feeding opportunities. Pilotfish (*Naucrates ductor*) often accompany sharks, probably as a means of grabbing a free meal with little expenditure of effort. Such behavior on the part of pilotfish is so inborn that they tend to accompany any large moving object, such as turtles or boats. Remoras and suckerfish (Echeneidae) will attach themselves to their host by means of powerful adhesive disks that are, in fact, highly modified dorsal fins. Remoras get a free ride from their host, usually a shark, other large fish, or a turtle, in addition to the free meal, which usually takes the form of scraps from the jaws of their host.

Some fishes exploit the foraging behavior of other species even if their feeding behaviors and food choices are quite different. Sea basses (Serranidae) follow herbivorous reef dwellers as they forage and snatch the inevitable small fish and crustaceans that become dislodged as a result of algal grazing by the herbivores. Goatfishes are bottom feeders among the coral reef community, and their disturbance of the sandy sediment over which they feed attracts a number of other fish species. Less assertive species, such as sea basses, may follow more fearsome predators, such as octopi and morays, and catch small fishes as they attempt to flee (Karplus, 1978; Montgomery, 1975). In Lake Malawi, the explosive radiation of cichlid species has resulted in similar feeding relationships among different species. The cichlid, *Cyrtocara moori*, follows several different species of bottom-feeding cichlids and will attack conspecifics and other species if they attempt to approach its feeding partner (Kocher and McKaye, 1983).

In many cases, fishes will develop associations with different classes of organisms. Some gobies seek shelter in burrows excavated by shrimp or in the cavities created by the siphon tubes of clams. The inhospitable array of spines presented by sea urchins makes them a safe haven for shrimpfishes (Centriscidae) and clingfishes (Gobiesocidae). Clownfishes (*Amphiprion* spp.) seek shelter among stinging sea anemones. The relationship is a mutualistic one; in exchange for shelter, the clownfish will provide food for the anemone either in the form of scraps from its own feeding or, in some cases, by acting as a lure to entice other fishes to the stinging tentacles of its host. This constant and intimate association has resulted in the clownfish being able to suppress nematocyst discharge against itself. Just how this is accomplished is unclear. It may be that the anemone becomes habituated to the constant presence of the clownfish. Although it is possible that clownfishes secrete substances that actively suppress nematocyst discharge, behavioral and other studies suggest that the relationship is possible because clownfishes transfer mucous secretions from their anemone host to their own bodies, resulting in inhibition of nematocyst discharge. Researchers have also discovered that certain compounds are secreted by the anemone that induce clownfishes to seek them out (Murata et al., 1986). The association of dolphins (particularly the subtropical genus *Stenella*) with tunas is well known as the reason why tuna harvesting operations can inflict high mortalities on the associated dolphins. Although they may be parasitized by a bewildering array of organisms, fishes themselves rarely engage in parasitic associations with other species. The small pearlfish (*Carapus*) seeks refuge within the body cavity of echinoderms, especially sea cucumbers. Some carapid species parasitize their hosts, consuming gonadal or respiratory tissues. The candiru (*Vandellia*), a tiny catfish known to enter the gill cavities of larger fish and feed on gill tissue, is apparently attracted to urinary streams and has been known to enter the urethra of humans, with painful consequences.

The best-known and most interesting relationships among fishes involve cleaning symbiosis. In this case, the relationship is definitely a mutualistic one, and both members of the association benefit. These relationships can occur among different species of fishes but may also involve invertebrates, particularly decapod shrimp (Limbaugh, 1961). Cleanerfishes are allowed by larger fishes to explore the surface of the body and even the mouth and gill cavities in search of parasites and loose or injured skin, which the cleaner devours. Although cleaning symbiosis is most frequently associated with coral reef environments, cleaning relationships are known from freshwater environments as well. Juvenile bluegill (*Lepomis macrochirus*) often clean larger adults, and adults will clean largemouth bass. This is a precarious relationship because bluegill are among the preferred food items for bass. Successful management of a farm pond depends on stocking the right proportion of each species in order to optimize the production of bass. In cleaning relationships, communication is essential. A fish may "request" cleaning by a characteristic head-down posture or by gaping the mouth. The cleaner then approaches, engaging in behavior that indicates cooperation. Bluegill will darken when approaching a bass to be cleaned; this color is thought to be an indication of submissiveness (Sulak, 1975).

In marine waters, cleanerfishes are known from a number of families, mostly tropical, although there are a few cleaners in temperate waters (Hobson, 1971). An example is the kelp perch (*Brachyistius frenatus*), a member of the temperate marine family Embiotocidae. This species does not seem to enjoy the relative immunity from predation seen in most tropical cleaners ("Temperate seas make for intemperate behavior" might be an appropriate motto here). Most cleaners in warm oceans are in the families of wrasses (Labridae), gobies (Gobiidae), and butterflyfishes (Chaetodontidae). It is no coincidence that fishes in these families characteristically have small, terminal mouths—ideal for picking up small morsels.

In a typical cleaning relationship, the cleaner, which is usually distinctively colored, will display itself at a prominent coral head, rock outcrop, sponge, or other conspicuous place, where cleaning activity will be carried out. This "focal point" may serve as a base from which the cleaner will range out over a larger territory that it has established, to the point that it may aggressively defend against potential competitors. Such intraspecific aggression will result in the partitioning of the reef into a number of non-overlapping, exclusively maintained cleaning territories.

The "cleanee" will usually present itself for cleaning singly, signaling the cleaner with a particular posture, opening the mouth and opercula, and erecting the fins. In some instances, entire schools will mill around a cleaning station, each individual patiently awaiting its turn. Cleaners, usually operating alone but sometimes in pairs or small groups, will proceed to remove any loose skin, parasites, or other fouling organisms that they come across. There are recorded instances in which two different species have been observed cleaning the same individual. Although visual cues seem most important in initiating the cleaner/cleanee relationship, tactile cues are important in maintaining it (Losey, 1978). The quiescent demeanor of the cleanee often serves as an invitation for other small fishes to approach and feed on its surface. Remarkable instances of cleaner mimicry are known among the wrasses, with the mimics stealing small bites of skin from the host. As might be expected, cleaner mimics are not tolerated by their models because they jeopardize the complex relationship that has evolved among the cleaner and its hosts.

About a half dozen species of decapod shrimp are also known to behave as cleaners of fishes. These tiny crustaceans, which would otherwise be a delectable morsel of food, are permitted to climb aboard a host fish and explore its mouth and gill cavities in search of food. Although early studies on cleaning symbiosis by Limbaugh (1961) demonstrated that cleaners are essential in maintaining the health of their hosts, similar studies by G. S. Losey (1972) on the cleaning wrasses inhabiting Hawaiian reefs have shown that their presence may not be as significant. When Losey excluded cleaners from his study site for six months, no increase in parasite infestation was observed in the host species. Consequently, the extent to which cleaning symbiosis may be a mutualistic relationship may vary significantly. Earlier suggestions that cleaning activity may promote fish diversity on coral reefs do not appear to be substantiated. Removal of cleaner wrasses (*Labroides*) from their cleaning stations on Indo-Pacific patch reefs had no impact on local fish diversity (Gorlick et al., 1987; Losey, 1972).

Social Behavior: Schools and Other Aggregations

It is important, at the outset, to define the distinctions between shoals and schools because the two terms have been used interchangeably, causing some confusion. A group of fishes brought together such that they constitute a social group can be termed a shoal in the same manner that the term *flock* defines the same situation for birds. As such, the term makes no implications about the characteristic structure or function of the social grouping. This is explicit in the definition of schools as entities in which the participants are defined by their orienting and swimming with a particular polarity and synchrony (Pitcher, 1986). The terms are unavoidably interchangeble because fish species themselves modify their social organization (Fig. 28–7).

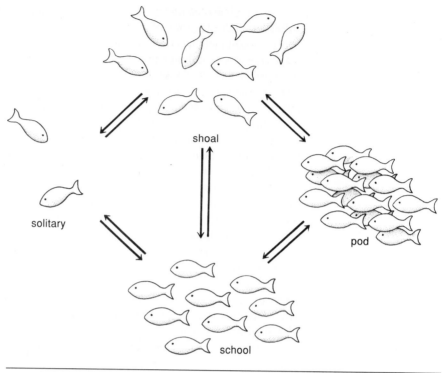

FIGURE 28–7   Relationships between various forms of fish grouping. Fishes may range from solitary existence at one extreme to the formation of pods or clumps at the other extreme. Intermediate associations include shoals and schools with varying amounts of social cohesion and communication implied. (Adapted from Breder, 1959.)

The tendency to approach and orient to other members of the species appears early in many fishes. For example, in the ontogeny of schooling behavior of the European minnow, *Phoxinus phoxinus*, individuals become attracted to each other as soon as the larvae become free swimming. It is only after three weeks, however, that schooling response occurs as a reaction to the threat of predation (Magurran, 1990). This behavior, often termed biotaxis, is largely dependent on visual stimuli. Size, shape, color, and patterning of conspecifics attract the very young of schooling species to one another. Although fishes are well equipped with physical senses other than vision and have keen chemical senses, vision appears to be vital in the formation and maintenance of schools. Cues to speed and directional changes come especially from the lateral visual fields. Several species have been shown to cease schooling at certain light intensities (Whitney, 1969). Blinded individuals of most tested species school poorly if at all. It becomes readily apparent to anyone watching a group of fish schooling in a small stream pool which individual is afflicted with a severe case of eye fungus, because this individual is unable to maintain proper orientation relative to its schoolmates. Other physical senses are useful in schooling and may compensate for the lack of visual cues. Pollock (*Pollachius virens*), for example, still demonstrate schooling while temporarily blindfolded (Pitcher et al., 1976). Sound production may be instrumental in reinforcing the integrity of the school in the dark, and electric fishes may rely on emission and detection of electromagnetic information as a way of keeping in contact in dark or turbid waters.

It is obvious that shoaling and schooling behavior confers some adaptive advantage to the species—the nature and extent of this has been the subject of much debate (Cushing and Harden-Jones, 1968; Pitcher, 1986; Radakov, 1973). The most obvious advantage to schooling in prey species is that it affords a greater degree of protection from predation (cf. Burgess and Shaw, 1979; Magurran, 1990; Shaw, 1978). Fishes in schools are more vigilant in their monitoring for the presence of predators (Magurran, 1990). Schools show predators multiple and shifting targets, and they may engage in a particular avoidance pattern that reduces their vulnerability (Fig. 28–8). Prey species, such as herring, form dense, writhing pods or "balls" that appear to confound a predator such that it will only strike at individuals when they become detached from the mass. In experiments, predatory fishes have been noted to consume more prey fishes when these are presented singly or in small numbers rather than as a large group (Lim, 1981; Major, 1978).

Schooling can bring advantages to fishes seeking food. In tropical reef communities, herbivores such as parrotfishes (Scaridae) form schools to facilitate the invasion of the territories of competitors. Surgeonfishes (Acanthuridae), a family that normally does not form schools, will do so as a means of invading feeding areas of competitors (Barlow, 1974; Robertson et al., 1976; Vine, 1974). The constant communication among members of a foraging school facilitates the localization of prey for a greater number of individuals. Cooperative foraging is apparent in schooling planktivores, which appear to be more efficient at feeding among dense patches of plankton than they are as individuals. Schooling predators can engage in "wolfpack" behavior to cooperate in encircling prey or, as in the case of voracious bluefish, driving prey schools into the shallows and even up onto the shore.

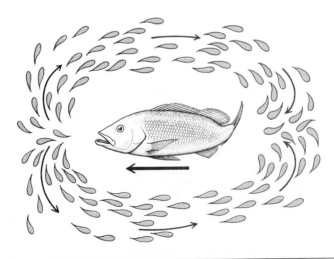

FIGURE 28–8   Illustration of the response of a schooling prey fish, *Lutianus monostigma*, to a predator, *Lutianus bohar*. The school retreats from in front of the predator and cascades around it while maintaining a fixed distance. The school then reforms behind the predator. (From Potts, 1970.)

Laboratory experiments demonstrate that young fishes allowed to feed in a group tend to grow faster than those eating by themselves. The advantages of close social cohesion as a means of improving foraging efficiency do have their limits, however. In some cases, as in those in which food resources are limiting, solitary foraging may be most effective. Some species—medaka (*Oryzias latipes*), for example—display aggression in competition for food and develop hierarchies when food is limited (Magnuson, 1962). Recent studies suggest that this species can rapidly evolve modifications in social behavior in response to changes in resource availability (Ruzzanti and Doyle, 1991).

Reproduction can be facilitated by schooling and shoaling behavior. Many fishes are communal spawners, and both sexes are represented in a single school so that, at the proper season and place, they have only to release gametes en masse in order to achieve a high rate of fertilization. Single-sex schools must, of course, commingle at spawning areas. In some species, the competitive aspect of reproductive behavior subverts the cooperation inherent in the school. During spawning season, individual white cloud minnow males quickly defect from the school and establish territories among vegetation. If available spawning sites are limited, conflict ensues within members of the school (Magurran and Bendelow, 1990). For pelagic species, showing up as a school and shedding gametes in the vicinity of members of the opposite sex may be the extent of their reproductive behavior. For other species, as we have seen, the act of reproduction involves a much greater array of behaviors.

Researchers have demonstrated that schools and shoals do not consist of randomly distributed individuals. Rather, fishes will segregate according to size in the formation of schools and shoals. Minnows and sardines exhibit size-segregative shoaling behavior. The threat of predation often results in different size groupings coming together in order to form a larger body of individuals (Pitcher, 1986). Fishes do not appear to move randomly among the school but have defined fields of activity such that some individuals may tend to be near the edge of the school while others concentrate near the center (Healey and Prieston, 1973). Change of direction by a school seems to be a concerted, instantaneous action on the part of all members so that it seems that they are of the same mind. By means of high-speed cinematography and other techniques, researchers have discovered that direction change can be initiated by individuals well within the mass of fish. Response to the initiator's directional change comes about within a fraction of a second.

Many observations of schooling behavior suggest that there may be some hydrodynamic advantage to fishes moving in schools. The ease with which a school seems to glide through the water seems to support such a contention, as does the observation that the role of "lead" fish in a school changes with every change in direction. Researchers studying the hydrodynamics of formation flying in birds claim that the lead position is traded within the formation as a means of sharing the arduous duty of trailbreaking because the lead individual gains no hydrodynamic advantage that formation flying affords to other members of the group. The mucous secretions of fishes have been suggested to enhance the movement of a school through the water because they surely assist in drag reduction for the individual fish. Hydrodynamic advantages to schooling, intuitively appealing as they may seem, have not been experimentally demonstrated. The experimental evidence that does exist seems to contradict the notion that fishes move easier as a group through the water (Pitcher, 1986).

The advantages of shoaling and schooling behavior do not necessarily constrain these activities to single-species aggregations. There are many instances in which fishes form mixed-species shoals. In fresh water, cyprinids (minnows) and catostomids (suckers) often form foraging-associated shoals among the rocks and cobble of stream bottoms. In the Columbia drainage, as many as five different species of suckers and minnows may comprise these shoals. Such formations are more common among juveniles, in which morphological and behavioral specializations for different kinds of resources have yet to develop. In highly structured benthic environments, mixed-species shoaling is relatively common. It has been most intensively studied among coral reef species (Wolf, 1983, 1985), where snappers (Lutjanidae), goatfishes (Mullidae), and grunts (Haemoelidae) have been reported to form multispecies schools when not actively feeding. These species tend to be nocturnal feeders thought to school together during the day as a means of increasing security against diurnal piscivores. Mixed-species shoaling appears to be relatively rare in open water pelagic species, however. Commercially important fishes, such as clupeids, gadoids, and scombroids, usually form single-species schools, although mixed-species schools of hake (Merluciidae) are known to occur (Pitcher, 1986).

Although schooling behavior has numerous adaptive advantages, some obvious and others more subtle, one must consider one of its chief liabilities. Sociality brings with it the increased likelihood of transmission of disease and parasites. Because of this, it would be advantageous if organisms were able to evaluate the level of infirmity or infestation of group members. Juvenile sticklebacks (*Gasterosteus aculeatus*) have been demonstrated to avoid schools in which conspecifics have been infected with crustacean ectoparasites (Dugatkin et al., 1994). Fishes probably are relying chiefly on the visual sense to make such evaluations, although stress-related secretions may also be detected by group members.

Modification of Behavior

Modification of behavior through learning has been mentioned several times in previous sections of this chapter. Obviously, learning is an integral component of the total adaptive response of an organism to its environment, and fishes have proven themselves adept at modifying their behavior in response to environmental constraints. Obviously, learning is involved in the return of migrating or displaced fish to spawning areas or home ranges. Recognition of mates, young, cleaners, or predators requires some behavior modification. Even general observation of the lives of fishes leads to the conclusion that fishes learn quickly and have reasonably good memories. Kieffer and Colgan (1992) have recently contributed a very good overview of the role learning plays in the lives of fishes. The appreciation of this aspect of fish behavior gives us insight into its application in the natural world and facilitates our manipulation of fishes in practical situations.

Behavioral Modification and Adaptation to the Environment

Recent studies on the ecology of fishes underscore the adaptive significance of behavioral plasticity in optimizing the utilization of aquatic resources by fishes. Studies on the feeding dynamics of fishes inhabiting small streams demonstrate the ability of prey species to detect the presence of predators and adjust their feeding regimen either spatially or temporally. Stoneroller minnows confine their algal grazing activities to pools in which predatory bass are absent (Power, 1987). The feeding behavior of the longnose dace (*Rhinichthys cataractae*), a nocturnally foraging species, can be experimentally modified using different light intensities (Beers and Culp, 1990).

Psychobiologists investigate the behavioral adaptations of fishes, not only because of an intrinsic interest in them, but because they make convenient subjects for the study of certain brain functions. The conditioned response is well known in fishes and is employed widely in a variety of studies. An experimental animal can be conditioned, through application of reward or punishment, to respond to a variety of stimuli. This response can take the form of obvious ones, such as flight or initiation of feeding behavior, or more subtle responses, such as changes in heart or respiratory rate. Conditioning paves the way for study of the ability of the fish to discriminate among colors, visual patterns, and small differences in sounds, odors, or temperature. Fishes can be trained to carry out tasks that require a series of actions. Once a subject is conditioned, memory can be investigated; and once memory is understood, the researcher can study the effects of chemicals, elapsed time, or other factors, such as removal of parts of the brain, on retention. Removal of the forebrain does not make experimental fishes incapable of learning or remembering, although some functions are slowed or altered. Certain chemicals appear to prolong memory in goldfish, but others are known to interfere with retention. For instance, DDT seems to reduce retention time in salmonids (Anderson and Peterson, 1969).

Applications of Behavioral Modification

The knowledge that anadromous salmonids return to the stream from which they migrated to the ocean, and not the stream to which their parents migrated, is of great importance to fishery management. Knowing that the young migrants imprint on the migration route by storing away olfactory and other cues as they move downstream, and then react to these cues on their subsequent upstream travels instead of following some genetically fixed urge to go to the same place their parents went, has made possible the use of a few strategically located hatcheries to supply downstream migrants for many streams, instead of having one hatchery for each stream's stock. Salmonids and many other species can be taught to eat artificially formulated diets instead of natural foods, which might be prohibitively expensive to furnish otherwise. Because marine species such as eels (*Anguilla japonica*) and buri (*Seriola quinqueradiata*) cannot be spawned in captivity, young must be rounded up from wild stocks. It is possible to pen-rear them in high concentrations, however, because they can be trained to consume commercially prepared diets. Not only do the fishes learn to accept unfamiliar foods, but they learn rapidly where and at what time the foods will be given. The ayu (*Plecoglossus altivelis*) normally feeds by scraping diatoms off submerged rocks but readily learns to take dry food from the surface. Channel catfish (*Ictalurus punctatus*) are normally subsurface feeders but are fed floating pelletized food in fish farms. This is advantageous to the fish farmer in that little food sinks to the bottom, where it might be wasted, and overfeeding and underfeeding can be avoided by balancing the food given against the feeding activity of the fishes. In some experimental aquacultural situations, salmon have been reared in open waters after training them to congregate for feeding in response to underwater sonic emissions.

It was mentioned earlier that studies have demonstrated the evolution of behavioral modifications in certain species of fishes. One of the most obvious ways to modify behavior in species of fish is through selective breeding, and the selection can be deliberate or incidental to the maintenance of fishes in an artificial hatchery or aquarium environment. We have already considered the potential for behavior modification by selective processes in the case of medaka that have been demonstrated to evolve modifications in social behavior as resource availability changes. The mutability of fish behavior enables fish culturists to breed out traits that would render a certain species unsuitable for high-density rearing situations. An example from one of the most intensively cultured species, the trout, is a case in point. In the wild, trout that are unwary of objects moving on the bank or overhead have a greater chance of falling victim to predators, whereas the wary ones that retreat from such disturbances are more likely to grow to reproductive age. In the hatchery, however, there is little premium placed on wariness. In fact, the wariest trout may not get its fair share of food, will grow poorly, and will be subject to greater stress. The least wary individuals are better fed and less subject to panic and injury brought on by reaction to disturbances. Hatchery strains of trout that have been under domestication for 40 or 50 years have behavior patterns quite different from those of wild trout of the same species. Fishes only one generation removed from the wild have been seen to exhibit changes in behavior.

The behavior of fishes is indeed fascinating in its diversity. It is a challenging topic, if only for the difficulty in extracting reliable information from a realm entirely different from our own. The challenge also lies in discerning the significance of behavioral adaptations to the environment in such a diverse and successful group of vertebrates. As such, studies of comparative ethology of fishes make a valuable contribution to understanding the basic principles of ecology and evolution.

Part Four

Populations, Species, and Communities

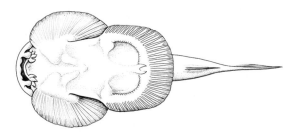

Hill-stream loach of the family Homalopteridae.

Ecology: Environments, Habitats, and Adaptations

Chapter
29

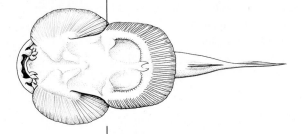

When queried by a group of clerics as to what he might infer about the nature of the Almighty from the scope of His creation, the esteemed evolutionary biologist J. B. S. Haldane is said to have simply responded, "an inordinate fondness for beetles." Were Haldane a vertebrate biologist, his reply would surely have been "an inordinate fondness for fishes." Such is the diversity of this group of animals, estimated at more than 24,000 species (Nelson, 1994), that it encompasses more species than all of the other vertebrate classes combined, and it remains, by far, the most poorly understood vertebrate group. It is an immense task, indeed, to attempt to understand the scope of ecological adaptations of fishes, a task made even more daunting by the fact that they exist in a medium so foreign to our own. In this chapter, we will consider the seemingly endless variety of fishes in the context of the adaptations they possess for the vast array of habitats that exist in the aquatic realm. Due to the scope of the topic, our coverage amounts to only a brief introduction to the ecology of fishes.

As might be expected, those species of economic value are the best understood from an ecological perspective (see, for instance, Mills's 1989 detailed account of the Atlantic salmon or Groot and Margolis, 1991, on Pacific salmon). Other species become well known because their life history and ecology are instructive in addressing fundamental biological problems—especially true in cases in which the species can be easily kept in captivity. Such is the case with the cyprinodontiform fishes, especially the family Poeciliidae. Meffe and Snelson's (1989) *Ecology and Evolution of Livebearing Fishes (Poeciliidae)* serves as an excellent introduction to one of the better understood fish families. This diverse group of small livebearers has become instrumental in advancing our knowledge of ecology, genetics, and evolution.

The Aquatic Realm

Before discussing the adaptations of fishes to aquatic environments, it is necessary to consider briefly some basic properties of the medium in which fishes have originated and diversified. The relationships between the structure and function of fishes and physical aspects of their aquatic environment are so fundamental that it should be intuitively obvious that the essence of this group of vertebrates derives from its origins in and adaptations to water. Yet, as we shall see, the diversity of fishes in the aquatic realm is so comprehensive that it includes many species that have escaped the confinements of a watery existence. These species have become relatively successful as amphibious creatures participating, to a greater or lesser extent, in terrestrial ecosystems.

Because water is about 800 times more dense than air, successful locomotion is dependent on development of effective streamlining. This is best exemplified in the fusiform shape of such species as tunas and billfishes. For benthic (bottom-dwelling) species, this is less of a constraint except for species that live in areas where water is moving at high velocity. In such places, species must present a low profile or smooth contour in order to maintain their position in torrents or riffles. Because the specific gravity of water (1 at 4°C) is slightly less than fish tissue, neutral buoyancy is achieved by the incorporation of low-density fatty tissue or gas bladders. This relieves fishes of the necessity to expend excessive amounts of energy in maintaining position. As might be expected, benthic fishes have reduced gas bladders or have lost them.

Pressure, which increases about 1 atm for every 10 m of depth, has profound effects on the life and structure of deep water fishes (witness the general body consistency of deep water marine species, which have very flabby, gelatinous tissue and skeletons that are weakly ossified). The absence of gas bladders in the deepest dwelling marine species is not apparently a consequence of the extremely high ambient pressures at that depth, but rather it is a consequence of their extremely reduced musculoskeletal systems—their body tissue density is so low that they are able to maintain neutral buoyancy even without the contributions of a gas bladder (Marshall, 1971).

Of great significance in the lives of fishes is the fact that the maximum density of fresh water is at 4°C, well above the freezing point. This means that ice forms on the surface, thus acting as a barrier between fishes and the overlying atmosphere during the winter. The freezing point of sea water is depressed relative to fresh water due to the concentration of dissolved substances. Because the tissue osmoconcentrations of marine fishes fall between that of fresh water and full-strength sea water, those at extremely high latitudes are faced with the threat of tissue destruction by freezing. This is counteracted by the presence of high-molecular-weight molecules, chiefly glycoproteins, in the body tissues, which act as a natural antifreeze by retarding the formation of ice crystals. The high specific heat of water means that warm or cold currents can be conveyed far beyond the latitudes where they form, thus affecting both the dispersal and distribution of fishes.

Although pure water is highly transparent, light absorption is rapid and differential. At depths of more than 100 m, only a fraction of the blue end of the spectrum is present. It is these wavelengths that, when reflected back to our retinas, give oceans their characteristic blue color. Most red wavelengths are absorbed in the upper 5 m, and orange is mostly gone at 15 m. Green and yellow penetrate to about 20 m. Although the depth record for photosynthetic activity in the ocean belongs to a species of red algae that was discovered at 268 m in the Bahamas (Littler et al., 1985), the vast majority of photosynthetic activity, especially that by chlorophyll-containing species, occurs at depths of less than 50 m. In highly turbid waters, as often occurs on wave-washed shores or muddy turbulent rivers, photosynthetic activity may be restricted to the upper few meters. An interesting consequence of differential light penetration is the effect it has on fish coloration. Rockfishes that appear bright red when they are hauled up on the deck of a trawler may, in fact, appear as a drab gray at the depth that they normally inhabit. Fishes living in oceanic realms not penetrated by light of any wavelength are uniformly black or dark brown but may possess highly reflective surfaces to facilitate the emission of light from photophores.

The dissolving capabilities of water are such that chemists term it the universal solvent. As it falls as rain, percolates through the ground, or runs along the surface of the earth, water collects numerous substances, including chlorides, sulfates, and carbonates of calcium, sodium, magnesium, and potassium. Silicon and phosphorous compounds, as well as many organic substances, are among the biologically important materials carried by water. The chemistry of water is indeed complex and has a significant impact on the adaptations of fishes to specific aquatic environments.

Water falling as rain collects both oxygen and carbon dioxide from the atmosphere. Although the solubility of oxygen is relatively low, it is of sufficient concentration that it can be successfully extracted by the respiratory apparatus of fishes. It is important to remember, though, that there is an inverse correlation between oxygen solubility and ionic concentration of the solvent. Whereas air contains nearly 21 percent oxygen by volume, fresh water can dissolve only 10.23 cc/l at 0°C, only about one twentieth that of air, and less as the temperature rises. Sea water of 30 ppt (parts per thousand or grams/liter) salinity holds about 8.8 cc/liter at 0°C. A relatively diverse freshwater fish fauna can exist at a dissolved oxygen concentration of 3.5 cc/l, or about 55 percent saturation at 20°C. Cool water species do better if maintained at concentrations nearer saturation (about 6.4 cc/l at 20°C). The evolution of air-breathing and ensuing terrestriality in several groups of fishes, including perhaps those lineages that gave rise to the first tetrapods, is believed to be largely the consequence of extremely low concentrations of dissolved oxygen, as can occur in swamps or even in some marine environments (Packard, 1974).

Carbon dioxide has a much higher solubility in water, causing rain to be naturally slightly acidic. An unfortunate consequence of industrial emissions, particularly sulfuric and nitric acids, is the increased acidification of the world's water supplies. In northeast America and Canada, with a largely granitic and insoluble substratum, bodies of water are soft and thus lacking in the buffering capacity of other regions. In those places, acid precipitation exerts its greatest impact. As much as 50 percent of the species in some taxonomic groups have been eliminated from bodies of water in the Adirondacks, the Poconos and Catskills, and in southern New England. Among the most severely impacted fish families are the Cyprinidae, Salmonidae, and Centrarchidae (Schindler et al., 1989).

Although nitrogen is inert, it can reach conditions of supersaturation under certain conditions, with harmful effects. Where spillways of large dams allow air to be carried deep into plunge pools, fishes may become subject to "gas bubble disease," a debilitating and often lethal condition caused by emboli that form in the tissues. Carbon dioxide dissolved in the free state—not bound as carbonate or bicarbonate—can affect the ability of fish blood to take up oxygen under conditions of supersaturation as well.

Dissolved salts, such as those encountered in the ocean or in various lakes and mineral springs, affect the development of osmoregulatory capacity in fishes. Although some species may be termed stenohaline due to their limited tolerance to fluctuations in medium osmoconcentration, others are euryhaline in that they can tolerate such fluctuations. The all-time champion osmoregulator among the vertebrates is the sheepshead minnow (*Cyprinodon variegatus*). Populations of this species have been found in bodies of fresh water if the level of dissolved calcium is sufficiently elevated and they have also become established in hypersaline lagoons, where the salinity may exceed twice that of full-strength sea water (Guillory and Johnson, 1986; Martin, 1972; Nordlie et al., 1991). In general, the osmoregulatory capacity of fishes is such that freshwater species regulate salt concentration at levels much higher than that of the surrounding medium, whereas marine fishes regulate at levels well below that of the surrounding medium (Table 29–1).

Freshwater Habitats and Adaptations

In discussing the adaptations of fishes to their respective environments, we must distinguish the terms habitat and niche. By habitat, we mean the place in which an organism normally lives. Physical characteristics as well as coexisting species would

TABLE 29–1

| Osmotic and Major Solute Concentrations in Selected Bodies of Water and in Representative Species of Teleosts | | | | | | | |

Body of Water	Average Osmotic Concentration (mOsm/liter)	Major Ions (mmol/l)					
		Na^+	K^+	Ca^{+2}	Mg^{+2}	Cl^-	So_4^{-2}
Average ocean	1000	470	10	10	54	548	38
Average river	~1	~.08	~.01	~.3	~.09	~.05	~.08
Little Manitou Lake, Canada	~2000	780	28	14	500	660	540
Great Salt Lake, Utah	~6000	3000	90	9	230	3100	150
Fish Species							
Cyprinus carpio (freshwater)	274	130	2.9	2.1	1.2	125	—
Lophius piscatorius (marine)	452	180	5.1	2.8	2.5	196	2.7
Oncorhynchus tshawytscha (freshwater phase)	288	161	.3	2.7	1.2	114	.5
O. tshawytscha (marine phase)	332	179	1.0	1.0	.9	139	.3

Sources: Sverdrup, H., et al., 1942, *The Oceans,* Prentice Hall; Hutchinson, G., 1957, *A Treatise on Limnology,* John Wiley & Sons; Holmes, W., & E. Donaldson, in W. Hoar & D. Randall (eds.), 1969, *Fish Physiology,* Academic Press; Evans, D., in D. Evans (ed.), 1993, *The Physiology of Fishes,* CRC Press.

define the habitat. It is the "address" of the species, so to speak. The niche is defined by the range of physical and biological conditions the species might encounter and with which it might interact. The niche thus includes the role the species might play in its respective community or ecosystem. Therefore, it might be construed to be the "occupation" of the species.

Fresh waters, in the form of habitable lakes and rivers, comprise an almost negligible proportion of the available surface waters of the earth (about 0.01 percent), yet they harbor about 41 percent of the known fish species (Horn, 1972). Habitat variability over the broad range of latitudes, as well as the increased opportunity for geographic isolation, probably account for this remarkable biogeographic phenomenon. Fresh waters may be still or moving; they differ greatly in temperature, depth, dissolved oxygen and other materials, suspended matter (including available food), and substratum type. Compared to sea water, the chemistry of inland waters is extremely variable (Table 29–1). Freshwater environments are subject to seasonal variation and, over a greater expanse of time, have experienced dramatic transformation by geological and biological processes. Changes in small streams and ponds may be readily observable over the course of a few years. Larger bodies of water change much more slowly but, from a geological perspective, are relatively young and short-lived phenomena. The Great Lakes, for example, can be dated only from the most recent glaciation events of North America—approximately 10,000 to 15,000 years ago. Ancient bodies of fresh water exceeding 1 million years of age (such as Lake Baikal, the world's deepest freshwater body, or the African Rift Valley lakes) are most unusual. Because the processes and consequences in running and still water differ considerably, the two series of environments will be discussed separately.

Running Water (Lotic and Fluviatile Environments)

As water drains off a watershed, a natural progression of flowing water habitat develops. Small rivulets join to form brooks that may be seasonally intermittent in their flow. These join to form larger streams that, themselves, combine to form rivers and river systems, which ultimately empty into interior lakes or the world's oceans. This progression is one of decreasing altitude and is generally marked by changes in velocity as water progresses from high-gradient torrents and riffles at higher altitudes to slower, meandering streams and rivers at lower altitudes near sea level. A popular ecological classification scheme, one that is especially useful in correlating fish abundance and distribution with stream size, is to designate flowing waters along an ordinal gradient, with the smallest headwater streams as order I streams. These join to produce an order II stream. Two order II streams merge to form an order III stream and so on. Order V streams and above would constitute the major river systems of a continent. Although somewhat simplistic in terms of ecological context, the stream order concept is a useful means of categorizing the continuum of stream types and associated fauna (Horton, 1945; Kuehne, 1962).

Upstream environments, referred to as the "rhithron" by Wootton (1992), are characterized by swift current and a range of substrate types. Oxygen levels vary with the velocity of the water. A typical stretch of upland stream will consist of riffles, in which the water moves swiftly and is saturated with dissolved oxygen, and shallow pools, in which biological activity results in the depletion of oxygen. The scouring action of swift water in the riffles results in a substrate that is typically rocks and cobble of varying texture, depending on local geological conditions, whereas fine-grain sediments and detritus will collect in the pools. The biological communities of stony streams tend to be quite similar worldwide; the greatest differences are seen as one moves from the tropics to the poles (Hynes, 1970). This enables the broadest application of faunal zonation patterns determined from the study of selected stream systems. Temperatures tend to be lower in upstream sections compared with downstream environments of the same system. Exposure to sunlight may be great in small streams above timberline, but brooks in forested areas can be covered by a canopy of vegetation, which will effectively limit the light reaching the water and thus hold temperatures down. Depending on latitude, altitude, and exposure, upland streams may be subject to ice and snow cover for part of the year. Seasonality in precipitation patterns can lead to intermittency of flow in these streams as well.

Some of the most remarkable of freshwater fishes are those species that are adapted to conditions of torrential flow. Asiatic hill-stream loaches (family Balitoridae, Fig. 29–1) have flattened ventral surfaces, with the pectoral and pelvic fins expanded to form a broad adhesive disc that enables the fish to hold its position in extremely fast-flowing water. The family Gyrinocheilidae (the popular "algae eaters" of the tropical fish trade) have a more conventional body form but possess a mouth that can act as a sucker for holding onto rocks. Other torrent fishes are in the South American catfish family Loricariidae (which includes species popular with aquarists), *Cheimarrichthyes* (Cheimarrhichthyidae) of New Zealand, and the Rhyacichthyidae of Asia. Lampreys use their primitive mouths in much the same fashion, gripping stones as they advance on their reproductive journey upstream (the family name Petromyzontidae literally means "stone sucker").

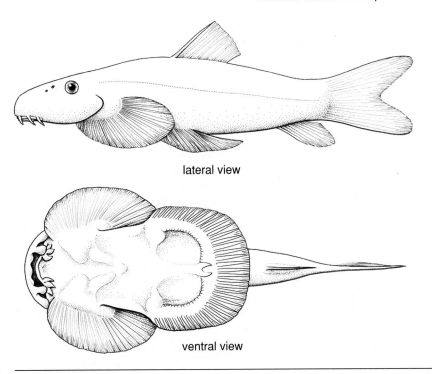

lateral view

ventral view

FIGURE 29–1   Hill-stream loach of the family Homalopteridae. Note flattened profile and pectoral and pelvic fins modified to form a sucking disc.

Of foremost importance in structuring any aquatic community is its productivity, its capacity to produce food for its component species. In water, as in terrestrial habitats, the bulk of food production depends on the photosynthetic capacity of the system. In high-gradient streams, there is little opportunity for rooted plants to develop, so primary productivity rests largely with algae, especially diatoms, that coat the bottom. These form a food base on which the primary and secondary consumers of the aquatic system may subsist. Although most fishes are secondary consumers, feeding on those creatures that feed on the diatoms, herbivory is known among some stream-dwelling species. The aforementioned hill-stream loaches and gyrinocheilids are vegetation scrapers. Stoneroller minnows (*Campostoma* spp.) and bluntnose minnows (*Pimephales notatus*) are among the most abundant minnows in upland streams of eastern North America. These two genera graze on diatoms and other attached material (*aufwuchs*) and can outnumber all of the other species combined in some especially productive stream stretches. As the gradient of the stream diminishes, there is a greater opportunity for biotic diversification, and the food base consequently becomes enhanced. Fish diversity increases in upstream sections, where there is a gradient low enough to permit the establishment of alternating pools and riffles. The accumulation of soft sediments in pools and their characteristically lower dissolved oxygen levels means that fewer feeding opportunities exist when compared to adjacent riffles. Yet, as we shall see, many species of fishes have become adapted to upland pool environments.

Terrestrial organisms constitute a significant nutritional input in upland as well as lowland streams. The abundant and diverse invertebrate population commonly termed "shredders" or "detritivores" may subsist on leaf litter that falls into streams. These then are an important food source for fishes. Perhaps the most remarkable examples of herbivory in fishes are those species, especially in the Amazonian basin, that subsist on seeds and fruits dropped from overhanging trees. The diet of some of the large characins, commonly referred to as pacu, of the Amazon River and its tributaries may consist almost entirely of seeds and fruits of several species of trees that form the streamside canopy (Goulding, 1980).

In higher latitudes of both the old and new world, salmonids are a conspicuous family of upland stream fishes but can range further downstream if temperature and oxygen conditions permit. These are active, streamlined fishes with strong locomotor capabilities. Species such as the rainbow trout (*Oncorhynchus mykiss*), the cutthroat trout (*O. clarki*), the brown trout (*Salmo trutta*), and chars of the genus *Salvelinus* can enter streams that are essentially all rapids with gradients up to 75 m/km. Typically, they will reside in stream pools, catching food as it drifts by. Salmonids also make occasional feeding forays into rapids and will invade intermittent streams on a seasonal basis. Salmonids will share this habitat with an assortment of other species. In North America, these will include dace of the genus *Rhinichthys*, creek chubs (*Semotilus*), as well as several species of suckers, such as the hog sucker (*Hypentelium nigricans*), the torrent sucker (*Moxostoma rhothoecum*), and the mountain sucker (*Catostomus platyrhynchus*). The two most abundant and diverse groups of benthic fishes in upland streams are sculpins (Cottidae) and darters (Percidae, subfam. Etheostomatinae). These are small, bottom-dwelling fishes that typically have dorsoventrally compressed bodies with flattened heads, broad pectoral fins, and reduced or absent gas bladders—all adaptations enabling them to hold position on the bottoms of upland streams, habitats that may be periodically subjected to torrential conditions. In western North America, sculpins, especially of the genus *Cottus*, are especially abundant, with species such as the Paiute sculpin (*Cottus beldingi*) and the shorthead sculpin (*C. confusus*) invading headwaters of streams. In eastern North America, however, sculpins are very poorly represented. Rather, it is the darters that are the most conspicuous of the benthic stream fish fauna. Geographic isolation, coupled with their limited mobility, has resulted in a high degree of endemism among the benthic fish fauna inhabiting upland streams—especially in the case of the darters, with over 140 species described so far. Studies have demonstrated considerable microhabitat partitioning among coexisting species of sculpins and darters. When two or more species are present, they will segregate according to current speed, substrate type, or depth (Daniels, 1987; Finger, 1982; Fisher and Pearson, 1987; Hlohowskyj and Wissing, 1986; Matthews, 1985; Ultsch et al., 1978). The reproductive adaptations of fishes in high-gradient streams include the burying of eggs in gravel or adhering them to suitable substrates in order to assure that they will not be swept away in the current. Trouts, chars, and salmons seek out gravelly bottom at the heads of riffles or in springs, where a nest is dug by the female. The demersal eggs are temporarily adhesive so that they remain in place until they are covered. Whitefish, grayling, and suckers spawn with vigorous activity that disturbs the gravel sufficiently to ensure that most of the eggs become lodged in interstices among the rocks. Sculpins and darters attach adhesive clumps of eggs to the underside of stones and the eggs may be vigorously guarded by male parents in some species.

The lower reaches of streams, termed the "potamon" by Wootton (1992), possess a greater variety of habitats less subject to seasonal fluctuation in physicochemical parameters. Consequently, they tend to exhibit a greater diversity of fish species (cf. Ebert and Filipek, 1988). The lower stretches of streams may be characterized by more pools and runs than riffles, and the current will be more variable but slower than upstream. Substrate size and texture become more variable, as do the kinds of plants. The increased amount of finer sediments permits the establishment of rooted plants to a greater degree. These provide greater microhabitat and feeding opportunities for the resident fish fauna. Members of the family Cyprinidae abound in these lower-gradient streams. In the Mississippi drainage, genera such as *Notropis* (the most diverse genus of North American cyprinids), *Hybopsis, Semotilus,* and *Pimephales* are representative. In various drainages of western North America, the squawfish (*Ptychocheilus* spp., predators and among the largest known cyprinids), the roach (*Hesperoleucus*), chubs (*Gila* spp.), the peamouth (*Mylocheilus*), and the redsides (*Richardsonius* spp.) can be encountered in running water of moderate to slow speed. Several species of suckers, such as *Catostomus commersoni* and *C. macrocheilus,* abound in the middle to lower stretches of streams, where greater opportunities for benthic feeding exist. North American streams are typically classified as warm water streams, occurring in lower latitudes and elevations, or cool water streams, seen at higher latitudes or elevations. Although the top carnivores in cool water streams are usually salmonids or larger members of the perch family, such as the walleye (genus *Stizostedion*), warm water streams will have a good representation of members of the sunfish family Centrarchidae. This relatively diverse family includes planktivores and microcarnivores, such as crappie (genus *Pomoxis*) and bluegill (*Lepomis macrochirus*), as well as large piscivorous game fishes, such as the basses of the genus *Micropterus.*

As stream order increases, currents become more sluggish, depth increases, bottom materials become finer, and water correspondingly becomes more turbid. Several species of large bottom feeders do well in such places. Suckers, including the genera *Carpoides* and *Ictiobus,* and catfishes of the genus *Ictalurus* forage among the fine-grained sediments and rooted plants of lowland streams and rivers. Ictalurids (family Ictaluridae) are the only catfish family native to North America and include some of the largest fishes found in fresh water. Although smaller ictalurids, such as the bullhead catfishes (*Ameirus*) and the madtoms (*Noturus*), abound in pools and riffles in smaller streams, some species, such as the blue catfish (*Ictalurus furcatus*) and the flathead catfish (*Pylodictes olivaris*), prefer large pools or channels of large river systems. Here, they may reach sizes in excess of 50 kg. The numerous small cyprinid species in the lower reaches of streams serve as forage for predatory pikes (*Esox* spp.) and centrarchids.

Patterns of distribution and abundance of freshwater fish species in rivers and streams in the eastern part of North America differ significantly from those seen in the western part of the continent due to the profound differences in the geologic history of the two regions. The freshwater fish fauna of the west does not approach the diversity of the geologically older eastern part of the continent, with only about one fourth as many species present (Smith, 1981a). Although the fish fauna of the Great Basin, Mojave, Sonoran, and Chihuahuan deserts of the western United States may not be as diverse as in regions experiencing a greater amount of precipitation, they are remarkable in their scope of adaptations to extremely arid conditions. They are also in possession of a unique evolutionary history, which is characterized by a high degree of endemism. Considering their arid nature, North American deserts are still surprisingly rich in aquatic biota. Desert waters consist usually of two types: rivers and streams, often with intermittent flow, and springs, which represent surface extrusions of groundwater rising up from geological faults. Marshes sometimes form in association with these sources of water, providing important habitat for aquatic organisms as well as a destination for many terrestrial animals. Those located along the flyways of migratory birds are especially critical because they provide a source of water and food en route to and from nesting and wintering areas. Only a few lakes are present, exclusively in the Great Basin (Soltz and Naiman, 1981). The modern desert environments are all of relatively recent origin. Basically, they are relicts of a much more extensive aquatic ecosystem that was present during the last great pluvial period, which ended 10,000 to 12,000 years ago (Soltz and Naiman, 1981). Consequently, the evolution of the resident fish fauna isolated in the various remnant springs and rivers is a relatively recent phenomenon.

In comparing spring-dwelling species with those in larger bodies of water, a few generalizations can be made. Spring dwellers tend to have a shorter, stubbier body profile; breeding season is more protracted, with a shorter time to first breeding; and there is a low fecundity, with a greater parental investment per offspring. A greater degree of territorial behavior, bolder reproductive coloration, and increased presence of alternative mating strategies also seem to be the case (Constantz, 1981). Killifishes of the family Cyprinodontidae are most representative of these traits. Over 20 species of the genus *Cyprinodon* have evolved in the North American deserts, including highly endemic species such as the Devil's Hole pupfish (*Cyprinodon diabolis*) and the Owens pupfish (*C. radiosis*). In these two species, their entire natural range is restricted to one or two isolated springs (Sigler and Sigler, 1987). The habitats of desert fishes may experience extremes in temperature, dissolved oxygen, and salinity. In this respect, cyprinodonts are well adapted to desert environments—their tolerance of environmental extremes is well known (cf. Hillyard, 1981). Many other North American families of freshwater fishes have representative species in the western deserts. These include salmonids (including the landlocked sockeye salmon variant known as the kokanee; whitefish, *Prosopium* spp.; and trout); cyprinids (especially the genus *Gila*); catostomids; ictalurids; and centrarchids (Sigler and Sigler, 1987). Cichlids have been introduced in some areas and are species of some concern because of their potentially adverse impact on the native biota.

Moyle and Herbold (1987) have made some interesting comparisons of the life history patterns of new and old world temperate fish communities. Communities of cold headwater streams in Europe and North America are quite similar, each consisting of salmonids, sculpins, and a few species of cyprinids (catostomids abound in upland streams in North America, but their old world distribution is restricted to western Asia), while those at lower elevations and latitudes tend to vary. In fact, European fish communities bear a greater resemblance to those of western North America than they do to eastern North America.

In Europe, typical running water cyprinids are the barbel (*Barbus barbus*), the chub (*Squalius cephalus*), and the dace (*Leuciscus leuciscus*). Typical downstream forms include the bream (*Abramis brama*), the carp (*Cyprinus carpio*—a species that had a dramatic impact on the native biota of North America when it was introduced in the 19th century), and the tench (*Tinca tinca*—another, more recent, introduction into North America).

Stream ecology has advanced to the point that zonation patterns can be derived as a means of classifying the habitats and associated fish fauna. Variation exists in the ichthyofauna due to differences in physical factors, such as altitude, latitude, and other factors that would bring about different temperature regimes in streams. In addition, centers of biogeographic origin must be considered. For example, within the family Percidae, why is the tribe Percini (yellow perches and related forms) found almost throughout holarctic waters, while darters of the tribe Etheostomatine are native only to North America? The diversity of fishes in some areas and the ecological versatility of many species add to the difficulty of classifying streams by fish habitat.

Most studies of stream fishes and their ecology have been reported in terms of associations of fishes, or stream zones have been named for the dominant species or associations found there. Often, these zones or associations have mainly local applications, but in Europe, where climatic and physiographic features allow the spread of an essentially similar fish fauna over a wide area, a zonation of streams proposed by Huet (1959) seems to have a broad application (Fig. 29–2) (Wootton, 1992). Li et al. (1987) have developed a similar zonation pattern for the salmonid- and cottid-rich rivers of the Pacific northwest, and similar ichthyofaunal zonations have been developed for other regions of the country.

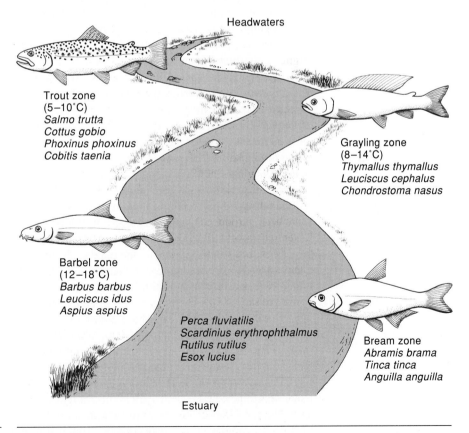

Headwaters

Trout zone
(5–10°C)
Salmo trutta
Cottus gobio
Phoxinus phoxinus
Cobitis taenia

Grayling zone
(8–14°C)
Thymallus thymallus
Leuciscus cephalus
Chondrostoma nasus

Barbel zone
(12–18°C)
Barbus barbus
Leuciscus idus
Aspius aspius

Perca fluviatilis
Scardinius erythrophthalmus
Rutilus rutilus
Esox lucius

Bream zone
Abramis brama
Tinca tinca
Anguilla anguilla

Estuary

FIGURE 29–2   Huet's zonation of fish assemblages of European rivers. Species representative of each zone are given as is the prevailing thermal regime of each zone. (*Source:* Huet, 1959; Wootton, 1992.)

Although the "typical" temperate stream system can be conveniently divided into high gradient, torrential upper reaches, moderate gradient regions with a flatter profile, and slow-moving rivers and streams of the lowlands, no such simple classification system exists for rivers in tropical regions. Variations in topography have led to some rivers originating in swampy regions, with large falls and long stretches of rapids occurring at widely spaced intervals (Lowe-McConnell, 1987). Seasonal patterns of precipitation result in vast stretches of watershed being a component of the aquatic ecosystem during the rainy season and a part of the terrestrial ecosystem at other times of the year. In the Amazon system, lowland forests may be flooded for up to half of the year, permitting many species of fish to forage over a broader range—this especially facilitates the feeding habits of those species that subsist on seeds, fruits, and other terrestrial vegetation. Reproductive migrations may also coincide with peak flooding such that young may be reared in the relative safety of the shallow reaches of flooded areas. These factors, in conjunction with the characteristically high rate of speciation in the tropics, have resulted in an exceedingly diverse ichthyofauna in tropical rivers and streams of the old and new world.

Where extensive flooding occurs, as in the neotropical rain forests, water chemistry is profoundly altered. Decomposing plant material makes the flooded forest waters acidic with low levels of dissolved oxygen. The forest canopy prevents wind mixing, resulting in stagnant waters. Characins and other species have developed a remarkable respiratory adaptation to these conditions. Under conditions of severe depletion of dissolved oxygen, they develop protrusions of the lower lip that appear to facilitate aquatic surface respiration, the uptake of oxygen from the thin layer of water at the surface (Saint-Paul and Bernardinho, 1988; Winemiller, 1989). This may permit herbivores to remain in the flooded forests as the water stagnates, while other species, including predators, are forced to return to the more richly oxygenated river channel. If flooding occurs in tropical savanna regions, conditions are quite different. Flooded savannas rapidly heat up in the absence of canopy cover. There is abundant light available for photosynthetic activity, and oxygen is replenished with wind mixing. In many tropical regions, these types of floodplains support important fisheries (Lowe-McConnell, 1987).

The tropical fish fauna of the old world includes a diversity of species mainly centered in Africa and Asia. Africa has over 2000 species of indigenous freshwater fishes. The most striking feature of the African freshwater fish fauna is its high degree of endemism, especially in the cichlids of the Rift Valley Lake system. The Cyprinidae are also well represented, especially in the four large river systems—the Niger, Nile, Zaire, and Zambesi—that drain the continent. The Zaire typifies the complexity of equatorial African river systems, consisting of large rapids, swamps, main river channels, shallow stretches, and seasonally flooded areas. Large lateral lakes are also part of the Zaire system. With such a diversity of habitat, the fish fauna is correspondingly rich and varied. In addition to the aforementioned cichlids and cyprinids, rivers may be populated with bichirs (Polypteridae), elephantnose fish (Mormyridae), characins (Characidae), and a variety of catfish families. Lungfishes (Protopteridae) occur in areas subject to seasonal drought; their remarkable aestivating abilities enable them to survive several months of desiccation. The Far East is a region that is extremely complex from a biogeographic perspective. For example, Borneo has over 300 species of primary freshwater fishes (primary freshwater fishes are those with virtually no tolerance to sea water and hence have dispersed entirely through freshwater corridors, while secondary fishes display limited tolerance to sea water and may include species with marine affinities) in 17 families. In contrast, just 140 km to the east, Sulawesi has but two primary freshwater species (Lowe-McConnell, 1987). In areas such as this, where primary freshwater fishes are conspicuous by their absence, atherinomorph fishes are particularly well established. Cyprinids dominate the tropical Asian fish fauna, as they do many other biogeographic regions. Typical of the cyprinids is the genus *Barbus,* with 11 species recorded from Sri Lanka (Kortmulder, 1987). Many species of catfishes, including some of the largest in the world, are also found in the main channels of river systems that drain tropical Asia. For example, *Pangasianodon gigas,* an enormous herbivore measuring up to 2 m in length, is found in the main channel of the Mekong River. Another remarkable species native to the lowland swamps is the archerfish (Toxotidae). Members of this family are able to obtain insect prey from overhanging vegetation by shooting them with jets of water.

The fish fauna of neotropical river systems is the richest and most diversified in the world, with over 2400 species described. It is also the most poorly understood. Compared to the old world tropical ichthyofauna, that of the new world is derived from fewer basic stocks. For example, primitive families, such as the osteoglossomorphs, are much more poorly represented, whereas characoids and siluroids have experienced explosive adaptive radiations. Fully 85 percent of the fishes of the Amazon Basin are ostariophysans, compared to 54 percent in the Zaire River system (Lowe-McConnell, 1987). Because so many South American river systems are yet to be thoroughly explored, few generalizations concerning the ecology of the ichthyofauna can be made. The fishes that have been the subject of extensive life history investigation have usually been the larger species, which may be commercially exploitable, or those species that are remarkable for some aspect of their life history, such as the fruit- and seed-eating characoids (Goulding, 1980).

A few river systems have been subjected to intensive collection efforts such that patterns of longitudinal zonation and habitat preferences can be discerned (cf. Ibarra and Stewart, 1989; Fig. 29–3). River conditions in the Amazon range from Andean headwaters carrying large quantities of snowmelt and associated sediments into the Amazon Basin, to blackwater rivers characterized by highly acidic waters deficient in dissolved inorganic ions, to clearwater rivers carrying only small amounts of suspended matter. The transparency of the latter permits blooms of phytoplankton and therefore supports abundant populations of planktivorous species (Lowe-McConnell, 1987). Fishes tend to be relatively sparse in the main channels but congregate near banks and beaches, where productivity and terrestrial nutrient input are greater. Those young fish ecologists adventurous enough to tackle the neotropical ichthyofauna are bound to reach the same conclusion: "so many fishes, so little time."

Still Water (Lentic Environments)

Once a lake has been created by one of the many geological processes that can cause basins to form in the surface of the earth, it begins to fill up with materials washed in by streams, blown in by winds, or produced in the lake itself. The geological processes most important in the formation of lakes in the northern hemisphere are

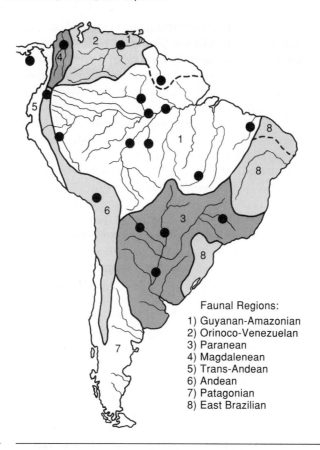

Faunal Regions:
1) Guyanan-Amazonian
2) Orinoco-Venezuelan
3) Paranean
4) Magdalenean
5) Trans-Andean
6) Andean
7) Patagonian
8) East Brazilian

FIGURE 29–3   Faunal regions of South America. Locations of extensive ichthyofaunal study are indicated with closed circles. (*Source:* Lowe-McConnell, 1987.)

those associated with glaciation during the Pleistocene. Usually, lakes have a life span measured in thousands of years, a mere instant in geologic time. Small shallow lakes, such as might form in irregularities in glacial moraines or be caused by a landslide damming a small valley, may become extinct in a few centuries. The natural progression is from lake to pond to swamp and finally to dry ground, if the processes of lake senescence are allowed to continue unabated. Lakes that owe their creation to major geologic phenomena on the earth's surface have a greater degree of permanence. Examples such as the aforementioned Baikal in Siberia or the deep rift lakes of Africa have life spans measured in the millions of years. Yet in them, too, the inexorable forces of environmental and hence biotic transformation are still occurring, albeit at a much slower rate.

Young lakes in temperate regions are low in nutrients and organic materials. These oligotrophic lakes, as they are called, are characterized by an abundant supply of dissolved oxygen at all depths such that the resident fish fauna has access to the entire body of water. Thermal stratification, which, in summer, layers the lake into a warm surface epilimnion, a middle mesolimnion characterized by the presence of the thermocline (a zone of rapidly changing temperature), and a lower, cool hypolimnion, permits the year-round existence of cold water fishes (Fig. 29–4). White-fishes, chars, and trout are the usual inhabitants of oligotrophic lakes in the northern hemisphere. The resident ichthyofauna may seasonally adjust its vertical range in order to find the optimum conditions of temperature, oxygen, and available food.

Fishes of the open water in lakes are typically strong swimmers that can seek out and capture the crustacean or insect prey that forms most of the food base. Some of the whitefishes and the sockeye salmon (*Oncorhynchus nerka*) are equipped with long, fine gill rakers for capturing small planktonic organisms. The young of the latter species spends its early life in lakes before migrating to the ocean. A freshwater variant called the kokanee completes its life cycle entirely in fresh water. Large, often piscivorous, predators are frequently the same species seen in larger rivers and streams. These include the rainbow trout (*Oncorhynchus mykiss*) and the lake trout (*Salvelinus namaycush*). The northern pike (*Esox lucius*), saugers and walleye (genus *Stizostedion*), and yellow perch (*Perca flavescens*) are typical of shallow water fishes in oligotrophic lakes but can be found in warmer water than salmonids.

Oligotrophic lakes support a relatively low biomass per unit of area or volume. Steep rocky banks support a sparse representation of rooted aquatic plants, and the nutrient-poor water supports a limited supply of phytoplankton such that the primary productivity is generally low. Because the recycling of nutrients proceeds slowly, growth of fish can be slow in this situation, and replacement of fish biomass removed may take considerable time. One of the most extreme examples of this was seen in populations of brook trout stocked in an extremely oligotrophic lake in the Sierra Nevada Mountains of eastern California. The combination of low temperatures and the paucity of food sources resulted in stunted populations that lived far longer than the normal life span for the species, as long as 24 years (Reimers, 1979).

As a basin ages, nutrients accumulate by minerals entering in solution, by organic matter washing or falling in, and through various other processes. Nutrients can be trapped in the lake by incorporation into the biomass of the lake. These are then recycled through the actions of scavengers and microbial decomposers. Any

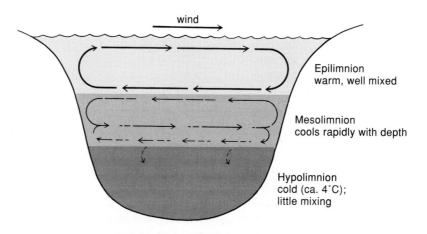

FIGURE 29–4   Zonation of a temperate oligotrophic lake, showing thermal stratification.

nutrients reaching a soluble state in the hypolimnion are not likely to be redistributed throughout the lake until there is a general recirculation of the water. Turnover of the lake occurs upon cooling in the autumn and warming in the spring, the result being that a generally uniform temperature is reached throughout the lake. This biannual circulation pattern redistributes the nutrients, dissolved oxygen, and other materials in the lake but can be modified depending on such factors as depth, altitude, latitude, and exposure to wind.

As bottom deposits increase and the nutrient content (and consequently the productivity) of the lake grows, it is generally said to be passing from the oligotrophic to the eutrophic condition, and its capacity to support a fish fauna and the species composition of that fauna change accordingly. Accumulation of soft bottom deposits, including much organic material, can increase the amounts of rooted vegetation in the shallow waters and can provide niches for many burrowing organisms, which may become available as forage for fishes. The rich water can support denser populations of phytoplankton, and the zooplankton community is thus provided with a greater food resource. Rapid recycling of nutrients can result in production of a large fish biomass. Because eutrophic lakes are capable of rapid replacement of biomass that is removed, they are suitable environments in which to develop sustainable fisheries for desirable sport and commercial fishes.

There are some trade-offs involved in the eutrophication process. Decomposition of organic material requires much oxygen, and the hypolimnetic waters of a rich lake can become anoxic. This can put some fishes in an uncomfortable squeeze during the warmer months. As surface waters warm, they may descend into the cooler depths only to encounter insufficient concentrations of dissolved oxygen. Massive fish kills can result. Salmonid populations in some eutrophic lakes become confined to a narrow stratum in the mesolimnion, thus restricting their feeding opportunities. Color and transparency can be affected during eutrophication due to organic compounds in solution and suspended materials, including plankton. The amount of carbon dioxide can increase significantly, as can that of ammonia and ammonium compounds. All these changes can be intensified and accelerated artificially by fertilization or organic pollution.

Fishes found in lakes that have evolved beyond the oligotrophic, salmonid-supporting situation vary with climate and faunal region. In northern areas, pike, perch, sauger, and walleye can coexist with or succeed salmonids. In warmer waters of North America, many centrarchid species that are also characteristic of river systems are encountered. These include black basses (*Micropterus*), crappies (*Pomoxis*), sunfishes (*Lepomis*), and rock basses (*Ambloplites*). Most centrarchids are compressed, deep-bodied fishes adapted to life around cover, such as vegetation or submerged logs and brush. Some tend to be solitary, whereas other species move in schools, but they tend not to be wide ranging in habit. Typical bottom feeders in warm lakes and ponds are those species of ictalurid catfishes, buffalofish, or other catostomid suckers that also frequent larger rivers and streams. Some species of small benthic families, such as the darters, also invade lake bottoms if suitable substrate is available. The shads, herrings, and alewives (genera *Alosa* and *Dorosoma)* of the primarily marine family Clupiedae and numerous species of cyprinids form an important part of the food base for predatory species. Introduction of clupeids as a food source for larger sportfishes, such as the basses of the genus *Micropterus,* is somewhat controversial, however. Some studies have suggested that introduced clupeids, in addition to providing forage for larger piscivores, will compete with the planktivorous young of these species, thus stunting their growth and diminishing the production capacity for desirable species in certain impoundments. This is especially true for the gizzard shad (*Dorosoma cepedianum*) (Lagler, 1956; Pflieger, 1975). The introduced carp can be very numerous in North American lakes and often disrupts the natural ecological relationships of native species.

As mentioned earlier, the natural succession of still-water habitats leads to the extinction of open water, with the final stages being bog ponds and swamps. These are characterized by shallow acid water, emergent vegetation, and soft, organically enriched bottoms. These waters are usually well past the peak of productivity as far as fish are concerned, and in North America they do not usually support a varied fish fauna. In the far north, Alaska blackfish (*Dallia pectoralis*) are found in shallow tundra ponds that freeze to the bottom, entrapping the fish, which survive if their body temperature does not drop much below 0°C. Sticklebacks (family Gasterosteidae) are another example of northern swamp-dwelling fishes. At lower latitudes, mudminnows (*Umbra* spp.), bullheads, some sunfish species, and, especially in warmer regions, gars (*Lepisosteus*) and the bowfin (*Amia calva*) inhabit swamps and backwaters.

The preceding examples are composites of conditions and faunas of temperate and cold North American lakes and actually represent a great simplification. There are many kinds of lakes classified by limnologists, according to such features as temperature, circulation patterns, productivity, and salinity. There are many inland bodies of water worldwide that, because of elevated salinities, support fish faunas more representative of marine environments. Such waters are termed athalassic if their saline condition does not arise from any preexisting association with the sea (Bayly, 1972). Fishes of the families Cyprinodontidae and Atherinidae are especially noteworthy for their osmoregulatory capacity and are conspicuous members of the ichthyofauna of inland saline lakes. Sticklebacks (Gasterosteidae) and even centrarchids are known to enter saline waters. Largemouth bass can tolerate exposure to moderately elevated levels of salinity and enter brackish marshes of the Gulf Coast of Texas and Louisiana (Meador and Kelso, 1990). The Salton Sea of southern California, an ancient pluvial basin in the Imperial Valley most recently reconstituted when floodwaters from the Colorado River refilled it early in this century, was so similar to coastal marine environments over 100 km to the west and south that marine species, such as bairdella (*Bairdella icistia*) and the orangemouth corvina (*Cynoscion xanthulus*), were introduced and formed the basis of a sport fishery. Steadily increasing salinities have eliminated this fishery in recent years, however (Moyle, 1976).

The age of such ancient lakes as Baikal and the African Rift Valley system has permitted the adaptive radiation of an amazingly diverse assemblage of fishes, based on the Cichlidae in the African lakes and on cottoids in Baikal. The most spectacular adaptive radiation of any vertebrate group has been the cichlids of the Rift Valley lakes, such as Malawi (Nyasa), Victoria, and Tanganyika. Over 500 species of cichlids, most of them endemic, are present in these three lakes. In Lakes Victoria and Malawi, over 80 percent of the lacustrine fish fauna are cichlids. Fully 60 percent of the nearly 300 species of Lake Tanganyika are members of the family Cichlidae. Cichlids have radiated to fill just about any ecological niche present in the lakes.

Examples are phytoplankton eaters of the genus *Tilapia,* zooplankton feeders (*Haplochromis intermedius* and *Limnochromis permaxillaris*), mollusk eaters (*Macropleurodes bicolor*), and piscivores (*Rhamphochromis* spp.). Others feed on algae scraped from rocks, some eat higher plants, and others filter organic matter from bottom deposits. Body form and feeding structures, including dentition and gill rakers, have become modified in a myriad of ways (Fig. 29–5). One group of hap-

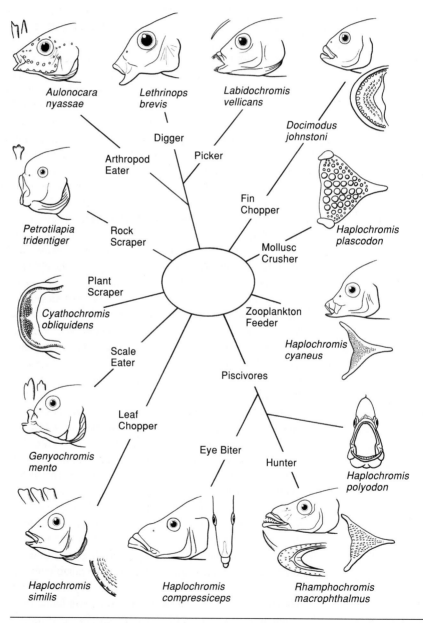

FIGURE 29–5 Explosive adaptive radiation in body form and dentition in African Rift Valley lakes. (*Source:* Fryer and Iles, 1972, in Lowe-McConnell, 1987.)

lochromine cichlids displays a remarkable specialization on different stages of development of other species of cichlids—different species consume recently fertilized eggs, embryos, early larvae, late larvae, shoaling groups of young, and older individuals of different sizes (Greenwood, 1974; Kortmulder, 1987; Witte, 1981). Studies on the community ecology and trophic structure of these complex communities reveal them to be consistently stable, characterized by long-term persistence and resilience of structure (Hori, 1991).

The sculpin subfamily Cottocomephorinae of Lake Baikal has diversified to include both benthic and open water members. The former are adapted to various bottom types and include some of the deepest-occurring freshwater fishes. They divide a food resource consisting largely of midge and caddisfly larvae and numerous species of copepods. Pelagic members of the family are said to specialize in feeding on planktonic copepods, with each fish species having a preferred prey species. The related Comephoridae are pelagic fishes that undertake diurnal vertical migrations from depths up to 1000 m. They are colorless and are distinguished by large mouths, extremely thin bones, and broad pectoral fins. Being viviparous and feeding on pelagic crustaceans, they have adapted completely to an open water existence (Berg, 1949; Nelson, 1976).

In tropical regions, where wet and dry seasons alternate, numerous kinds of swamp-dwelling fishes have evolved habits and structures that permit them to occupy environments that are often lacking in dissolved oxygen or even water. Several methods of air breathing have developed in swamp fishes; some are capable of moving overland to escape desiccation, and the African and South American lungfishes burrow into the mud during the dry season. The African lungfishes even form a cocoon and enter into prolonged periods of aestivation at this time. So-called annual fishes of the family Aplocheilidae (the African genera *Fundulopanchax* spp. and *Nothobranchius* spp.) and the Rivulidae (the South American genera *Cynolebias, Pterolebias,* and others) respond to the periodic drying of their environments by laying large numbers of drought-resistant eggs, which develop over the dry season and hatch when the rains return (Berra, 1981; Breder and Rosen, 1966; Simpson, 1979). This unusual life history trait has enabled biological supply houses to market "instant fish" kits consisting of eggs from the genus *Nothobranchius* that can be held in a suspended state until needed.

Cave Systems: An Unusual Aquatic Habitat. Caves form very special habitats for some species of freshwater fishes. A gradation of aquatic habitats is present, ranging from conditions similar to noncave environments near the mouth, to a twilight area of decreasing light and minimal temperature fluctuations, to the lightless interior where temperatures show little, if any, seasonal fluctuation. Virtually all organic matter in the inner cave habitat must originate outside. Streams flowing through may bring debris and plants, which provide food for invertebrates directly or can nurture fungi upon which the invertebrates can feed. In some caves, bat droppings provide considerable organic material. Invertebrate cave inhabitants include collembolans, crickets, ants, millipedes, spiders, isopods, and crayfishes. Some fish species that normally live outside caves occasionally enter them for food or shelter. These are called trogloxenes and may include members of several families, including the Ictaluridae, Cyprinidae, and Cyprinodontidae. Fishes that normally spend part of their life cycle in, and part out, of caves are termed troglophiles and include the springfish

(*Chologaster agassizi*) of North America and species of the genus *Chondrostoma* in Europe. These, of course, have well-developed eyes. Those species that are confined to caves and live in constant darkness are known as troglobites. The Amblyopsidae of North America are well-known cavefishes. Included are the genera *Amblyopsis* and *Typhlichthys*. Other families with troglobitic members are the ictalurids (*Satan, Trogloglanis*) of North America (these two genera were recovered from deep artesian wells in Texas, up to 500 m beneath the surface); Clariidae of Africa; Pimelodidae (*Typhlobagrus*) of Brazil; Cyprinidae (*Aulopyge, Coecobarbus*) of Europe and Africa; Synbranchidae (*Pluto*) of the Yucatan; and Brotulidae (*Stygicola*) of Cuba. *Poecilia sphenops*, a live-bearer from Mexico, has been noted as having normally sighted forms outside a cave near Tabasco, but with some sightless populations and intergrades inside. Some of the blind individuals are born without sight, whereas others are secondarily blinded by the growth of circumorbital tissue over the eyes. Cephalic lateralis canals are modified in the cave dwellers and are enlarged and partially open. In addition, those living in the dark depend on tactile stimulation in the courtship behavior rather than the typical visual stimulation generally seen in the species (Barr, 1968; Culver, 1982; Lee et al., 1980). A blind cave characin popular with tropical fish hobbyists has been designated the genus *Anoptichthys,* yet it is, in fact, a variant of the Mexican species *Astyanax mexicanus.* Eyeless and eyed forms readily interbreed and produce fertile offspring (Avise and Selander, 1972; Schemmel, 1980). Anyone who has kept these interesting fishes in a home aquarium soon loses any initial feelings of sympathy for these sightless individuals because they are invariably the first to find the food when sprinkled in the tank.

True cavefishes are usually sightless and lack pigment in the skin. The lack of vision is compensated for by development of other senses, particularly the acousticolateralis system and olfaction. The cave-dwelling catfishes utilize the sense of touch (and possibly taste) concentrated on their barbels. The lateral-line organs of amblyopsids are set out on the surface.

Due to a paucity of predators in cave ecosystems, cavefishes tend not to display escape responses. Although no seasonality exists within the cave environment, some species show an annual reproductive cycle. Amblyopsids have developed the habit of incubating their eggs in the branchial chamber. Although predation may not exist on the adults, the evolution of such behavior suggests that eggs and possibly larvae are vulnerable.

Marine Habitats and Adaptations

The oceans are characterized by great size and depth, continuity in time and space, and diversity of bottom type, water motion, temperature, and salt content. Although the salinity may vary locally due to the relative contributions of fresh water versus evaporation, the relative ionic composition is uniform worldwide. Marine fishes can live on or near the bottom, in what we have termed the benthic realm, or in the open sea, the pelagic realm. Marine ecosystems encompass these two realms, with some communities linking them together. For example, benthopelagic fishes, such as cod, live in the water column itself but feed on the bottom. The usual framework within which marine ecologists work is one of zonation of the two realms mentioned. Zone boundaries are crossed by numerous animals, but because the zones are established according to such factors as light penetration, temperature, and extent of the continental shelf and slope, they have considerable biological significance.

The benthic realm is divided into the shelf zone, which extends down to about 200 m; the upper slope zone, which extends to about 1000 m; the lower slope, reaching to about 3000 m; the abyssal plain, which extends to about 6000 m; and the hadal zone, which includes the oceanic trenches as deep as 11,000 m. The pelagic realm is divided into the epipelagic zone, to about 200 m (this roughly corresponds to the depth of light penetration sufficient for photosynthetic activity and the edge of the continental shelf); the mesopelagic zone, to about 1000 m, which is the limit of all surface light; the bathypelagic zone, which is aphotic and reaches the 6000-m depth; and the hadopelagic zone, in the deep trenches below 6000 m. Some ecologists find it convenient to divide the ocean into the neritic system (on and above the continental shelf) and the oceanic system (beyond the shelf) (Fig. 29–6).

The Benthic Realm

The Intertidal Zone. The part of the shore that is periodically covered and uncovered by the tides, the intertidal or littoral environment, constitutes a very special section of the shelf zone. Depending on exposure to waves and currents, the shore can be the site of erosion or deposition. Eroding shores are typically rocky and rugged. This environment can be enriched by coastal upwelling and runoff from the adjacent landmass—this results in high levels of primary production. The combination of turbulent conditions, caused by currents and wave wash, and the regular cycle of emersion and immersion as the tides ebb and flow make this a harsh and stressful environment in which to live. Life flourishes here nonetheless; the abundance and diversity of flora and fauna (including a number of fish families) and their easy accessibility have made intertidal shores one of the most significant ecosystems in

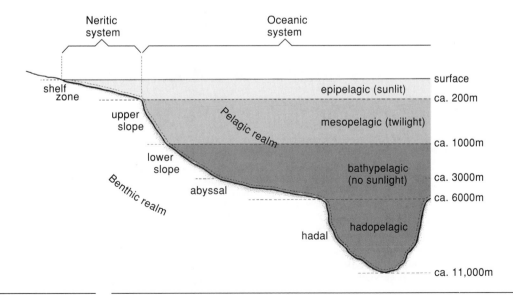

FIGURE 29–6 Ecological zonation of the ocean.

terms of general contribution to our understanding of ecological principles. The irregular substrate usually has depressions that retain water as the tide recedes so that fully marine conditions can be perpetuated there when the tide recedes. These tide pools are often strewn with boulders and harbor a variety of marine algae and invertebrates. Many species of fishes are specifically adapted to the intertidal zone, while other species enter the intertidal zone at high tide to forage but retreat when the tide recedes. Permanent residents of the intertidal zone seek refuge in the tide pools at low tide, although many species exist worldwide that can withstand periods of emersion and will wait out the low-tide interval secreted beneath damp boulders or in clumps of seaweed. Fishes found on exposed rocky intertidal shores must be able to withstand strong wave motion by swimming, hiding, or clinging (Fig. 29–7).

In temperate regions, intertidally occurring fishes must be tolerant of fluctuating temperature and salinity, especially if they frequent shallow pools close to the high-tide level. The terms *eurythermic* (having broad temperature tolerance) and *euryhaline* (broad salinity tolerance) have been used to describe the physiological adaptations of these fishes. Many intertidal fishes display anatomical features consistent with their benthic existence in a shallow, turbulent region. They are generally small, rarely longer than 30 cm, which enables them to occupy holes, crevices, and interstices among the rocks and vegetation. Distinctive fin structures are also an adaptation to intertidal life. Broad, thickened pectoral fins can be used to wedge the animal into crevices; highly terrestrial mudskippers can raise themselves on their pectorals in order to facilitate locomotion on land. In some families, the fins may become modified into adhesive discs further to facilitate maintaining position in turbulent conditions. The skin of intertidal fishes is generally tough so it can withstand the abrasive conditions normally encountered in its environment, yet many species, as an

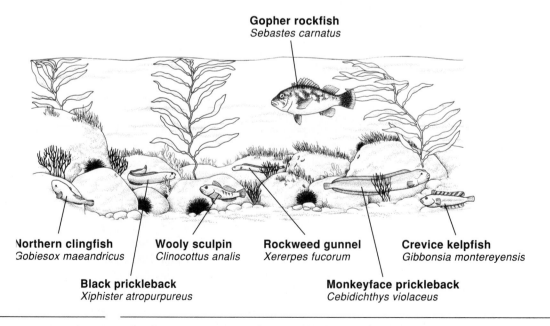

Gopher rockfish
Sebastes carnatus

Northern clingfish
Gobiesox maeandricus

Wooly sculpin
Clinocottus analis

Rockweed gunnel
Xererpes fucorum

Crevice kelpfish
Gibbonsia montereyensis

Black prickleback
Xiphister atropurpureus

Monkeyface prickleback
Cebidichthys violaceus

FIGURE 29–7   Representative intertidal fishes from the California coast. (*Source:* Horn and Gibson, 1988.)

adaptation to periodic deprivation of aquatic sources of oxygen, can apparently engage in cutaneous gas exchange (Horn and Gibson, 1988). Cryptic coloration is the rule in intertidal fishes—many are remarkable in the way that they can blend with the substrate and its associated vegetation. Those groups with the greatest representation in the intertidal zone worldwide include the Cottidae (sculpins), blennioids (families Blenniidae, Stichaeidae, Pholididae, and Clinidae), Gobiidae (including among the most terrestrially adapted of fishes, the mudskippers), and Gobiesocidae (the clingfishes—a family that is characterized by a ventral surface modified into a broad sucking disc—the vertebrate equivalent of a limpet). A few species from these families live in supratidal areas, some in semipermanent waters that are replenished by wave splash or spray. In rocky areas, sculpins, gobies, sleepers (gobioids—classified by some in a separate family, the Eleotridae), blennies, and clingfishes often inhabit the supratidal "spray zone." The preferred habitat of the large Chilean clingfish (*Sicyases sanguineus*) is out of the water and well above the water line, where it will remain as long as it can be occasionally doused by wave spray (Ebeling et al., 1970). In soft-bottomed areas of the warmer shores, the mudskippers (gobioids often classified in the separate family Periopthalmidae) commonly clamber about in mangrove roots or haul themselves along mudbanks with their armlike pectoral fins.

Depositing shores are characterized by fine bottom materials, usually sand along the open coast, and mud or mixtures of sand and mud in well-protected areas such as bays and mangrove swamps. Sandy beaches seldom provide hiding places for fishes to remain in between tides, but a few fishes are adapted for burrowing or hiding in the sand just below the tideline. The sandfish (*Trichodon*) has fringed lips that allow respiratory water to pass but that strain out the sand. Flounders (Pleuronectidae), sanddabs (Bothidae), and various skates (Rajidae) and rays (Dasyatidae) can cover themselves with sand. *Ammodytes,* the sandlance, is said to dive into the sand for protection. The venomous weeverfishes (Trachinidae) and stargazers (Uranoscopidae) habitually bury themselves in sand.

In general, sandy shores are not as rich in food resources as are the muddy areas, where a greater variety of vegetation can attach and grow and where accumulated detritus can foster the development of a diverse invertebrate fauna. These soft-bottomed areas may not host the diversity of intertidal fishes seen on rocky shores but may provide daily food for a much greater biomass of fish per unit area. Most fishes that feed in these areas move in from near shore as the tide covers the sand or mudflats and seek the variety of crustaceans, worms, molluscs, and other invertebrates found there. A few species, especially gobies, remain in the flats, hiding in clam holes or shrimp burrows. Sculpins, flounders, tonguefishes (Cynoglossidae), soles (Soleidae), sanddabs, surfperches (Embiotocidae), mullets (Mugilidae), and silversides (Atherinidae) are some examples of the many kinds of fishes that might be expected to move inshore at high tide. Some piscivorous species take advantage of the food resources in the intertidal zone. The striped bass (*Morone saxatilis*), a species native to the Atlantic Coast of North America, is such an adaptable inshore predator that it was successfully introduced on the Pacific Coast in the nineteenth century, where it has become a highly esteemed sport fish.

The Shallow Subtidal Zone. Below the limit of low tide, the environments are more stable than in the intertidal area. At shallow depths, however, there is water motion due to wave action and tidal currents, and seasonal fluctuations of light, temperature, and salinity are possible. Inshore areas with suitable substrates support great beds of marine algae, such as the kelp forests of the California coast. These kelp beds support a characteristic ichthyofauna ranging from large rockfishes (Scorpaenidae) and surfperches to minute clingfishes that adhere to the kelp blades. Exceptionally high productivity is encountered where upwelling brings nutrients from the deep ocean into nearshore areas. Upwelling often determines the productivity of commercial fishing grounds in shelf areas. Fish faunas are at their richest in the shelf zone, especially in the upper 50 m, the "inner sublittoral" part of the shelf. Scores of families and hundreds of genera have representatives in this biome. Included are the families encountered in the intertidal plus, for example, sharks (Carcarhinidae and others), guitarfishes (Rhinobatidae), many eels (Anguillidae, Ophichthyidae, Muraenidae), cods (Gadidae), pipefishes (Syngnathidae), numerous perchlike families (such as goatfishes, Mullidae; drums, Sciaenidae; nibblers, Girellidae; wrasses, Labridae), scorpionfishes (Scorpaenidae), and triggerfishes (Balistidae).

Coral Reefs. One of the most spectacular sections of the shallow shelf is the coral reef environment. The skeleton-forming capabilities of cnidarian corals, as well as some species of sponges and algae, result in prominent geological features in some shallow seas. Corals require consistently warm temperatures (23° to 25°C) and uniform salinities (they will not form near bays and estuaries, where fresh water mixes with ocean water). They require shallow water in order to support the photosynthetic activity of the symbiotic algae that reside in their tissues. The heterogeneity of the reef substrate provides a wealth of habitats and niches for biota, thus making coral reefs the richest biome in the world's oceans. Early estimates of the productivity of coral reefs suggested that they were up to 20 times more productive than the surrounding ocean (Rhyther, 1959). More recent studies suggest that the overwhelming complexity of interrelationships among coral reef associates makes reliable assessment of actual reef-based productivity extremely difficult, if not impossible, to obtain (Barnes and Hughes, 1982). Coral reefs are distributed widely in the Indo-Pacific between 30°N and 30°S and are well-developed around the West Indies and the Caribbean. The total number of fish species associated, either peripherally or entirely, with coral reef ecosystems has been estimated at between 6000 and 8000 species, or up to 40 percent of the known fish fauna (Ehrlich, 1975). The most characteristic groups can be broken down into four major assemblages: labroids (wrasses, Labridae; and parrotfishes, Scaridae), pomacentrids (damselfishes, Pomacentridae), acanthuroids (surgeonfishes, Acanthuridae; rabbitfishes, Siganidae; Moorish idols, Zanclidae), and chaetodontoids (butterflyfishes, Chaetodontidae; angelfishes, Pomacanthidae) (Sale, 1991). Conspicuous signaling coloration and deep, compressed bodies that allow rapid turns and access to narrow spaces are characteristic of the groups that forage in the water column above the coral. The tetraodontiform fishes, including the curious boxfishes (Ostraciidae), triggerfishes (Balistidae), and porcupinefishes (Diodontidae), hover about the reef using their highly modified fins as their primary propulsive organs. A host of other species, more cryptically colored to blend in with the bottom, make their home in the crevices and holes that abound in the reef environment. Anguilliform fishes, including the morays (Muraenidae), snake eels, and worm eels (Ophichthidae) are especially well adapted for such a cryptic existence. Some species avail themselves of the most unusual of habitats. The shrimpfish (Centriscidae, named for its crustacean-like carapace), for example, can insert itself among the formidable spines of sea urchins.

Modifications for food gathering range from the small mouths and long snouts of the longnose butterflyfishes (*Forcipiger* spp.) and the Moorish idol (*Zanclus*) to the coral-crushing beak of the parrotfishes. Wrasses usually have small to moderate-sized mouths, with anterior, canine-like teeth adapted for picking and grasping and posterior teeth modified for crushing. The numerous sea basses have wide mouths suited to predatory habits. Food resources of the reefs are tremendous, including not only the coral polyps and other benthic invertebrates, but also algae and plankton.

To catalog the myriad of adaptations encountered in the dozens of families inhabiting the reefs is beyond the scope of this discussion, but the fishes discussed here range from beautiful to bizarre, from clownlike to deadly dangerous. Some are adapted to life in the surf and in the channels that dissect the reefs. Others seek quieter water in the lagoons that are often encircled by the reef itself.

Although coral reef ecosystems are immensely complex, with a staggering diversity of associated organisms, their relative ease of accessibility, especially with the advent of SCUBA, has made them ideal environments to test and refine basic ecological concepts. For example, Doherty and Fowler (1994) have examined the role of larval recruitment in determining the structure of damselfish (Pomacentridae) populations inhabiting a series of patch reefs. Although many predictive models do not consider recruitment to be a potentially significant variable, Doherty and Fowler demonstrated that population abundance on localized patch reefs could be explained almost entirely through assessment of recruitment strength.

The Offshore Continental Shelf. The continental shelf, here restricted by definition as that part of the sea bottom shallower than 200 m, is variable in width. Where coastal mountains plunge directly into the depths of the ocean, no shelf may exist. In other places, the shelf may form a broad, shallow expanse up to 1200 km wide. The shelf averages 70 km worldwide. The shallowest parts of the shelf adjacent to the coastline are characterized by high productivity, many species of large algae, and a great quantity of fish species. Depending on the transparency of the water, the remainder of the shelf, the "outer sublittoral," retains some capacity to support plant life and therefore to produce organic matter. Much of the organic matter of the continental shelf has its origins in the overlying neritic waters, while a large proportion of it is terrigenous, having its origin from adjacent landmasses. Daily and seasonal fluctuations of temperature and light are modulated by the depth, and the bottom tends to be more uniform over larger areas, especially in sedimented regions. The shelf region typically consists of muddy bottoms, with sediments of varying composition interspersed with rock outcrops. Sediment composition often is determined by coastal geological processes. For example, off the coast of Maine, relatively recent glacial events have scoured much of the sedimentary strata, exposing a greater amount of bedrock. Freshwater runoff brings locally heavy deposits of terrigenous sediments as well.

This environment, although offering a considerable variety of habitats, can provide only a fraction of the niches available in inshore areas with greater heterogeneity of substrata. Instead of great numbers of species, each with a small or moderate biomass, shelf areas typically have fewer species, but these may exist in considerable abundance. Many of the world's great fisheries target benthic or benthopelagic species of the shelf regions, including soles and flounders, rockfishes (*Sebastes*), and the gadoids (*Gadus, Pollachius, Melanogrammus, Urophycis*).

Many shelf species reach lengths of 30 cm to 1 m. They are adapted to feed on the worms and molluscs of the bottom and on the crustaceans that are found both on or near the bottom. The abundance and diversity of such bottom fare often is the factor determining the productivity of a fishery in a given area. Fishes are usually visual in their feeding behavior on the bottom, with eyes that are well developed and somewhat enlarged due to the lower light levels. Benthic species usually lack gas bladders, and some, such as the poachers (Agonidae) and a few sculpins (such as *Radulinus* spp.), are armored. Benthopelagic species, such as the cods, are equipped with gas bladders and can thus hold positions above the bottom. Sensory barbels are present in the cods, poachers, and some of the sculpins, apparently to facilitate feeding on bottom fauna.

The Continental Slope and Abyssal Plain. Moving from the shelf onto the upper slope, temperature decreases steadily over the 200- to 1000-m span and the water movement is generally slight. The last weak rays of light from the surface are eventually extinguished at about 750 to 1000 m, but the greatly enlarged and extremely sensitive eyes of many fishes living at this depth can still make use of such low levels of ambient light intensity. Although bioluminescence is important in the pelagic realm at depths corresponding to the upper slope, it is relatively rare in the benthic fishes. A few rattails (Macrouridae) display luminescence at these depths. Lateral-line systems are also typically highly developed.

This zone depends, as do the deeper zones, on the shallower parts of the sea for virtually all the available food. Remains of plankton and other pelagic organisms accumulate on the bottom, where scavengers attack and recycle them. Many of the fish families from the upper slope are also encountered on the shelf. Flatfishes, including the halibuts (*Hippoglossus* spp.), are common, as are the hagfishes, cods, morids, scorpaenids, skates, squaloid sharks, eelpouts (Zoarcidae), seasnails (Liparidae), and cusk-eels (of Ophidiidae). The latter three families illustrate the relative success of the elongate body form among these deep-dwelling fishes. A common body shape is to have a large head, well-developed pectoral fins, and long dorsal and anal fins along a tapering afterbody. Lower shelf fishes also showing this body form include the chimaeras, rattails, Halosauridae, and Notacanthidae. It is thought that the combination of an elongate snout region with an anal fin that is longer than the dorsal fin aids in positioning the body in a head-down manner to facilitate bottom feeding. Many of these fishes reach lengths of 30 to 45 cm and more.

Reproductive adaptations in some species appear to prevent young from being removed too far from suitable habitat. Hagfishes, skates, and chimaeras deposit large eggs, from which hatch precocial juveniles essentially like the adults. Some brotulids and electric rays are ovoviviparous, and a number of eelpouts are viviparous. Agonids and some eelpouts deposit demersal eggs, and the latter guard the spawn. Cods, macrourids, and eels have pelagic eggs that can drift freely with the deep currents. These species typically have very high fecundities. The eggs are small and may hatch pelagic larvae that drift in the currents for a while. They are thus more vulnerable to predation and are more likely to drift away from optimum living areas. In many species, the eggs and larvae are hydrostatically balanced to float not at the surface but at some intermediate depth. In some flatfishes, eggs and larvae may drift inshore, where better nursery areas may exist. Juveniles subsequently return to deeper water. A few liparids place their eggs in the gill cavities of crabs.

In the lower slope and abyssal plain, physical conditions become quite uniform except for increasing pressure with depth. These zones are cold and dark, with slow bottom currents. The sediments contain less terrigenous materials and are predominantly composed of calcareous and siliceous deposits from the pelagic realm or, in very deep areas, of red clay. Food also originates from above. Particulate matter that reaches the bottom, if not eaten by scavengers, can be broken down by bacteria, which might then be eaten by mud-ingesting invertebrates, which in turn can be eaten by fishes.

Although the fish fauna of the benthic regions below 1000 m contains many elements of the upper slope, modifications for the deeper life are evident. Eyes are generally small, even vestigial in a few groups, and are so highly modified as to be considered absent in at least one genus (*Ipnops*). Pigmentation, though usually black or otherwise dark, may be lacking in some fishes. Lateral-line systems are greatly developed, with the neuromasts often set out on stalks. Many deep-ocean gadiform species have the ability to produce sounds that may be useful in locating mates. Bioluminescence is also common among the macrourids found in the deep ocean.

Gas bladders are present in about half of the species below 2000 m and are often of large size. Skeletons and scales are reduced greatly in weight, and many fishes are flabby and watery. The tripodfishes (*Bathypterois* spp.) have elongate rays on the pelvic and caudal fins. These are used to prop the animal above the soft ooze bottoms on which it lives.

In the late 1970s, oceanographers studying the deep ocean bottom about 280 km northeast of the Galapagos Islands discovered one of the most unusual ecosystems on earth, one that is apparently energetically based on warm, chemically enriched water escaping from a series of fissures on the ocean bottom at a depth of about 2500 m. Clustered around the vents were mussels, clams, and spectacular giant pogonophoran worms. Species of barnacles, crabs, and shrimp never before seen have been described as members of this community. A number of fish species have also been described as associated with these deep ocean thermal springs. Cohen and Haedrich (1983) have recorded about 20 species of fishes living in the vicinity of the thermal springs. The most common were three species of macrourid rattails (genus *Coryphaenoides*), an ophidiid (*Bassozetus*), and two or three species of zoarcid eelpouts. An unidentified member of the family Bythitidae was discovered to be actually living in the vents. It is obvious that the trophic enrichment afforded by these thermal vents serves as a magnet, attracting a number of fishes in the vicinity.

Not many fishes occur in the hadal zone. A brotulid (*Bassogigas profundissimus*) and a liparid (*Careproctus amblystomopsis*) have been collected deeper than 7000 m, and the deep sea explorers Piccard and Walsh observed a "flatfish" at 10,800 m in the Marianas Trench, the deepest known point of the world's oceans. The identity of this, possibly the deepest occurring of all fishes, is not known, although Wolff (1961) suspects that it may not be a fish at all but rather some species of holothurian (sea cucumber).

The Pelagic Realm

The pelagic realm is constituted of the great volume of the interconnected oceans consisting of almost all of the earth's water. A pelagic existence implies a life more or less independent of the ocean bottom. Fishes that are completely independent, carrying out their entire life cycle in open waters, are termed holopelagic. Others that swim in the water column but are bottom associated for purposes of feeding or reproduction are termed meropelagic or hemipelagic. Although most pelagic species are associated with the more productive shelf regions, others are much more wide ranging, with a life style that may entail annual migrations of several thousand kilometers over the open ocean.

With buoyancy of paramount importance for a pelagic existence, it is not surprising that the gas bladder is an almost universal organ among epipelagic fishes and is common among mesopelagic ones as well. The inclusion of fats and oils in muscles, body cavity, gas bladder, or liver is an effective flotation device as well but requires a greater mass of material than does gas. Pelagic sharks may have livers constituting a quarter of their total weight. The specific gravity of a fish can be brought close to that of water simply by including a large proportion of water somewhere within the confines of the skin, giving many pelagic fishes a gelatinous consistency. Many species of snailfishes and the ocean sunfish (*Mola mola*) are good examples of this. Reduction in the weight of the skeleton confers some hydrostatic advantage. Many deep-living species that lack gas bladders have very thin papery bones and lack scales.

One might expect that scombroids, which are powerfully built pelagic fishes, would typify fishes having well-developed gas bladders, yet this is not the case. Many scombroids regularly make rapid vertical ascents and descents in conjunction with feeding activities. To avoid the limitations on the rate of vertical movement that might be imposed by filling and emptying a gas bladder to maintain buoyancy, these species have simply divested themselves of the structure and remain aloft in the water from the lift generated by their rapid forward motion.

A high surface-to-volume ratio is useful to reduce the rate of sinking in pelagic organisms. This is probably best illustrated in planktonic crustaceans (copepods) that possess elaborate cirri on antennae and other appendages. Fishes as small as the larvae of pelagic species and as large as the manta ray probably also employ high body surface areas to retard the rate of sinking.

Pelagic life requires special adaptations for food gathering. The basis of the food web is phytoplankton, on which numerous invertebrates and larval fishes (the zooplankton) feed. Fishes such as the clupeoids may have fine sievelike gill rakers with which to strain out both phyto- and zooplankton, or they may be adapted to pick off the zooplankters one by one. Some larval fishes and invertebrates are predators, thus assuming the niche of top carnivores in the plankton community. Great swimming powers are common among epipelagic fishes and are necessary for capture of prey, avoidance of predators, and migration. One definition of a planktonic organism is that it does not disperse under its own power but rather is dependent on the ocean currents. If this is the case, the ocean sunfish (*Mola mola*) is probably the largest planktonic organism given its limited powers of locomotion. Ocean sunfishes also serve as a reference point in an open ocean that is virtually devoid of any substratum with which to orient. Consequently, they are host to a myriad of organisms, including a number of parasitic forms that use the sunfish's vast bulk as a source of food and shelter. Other large "planktonic" forms include the oarfish (*Regalecus*), while some of the smaller planktonic fishes would include the sauries and flying fish. The genus *Schindleria* is one of the smallest ichthyological members of the plankton community, achieving an adult size of only 2 cm. This enigmatic genus of minute, paedomorphic fishes has been classified recently as a gobioid (Johnson and Brothers, 1993).

The Epipelagic Zone. This thin surface layer of the oceans is generally considered to extend to about 200 m, although conditions typical of the epipelagic may disappear at shallower depths or extend deeper, depending on location and numerous interacting physical factors. Although the epipelagic zone is generally the warmest and best-lighted layer, it is subject to daily and seasonal changes in many physical features. Wherever the light and sufficient nutrient materials coincide, plants can photosynthesize, thus producing the organic material that supports life, not only in the upper layers, but indirectly in the lower layers as well. Dissolved oxygen is high in the epipelagic zone, near saturation in the upper layers. Salinity ranges from about 33 to 37 ppt.

Close to the surface, usually in the upper 10 to 25 m in high latitudes but sometimes reaching to 200 m or more in the subtropics, there is a stratum of uniform temperature. This may reach or exceed 20°C in the tropics and fluctuates only a few degrees during the year. Proceeding poleward, through temperate to polar latitudes, the upper temperatures become lower, with greater seasonal fluctuations in temperate areas than in colder regions. Below the surface layer there is a zone of rapidly falling temperature, the thermocline, in which the temperature can drop nearly 1° per 10 m of depth. This is a constant feature of warm oceans but develops only during the summer in high latitudes. Below the thermocline (and well below the epipelagic) the sea is cold, in some places reaching below 0°C. Stenothermic fishes are confined to certain depths, water masses, or latitudes by their narrow tolerances, whereas eurythermic species can range from one temperature to others, depending on their overall tolerance and requirements.

With 70 families represented in the epipelagic zone, this vast region of the world's oceans encompasses less than 2 percent of all known species of fishes. These range from the aforementioned *Schindleria* and dwarf sauries (Scomberesocidae) of 15 cm up to the gigantic whale shark of 18 m, but most representatives are from 30 cm to 1 m in length. The swift swimmers, such as tunas and other mackerel-like species, pelagic sharks, and some carangids, are fusiform in body shape with stiff, narrow, keeled caudal peduncles and crescent-shaped caudal fins. This design permits maximum power output with minimum resistance imposed on the rapidly beating caudal fin. The powerful billfishes (marlins, swordfish) probably derive a considerable streamlining advantage at high speeds from their elongate bills; these animals are reported to reach 130 km per hour. Many families have somewhat less of the classic fusiform shape, some retaining a lunate or forked caudal but having a more compressed body and others being more elongate, even arrow shaped, such as the barracudas.

Epipelagic fishes are remarkably uniform in coloration—a dark dorsal surface, either green or blue, giving way to silvery sides that reflect whatever the prevailing colors are in the immediate environment. This pattern, known as obliterative countershading, enables the fish to blend in with the background regardless of the angle from which it is viewed. Neritic fishes may sometimes add variations in the form of stripes or bars.

The richest parts of the pelagic realm are usually the neritic regions, close to mineral enrichment from terrestrial runoff, and subpolar areas, where vertical interchange of water occurs each winter. Furthermore, localized coastal upwelling can enrich the surface waters with nutrients liberated by the breakdown of organic matter in the depths. This rich water is brought up also in regions of current divergence, as along the equator. Great abundances of fishes, many of them of commercial significance, can be supported in these areas. The distribution of nutrients, and thus primary productivity is very patchy over the epipelagic realm, with much of the open ocean a "wet desert" with minimal organic production. In those patches where primary productivity is high, fisheries can be enormously productive. The waters off the coast of Peru and Ecuador, for example, are host to a fishery for the Peruvian anchovy (*Engraulis ringens*) that was until recent times, the single most productive fishery in the world. Overfishing, combined with a series of unusually severe El Niño events, have caused a drastic decline in this fishery.

A special fish community of the pelagic realm consists of essentially demersal species that live among the dense beds of Sargasso weed found far offshore in the Atlantic Ocean (a region known as the Sargasso Sea) and in the western Pacific. The Sargasso Sea is a vast region that is encircled by the main currents of the north Atlantic (Fig. 29–8). Although confined to a relatively nutrient-deficient water mass,

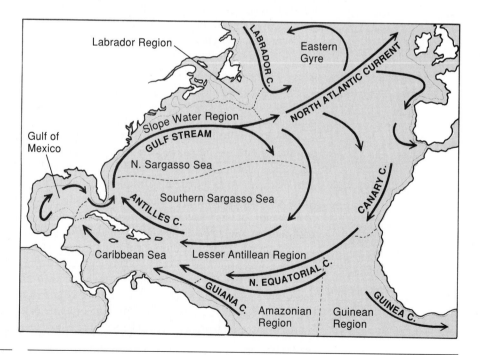

FIGURE 29–8   Map of the North Atlantic Ocean showing the location of the Sargasso Sea. (*Source:* Ingmanson and Wallace, 1989.)

the sargassum weed community is still home to about 50 species of fishes (from 23 families) and invertebrates. Included in the fish fauna are two endemic species—the sargassum fish (*Histrio histrio*) and the sargassum pipefish (*Syngnathus pelagicus*)—some species of filefishes (*Alutera*), and the young of several true pelagic species (Dooley, 1972). Flotsam, such as *Sargassum* and other algal species, or debris of a terrestrial origin attracts not only small animals that seek shelter but also larger fishes that feed on these organisms. Large floating jellyfishes attract certain small fishes that can live unharmed among the tentacles. The best example appears to be the man-of-war fish (*Nomeus gronovii*), which lives with the siphonophore *Physalia* (Mansueti, 1963). Unfortunately, the *Sargassum* community is but one of the many complex macroalgal communities that are threatened by commercial harvesting. Seaweeds are harvested for a number of uses, including meal products for human and animal consumption as well as alginates for the pharmaceutical and cosmetic industries. Controversy surrounds these operations because the magnitude of their impact on fish populations has yet to be determined (cf. Black and Miller, 1991, 1994; Rangeley, 1994). If large-scale harvesting of macroalgae is to continue, however, measures must be taken to minimize damage to the invertebrates, fishes, and turtles that comprise the intricate food web of these remarkable communities.

Reproductive adaptation of epipelagic bony fishes involves high potential fecundity, with most larger species releasing hundreds of thousands to millions of eggs. Tunas, depending on size, release from about 1 to 10 million eggs. The ocean sunfish is reported to spawn up to 300 million eggs. Such high fecundity among epipelagic species is obviously a compensation for the extremely high mortality experienced in their planktonic egg and larval stages. In holopelagic species, the eggs are released in the open ocean, mostly to drift freely with the winds and currents, but some sauries and flyingfishes attach eggs to flotsam by means of threadlike structures on the shell. Duration of incubation is short in most of the drifting eggs, especially in the tropics. Usually they hatch in from one to three days. Some species—especially sauries, halfbeaks, and flyingfishes—require an incubation time of one to two weeks. Some holopelagic species are intermittent spawners, extending their reproductive activity over several months.

Mesopelagic fishes that deposit demersal eggs must do so where the larvae can drift to proper nursery areas before transforming into juveniles. This is also a concern for some holopelagic species. They have therefore evolved spawning migration patterns that place a large proportion of the drifting eggs and larvae in favorable currents. The larvae are usually modified for flotation, with a great surface-to-volume ratio or with inclusions of oil. Epipelagic sharks, for the most part, are ovoviviparous or viviparous, giving birth to young well able to swim and forage. Consequently, broods are small. The giant whale shark is oviparous, depositing eggs surprisingly small for such a great animal (Breder and Rosen, 1966; Nikolsky, 1963). Fish families most frequently associated with the epipelagic realm are indicated in Table 29–2.

The Mesopelagic Zone. Although sunlight attenuates and disappears almost completely near the lower boundary of the mesopelagic zone (at approximately 1000 m), small populations of phytoplankton can still be found there, along with debris from epipelagic plankton. This forms the food base for a sometimes abundant and varied zooplankton community that includes some species with the habit of moving

TABLE 29–2   **Families and Representative Species Present in the Epipelagic Zone**

Family	Species	Remarks
Lamnidae	white shark, *Carcharodon carcharias*	Holopelagic, neritic
	salmon shark, *Lamna ditropis*	Holopelagic
Cetorhinidae	basking shark, *Cetorhinus maximus*	Holopelagic, neritic
Alopiidae	thresher shark, *Alopias vulpinus*	Holopelagic, neritic
Rhincodontidae	whale shark, *Rhincodon typus*	Meropelagic
Carcharinidae	tiger shark, *Galeocerdo cuvieri*	Holopelagic, neritic
Clupeidae	herring, *Clupea harengus*	Meropelagic
	sardine, *Sardinops sagax*	Holopelagic, neritic
Engraulidae	anchovy, *Engraulis encrasicolus*	Holoepipelagic, neritic
	northern anchovy, *E. mordax*	Holoepipelagic, neritic
Salmonidae	Atlantic salmon, *Salmo salar*	Meropelagic
	chinook salmon, *Oncorhynchus tshawytscha*	Meropelagic
Myctophidae	(Numerous species of lanternfishes migrate into epipelagic at night from below)	
Exocoetidae	oceanic flyingfish, *Exocoetus obtusirostris*	Holopelagic
	California flyingfish, *Cypselurus californicus*	Holopelagic, neritic
Scomberesocidae	Pacific saury, *Cololabis saira*	Holopelagic
	Atlantic saury, *Scomberesox saurus*	Holopelagic
Lampridae	opah, *Lampris guttatus*	Holopelagic
Echeneidae	remora, *Remora remora*	Holopelagic
Carangidae	pilotfish, *Naucrates ductor*	Holopelagic
	rough scad, *Trachurus symmetricus*	Holopelagic, neritic
Coryphaenidae	dolphin, *Coryphaenus hippurus*	Holopelagic, neritic
Bramidae	bigscale pomfret, *Taractes longipinnis*	Holopelagic
Luvaridae	louvar, *Luvarus imperialis*	Holopelagic
Gempylidae	escolar, *Lepidocybium flavobrunneum*	Holopelagic
Scombridae	frigate mackerel, *Auxis thazard*	Holopelagic, neritic
	albacore, *Thunnus alalunga*	Holopelagic, neritic
Xiphiidae	swordfish, *Xiphias gladius*	Holopelagic
Istiophorinae	white marlin, *Tetrapturus albidus*	Holopelagic
Centrolophidae	medusafish, *Icichthys lockingtoni*	Holopelagic
Nomeidae	silver driftfish, *Psenes maculatus*	Holopelagic
Tetragonuridae	bigeye squaretail, *Tetragonurus atlanticus*	Holopelagic
Stromateidae	butterfish, *Peprilus triacanthus*	Meropelagic
Molidae	ocean sunfish, *Mola mola*	Holopelagic
	slender mola, *Ranzania laevis*	Holopelagic

upward into shallower depths during the night. Larger mesopelagic organisms, including many species of fishes, also migrate vertically. The lanternfishes (Myctophidae) are among the best-known family of mesopelagic vertical migrators. In terms of numerical abundance and diversity, the myctophids, as well as the bristlemouths (Gonostomidae), dominate the meso- and bathypelagic realms. Like many fishes inhabiting these depths, they are generally small (less than 20 cm long), dark in color (black and red are the most common colors in mesopelagic animals), and supplied with photophores arranged in species-specific, and sometimes gender-specific, patterns on the body (Marshall, 1970, 1971, 1980).

Rapid vertical migrations of 200 m or more—one round trip each night—subject the fish to tremendous changes of pressure. Because of the relationship between pressure and volume of gases, filling and emptying the gas bladder in order to maintain a constant volume and hence neutral buoyancy becomes a physiologically challenging task. About half of the mesopelagic species have divested themselves of a gas bladder and depend on fatty deposits or other means for hydrostatic balance. Fats, which are buoyant but much less compressible, often replace gases in the gas bladder. Not all vertical migrants move up into the surface layer of isothermic water. Some migrate from below the thermocline into the thermocline. These form a sometimes dense concentration of life that is readily detectable by underwater SONAR—hence the name "deep scattering layer" (Farquhar, 1970).

Mesopelagic fishes tend to have relatively large, well-developed eyes. Some species have large light-gathering areas associated with the eyes. Sunlight, however dim, still plays an important role in the lives of mesopelagic organisms. Visual pigments are those sensitive to light in the blue portion of the spectrum, that wavelength that penetrates the deepest and hence imparts the characteristic blue color of the oceans. Visual cells consist largely of rods—exclusively rods in those species that are confined to the mesopelagic. Bioluminescence is another source of light and plays a significant role in predator–prey relationships and reproductive mechanisms (Marshall, 1971).

Predators in the mesopelagic range from small stomiatoids of less than 20 cm to the daggertooth (*Anotopterus*) and lancetfishes (*Alepisaurus* spp.) of 1 m or more. Although most species are small, they are fearsome in appearance, with large eyes and large mouths lined with dagger-like or barbed teeth. Some species, such as the swallowers (Chiasmodontidae), have greatly distensible stomachs and can swallow just about anything that comes their way (Fig. 29–9).

Modifications of the lateralis system, apparently for greater sensitivity, appear in some mesopelagic fishes. These involve placement of the neuromasts on the surface of the body or actually on papillae or pedicels, as in snipe eels and some of the ceratioid anglers. Other groups, such as whalefishes, have enlarged pores opening from large canals. The auditory organs of deep sea fishes differ greatly in the relative size of the sacculus and lagena, those parts concerned with hearing. Some investigators believe that size is correlated with sensitivity to different frequencies. There is probably abundant sound in the deep waters for the ears to respond to, because many fishes and crustaceans have what are obviously sound-producing structures.

Reproductive adaptations of mesopelagic fishes are poorly understood. Most apparently release pelagic eggs, some of which float at the surface and some at density layers below the thermocline. The larvae are also pelagic and usually drift at the same level as the eggs.

There is a great diversity of life in the mesopelagic zone. Zooplankton species there usually outnumber those of the epipelagic, and there are about 1000 fish species represented, some of which are interzonal. Resources are available for the establishment of many niches, and most fish species are specialized in some way. Soft-rayed fishes dominate; there are about 200 species of Myctophidae and numerous salmoniform species. Table 29–3 shows examples of species encountered in the mesopelagic zone, and a few representative forms are shown in Fig. 29–9.

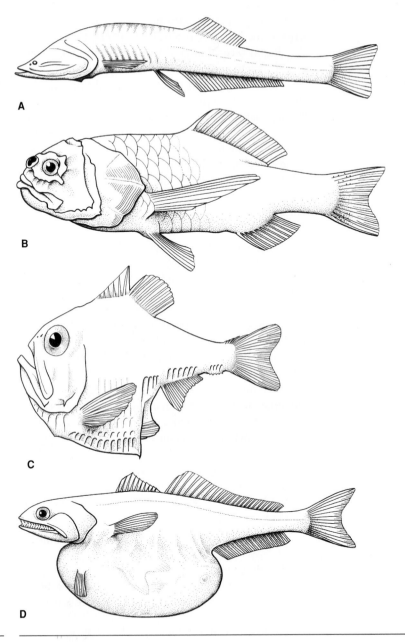

FIGURE 29–9   Examples of mesopelagic fishes. **A**, bristlemouth (*Cyclothone*); **B**, bigscale (*Poromitrus*); **C**, marine hatchetfish (*Argyropelecus*); **D**, black swallower (*Chiasmodon*).

The Bathypelagic and Hadopelagic Zones. Below the limit of light, the environment is uniformly dark and cold. Generally, food is scarcer than in the upper zones, and fish populations are sparser and consist of fewer species. The fishes living in these cold temperatures and great pressures typically have flaccid bodies with reduced skeletal deposition. Gas bladders are usually absent in species below 1000 m. The

TABLE 29–3	**Families and Representative Species Present in the Mesopelagic Zone**	

Family	*Species*
Squalidae	broadband dogfish, *Etmopterus gracilospinous*
	collared dogfish, *Isistius brasiliensis*[1]
Synaphobranchidae	Atlantic deep sea eel, *Synaphobranchus infernalis*
Nemichthyidae	slender snipe eel, *Nemichthys scolopaceus*
Argentinidae	Pacific argentine, *Argentina sialis*
Bathylagidae	California smoothtongue, *Leuroglossus stilbius*
Opisthoproctidae	spookfish, *Macropinna microstoma*
Alepocephalidae	slickhead, *Alepocephalus bairdi*
Gonostomatidae	lightfish, *Gonostoma denudotum*
	anglemouth, *Cyclothone microdon*[1]
Sternoptychidae	hatchetfish, *Argyropelecus olfersi*[1]
Stomiatidae	boafish, *Ichthyococcus ovatus*
Chauliodontinae	Pacific viperfish, *Chauliodus macouni*
Melanostomiatinae	longfin dragonfish, *Tactostoma macropus*[1]
Idiacanthinae	black dragonfish, *Idiacanthus fasciola*
Chlorophthalmidae	shortnose greeneye, *Chlorophthalmus agassizi*
Scopelarchidae	northern pearleye, *Benthalbella dentata*
Paralepididae	duckbilled barracudina, *Paralepis atlantica*[1]
Alepisauridae	longnose lancetfish, *Alepisaurus ferox*[1]
Anotopteridae	daggertooth, *Anotopterus pharao*[1]
Myctophidae	lanternfish, *Myctophum punctatum*[1]
	northern lampfish, *Stenobrachius leucopsaurus*[1] (numerous additional genera and species)
Trachipteridae	dealfish, *Trachipterus arcticus*
Bregmacerotidae	antenna codlet, *Bregmaceros atlanticus*
Gempylidae	oilfish, *Ruvettus pretiosus*

[1]Interzonal.

only visual cues are those associated with the luminous organs present in many deep pelagic fishes. By far the most common fishes in the bathypelagic are bristlemouths (Gonostomidae) of the genus *Cyclothone* (Marshall, 1971).

Modifications for feeding are remarkable in these zones. The mouths of fishes such as the gulpers and anglerfishes are unbelievably capacious. The gulpers (families Saccopharyngidae and Eurypharyngidae) also possess distensible stomachs that enable them to engulf prey larger than themselves (Fig. 29–10). Anglers apparently attract prey by displaying a luminous lure called the esca, borne on a "fishing rod" or illicium on the head. This is a feature characteristic of the order Lophiiformes and, as such, is seen on many shallow water relatives, such as the frogfishes and sargassumfish (Antennariidae) and the goosefishes (Lophiidae). Among the most bizarre of the deep water anglers are the members of the family Linophrynidae. These possess an enormous set of luminous barbels below the mouth as well (Fig. 29–10). The viperfishes (*Chauliodus*) have an elongate first ray on the dorsal fin that appears to function as a lure. Any of these unusual devices will assist in the capture of prey in an environment that is characteristically so low in biomass that encounters between predator and prey are relatively rare.

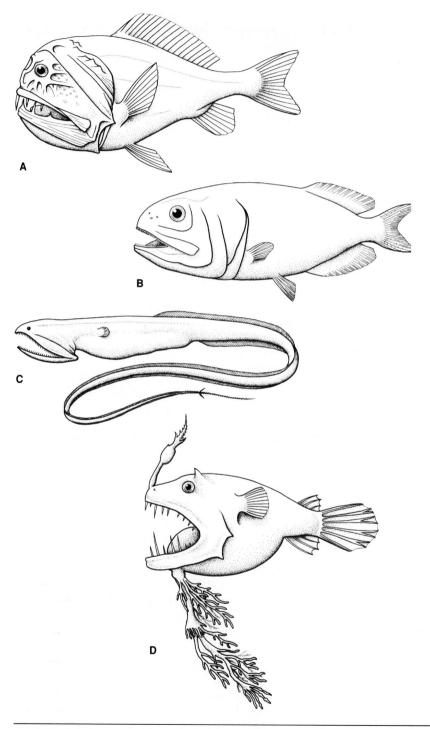

FIGURE 29–10 Examples of bathy- and hadopelagic fishes. **A,** fangtooth (*Anoplogaster*); **B,** whalefish (*Rondeletia*); **C,** swallower (*Saccopharynx*); **D,** ceratioid anglerfish (*Linophyrne*).

TABLE 29–4	**Representative Families and Species in Pelagic Zones Below 1000 Meters**
Family	**Species**
Nemichthyidae	snipe eel, *Nemichthys scolopaceus*
Saccopharyngidae	whiptail gulper, *Saccopharynx ampullaceus*
Eurypharyngidae	pelican gulper, *Eurypharynx pelecanoides*
Bathylagidae	blacksmelt, *Bathylagus antarcticus*
Gonostomatidae	bristlemouth, *Gonostoma bathyphilum*
Stomiidae	loosejaw, *Malacosteus niger*
	ribbon sawtail fish, *Idiacanthus fasciola*
Giganturidae	gianttail, *Gigantura vorax*
Paralepididae	slender barracudina, *Lestidium ringens*
Evermannellidae	sabertooth, *Evermannella atrata*
Himantolophidae	footballfish, *Himantolophus groenlandicus*
Ceratiidae	seadevil, *Cryptopsaras couesi*
Chiasmodontidae	black swallower, *Chiasmodon niger*

The ceratioid anglerfishes are well known for their obviously efficient method of ensuring that the sexes will be together at spawning time—a chancy situation given the comparatively low numbers of individuals in the bathypelagic realms. The tiny males of many species are parasitic on the females. They grasp the female with their jaws and subsequently become nothing more than testes-bearing lobes on her body, depending on her circulatory system for nourishment. Little is known of reproduction in the deep sea except that many species have pelagic eggs and larvae and that males of many species are smaller than the females.

Table 29–4 lists some families and representative species found in the deep pelagic zones below 1000 m. Examples of deep pelagic fishes, including the aforementioned gulper and linophrynid anglerfish, are shown in Figure 29–10.

Life History Variation

Fishes are nothing if not adaptable to changes in their environments. The success of transplantation into new environments, as in the case of carp from the old world into the new world or striped bass from the Atlantic Coast to the Pacific Coast, is testament to the adaptability of fishes to novel environmental situations. Such adaptation surely involved modification in the ways that fishes live—changes in foods and feeding behavior, reproductive habits, and other life history traits. In some cases, changes in the environment have been documented in influencing the course of evolution by selecting for certain life history attributes that are optimal for certain environmental situations (cf. Bruton, 1990).

It is expected that evolutionary history is of overriding importance in dictating life history patterns. For example, although elasmobranchs and bony fishes share similar marine habitats, the former typically practice internal fertilization and produce few large eggs, whereas the latter typically produce large numbers of small eggs that are usually fertilized externally. Even closely related species may differ profoundly in life history patterns. The genus *Oncorhynchus* includes Pacific salmon, which die after breeding once (a condition known as semelparity), as well as the largely freshwater trouts, which are repeat spawners (a condition known as iteroparity). The Atlantic salmon of the closely related genus *Salmo* is also an iteroparous species, with many adults surviving to breed again.

Life history traits can be conceived of as conforming to one of two strategies, termed r-selection or K-selection. An r-selected species is adapted to an environment of high unpredictability in which resources are less limited. The adaptive response to this situation is to produce the largest possible number of offspring but invest little care to ensure their survival due to the abundance of resources. A K-selected individual has a life history pattern that is adapted to a more constant environment in which competition for resources is more intense. In such cases, it is evolutionarily more prudent to produce fewer young and invest a greater amount of energy in parental care to ensure the survival of those that are produced. For fishes, larger eggs give rise to larger larvae. Because mouth size, swimming capacity, and sensory abilities increase with size, larger larvae would seem to have a greater chance of survival. This implies a trade-off since fecundity is inversely correlated with egg size in fishes (Wootton, 1992). Although cod and salmon are of approximately the same size when sexually mature, the cod produces several hundred thousand small-diameter eggs that are broadcast into the ocean waters, whereas the salmon produces far fewer larger eggs that are buried in a nest. For the salmon, the evolution of larger egg size and greater egg care ensures an increased survival of early life history stages. Variations in life history patterns within a particular genus of fishes can occur along a continuum ranging from r-strategists to K-strategists. The darters (genus *Etheostoma*), for example, have been demonstrated to exhibit a range of reproductive capacities (the measurable parameters being number and size of eggs produced) that correlate with body size, latitude of occurrence, and spawning habits (Paine, 1990). These kinds of studies may enable us to determine why some species of a given genus are so rare in some habitats while others may be exceedingly common.

As one might expect, life history strategies are not so simple that they can easily fit the dictates of the r- and K-selection model exclusively. A key feature that this model does not consider is the possibility of differential mortality of young versus adult stages. An alternative model, termed "bet hedging," assumes a survival rate for immature individuals that is much different from that for adults. If the optimal number of offspring produced varies from year to year, selection would favor a bet-hedging strategy in which a smaller than optimal number of offspring would be produced. This model may be more applicable in that it considers the consequences of spatial and temporal variation in environmental factors influencing survival, growth, and reproduction of individuals (Stearns, 1976; Stearns and Crandall, 1984; Winemiller and Rose, 1993; Wootton, 1990).

Perhaps the most intriguing life history studies are those in which we can detect the specific environmental factors that are influencing the evolutionary modification of a particular life history trait. Streamflow variations may be a selective factor in the determination of life history traits such as fecundity and egg size (Heins, 1979). Other physicochemical features of the environment may exert an impact on the evolution of life history patterns (Garrett, 1982; Heins and Baker, 1987). Field and laboratory studies by Reznick and Endler (1982) have revealed remarkable adaptive responses by Trinidad guppies to different levels of predation. In some streams, killifish are the chief predators, and they focus on immature guppies but are unable to consume mature adults. In other locations, predatory pike cichlids feed on adults but seldom expend energy on the young. Reznick and Endler (1982) discovered that guppy populations in areas where the pike cichlids were present matured sooner, were smaller at maturity, and mated at an earlier age. Larger numbers of offspring were produced at a greater frequency than were produced in areas experiencing only killifish predation. For male guppies, the situation is especially precarious. Males rely on bold coloration and courtship displays to attract the larger, but more drably colored, females. For them, conspicuous behavior can result in one of two outcomes—successful mating or a quick snack for a passing pike cichlid.

Trophic and Other Ecological Relationships of Fishes

Who Consumes Whom?

The trophic interrelationships of organisms are often conceived of as food chains in which the roles of component organisms as either predators or prey are indicated. In the aquatic realm, fishes are both predators and prey. A food chain usually has no more than five or six links, starting with the primary production of photosynthetic organisms and ending with the top carnivores. Fishes may comprise several of these links. For example, a pelagic food chain may consist of clupeoids consuming both phyto- and zooplankton, codfish consuming the herring, and sharks consuming the codfish. But some sharks, such as the whale shark, basking shark, or "megamouth," feed lower on the food chain by being essentially planktivorous. This enables these species to gain access to a much larger biomass of available prey and may have contributed to their being among the largest of all fishes.

Trophic relationships of fishes can be exceedingly complex. Fishes may occupy different levels of the food chain at different stages of their life histories. Comparatively few fish species are herbivorous as adults. Most of these consume planktonic or filamentous algae, but a few can live on larger plants. Examples of algae eaters are several cichlids, *Tilapia mossambica, T. nilotica, T. galilaea,* and *T. macrochir.* Others that feed largely on algae are loricariid catfishes, the silver carp, *Hypophthalmichthys molotrix,* and species of the cyprinid genera *Campostoma* and *Discognathichthys.* The grass carp, *Ctenopharyngodon idella,* can feed exclusively on rooted aquatic plants. *Tilapia melanopleura* eats both algae and higher plants (Bardach et al., 1972). As mentioned earlier, some characins, such as *Metynnis* spp. and *Colossoma,* are vegetarians, and some species include terrestrial fruits and nuts in their diets. Herbivory is especially rare in the marine environment, especially in temperate seas (Horn, 1989). One of the largest blennioid species, the monkeyface prickleback *Cebidichthys violaceus,* grazes on leafy benthic algae that abounds in its north Pacific nearshore habitat. Herbivores also obtain significant nourishment from the small, vegetation-associated animals that are also consumed.

Many fish species are termed microphagous because of their habit of feeding on tiny organisms or detritus in bottom deposits. These species usually have some modification for filtering out desired materials, but some, living in rich surroundings, devour the bottom mud itself and digest the living and dead organic matter contained in it. The larvae of lampreys are well known for their microphagous habits, living in organically enriched bottoms and straining out their food at the mud-water interface by means of an intricate oral sieve. Other microphages are the gray mullet, *Mugil auratus,* the South American hemiodontines and anodontines, and the mud carp, *Cirrhinus molitorella. Tilapia leucosticta* and *Trichogaster trichopterus* eat plankton and decomposing plant materials. Given the diet of these bottom-feeding species, they may be fulfilling the combined roles of herbivore, carnivore, and scavenger. The common carp (*Cyprinus carpio*) is another example of a truly omnivorous fish because it can feed on benthic invertebrates, vegetation, and occasionally other fishes. Its hardy constitution and omnivorous nature have enabled it to adapt to a wide variety of aquatic environments in which it has been introduced—often to the detriment of native species. Most fishes, feeding further up on the food chain, consume, in addition to insects, other aquatic invertebrates, such as molluscs and crustaceans. Piscivory is, as expected, common among both marine and freshwater fishes; but frogs, snakes, and even birds and mammals may figure in the diet of larger species. The largest of the top carnivores is the dreaded white shark (*Carcharodon carcharias*), which reaches such a size (up to 10 m) that it can feed with impunity on just about anything that can be engulfed by its cavernous mouth, including large fishes, seals, and the occasional unfortunate human. The mako (*Isurus* spp.), tiger (*Galeocerdo*), and hammerhead (*Sphyrna* spp.) sharks are other fearsome predators that account for most attacks on humans (Gruber et al., 1984). Barracudas, particularly *Sphyraena barracuda,* strike an especially threatening pose as they cruise the margins of coral reefs. Although their habit of approaching divers and following them is unnerving, they pose no real threat unless molested.

Perhaps the best way to envision the ecological relationships of organisms is through the construction of food webs (Fig. 29–11). Such a model of the trophic interrelationships of members of an aquatic community makes it easier to perceive the impact that changes in abundance or distribution of one member of the food web can have on other members. The situation is complicated by the fact that the feeding habits of fishes change as they get older and larger. Consequently, their "niche" in the food web may be only a temporary one that is defined by life history stages. Overall control of abundances of individual members of the food web can be perceived as being from the top down or from the bottom up. In a pelagic community, it has been suggested that species composition is largely controlled by higher trophic levels, such as piscivorous species—an example of top-down regulation. If abiotic factors, such as light, temperature, salinity, or nutrient availability, are more important, the community is said to be controlled from the bottom up (Wootton, 1990).

Impact of Trophic Relationships on the Structure of Ecological Communities

Because the majority of fishes are carnivores, we might assume that they are especially important as top-down regulators of community composition. Fish predation can alter the scope of available habitat by prey species—often including smaller individuals of those same predators. The concept of a "keystone species"—one whose presence or absence determines the composition of the community—is valid especially in relatively confined systems such as lakes and streams. For example, the re-

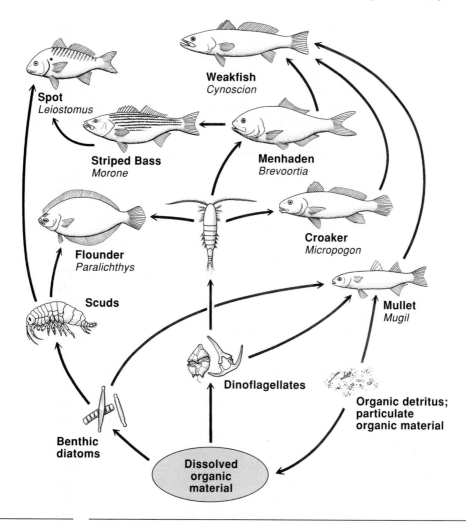

FIGURE 29–11 A simplified estuarine food web indicating the feeding niches of some fish species common to coastal waters of the southeastern United States. (*Source:* Smith, R. L. 1980.)

cent introduction of the piscivorous Sacramento squawfish (*Ptychocheilus grandis*) into the Eel River in northern California caused significant shifts in microhabitat preferences of the juveniles of several co-occurring species (Brown and Moyle, 1991). The structuring of a fish community may be the result of a complex interaction of physical and biological factors, including predation. Recent studies on ecological relationships in communities defined by a high degree of physical instability, such as the headwaters of small streams or rocky intertidal shores, have produced conflicting schools of thought as to how these communities achieve their characteristic composition, if in fact they ever do achieve a characteristic composition. Grossman (1982; Grossman et al., 1990) and others have proposed that some fish communities are "stochastic" in the sense that variation in the physical environment results in large variation in species composition and abundance. These communities never reach an equilibrium state because they are constantly experiencing perturbation. Consequently, competitive exclusion and resource limitation rarely figure in the structuring of such communities. Such bold assertions have not gone unchallenged, however (cf. Herbold, 1984). What is apparent after decades of study is that no single variable or process can account for the diversity and complexity encountered in the structure of fish communities. The subtle interplay of a number of variables in both time and space is something we are only beginning to appreciate.

Symbiotic Associations Among Fishes

In the aquatic realm, there exist many instances of fishes establishing specific relationships among themselves or with other organisms. Such symbiotic relationships may be mutually beneficial (mutualism), beneficial to only one member of the association (commensalism), or beneficial to one member to the detriment of its associate (parasitism). Among the most conspicuous associations are those involving large fishes, such as sharks and manta rays, and smaller individuals, such as the pilotfish (*Naucrates ductor*). By swimming very close to the shark, the pilotfish's own swimming efforts may be reduced and, in addition, it has the opportunity to feed on bits and pieces of the shark's meals. Sharksuckers or remoras (family Echeneidae) pursue an even more intimate association with the aid of an oval adhesive disc on their dorsal surface, which is used to attach to large swimming creatures. They are thus carried about with little effort of their own, and they can detach simply by swimming forward when an opportune time for feeding arises.

Some remarkable symbiotic relationships have evolved between coelenterates and fish. Members of the stromateoid genus *Nomeus* are known as man-of-war fish because they live with the large siphonophore *Physalia,* swimming among the tentacles without injury. The medusafish (*Icichthys lockingtoni*—another stromateoid genus) has similar habits. Larvae and small juveniles of other marine families, such as Carangidae and Gadidae, are known to shelter under jellyfish (Mansueti, 1963).

Species of *Amphiprion* (Pomacentridae) are able to associate closely with sea anemones. Anemonefish are not immune to the stinging nematocysts of their hosts, but they apparently suppress their discharge by mouthing and nibbling on the tentacles, thus transferring much of the anemone's mucous coat to the anemonefish. Apparently, the anemone senses this mucus on the fish, and nematocyst discharge is suppressed. Similar immunity has also been noted in juveniles of the sculpin genus *Artedius.* The pearlfishes (Carapidae) are often associated with echinoderms and molluscs. Most commonly, certain species hide inside sea cucumbers, entering through the anus tail first in a corkscrew fashion. A scorpionfish of the genus *Minous* carries a partial covering of hydroids on its skin, apparently to conceal the fish. The blind goby, *Typhlogobius californiensis,* usually lives with the ghost shrimp, *Calianassa,* in the shrimp's burrows. A closer relationship is seen between crustacea and the sea snail genus *Careproctus,* members of which deposit eggs on, or in, the branchial region of certain large crabs. The habit of the bitterling, *Rhodeus,* of ovipositing in the incurrent siphon of freshwater clams is well known (Fig. 29–12).

One of the most remarkable forms of symbiosis is cleaning, in which certain marine species remove parasites, bacteria, and infected tissue from other, usually larger, species. Most cleaners are tropical and represent evolutionarily more recent groups, such as the gobies, angelfishes, butterflyfishes, and wrasses. A surfperch,

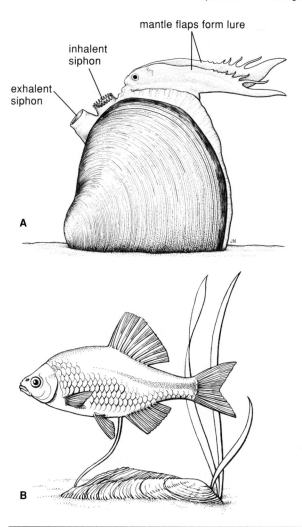

mantle flaps form lure

inhalent
siphon

exhalent
siphon

A

B

FIGURE 29–12   Bivalve mollusks exploiting fishes, and vice versa. **A,** The mantle tissue of the freshwater clam *Lampsilis ovata* mimics a small fish for the purpose of attracting a larger predator, to which the ejected glochidia larvae may attach. **B,** Bitterling (*Rhodeus*) depositing eggs in mantle cavity of a mussel. (*Source:* Wickler in Barth and Broshears, 1982.)

Brachyistius frenatus, is an unusual example of a cleaner that inhabits temperate latitudes. The cleaners may not engage in this activity on a full-time basis and probably do not depend on it for a complete livelihood; some clean other fishes only as juveniles and forsake the habit as adults. In those species in which cleaning provides a significant source of food, special colors and patterns have evolved to advertise their services. They may maintain cleaning stations, where other fishes are often seen patiently queueing up and waiting for their turn. Fishes coming to the stations to be cleaned cooperate fully with their cleaners, remaining still, opening the mouth and gill covers when necessary, and, most important, refraining from eating the cleaner. Large barracudas and groupers allow the smaller fishes to swim around their mouths and gills with impunity. Although the cleaner performs a service for the larger fishes, there are a few "con artists" among the wrasses that mimic closely the color and behavior of the cleaners but actually feed on bits of skin from the unsuspecting customer.

Another type of relationship is parasitism, in which one species gains its food from the body of another while closely associated with it. In the strictest definition of the term, parasites are prudent in that they do not kill their hosts because this would result in their own demise. Lampreys are often seen as one of the few cases of nutritional parasitism among the vertebrates, yet their actions, consisting of rasping a wound on the host and sucking out its body fluids, actually seem more like a specialized form of predation. Hagfishes sometimes bore into the body cavities of large fish and feed on them; probably the only fishes that would permit this kind of action have already been incapacitated. The deep-living snubnose eel, *Simenchelys parasitica,* has been known to "parasitize" large halibut, however. As mentioned in Chapter 28, some trichomycterid catfishes of South America are parasitic upon larger species, usually living in the branchial chamber of the host.

A kind of "nest parasitism" occurs among some North American minnows, which deposit their eggs in nests made by other species of minnows or by centrarchids. Survival of the brood appears to be enhanced by having another, perhaps larger, individual guard the eggs.

There are numerous examples of close relationships between fishes and other animals, one of the commonest being the hosting of numerous kinds of parasites. Fish may harbor protozoans, coelenterates, acanthocephalans, flukes, tapeworms, roundworms, isopods, amphipods, and the glochidia larvae of freshwater mussels. In females of the freshwater clam *Lampsilis ovata,* the mantle edges are modified to mimic a fish. When a predatory fish becomes attracted to the lure, the clam blows glochidia larvae out its excurrent siphon. These then anchor in the gill cavity of the host fish (Fig. 29–12). Some individual fishes can qualify as swimming communities, with many of the aforementioned groups represented in or on their bodies. The life cycles of many parasitic worms involve several hosts, and the fish may serve as an intermediate or as the definitive host; in these cases, the life of the fish may be linked with the lives of copepods, snails, birds, or other vertebrates.

Studies on the ecology of fishes become more critical as increasing amounts of the earth's surface are destroyed and the associated species are irretrievably lost. The greatest irony of this situation is that habitat destruction is most extensive in the tropics, in those regions where the biota is least understood. Although state and federal legislation affords some degree of protection to rare and endangered species in developed countries, the same cannot be said for third world countries, where considerations of habitat quality and species diversity are secondary to the more pressing problems of human survival, a survival that regrettably is dependent on exploitation of natural resources.

Distribution and Migrations

Tropical Eastern Pacific Region
Indo-Pacific Warm Temperate: South Australian Region
Northern New Zealand Region
Western South American Region
Southern Cold Temperate Region
Antarctic Region
Western Atlantic Boreal Region
Eastern Atlantic Boreal Region
Atlantic Warm Temperate: Carolina Region
Mediterranean-Atlantic Region
Tropical Atlantic–Western Atlantic Region
Eastern Atlantic Tropical Region
Eastern South American Warm Temperate Region
Southern African Warm Temperate Region
Worldwide

Introduction

An armchair theoretician could not dream up a better suite of organisms than fishes to study the convolutions of earth history. Evolutionarily older and much more numerous than all the other vertebrates combined, fishes are distributed literally everywhere that fresh or sea water exists on the globe. Since the vast majority of fishes are physiologically restricted to either one medium or the other, many are isolated in one or a few freshwater drainages. About 42 percent of the world's fish species are restricted to fresh water (Cohen, 1970). Another 39 percent are continental (i.e., they live in shallow cold or warm waters on the continental shelves) (Cohen, 1970). This means that the vast majority of fishes are, for biogeographical purposes, as continental as the land vertebrates (Rosen, 1975). Exclusively oceanic species, deep or shallow, constitute only the remaining 19 percent of the fishes, but they may have tremendous biomass because of the vast ocean areas they occupy. Only a small percentage (less than 5 percent) of the world fishes are specialized for intermediate salinities or are diadromous, migrating back and forth between fresh and salt water (Cohen, 1970; McDowall, 1988; Nelson, 1994).

In addition to the physiological restriction to fresh or marine water, the size of fishes and their lack of protection against drying out make it unlikely that they will be transported long distances by the wind or be carried by birds or other animals. Fishes must rely on the passage of geological time and concomitant physiographic changes to disperse to new areas. For example, the size of many island faunas, or on habitat islands on land, is closely related to the size, elevation, or vegetative cover of the island. The application of these analyses to the fish faunas of lakes or rivers often finds poorer relationships than those using birds or insects on isolated habitat patches. This is because the birds and insects disperse in ecological time (days to months) rather than over hundreds of years or more, which is more like the natural dispersal rates in most fishes. Thus many freshwater rivers and lakes are undersaturated with fishes, at least in relation to typical parameters of lakes and streams. This is because fish usually colonize areas over geological time scales and have not had time to fill these aquatic "islands" (Barbour and Brown, 1974; Brown and Gibson, 1991; Swift et al., 1986; Welcomme, 1979).

The economic importance of fishes has generated considerable support for faunal and taxonomic work, and surveys of fishes were accomplished early (compared to other organisms) for many areas. Because the distributions of fishes are keyed to geological time scales, ichthyologists figured prominently in the classical analyses of the global distributions of animals (Boulenger, 1905; Briggs, 1974; Darlington, 1957; de Beaufort, 1951; Gunther, 1880; Jordan, 1901, 1928; Myers, 1938, 1941, 1949, 1951). Many of the biological or biogeographical regions of the freshwater (terrestrial) and marine environment have had a strong ichthyological component in their definition. Naturally, fishes often show close correspondence to the distributions of other groups of organisms. The general correspondence of geographic distributions was used to define areas of importance, such as areas of endemism or of faunal interchange. Inevitably, correspondence varies with the biology of the species, and the boundaries can often be defined in various ways depending on interpretation and the fineness of the topographic scale used. Beginning in the early 1960s, more mathematically rigorous methods were applied, such as regression and cluster analysis, to quantify the relationships among and between various areas and faunas. Most recently, in the 1970s, congruence again became important. However, this time it was congruence of cladograms or genealogies that was considered important. Cladistic methodology relies on congruence among the cladograms of various taxonomic groups that overlap in their distributions. The cladograms are hypotheses of the genealogical relationships of each taxonomic group in question and reflect the speciation events during the evolution of the group. At least three separate areas with separate diagnosable and interrelated taxa are required. Simple cases can be done by hand, but more than a few cladograms or taxa require computer algorithms. The program can calculate a cladogram of areas, basically a single cladogram, or tree, that most parsimoniously combines the features of the genealogies (also cladograms) of the pertinent taxa. The three historically sequential but overlapping methods of classifying organisms (and doing biogeography) are usually called evolutionary (the earliest and most intuitive), numerical (the most mathematical and quantitative, often informative only at ecological time scales), and cladistic (using only shared, derived characters to define branching points in the classification), usually used at geological time scales. The features and application of these methods are fully explored with an ichthyological emphasis by Lundberg and McDade (1990), Mayden and Wiley (1992), and Stiassny and de Pinna (1994).

The distribution of fishes can be analyzed at many levels, from the short-term colonization of new habitat (Matthews and Heins, 1987) to the worldwide distribution patterns of living and fossil fishes (Grande and Bemis, 1991; Lundberg, 1993; Parenti, 1981; Patterson, 1981; Springer, 1982; Wiley, 1976). Before the 1970s, considerable reliance was placed on the use of ecologically defined categories (Table 30–1) of fishes based on their perceived salinity tolerance and migration patterns. Those most restricted to fresh water (primary and secondary) were considered better (today we say more predictive) for analyzing distributions because they were presumed not to cross marine barriers, today or in the past (McDowall, 1988; Myers, 1949; Patterson, 1975). Several scientific factors reduced the importance of such

TABLE 30–1	**Divisions of Freshwater Fish Groups***
Division	*Remarks*
I. Primary	Groups with little or no tolerance for sea water; such as lungfishes, paddlefishes, bichirs, pikes, minnows, catfishes (except plotosids and ariids), characins, centrarchids, living osteoglossids.
	Archaeolimnic: originating in fresh waters and always so confined (Patterson, 1975)
	Telolimnic: confined to fresh waters at present, but less so in the past.
II. Secondary	Groups usually restricted to fresh water but with enough salt tolerance so that members can enter the ocean and sometimes cross narrow saltwater barriers; garpikes, killifishes, live bearers, cichlids.
III. Diadromous	Migratory between fresh water and sea for the purpose of breeding (McDowall, 1988).
	1. Anadromous: diadromous fishes which spend most of their life in the sea and mature. When fully grown they return to fresh water to breed.
	2. Catadromous: diadromous fishes which spend most of their life in fresh water and go to the sea as adults to breed.
	3. Amphidromous: regularly migrating between fresh water and the sea for purposes other than breeding.
	A. Marine: spawning in marine water, with larvae and juvenile stages briefly in fresh water.
	B. Freshwater: spawning in fresh water with larvae and juvenile stages; temporarily marine before returning to fresh water.
IV. Vicarious	Non-migratory species of otherwise marine groups living in fresh water, such burbot, brook silverside, tule perch.
V. Complementary	Groups with close marine relatives that dominate fresh waters in the absence of primary and/or secondary fishes, such as the melanotaeniids of Australia–New Guinea.
VI. Sporadic	Fish that go back and forth between each medium and can breed in either medium, some anchovies, mullets, snappers, and gobies.
VII. Peripheral	A term coined to include categories III-VI above, namely all species that move between fresh and salt water at some or many life stages.
VIII. Euryhaline	Estuarine fishes that often and freely go between marine and fresh water, differing from most of the above categories which usually are capable of changing mediums only at particular life stages.

*After Myers, 1938, 1949, 1951; McHugh, 1967; Patterson, 1975; McDowall, 1988.

ecological classifications. First, whole families of fishes were either included or excluded, even when individual species in the families might be as primary or secondary as members of the designated freshwater families. This led some workers (Swift et al., 1986) to classify individual fish species ecologically. Second, other workers, particularly in the western United States, with a high proportion of secondary freshwater, diadromous, and peripheral fishes, referred to all fishes in fresh water as inland fishes to emphasize their role in freshwater ecosystems (Moyle, 1976; Wydowski and Whitney, 1979), regardless of their known or hypothesized salinity tolerance. Third, the realization that many additional ecological factors could be equally restrictive to the distribution of fishes diminished the importance of salinity tolerance per se as a restriction to movement. Fourth, the analysis of Cohen (1970) showed that most marine fishes lived on, and were restricted to, the continental shelf. As such, they could be as informative as the freshwater fishes. Only a few euryhaline and diadromous species were widespread, and even these were restricted to the offshore estuarine zone (Haedrich, 1983; McHugh, 1967). Finally, the verification by geologists of the reality of continental drift forced the realization that fishes could be carried long distances as passengers on pieces of terranes that had moved all over the globe. Long-distance dispersal was no longer the only possible way for fish faunas to be found on more than one continent. Such reasoning led Rosen (1975) to declare that fish species need only be considered continental (rather than not) to be useful for biogeographical studies. In addition, even oceanic fishes can still be informative in analyzing the formation of the ocean basins and water masses (White, 1994).

It is still important to study the size and diversity of fish faunas in a descriptive sense. Many of the described faunal regions coincide with major drainages and help identify areas of past speciation events (cladogenesis) (Banarescu, 1990, 1992, 1994; Greenwood, 1983; Hocutt and Wiley, 1986). This is particularly important where faunas are being heavily impacted by humans and may be partially or completely lost. Rapid changes are taking place both in poorly known areas, such as South America and Indonesia, and the best-studied areas, such as the United States, Japan, and Europe (Moyle and Leidy, 1992; Moyle and Williams, 1990; Stiassny and de Pinna, 1994). Any scientific assessment of the gain and loss of biodiversity is dependent on accurate historical records, and many faunal works gain in value with time. Unfortunately, they also are often nostalgic reminders of the degree to which humans have affected the environment. With a large and diverse vertebrate group such as the fishes, these reminders are everywhere. This pessimism is balanced, for ichthyologists, through the faunas of remote areas. Here one can still discover new species with unique biological features, a seldom, unlikely event with other better known and well-described vertebrate groups. A new journal was recently established for the volume of freshwater fish papers being generated worldwide: *Ichthyological Explorations in Freshwaters.*

Migrations of Fishes

Many fishes undergo long migrations on an annual or seasonal basis for a variety of reasons. Here we adopt a broad definition of *migration* that will include any mass movement from one habitat to another with characteristic regularity in time or according to life history stage. This broad view allows the inclusion of both active and passive mass movements, whether they are extensive seasonal or annual changes of habitat or short-term, short-distance travels. Usually these migrations are round trip.

Most migrations are attuned to food gathering, adjustment to temperature, or reproduction and are sometimes referred to simply as feeding, breeding, or wintering. In many fishes, these take place on an annual basis but, in others, individuals are involved in certain changes of habitat only at certain developmental stages. Others are daily, such as the vertical migrations of fishes in the deep scattering layer. The most accepted descriptive scheme for classifying migratory fishes is that of Myers (1949), as discussed by McDowall (1988):

Diadromous. These are truly migratory fishes that travel between the sea and fresh water. The three following types are included:

Anadromous. Diadromous fishes that spend most of their lives in the sea and migrate to fresh water to breed.

Catadromous. Diadromous fishes that spend most of their lives in fresh water and migrate to the sea to breed.

Amphidromous. Diadromous fishes in which migration from fresh water to the sea, or vice versa, is not for the purpose of breeding, but occurs regularly at some other definite stage of the life cycle. Both marine and freshwater amphidromy exist, as first noted by McDowall (1988).

Potamodromous. These are truly migratory fishes that migrate wholly within fresh water (non-diadromous).

Oceanodromous. These are truly migratory fishes that live and migrate wholly in the sea (non-diadromous).

The first three terms, diadromous, anadromous, and catadromous, have gained considerable usage among ichthyologists and fish biologists, whereas the latter two are not used as extensively. Several terminologies preceded these, overlap with them, or further refine the meanings. Other terms were introduced to classify fishes inhabiting various salinities without regard to migration. Many of these have not been adopted by ichthyologists because fishes are usually more vagile and not restricted to small areas (Haedrich, 1983; McHugh, 1967). Much of this history is discussed by McDowall (1988), who accepts the definitions as presented in Table 30–1 and points out the distinction between marine and freshwater amphidromy.

Migrations enable organisms to take advantage of additional habitat and to avoid adverse conditions. Disadvantages include not only the expenditure of energy, but exposure to predators along the migration route, losses of passively drifting eggs and larvae that end up in unsuitable regions, and adjustment to changes in habitat parameters. Migratory species store extra energy in the form of fat prior to the onset of migrations. Some species, such as Pacific salmons and lampreys, mobilize all their resources and commit them irreversibly to the acts of migration and spawning, resulting in death following spawning.

Advantages of migration, of course, are concerned with placing the individuals of a species in the best or necessary location for given biological activities at the appropriate time. Nocturnal feeding near the sea surface by species that spend the day in the twilight of the depths places these species in contact with a high concentration of food organisms and allows them to feed when some avian and other diurnal predators are inactive. As might be expected, several dark-colored species of sea birds also feed at night. Feeding migrations of the far-ranging species allow them to reach or to follow food resources that can support far larger populations than the resources of the spawning or nursery grounds. Spawning migrations are important to species with restrictive requirements for reproduction and those whose eggs and larvae must drift to specific nursery grounds. In tropical freshwater rivers, migrations of some species are closely attuned to the seasonal changes in river flow. Opportunities for feeding and breeding are presented by flooding during the rainy season in African, South American, and Asian floodplain rivers (Goulding, 1980; Lowe-McConnell, 1975; Welcomme, 1979). Shallow flooded expanses lateral to the river provide fruits and small animals that greatly increase the food supply for these fishes.

Examples of migrations range from the insignificant to the herculean. Many stream fishes, such as minnows, suckers, trout, and darters, may move only a few yards from feeding to spawning grounds. Many temperate fishes move downstream to deeper holes or larger streams for the winter. Often the migrations can be any direction as long as the desired substrate, depth, or temperature is present. Large potadromous characoids and catfishes in large tropical rivers or diadromous salmonids in large northern hemisphere rivers may move hundreds of miles upstream to spawn.

Short, daily nocturnal feeding migrations to the surface layers by many mesopelagic fishes enable them to take advantage of food organisms in the epipelagic zone, where photosynthesis takes place during the day. These fishes (and many other marine organisms) form as a dense layer at 500 to 600 feet down during the day. They collectively are known as the deep scattering layer, which was discovered with the advent of sonar during World War II. Many stream, lake, and marine shore fishes make similar, but much less spectacular, onshore movements at night only to return to deeper water during the day. During the reproductive season, similar nocturnal movements occur to concentrate the spawning adults on gravel bars, rocks, or sandy areas, whichever is appropriate for the species in question. Species with pelagic eggs, such as clupeids, hiodontids, and percichthyid basses, move upstream and the fertilized eggs and subsequent larvae drift passively downstream for a certain period before taking up their juvenile habitat.

Many large marine fishes move in schools north and south on an annual basis, following annual temperature variations, usually at the opposite times of the year in each hemisphere. The long north–south continental margins span wide temperature ranges and often induce cool upwelling currents. The large pelagic fishes have adapted their life cycles to these features. Even more wide ranging are the tunas, which may travel through much of an ocean basin (Nakamura, 1969). In the north Pacific, the albacore that have moved north along the California-Oregon coast during the summer may be in mid-ocean during the winter and back to the coast in June. Not all return, because there is another pattern sometimes followed by a portion of the population. This takes them from the mid-Pacific to Japan, where they move northeast along the coast in May and June, returning to mid-ocean by fall (Fig. 30–1). The spawning migration of mature fishes occurs as a divergence from the western Pacific pattern. The bluefin tunas of the Atlantic cross temperature boundaries on their known migrations moving between Florida and Norway and between Norway and Spain. Returns of tags from tunas indicate that some move between Baja California and Tokyo and between West Australia and northern New Zealand. Tagged and recaptured marlin have been logged at a minimum known speed of about 1850 km per month while traveling from California to Hawaii. Blue sharks tagged off New York and New Jersey were recovered two years later off South America and off Africa, distances of 3200 and 4800 km, respectively. Casey et al. (1980) report a blue shark tagged off Montauk, New York, that was recaptured 5844 km away on the coast of Liberia nine months later. The minimum speed for such a trip (over the curve of the earth) would be about 21 km per day.

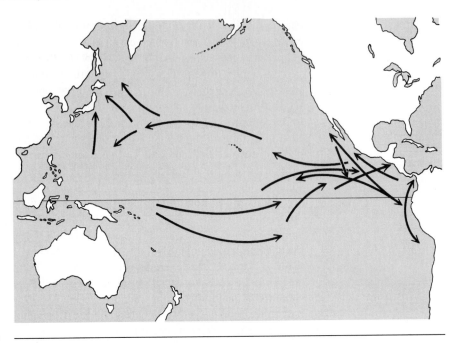

FIGURE 30–1 Most economical path of travel between sites of tagging and subsequent sites of recapture (arrows), reflecting the circular, trans-Pacific migration patterns of some oceanic fishes, particularly yellowfin tuna.

Diadromous migrations are remarkable because a change from one medium to another is necessary, and because the distances involved often require amazing feats of orientation, navigation, and precise recognition of home spawning areas. Catadromous migrations range from the short downstream movements of prickly sculpins in creeks tributary to the north Pacific Ocean to the mid-ocean sojourns of the eels of the genus *Anguilla*. *Galaxias maculatus* of New Zealand resembles the prickly sculpin in descending just to the estuarine areas, whereas the species of the tropical mullet genera *Joturus* and *Agonostomus* move long distances down large, tumultuous rivers to spawn near the coasts of Middle America.

The longest catadromous migrations are those of the eels of the genus *Anguilla*. They occur in the Indo–West Pacific and on both sides of the north Atlantic; their biology is best known in the Atlantic. Two species—the American eel, *A. rostrata*, and the European eel, *A. anguilla*—ascend freshwater streams as small elvers, 100 to 150 mm long. The males remain mostly in the lower reaches or even in the estuaries. The females move farther upstream and can move hundreds of miles inland. They can wriggle through wet grass or in crevices in obstructions to surmount barriers impassable to many other fishes. They spend several years in fresh water, and females reach about 1 m in length. The males spend less time in fresh water and are somewhat smaller. They have small eyes and a greenish or yellowish color. Before beginning the seaward migration, this color changes to blackish or silvery and the eyes get much larger. After emigrating to the sea, the eels virtually disappear; only a few marine records exist for these fishes, and these are mostly in the shallows of the continental shelf. One sighting was made of two fishes at 2000 m deep near the Bahamas (Nelson, 1994).

Subsequent captures of the smallest larvae indicate that spawning takes place in deep water in the Sargasso Sea (Figure 30–2) at depths of 400 to 700 m. This is where, in 1929, Johannes Schmidt discovered the smallest larvae coming to the surface. As in most eels, the larvae are delicate, leaf shaped, and transparent and are called leptocephali. They drift months or years before encountering the coastal areas, metamorphosing into elvers, and repeating the cycle. Since the Atlantic Ocean floor has been spreading since the mid-Mesozoic, the eels have had farther and farther to go to get to the opposite shores. European larvae have to go much farther and defer metamorphosis much longer than American eels. In fact, some ichthyologists proposed that the European eels may not spawn at all. It might be too far for them to travel to the Sargasso Sea. If adult European fishes do not or cannot make it to the spawning grounds, the elvers that reach Europe must be derived from spawning American eels. Known differences in vertebral numbers could be due to cooler water and the longer time in the plankton. Both of these factors are known to increase the vertebral count in larvae of other fish species. Up to four or five species of anguillids occur together in the Indo–West Pacific, and the relationships must be more complicated. Much remains to be discovered about the life history of these remarkable eels (McDowall, 1988).

The species of Galaxiidae, Retropinnidae, and Aplochitonidae in New Zealand, Australia, and South America have long been thought to have a diadromous life style. However, McDowall (1988) believes that only *Galaxias maculatus* is truly catadromous. Most other species do not actually enter marine water enough to be

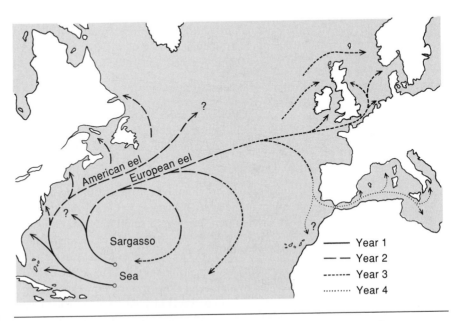

FIGURE 30–2   Movement of developing larvae of American and European eels from closely adjacent areas of hatching in the Sargasso Sea. Note differences in direction and longevity, mostly due to the largely passive transport of the larvae by oceanic currents.

considered diadromous. McDowall considers most of the other galaxiids, the retropinnids, and the aplochitonids to be amphidromous. However, one aplochitonid, *Lovettia seali,* the Tasmanian whitebait, is well known enough biologically to be called anadromous. The aplochitonids from southern South America are considered amphidromous but have not been well studied. McDowall (1988) also considers that they may not be closely related to the other members of the family in Tasmania.

In Japan, the ayu, *Plecoglossus altivelis,* an annual fish, descends to lay its eggs in the lower reaches of the rivers. The larvae are carried out to sea, where they remain for a few months. Then they move back into the middle and upper reaches of the rivers to grow and then migrate downstream to spawn and die soon thereafter. This life cycle is best described as freshwater amphidromy (McDowall, 1988). Marine amphidromy is seen in the mullets, *Mugil cephalus,* which spawn at sea and whose young spend a short period in fresh or low-salinity water before becoming more marine.

Variation also occurs in the degree of anadromy in North American salmonids and clupeids. Cutthroat trout in the Pacific Northwest often descend only into coastal estuaries before moving upstream to spawn in late fall or early spring. Steelhead range over much of the north Pacific Ocean and migrate hundreds of miles in the larger rivers. They often return for two or three successive years to spawn. The species of Pacific salmon spawn from the coastal estuaries just above salt water to hundreds of miles upstream, depending on the species. The species of Pacific salmon all die after spawning. In the Sacramento River of California, four distinct runs of chinook salmon occur, one for each season (Moyle and Williams, 1990).

Although many sturgeon of the genus *Acipenser* are considered anadromous, many only descend to estuarine areas and seldom, if ever, reach fully saline water. In the eastern United States, the hickory shad, *Alosa chrysochloris,* migrates downstream to coastal areas of low salinity but has not been taken at sea. The Alabama shad, *A. alabamae,* ranges widely at sea in the Gulf of Mexico. As more species become better known, more refinements will be necessary in our understanding of fish migration.

Active migration requires adjustment to environmental conditions, including temperature, light, water current, salinity, alkalinity, and other sensory modalities. Major physiological changes accompany the transition from marine to fresh water and occur only at certain stages in the life cycles of diadromous fishes, so they are not euryhaline fishes, as is often assumed (McDowall, 1988). It is well known that Pacific salmon orient to their natal stream largely by smell. Other fishes are known to orient to the angle of the sun in the sky, and this provides an orientation mechanism in the open sea, where smell may not be usable and other cues are absent. Tunas have a well-developed pineal eye on top of the skull that may function in orientation. Some sharks are known to detect weak electrical fields. It is now known to be physiologically possible for sharks to detect and orient to the weak electrical field generated by salt water flowing within the earth's magnetic field. This could explain migrations in deep or turbid water, where vision cannot be used. This seems to be the kind of mechanism required by the anguillid eels. Several organisms are known to respond directly to the earth's magnetic field. These possibilities seem to be required by known migrations that appear to take place in a clueless environment.

Just as we are beginning to understand the mechanisms of migration, many of our most spectacular migrating fishes are rapidly declining in numbers. Concern for their survival is stimulating drastic actions to conserve the fishes that are left and providing an impetus to study them more thoroughly and thus help them survive (Moyle and Leidy, 1992; Moyle and Williams, 1990; Stiassny and de Pinna, 1994).

Continental Distribution of Fishes

Each continent has a distinctive fish fauna with a unique combination of families, genera, and species found there exclusively and others shared with other continents. Strong resemblances between certain faunas may be the result of past land connections, early geological proximity, ancestral marine habits lost in today's descendants, or simply the presence of many freshwater representatives of salt-tolerant groups. Climatic, altitudinal, and other habitat zonation is important in the distribution of fishes, and earlier ichthyologists began their treatment of the subject with the recognition of a northern zone, an equatorial zone, and a southern zone (Günther, 1880; Boulenger, 1905). These zones were defined by present-day boundaries in the ranges of the fishes involved.

Günther's northern zone includes the area roughly north of the Tropic of Cancer, except that its boundary crosses Asia at 30 to 35 degrees north latitude. The southern zone takes in only part of Victoria in Australia, Tasmania, New Zealand, and a strip of Chile, plus Patagonia. In between the two is the equatorial zone, divided into cyprinoid (Africa and Asia) and acyprinoid (South America and Australia) regions. The utility of this system in describing general and ecological resemblance between fish faunas is evident in that the fresh waters of the north are characterized by the salt-tolerant families of salmonids, sturgeons, smelts, northern lampreys, and several freshwater families, including pikes, leuciscine cyprinids, and perches. The warm water zone contains cichlids, characins, lungfishes, osteoglossids, etc., and the southern bits of land have a sparse fauna of salt-tolerant galaxiids, haplochitonids, retropinnids, and southern lampreys. These three zones correspond roughly to what are today called the cold or boreal, the temperate, and the tropical divisions in each hemisphere.

Further knowledge of the fishes led to a refinement of the zoogeographical divisions from which to consider both the descriptive and historical aspects of fish distributions and dispersal. Fish distributions align themselves into the world divisions usually accepted by biogeographers, because fishes often were part of the defining elements. Fishes obviously have many of the same limiting factors as other marine and freshwater ectothermic organisms. The world is divided into three faunal realms. The first, historically called Arctogea, will be called by Darlington's (1957) term Megagea. It is subdivided as follows: (1) the Ethiopian Region, which consists of Africa except for the Atlas Mountain area of northwestern Africa (part of southern Arabia is included); (2) the Oriental Region, which includes tropical Asia plus some of the continental islands to the southeast; (3) the Palearctic Region, which includes temperate and cold Eurasia plus northwestern Africa; and (4) the Nearctic Region, which consists of North America south to the tropical areas of Mexico. The second realm is Neogea, which consists of the Neotropical Region (South and Central America south to the tropical areas of Mexico). The third realm is Notogea, which contains the Australian Region (Fig. 30–3). Sometimes the Nearctic and Palearctic together can be called the larger Holarctic Region, which is useful in describing fishes distributed across the northern parts of both the old and new worlds. The borders of these regions usually coincide with prominent physical or climatic changes. The ranges of all the living families and often the larger subdivisions of the families can be found in Nelson (1994), and distribution maps of most of the freshwater families are given in Berra (1981).

The Holarctic Region

The fishes of the Nearctic and Palearctic regions show close relationships, and several fish families are distributed across most of both of these areas. The similarity is greatest in the northern parts and becomes less as one approaches the transitional areas farther south. In both North America and Eurasia, the fish faunas are much richer in the southeastern parts than in the western, northern, and northeastern areas. The Rocky Mountains are a boundary of importance in North America, and the Urals also stand in analogous fashion between eastern and western Eurasia. Both continents are divided into a richer eastern and a depauperate western half.

Six freshwater families are found only in the Holarctic region: Polydontidae, Esocidae, Umbridae, Dalliidae, Catostomidae, and Percidae. The two living members of the Polyodontidae, or paddlefishes, are in the Yangtze River of China and the Mississippi River and associate drainages in eastern North America (Grande and Bemis, 1991). These and several fossils from North America support an early Tertiary transpacific biogeographic relationship (Grande and Bemis, 1991; Fig. 30–4). The Esocidae has a wide distribution, especially in the northern part of the Holarctic

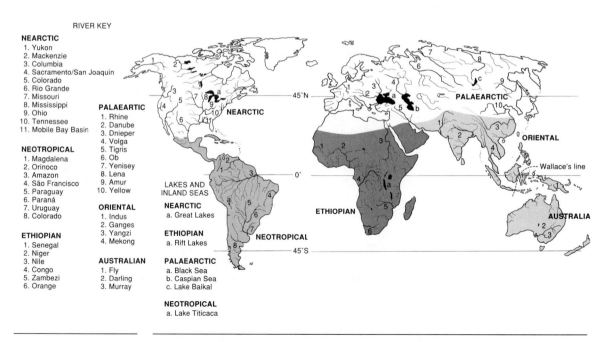

RIVER KEY

NEARCTIC
1. Yukon
2. Mackenzie
3. Columbia
4. Sacramento/San Joaquin
5. Colorado
6. Rio Grande
7. Missouri
8. Mississippi
9. Ohio
10. Tennessee
11. Mobile Bay Basin

NEOTROPICAL
1. Magdalena
2. Orinoco
3. Amazon
4. São Francisco
5. Paraguay
6. Paraná
7. Uruguay
8. Colorado

ETHIOPIAN
1. Senegal
2. Niger
3. Nile
4. Congo
5. Zambezi
6. Orange

PALAEARTIC
1. Rhine
2. Danube
3. Dnieper
4. Volga
5. Tigris
6. Ob
7. Yenisey
8. Lena
9. Amur
10. Yellow

ORIENTAL
1. Indus
2. Ganges
3. Yangzi
4. Mekong

AUSTRALIAN
1. Fly
2. Darling
3. Murray

LAKES AND INLAND SEAS

NEARCTIC
a. Great Lakes

ETHIOPIAN
a. Rift Lakes

PALAEARCTIC
a. Black Sea
b. Caspian Sea
c. Lake Baikal

NEOTROPICAL
a. Lake Titicaca

FIGURE 30–3 Major biogeographic realms, rivers, and larger inland water masses of the world.

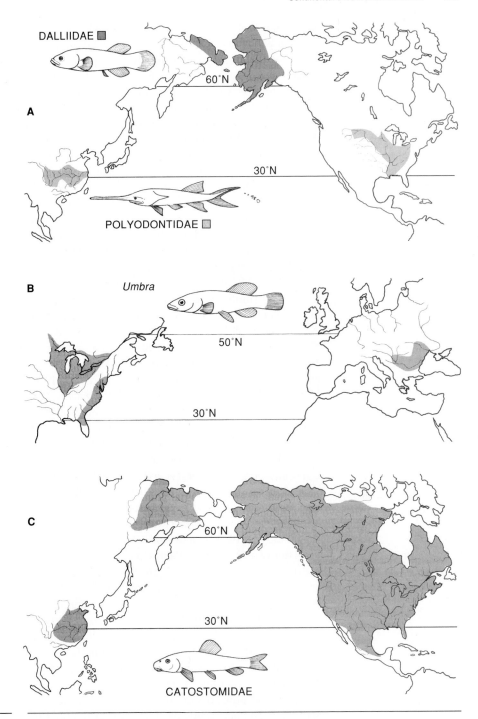

FIGURE 30–4   Representative Holarctic distributions of freshwater fishes. Recent trans-Pacific distribution of the Dalliidae and Siberian Catostomidae due to the elevation of Bering Straits, several tens of thousands of years ago. Older relationships are demonstrated by the Polyodontidae (trans-Pacific) and *Umbra* (trans-Atlantic), both tens of millions of years old.

area, although members range south to tributaries of the Gulf of Mexico and into the Caspian Sea area. *Esox lucius* has a truly circumpolar distribution and is the only pike found in many areas of North America and Europe. Otherwise, only diadromous or salt-tolerant fishes are holarctic in distribution, such as rainbow smelt, *Osmerus mordax;* fourhorn sculpin, *Myxocephalus quadricornis;* and threespine stickleback, *Gasterosteus aculeatus.* The mud minnows, Umbridae, have a disjunct distribution. *Umbra krameri* lives in the Danube of southern Europe, *U. pygmaea* on the Atlantic and Gulf of Mexico coasts of North America, and *U. limi* in the Great Lakes and upper Mississippi River drainages. These probably date from a trans-Atlantic connection of Europe and North America near the end of the Mesozoic (Patterson, 1981).

The Dalliidae, long considered monotypic, contains one species, *Dallia pectoralis,* in Alaska, and two or three others in Russian territory across the Bering Straits (Nelson, 1994). These fishes occur in fresh water on both sides of the straits, an area known to have been above the sea repeatedly during the Pleistocene ice ages. The Catostomidae show two trans-Pacific relationships. An old one (early Tertiary) is represented by the Chinese genus *Myxocyprinus* and the North American genus *Cycleptus* (Smith, 1992). More recently, the species *Catostomus catostomus,* like the genus *Dallia,* occurs on both sides of the Bering Straits, as another example of Pleistocene exchange.

The Percidae are represented in the two subregions by two shared genera, *Stizostedion* and *Perca,* plus three other genera in Eurasia and three rather speciose genera in eastern North America with more than 150 species of darters (Burr and Mayden, 1992; Wiley, 1992). One species reaches Pacific coastal drainages in the mountains of northern Mexico. It is not well established whether the percids are a trans-Pacific or trans-Atlantic relationship; it is possible that both relationships are involved (Collette and Banarescu, 1977; Page, 1983; Wiley, 1992).

The remaining primary (and archaeolimnic) family found in the Holarctic is Cyprinidae, one of the largest and most widespread of the freshwater groups worldwide. It forms an important component of all continental faunas except those of Australia, South America, and, of course, Antarctica and Greenland, which harbor no living freshwater fishes (Fig. 30–5). The subfamily Leuciscinae is well represented in Asia, Europe, and North America; and one genus, *Phoxinus,* is shared by the latter two. Both trans-Pacific and trans-Atlantic relationships exist among the cyprinids (Cavender and Coburn, 1992). The intra–North American relationships are described in Coburn and Cavender (1992), and the subdivisions of whole freshwater fish fauna of North America are detailed in Hocutt and Wiley (1986).

The killifish family, Fundulidae, is exclusively new world and is found from the Great Lakes region to central Mexico on the Atlantic Coast and from southern California to the southern tip of Baja California on the Pacific. The related Cyprinodontidae range from North America and South America to north Africa, including the eastern Mediterranean. The lampreys, Petromyzontidae, and the eels, Anguillidae, are widespread diadromous species in the Holarctic, but the anguillids are absent from the Pacific side of North America. Other salt-tolerant families that are often estuarine or anadromous are the Acipenseridae, Osmeridae, Salmonidae, Gasterosteidae, and Cottidae. The cod family, Gadidae, has a freshwater genus, *Lota,* distributed circumglobally in the north, somewhat like northern pike.

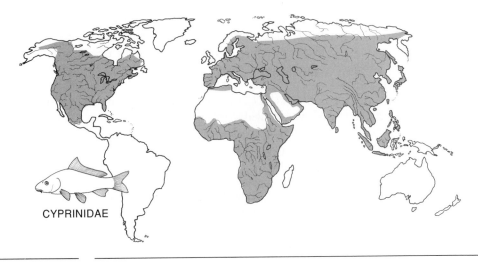

CYPRINIDAE

FIGURE 30–5 The northern hemisphere distribution of a primary or archaeolimnic freshwater minnow family, the Cyprinidae, that is not present in South America, Australia, or Antarctica.

The Palearctic region shares all strictly freshwater groups with other continental landmasses. Three peculiar endemic salt-tolerant families occur there. The amphidromous ayu (*Plecoglossus altivelis*) of Japan and Taiwan once constituted the monotypic family Plecoglossidae, but is now placed in Osmeridae. The Salangidae is another family found along the east coast of Asia; the neotenic icefish, *Salangichthys microdon,* enters the rivers of China (Nelson, 1994). Two endemic families, the Comephoridae and Abyssocottoidae, are found in the Lake Baikal region. A species flock of abyssocottids, and including a few species of cottids and comephorids, has evolved, apparently in situ, in this old rift lake (Sideleva, 1982).

The Palearctic shares a number of freshwater families with tropical Africa and the Asian tropics of the Oriental region. It is richest in cypriniforms, the Cyprinidae, and related families, many of which show trans-Pacific relationships or have relatives in Africa. Loaches (Cobitidae) occur in all three areas, including Europe. Climbing perches (Anabantidae), snakeheads (Channidae), the spring swamp eels (Mastacembelidae), and the bagrid catfishes range from tropical Asia throughout the transitional area into temperate Asia. The hill-stream catfishes (Sisoridae), absent from Africa, are found in both tropical and temperate Asia. The old world catfishes, or sheatfishes (Siluridae), are shared by the Oriental and Palearctic. Homalopterids are found both north and south of the Himalayas and are more adapted to steep mountain streams than to particular Palearctic or Oriental drainages per se (Banarescu, 1992).

The Nearctic region is characterized not only by the holarctic fishes, but by several archaic living fish groups that are relics of formerly wider distributions, a few endemic freshwater families, and a few southern species that just enter the southern portion of the region. The biogeography of North American fishes in relation to the other continents was discussed by Patterson (1981). Hocutt and Wiley (1986) and Mayden (1992) provided many details of relationships within the continent.

The bowfin family Amiidae is represented in Jurassic and Cretaceous seas by fossils in Europe, Asia, and South America. Only the single living species, *Amia calva,* of eastern North America seems to have lived only in fresh water, judging from Eocene and Miocene fossils from North America, Europe, and Asia. Thus the Amiidae show a relationship with Europe originally but are now restricted to North America. A similar situation prevails for the gars, Lepisosteidae. Wiley (1976) found fossils of this group in India, Europe, and Africa, but they are restricted to North America, Mexico, Cuba, and the Pacific side of Middle America today. Some of the species are restricted to fresh water, whereas others are salt tolerant and have been seen swimming in coastal sea water. However, this is probably not relevant to their division into species over geological time. The fossil record clearly indicates that they were present when the continents were close together or confluent, and the ancestral gars had no expanses of salt water to cross.

Another endemic eastern North American family, the Hiodontidae, are related to the living Notopteridae of Asia and Africa. However, like the paddlefishes, their closest phylogenetic relatives are in eastern Asia, the fossil Lycopteridae (Patterson, 1975). Several extinct forms also occur in North America (Cavender, 1986; Wilson and Williams, 1992). In the absence of European fossils, a trans-Pacific relationship is indicated.

Three small freshwater families—Percopsidae, Aphredoderidae, and Amblyopsidae—comprise nine species of small, primitive, spiny-rayed fishes whose fossil record is virtually all North American (Patterson and Rosen, 1989). Two additional families of North American freshwater fishes are the catfishes, Ictaluridae (48 species), and sunfishes, Centrarchidae (32 species). Both are almost exclusively eastern North American but have extensive fossil records in the western United States. One centrarchid, *Archoplites interruptus,* is found in central California (Cavender, 1986; Minckley et al., 1986).

Middle America (Central America plus Mexico) has been recognized as an area of interchange of the North and South American fauna, as well as an area of endemism, for a long time (Bussing, 1985; Regan, 1906-1908). The tropical and largely South American Cichlidae, Characidae, and Pimelodidae extend from Panama to southern Texas and become progressively fewer in more northern drainages. In a similar manner, the Ictalurids become scarce southward into southern Mexico. Large numbers of poeciliids, cyprinodontids, fundulids, profundulids, pimelodids, cichlids, and atherinids are endemic to subdivisions of Middle America (Bussing, 1985) (Fig. 30–6).

Pleistocene glaciation, which covered much of northern North America (and Eurasia) several times in the last 1.7 mya, drove most aquatic life southward (Fig. 30–7). Many interesting faunal patterns were established as fishes and other aquatic organisms reinvaded as the glaciers retreated northward (Crossman and McAllister, 1986; Trautman, 1981; Underhill, 1986). Small unglaciated areas in the north, such as the driftless area of Wisconsin and Pacific coastal areas moderated by the oceanic influence, served as refuges. Most species probably were driven southward into southeastern North America. Populations of rainbow trout, now isolated in the high mountains of Mexico (and brown trout in the Atlas Mountains of northern Africa), are relics of colder climates created during glacial advances. The richness of the faunas of the southern areas is due to the fact that the glaciers did not devastate the terrain. During these glacial advances, arid areas like the American Southwest received much more rainfall, had many lakes and streams, and had larger fish faunas.

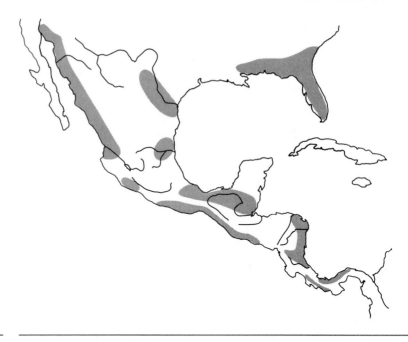

FIGURE 30–6   The several areas of endemism at the genus and species level in Middle American freshwater fish families such as Poecilidae, Lepisosteidae, Fundulidae, Cichlidae, and Characidae.

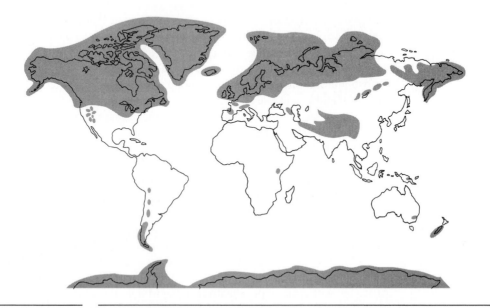

FIGURE 30–7   Extent of Pleistocene glaciation on a worldwide basis. Areas covered with glaciers were uninhabitable by fishes, and unstable conditions also existed some distance from the glacial margins. This occurred several times in the last 1.6 million years.

However, the faunas were still much smaller than those from passive continental margins, and today they are among the most depauperate known (Minckley and Deacon, 1991; Naiman and Soltz, 1981; Swift et al., 1993). The water tied up in glacial ice caused a fall in sea level of 300 m or more and exposed the Bering Straits, connecting Alaska and Russia. This allowed the exchange of species like *Dallia pectoralis, Catostomus catostomus,* and *Lota lota.* Continental shelves were often much wider, such as along the Florida peninsula and along the Atlantic coast. Long Island (New York) and Vancouver Island (British Columbia) were connected to the mainland, but areas separated by deep water, such as the Bahama Banks and the southern California Channel Islands, were never connected.

The Oriental Region

The Oriental region is less definable because it has broad land connections with the Palearctic and the Ethiopian, with many overlapping ranges of fishes. The Himalayas form a definite boundary to the north, and dry areas of the Middle East form a partial barrier in the west. This fish fauna generally ranges from eastern Iran through tropical Asia, with a transitional zone just south of the Yangtze River in China. Taiwan, the Philippines, and Indonesia are included as far as (or eastward to) Wallace's Line that separates Bali from Lombok and Borneo from Celebes. A few freshwater fishes cross this line, which was originally based on terrestrial faunas. The large shallow seas surrounding most of the Indonesian islands mean that they were all connected by one large river drainage 10,000 years ago. Today each island has a subset of the fishes and shares many species with the other islands and with mainland Indochina (Roberts, 1989).

The Oriental region is rich in ostariophysian fishes, both Cypriniformes and Siluriformes. This richness led early workers to postulate that the Oriental region is the center of origin of the cyprinid fishes, which comprise several endemic subfamilies and endemic species in this region (Winfield and Nelson, 1991). Some genera from the cyprinid subfamilies Barilinae, Barbinae, and Garrinae are shared with Africa. Catfish families endemic to the Oriental region are the Amblycepidae, Chacidae, Cranoglanididae, and Pangasiidae. Catfishes found in both the Oriental and Palearctic are Bagridae, Siluridae, and Sisoridae; those shared with Africa are Bagidae, Clariidae, and Schilbeidae. The hill-stream Homalopteridae are essentially confined to the Oriental region, but their relatives, the Cobitidae, are present in temperate Eurasia and Africa as well.

Other groups in the Oriental region are Osteoglossidae (Figure 30–8) (shared with Africa, Australia, and South America), Notopteridae (shared with Africa), Nandidae (shared with Africa and South America), Mastacembelidae (shared with Africa and temperate Asia), Chaudhuriidae and Channidae (shared with Africa and temperate Asia), Luciocephalidae, and the five or six anabantoid families.

Other groups, formerly not considered because they were more salt tolerant, include the widespread Rivulidae and Cichlidae, mainly found in South America and Africa but with a few representatives in the Middle East, India, and Sri Lanka (formerly Ceylon). The Toxotidae, the archerfishes, range from India east and south throughout the islands to Australia. This distribution is common for many marine shore species and species groups as well. The priapium fishes, Phallostethidae, occur from Thailand to Borneo and the northern Philippine Islands. The Syn-

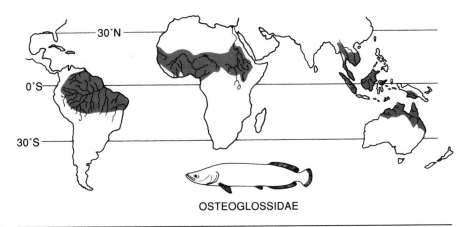

OSTEOGLOSSIDAE

FIGURE 30–8   The southern hemisphere, or Gondwana, distribution of the freshwater fish family Osteoglossidae.

branchidae, or swamp eels, contains a small number of species that occur worldwide in the tropical areas and only rarely occur, if ever, in marine water. The peculiar Indostomidae is endemic to Burmese fresh waters and is possibly related to the Gasterosteiformes, many of which are also freshwater, but in the Holarctic. The recently discovered *Sundasalanx,* in Borneo and Thailand (Roberts, 1984), is now placed in the Palearctic Salangidae (Nelson, 1994).

The Ethiopian Region

The continent of Africa, excluding the Atlas mountains in the northwest, is almost continuous with the Ethiopian Region and is bounded by ocean, except in the Suez Canal area. The mountains are part of the Palearctic because of the northern fauna found there (Greenwood, 1983; Roberts, 1975). The small fauna of Madagascar is African or endemic complementary fishes.

The equator bisects Africa, so the climate is hot to warm in the central portion and cooler but moderate at the northern and southern extremities and at higher elevations. Large areas of desert constitute nearly 40 percent of the landmass and are growing. These areas have little or no permanent water and few native fishes. Tropical rainforests lie near the equator and the Zaire and upper Nile river systems.

In earlier times, great inland drainage basins held extensive lakes. These facilitated fish dispersal across the continent through formerly connected drainages. Several rift lakes contain large species flocks that are believed to have evolved in situ, without much of the usual geographic allopatry often considered a requirement for speciation in fishes (Greenwood, 1983).

Several families are restricted, or endemic, to Africa. Two of these are archaic groups of lower bony fishes, including the lungfishes (Protopteridae) and the bichirs (Polypteridae). Archaic teleost families endemic to Africa are the Denticipitidae, Pantodontidae, Gymnarchidae, Mormyridae, Phractolaemidae, and Kneriidae. The remainder of the endemic families are Characiformes (Distichodonditae, Citharinidae, Ichthyoboridae, and Hepsetidae) or Siluriformes (Amphiliidae, Malapteruridae, and Mochokidae).

Several families are shared by Asia and Africa. The Notopteridae, Channidae, Mastacembelidae, Bagridae, Anabatantidae, and Clariidae are found only in Africa and Asia. A few genera of Cyprinidae, noted earlier, are shared with Asia. A much older relationship with South America is indicated by the sharing of the characiform fishes, one of the few unequivocal indications of such a connection. In addition, the following fishes are distributed in Asia, South America, and Africa: Osteoglossidae, Aplocheilidae, Synbranchidae, Nandidae, and Cichlidae. Another link with South America is the Protopteridae, closely related to the Lepidosirenidae in South America. As noted earlier, this relationship must accommodate fossils from Bolivia that indicate both types of lungfishes in South America in the Paleocene. Many other divisions between taxa shared by southern continents are difficult to define exactly (Lundberg, 1993), and may predate the split-up of these two continents.

Disregarding the desert areas, the fish fauna is poorest in the highlands of Ethiopia and the southern tip of Africa. Cyprinidae, Cichlidae, and some catfish families are most typical, but Ethiopia has some species of cobitids, and the Cape area in the southwest has the Anabantidae and Aplocheilidae. Tropical West Africa, especially the Zaire (Congo) basin and the rain forests along the Gulf of Guinea, has the richest fish fauna, with all the freshwater fish families of the continent represented, except the Cobitidae and Gymnarchidae. The Nile and the faunally related drainages that stretch south of the Sahara to Senegal are also rich in fishes. Other more peripheral areas have smaller faunas (Roberts, 1975; Skelton, 1993).

Many of Africa's freshwater fish species have arisen in a few large, old rift lakes, which have been great historical experiments in fish diversity and the packing of numerous fish species into a relatively restricted area. Lakes Tanganyika, Victoria, and Malawi all have 100 to 300 fish species each, dominated by cichlids, which have diversified morphologically to exploit many niches nearly analogous to those exploited by many different families of fishes on the marine tropical coral reefs (Fryer and Iles, 1972; Greenwood, 1983). Several species of mastacembelids and clariid catfishes have evolved in these systems as well. Large predatory fishes, genus *Lates,* have recently been introduced into some of these lakes and threaten to extirpate much of this diversity. Africa has been divided into about 12 zoogeographic regions by Roberts (1975) and Greenwood (1983), and southern Africa is divided into five regions by Skelton (1993).

The Neotropical Region

South America is also completely bounded by ocean, except for a narrow connection with the Isthmus of Panama. This connection is also significant for marine faunas since it separates the Atlantic and Pacific oceans. This connection is only about 3.5 million years old. Relatively few freshwater fishes have invaded northward into Middle America or farther, or entered South America from the north, as noted earlier. In contrast, a great interchange took place among terrestrial mammals and is well documented in the fossil record. The vast majority of the continent is dominated by the tropical rainforest of the Amazon and related basins, the Orinoco to the north and the Parana–La Plata to the south. Only the southern tip—namely, the Patagonian regions of Chile and Argentina—and higher elevations of the Andes mountains bordering the continent on the west have temperate climatic conditions. Parts of temperate areas were also glaciated, much like the northern hemisphere (Fig. 30–7).

The fish fauna of South America is rich in Characiformes and Siluriformes; estimates of the total freshwater fauna range as high as 5000 species (Böhlke et al., 1979; Fink and Fink, 1979). Often closely related, or cognate, species occur, one each in the Amazon, Orinoco, Parana, and Magdalena river systems. Additional species are endemic to particular tributaries of each system, multiplying the numbers. The characiforms are so different from those of Africa that only one very old ancestral group must have been shared between the continents. The weakly electric gymnotiforms, divided into five or six families, are also rich, numbering about 150 species. The electric eel, that most famous of gymnotiforms, is endemic to South America. Fourteen families of catfishes are endemic to the continent, including the armored loricariid catfishes and the elongate, naked Trichmycteridae, famous for the habit of invading the urethra of mammals. Other freshwater fishes, probably with marine ancestors (and thus telolimnic), are the freshwater stingrays, Potamotrygonidae, and the Jensyiidae and Anablepidae. Freshwater telolimnic representatives of the Clupeidae, Engraulidae, Belonidae, Sciaenidae, and Soleidae also occur, usually ten species or less in each family. Since the Amazon is the largest drainage in the world, it is fitting that the largest freshwater fish fauna should be found there. Past geological events have precipitated much of the speciation, including large periods of lacustrine habitats occupying much of the drainage and an outlet to the Pacific Ocean for much of the basin's history. In addition, evidence points to considerable variation in flow during glacial periods and repeated fragmentation of the tropical habitat. All of these changes provided opportunity for speciation in fishes (Fink and Fink, 1979; Gery, 1969; Lowe-McConnell, 1975; Welcomme, 1979). The fauna is just beginning to be reasonably well surveyed, and as more cladograms become available, interpretation of the historical evolution of the fish fauna will be an active area of research.

The Orinoco–Venezuelan fish fauna to the north of the Amazon, including the island of Trinidad, holds 400 or more species. The drainages farther along the Caribbean toward Panama have fewer fish species, but fossil evidence indicates that these drainages had a richer fauna in the past (Lundberg et al., 1986). Thus extinction, as well as distance from the Amazon, has played a part in the composition of the fauna. Pacific drainages in Colombia and Ecuador have what is called a trans-Andean fauna, much reduced in number of species but with relationships to the Amazonian fauna. These relationships are probably due both to lateral lowland movements by fishes around the northern edge of the continent and stream captures across the Andean divide.

The spine of the high Andes is inhabited by a small fauna of cold-tolerant fishes, primarily Trichomycteridae, Loricariidae, and Cyprinodontidae, genus *Orestias.* Ancient high-altitude lakes in Bolivia, including Lake Titicaca, have a large species flock of orestine cyprinodontids that evolved in the absence of predatory fish species (Parenti, 1984b). These are examples of upland fishes distributed more in relation to a physiographic region than to particular drainages. Southward in the temperate region, a small fauna of primitive trichomycterids, the most primitive living catfish, genus *Diplomystes,* and several species of percicthyid percoids occur. The introduction of rainbow trout has adversely impacted the native fish faunas of this area.

Across Amazonian drainage divides to the south are the Parana, La Plata, Uruguay, and Paraguay systems. The Parana system, broadly connected to the

Amazon by the vast marshes of the Gran Chaco, shares many relationships with the Amazon; the others are separated by much higher elevations and have much smaller faunas (Smith, 1981).

The Australian Region

Most of the freshwater fishes of Australia and New Guinea are freshwater fishes related to, or part of, families that are estuarine or marine elsewhere. The only classically defined freshwater fishes are the Australian lungfish, *Neoceratodus,* and an osteogolossid, *Scleropages leichhardti.* The lungfish has widespread fossil relatives, and the genus *Scleropages* is represented in Borneo and southeast Asia by *S. formosus* and by fossils on Sumatra. These archaic freshwater fishes are restricted to the wetter parts of Queensland in eastern Australia. The freshwater species with tropical marine relatives include the Melanotaeniidae, Gadopsidae, Clupeidae, Ariidae, Plotosidae, Atherinidae, Mugilidae, Gobiidae, Eleotridae, Kuhliidae, and Teraponidae. Other salt-tolerant or diadromous species have relatives in the temperate northern hemisphere, a pattern called bipolar, antitropical, or (if narrowly separated, like many marine species) antiequatorial (Randall, 1982). Families included are Geotridae (southern lampreys), Percichthyidae, Galaxiidae, Aplochitonidae, and Retropinnidae. Two very different tropical eel-like fishes also occur; the Anguillidae are catadromous, and several species occur in Australia. Three species of synbranchid eels occur in northern tropical drainages. The fauna contains about 185 species of freshwater fishes, almost all found in small drainages on the periphery of this large, arid continent (Allen, 1991). Several of these are shared with New Guinea to the north and Tasmania to the south. The shallow Bass Straits separating Tasmania, and the Arafura Sea, separating New Guinea, became dry land during Pleistocene low-sea stands. This allowed an exchange of freshwater fishes between Australia and these smaller landmasses.

New Zealand, about 1600 km east of Australia, has about 30 species of native freshwater fishes. It has no archaeolimnic freshwater fishes and apparently has not been close to or connected to larger continental landmasses for 60 to 70 million years (McDowall, 1990). Only four diadromous species are not endemic to this island country, indicating a long isolation.

Land Bridges and Moving Continents

When the taxa of fishes of the various continents have a genealogical relationship, this indicates that a relationship between the geographic areas is involved. These areas can be adjacent drainages, springs, and/or lakes a few thousand years old. They can be large islands, like New Guinea or Borneo, that were connected by land bridges several thousand years ago, or far-away continents separated by millions of years. Such relationships provide a measure of closeness of association and can provide an estimate of the chronological order of past events, whether fossils are included or not (Wilson and Williams, 1992).

Plate tectonics, sea-floor spreading, and the sequential changes in the paleomagnetism of the earth's crust have provided the evidence placing the concept of continental drift on a firm logical base. The earth's crust, or lithosphere, is about 100 km thick and consists of a few major plates and several smaller ones, all of which ride on part of a mantle that is thickly plastic. Convection currents in the mantle send new rock to the surface along ridges in the oceans. The sea floor spreads in both directions at right angles to the ridges. Continental masses, such as Africa and South America or North America and Europe, are moving in opposite directions from each other. The east coast of the United States is a passive margin, with stable drainages and concentric rings of younger and younger sediments from the mountains to the coast. At the great oceanic trenches and along the west coast of North and South America, plates meet and one slides down at a steep angle below the other. This pressure is largely responsible for the mountain building along the western margins of the New World; these margins are called active margins. The high level of tectonic activity has kept aquatic systems disrupted and has prevented the evolution of large, diverse freshwater fish faunas. The largest faunas are found on the areas that have been relatively stable for long periods of time geologically.

The continents combined and recombined several times before the most recent coming together about 225 million years ago (Triassic). This combined area is called Pangaea, and it began to split apart in the Jurassic, about 180 million years ago (Owen, 1981; Fig. 30–9). This first separation was between Africa and North America. Soon two tiers of continents were formed, a northern group called Laurasia and the southern ones combined into Gondwanaland (Fig. 30–9). North America and Europe were connected until about 50 million years ago (Late Cretaceous–Paleocene), and North America and Asia were connected from 50 to 30 million years ago. They separated about Miocene time (about 15 million years ago) and were repeatedly connected and separated by the Pleistocene glaciations in the last 2 or 3 million years. North and South America first separated about 70 million years ago, and a continuous land connection was re-established only 3.7 million years ago (Pliocene). Approximate times for separation (in millions of years) of other landmasses are as follows: South America–Africa, 100; Africa–India, 65; India–East Antarctica, 100; East Antarctica–Australia, 55; Australia–New Zealand–West Antarctica, 80. South America and West Antarctica were possibly connected until about 10 to 12 million years ago. The splitting of Africa and East Antarctica is estimated to have been from 90 to 180 million years ago. Madagascar and Africa have been separated for more than 150 million years. It is much more difficult to tell when continents were effectively separated enough for isolation of the faunas. It is easy to imagine large, shallow, perhaps even brackish seas separating the continents initially and that these were easy for freshwater fishes to cross.

Early biogeographers tried to deduce centers of origin for fishes before continents were understood to have moved considerably. This was more plausible when dispersal was considered the only means for fishes to get from one continent to another. The distribution of freshwater fishes is the result of numerous geological, climatological, and biological processes, some of which are obvious, some disguised, some hidden, and some possibly not yet considered. All of these processes can be studied against the backdrop of the genealogies of the fishes, to interpret the way conditions evolved and arrive at the patterns of distribution that we have today. This kind of detective work makes the work of the ichthyogeographer interesting and challenging.

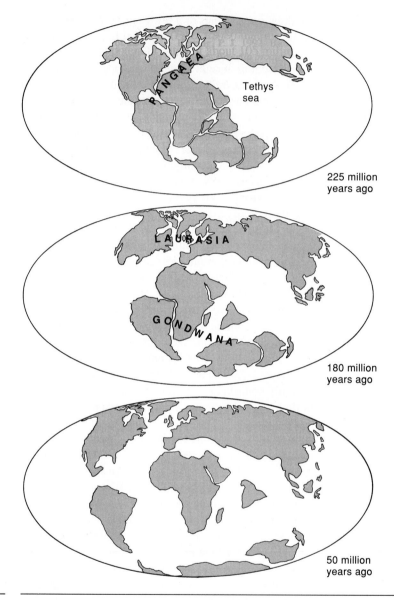

FIGURE 30–9 Approximate changes in position of the continents from the early Mesozoic to the early Cenozoic, demonstrating the increasing isolation of the continents and intervening ocean basins. This fragmentation subdivided the freshwater and marine fish faunas of the world, leading to a greater diversity today than at any time in the past.

Distribution of Marine Fishes

The break-up of the continents also fragmented the ocean basins, which must have been more continuous around the combined continents. In addition, for a long period of continental separation, a continuous, east–west trending seaway, the Tethys

Sea, separated the northern and southern suites of continents. Only for the last 3.7 million years have both the Isthmus of Panama and the Isthmus of Suez (the latter closing in the Miocene) been closed off to interchange of the Atlantic and Pacific Oceans and the Indian Ocean and Mediterranean Sea, respectively. Ichthyologists were prominent in protesting the construction of a sea-level canal in Panama, where two great marine faunas would have been thrown together with unpredictable consequences. The Suez Canal has resulted in a minor invasion of marine fishes into the less rich eastern Mediterranean from the much more diverse Red Sea; few, if any, fishes have gone in the opposite direction.

Many cold water marine fishes occur poleward in each hemisphere and are absent in the intervening tropics; such distributions are called antitropical or, if separated more narrowly, antiequatorial (Randall, 1982). Fishes of the families Scorpaenidae, Cottidae, Zoarcidae, Agonidae, Oplegnathidae, Girellidae, Clinidae, Heterodontidae, and some Pomacentridae, Atherinidae, and Labridae, among others, figure prominently in these distributions. Temperature restricts some of these species; others may bridge the tropics in deeper, cooler water, such as the Heterodontidae. Most are permanently and unquestionably separated today and had to rely on past climatic cooling to allow them or their ancestors to disperse across the tropics (White, 1986). Other fishes, particularly intertidal species and mesopelagic species that migrate vertically, traverse wide temperature ranges on a daily basis.

Salinity can also affect distributions; obviously, the vast majority of fishes are either fresh or saltwater species. Only a few are diadromous or euryhaline and pass back and forth between both media, and fewer are adapted to intermediate salinities. The eastern Mediterranean, Black, and Caspian seas are exceptional in their brackish water adapted faunas, which combine derivative species from both freshwater, marine, and diadromous taxa. Otherwise, such taxa are rare and occur in narrow strips or isolated patches at the fresh–saltwater interface (Haedrich, 1983; Swift et al., 1989). McHugh (1967) noted that larger, vagile fishes inhabit an expanded estuarine zone that extends outside the usual definitions of small, semienclosed bodies of coastal water. He included oceanic areas with surface salinities usually under 33 parts per thousand, including the north Pacific, Panama Bay, much of coastal Indo-China, and the western Pacific between northern Japan and Korea. These areas support large populations of estuarine fishes like salmonids, clupeids, engraulids, sciaenids, and others. Many of these species do not occur or are scarce in fully marine situations away from the influence of fresh water.

Land is an obvious barrier to marine fishes and, more important, often changes through geological time to provide new mixes of the faunal components. Until 3.7 million years ago, warm Atlantic and Caribbean waters flowed west into the Pacific over a submerged Isthmus of Panama. When the gap closed, warm water was deflected northward, giving rise to the Gulf Stream, which affected the whole north Atlantic Ocean. Thus much of the oceanic circulation of the Atlantic is relatively new, and the arrangement of the fauna must be interpreted in light of these facts. Underwater land and ridges that separate ocean basins effectively isolate populations of mesopelagic and benthic fishes, so that some speciation has taken place (Ebeling, 1967; White, 1994), albeit much less than among marine shore or freshwater fishes. As far as is known, a deep sea environment did not come into existence until the end of the Mesozoic, coincident with the evolution of many of the teleosts. Several basal euteleost groups are almost entirely deep sea forms, such as stomiiforms, myctophiforms, and aulopiforms (Nelson, 1994).

Pelagic Fishes

Pelagic fishes have wide expanses of the seas in which to feed and live, and little to isolate them into different groups. Since the divisions of today's oceans have been developing throughout the Cenozoic, today's configuration is relatively recent and the ocean basins were more confluent in the past. Only about 15 percent of the world's species of fishes are pelagic (Cohen, 1970). Often, the division of epipelagic species matches closely the divisions among mesopelagic species, indicating considerable vertical cohesion in the pelagic faunas. There are so few species that they can be singled out individually, whereas in the other areas, at most, genera, and more usually families, are noted.

Polar Waters. In the Arctic and Antarctic, waters range from nearly –2°C in winter to 5° or 6°C in summer. Parts of these areas can be covered much (or all) the year with ice and have no pelagic fauna. In the Arctic, the cold water extends south into the northern Bering Sea, reaching approximately to the 60th parallel. On the Atlantic side, the boundary of the Arctic water begins at the southeastern corner of Labrador at about 50° N, extends up to Iceland, and then continues northeast over Norway to a bit east of North Cape, at about 70° N. This displacement is due to the southward flow of the Labrador current in the west and the northeastward drift of the warm currents of the Atlantic (Fig. 30–10).

Pelagic fishes of the Arctic waters include some anadromous salmonids, such as the arctic char, *Salvelinus fontinalis,* and the pink salmon, *Oncorhynchus gorbuscha;* the gadids, *Arctogadus* and *Boreogadus;* and the herring, *Clupea harengus.*

Around the Antarctic, south of the Antarctic convergence, located at about 60° S in the Atlantic, there is extremely cold water that remains at less than 5°C even in summer (Fig. 30–10). This frigid area supports a very small pelagic fish fauna, including representatives of many of the typical mesopelagic families and some of the nototheniiforms that leave the bottom to eat the abundant krill. The chaenichthyid *Champsocephalus gunnari* is an example.

North Pacific: Cold Temperate Waters. In the North Pacific, cold temperate waters are found between the Arctic and a line roughly extending from Tokyo to southern California, the southern limit being along the winter 13° isotherm (about 34° to 35° N) (Fig. 30–11). The northern part (from about 42° N to the Arctic) is characteristically colder than 10°C in winter and has a surface salinity of less than 34 parts per thousand (ppt). This is called the Pacific subarctic or north boreal region. Typical pelagic fishes of the region are anadromous Pacific salmons and steelhead of the genus *Oncorhynchus.* Two other salmonid genera, *Stenodus* (one species) and *Salvelinus* (two species), also can be anadromous but do not range as widely offshore. One hexagrammid, the atka mackerel, *Pleurogrammus monopterygius,* lives a pelagic life; and the young of the other greenlings, especially *Hexagrammos decagrammus,* can be found in the open waters of the Gulf of Alaska. Relatives of the greenlings in open waters are the anoplopomatids, the skilfish, *Erilepis zonifer,* and the young of the sablefish, *Anolopoma fimbria. Theragra chalcogramma,* the

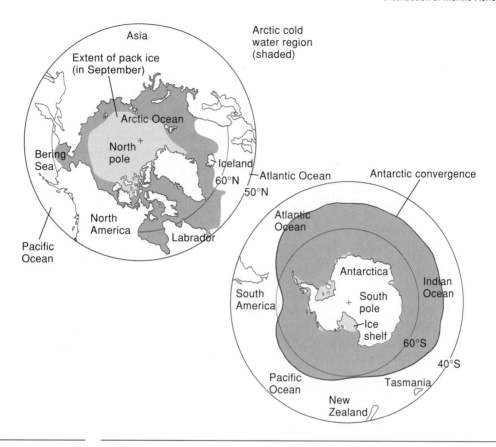

FIGURE 30–10   Oceanic subdivisions of the polar regions.

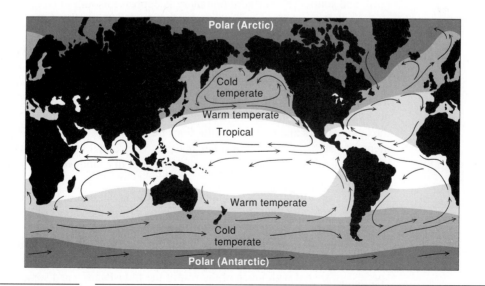

FIGURE 30–11   Major current and temperature patterns at the surface of the world's oceans.

walleye pollack, is another typical fish of the region, as is the Pacific herring, *Clupea harengus pallasi.*

Other common fishes that also range farther south than the aforementioned species include the wide-ranging Pacific saury, *Cololabis saira,* jack mackerel, *Trachurus symmetricus,* salmon shark, *Lamna nasus,* basking shark, *Cetorhinus maximus,* and the Pacific pomfret, *Brama japonica.*

The Pacific south boreal extends from about 34 to 42° N and is characterized by pelagic fishes that often cross regional boundaries. In fact, some fishery workers regard the area as a transitional one. The pomfret and saury are much more typical of the south than of the north boreal, as are the sardine, *Sardinops sagax,* the northern anchovy, *Engraulis mordax,* the medusafish, *Icichthys lockingtoni,* and the blue shark, *Prionace glauca.* Various tunas, including the albacore, *Thunnus alalunga,* and the bluefin, *T. thynnus,* are common in the area during certain times in their migrations.

North Pacific: Warm Temperate Waters. The southern boundary of the warm temperate region in the North Pacific is the winter 20° isotherm, or about 23° N. The northern boundary is at about 34° N, although in the mid-Pacific the regional boundaries trend to the north. This region has a well-developed pelagic fish fauna, with many species shared with adjoining regions. Many tunas and flying fishes live in these waters, along with marlins, sharks, and others. Typical species are the flyingfishes, *Prognichthys rondeleti* and *Cypselurus pinnatibarbatus,* the albacore, *Thunnus alalunga,* the skipjack, *Euthynnus pelamis,* the squaretail, *Tetragonurus cuvieri,* and the striped marlin, *Tetrapterus audax.* The rather uncommon opah, *Lampris regius,* and the louvar, *Luvarus imperialis,* appear to prefer the warm temperate water.

Tropical Indo-Pacific. The 20° winter isotherm limits the tropical waters both to the north and south. The northern boundary is depressed south to Hainan in the west, rises north through Taiwan, and runs across the Pacific at about 23° N before dipping south nearly to the tip of Baja California. The southern boundary extends from the eastern coast of South Africa to about mid-Australia, and from the east coast of mid-Australia to near the coast of South America, where it turns north along the colder Peru current to about 3° S. In the northern Indian Ocean, only the northern part of the Persian Gulf is excluded from the tropics. Although there is considerable similarity among the tropical marine fish faunas of the Atlantic and Pacific, the land barriers and previous extinctions have given each area its distinctive elements. In addition, tidal and other features vary considerably between various areas, some only separated by a short distance of dry land, such as the Isthmus of Panama.

The eastern Pacific province occupies the area adjacent to the coast of the Americas, extending out about 3500 kilometers from the shores of South America (Briggs, 1961). It is characterized by a number of endemic species of small anchovies, genus *Cetengraulis,* thread herrings, *Opisthonema* spp., flyingfishes, *Cyselurus* and *Prognichthys,* and several carangids or jacks. Many large pelagic species are shared with much of the rest of the Indo-Pacific, and some species, like the yellowfin tunas, circumnavigate much of the tropical Pacific on an annual basis.

The Indo–West Pacific pelagic fauna is relatively similar and is dominated by blue sharks, whale sharks, thresher sharks, and silky sharks, along with tunas and flying fishes. Several large species of billfishes are also concentrated in this zone, including marlins and sailfishes and several of the tunas.

Southern Warm Temperate Region. Because the warm temperate waters are continuous around southern Africa and barely interrupted by southeastern Australia, a circumglobal fauna has developed. The major land barrier is the southern part of South America, from about 41° S on the west side and from the region of Buenos Aires on the east. Along the west coast of South America, a couple of endemic species occur, the Inca scad, *Trachurus murphyi,* and chub mackerel, *Scomber japonicus.* Most of the other pelagic species are circumglobal in the warmer parts of the oceans: southern bluefin tuna, *Thunnus maccoyi,* the saury, *Scomberesox saurus,* the pomfret, *Palinurichthys antarcticus,* and the squaretail, *Tetragonurus cuvieri.*

Southern Cold Temperate Region. The cold temperate area between the Antarctic and southern subtropical convergences does not seem to have a distinct pelagic fish fauna, sharing most species with the warm temperate waters to the north. Examples are the basking shark, *Cetorhinus maximus,* the saury, albacore, and butterfly mackerel, *Gasterochisma melampus,* and the gempylid, *Thrysites atun.*

Atlantic Ocean: Boreal Waters. The Atlantic boreal includes the ocean between southern Labrador and northern Norway on the north and between Cape Hatteras and the English Channel on the south. The pelagic fish fauna of the region is mixed, made up of eurythermal species that can range to the north or the south. Examples are the mackerel shark, *Lamna nasus,* basking shark, opah, *Lampris regius*; Atlantic mackerel, *Scomber scombrus*; bluefin tuna; swordfish, *Xiphias gladius*; and mola, *Mola mola.* Using mid-water fishes (Backus, 1986) delimits this zone near the 9° isotherm on the north and the 15° isotherm on the south.

Atlantic Ocean: Warm Temperate Waters. This region is narrow on the west, from Cape Hatteras to mid-Florida, and much wider in the east, from the English Channel to Senegal. Again we have skipjack; saury; blue shark; whitetip shark, *Carcharhinus longimanus*; pilotfish, *Naucrates ductor*; the dolphin, *Coryphaena hippurus*; and the albacore, *Thunnus alalunga.* The epipelagic fauna of the Mediterranean is a subset of the Atlantic, probably due to extinctions and probably even large-scale drying up of the Mediterranean during the last 5 or 6 million years.

Atlantic Tropical Region. This region is broad to the west, from Florida to Rio de Janiero, and constricted on the east by the cool currents (Senegal to Angola). The pelagic fishes of the region are eurythermal and range into the adjacent warm temperate regions as well. A notable feature is a large circular gyre of ocean water in the west off the Bahamas, the Sargasso Sea. Large masses of these drifting algae of the genus *Sargassum* create a whole community of small fishes, including the Sargassum fish, *Histrio histrio,* and many juvenile pipefishes, jacks, and stromateids. In very deep water, the two Atlantic species of *Anguilla* are believed to spawn at depths as yet poorly known. Backus (1986) divides this into a tropical and another more northern subtropical region.

Continental Shelf Fishes

Shelf fishes generally live shallower than about 100 m depth, which usually corresponds more or less to the edge of the continental shelves. Most of the fishes are benthic or epibenthic and have much narrower distributions than many pelagic species. Movement or dispersal is restricted to continental margins, and large expanses of deep water, strong inflows of fresh water (such as river deltas), and temperature and substrate changes can cause breaks in distribution. Long-lived larval stages provide the potential for colonizing across barriers, but studies on the Great Barrier Reef found little correlation between length of larval life and extent of geographic distribution for a large number of reef species (Sale, 1991). Obviously, many ecological factors enter into the ability of shore fishes to extend their range and colonize new and old habitats. As many, if not more, of these must involve mechanisms to keep the offspring near suitable habitat instead of casting them to the uncertainty of the open ocean (Sale, 1991). Many temperature-associated shelf fishes move with the seasons; warm water fishes move northward into southern California or southward along the New South Wales coast of Australia during spring and summer, only to retreat again in the fall. Many tropical species colonize the inshore areas of the northern Gulf of Mexico in the spring and summer and in the fall move southward, into deeper water, or are extirpated. Cool water species do the reverse, so that in winter skates, gadids, and the spiny dogfish, *Squalus acanthias,* are present in shallow water in the winter and retreat to deep cool water in the summer.

On a longer time scale, cold water fishes can be distributed in pockets of cold water upwelling along north–south, western-facing coasts, such as North and South America. In areas of cold water upwelling, the upper ends of larger bays are often much warmer and hold populations of warmer water fishes that are relics of earlier, warmer climatic conditions. From Point Conception in southern California to the southern tip of Baja California, areas of cold upwelling alternate with warm bays, creating a complex intertwining of northern and southern faunal elements (Hubbs, 1961; Rosenblatt, 1967). These idiosyncracies make it difficult to define areas by their fish faunas; the general subdivisions of Briggs (1974) are followed here as a rough guide to the way marine fishes are arranged on earth.

Arctic Region. The Arctic region includes the shores of North America from southeast Labrador to Nunivak Island, Alaska, on the Bering Sea. In Eurasia, it includes the shore from the Murmansk Peninsula around to the northern part of the Bering Sea at about 62° N. All of Greenland is included, but only the north coast of Iceland. The Arctic fish fauna consists of a relatively few species that are restricted to the region and some that are shared with the boreal shelf. Typical cold water forms are *Boreogadus saida*; *Arctogadus borisovi*; the Arctic flounder, *Liopsetta gracilis*; the fourhorn sculpin, *Myxocephalus quadricornus*; the Arctic sculpin, *M. scorpioides*; and the eelpout, *Lycodes fragilis.* Boreal species entering the Arctic from the Atlantic include the sleeper shark, *Sominosus microcephalus,* the skate, *Raja radiata,* the wolf fish, *Anarhichas lupus,* the snailfish, *Liparis liparis,* and the sand lance, *Ammodytes hexapterus.* Boreal species from the Pacific include the capelin, *Mallotus villosus,* the pond smelt, *Hypomesus olidus,* the eelpout, *Gymnelis viridis,* and the sculpin, *Artediellus scaber.* Fourhorn sculpin, pond smelt, and capelin can invade coastal fresh water as well; much of the surface water in the Arctic is below 33 parts per thousand. Andriashev and Chernova (1994) tabulated 415 species from Arctic seas and adjacent waters, including diadromous and freshwater species.

Pacific: Eastern Boreal Region. This region includes the Aleutian Islands and the coast of North America from the southern part of the Bering Sea to Point Conception in southern California. The North American coast is dominated by the southward-flowing California current, which is deflected offshore by Point Conception. A northern Aleutian province is separated at southeastern Alaska from a southern Oregon province. The fauna is similar to that across the Pacific in the western boreal region, with a rich fauna of sculpins, rockfishes, snailfishes, right-eyed flounders, gunnels, and pricklebacks. The boreal faunas are considerably richer than the Arctic, but less so than the warm temperate region just to the south. Horn and Allen (1978) divide this region into two, a northern Alaska region and a southern Vancouverian one, based on a cluster analysis of the coastal marine fishes (excluding bay-inhabiting species).

Pacific: Western Boreal Region. This region includes the coast of Asia north of the Formosa Strait to Cape Olyutorsky in the Bering Sea, except for the tip of Korea and southern Japan, which are influenced by warm currents. Three subdivisions or provinces have been recognized in the Oriental to the south, the Kurile to the north, and the Okhotsk in the northwest. In the northerly areas are migratory salmonids; icefish, *Salangichthys*; the ayu, *Plecoglossus*; and the Japanese huchen "chevitza," *Hucho perryi*. In addition, species of the families Liparidae, Scorpaenidae, Cottidae, Osmeridae, Hexagrammidae, and Pleuronectidae are common.

North Pacific Warm Temperate: Japan. This region includes the southern tip of Korea; the coast of Honshu (Japan) south of about 35° N Shikoku, Kyushu; the west coast of Taiwan; and the coast of China from about Wenchou to Hong Kong. Like the southern California fauna, a strong mixture of northern and southern species, often fluctuating seasonally, in addition to endemic antitropical species, dominate the fauna. Some prominent families are lizardfishes (Synodontidae), flyingfishes, mullets (Mugilidae), jacks (Carangidae), sea basses (Serranidae), and croakers (Sciaenidae).

North Pacific Warm Temperate: California Region. As noted earlier, a mixture of warm, cold, and endemic species characterized the region from southern California to southern Baja California. Some of the endemics are atherinids, including the grunion, *Leuresthes tenuis,* and its relatives the topsmelt and jacksmelt; the California black sea bass; the California white sea bass (a croaker); the garibaldi (a pomacentrid); the senorita (a labrid); the opaleye (a girellid); and the California halibut (a bothid). Many groups are shared with the western Pacific, such as the embiotocids; the *Chasmichthys* group of gobies (Birdsong et al., 1988); many rockfishes, genus *Sebastes;* and many cottids. A subset of these fishes occurs in the upper Gulf of California, separated from those of the outer coast by peninsular Baja California and the tropical waters around its southern extremity (Thomson et al., 1979). These species are conspecific and at most slightly differentiated, indicating a relatively recent separation, probably 1 or 2 million years at most. Otherwise, clinids, cottids, embiotocids, bothids, gobies, croakers, and rockfishes dominate this fauna. Horn and Allen (1978) found seven principal faunal discontinuities from Alaska to Baja California in their cluster analysis of shore fishes. Three distinct regions were found in California: northern, central, and southern. These were generally congruent with those based on other organisms and largely reflected marine climate patterns.

Tropical Indo–West Pacific Region. The richest marine fish fauna on earth is encompassed by the tropical waters of the western Pacific and Indian oceans. A recent book on the large visible fishes of the Coral Sea (Randall et al., 1992) near Australia lists over 1000 species. If one includes the as yet untabulated small species, probably 2500 to 3000 species are present. The fish fauna farther north, nearer to the Philippines, and west toward the Indo-Malayan area is considered even richer than the Coral Sea (Sale, 1991). Families in these coral reef habitats are the Muraenidae, Ophichthyidae, Syngnathidae, Labridae, Scaridae, Pomadasyidae and related families, Pomacentridae, Blenniidae, Apogonidae, Gobiidae, Serranidae, Lutjanidae, Acanthuridae, Chaetodontidae, Pomacanthidae, and Eleotridae. When these faunas are more completely known, it will be interesting to see the final tallies. The largest fish faunas may still fall in countries like Japan, Australia, or South Africa, which have large areas of both the Indo-Pacific fauna and cold temperate faunas to augment their numbers of fishes. Smith and Heemstra (1986) noted that the 2200 marine species of South Africa represent about 15 percent of the world total of marine fishes, including 83 percent of the families. In contrast, Japan has 86 percent, Australia 68 percent, and the Philippines 60 percent of the world's families (Smith and Heemstra, 1986). Nakabo (1993) gives the figure 3,600 species of marine and freshwater fishes in Japan.

Springer (1982) pointed out a strong disjunction in fish distributions between the Indo–West Pacific and the Pacific Plate, the latter occupying much of the open Pacific north and west of Japan, the Marianas Islands, the Solomon Islands, New Hebrides, Samoa, and New Zealand to the west coasts of North, Middle, and South America (Figs. 30–12 and 30–13). Thus his Indo–West Pacific included only those areas on the Philippine, Eurasian, and Indo-Australian plates. Genera and species groups of many fishes occur in the Indo–West Pacific or on the Pacific Plate and not on both. Springer's exhaustive survey argues strongly for the distinctness of each area, but the reasons for the dichotomy are not clear.

Tropical Eastern Pacific Region. This region extends from the southern one third of the Baja California peninsula south to about the Gulf of Guayaquil. A few offshore islands are included, such as the Revillagigedos, Malpelo, Cocos, Clipperton, and the northernmost Galapagos Islands. Most of the Galapagos Islands are farther south, influenced by cold currents, and have a temperate fauna. The fauna is closely related to that of the western Atlantic and Caribbean, but with considerably fewer species among the reef-dwelling fishes, largely because coral reefs are poorly developed (Allen and Robertson, 1994). Bay and estuarine species of sea catfishes, croakers, and clupeoids are about equally rich on both sides of the Isthmus of Panama. Some of both reef and estuarine species are barely, or not at all, differentiated between Atlantic and Pacific populations. Only a few western Pacific species occur in the eastern Pacific; the vast expanse of water separating these two areas is called the

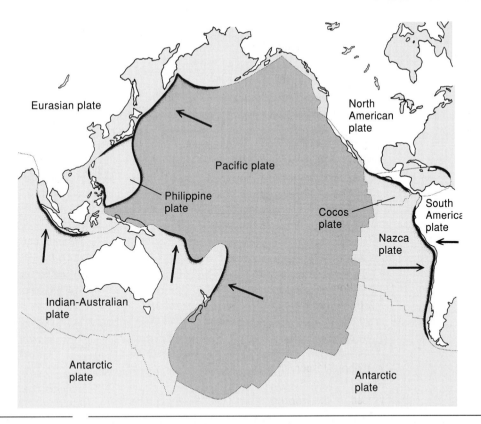

FIGURE 30–12 The Pacific Plate and surrounding plates, a causal element in marine shore fish distributions. Darker lines represent deep oceanic trenches.

eastern Pacific barrier (Allen and Robertson, 1994). Many reef-dwelling species extend virtually the whole length of this area but are interrupted by the estuarine influence of the Gulf of Nicoya and Panama Bay. Both reef and estuarine forms often have distinct endemics in the Gulf of California. Many genera and subfamilies of marine fishes are endemic to the new world, but only one family, Dactyloscopidae, is endemic.

Indo-Pacific Warm Temperate: South Australian Region. The southwestern and southeastern coasts of Australia hold a temperate fauna with some tropical species and at least three endemic families, Peronedysidae, Pataecidae, and Gnathanacanthidae. Many other genera are endemic to these two areas.

Northern New Zealand Region. The northern island of New Zealand and the Kermadec Islands north of it make up a region that shares several families with temperate Australia—namely, Odacidae, Latridae, Leptoscopidae, Arripidae, and Chironemidae.

Western South American Region. The northward-flowing cold Peru current somewhat mirrors the southward-flowing California current in California, carrying cold water conditions to lower latitudes than expected. In both cases, during occasional

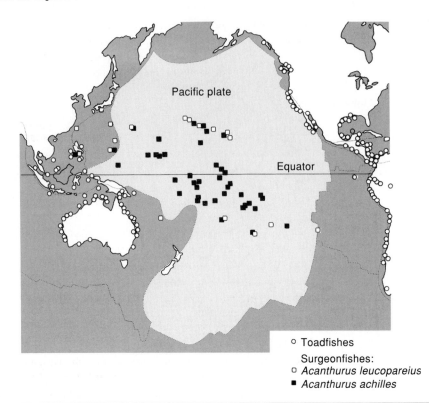

FIGURE 30–13 Pacific shore fish distributions: non-Pacific Plate (toadfishes, circles), Pacific Plate endemics (both *Acanthurus* species, open and closed squares); and an anti-equatorial distribution (*Acanthurus leucopareius,* open squares).

El Niño conditions (usually every few years), warm equatorial water pushed east across the central Pacific is deflected north and south by Middle America, and warm water and its associated fishes are pushed to higher latitudes than usual. An influx of tropical fishes is matched by a decline in many resident temperate fishes. Clearly, the faunal zones are blurred by these variations over ecological and geological time. Under more typical cooler conditions, both areas can enjoy tremendous fisheries based on clupeoids that flourish in the cool, rich, upwelling waters. In the north, these have been based on *Sardinops sagax,* and, more recently, *Engraulis mordax.* In the south it is *Engraulis ringens.*

Southern Cold Temperate Region. Four regions of southern cool water are recognized: southern Victoria and Tasmania (Australia), the southern island of New Zealand, southern South America, and the sub-Antarctic islands such as Keruelen and Macquarie. The families Bovichthyidae and Nototheniidae are found in all these areas. Nototheniidae is also in the Antarctic shore waters. In addition, a few more temperate species of blennies, gobies, clingfishes, and pipefishes occur here. This region and the following Antarctic one are overwhelmingly dominated by the nototheoids (Eastman, 1993), numbering about 190 shore species.

Antarctic Region. Cold Antarctic water sinks below the warmer ocean water farther north, creating a region in which water temperatures hover near 1°C. A small fauna of cold-adapted (cryophilic) fishes lives in the shore areas, often under the ice, including the families Chaenichthyidae, Nototheniidae, Bathydraconidae, and Harpagiferidae (Eastman, 1993).

Western Atlantic Boreal Region. Between Newfoundland and Cape Hatteras, the coast of North America is influenced by the cold, southward-flowing Labrador current. This prevents the establishment of many tropical and subtropical species, which invade in the summer at least as far as Cape Cod with the warmer Gulf Stream. This cooler region supports a small fauna of salmonids, sturgeons, sticklebacks, poachers, sculpins, wolf fishes, cods, right-eyed flounders, and some others characteristic of the north Pacific and northeast Atlantic as well.

Eastern Atlantic Boreal Region. This region includes the southern coast of Ireland, all of the British Isles, and the coasts of northern Europe eastward through the Baltic Sea and Scandinavia. Many of the same Arctic and boreal families are present, and many genera and a few species are shared with the western Atlantic. Southern species often stray north in the warm months.

Atlantic Warm Temperate: Carolina Region. Often called the Carolinian region, this area is split into two by the Florida Peninsula, which extends southward into fully tropical seas and serves today as a barrier between the similar fishes inhabiting both the southeastern Atlantic Coast and the northern Gulf of Mexico. Like the Baja California Peninsula on the western coast of North America, this pattern probably derives both from a former Late Miocene submergence of the peninsula and cooler climatic conditions in the Pleistocene, which permitted some taxa to circumnavigate the southern end. In the Carolinian region, the speciose taxa are the sciaenids, clupeids, gobiids, pipefishes, atherinids, belonids, fundulids, cyprinodontids, left-eyed flounders, sea robins, and puffers. The seasonal cycle in the northern Gulf was noted earlier.

Mediterranean-Atlantic Region. This region includes the Atlantic and Mediterranean coasts of Europe and Africa, from the English channel to about 14° at Cape Verde. The Black, Caspian, and Aral seas are included because of the similarities of their vertebrate faunas (Zenkevitch, 1963). The Mediterranean has many endemic species closely related to taxa that occur in the eastern Atlantic outside of it. In addition, the Black and Caspian seas have endemic species or subspecies, particularly in taxa that tolerate low salinity, like gobies, clupeids, and anchovies. In addition, the Caspian Sea has some northern species of cyprinids and diadromous species like lampreys, the whitefish genus *Stenodus,* and sturgeons that are relics of cooler glacial conditions in the past. The Aral Sea has lower salinity and the smallest fish fauna of the three. Today these distinctive faunas are threatened with almost complete extirpation by appropriation of the water from their tributary rivers. In the Pleistocene, glacial advances pushed brackish-water fishes out into the eastern Mediterranean, which was much fresher, at least at the surface. Many of these brackish-water species are left as relics in coastal lagoons and estuaries, isolated by the return of fully marine conditions. The marine fauna of the Mediterranean is depauperate for such a large body of sea water and probably suffered extinctions from the wide climate fluctuations in the last few million years. Interestingly, since the Suez Canal opened a sea-level connection with the Red Sea, 40 or more Red Sea species have colonized the eastern Mediterranean, but few, if any, fish species have gone the opposite direction.

Tropical Atlantic–Western Atlantic Region. This region includes the Caribbean Sea and surrounding areas of southern Florida, the Bahamas, and parts of the coast of Brazil. It is broadly interrupted by the Amazon delta but exists south of the delta along parts of the southern coasts of Brazil. It is the sixth or seventh richest tropical coral reef area after the Indo–West Pacific, harboring about 1000 species and somewhat fewer at any one specific locality (Sale, 1991). The north coasts of South America also harbor large numbers of estuarine ariids, sciaenids, and clupeoids. A few of the same or closely related species occur across the Atlantic in the Gulf of Guinea; thus the Atlantic Ocean has not been a complete barrier to dispersal of the marine shelf fauna (Robins, 1971).

Eastern Atlantic Tropical Region. This region extends from Cape Verde on the African coast south to Mossamedes at about 15° S and is constricted by the cool, southward-flowing Canary current and the northward-flowing Benguela current. The tropical fauna is only about half the number of the western Atlantic, and only a few species are shared with the Indo–West Pacific fauna, which occurs almost entirely on the opposite side of the African continent.

Eastern South American Warm Temperate Region. From Rio de Janiero south to the Rio de la Plata, a small temperate fauna exists, with some cognate species of the north temperate and tropical regions. The much smaller land area and lack of major physiographic divisions has left a less differentiated fauna than in the northwestern Atlantic. It consists of some of the same dominant fishes, such as Ariidae, Sciaenidae, clupeoids, Gobiidae, Scorpaenidae, and Serranidae.

Southern African Warm Temperate Region. The southern end of Africa from southern Angola around to Durban is warm temperate and shares some species with the more tropical Indian Ocean farther east and north. It is dominated by cold upwelling that contrasts sharply with the warm waters farther east along the Indian Ocean side of South Africa (Smith and Heemstra, 1986).

Worldwide. About 100 marine species have a worldwide distribution (Briggs, 1960). These are mostly pelagic or deep sea species and include fishes like the white shark, tiger shark, ocean sunfish, and swordfish. In many cases the continents effectively isolate them into two or more non-interbreeding populations. For at least two or three million years these isolates have differentiated very little from each other.

Chapter 31

Fishes and Human Endeavor

In considering the relationships between fishes and humans, one automatically thinks of fishes as food, for this is the greatest and most obvious relationship. This section will devote attention to the widespread fisheries and products made from fishes, but other relationships will also be examined. Most of these will emphasize positive aspects of interaction between fishes and humans, such as recreational fisheries and the use of fishes for aquarium display, but attention will first be given to some negative or more unpleasant relationships. Fortunately, the incidence of these negative experiences is minor in comparison to the many important positive associations between humans and fishes. However, some may receive undue attention; one nonfatal shark attack is sure to get more newspaper coverage than the capture of enough fish to feed a large city.

Negative Aspects

Poisonous and Venomous Fishes

Throughout the many orders of fishes, there are species known to be poisonous to humans. Russell (1969) stated that at least 1000 species may be implicated. Some can cause illness or death when eaten; others have stinging spines that introduce venoms. Together they constitute genuine hazards to divers, fishermen, and bathers in the natural habitat of the fishes, and to aquarists and diners who might come into contact with fishes or fish flesh in areas far removed from their origin.

A considerable vocabulary of terms designates the toxins of the fishes and the conditions they cause in human beings. Some of the important terms are the following (see Halstead 1970, 1992):

Ichthyotoxin: Generally, any poison originating from fishes

Ichthyosarcotoxin: Poison found in flesh of fishes, excluding poisons due to bacterial action

Ichthyohemotoxin: Poison found in blood of fishes

Ichthyootoxin: Poison found only in roe of fishes

Ichthyocanthotoxin: Poison secreted at the site of a venom apparatus such as spines, stings, or teeth of fishes

Ciguatera: A particular ichthyosarcotoxism caused by eating various marine fishes of tropical and subtropical areas

Poison: A substance that can cause structural damage or functional disturbance by its chemical activity in a biological system when eaten, injected, or otherwise applied to a body in small quantities

Scombroid poisoning: An ichthyosarcotoxism caused by eating improperly preserved scombroid fishes

Tetrodotoxin: The poison in viscera of puffer fishes

Toxin: In modern usage, any poisonous substance from microbic, mineral, animal, or vegetable origin

Venom: Poison originating in the venom-secreting apparatus of an animal

In this treatment, the adjective *poisonous* will be used for fishes containing poisons (ichthyosarcotoxins, etc.) that affect humans after ingestion of the fish. *Venomous* will be used for those fishes that introduce toxins by means of stings, spines, or teeth. This usage roughly follows that of Halstead (1970).

Poisonous Fishes

Ciguatera. The most widely known type of poisoning caused by fishes is ciguatera, a type of intoxication that causes a variety of symptoms such as nausea, vomiting, abdominal pain, reversal of hot and cold sensation, and numbness of the mouth. Various other symptoms may include headache, muscular aches, dizziness, and, occasionally, blistering and loss of skin on hands and feet. A great variety of tropical marine fishes cause this type of poisoning, although some species may be toxic in some geographical areas and not in others. About 20 families, several of which contain normally safe and popular food fishes, are mentioned as being at times ciguatoxic by Halstead et al. (1990). Usually large individuals of a species are poisonous whereas small ones are not. The toxin appears to be obtained through the food chain, originating in certain dinoflagellates living on macroalgae that grow around coral reefs. Among the several toxic genera of dinoflagellates, two, *Amphidinium* and *Gambierdiscus,* have been implicated as producers of especially potent toxins (Withers, 1988; Yasumoto et al., 1987). An important toxic agent of ciguatera is ciguatoxin, which is very potent; a lethal dose occurs at 50 percent (LD^{50}) of 0.45 g/kg in mice when injected intraperitoneally. Maitotoxin, another toxin of ciguatera produced especially by *Gambierdiscus toxicus,* is even more potent, with an LD^{50} of 0.13 g/kg in mice. This is known as the most potent toxin of marine origin (Ohizumi and Kobayashi, 1990). Other less well known toxins, such as scaritoxin, are involved in ciguatera (Withers, 1988). Hashimoto (1979) mentions two genera of blue-green algae *Schizothrix* and *Microcoleus* as being implicated in ciguatera. Dinoflagellates with the contained toxins are thought to be more prevalent on dead coral reef areas than on undisturbed reefs (Bagnis, 1981). The toxic substances are eaten by herbivorous fishes, which then become the food of carnivores. Large carnivores tend to concentrate the poison to the point that they become dangerous to eat.

Fishes most often implicated in ciguatera are morays (Muraenidae), barracuda (Sphyraenidae), snappers (Lutjanidae), groupers (Serranidae), and jacks and their close relatives (Carangidae). All these families contain food fishes that are habitually eaten in tropical areas. There have been over 400 species known to cause ciguatera (Halstead and Vinci, 1988). Bagnis (1981) identified 53 species (13 families) from the Caribbean and 74 species (16 families) from the Indo-Pacific as being ciguatoxic. Randall (1980) found that over one third of the specimens examined from Enewetok and one fifth of the specimens from Bikini showed ciguatoxicity.

Additional families of fishes, some of which are herbivores or plankton feeders, that have been implicated in one or more poisonings include bonefishes (Albulidae), milkfish (Chanidae), tarpons (Elopidae), herrings (Clupeidae), anchovies (Engraulidae), lizardfishes (Synodontidae), conger eels (Congridae), flyingfishes (Exocoetidae), squirrelfishes (Holocentridae), surgeonfishes (Acanthuridae), butterflyfishes (Chaetodontidae), mackerels and tunas (Scombridae), plus a number of other perciform fishes and some of the Tetradontiformes (filefishes and relatives).

There have been mentions of poisonings from ciguatera dating back to the seventh century A.D. in China and the seventeenth century in Europe. The explorer Captain James Cook was apparently affected by ciguatera on his voyage of 1774 (Halstead, 1978). There are stories that usually tell of incidents involving a ship's crew or a family of islanders, but some mention 50 or 60 persons taken ill and at least one report tells of 1500 persons affected. Relatively few deaths (less than 10 percent) are reported in most instances, but there is a report of more than 400 deaths among Marshall and Caroline islanders during 1940 and 1941. In the United States, in the years 1973 to 1980, there was a total of 2841 outbreaks (about 125,000 cases) of acute food-borne disease, according to Hughes and Potter (1991). The causative agent was identified in only 38 percent of the outbreaks, and of these 697 were caused by chemical agents. About 240 were caused by ciguatera. The annual number of ciguatera cases worldwide is estimated at from 10,000 to 50,000 (Hilgerd, 1983).

There is no effective remedy for ciguatera at this time. Cooking does not destroy the toxin. No immunity is imparted by a prior attack, according to Halstead (1978, 1988).

Scombroid Poisoning. Another common poisoning results from eating the flesh of tunas, their relatives, and certain other fishes if the fishes have not been properly stored. Hughes and Potter (1991) indicate that the problem is about as common in the United States as ciguatera. Several fishes other than scombroids have been implicated in this type of poisoning, including members of the families Clupeidae, Engraulidae, Scombresocidae, Carangidae, Coryphaenidae, Arripidae, and Pomatomidae (Auerbach, 1988). Taylor and Bush (1988) list 45 species (in seven families) that may be implicated in scombrotoxism. Many of the species involved have a large proportion of dark muscle and maintain a temperature higher than ambient. The flesh of these fishes can undergo rapid bacterial decomposition if stored without refrigeration for a significant period. Tuna held at 20° to 25°C for several hours acquire toxicity (see Halstead, 1978). The bacterial action converts free histidine, which is common in scombroid muscle, to histamine and derivatives such as saurine, the phosphate salt of histamine (Taylor and Bush, 1988). Convincing evidence that histamine is the toxic agent causing scombroid poisoning is presented by Morrow et al. (1991).

Strong allergic reactions can be caused by ingestion of histamine-rich fish. These include flushing of the skin, dizziness, nausea, diarrhea, vomiting, thirst, and palpitations (Auerbach, 1988; Halstead, 1978, 1988). Because the problem of histamine in fish flesh is not limited to scombroids and the symptoms are allergy-like, the poisoning is referred to as histamine (scombroid fish) poisoning by Taylor and Bush (1988) and as pseudo-allergic fish poisoning by Prescott (1984).

Poisonous and venomous fishes may have some ecological advantage in that predators may be injured or killed by stings or poisons. Some species, such as the Moses sole (*Pardachirus*), have their toxin well spread from glands, so predators are warned away. Many venomous fishes exhibit warning coloration.

Tetrodotoxin Poisoning. Much has been written about puffer poisoning (tetrodotoxism) caused by eating the viscera of various tetraodontiform fishes, mostly of the genus *Takifugu* (= *Fugu*), which are common table fare in Japan (Halstead, 1978, 1988). The toxin, tetrodotoxin, is particularly strong, and the fatality rate is usually over 50 percent of those intoxicated. Nonetheless, the flesh is of such high

quality that the Japanese maintain a fishery for the species and license chefs to prepare them so that only the nontoxic portions reach the table. The ovary and liver are
the most toxic parts of the fish, with the stomach and intestines being nearly as virulent, especially in *Takifugu niphlobles.* The eyes and kidneys are toxic as well. The
skin, subcutaneous tissue, and testes are moderately poisonous in some species (see
Halstead, 1978; Hashimoto, 1979). Other poisonous genera include *Arothron, Lago-
cephalus, Sphaeroides,* and *Tetraodon* (Halstead, 1992).

   Puffers (Tetraodontidae), porcupinefishes (Diodontidae), molas (Molidae; Fig.
16–2**B**), and a goby *Gobius criniger* (= *Rhinogobius nebulosus*) (Hashimoto, 1979)
have been implicated in tetrodotoxism. These are widely distributed in warm and
warm-temperate waters and have been the cause of illness and death throughout
their ranges, especially in the Orient. The poison found in puffers and their relatives
has been isolated and its chemical structure determined. It is identical to tarichatoxin, a poison found in newts of the genus *Taricha* (see Halstead, 1978).

Other Fish Poisons. Other types of fish poisoning are known but do not appear to be
as common as the foregoing. Lampreys and hagfishes have been implicated in what is
called cyclostome poisoning, which in the case of the lampreys might be due to an
excess of bile salts in the flesh (MacDowall, 1988). Several families of sharks have
toxic livers or flesh. Elasmobranch poisoning might be due to three different kinds of
toxins (including ciguatera). The flesh of the Greenland shark *Somniosus macrocephalus* often contains a toxin that can be removed by drying or thoroughly washing
strips of the meat. Many tropical sharks have toxic livers, a condition apparently not
related to hypervitaminosis A, which could result from eating the vitamin-rich livers.

   Oilfishes (Gempylidae) have been called "purgativefishes" by seafarers because of the diarrhea caused by eating their oily flesh. A few fishes are known to
produce toxins from skin cells or glands not associated with any kind of a stinging
structure. These poisons are called ichthyocrinotoxins and are harmful if ingested by
humans. Fishes known to have poisonous skin or slime or to secrete toxins upon
being disturbed include hagfishes, lampreys, morays, soapfishes (Grammistidae),
puffers, and their near relatives.

   Tunas and swordfishes are known to concentrate mercury through the food
chain, and unacceptable levels have been noted by the U.S. Food and Drug Administration. Landed fish are routinely tested, and those with excessive mercury content
are not used for human consumption. Pollution of waters by various heavy metals
and organic compounds has become a serious problem in many areas of the world
because of agricultural and manufacturing practices. For instance, pesticides such as
DDT have accumulated in fish to the point that in the case of this compound, advisories were issued at one time concerning the danger of eating the fatter portions of
fish from the Great Lakes (see Foran et al., 1989). Kepone bioaccumulation in fishes
and shellfishes has led to the closing of certain fisheries (Huggins, 1989). Industrial
chemicals, such as the polychlorinated biphenyls (PCBs), have exceeded U.S. Food
and Drug Administration tolerance levels in fishes in many waters. In 1989 in the
United States, 37 of the states had advisory programs of some kind to inform people
of risks of consuming fish from designated polluted waters (Reinert et al., 1991).

   The worldwide nature of the problem of bioaccumulation of harmful chemicals
is illustrated by the presence in *Latimeria* of organochlorines, including PCBs and
DDT (Hale et al., 1991). There is international concern involving the levels of these

contaminants, from both ecological and public health standpoints (Borlakoglu and Dils, 1990).

Venomous Fishes. A number of fish families contain species capable of inflicting painful stings that combine mechanical injury with release of venom (Halstead, 1978; Maretic, 1988). The stingrays of Dasyatidae, Potamotrygonidae, Urolophidae, and Gymnuridae are among the best known of stinging fishes. Their generally larger relatives, eagle rays (Myliobatidae) and cownose rays (Rhinopteridae), have stings and venom-producing tissue but are not often implicated in injury to humans, probably because the rays spend much of the time off the bottom, and the stings are very close to the base of the tail. The stings of the rays are stiff, dagger-like spines with recurved teeth on the edges. In the integumentary sheath, especially along the ventrolateral grooves of the spine, there is a layer of venom-producing glands. When the spine is thrust into the flesh of a victim, the skin sheath is broken and the venom released into the wound.

Spines of stingrays are usually situated close to the thick muscular base of the tail (Fig. 5–6**C**). When the ray is stimulated to defend itself—for example, by a person stepping on the disc—the tail is curled quickly over the back and the spines thrust at the offender. The wounds resulting from these stings are dangerous and painful. The victim is usually in danger from secondary infections, tetanus, and gangrene as well as the effects of the venom.

Other cartilaginous fishes with venomous spines are the dogfish shark (*Squalus*) (Fig. 5–5**C**) and the chimaeras (Fig. 5–2**B**). In these the venom glands are along the dorsal spines, and the danger of being stung is not as great as with the stingrays. Most punctures come from careless handling. The venom of sharks and chimaeras is not as dangerous as that of stingrays.

Among the bony fishes certain catfishes, weevers, surgeonfishes, scorpionfishes, stargazers, rabbitfishes, toadfishes, and the sabertooth blennies are known to be venomous, and several other groups, including Carangidae and Scatophagidae, are reported to have venom associated with spines.

Catfishes have venom glands in the skin sheathing the dorsal and pectoral spines, and some groups have axillary venom glands that supply their secretions to the exterior of the pectoral spine. Most catfish stings are painful but not dangerous, but some species cause edematous swelling and gangrene at the wound. The family Plotosidae, an Indo-Pacific group containing both fresh- and saltwater members, contains the most dangerous venomous catfishes, some capable of causing death in humans (Halstead, 1978). Some of the catfishes commonly kept in home aquaria can inflict severe stings that occasionally result in numbness and shock. Included are the shovelhead catfishes of Pimelodidae, a South American family; the air-breathing Clariidae and Heteropneustidae, which is an aggressive stinger; and the armored Doradidae. Other venomous catfishes are the Bagridae, Siluridae, the marine Ariidae, and the familiar North American Ictaluridae.

Catfishes react to being grabbed or restrained by lashing violently from side to side. Usually the pectoral and dorsal spines are locked in an erect position during this activity so they can pierce the attacker. Care should be exercised when handling live catfishes of any kind. Usually they can be grasped directly behind the pectoral fins with reduced danger of being stung.

Fishes of the family Scorpaenidae are widespread along tropical and temperate shores. Members of the family generally have venom glands in grooves along the dorsal, anal, and pelvic spines (Roche and Halstead, 1972). In some genera, such as *Sebastes,* the venom is not virulent and the spines mainly inflict a painful wound. In other genera, such as *Scorpaena* and the tropical *Pterois,* the venom is more powerful (that of *Pterois* sometimes kills human beings). Species of *Pterois,* called turkeyfish, lionfish, etc., are extremely colorful and are sometimes kept in home aquaria.

Fishes of the family Synanceidae (stonefishes) have large venom glands associated with the dorsal, anal, and pelvic spines. The ducts of the spines run in a groove to a point near the tip of the spine. Stonefish venom is extremely dangerous, and because the fishes are so well camouflaged, there are many recorded instances of people stepping on them and being severely affected by the venom. There are records of several deaths caused by stonefishes. The family lives in the warm Indo-Pacific region.

Weevers (Trachinidae; Fig. 15–1**B**) are among the fishes that have venomous opercular spines as well as venomous fin spines. Weevers are found from the North Sea south into the Mediterranean and have considerable contact with fishermen and divers throughout their range. They are said to attack when disturbed, or even without provocation, striking with the bladelike opercular spine (Maretic, 1988). Both fin and opercular spines are deeply grooved, as in most venomous fishes (Fig. 31–1). The venom-producing tissue is in the grooves. Few fatalities result from weever stings, but permanent damage to joints can result (Maretic, 1988).

Toadfishes (Batrachoididae; Fig. 12–2**C**) are found mainly in warm coastal waters of most seas. One subfamily, the Thalassophryninae, has members with hollow opercular and dorsal fin spines surrounded by venom glands (Halstead, 1970). Their venom is not regarded as being as dangerous as that of the weevers. Venomous toadfishes occur in the tropical eastern Pacific and western Atlantic. Two freshwater species are known from South America.

FIGURE 31–1   Grooved dorsal spine of venomous toadfish.

Very few fishes appear to have a venomous bite. Morays have been reported to have poison fangs, but that has been disproved. There may be some venom-producing tissue in the skin of the palate, according to some authorities. The sabertooth blennies, *Meiacanthus,* have large canine teeth in the lower jaw. The teeth are grooved and are associated with venom glands, so the bite is very painful (Fishelson, 1974).

Fishes with strong venoms or exceptionally strong poisons in their flesh probably obtain some ecological advantage from their toxins. Little is known about the subject, although laboratory tests have shown that some species are toxic to other fishes.

Traumatogenic Fishes

Among the fishes that are injurious to humans, there are electrogenic species such as *Torpedo*, *Electrophorus,* and *Malapterurus* (electric ray, eel, and catfish; Fig. 19–8). These can all cause pain and temporary numbness, and the electric eel is large and powerful enough to produce fatal shocks. Other nonbiting but potentially traumatogenic fishes are the billfishes and sawfishes. Both types are large and armed with a long extension of the rostrum. Occasionally billfishes ram boats, doing considerable damage to the small craft; swordfishes have been recorded as penetrating small boats with the bill (Schultz and Stern, 1948). Sawfishes inhabit shallow water and habitually enter tropical or subtropical rivers, so they often come in contact with bathers or fishermen. Serious injuries and some deaths have been caused by sawfishes in various areas, especially in India, where the Ganges can be crowded with pilgrims. A sawfish 4 or 5 m long blundering into a crowd of bathers and turning swiftly is certain to do damage. Helm (1976) tells of some encounters, one fatal, by fishermen and divers with sawfishes in Central America.

Needlefishes (Belonidae; Fig. 13–1C) can leap out of the water at great speed, their sharp-pointed beaks and arrow-shaped bodies combining to make them dangerous projectiles. They occasionally strike boaters, swimmers, or surfboard riders and cause serious effects. In Hawaii during the summer of 1977, at least three persons were struck. Two were hit in the leg, and the beak passed through the calf of one. The third, a ten-year-old boy, was fatally injured when a meter-long needlefish struck him in the eye.

The candiru (*Vandellia*) is a tiny South American catfish that parasitizes larger fishes, living in their gill cavities. It is apparently attracted to urine and is reputed to enter the urethra of humans (Nelson, 1994). Surgery is required to remove it because of its recurved spines.

Fishes that bite are probably feared more than all others. Most predatory fishes and some others, such as the coral-eating parrotfishes, can deliver painful or injurious bites when mishandled after capture. There are stories of the wolf-herring (*Chirocentrus dorab;* Fig. 8–2D) attempting to reach and bite its captors when boated, but most bites from captured fish are probably accidental. The pike eels (*Muraenesox*) are also dangerous to capture. There are some fishes, however, that have made direct biting attacks on bathers and shipwrecked individuals. The deep-set, knifelike teeth of the barracuda are capable of tearing out great chunks of flesh. Morays occasionally make unprovoked attacks on swimmers. One of the most feared groups of fish are the piranhas (*Serrasalmus, Pygocentrus*). These South American characins have strong jaws and sharp triangular teeth and commonly feed on animals larger than themselves. They can attack and severely wound or kill wading or swimming mammals, including humans. The wounds from piranha bites seen on other fishes are neat hemispherical hollows.

Rarely, carnivorous fishes may attack hands or feet of swimmers, boaters, or fishermen. There are documented records of attacks by the bluefish, *Pomatomus saltatrix,* in Florida (deSylva, 1976), and occasional newspaper articles originating in the Great Lakes states report bites by muskellunge, *Esox masquinongy.*

Some nest-guarding species may threaten human intruders, and at least two are confirmed as carrying out the threat to bite (Randall and Millington, 1990). Two large Indo-Pacific triggerfishes, the blue triggerfish (*Pseudobalistes fuscus*) and the titan triggerfish (*Balistoides viridescens*), are known to have inflicted painful wounds on divers that have ventured too close to nests. At least two divers bitten by the blue triggerfish, which may reach a length of 75 cm and a weight of 7 kg, required hospitalization.

Sharks, of course, are the fishes that draw most attention as "maneaters." Some are of large size and have voracious appetites, seeking large prey. These apparently attack humans as a source of food, as they would attack a seal or a large fish. Smaller sharks, too, have been known to feed on humans, but some authorities believe that one half or more of known shark attacks have not involved feeding sharks.

The Office of Naval Research sponsored the accumulation of data on more than 1600 shark attacks in a file at the Smithsonian Institution. Data from over 1100 attacks were subjected to computer analysis at the Mote Marine Laboratory, Sarasota, Florida, in an attempt to discover important factors related to shark attacks (Baldridge, 1974). Some of that laboratory's findings are summarized here. After support by the Navy was lost, the shark attack file was transferred to the University of Rhode Island, and then in 1988 to the Florida Museum of Natural History, where it is under the auspices of the American Elasmobranch Society (Burgess, 1990).

In general, shark attacks occur where large numbers of swimmers and divers frequent waters containing large numbers of sharks. Usually swimmers are found in water over 20°C and sharks are seldom found in water over 30°C, so most attacks occur between those temperatures, with the peak range between 21° and 24°C. The white shark has attacked humans at temperatures as low as 10.5°C.

That male victims of shark attacks outnumber females by about 9 to 1 is probably a reflection of the relative swimming and diving activities of the sexes rather than the taste preference of the sharks. Nearly a fifth of the more than 1000 victims for which such information was available were spearfishermen, with many known to have been carrying captured fish. About two thirds of the attacks recorded took place within 200 feet of shore, which, of course, is the area where the greatest numbers of people are found. Actual possibility of attack is probably greater farther from shore. The fatality rate among shark victims is about 35 percent. There have been several shark attacks on the coasts of Oregon and California, mainly involving surfboarders (Lea and Miller, 1985). Some victims have been severely injured by white sharks, which seem to be mainly motivated to bite and shake the surfboard. Even though the swimmers have bled profusely, the sharks have not pressed the attacks, and some biologists have thought that the attacks might be some kind of territorial behavior. However, Tricas and McCosker (1984) and McCosker (1985) call attention to the similarity, when viewed from below, of the silhouette of a surfer lying supine on a short surfboard and that of an adult seal, a common prey of white sharks.

Information on white shark attacks in Chile indicates that spearfishermen were attacked as prey. Two fatal attacks reported by Engana and McCosker (1984) have a "Jaws"-like aspect. In one, a shark of about 4 m was observed with a speargun trailing from its mouth. A search for the fisherman disclosed only a torn wetsuit jacket and a diving fin. In the other, a shark of more than 7 m long and about 1.4 m broad decapitated and mangled a diver. Those attacks, plus others, occurred near pinniped haulouts. Marine mammals are natural prey of sharks of many kinds, including the "cookiecutter" species of *Isistius* (LeBoeuf et al., 1987), as well as white sharks (LeBoeuf et al., 1982).

In about 270 human attack cases, sharks of eight families were identified. The worst offender appeared to be the white shark (*Carcharodon carcharias*), a giant that wanders well into temperate waters. The tiger shark (*Galeocerdo cuvieri*) was a close second. The mako shark and its relatives (*Isurus*) have been implicated in many attacks, as have the hammerheads (Sphyrnidae). Sharks of the genus *Carcharinus*, some of which enter tropical and subtropical rivers, including the Mississippi, often attack humans. *C. leucas*, the bull shark, is widespread in warm seas and seems to be the species most often implicated in attacks on humans in fresh water. Coad and Papahn (1988) show that sharks ascend the rivers at the head of the Persian Gulf to a distance of 850 river km and that attacks have occurred as far as 420 km upstream in southern Iran. Eleven attacks, probably by the bull shark, were documented for the period 1958 to 1985. Although earlier records were not complete, 34 attacks are reported from 1941 to 1985. About half of these attacks were fatal.

Fishes as Carriers of Parasites and Diseases

One negative aspect of our relationship with fishes that is important in some geographical areas is the ability of various species to harbor parasites that can affect us directly. The parasites are usually of concern in areas where freshwater fishes are eaten raw or without sufficient processing, although there are some parasites that can be transmitted to humans by marine fishes.

Most of the parasites involved are worms—nematodes (roundworms), cestodes (tapeworms), and trematodes (flukes). One potentially dangerous nematode is the kidney worm, *Dioctophyma renale*, which is known mainly from the Orient. Another type (marine) occasionally found in humans, especially in regions where raw herring is consumed, is represented by *Anisakis* and its relatives. These worms can cause illness if a person becomes infested with enough of them. The practice of swallowing live minnows has led to severe consequences caused by perforation of the intestinal wall by the larvae of nematodes of the genus *Eustrongyloides* (Centers for Disease Control, 1982).

Probably the best known fish-borne cestode affecting humans is the broad tapeworm, *Diphyllobothrium latum*, common in some freshwater fishes in northern European countries and parts of Asia and North America. A relative, *D. pacificum*, carried by marine fishes, also can affect humans (Higashi, 1985). In North America, *D. latum* is known primarily from the Great Lakes area. Those infested with this worm generally suffer from anemia. The worm is transmitted to humans through uncooked freshwater fishes.

Among the trematodes are a few that are known to be transmitted from fish to humans. Some are intestinal parasites, but one family, Opisthorchidae, contains liver parasites that can cause serious effects. Infestations by *Clonorchis sinensis* are known mainly from Asia, where 20 million persons are estimated to be affected (Higashi, 1985). An interesting relationship involving a snail, a fluke, and one of our most highly prized domestic animals, the dog, is found in coastal portions of the Pacific Northwest. A snail, *Oxytrema silicula,* harbors the early stages of the fluke, *Nanophyetus salmonis,* the cercariae of which are carried mainly by salmon and trout. The adult fluke is found in various carnivores, including skunks, raccoons, and others. Dogs and other canids are susceptible to the fluke, which of itself is not a dangerous parasite but carries a rickettsial disease that is extremely dangerous to members of the dog family. The fluke's cercariae are so common in salmonids of the region that before the true nature of the disease was discovered, it was generally believed that salmon were poisonous to dogs. Methods of prevention and treatment are now known, so the mortality rate of dogs that eat raw salmon has been considerably reduced (Baldwin et al., 1967). This fluke has now been shown to cause illness in humans (Fritsche et al., 1989). The cercaria can be transferred by ingestion of undercooked salmonids or, as in one reported case, even by handling the fish (Harrell and Deardorf, 1990).

Under certain circumstances, fish may be sources of microorganisms that cause infections and disease in humans. Bites, punctures by spines, or scratches by fins or scales can introduce infective agents. Various species of potentially infective *Vibrio* have been isolated from white shark teeth (Buck et al., 1984). Other species of *Vibrio* include *V. damsa* (Fouz et al., 1992), *V. parahaemolyticus* (Ghittino, 1972), and a toxigenic strain of *V. cholerae* that was discovered in fish and shellfish from Mobile Bay, Alabama, in 1991 (Anonymous, 1991). Other kinds of microorganisms that are carried by fish and are potentially infective to humans include *Aeromonas, Mycobacterium, Shigella, Salmonella, Clostridium,* and *Erysipalothrix* (Ghittino, 1972; Janssen, 1970). A painful skin disease called "fish handler's disease" can be caused by *E. rhusiopathiae* (Sonnenworth et al., 1980). *M. marinum* can cause subcutaneous abscesses after gaining entry through broken skin (Wolinsky, 1980).

Home aquaria and fish culture facilities can be sources of infection. For instance, *Edwardsiella tarda* has been implicated in protracted diarrhea in a situation wherein the only source of the infective agent appeared to be an aquarium (Vandepitte et al., 1983). *Salmonella* has been found in eel culture ponds in Japan (Saheki et al., 1989).

Fishes Out of Place

One definition of a weed is that it is a plant out of place, and the same might be said of some fishes. Regardless of their adaptation to the environment in which they evolved, fishes, when introduced to other areas, can have adverse effects on the new environment or on other species there. Adverse effects may materialize through competition, predation, or hybridization. In some instances, such as when fish managers wish to enhance populations of selected species, the presence of even transplanted native fishes can be undesirable.

Nearly 50 exotic species have been introduced and established in the United States and Canada (Courtenay et al., 1991). Some of them, such as common carp and brown trout, were deliberately released by agencies expecting beneficial use, but most represent unauthorized introduction and release. Many are ornamental fishes dumped by aquarists without regard for the native fauna. There are 17 cichlids, eight cyprinids, and seven live-bearers on the list of introductions.

A good example of an introduced fish considered a "weed" or "trash fish" in many areas is the carp (*Cyprinus carpio*). In Europe and in Asia, where this species is native, the carp is esteemed as food, is subject to fisheries, and apparently has ecological checks and balances, so it does not dominate. In North America, there are only minor fisheries for carp, and populations in many places run unchecked. The carp is such an efficient forager that few native fishes can compete with it, especially as the carp can uproot and destroy aquatic vegetation, which serves as cover for fish and fish food organisms. Some of the greatest difficulty with carp is on certain migratory waterfowl refuges, where the fish destroys the aquatic vegetation upon which the waterbirds feed. Another introduced cyprinid, the grass carp, *Ctenopharyngodon idella,* although it can be useful in restricted situations, is certain to bring about undesired environmental changes because its main food is aquatic vegetation. Its potential effects on water plants and the fauna dependent on them are great, even though the fish might be useful in some applications (see the following section).

Some native North American minnows have bad reputations among trout managers, especially when they are introduced to lakes in which only trout are desired. Tui chub (*Gila bicolor*) and redside shiner (*Richardsonius balteatus*) are two minnows that seem to compete with trout, in some situations to the point of near exclusion. Often management agencies resort to poisoning the entire lake with rotenone or other ichthyocides and begin the management over again. In some instances, such control operations are not detrimental, but in others there may be some danger to rare fishes or to susceptible invertebrates. The Miller Lake lamprey is extinct because of such an operation (Bond and Kan, 1973).

Perhaps the greatest fish control campaign of all was the effort directed at the sea lamprey (*Petromyzon marinus*) after it invaded the upper Great Lakes, contributing to the decline of the commercial and sport fisheries. While a specific toxin for the lamprey was being sought, electrical barriers were devised that decreased the numbers of adult lampreys reaching the spawning grounds. After a specific poison was discovered and developed for use, a systematic effort was made to eradicate the larval lampreys in all the tributary streams. The effort was sufficiently successful that salmonid fishes, especially the introduced coho salmon (*Oncorhynchus kisutch*), are now the objects of fisheries.

Wherever fishes or other aquatic organisms are grown in monoculture, species other than the chosen object of culture will be removed, even though they might be prized for culture in another application.

Positive Aspects

Fishes as Biological Control Agents

The varied feeding habits of fishes sometimes lead to the use of certain kinds of fishes to control organisms considered undesirable. Among the insect-eating species are various small species of Cyprinidae and members of the family Poeciliidae. One poeciliid, *Gambusia affinis,* called the mosquitofish although mosquitos are not a preferred food, has been introduced into many areas outside its native range in the Mississippi, adjacent drainages, and the east coast of North America. Its undeserved reputation as a mosquito eater is so great that it even has been taken to other continents, where there probably are native top-feeding fishes of equal or greater efficiency.

There are several fishes that feed on freshwater snails, thereby destroying intermediate hosts of parasites of humans. Of special note are certain African cichlids and the black or mud carp of Asia, *Mylopharyngodon piceus.*

Another Asiatic carp, the grass carp (*Ctenopharyngodon idella*), feeds almost exclusively on aquatic plants, ingesting prodigious amounts daily. Although only a portion is digested, the vegetation is shredded by the carp's pharyngeal teeth and passed through the alimentary canal, becoming, in effect, a fertilizer for other plant growth. Usually phytoplankton growth is promoted by the activity of the carp, so rooted plants are destroyed not only by being eaten but by the shading due to phytoplankton blooms. The grass carp is an excellent table fish and has been introduced to many countries, where it serves double duty as a food fish and an aquatic weed control agent. In the United States, its best use could be in the control of plants in small ponds or lakes, where growth of vegetation conflicts with recreational or other primary uses. Unfortunately, the grass carp has been liberated into the Mississippi and other river systems, where it may have undesirable effects on the environment. Various members of Cichlidae, especially species of *Tilapia* and *Sarotherodon,* feed on filamentous algae and soft vascular plants. Some are efficient at removing vegetation from irrigation systems and are used for weed control in ditches in the southwestern United States. In some areas, they cannot survive the cool winters but can be planted annually in the spring. Reproduction during the summer increases populations sufficiently to control the problem plants.

In some instances, predatory fishes are used to control populations of other fishes. In tilapia culture in rice fields, for example, a few snakeheads (*Channa*) are often added to crop some of the excess juveniles. The walleye (*Stizostedion vitreum*) is sometimes used in managed lakes as a control for the yellow perch (*Perca flavescens*), but success is not always achieved. The northern pike (*Esox lucius*) has been successfully employed as a control for sunfishes.

There is a possibility that chemicals capable of repelling sharks can be found in fishes (Zahuranec, 1983). The best possibility appears to be a toxic substance called "pardaxin" found in the Moses sole, *Pardachirus marmoratus,* of the Red Sea. The species produces the toxin through pores from glands that lie along the bases of the dorsal and anal fins. The toxin is fatal to other fishes and appears to repel sharks effectively (Clark, 1983).

Scientific Uses of Fishes

In general, fishes are excellent subjects for the study and demonstration of anatomy, physiology, ecology, evolution, and other aspects of science. Because of their ready availability and their representation of typical structure of lower vertebrates, the dogfish (*Squalus acanthias*) and yellow perch (*Perca flavescens*) have become standard dissection laboratory animals. These two species are the basis of an important trade and are the subjects of numerous laboratory manuals. Lampreys and their larvae are dissected routinely in college laboratories as representatives of Agnatha, and there is a good trade in gars and bowfins as primitive bony fishes.

The hagfishes, with their extra hearts, have aided physiologists in studies of cardiac pacemaker cells. Rainbow trout have been used extensively in the study of relationships of various food components to hepatoma and other cancers. Many species, including the rainbow trout, are used in studies of water pollution and the effects of various pesticides and waste products on aquatic life. The fathead minnow (*Pimephales promelas*), the bluegill (*Lepomis macrochirus*), and the goldfish (*Carassius auratus*) are favorite bioassay animals and, along with the mummichog (*Fundulus heteroclitus*), are also used in physiological experiments. Many species have been used in research in ethology and experimental psychology, including the goldfish, various sticklebacks, and cichlids. The medaka (*Oryzias*) and the zebra danio (*Danio rerio*), both of which reproduce well in captivity, are among fishes that are useful in the study of embryology. In addition, zebra danio clones are of growing importance in physiological studies.

The professional societies that work with fish in research have strong interests in the proper care and use of experimental animals. The American Society of Ichthyologists and Herpetologists, the American Fisheries Society, and the American Institute of Fisheries Research Biologists together produce a publication called *Guidelines for Use of Fishes in Field Research*. Humane treatment of experimental animals is a responsible goal of these scientists.

The discovery of squalamine, a potent aminosterol antibiotic, in the dogfish shark (Moore et al., 1993) was mentioned in Chapter 5.

Recreational Fisheries

It is easy to imagine how fishing for fun originated from hook-and-line fishing for food fishes. Many fishes have qualities of speed, stamina, and leaping abilities that allow them to put up noble struggles to prevent capture, and the lightening of tackle to give the fishes a fighting chance can be seen as a natural development transforming a serious matter of food gathering into a form of play. Usually, anglers utilize their catches as food, so for many sport fishing serves a dual purpose. There are, however, some sport fisheries, based on the use of artificial flies as lures, in which no fish are killed, but they are returned to the water to be caught and released another day. Types of fishes sought by anglers range from the sharks to the lower bony fishes to the higher bony fishes. The size range runs from trouts and others of a few centimeters in length to the giant tunas, swordfishes, and sharks. Sport fishing is geographically widespread, but it reaches its greatest importance in affluent societies, where the demand for outdoor recreation transcends the need for commercial fisheries.

Freshwater angling in the Northern Hemisphere is based on a variety of fish families, with the Salmonidae ranking high in popularity and in money spent in pursuit of sport. Salmon and trout are found in abundance in the northern part of the hemisphere and are represented at high altitudes south to Turkey, Morocco, and Pakistan. Many species support recreational fisheries, although salmon fisheries are declining in the northwestern United States. The arctic char and its relatives of the genus *Salvelinus* are limited to the coldest waters and so are best distributed in the far north and high altitudes. Two of the best-known sport fishes—the rainbow trout (*Oncorhynchus mykiss*) and the brown trout (*Salmo trutta*)—have been introduced to many areas outside their original distributions. Other famous salmonids are the Atlantic salmon (*S. salar*) and two of the Pacific salmons, the chinook (*O.*

tshawytscha) and the coho (*O. kisutch*). Various whitefishes (Coregoninae), the grayling (*Thymallus*), the huchen (*Hucho*) of Europe and Asia, and others are locally important as game fishes.

Pikes (Esocidae) and perches (Percidae) are other popular northern game fishes. Members of these families range from cold to temperate regions and include both large and small species. The pike (*Esox lucius*) is holarctic in distribution and provides sport throughout its wide range. Other large pike are the muskellunge (*E. masquinongy*) and the Amur pike (*E. reicherti*). The pikeperches (*Stizostedion* spp.), the largest members of Percidae, are found in both North America and Eurasia. The walleye (*S. vitreum*) is a noted game fish in the United States and Canada. The perch of Europe (*Perca fluviatilis*) and the yellow perch (*P. flavescens*) of North America are locally numerous in rivers, lakes, and reservoirs and have good reputations as panfishes.

Minnows (Cyprinidae) are distributed in both cool and warm waters of the northern hemisphere. In Europe, several species are important for recreation, including many smaller species as well as large ones such as the carp (*C. carpio*) and the tench (*Tinca tinca*). In India the very large minnow, the mahseer (*Barbus tor*), is a favorite of sport fishermen. The mahseer and related species of the Tigris (*B. schejki*) may reach about 2 m and nearly 100 kg. A few North American anglers fish for the introduced carp, but other minnows are not generally sought. Warm-water sport fishes of North America include many of the catfish family, Ictaluridae. Three of these are large fishes—the flathead catfish (*Pylodictus olivaris*), the blue catfish (*Ictalurus furcatus*), and the channel catfish (*I. punctatus*). Members of *Ameiurus* are smaller but are popular panfishes. In eastern Europe, the huge catfish *Silurus glanis* is sought by anglers. Other favorite warm-water game fishes are included in the North American sunfish family, Centrarchidae. Chief among these are the black basses of the genus *Micropterus,* but the crappies, *Pomoxis,* and the sunfishes, *Lepomis,* have many devotees.

Many of the game fishes and panfishes of the northern hemisphere have been introduced to areas where they were not natively found. For instance, centrarchids, ictalurids, and the rainbow trout have been established in Europe, and the carp and brown trout have been brought to North America. There has been a strong tendency for anglers to take their favorite sport fishes with them wherever they go, and many northern fishes are now established in the southern hemisphere. Rainbow and brown trout are now present in suitable cold streams and lakes in Africa, South America, Australia, and New Zealand. The carp is now present in many places in the southern hemisphere, and a few centrarchids, notably the largemouth bass, have been introduced to Africa.

Native southern hemisphere sport fishes include some large and interesting species. The arapaima (*Arapaima gigas*) of the Amazon is sought with hook and line. This species is reported to reach a weight of over 200 kg at 2.4 m, and specimens of about 100 kg have been taken by anglers. The dorados[1] of South America (*Salminus* spp.) are large, colorful fishes reputed to be among the world's best game species. *S. maxillosus* reaches a weight of about 23 kg. Some of the cichlids of South America are colorful and sporting. One, the peacock cichlid, *Cichla ocellaris,* has been tried as a game fish in Florida.

---

[1]Not to be confused with the marine genus *Coryphaena*, which is sometimes called dorado.

African game fishes include the huge Nile perch (*Lates niloticus*) and the voracious tigerfish (*Hydrocynus goliath;* Fig. 9–3C). The former reaches a length of more than 2 m and 270 kg, the latter slightly less. In Australia a large member of the Percichthyidae, the Murray cod (*Maccullochella macquariensis*), is a favored game fish. It reaches about 1.8 m long.

Just as there are hundreds of species of freshwater fishes sought by anglers, there is also a tremendous variety of saltwater species that provide sport fishing. The so-called big game fishing is pursued mainly in tropical and subtropical oceans, with large scombroid fishes as the targets. The marlins (*Makaira, Tetrapturus*), swordfish (*Xiphias*), and bluefin tuna (*Thunnus thynnus*) are some of the most highly prized big game fishes; but others, such as the white shark (*Carcharodon*) and the mako shark (*Isurus*), are also popular. (In some areas, the white shark is now protected.) Other large marine game fishes include the tarpon (*Megalops atlanticus*), the sailfishes (*Istiophorus*), jewfish (*Epinephelus itajara*), giant sea bass (*Stereolepis gigas*), barracuda (*Sphyraena*), and cobia (*Rachycentron*).

Most saltwater anglers seek smaller quarry, including members of several common families such as temperate basses (Percichthyidae), cods (Gadidae), sea basses (Serranidae), jacks (Carangidae), snappers (Lutjanidae), dolphins (Coryphaenidae), porgies (Sparidae), drums (Sciaenidae), surfperches (Embiotocidae), rockfishes (Scorpaenidae), greenlings (Hexagrammidae), and flounders (Pleuronectidae).

The importance of sport fishing is growing in many areas. In North America, the sport catch of some species exceeds the commercial take, and some species and fishing areas are reserved exclusively for sport angling. Considerable economic value is placed on recreational fisheries in developed countries. Besides being an excellent use of leisure time, angling promotes a great circulation of money. Anglers purchase such items as fishing tackle, boats, outboard motors, and camping gear. They expend money for travel, food, and lodging to the extent that some small coastal communities depend on good fishing for continued prosperity. In addition, there is great nutritional value in the fish caught, even if the angler's cost per pound in catching them greatly exceeds the price per pound in the fish market.

The Aquarium Trade

The color, form, motion, and habits of many fish species are enjoyed by aquarists in many parts of the world. For instance, home aquaria and ornamental pond enthusiasts are estimated to number more than 20 million in the United States and about 2 million each in Canada and Japan. There are records of imports of ornamental fishes by most of the nations of Europe as well as some Asian nations (Table 31–1). The aquarium hobby probably rivals photography as the most popular worldwide hobby.

Freshwater species are used much more than marine species because of the relative ease of culture, care, and shipping, but marine species are becoming more popular as the technology of maintaining small saltwater systems advances. There are probably over 1000 species of freshwater fishes available to aquarists in the United States. Most belong to the following few families: the characins, tetras, etc. (Characidae); the minnows, carps, and relatives (Cyprinidae); the loaches (Cobitididae); the topminnows (Cyprinodontidae); the live bearers (Poeciliidae); various South American catfishes (Callichthyidae, Loricariidae, and Doradidae); the cichlids (Cichlidae); and the gouramis (Belontiidae). Fishes of these and many other

TABLE 31–1	**Leading Nations in Imports of Ornamental Fishes, 1991***

Country	Value in US$
World Total	214,892,000
U.S.A.	56,654,000
Japan	31,562,000
United Kingdom	21,627,000
F. R. Germany	19,565,000
France	15,446,000
Singapore	10,802,000
Hong Kong	7,808,000
Netherlands	7,808,000
Italy	7,246,000
Belgium	6,122,000
Spain	4,861,000
Canada	4,446,000

*Adapted from *FAO Yearbook of Fishery Statistics,* vol. 73.

tropical freshwater families are imported to temperate climates in what has been until recently a relatively unrestricted trade.

Saltwater species constitute only a small percentage of the commerce in aquarium fishes but represent a growing segment of the trade and generally command high prices. Usually, reef fishes such as butterflyfishes (Chaetodontidae), damselfishes (Pomacentridae), surgeonfishes (Acanthuridae), triggerfishes (Balistidae), squirrelfishes (Holocentridae), and cardinalfishes (Apogonidae) are the most popular. Over 40 families of marine fishes used in home aquaria are covered by Axelrod and Burgess (1987).

Because most marine aquarium fishes are collected from the wild, and must be held for some time before shipment from tropical areas to the mostly temperate areas for retail sales, mortality prior to final sale may be around 30 percent (Wood, 1985). Wood found that about half of the specimens survived six months in home aquaria and that survival to one year was 34 percent, and he recommended that measures be taken to protect and manage the wild populations of the popular reef fishes.

World export trade in ornamental fishes in 1991 amounted to over 114,000,000 U.S. dollars (FAO, 1993). The major exporter of aquarium fishes is Singapore, with about 40 million U.S. dollars in trade annually. Other important exporters are Hong Kong and the United States, both with a trade of over 13 million dollars. The Netherlands, Germany, Japan, and Indonesia all export from 5 to 7 million dollars worth per year. Colombia had an export trade of nearly 7 million dollars in 1982, but that has diminished to about 1.5 million, in part because of restrictions.

Usually fishes for export are captured by means of traps, seines, or dip nets by individuals, who then sell them at a collecting point from which they are transported to the exporters. For instance, in Thailand they are trapped from creeks flowing into the Chao Phraya River and then taken by small boat to a buyer, who sorts the species into liveboxes attached to a barge. When enough are accumulated, the barge is taken downriver to Bangkok, where the fishes are sold to an exporter, who ships to Singapore or Hong Kong for shipment to other countries.

Aquarium-related trade in the United States, including fishes and accessories, was estimated to have an annual value of about 700 million dollars (Conroy, 1975). This included sales of fishes and other aquarium animals, fish food, fish health aids, aquaria, pumps, filters and other accessories, books, and magazines.

Restrictions are being placed on the transport of aquarium fishes by various countries and states for a variety of reasons. Some countries are realizing that unrestricted export of wild native fishes will eventually deplete the supplies of certain species. Other countries prohibit export or import of species declared endangered or especially rare. In addition, there are species considered potentially harmful to humans or to the environment of a country where they might be introduced, and these are therefore restricted from free trade. Certain stingrays, stonefishes, lionfishes, and weevers are listed as undesirable in the United States. Various states prohibit piranhas, walking catfishes, snakeheads, parasitic catfishes, the electric eel, and others.

In addition to the enjoyment afforded by private aquaria, there is the pleasure of public aquaria, where both native and exotic fishes of all sizes can be displayed. Many large aquaria have facilities for keeping both fresh- and saltwater species. Some feature giant circular tanks that accommodate large sharks and rays and active fishes such as tunas. Such aquaria are often set up in conjunction with museums or public parks, or they may be attached to commercial ventures where marine shows are staged.

Fishes as Items of Commerce

Although fishes and fisheries may be of considerable importance and value from the standpoints of hobbies, recreation, biological control, and other interests, the greatest value from a worldwide standpoint arises from the commercial use of fishes as industrial products and food. The industrial uses to which fishes can be put are numerous and varied, but the most important appears to be the production of fish meal, mainly for agriculture. Protein-rich fish meal is a basic ingredient in the food of poultry, trout, catfish, pigs, and other domestic animals. Some grades of meal have been used in the fertilizer industry. Fish oil is another important product of reduction plants and is used in many manufacturing processes and in various foodstuffs.

Other than meal and oil, fishes are used in the manufacture of glue, as a source of leather, as a source of silver pigment (guanine) for certain paints, and as a source of many other minor items or products. Fishes have been especially important in some primitive societies, furnishing, for instance, spines for needles and awls; skin for leather; and, in some areas, large stinging spines or large teeth for spear or arrow points or for making club edges more formidable.

The greatest use of fishes is as food. Species of all sizes and habitats contribute to the table fare of people in most parts of the world. Small species may be of great local importance. For example, larval fishes are harvested in New Zealand and Asia as they make their way upstream. In the Philippines and Southeast Asia, tiny gobies and larvae of other fishes are used in the manufacture of bagoong or fish sauces of various types. Sardines, sprats, and other small marine fishes are harvested en masse

TABLE 31–2

"Species Group"	Landings (Metric Tons)
"Herrings, sardines, anchovies" plus "shads"	22,069,238
"Cods, hakes, haddocks"	10,467,234
"Jacks, mullets, sauries"	10,076,284
"Redfishes, basses, congers"	5,742,025
"Carps, barbels and other cyprinids"	5,394,216
"Tunas, bonitos, billfishes"	4,478,296
"Mackerels, snoeks, cutlassfishes"	3,479,642
"Salmons, trouts, smelts"	1,637,819
"Miscellaneous marine fishes"	10,390,521
"Miscellaneous freshwater fishes"	6,319,521

World Landings of Fish by Selected FAO "Species Groups" 1991*

*Data from *FAO Yearbook of Fishery Statistics,* vol. 72.

for use as food. Although the relatively small herrings (Clupeidae) and anchovies (Engraulidae) account for a great share of the world's catch—historically 20 to 30 percent—most of the tonnage landed of those families is used for industrial purposes.

Most food fishes are medium sized, weighing a kilogram or more, but a few species are giants, such as the great tunas and the swordfishes. Some of the preferred groups of food fishes on the world markets other than the herrings and allies are the cods and relatives; tunas and other mackerel-like fishes; salmon, trout and smelt; and flatfishes. Table 31–2 shows world landings of fishes by the groups used for statistical purposes by the Food and Agricultural Organization of the United Nations.

The impact of fisheries on the world food supply can be appreciated from the total landings. The estimated world catch of fish and shellfish in 1989 was over 100 million metric tons (one metric ton equals 1.1 U.S. short tons) but has declined since that year. By 1991, the catch was estimated at about 97 million metric tons, of which more than 82 million metric tons were made up of finfish. About one third was used for industrial purposes (and in part cycled through farm animals to produce protein from farm animals). Fish provides about 50 percent of all animal protein consumed in Southeast Asia and is known to be extremely important in the protein supply in many other areas.

The World Fisheries

Fisheries are pursued in both fresh and salt water, with the great bulk of landings near shore on the continental shelf and slope. Freshwater fishes account for a catch of about 13 million metric tons annually, according to best estimates. The remainder of the finfish catch, about 73.6 million metric tons, is taken from marine waters. Landings of shellfish amount to about 12.7 million metric tons annually.

The catch of marine fishes largely reflects the availability of stocks that can be harvested economically under prevailing conditions of technology and demand for fishery products. Ryther (1969) estimated that the annual production of fish of all kinds in the ocean is about 242 million metric tons, about 3.3 times the annual marine catch in the late 1980s. Not all of that production is potentially catchable or useful. There is an estimate (FAO, 1979) that the fish stocks conventionally fished should provide a potential catch of about 105 million metric tons. Earlier estimates put the potential sustainable yield at up to 290 million metric tons (see Rounsefell, 1975). Rothschild (1981) reviewed estimates of future demand and production. The estimated demand for fishery products for the year 2000 of 97 million metric tons was exceeded in 1988, when the estimated world catch of fish and shellfish was 99 million metric tons. That figure was exceeded in 1989, when the landings reached 100.2 million metric tons (FAO, 1991). Greater use of species and stocks that were not used to their greatest potential earlier helped increase the catch. Exploitation of the great stocks of walleye pollock (*Theragra chalcogramma*) and the Inca scad or "Chilean jack mackerel" (*Trachurus murphyi*) contributed to the upswing. In addition, miscellaneous fishes, including sharks, skates, and various mesopelagic stocks (Gjosaeter and Kawaguchi, 1980), were fished more as proper gear and manufacturing processes were developed.

The landings of walleye pollock constituted the largest single-species fishery in the world in 1989, with about 6.25 million metric tons landed. This exceeded the catch of Peruvian anchoveta. Catch of the pollock by the United States for the years 1988 to 1991 averaged 0.91 million metric tons annually, somewhat less than the menhaden catch, which averaged 0.97 million metric tons for those years (Table 31–3).

The marine catch of finfish, which was nearly 68 million metric tons in 1991, comes mainly from areas where upwellings of cold, nutrient-rich waters or other rich currents (usually cool) promote growth of plankton, which forms the basis for productive fish communities. Great fishing areas where cold and warmer currents meet at the surface include grounds off New England and the maritime provinces of Canada; the North Sea; the Bering Sea; the continental shelf and slope along Alaska, British Columbia, and south to Mexico; and the waters from the East China Sea north past the Japanese islands to Kamchatka and the Kuriles. There is similar upwelling on the Atlantic coast of Africa and to some extent on the west coast of India.

A notable example is the Pacific Ocean off Peru, where the impingement of the cold Humboldt current on warmer water in an area of high illumination is the basis of the tremendous fishery of the Peruvian anchoveta, *Engraulis ringens*. Because of this species, Peru reached prominence among the leading fishing nations, landing over 12 million metric tons in 1970 and over 10 in 1971. (Table 31–4 shows some of the prominent fishing nations.) Peru's fishery suffered a collapse to a low of 0.5 million metric tons in the late 1970s (Bardach and Santerre, 1981). This was caused by anomalous wind patterns along the equator in the Pacific. This short-term climatic change allowed warm water to move inshore along the coast of South America. The phenomenon is called El Niño (the Christ child) because its usual manifestation is at Christmas time.

In 1982 and 1985, the far-reaching effects of an El Niño event depressed the salmon fisheries in Oregon and Washington. Advection of warm water northward along the Northwest coast displaced the salmon and allowed subtropical fishes such as lizardfish, triggerfish, and catalufas to move up to Oregon. Barracudas were taken in Alaskan waters.

Most of the world's fisheries involve capture of wild stocks of fishes wherever they can be found. This, of course, is a form of hunting, for finding the fishes in

TABLE 31-3	Average U.S. Commercial Landings of Selected Fish Groups, 1988–1991*	
	Group	**Pounds**
	Herring	220,563,000
	Menhaden	2,003,503,000
	Pacific Salmon	727,112,000
	Pollock	2,419,318,000
	Tuna	52,212,000
	Flounder	272,653,000
	All others	1,565,908,000
	Total landings	7,283,764,000

*Data from USDA. *Aquaculture Situation and Outlook.* Sept. 1992.

TABLE 31-4	Leading Fishing Nations with 1991 Landings (Metric Tons)*	
	Nation	**Metric Tons**
	World Total	96,925,900
	China	13,134,967
	Japan	9,306,827
	Former U.S.S.R.	9,216,927
	Peru	6,944,172
	Chile	6,002,867
	U.S.A.	5,473,321
	India	4,036,931
	Indonesia	3,186,000
	Thailand	3,065,170
	Rep. Korea	2,515,305
	Philippines	2,311,797

*Data from *FAO Yearbook of Fishery Statistics,* vol. 72 (includes shellfish).

large lakes, rivers, and the seas may account for much of the effort expended by fishermen. Methods of hunting range from subsistence fishermen groping along the bottom of a swamp with a plunge basket to the employment of aircraft or the use of sophisticated sonic detectors. Few modern fishing vessels of any size are without some means of detecting fish.

Methods of fishing have been evolving from some unknown time in prehistory. Bone and shell hooks, bone gorges, and harpoons have been recovered from various paleolithic archeological sites and also with net fragments from neolithic remains. Cushing (1988) gives a short review of fisheries in historical antiquity. Gear now employed in fisheries includes the entire arsenal from the spears, traps, hand lines, and small nets of the traditional subsistence fisherman to the tremendous trawls, purse seines, and 30-mile-long drift nets of the modern, high-seas fishing fleets. The high-seas drift nets, developed especially for the capture of squid in the Pacific, are indiscriminate and capture almost anything they encounter, including marine mammals, birds, and any pelagic fishes whether protected by international agreements or not. Their use is controlled to some extent by international conventions.

Usually, fishing gear is devised and employed to take advantage of the habits, behavior, and movements of the target species (see Rounsefell, 1975). Ground fishes of the families Gadidae (cods), Merlucciidae (hakes), Pleuronectidae (flounders), Scorpaenidae (rockfishes), and others are taken by means of various trawls, which are dragged along the bottom at shallow and moderate depths. Some slow-moving schooling species that live well off the bottom can be taken with midwater trawls. Pelagic schooling species, such as the families Engraulidae (anchovies), Clupeidae (herrings), Salmonidae (salmons), and Scombridae (tunas), are vulnerable to capture by purse seines or other encircling nets.

Migratory species or others having definite patterns of movement can be captured by set or drifting entangling gear, such as trammel nets or gill nets, or can be led by long weirs or fences into pound nets or traps. Beach seines of various types can be utilized to catch species that migrate close to shore.

Hooks and lines have their places in modern commercial fisheries. Trollers rigged to tow several lures at various depths seek salmon and tuna. Halibut are caught by long lines stretched along the bottom; tuna are taken by long lines at the surface or by lures on short poles and lines. In this type of fishing, live bait is released into a school of tuna, and the lures are cast among the frenzied feeding fish. The hooked tunas are heaved onto the deck of the boat, where they are easily freed from the barbless hook. If the tuna are large, two or more poles are used to each lure, and the fishermen work together to throw the fish onto the deck.

Modern fishing vessels, in what Cushing (1988) terms the second industrialization of fisheries, are equipped with cold storage or quick-freezing facilities so that catches from trips of many weeks or months can be kept until delivered ashore. Fleets of some countries feature factory ships or "mother ships" that receive and process fish captured by trawlers. Other ships bring supplies, transport products, and even shuttle crewmen on rotation, so that the fleet can stay on the fishing grounds for long periods of time and high-quality fish products can reach the home country. The upswing in the landings of the modernized fleets and the success of better processing and distribution systems brought about a tripling of world fisheries production during the 1950s and 1960s (Royce, 1987).

Aquaculture

For thousands of years, fishes have been kept in enclosures or impoundments, either to hold them alive until needed for food or to allow them to grow, thus producing more food. The practice has been traditional in China and other parts of Asia and was used by the ancient Romans. Fish culture of several kinds probably predates history, and many traditional methods practiced today have no doubt persisted for a thousand years or more (Bardach et al., 1972).

Aquaculture, which can be defined simply as rearing aquatic organisms under more or less controlled conditions, is taking on a new importance in modern times

(Stickney, 1979), supplying probably 8 to 10 million metric tons annually. As the capture fisheries have reached the limit of their production in some areas, and because the expansion of fisheries to other areas farther from population centers or to areas with lower fish production tends to raise prices, more attention has been given to culture of desirable fishes at competing prices. There are, however, many other reasons for the growing popularity of fish culture in addition to the obvious advantages of producing protein for subsistence or high-quality fish for luxury prices. For instance, fish culture can make better use of some lands than the farming of terrestrial animals, or it can be a means of recycling farm wastes to produce a useful product. Sometimes pig, chicken, or duck farms are built over fish ponds in Asia so fertilization is continuous and direct. Fish culture can be used to replace ruined or inaccessible spawning grounds or to supplement natural populations. Obviously, depending on the worth of the species cultured, some types of culture will warrant greater effort and financial expenditure than others, so various levels of intensity can be recognized. A low level of intensity might consist of transferring young fish from the wild into ponds, where they could be left without care until harvested. In high-intensity culture, the entire life cycle of the stock is controlled. Brood fish are kept in special ponds and are induced to spawn or spawned artificially; the eggs are cared for, and the young are provided with prepared food until they reach harvestable size.

Fishes of many families are cultured—from sturgeons to puffers—because of the combination of availability, demand, and adaptability to artificial surroundings. A high demand can justify, economically at least, the culture of high-cost carnivorous fishes such as trout and salmon (Salmonidae), porgies (Sparidae), the buri (*Seriola quinquiradiata*), eels (Anguillidae), and others, even though they must be fed high-protein food made in part from other fishes. The channel catfish (*Ictalurus punctatus*) is another fairly high-priced fish fed high-protein food.

Greater bioenergetic efficiency, but not necessarily greater profits, can be realized by rearing omnivores, planktivores, or herbivores. These are closer to the base of the food web than the carnivores and can be cultured with less outlay for food. Fishes of the family Cyprinidae are extremely popular for fish culture in Asia, Europe, and parts of Africa. Culture of more than one species in the same pond at the same time (polyculture) can be set up with a number of Chinese carps so that fresh vegetation can be fed the grass carp (*Ctenopharyngodon idella*), which passes a large proportion of the vegetation through its digestive tract shredded but undigested, thus fertilizing the water. Plankton resulting from this fertilization is eaten by such fishes as the silver carp (*Hypophthalmichthys molotrix*) and bighead carp (*H. nobilis*). Bottom organisms can be eaten by the common carp (*Cyprinus carpio*) and others. Several ecological niches are filled in such a pond, and greater production is realized than if only one species were used. Polyculture systems are used in India, utilizing fishes of the genera *Gibelion, Cirrhina,* and *Labeo.*

The common carp is one of the most important cultured fishes in the world. It is an omnivorous species that has been bred for rapid growth, great efficiency in utilization of food, good body conformation, and other desirable attributes. At present it does not enjoy great popularity in North America, but the time may come when this efficient animal will be instrumental in providing protein to an ever-growing population of Americans.

Another important family containing some herbivorous fishes used in fish culture is Cichlidae. African species of the genus *Tilapia* (including *Sarotherodon* and *Oreochromis*) are used in subsistence fish culture both in Africa and Asia and are subject to culture in other areas, including North America. *Tilapia* (*Oreochromis*) *mossambica* has been widely introduced and used with varying success in many warm countries. One of the greatest disadvantages of that species is its fast reproductive rate, which often leads to overpopulation and stunting, with consequent production of fish unattractively small for table use. Other species, such as *T. nilotica, T. melanopleura,* and *T. zillii,* are also cultured. The native cichlids of India (*Etroplus* spp.) are of minor use in fish culture. In South and Central America, native fishes of the genera *Cichlasoma, Astronotus,* and *Cichla* are subject to culture.

Important marine fishes from the standpoint of fish culture are the herbivorous *Chanos chanos* or milkfish and mullets of the genus *Mugil,* which are generally detritivores. The milkfish is cultured in brackish water ponds in the Indo-Pacific region, especially the Philippines, Indonesia, and southern Taiwan. Mullets are cultured in brackish or marine waters in many regions that have tropical to warm-temperate climates.

Although many other species and families are used in fish culture, those mentioned are among the most prominent. Research efforts are constantly being directed at discovering additional species to culture and developing new and more efficient methods, better food, and so on.

A recent development in the North Pacific is the concept of "ocean ranching," in which Pacific salmon are reared to migratory age and size in a hatchery and then released to the ocean after being held for a time at a place that serves both as the release and capture site. While being held prior to release, they are behaviorally imprinted with the odor of the release site, so that after they have grown to maturity in the ocean they return to the same site and are harvested. In some instances, that type of aquaculture has resulted in oversupply and depressed prices. In 1991, the return of pink salmon to a hatchery site in Prince William Sound was so great that prices fell to so low a level that fishing was not economical. This resulted in the capture and dumping at sea of tons of fish by the corporation operating the facilities to prevent organic pollution on the bay in which the hatchery was located.

Net-pen rearing of salmon is gaining popularity in several areas. Coho salmon are cultured in Chile and in British Columbia, and the Atlantic salmon is cultured in many countries, especially in Scandinavia (see Heen et al., 1993).

Production of fish from aquaculture is significant, but exact figures are difficult to obtain because much of the culture is carried out in village or family ponds, and not all large-scale culturists maintain accurate records. Estimates for 1975 by Ryther (1981) include more than 2.5 million metric tons of freshwater fish and 0.66 million tons of marine fish. Estimates of mollusc culture for that year give a total of about 1.4 million tons. Crustacean culture was estimated at 0.016 million tons, and seaweed culture was 1.06 million metric tons. Fish accounted for nearly 58 percent of the total production of 5.54 million metric tons. Arrignon (1982) estimated that aquacultural production in 1975 was 6 million metric tons, of which 4 million was fish. He pointed out that aquaculture provided 4 percent of the animal protein (excluding milk) for 1975.

Estimates from various sources for 1980 are near 8 million metric tons for all aquaculture. Nash (1988) reviewed estimates for 1985, when total production was about 10.6 million metric tons, 4.7 million of which was fish. China was the leading nation, with total aquacultural production of 5.2 million metric tons, 2.4 of which was fish. The United States was fifth, with 0.35 million metric tons total production and about 0.19 million metric tons of fish.

Cyprinidae accounted for one half of the production in 1985, with the bighead carp (*Hypophthalmichthys nobilis*) and silver carp (*H. molotrix*) the most important species (Nash, 1988). Inland aquaculture in the United States relies heavily on the channel catfish and rainbow trout. The annual production is around 400 million pounds of catfish and possibly 150 million pounds of trout.

Management of Fisheries

When only a few fishermen were fishing, with primitive and inefficient gear, the waters must have seemed inexhaustible. Subsistence fisheries in fresh waters and near the shores of the ocean traditionally have been pursued with the idea of harvesting as much usable protein as possible with the least amount of effort. Generally, the history of such a fishery is that as more fishermen enter the activity and as better gear is devised, the capacity of the fish populations to produce large numbers of large fish is diminished. This becomes evident in the smaller average size of the fish landed, in the increased amount of effort needed to harvest the catch, and possibly in an absolute decrease in the weight of the annual harvest.

These are symptoms of overfishing, and they can be recognized even in the absence of environmental degradation—not only in traditional fisheries in small or confined fishing grounds but also in modern, large-scale commercial and recreational fisheries. An unexploited or lightly used fish stock often has a significant proportion of its biomass in the slow-growing older brood stock. These large individuals may represent, because of the peculiarities of fish population dynamics, a deterrent to the recruitment of great numbers of smaller fish into the usable size classes. In addition, these mature fishes constitute a bonanza for the fishermen who first exploit the stock and may attract the fishing pressure that can eventually result in reducing the number of spawners to the point at which optimum numbers of recruits cannot be produced. Fishery scientists have determined a hypothetical relationship of numbers of spawners to numbers of young fishes entering the fishery (Fig. 31–2).

Fishermen and scientists generally recognize that the aquatic ecosystems are not inexhaustible, because they now recognize factors that limit biological productivity. They also recognize that human activities can alter natural processes either through impact on the populations themselves or on the environment upon which the fish populations depend. Most alterations in the environment that attend advancing civilization can have deleterious effects on fisheries. Such activities as land clearing, logging, grazing, land drainage, diversion of water for irrigation, obstruction of streams by dams, and industrial and domestic pollution can all diminish the productivity of water. Because productivity of future fisheries may be contingent on current treatment of fish populations and the ecosystems that produce them, management of the fisheries to protect their capacity for producing a sustained yield is highly desirable. What constitutes the yield will differ among fisheries, depending on what a community or nation expects from a particular fishery. There can be considerable differences in what fishery scientists believe should be taken from a given fishery. Two concepts of yield are maximum sustainable yield and optimum sustainable yield. The maximum sustainable yield is basically the weight of fish harvested (Weithman,

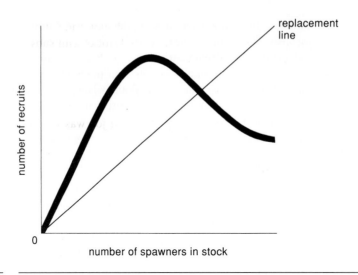

FIGURE 31–2   Hypothetical relationship between a fish stock and the young fish recruited to the stock.

1993). Optimum sustainable yield includes not only harvest but can include other benefits of a biological and socioeconomic nature. In recreational fisheries, the yield sought may be different. Numbers of fish caught may be important in some sport fisheries, but in others merely the opportunity to try for trophy-sized fish may be the return that devotees wish to perpetuate.

Regardless of the beneficial product expected, fishery managers are devoted to using all available knowledge and technology to perpetuate, distribute, and enhance the fisheries for our greatest advantage. The managers are backed by scientists in many disciplines—mathematicians involved in studies of population dynamics, physiologists, pathologists, ecologists, statisticians, and many more—as well as that combination of many of these disciplines known as the fishery biologist. There are, however, many problems in maintaining the backing of communities and industries whose livelihoods or profits may be threatened by measures designed to perpetuate certain fisheries. Even when facing a declining catch that seems destined to disappear, fishermen find it difficult to support regulations that diminish their catch.

Historically, fishery management seems to have begun with laws and regulations (Rounsefell, 1975; Sigler and Sigler, 1990). Artificial propagation may have been the next method, followed by control and improvement of environmental factors. Laws and regulations usually were instituted to protect breeding populations or undersized fish in order to ensure recruitment of adequate numbers of usable-sized fish. These regulations restricted length of season or size of fish taken, or limited the type of fishing gear used. In recent times, the amount of gear entering a fishery may be limited, or overall catch quotas for an entire fleet may be set. In recreational fisheries, there may be special applications of regulations. There may be both upper and lower size limits to protect both undersized fish and the large brood stock, or some sport fisheries may operate on a "catch and release" basis so that a single fish might produce sport for a number of different fishermen during its life.

Artificial propagation, the incubation and hatching of eggs stripped from female fish from wild or domesticated stocks with subsequent release of young fish, was instituted with the idea that it would improve on natural processes and provide more recruits to the fisheries. Much early effort in this field was predicated on the false notion that the fertilization rate in the natural situation was poor and that only a small percentage of the eggs survived and hatched. Consequently, the old-time hatchery owners considered that their job was well done when eggs in their care hatched and the unfed fry were released. Most often, the eggs were robbed from a wild run of fish trapped at a weir and not allowed to spawn naturally.

The great successes predicted for artificial propagation in enhancing the fisheries were not evident for decades. The practice was useful in transplanting stocks, and occasional success was experienced in lakes and in streams, where natural runs were cut off by dams, but the real utility of artificial propagation of salmonids was not realized until better knowledge of the life history and ecology of cultured species was obtained and advances in hatchery techniques were made. Better nutrition, disease control, and selective breeding of hatchery-generated brood stock have allowed fishes to be reared economically for several months to a year before release. Fishes may be released for several purposes. Trout may be released in streams or lakes for immediate capture, with those placed in lakes usually giving a much better return to the angler. Others may be released in autumn into lakes rich in food, where good winter survival is expected. Anadromous species may be reared to the smolt phase and released with the expectation that they will immediately descend to feeding grounds in the ocean or in large lakes. Although early promise for success in maintaining coho salmon runs by means of hatcheries has not held up, there has been some success with artificial propagation of the chum and pink salmon, and research in progress promises better culture of these species in the future.

In freshwater fisheries, protection and improvement of the environment has become important. Many streams in their natural state have falls or other barriers (including certain types of log jams) that prevent or delay migration of fishes, or these streams may not have a distribution of riffles and pools, which provide for a viable combination of optimum food production and nesting and hiding places. Such situations can often be improved by construction of fish ladders around falls and barriers, removal of unwanted log jams, and placement of check dams and other devices to alter the character of the stream.

Often the capacity of streams to support normal populations of fishes is degraded by activities such as logging, farming, construction of large dams, and release of pollutants into the waters. Some productivity can be restored through proper watershed management, pollution control and abatement, and the provision of passage facilities at the dams.

References for management of fisheries are numerous and diverse, including classics such as Cushing (1968), *Fisheries Biology, A Study on Population Dynamics;* and Ricker (1954), *Stock and Recruitment.* Others include Coull (1993), *World Fisheries Resources;* Cowx (1991), *Catch Effort Sampling Strategies;* Gulland (1983), *Fish Stock Assessment,* Volume 2; Gulland (1988), *Fish Population Dynamics: Implications for Management;* Gunderson (1993), *Surveys of Fisheries Resources;* and Walters (1986), *Adaptive Management of Renewable Resources.*

Threatened and Endangered Species

In many places, there are fish species in danger of extinction and, unfortunately, a number that have become extinct. The International Union for the Conservation of Nature (IUCN, 1988) lists 77 species as endangered worldwide. Of these, 30 were listed for the United States. The list contains 24 recently extinct fishes, 15 from the United States, seven from Mexico, and one shared by those two countries. According to the criteria of the Endangered Species Committee of the American Fisheries Society (Williams et al., 1989), there are over 70 endangered North American fish species and over 80 that are threatened with endangerment. In addition, Williams et al. (1989) consider numerous subspecies to be in those two categories or of special concern, so they listed over 360 taxa worthy of protection. Miller et al. (1989) listed 40 taxa (species and subspecies combined) that have become extinct in North America during the past 100 years. The lack of extinct category listings from other parts of the world may be more a matter of maintenance of adequate records than better management of the faunas.

Many species or subspecies of fishes have restricted ranges or exacting ecological requirements. Involved may be narrow temperature limits, high oxygen needs, necessary migration routes, limited types of substrate for spawning, specific foods for various life history stages, and other requirements. Some species may have no natural defense against competition or predation by species with which they have not coevolved. Restricted ranges may be due to desiccation of large bodies of water, as in western basins, where marvelously large Pleistocene lakes have dwindled to small alkaline remnants and the toughest of the fauna remain in these remnants, or the luckiest survive in the permanent headwaters of intermittent streams or in constant springs (see Naiman and Soltz, 1981). Some taxa may have restricted ranges above impassable falls or, for instance, in coastal streams, where their ancestors were isolated by stream capture. Some may be glacial relicts, holding to a fragment of a once mighty distribution now diminished by consequences of the advance and retreat of the ice sheet. For whatever reason the ranges are restricted, some of these survivors of glacial advance and pluvial and altithermal periods have little leeway in their fight for survival, especially if the slow natural processes are disturbed by human interference.

Typically fishes that are threatened with population decline or extinction are those with small geographical ranges, but others with wider distribution may become endangered, usually because of ecological changes. Species have disappeared from some of the Great Lakes and some large river systems of the United States.

Water is a natural resource with a high demand and value, and many of the woes of fishes are direct results of human use of waterways. Even in areas with adequate rainfall, manipulation of streams to supply water for domestic or industrial purposes, or damming them to produce water wheels or hydroelectric plants with energy, can upset the normal ecology of fishes. Interconnection of waterways by canals or ditches can allow the spread of predators or competitors into drainages in which sensitive species live. There are many kinds of organic and chemical pollutants that, along with siltation, change habitats for the worse. Channelization of rivers can destroy habitat. Civilization and industrialization have, overall, been unkind to fishes.

Demand for irrigation water in the arid western United States has had strong influences on the desert fishes. Springs have been pumped dry on occasion, and overuse of limited aquifers has caused failure of springs at some distance from the wells being pumped (Minckley and Deacon, 1991; Pister, 1981; Williams et al., 1985). Other aspects of agriculture, including land clearing, drainage of wetlands, and use of certain pesticides, though necessary for production of food and fiber, can impact fishes. Disruption of natural watersheds by logging or certain grazing practices can have deleterious effects.

Other conditions mentioned by Williams et al. (1989, p. 3) as being threats to sensitive species are hybridization (which can follow introduction of congeners or may occur following disruption of migration routes or spawning habitat) and "overuse for commercial, recreational, scientific or educational purposes." To the foregoing should be added well-meant but destructive management practices, including (especially in former times) ill-advised hatchery practices and the use of poisons in the habitats of rare species.

The perception of need for protection of wildlife species of all kinds has grown over the years and has brought about the passage by the U.S. Congress of the Endangered Species Preservation Act of 1966 and the more comprehensive Endangered Species Act (ESA) of 1973, plus its subsequent modifications. (See Williams and Deacon, 1991, for a review of the development of ethics and legislation on protection of species.) The ESA is a powerful statute, capable of far-reaching protection of species and their habitats regardless, in large part, of economic consequences. In essence, according to DiSilvestro (1989, p. 160), it has "outlawed extinction." The impact of the ESA on the construction of the Tellico Dam in Tennessee showed that economic considerations can win, through congressional action, over an endangered species (DiSilvestro, 1989). What the consequences of listing the northern spotted owl and various populations of Pacific salmon as endangered or threatened will be is yet to be seen. Listing of the owl had immediate impact on the lumber industry in the Northwest, and listing of the Pacific salmon will affect major irrigation and hydroelectric systems, as well as commercial fishing.

Values of species with no current economic importance are difficult to explain and defend, but there may be undiscovered importance (including economic value) and the decline of certain species may be symptomatic of irreversible and detrimental changes in the landscape. Economics will enter deliberations on endangered species more and more. As Decker and Goff (1987, p. xx) put it, "Natural resource professionals who believe that they can sequester themselves and concentrate on biological study apart from social and economic influences are naive, misled, or both." Many emphasize that the ecological value of fish and wildlife is part of the processes on which human survival depends. Other values included are economic use, education, and recreation as well as ethical, philosophical, and aesthetic values.

The continued survival of endangered fishes depends on such methods of protection as laws and regulations and proper management, including restoration of habitat, transplantation, and artificial propagation. Above all, increased awareness of the interrelationships of humans and the natural environment must be accomplished through education.

References

Abbot, F. S. 1973. Endocrine regulation of pigmentation in fish. Am. Zool. 13:885–894.

Abrahamson, T., and S. Nilsson. 1976. Phenylethanolamine-N-methyl transferase (PNMT) activity and catecholamine content in chromaffin tissue and sympathetic neurons in the cod, *Gadus morhua.* Acta Physiol. Scand. 96(1):94–99.

Adam, H., and R. Strahan. 1963. Systematics and geographical distribution of myxinoids, p. 1–8. *In*: The biology of *Myxine,* A. Brodal and R. Fänge (eds.). Universitetsforlaget, Oslo.

Adelman, I. R. 1987. Uptake of radioactive amino acids as indices of current growth rate of fish: A review, p. 65–79. *In*: Age and growth of fish, R. C. Summerfelt and G. E. Hall (eds.). Iowa State University Press, Ames, IA.

Agarkov, G. B., Yu. N. Varich, and K. A. Snezhina. 1976. Innervation of the locomotor organs in some fish. J. Hydrobiol. 12(2): 57–59.

Agarwal, S., and P. A. John. 1975. Functional morphology of the urinary bladder in some teleostean fishes. Forma Functio 8(2): 19–26.

Ahlberg, P. E. 1989. Paired fin skeletons and relations of the fossil group Porolepiformes (Osteichthytes: Sarcopterygii). Zool. Journ. Linn. Soc. 96:119–166.

Ahlstrom, E. H. 1968. Review of "Development of fishes of the Chesapeake Bay region, an atlas of egg, larval, and juvenile stages, Part 1." Copeia 1968:648–651.

Ahlstrom, E. H., J. Richards, and S. H. Weitzman. 1984. Families Gonostomatidae, Sternoptychidae, and associated stomiiform groups: Development and relationships, p. 184–198. *In*: Ontogeny and systematics of fishes, H. G. Moser, W. J. Richards, D. M. Cohen, M. P. Fahay, A. W. Kendall, Jr. and S. L. Richardson (eds.). Am. Soc. Ichthyol. and Herpetol. Sp. Pub. 1.

Ahlstrom, E. H., K. Amaoka, D. A. Hensley, H. G. Moser, and B. Y. Sumida. 1984. Pleuronectiformes: Development, p. 640–670. *In*: Ontogeny and systematics of fishes, H. G. Moser, W. J. Richards, D. M. Cohen, M. P. Fahay, A. W. Kendall, Jr. and S. L. Richardson (eds.). Am. Soc. Ichthyol. and Herpetol. Sp. Pub. 1.

Ahsan-ul-Islam. 1949. The comparative histology of the alimentary tract of certain freshwater teleost fishes. Proc. Ind. Acad. Sci. B33:297–321.

Aldridge, R. J., D. E. G. Briggs, M. P. Smith, E. N. K. Clarkson, and N. D. L. Clark. 1993. The anatomy of conodonts. Phil. Trans. Royal Soc., London (B). Biol. Sci. 340(1294):405–421.

Alexander, R. McN. 1964. Adaptation in the skulls and cranial muscles of South American characinoid fish. J. Linn. Soc. Zool. 45(305):169–190.

Alexander, R. McN. 1969. The orientation of muscle fibers in the myomeres of fishes. J. Mar. Biol. Assoc. U.K. 49:263–190.

Alexander, R. McN. 1970. Mechanics in the feeding action of various teleost fishes. J. Zool. Lond. 162:145–156.

Alexander, R. McN. 1972. The energetics of vertical migration by fishes. Symp. Soc. Exp. Biol. 26:273.

Alexander, R. McN. 1975. The chordates. Cambridge University Press, London, New York.

Aleyev, U. G. 1977. Nekton. Junk, The Hague.

Algranati, F. D., and A. Perlmutter. 1981. Attraction of zebrafish, *Brachidanio rerio,* to isolated and partially purified chromatographic fractions, p. 31–38. *In*: Ecology and ethology of fishes, D. L. G. Noakes and J. A. Ward (eds.). Junk, The Hague.

Alheit, J. 1989. Comparative spawning biology of anchovies, sardines, and sprats, p. 7–14. *In*: The early life history of fish, J. H. S. Blaxter, J. C. Gamble and H. v. Westernhagen (eds.), Rapp. P. -v. Reun. Cons. Int. Explor. Mer, 191.

Al-Hussaini, A. H. 1946. The anatomy and histology of the alimentary tract of the bottom feeder, *Mulloides auriflamma* (Forsk.). J. Morphol. 28:121–154.

Al-Hussaini, A. H. 1949. On the functional morphology of the alimentary tract of some fish in relation to differences in their feeding habits: Anatomy and histology. Q. J. Microsc. 90:109–139.

Ali, M. A. (ed.). 1975. Vision in fishes: New approaches in research. Plenum Press, New York.

Ali, M. A., and M. Anctil. 1976. Retinas of fishes. Springer-Verlag, Berlin.

Ali, M. A., and M. A. Klyne. 1985. Vision in vertebrates. Plenum Press, New York and London.

Allen, G. R. 1991. Freshwater fishes of Australia. T. F. H. Publications, Neptune, NJ.

Allen, G. R., and D. R. Robertson. 1994. Fishes of the tropical eastern Pacific. University of Hawaii Press, Honolulu.

Allen, L. G. 1984. Gobiesociformes: Development and relationships, p. 629–636. *In*: Ontogeny and systematics of fishes, H. G. Moser, W. J. Richards, D. M. Cohen, M. P. Fahay, A. W. Kendall, Jr. and S. L. Richardson (eds.). Am. Soc. Ichthyol. and Herpetol. Sp. Pub. 1.

Allendorf, F. W., and G. A. Thorgaard. 1984. Tetraploidy and the evolution of salmonid fishes, p. 1–53. *In*: Evolutionary genetics of fishes, B. J. Turner (ed.). Plenum Press, New York.

American Society of Zoologists. 1973. Symposium: The current status of fish endocrine systems. Am. Zool. 13:710–936.

American Society of Zoologists. 1977. Symposium: Recent advances in the biology of sharks. Am. Zool. 17:287–515.

Amoroso, E. C. 1960. Viviparity in fishes. Symp. Zool. Soc. Lond. 1:153–181.

Anderson, A. J. 1986. Nutrition and feeding of fish, p. 25–31. *In*: Aquaculture in Australia, P. Owen and J. Bowden (eds.). Rural Press Queensland, Brisbane.

Anderson, B. G., and R. D. Loewen. 1975. Renal morphology of freshwater trout. Am. J. Anat. 143(1):93–114.

Anderson, J. M., and M. R. Peterson. 1969. DDT: Sublethal effects on brook trout nervous system. Science 164:440–441.

Anderson, M. E. Symp. 1983; publ. 1984. Zoarcidae: Development and relationships, p. 578–582. *In*: Ontogeny and systematics of fishes, H. G. Moser, W. J. Richards, D. M. Cohen, M. P. Fahay, A. W. Kendall, Jr. and S. L. Richardson (eds.). Am. Soc. Ichthyol. and Herpetol. Sp. Pub. 1.

Andrews, S. M. 1973. Interrelationships of crossopterygians, p. 137–177. *In*: Interrelationships of fishes, P. H. Greenwood, R. S. Miles and C. Patterson (eds.). Academic Press, New York.

Andriashev, A. P., and N. V. Chernova. 1994. Annotated list of fish-like vertebrates and fishes of the Arctic Seas and adjacent waters. Vop. Ichthol. 34:435–456. (In Russian)

Anonymous. 1991. Cholera found in Mobile Bay. Fisheries News 2(11):1.

Antony, A. C. 1952. Use of fish slime in structural engineering. J. Bombay Nat. Hist. Soc. 50(3):682.

Aprieto, V. L. 1974. Early development of five carangid fishes of the Gulf of Mexico and the south Atlantic coast of the United States. Fish. Bull. U. S. 72:415–443.

Arai, K., K. Matsubara, and R. Suzuki. 1991. Karyotype and erythrocyte size of spontaneous tetraploidy and triploidy in the loach *Misgurnus anguillicaudatus*. Nippon Suisan Gakkaishi 57:2167–2172.

Arnold, L. R. 1974. Rheotropism in fishes. Biol. Rev. 49:515–576.

Arnott, H. J., N. J. Maciolek, and J. A. C. Nicol. 1970. Retinal tapetum lucidum: A novel reflecting system in the eye of teleosts. Science 169:478–480.

Arratia, G. 1982. *Chongichthys dentatus*, new genus and species, from the late Jurassic of Chile (Pisces: Teleostei: Chongichthy-diae, new family). J. Vertebrate Paleontology 2(2):133–149.

Arratia, G., and H. P. Schultze. 1990. The urophyal: Development and homology within osteichthyans. J. Morph. 203:247–282.

Arrigon, J. 1982. Amenagement ecologique et piscicole des eaux douces. Gauthier-Villars, Paris.

Arthur, D. K. 1976. Food and feeding of larvae of three fishes occurring in the California current, *Sardinops sagax, Engraulis mordax,* and *Trachurus symmetricus*. Fish Bull. U. S. 74:517–530.

Atema, J. 1971. Structures and functions of the sense of taste in the catfish (*Ictalurus natalis*). Brain, Behav. Evol. 4:273–294.

Atema, J. 1980. Chemical senses, chemical signals, and feeding behavior in fishes, p. 57–101. *In*: Fish behavior and its use in the capture and culture of fishes, J. E. Bardach, J. J. Magnuson, R. C. May and J. M. Reinhart (eds.). ICLARM Conf. Proc. No. 5, Manila.

Atkins, D. L., and W. L. Fink. 1979. Morphology and histochemistry of the caudal gland of *Corynopoma riisei* Gill. J. Fish Biol. 14: 465–469.

Atz, J. W. 1952a. Internal nares in the teleost *Astroscopus*. Anat. Rec. 113:105–116.

Atz, J. W. 1952b. Narial breathing in fishes and the evolution of internal nares. Q. Rev. Biol. 27:366–376.

Atz, J. W. 1964. Intersexuality in fishes, p. 145–232. *In*: Intersexuality in vertebrates including man, C. N. Armstrong and A. J. Marshall (eds.). Academic Press, New York.

Auerbach, P. S. 1988. Clinical therapy of marine envenomation and poisoning, p. 493–565. *In*: Handbook of natural toxins, Vol. 3, Marine toxins and venoms, A. T. Tu (ed.). Marcel Dekker Inc., New York and Basel.

Avella, M., C. B. Schreck, and P. Prunet. 1991. Plasma prolactin

and cortisol concentrations of stressed coho salmon, *Oncorhynchus kisutch*) in fresh water or salt water. Gen. Comp. Endocrinol. 81:21–27.

Avella, M., G. Young, P. Prunet, and C. B. Schreck. 1990. Plasma prolactin and cortisol concentrations during salinity challenges of coho salmon (*Oncorhynchus kisutch*) at smolt and post-smolt stages. Aquaculture 91(4):359–372.

Avise, J. C., and R. K. Selander. 1972. Evolutionary genetics of cave-dwelling fishes of the genus *Astyanax*. Evol. 26:1–19.

Avise, J. C., and M. H. Smith. 1974. Biochemical genetics of sunfish. I. Geographic variation and subspecific intergradation in the bluegill, *Lepomis machrochirus*. Evol. 28:42–56.

Avise, J. C., G. S. Helfman, N. C. Saunders, and L. S. Hales. 1986. Mitochondrial DNA differentiation in North Atlantic eels: Population genetic consequences of an unusual life history pattern. Proc. Nat. Acad. Sci. 83:4350–4354.

Avtalion, R. R., and L. Reich. 1989. Chromatophore inheritance in red tilapias. Bamidgeh (Isr. J. Aquacult.) 41(3):98–104.

Axelrod, H. R., and W. E. Burgess. 1987. Saltwater aquarium fishes (Third edition). T. F. H. Publications, Neptune City, NJ.

Backus, R. H. 1986. Biogeographic boundaries in the open ocean, p. 9–13. *In*: Pelagic biogeography—proceedings of an international conference, A. C. Pierrot-Bults, S. van der Spoel, B. J. Zahurance and R. K. Johnson (eds.). UNESCO Tech. Pap. Mar. Sci. 19.

Backus, R. H., J. E. Craddock, E. L. Haedrich, D. L. Shores, J. M. Teal, A. S. Wong, and G. W. Mead. 1968. *Ceratoscopelus maderensis*: Peculiar sound-scattering layer identified with this myctophid fish. Science 160:991–993.

Baglioni, S. 1908. Der Attmungmechanismus der Fische. Z. allgem. Physiol. 7:177–282.

Bagnara, J. T. 1983. Developmental aspects of vertebrate chromatophores. Amer. Zool. 23:465–478.

Bagnis, R. 1981. L'ichtyosarcotoxisme de type ciguatera: Phenomene complexe de biologie marine et lumaine. Oceanol. Acta 4(3):375–387.

Bailey, R. M., and T. M. Cavender. 1971. Fishes. *In*: McGraw-Hill Encyclopedia of Science and Technology (Third Edition). McGraw-Hill, New York.

Bainbridge, R. 1960. Speed and stamina in three fish. J. Exp. Biol. 35:109–133.

Baldridge, H. D. 1974. Shark attack: A program of data reduction and analyses. Cont. Mote Mar. Lab., Vol. 1, No. 2.

Baldwin, N. L., R. E. Millemann, and S. E. Knapp. 1967. "Salmon poisoning disease." III. Effect of experimental *Nanophyetus salmincola* infection on the fish host. J. Parasitol. 53:556–564.

Balinsky, B. I. 1970. An introduction to embryology. W. B. Saunders, Philadelphia, PA.

Ball, J. N., and B. I. Baker. 1969. The pituitary gland: Anatomy and histophysiology, p. 1–110. *In*: Fish physiology, Vol. II, W. S. Hoar and D. J. Randall (eds.). Academic Press, New York.

Balon, E. K. 1975a. Reproductive guilds of fishes: A proposal and definition. J. Fish. Res. Bd. Can. 32:821–864.

Balon, E. K. 1975b. Terminology of intervals in fish development. J. Fish. Res. Bd. Can. 32:1663–1670.

Balon, E. K. 1979. The theory of saltation and its application to the ontogeny of fishes: Steps and thresholds. Env. Biol. Fish. 4:97–101.

Balon, E. K. 1981. Saltatory processes and altricial to precocial forms in the ontogeny of fishes. Am. Zool. 21(2):573–596.

Balon, E. K. 1984. Patterns in the evolution of reproduction styles in fishes, p. 35–53. In: Fish reproduction: Strategies and tactics, G. W. Potts and R. J. Woodtton (eds.). Academic Press, London, Orlando.

Balon, E. K. 1985a. Early life histories of fishes: New developmental, ecological and evolutionary perspectives. Junk, Dordrecht.

Balon, E. K. 1985b. The theory of saltatory ontogeny and life history models revisited, p. 18–28. In: Early life histories of fishes, E. K. Balon (ed.). Junk, Dordrecht.

Balon, E. K. 1986. Saltatory ontogeny and evolution. Riv. Biol.—Biol. Forum. 79:151–190.

Balon, E. K. 1991. Prelude: The mystery of a persistent life form, p. 9–13. In: The biology of *Latimeria chalumnae* and evolution of coelacanths, J. A. Musick, M. N. Bruton, and E. K. Balon (eds.). Kluwer Academic Publishers, Dordrecht, Boston, London.

Banarescu, P. 1990. Zoogeography of freshwaters, Vol. I. Aula Verlag, Wiesbaden.

Banarescu, P. 1992. Zoogeography of freshwaters, Vol. II. Aula Verlag, Wiesbaden.

Banarescu, P. 1994. Zoogeography of freshwaters, Vol. III, Distribution and dispersal of freshwater animals in Africa, Pacific areas, and South America. Aula Verlag, Wiesbaden.

Banister, K. E. 1970. The anatomy and taxonomy of *Indostomus paradoxus* Prashad and Mukerji. Bull. Br. Mus. (Nat. Hist.) Zool. 19(5):179–209.

Barbour, C. D., and J. H. Brown. 1974. Fish species diversity in lakes. Amer. Nat. 108:473–489.

Bardach, J. E., and R. M. Santerre. 1981. Climate and the fish in the sea. Bio-Science 31(3):206–215.

Bardach, J. E., M. Fujiya, and A. Holl. 1967. Investigation of external chemoreceptors of fishes, p. 641–665. In: Olfaction and taste II, T. Hayashi (eds.). Pergamon Press, Oxford.

Bardach, J. E., J. H. Ryther, and W. O. McLarney. 1972. Aquaculture: The farming and husbandry of freshwater and marine organisms. Wiley-Interscience, New York.

Bardach, J. E., J. H. Todd, and R. Crickmer. 1967. Orientation by taste in fish of the genus *Ictalurus*. Science 155: 1276–1278.

Bardack, D. 1985. Les premiers fossiles de hagfish (Myxiniformes) et Enteropneusta (Hemichordata) depots de la faune (Pennsylvanienne) du Mazon Creek dans l'Illinois, U.S.A. Bulletin Trimest. Soc. Hist. Nat. Amis. Mus. Autun No. 116.

Bardack, D. 1991. First fossil hagfish (Myxinoidea): A record from the Pennsylvanian of Illinois. Science 254:701–703.

Bardack, D., and E. S. Richardson, Jr. 1977. New agnathous fishes from Pennsylvanian of Illinois. Fieldiana. Geol. 33(26):489–510.

Barham, E. G. 1971. Deep-sea fishes—lethargy and vertical orientation, p. 100–118. In: Proceedings of an International Symposium on biological sound scattering in the ocean, G. B. Faraguhar (ed). Maury Center for Ocean Science. Dept of the Navy, Washington, DC.

Barlow, G. M. 1961. Causes and significance of morphological variation in fishes. Syst. Zool. 10:105–117.

Barlow, G. M. 1963. Ethology of the Asian teleost *Badis badis*. II. Motivation and signal value of the colour patterns. Anim. Behav. 11:97–105.

Barlow, G. M. 1972. The attitude of fish eyelines in relation to body shape and to stripes and bars. Copeia 1972:4–12.

Barlow, G. M. 1974. Extraspecific imposition of social grouping among surgeonfishes. J. Zool. Lond. 174:333–340.

Barlow, G. M. 1981. Patterns of parental investment, dispersal and size among coral-reef fishes, p. 65–85. In: Ecology and ethology of fishes, D. L. G. Noakes and J. A. Ward (eds.). Junk, The Hague.

Barlow, G. M. 1991. Mating systems among cichlid fishes, p. 173–190. In: Cichlid fishes: Behavior, ecology and evolution, M. H. A. Keenleyside (ed.). Chapman and Hall, London, New York.

Barnes, R. S. K., and R. N. Hughes. 1982. An introduction to marine ecology. Blackwell Scientific Publications, Oxford.

Barr, T. C., Jr. 1968. Cave ecology and the evolution of troglobites, p. 35–102. In: Evolutionary biology, Vol. 2, T. Dobzhansky, M. K. Hecht, and W. C. Steere (eds.). Appleton-Century-Crofts, New York.

Barrington, E. J. W. 1957. The alimentary canal and digestion, p. 109–161. In: The physiology of fishes, Vol. 1, M. E. Brown (ed.). Academic Press, New York.

Barrington, E. J. W. 1972. The pancreas and intestine, p. 135–169. In: The biology of lampreys, Vol. 2, M. W. Hardisty and I. C. Potter (eds.). Academic Press, London.

Barry, M. A., and M. V. L. Bennett. 1989. Specialized lateral line receptor systems in elasmobranchs: The spiracular organs and vescicles of Savi, p. 591–606. In: The mechanosensory lateral line, S. Coombs, P. Görner and H. Münz (eds.). Springer-Verlag, New York.

Bass, A. H. 1986. Electric organs revisited: Evolution of a vertebrate communication and orientation organ, p. 13–70. In: Electroreception, T. H. Bullock and W. Heiligenberg (eds.). Wiley, New York.

Bass, A. H., and J. A. Ballard. 1972. Buoyancy control in the shark *Odontaspis taurus* (Rafinesque). Copeia 1972:594–595.

Bass, A. H., and M. A. Marchaterre. 1989. Sound generating (sonic) motor system in a teleost fish (*Porichthys notatus*): Sexual dimorphism in the ultrastructure of myofibrils. J. Comp. Neurol. 286(2):141–153.

Batty, R. S., and C. S. Wardle. 1979. Restoration of glycogen from lactic acid in the anaerobic swimming muscle of plaice, *Pleuronectes platessa* L. J. Fish Biol. 15:509–519.

Bauchot, R., A. Thomot, and M. L. Bauchot. 1989. The eye muscles and their innervation in *Chaetodon trifasciatus* (Pisces, Teleostei, Chaetodontidae). Env. Biol. Fishes 25(1–3):221–233.

Bayly, I. A. E. 1972. Salinity tolerance and osmotic behavior of animals in athalassic saline and marine hypersaline waters. Ann. Rev. Ecol. Syst. 3:233–268.

Beacham, T. D., R. E. Withler, C. B. Murray, and L. W. Barner. 1988. Variation in body size, morphology, egg size, and biochemical genetics of pink salmon in British Columbia. Trans. Am. Fish. Soc. 117:109–126.

Beamish, F. W. H. 1964. Respiration of fishes with special emphasis on standard oxygen consumption. II. Influence of weight and temperature on respiration of several species. Can. J. Zool. 42:177–188.

Beamish, F. W. H. 1978. Swimming capacity, p. 101–187. In: Fish physiology, Vol. VII, W. S. Hoar and D. J. Randall (eds.). Academic Press, New York.

Beamish, R. J. 1980. Adult biology of the river lamprey (*Lampetra ayresi*) and the Pacific lamprey (*Lampetra tridentata*) from the Pacific coast of Canada. Can. J. Fish. and Aquat. Sci. 37(11): 1906–1923.

Beamish, R. J., and G. A. McFarlane. 1987. Current trends in age determination methodology, p. 15–42. *In*: Age and growth of fish, R. C. Summerfelt and G. E. Hall (eds.). Iowa State University Press, Ames, IA.

Beamish, R. J., and P. S. Mookherjii. 1964. Respiration of fishes with special emphasis on standard oxygen consumption. I. Influence of weight and temperature on respiration of goldfish, *Carrasius auratus* L. Can. J. Zool. 42:161–175.

Beatty, D. D. 1975. The role of the pseudobranch and choroid rete mirabile in fish vision, p. 673–678. *In*: Vision in fishes: New approaches in research, M. A. Ali (ed.). Plenum Press, New York.

Beatty, D. D. 1984. Visual pigments and the labile scotopic visual system of fish. Visual Pigment Biochem. 24(11):1563–1573.

Beers, C. E., and J. M. Culp. 1990. Plasticity in foraging of a lotic minnow (*Rhinichthys cataractae*) in response to different light intensities. Can. J. Zool. 68:101–105.

Begle, D. P. 1991. Relationships of the osmeroid fishes and the use of reductive characters in phylogenetic analysis. Syst. Zool. 40(1):33–53.

Begle, D. P. 1992. Monophyly and relationships of the argentinoid fishes. Copeia 1992:350–366.

Beldridge, H. D., Jr. 1972. Accumulation and function of liver oil in Florida sharks. Copeia 1972:306–325.

Bell, M. 1993. Convergent evolution of nasal structure in sedentary elasmobranchs. Copeia 1993:144–158.

Bell, M. 1994. Paleobiology of sticklebacks, p. 438–471. *In*: Evolutionary biology of the threespine stickleback, M. Bell and S. Foster (eds.). Oxford University Press, Oxford.

Belman, B. W., and M. E. Anderson. 1979. Aquarium observations on feeding by *Melanostigma pammelas* (Pisces: Zoarcidae). Copeia 1979:366–369.

Belokopytkin, Yu. S., and G. E. Shul'man. 1989. On energy cost of fish swimming. Zh. Obshch. Biol., J. Gen. Biol. 50(6):836–840.

Belsare, D. K. 1973. Comparative anatomy and histology of the corpuscles of Stannius in teleosts. Z. Mikrosk. Anat. Forsch. 87(4):445–456.

Belsare, D. K. 1974. Morphology of the pineal organ in some carps. Zool. Beitr. 20(1):47–54.

Bemis, W. E. 1987. Feeding systems of living Dipnoi: Anatomy and function, p. 249–275. *In*: The biology and evolution of living fishes, W. E. Bemis, W. W. Burggren, and N. E. Kemp (eds.). Liss, New York.

Bemis, W. E., and T. E. Hetherington. 1982. The rostral organo of *Latimeria chalumnae*: Morphological evidence of an electroreceptive function. Copeia 1982:467–471.

Bendy, P. J. 1976. Comparative vertebrate endocrinology. Cambridge University Press, Cambridge.

Bennett, G. W. 1971. Management of lakes and ponds. Van Nostrand Reinhold, New York.

Bennett, M. V. L. 1970. Comparative physiology: Electric organs. Ann. Rev. Physiol. 32:471–528.

Bennett, M. V. L. 1971. Electric organs, p. 347–491. *In*: Fish physiology, Vol. V, W. S. Hoar and D. J. Randall (eds.). Academic Press, New York.

Bentley, P. J. 1971. Zoophysiology and ecology, Vol. 1. Endocrines and osmoregulation: A comparative account of the regulation of water and salt in vertebrates. Springer-Verlag, Berlin.

Benton, M. 1990. Vertebrate palaeontology. Chapman and Hall, London.

Bentzen, P. A., S. Harris, and J. M. Wright. 1991. Cloning of hypervariable minisatellite and simple sequence microsatellite repeats for DNA fingerprinting of important aquacultural species of salmonids and tilapia, p. 243–262. *In*: DNA fingerprinting: Approaches and application, T. Burke, G. Dolf, A. J. Jeffreys and R. Wolff (eds.). Birkheuser Verlag, Basel.

Bentzen, P., W. C. Leggett, and G. G. Brown. 1988. Length and restriction site heteroplasmy in the mitochondrial DNA of American shad (*Alosa sapidissima*). Genetics 118:509–518.

Berg, L. S. 1940 (Reprint 1947). Classification of fishes both recent and fossil. Trav. Inst. Zool. Acad. Sci. URSS 5:87–517. Reprint 1947, J. W. Edwards, Ann Arbor, Michigan.

Berg, L. S. 1949. Freshwater fishes of the U.S.S.R. and adjacent countries (Fourth Edition), Vol. 3. Israel Program for Scientific Translations, Jerusalem.

Berg, T., and J. B. Steen. 1966. Regulation of ventilation in eels exposed to air. Comp. Biochem. Physiol. 18:511–516.

Berlind, A. 1973. Caudal neurosecretory system: A physiologist's view. Am. Zool. 13:759–770.

Bermingham, E., T. Lamb, and J. C. Avise. 1986. Size polymorphism and heteroplasmy in the mitochondrial DNA of lower vertebrates. J. Hered. 77:249–252.

Bern, H. A. 1967. Hormones and endocrine glands of fishes. Science 158:455–462.

Bern, H. A., and R. S. Nishioka. 1985. Endocrine control of salmonid development and seawater adaptation, p. 7. *In*: Advances in aquaculture and fisheries research: A California sea grant symposium, May 1983, K. E. Anderson (ed.). California Sea Grant College Program, Davis, CA.

Bernstein, J. J. 1970. Anatomy and physiology of the central nervous system, p. 1–90. *In*: Fish physiology, Vol. IV, W. S. Hoar and D. J. Randall (eds.). Academic Press, New York.

Berra, T. M. 1981. An atlas of distribution of the freshwater fish families of the world. University of Nebraska Press, Lincoln, NE.

Berra, T. M., D. M. Sever, and G. R. Allen. 1989. Gross and histological morphology of the swimbladder and lack of respiratory structures in *Lepidogalaxias salamandroides*, an aestivating fish from Western Australia. Copeia 1989:850–856.

Bertelsen, E. 1951. The ceratioid fishes. Dana Rep. 39:1–276.

Bertelsen, E. 1984. Ceratoidei development and relationships, p. 325–335. *In*: Ontogeny and systematics of fishes, H. G. Moser, W. J. Richards, D. M. Cohen, M. P. Fahay, A. W. Kendall, Jr. and S. L. Richardson (eds.). Am. Soc. Ichthyol. and Herpetol. Sp. Pub. 1.

Bertelsen, E., and N. B. Marshall. 1956. The Miripinnati, a new order of teleost fishes. Dana Rep. 42:1–34.

Bertelsen E., and N. B. Marshall. 1958. Notes on Miripinnati: (addend. to Dana Rep. 42). Dana Rep. 45:9–10.

Bertelsen, E., J. G. Nielsen, and D. G. Smith. 1989. Suborder Saccopharyngoidei: Families Saccopharyngidae, Eurypharyngidae and Monognathidae, p. 636–665. *In*: Fishes of the western North Atlantic, Part 9, Vol. 1, Orders Anguilliformes and Saccopharyngiformes, E. B. Böhlke (ed.). Mem. Sears Found. for Marine Res., Memoir, New Haven, CT.

Bertelsen, E., T. W. Pietsch, and R. J. Lavenberg. 1981. Ceratioid anglerfishes of the family Gigantactinidae: Morphology, systematics, and distribution. Cont. in Science, Nat. Hist. Mus., Los Angeles Co. (332):1–74.

Bertin, L., and C. Arambourg. 1958. Super-ordre des Teleosteens. Trait. Zool. 13(3):2204–2500.

Bertmar, G. 1968. Lungfish phylogeny, p. 259–283. *In*: Current problems of lower vertebrate phylogeny, T. Orvig (ed.). Wiley-Interscience, New York.

Best, A. C. G., and J. A. C. Nicol. 1980. Eyeshine in fishes. A review of ocular reflectors. Can. J. Zool. 58:945–956.

Billard, R. 1981. The reproductive cycle in teleost fish. (Le cycle reproducteur chez les poissons Teleosteens). Cah. Lab. Hydrobiol. Montereau (12):43–56.

Billard, R., and B. Breton. 1978. Rhythms of reproduction in teleost fish, p. 31–53. *In*: Rhythmic activity in fishes, J. E. Thorpe (ed.). Academic Press, New York.

Billard, R., A. Fustier, C. Weil, and B. Breton. 1982. Endocrine control of spermatogenesis in teleost fish. Can. J. Fish. Aquat. Sci. 39:65–79.

Birdsong, R. S., E. O. Murdy, and F. L. Pezold. 1988. A study of the vertebral column and median fin osteology in gobioid fishes with comments on gobioid relationships. Bull. Mar. Science 42(2):174–214.

Birkhoff, G. 1950. Hydrodynamics. Princeton University Press, Princeton, NJ.

Bishop, C., and P. H. Odense. 1966. Morphology of the digestive tract of the cod, *Gadus morhua*. J. Fish. Res. Bd. Can. 23:1607–1615.

Bjerring, H. C. 1973. Relationships of coelacanthiforms, p. 179–205. *In*: Interrelationships of fishes, P. H. Greenwood, R. S. Miles and C. Patterson (eds.). Academic Press, New York.

Bjoesaeter, J., and K. Kawaguchi. 1980. A review of the world resources of mesopelagic fish. FAO Fish. Tech. Pap. No. 193.

Black, R., and R. J. Miller. 1991. Use of the intertidal zone by fish in Nova Scotia. Env. Biol. Fishes 31:109–121.

Black, V. S. 1957. Excretion and osmoregulation, p. 163–205. *In*: The physiology of fishes, Vol. 1, M. E. Brown (ed.). Academic Press, New York.

Blackett, R. F. 1962. Some phases in the life history of the Alaskan blackfish, *Dallia pectoralis*. Copeia 1962:124–130.

Blake, I. H. 1930. Studies on the comparative histology of the digestive tube of certain teleost fishes. I. A predaceous fish, the sea bass (*Centropristes striatus*). J. Morph. 50:39–70.

Blake, I. H. 1936. Studies on the comparative histology of the digestive tube of certain teleost fishes. III. A bottom-feeding fish, the sea robin (*Prionatus carolinus*). J. Morph. 60:77–120.

Blake, R. W. 1981a. Mechanics of ostraciform propulsion. Can. J. Zool. 59(6):1067–1071.

Blake, R. W. 1981b. Mechanics of drag-based mechanisms of propulsion in aquatic vertebrates, p. 29–52. *In*: Vertebrate locomotion, Symp. Zool. Soc. Lond. 48, M. H. Day (ed.). Academic Press, New York.

Blake, R. W. 1981c. Pseudomacrostomism in coral reef fishes. Can. J. Zool. 59(6):1183–1185,

Blake, R. W. 1983. Fish locomotion. Cambridge University Press, Cambridge.

Blaxter, J. H. S. 1968. Rearing herring larvae to metamorphosis and beyond. J. Mar. Biol. Assoc. U.K. 43:17–28.

Blaxter, J. H. S. 1969. Development: Eggs and larvae, p. 177–252. *In*: Fish physiology, Vol. III, W. S. Hoar and D. J. Randall (eds.). Academic Press, London.

Blaxter, J. H. S. (ed.). 1974. The early life history of fish: the proceedings of an international symposium held at the Dunstaffnage Marine Research Laboratory of the Scottish Marine Biological Association at Oban, Scotland from May 17-23, 1973. Springer-Verlag, Berlin, New York.

Blaxter, J. H. S. 1980. Vision and the feeding of fishes, p. 32–56. *In*: Fish behavior and its use in the capture and culture of fishes, J. E. Bardach, J. J. Magnuson, R. C. May, and J. M. Reinhart (eds.). ICLARM Conf. Proc. No. 5, Manila.

Blaxter, J. H. S. 1981. The swimbladder and hearing, p. 61–71. *In*: Hearing and sound communication in fishes, W. N. Tavolga, A. N. Popper, and R. R. Fay (eds.). Springer-Verlag, New York.

Blaxter, J. H. S., and R. S. Batty. 1984. The herring swimbladder: Loss and gain of gas. J. Mar. Biol. Assoc. U.K. 64(2):441–459.

Blaxter, J. H. S., J. C. Gamble, and H. v. Westernhagen. 1989. The early life history of fish. Rapports et Proces-Verbaux des Reunions. Cons. Int. Explor. Mer 191.

Bleckmann, H. 1988. Prey identification and prey localization in surface-feeding fish and fishing spiders, p. 619–641. *In*: Sensory biology of aquatic animals, J. Atema, R. R. Fay, A. N. Popper and W. N. Tavolga (eds.). Springer-Verlag, New York.

Block, B. 1987. Billfish brain and eye heater: A new look at non-shivering heat production. News in Physiol. Sci. 2:208–214.

Blot, J. 1966. Holocephales et elasmobranches systematique, p. 702–776. *In*: Traite de paleontologie, Tome 4, (2), J. Piveteau (ed.). Masson et Cie, Paris.

Blumel, L. 1979. Paternal care in the bony fishes. Quart. Rev. Biol. 54:149–161.

Boehlert, G. W. 1978. Intraspecific evidence for the function of single and double cones in the teleost retina. Science 202:309–311.

Boehlert, G. W. 1979. Retinal development in postlarval through juvenile *Sebastes diploproa*: Adaptations to a changing photic environment. Rev. Can. Bi. 38(4):265–280.

Bohensky, F. 1981. Photo manual and dissection guide of the shark. Avery, Wayne, NJ.

Böhlke, J. E. 1966. Lyomeri, Eurypharyngidae, Saccopharyngidae, p. 603–628. *In*: Fishes of the western North Atlantic. Mem. Sears Found. Mar. Res. 1(5).

Böhlke, J. E., and V. G. Springer. 1961. A review of the Atlantic species of the clinid fish genus *Starksia*. Proc. Acad. Nat. Sci. Phila. 113(3):29–60.

Böhlke, J. E., S. H. Weitzmann, and N. Menezes. 1979. Estado actuale da sistemic dos peixes de agua doce da America do Sul. Acta Amazonia 8 (1979) (4):657–677.

Bolsano, J. S., K. Kucharski, E. J. Randle, and P. J. Monaco. 1985. Reduction of competition between bisexual and unisexual females of *Poecilia* in northeastern Mexico. Env. Biol. Fish. 12(4): 251–263.

Bolton, J. P., N. L. Collie, H. Kawauchi, and T. Hirano. 1987. Osmoregulatory actions of growth hormone in rainbow trout (*Salmo gairdneri*). J. Endocr. 112:63–68.

Bond, C. E., and T. T. Kan. 1973. *Lampetra* (*Entosphenus*) *minima* n. sp., a dwarfed parasitic lamprey from Oregon. Copeia 1973: 568–574.

Bond, C. E., and D. L. Stein. 1984. *Opaeophacus acrogeneius*, a new genus and species of Zoarcidae (Pisces: Osteichthyes) from the Bering Sea. Proc. Biol. Soc. Wash. 97(3):522–525.

Bond, C. E., and T. Uyeno. 1981. Remarkable changes in the vertebrae of (the) perciform fish *Scombrolabrax* with notes on its anatomy and systematics. Jpn. J. Ich. 28(3):259–262.

Bond, C. E., T. T. Kan, and K W. Myers. 1983. Notes on the marine life of the river lamprey, *Lampetra ayersi,* in Yaquina Bay, Oregon and the Columbia River estuary. Fish. Bull. U. S. 81(1):165–167.

Bone, Q. 1963. The central nervous system, p. 50–91. *In*: The biology of *Myxine,* A Brodal and R. Fänge (eds.). Universitets Forlaget, Oslo.

Bone, Q. 1971. On the scabbard fish *Aphanopus carbo.* J. Mar. Biol. Assoc. U.K. 51:219–255.

Bone, Q. 1972. Buoyancy and hydrodynamic functions in the caster oil fish, *Ruvettus pretiosus* (Pisces: Gempylidae). Copeia 1972:78–87.

Bone, Q. 1975. Muscular and energetic aspects of fish swimming, p. 493–528. *In*: Swimming and flying in nature, T. Y.-T. Wu, C. J. Brokaw, and C. Brennan (eds.). Plenum Press, New York.

Bone, Q. 1978. Locomotor muscle, p. 361–424. *In*: Fish physiology, Vol. VII. W. S. Hoar and D. J. Randall (eds.). Academic Press, New York.

Bone, Q. 1988. Muscles and locomotion, p. 99–141. *In*: Physiology of elasmobranch fishes, T. J. Shuttleworth (ed.). Springer-Verlag, Berlin.

Bone, Q. 1989. Evolutionary patterns of axial muscle systems in some invertebrates and fish. Amer. Zool. 29:5–18.

Bone, Q., and N. B. Marshall. 1982. Biology of fishes. Blackie, Glasgow and London.

Bone, Q., J. Kiceniuk, and D. R. Jones. 1978. On the role of the different fibre types in fish myotomes at intermediate swimming speeds. Fish. Bull. U. S. 76:691–699.

Bone, Q., N. B. Marshall, and J. H. S. Blaxter. 1995. Biology of fishes (Second Edition). Blackie Academic and Professional, Glasgow.

Borlakoglu, J., and R. Dils. 1990. Polychlorinated biphenyls (PCBs) and marine food chains. Biologist 37(5):145–147.

Borwein, B., and M. J. Hollenberg. 1973. The photoreceptors of the "four-eyed" fish, *Anableps anableps.* J. Morphol. 140:405–439.

Boulenger, G. A. 1904. Teleostei (systematic part), p. 539–727. *In*: Fishes, S. F. Harmer and A. E. Shipley (eds.). Cambridge Natural History, Vol. VII.

Boulenger, G. A. 1905. The distribution of African freshwater fishes. Nature 72:413–421.

Boulenger, G. A. 1910. Fishes, p. 140–727. *In*: Cambridge Natural History, Vol. 7, S. F. Harmer and A. E. Shipley (eds.). MacMillan, London.

Bourne, G. H. 1980. Hearts and heart-like organs. Academic Press, New York.

Bowen, S. H. 1987. Dietary protein requirements of fishes—a reassessment. Can. J. Fish. Aquat. Sci. 44:1995–2001.

Bowmaker, J. K. 1990. Visual pigments of fishes, p. 81–107. *In*: The visual system of fish, R. Douglas and M. Djamgoz (eds.). Chapman and Hall, London.

Bowmaker, J. K., and Y. W. Kunz. 1987. Ultraviolet receptors, tetrachromatic colour vision and retinal mosaics in brown trout (*Salmo trutta*): Age-dependent changes. Vision Research 27: 2101–2108.

Bowmaker, J. K., H. J. A. Dartnall, and P. J. Herring. 1988. Longwave-sensitive visual pigments in some deep-sea fishes: Segregation of "paired" rhodopsins and porphyropsins. J. Comp. Phys. A 163(5):685–698.

Bradbury, M. G. 1967. The genera of batfishes (family Igcoelphaslidae). Copeia 1967:399–422.

Bradbury, M. G. 1988. Rare fishes of the deep-sea genus *Halieutopsis*: A review with descriptions of four new species (Lophiiformes: Ogcocephalidae). Fieldiana Zoology (n. s.) 44:1–22.

Braekevelt, C. R. 1974. Fine structure of the retinal pigment epithelium, Bruch's membrane, and choriocapillaris in the northern pike *Esox lucius*. J. Fish. Res. Bd. Can. 31(10):1601–1605.

Brannon, E. L. 1987. Mechanisms stabilizing salmonid fry emergency timing, p. 120–124. *In*: Sockeye salmon (*Oncorhynchus nerka*) population biology and future management, H. D. Smith, L. Margolis, and C. C. Wood (eds.). Can. Spec. Pub. Fish. Aquat. Sci. 96.

Branson, B. A., and G. A. Moore. 1962. The lateralis components of the acoustico-lateralis system in the sunfish family Centrarchidae. Copeia 1962:1–108.

Branson, B. A., and G. U. Ulrikson. 1967. Morphology and histology of the branchial apparatus in percid fishes of the genera *Percina, Etheostoma* and *Ammocrypta* (Percidae: Percinae: Etheostomatini). Trans. Am. Microsc. Soc. 86(4):371–389.

Bratton, B. O., and J. L. Ayers. 1987. Observations on the electric organ discharge of two skate species (Chondrichthyes: Rajidae) and its relationship to behaviour. Environ. Biol. Fish 20(4): 241–254.

Breder, C. M., Jr. 1926. The locomotion of fishes. Zoologica (N.Y.) 4:159–297.

Breder, C. M., Jr. 1946. An analysis of the deceptive resemblances of fishes to plant parts with critical remarks on protective coloration. Bull. Bingham Oceanographic Coll. 10(2):1–49.

Breder, C. M., Jr. 1959. Studies on social grouping in fishes. Bull. Am. Mus. Nat. Hist. 117:393–482.

Breder, C. M., Jr. 1972. On the relationship of teleost scales to pigment patterns. Contrib. Mote Mar. Lab. 1:1–79.

Breder, C. M., Jr. and P. M. Bird. 1975. Cave entry by schools and associated pigmentary changes of the marine clupeid, *Jenkinsia*. Bull. Mar. Sci. 25(3):377–386.

Breder, C. M., Jr. and P. Rasquin. 1950. A preliminary report on the role of the pineal organ in the control of pigment cells and light reactions in recent teleost fishes. Science 111:10–12.

Breder, C. M., Jr. and D. E. Rosen. 1966. Modes of reproduction in fishes. Natural History Press, Garden City, NY.

Brenner, S., G. Elgar, R. Sanford, A. Macrae, B. Venkatesh, and S. Aparicio. 1993. Characterization of the pufferfish (*Fugu*) genome as a compact model vertebrate genome. Nature 366:265–268.

Bres, M. 1993. The behavior of sharks. Rev. Fish Biol. Fisheries 3:133–159.

Brett, J. R., 1957. The eye, p. 121–154. *In*: The physiology of fishes, Vol. 2, M. E. Brown (ed.). Academic Press, New York.

Brett, J. R. 1971. Energetic response of salmon to temperature. A study of some thermal relations in the physiology and freshwater ecology of sockeye salmon (*Oncorhynchus nerka*). Am. Zool. 11:99–113.

Brett, J. R. 1979. Environmental factors and growth, p. 599–675. *In*: Fish physiology, Vol. VIII, Bioenergetics and growth, W. S. Hoar, D. J. Randall and J. R. Brett (eds.). Academic Press, New York.

Brett, J. R. and D. MacKinnon. 1954. Some aspects of olfactory perception in migrating adult coho and spring salmon. J. Fish. Res. Bd. Can. 11:310–318.

Bridge, T. W. 1910. Fishes, p. 140–727. *In*: Cambridge Natural History, Vol. 7, S. F. Harmer and A. E. Shipley (eds.). MacMillan, London.

Bridge, T. W., and G. A. Boulenger. 1904. Fishes. *In*: The Cambridge Natural History, Vol. 3, S. F. Harmer and A. E. Shipley (eds.). MacMillan, London. Reprint ed. 1958, Wheldon and Wesley Ltd., Codicote, England, and H. R. Engelmann (J. Cramer), Weinheim, Germany.

Briggs, D. E. G. 1992. Conodonts: A major extinct group added to the vertebrates. Science 256:1295–1296.

Briggs, J. C. 1955. A monograph of the clingfishes (order Xenopterygii). Stanford Ichthyol. Bull. 6:1–224.

Briggs, J. C. 1960. Fishes of worldwide (circumtropical) distribution. Copeia 1960:171–180.

Briggs, J. C. 1961. The East Pacific Barrier and the distribution of marine shore fishes. Evolution 15(4):545–554.

Briggs, J. C. 1974. Marine zoogeography. McGraw-Hill, New York.

Briggs, J. C. 1981. Do centers of origin have a center? Paleobiology 7(3):305–307.

Briggs, J. C. 1986. Order Gobiesociformes, p. 378–380. *In*: Smith's sea fishes, M. M. Smith and P. C. Heemstra (eds.). Springer-Verlag, Berlin, Heidelberg, New York, London, Paris, Tokyo.

Brill, R. W., and A. E. Dizon. 1979. Red and white muscle activity in swimming skipjack tuna, *Katsuwonus pelamis* (L.). J. Fish Biol. 15:679–685.

Brocksen, R. W., and J. P. Bugge. 1974. Preliminary investigations on the influence of temperature on food assimilation by rainbow trout *Salmo gairdneri* Richardson. J. Fish Biol. 6:93–97.

Brocksen, R. W., G. E. Davis, and C. E. Warren. 1968. Competition, food consumption and production of sculpins and trout in laboratory stream communities. J. Wildl. Mgt. 32:51–75.

Brodal, A., and R. Fänge (eds.). 1963. The biology of *Myxine*. Universitetsforaget, Oslo.

Brown, J. H., and A. C. Gibson. 1991. Biogeography (Second Edition). Mosby-Yearbook, Inc., St. Louis, MO.

Brown, L. R. 1989. Temperature preferences and oxygen consumption of three species of sculpin (*Cottus*) from the Pit River drainage, California. Env. Biol. Fish. 26(3):223–236.

Brown, L. R., and P. B. Moyle. 1991. Changes in habitat and microhabitat partitioning within an assemblage of stream fishes in response to predation by Sacramento squawfish (*Ptychocheilus grandis*). Can. J. Fish. Aquat. Sci. 48:849–856.

Brown, M. E. 1957. Experimental studies on growth, p. 351–400. *In*: The physiology of fishes, M. E. Brown (ed.). Academic Press, New York.

Brown, W. M. 1983. Evolution of animal mitochondrial DNA, p. 62–88. *In*: Evolution of genes and proteins, M. Nei and R. K. Koehn (ed.). Sinauer Assoc., Sunderland, MA.

Brown, W. M., M. George, Jr., and A. C. Wilson. 1979. Rapid evolution of animal mitochondrial DNA. Proc. Nat. Acad. Sci. U.S. 76:1967–1971.

Bruton, M. N. (ed.). 1990. Alternative life history styles in fishes. Kluwer Academic Publishers, Hingham.

Bruun, A. F. 1940. A study of the fish *Schindleria* from South Pacific waters. Dana Rep. 21:1–12.

Buck, J. D., S. Spotte, and J. J. Gadbaw, Jr. 1984. Bacteriology of the teeth from a great white shark: Potential medical implications for shark bite victims. J. Clin. Microbiol. 20(5):840–851.

Buhler, P. 1977. Comparative kinematics of the vertebrate jaw frame, p. 123–138. *In*: Physiology of movement—biomechanics, W. Nachtigall (ed.). Gustave Fischer Verlag, Stuttgart.

Bullock, T. H. 1981. Comparisons of the electric and acoustic senses and their central processing, p. 525–570. *In*: Hearing and sound communication in fishes, W. N. Tavolga, A. N. Popper and R. R. Fay (eds.). Springer-Verlag, New York.

Bullock, T. H., and W. Heiligenberg (eds.). 1986. Electroreception. Wiley, New York.

Bulow, F. J. 1987. RNA-DNA ratios as indicators of growth in fish: A review, p. 45–64. *In*: Age and growth in fishes, R. C. Summerfelt and G. E. Hall (eds.). Iowa State University Press, Ames, IA.

Burch, S. J., R. Lawson, and D. H. Davies. 1984. The relationships of cartilaginous fishes: An immunological study of serum transferrins of holocephalans and elasmobranchs. J. Zool. 203(3):303–310.

Burger, J. W. 1962. Further studies on the function of the rectal gland in the spiny dogfish. Physiol. Zool. 35:205–217.

Burger, J. W., and W. N. Hess. 1960. Function of the rectal gland in the spiny dogfish. Science 131:670–671.

Burgess, G. H. 1990. Shark attack and the international shark attack file, p. 101–105. *In*: Discovering sharks, S. H. Gruber (ed.). American Littoral Society, Highlands, NJ.

Burgess, J. W., and E. Shaw. 1979. Development and ecology of fish schooling. Oceanus 22:11–17.

Burggren, W. W., and K. Johansen. 1987. Circulation and respiration in lung fishes (Dipnoi), p. 217–236. *In*: The biology and evolution of lungfishes, W. E. Bemis, W. W. Burggren, and N. E. Kemp (eds.). Alan R. Liss, Inc., New York.

Burggren, W. W., and J. L. Roberts. 1991. Respiration and metabolism, p. 353–435. *In*: Comparative animal physiology (Fourth

Edition), Environmental and metabolic animal physiology, C. L. Prosser (ed.). Wiley-Liss, New York.

Burggren, W. W., B. McMahon, and D. Powers. 1991. Respiratory functions of blood, p. 437–508. *In*: Comparative animal physiology (Fourth Edition), Environmental and metabolic animal physiology, C. L. Prosser (ed.). Wiley-Liss, New York.

Burkhardt, D. A., J. Gottesman, J. S. Levine, and E. F. MacNichol, Jr. 1983. Cellular mechanisms for color-coding in holostean retinas and the evolution of color vision. Vision Res. 23:1031–1041.

Burns, J. R. 1976. The reproductive cycle and its environmental control in the pumpkinseed, *Lepomis gibbosus* (Pisces: Centrarchidae). Copeia 1976:449–455.

Burr, B. M., and R. L. Mayden. 1982. Life history of the brindled madtom *Noturus miurus* in Mill Creek, Illinois (Pisces: Ictaluridae). Am. Midl. Nat. 107:25–41.

Burr, B. M., and R. L. Mayden. 1992. Phylogenetics and North American freshwater fishes, p. 18–75. *In*: Systematics, historical ecology, and North American freshwater fishes, R. L. Mayden (ed.). Stanford University Press, Palo Alto.

Burton, R. F. 1968. Cell potassium and the significance of osmolarity in vertebrates. Comp. Biochem. Physiol. 27:763–773.

Busacker, G. P. and I. R. Adelman. 1987. Uptake of ¹⁴C-glycine by fish scales (in vitro) as an index to current growth rate, p. 355–357. *In*: Age and growth in fish, R. C. Summerfelt and G. E. Hall (eds.). Iowa State University Press, Ames, IA.

Bussing, W. A. 1985. Patterns of distribution of the Central American ichthyofauna, p. 453–473. *In*: The great American interchange, F. G. Stehli and S. D. Webb (eds.). Plenum Press, New York.

Buth, D. G. 1983. Duplicate isozyme loci in fishes: Origins, distribution, phyletic consequences, and locus nomenclature, p. 381–400. *In*: Isozymes: Current topics in biological and medical research. Vol. 10: Genetics and evolution, M. C. Rattazzi, J. G. Scandalios, and G. W. Whitt (eds.). Liss, New York.

Buth, D. G. 1984a. Allozymes of the cyprinid fishes: Variation and application, p. 561–590. *In*: Evolutionary genetics of fishes, B. J. Turner (ed.). Plenum Press, New York.

Buth, D. G. 1984b. The application of electrophoretic data in systematic studies. Ann. Rev. Ecol. Syst. 15:501–522.

Buth, D. G., and T. R. Haglund. 1994. Allozyme variation in the *Gasterosteus aculeatus* complex, p. 61–84. *In*: The evolutionary biology of the three spine stickleback, M. A. Bell and S. A. Foster (eds.). Oxford University Press, Oxford.

Buth, D. G., T. R. Haglund, and W. L. Minckley. 1992. Duplicate gene expression and allozyme divergence diagnosis for *Catostomus tahoensis* and the endangered *Chasmistes cujus* in Pyramid Lake, Nevada. Copeia 1992:935–941.

Butler, D. G. 1966. Effect of hypophysectomy on osmoregulation in the European eel (*Anguilla anguilla* L.). Comp. Biochem. Physiol. 18:773–781.

Butler, D. G. 1973. Structure and function of the adrenal gland of fishes. Am. Zool. 13:839–879.

Butler, P. J. 1986. Exercise, p. 102–118. *In*: Fish physiology: Recent advances, S. Nilsson and S. Holmgren (eds.). Croom Helm, London, Sydney, Dover, NH.

Butler, P. J., and J. D. Metcalfe. 1988. Cardiovascular and respiratory systems, p. 1–47. *In*: Physiology of elasmobranch fishes, T.

J. Shuttleworth (ed.). Springer-Verlag, Berlin, Heidelberg, New York, London, Paris.

Cagan, R. H. 1984. Olfactory recognition in rainbow trout, p. 285–299. *In*: Comparative physiology of sensory systems, L. Bolis, R. D. Keynes, and S. H. P. Maddrell (eds.). Cambridge University Press, Cambridge.

Cai, Z, and L. Curtis. 1989. Effects of diet on consumption, growth and fatty acid composition in young grass carp. Aquaculture 81:47–60.

Caillet, G. M., and R. L. Radtke. 1987. A progress report on the electron microprobe analysis technique for age determination and verification in elasmobranchs, p. 359–369. *In*: Age and growth in fishes, R. C. Summerfelt and G. E. Hall (eds.). Iowa State University Press, Ames, IA.

Caillet, G. M., R. L. Radtke, and B. A. Weldon. 1986. Elasmobranch age determination and verification: A review, p. 345–360. *In*: Proceedings of the Second International Conference on Indo-Pacific Fishes, Tokyo, T. Uyeno, R. Arai, T. Taniuchi, and K. Matsuura (eds.). The Ichthyol. Soc. of Japan, Tokyo.

Callard, G. V. 1988. Reproductive biology (Part B), p. 292–317. *In*: Physiology of elasmobranch fishes, T. J. Shuttleworth (ed.). Springer-Verlag, Berlin.

Calton, M. S., and T. E. Denton. 1974. Chromosomes of the chocolate gourami: A cytogenetic anomaly. Science 185:618–619.

Cameron, P. A., and E. N. Pugh. 1991. Double cones as a basis for a new type of polarization vision in vertebrates. Nature 353:161–164.

Campagno, L. J. V. 1991. The evolution and diversity of sharks, p. 15–22. *In*: Discovering sharks, S. H. Gruber (ed.). Sp. Pub. 14, Am. Littoral Soc., Highlands, NJ.

Campton, D. 1987. Natural hybridization and introgression in fishes: Methods of detection and genetic interpretations, p. 161–192. *In*: Population genetics in fishery management, N. Ryman and F. M. Utter (eds.). University of Washington Press, Seattle.

Campton, D., and B. Mahmoudi. 1991. Allozyme variation and population structure of striped mullet (*Mugil cephalus*) in Florida. Copeia 1991: 485–492.

Cappetta, H. 1987. Chondrichthyes II. Mesozoic and Cenozoic Elasmobranchii. *In*: Handbook of paleoichthyology, Vol. 3B, H.-P. Schultze (ed.). Gustav Fischer Verlag, Stuttgart.

Caprio, J. 1984. Olfaction and taste in fish, p. 257–284. *In*: Comparative physiology of sensory systems, L. Bolis, R. D. Keynes and S. H. P. Maddrell (eds.). Cambridge University Press, Cambridge.

Caprio, J. 1988. Peripheral filters and chemoreceptor cells in fishes, p. 313–338. *In*: Sensory biology of aquatic animals, J. Atema, R. R. Fay, A. N. Popper and W. N. Tavolga (eds.). Springer-Verlag, New York.

Carey, F. G., and Q. H. Gibson. 1983. Heat and oxygen exchange in the rete mirabile of the bluefin tuna, *Thunnus thynnus*. Comp. Biochem. Physiol. 74A(2):333–342.

Carey, F. G., J. M. Teal, J. W. Kanwisher, K. D. Lawson, and J. Beckett. 1971. Warm-bodied fish. Am. Zool. 11 (1):137–145.

Carlander, K. D. 1977. Handbook of freshwater fishery biology, Vol. 2. Iowa State University Press, Ames, IA.

Carlander, K. D. 1987. A history of scale age and growth studies of

North American freshwater fish, p. 3–161. *In*: Age and growth of fish, R. C. Summerfelt and G. E. Hall (eds.). Iowa State University Press, Ames, IA.

Carlson, H. R., and R. E. Haight. 1972. Evidence for a home site and homing behavior of adult yellowtail rockfish, *Sebastes flavidus*. J. Fish. Res. Bd. Can. 29:1011–1014.

Carpene, E., A Veggetti, and F. Mascarello. 1982. Histochemical fibre types in the lateral muscle of fishes in fresh, brackish and saltwater. J. Fish. Biol. 20(4):379–396.

Carr, S. M., and D. Marshall. 1991. DNA sequence variation in the mitochondrial cytochrome *b* gene of Atlantic cod (*Gadus morhua*) detected by the polymerase chain reaction. Can. J. Fish Aquat. Sci. 48:48–52.

Carr, W. E. S. 1982. Chemical stimulation of feeding behavior, p. 259–274. *In*: Chemoreception in fishes, T. J. Hara (ed.). Chapman and Hall, London.

Carroll, R. L. 1988. Vertebrate paleontology and evolution. Freeman, New York.

Carscadden, J. E., and W. C. Leggett. 1975. Life history variations in populations of American shad, *Alosa sapidissima* (Wilson), spawning in tributaries of the St. John River, New Brunswick, J. Fish Biol. 7(5):595–609.

Carter, G. S. 1967. Structure and habit in vertebrate evolution. University of Washington Press, Seattle.

Caruso, J. H. 1985. The systematics and distribution of the Lophiid anglerfishes: III. Intergeneric relationships. Copeia 1985:870–875.

Caruso, J. H. 1989. Systematics of and distribution of the Atlantic chaunacid anglerfishes (Pisces: Lophiiformes). Copeia 1989:153–165.

Castell, J. D., R. O. Sinnhuber, J. H. Wales, and D. J. Lee. 1972. Essential fatty acids in the diet of rainbow trout (*Salmo gairdneri*): Growth, feed conversion and some gross deficiency symptoms. J. Nutr. 102:77–86.

Casey, J. H., H. L. Pratt, Jr. and C. Stilwell. 1980. The shark tagger. Newsl. Coop. Shark Tagging Program. NOAA, NMFS Narragansett, RI.

Castle, P. H. J. 1976. Criteria of satisfactory fixation and preservation of eel larvae (Leptocephali), p. 319–320. *In*: Zooplankton fixation and preservation, H. P. Steedman (ed.). Unesco Press, Paris.

Castle, P. H. J. 1984. Notacanthiformes and Anguilliformes: Development, p. 62–102. *In*: Ontogeny and systematics of fishes, H. G. Moser, W. J. Richards, D. M. Cohen, M. P. Fahay, A. W. Kendall, Jr. and S. L. Richardson (eds.). Am. Soc. Ichthyol. and Herpetol. Sp. Pub. 1.

Cavender, T. M. 1970. A comparison of coregonines and other salmonids with the earliest known teleostean fishes, p. 1–32. *In*: Biology of coregonid fishes, C. C. Lindsey and C. S. Woods (eds.). University of Manitoba Press, Winnipeg.

Cavender, T. 1986. Review of the fossil history of North American freshwater fishes, p. 699–724. *In*: The zoogeography of North American freshwater fishes, C. H. Hocutt and E. O. Wiley (eds.). John Wiley and Sons, New York.

Cavender, T., and M. Coburn. 1992. Phylogenetic relationships of North American Cyprinidae, p. 293–327. *In*: Systematics, historical ecology, and North American freshwater fishes, R. L. Mayden (ed.). Stanford University Press, Palo Alto.

Centers for Disease Control. 1982. Intestinal perforation caused by larval *Eustrongyloides*-Maryland. Morbidity Mortality Weekly Rept. 32(28):383.

Chan, C. B., and E. Hale. 1992. Effects of somatostatin on intragastric pressure and smooth muscle contractility of the rainbow trout, *Oncorhynchus mykiss* Walbaum. J. Fish Biol. 40(4):545–556.

Chan, S. J., Q.-P. Cao, S. Nagamatsu, and D. F. Steiner. 1993. Insulin and insulin-like growth factor genes in fishes and other primitive chordates, p. 407–418. *In*: Biochemistry and molecular biology of fish, Vol. 2: Molecular biology frontiers, P. W. Hochachka and T. P. Mommsen (eds.). Elsevier, Amsterdam.

Chang, M. M. 1991. Rhipidistians, dipnoans and tetrapods, p. 3–28. *In*: Origins of major groups of tetrapods: Controversies and consensus, H.-P. Schultze and L. Trueb (eds.). Cornell University Press, Ithaca, NY.

Chang, W. 1971. Studies on feeding and protein digestibility of silver carp, *Hypophthalmichthys molotrix* (C. & V.). Chinese-American Joint Comm. on Rural Reconstr. Fish., Ser. 11:96–114.

Chang, Y.-S., F.-L. Huang, and T.-B. Lo. 1994. The complete nucleotide sequence and gene organization of carp (*Cyprinus carpio*) mitochondrial genome. J. Mol. Evol. 38:138–155.

Chao, L. N. 1973. Digestive system and feeding habits of the cunner, *Tautogolabrus adspersus,* a stomachless fish. Fish. Bull. U. S. 71(2):565–586.

Chapleau, F. 1988. Comparative osteology and intergeneric relationships of the tongue soles (Pisces: Pleuronectiformes: Cynoglossidae). Can. J. Zool. 66:1214–1232.

Chapleau, F. 1993. Pleuronectiform relationships: a cladistic reassessment. Bull. Mar. Sci. 52(1):516–540.

Chapleau, F., and A. Keast. 1988. A phylogenetic reassessment of the monophyletic status of the family Soleidae, with comments on the suborder Soleoidei (Pisces: Pleuronectiformes). Can. J. Zoo. 66:2797–2810.

Chapman, L. J., L. Kaufman, and C. A. Chapman. 1994. Why swim upside down? A comparative study of two mochokid catfishes. Copeia 1994:130–155.

Chardon, M. 1968. Anatomie comparee de l'appareil de Weber et des structures connexes chez les Siluriformes. Mus. R. Afr. Cent. Ann. (Ser. 8, Zool.) 1969:1–277.

Charman, W. N., and J. Tucker. 1973. The optical system of the goldfish eye. Vision Res. 13:1–8.

Chen, S.-J., N.-H. Chao, and I.-C. Liao. 1986. Diploid gynogenesis induced by cold shock in cyprinid loach. *Misgurnus anguillocaudatus*. Proc. First Asian Fisheries Forum, Manila, 16–31 May 1986:105–108.

Cherfas, B. I. (ed.) 1969. Genetics, selection, and hybridization of fish. Academy of Sciences of the U.S.S.R., Ministry of Fisheries of the U.S.S.R., Ichthyological Commission. Jerusalem, Israel Program for Scientific Translations.

Chessman, B. C., and W. D. Williams. 1975. Salinity tolerance and osmoregulatory ability of *Galaxias maculatus* (Jenyns) (Pisces, Salmoniformes, Galaxiidae). Freshwater Biol. 5(2):135–140.

Chevassus, B. 1983. Hybridization in fish. Aquaculture 33:245–262.

Chiarelli, A. B., and E. Capanna. 1973. Checklist of fish chromosomes, p. 206–232. *In*: Cytotaxonomy and vertebrate evolution,

A. B. Chiarelli and E. Capanna (eds.). Academic Press, New York.

Childers, W. F. 1967. Hybridization of four species of sunfishes (Centrarchidae). Illinois Nat. Hist. Surv. Bull. 27:159–214.

Childress, J. J., and R. P. Meek. 1973. Observations on the feeding behavior of a mesopelagic fish (*Anoplogaster cornuta*, Beryciformes). Copeia 1973: 602–603.

Chow, P. H., and D. K. O. Chen. 1975. The cardiac cycle and the effects of neurohumors on myocardial contractility in the Asiatic eel, *Anguilla japonica*, Timm and Schle. Comp. Biochem. Physiol. 52C:41–45.

Christophe, B., and F. Baguet. 1982. Luminescence of isolated photophores and supracaudal gland from *Myctophum punctatum*: Electrical stimulation. Comp. Biochem. Physiol. 71A:131–136.

Ciereszko, A., L. Ramseyer, and K. Dabrowski. 1993. Cryopreservation of yellow perch semen. Prog. Fish. Cult. 55(4):261–264.

Clack, J. A. 1989. Discovery of the oldest tetrapod stapes. Nature 342:425–427.

Clark, E. 1983. Shark repellent effect of the Red Sea Moses sole, p. 135–150. *In*: Shark repellents from the sea: New perspectives, B. J. Zahuranec (ed.). Westview Press, Boulder, CO.

Clarke, W. D., and R. L. Haedrich. 1968. Dive 218, p. 14–15. *In*: Gulfview-Diving Log, May 27–June 12, 1967, R. D. Gaul and W. D. Clarke (eds.). Gulf Universities Research Corp.

Clarke, W. D., and W. G. Pearcy. 1968. Dive 217, p. 10–14. *In*: Gulfview-Diving Log, May 27–June 12, 1967, R. D. Gaul and W. D. Clarke (eds.). Gulf Universities Research Corp.

Clarke, W. D., and R. H. Rosenblatt. 1968. Dive 219, p. 16–17. *In*: Gulfview-Diving Log, May 27–June 12, 1967, R. D. Gaul and W. D. Clarke (eds.). Gulf Universities Research Corp.

Clausen, H. S. 1959. Denticipitidae, a new family of primitive isospondylous teleosts from west African freshwater. Vidensk. Medderr. Dansk. Naturh. Foren. 121:141–156.

Cloud, J. G. 1990. Strategies for introducing foreign DNA into the germ line of fish. J. Reproduct. Fert., Suppl. 41:107–116.

Cloutier, R., and P. L. Forey. 1991. Diversity of extinct and living actinistian fishes (Sarcopterygii), p. 59–74. *In*: The biology of *Latimeria cl alumnae* and the evolution of coelacanths, J. A. Musick, M. N. Braton and E. K. Balon (eds.) Env. Biol. Fish. 32(1–4).

Coad, B. W., and F. Papahn. 1988. Shark attacks in the rivers of southern Iran, p. 131–134. *In*: On lampreys and fishes: a memorial anthology in honor of Vadim D. Vladykov, D. E. McAllister and E. Knott (eds.). Kluwer Academic Publications, Dordrecht, Boston, London.

Coates, C. W. 1937. Slowly the lungfish gives up its secrets. Bull. N.Y. Zool. Soc. 40:25–34.

Coates, M. I., and J. A. Clack. 1991. Fish-like gills and breathing in the earliest-known tetrapod. Nature 352:234–235.

Coburn, M. M., and T. M. Cavender. 1992. Interrelationships of North American cyprinids, p. 329–373. *In*: Systematics, historical ecology, and the North American freshwater fish fauna, R. L. Mayden (ed.). Stanford University Press, Palo Alto.

Cohen, D. M. 1970. How many recent fishes are there? Proc. Calif. Acad. Sci., 4th Series, 38:341–346.

Cohen, D. M. 1984. Gadiformes: Overview, p. 259–265. *In*: Ontogeny and systematics of fishes, H. G. Moser, W. J. Richards, D.

M. Cohen, M. P. Fahay, A. W. Kendall, Jr. and S. L. Richardson (eds.). Am. Soc. Ichthyol. and Herpetol. Sp. Pub. 1.

Cohen, D. M. (ed.). 1989. Papers on the systematics of gadiform fishes. Nat. Hist. Mus., Los Angeles Co. Sci. Ser. 32.

Cohen, D. M., and R. L. Haedrich. 1983. The fish fauna of the Galapagos thermal vent region. Deep-Sea Research 30(4A): 371–379.

Cohen, D. M., and J. G. Nielsen. 1978. Guide to the identification of genera of the fish order Ophidiiformes with a tentative classification of the order. NOAA Tech Rep., NMFS Circ. 147.

Cohen, D. M., T. Inada, T. Iwamoto, and N. Scialabba. 1990. Gadiform fishes of the world (Order Gadiformes). An annotated and illustrated catalogue of cods, hakes, grenadiers and other gadiform fishes known to date. FAO Species Catalog, Vol. 10. FAO Fish. Synop. 125.

Cohen, J. 1977. Reproduction. Butterworths, London.

Cohen, J. L. 1990. Vision in elasmobranchs, p. 463–490. *In*: The visual system of fish. R. Douglas and M. Djamgoz (eds.). Chapman and Hall, London, New York, Tokyo, Melbourne, Madras.

Colbert, E. H. 1980. Evolution of the vertebrates (Third Edition). Wiley, New York.

Colbert, E. H., and M. Morales. 1991. Evolution of the vertebrates (Fourth Edition). Wiley-Liss, New York.

Cole, L. C. 1954. The population consequences of life history phenomena. Q. Rev. Biol. 29:103–137.

Colgan, P. 1986. Motivational basis of fish behavior, p. 23–46. *In*: The behavior of teleost fishes, T. J. Pitcher (ed.). Johns Hopkins University Press, Baltimore.

Collette, B. B. 1966a. *Belonion*, a new genus of freshwater needlefishes from South America. Am. Mus. Novit. No. 2274:1–22.

Collette, B. B. 1966b. A review of the venomous toadfishes, subfamily Thalassophryninae. Copeia 1966:615–623.

Collette, B. B., and P. Banarescu. 1977. Systematics and zoogeography of the fishes of the family Percidae. J. Fish. Res. Bd. Can. 34:1450–1463.

Collette, B. B., and C. E. Nauen. 1983. FAO species catalogue, Vol. 2, Scombrids of the world. An annotated and illustrated catalogue of tunas, mackerels, bonitos and related species known to date. FAO Fish Synop. 125.

Collette, B. B., and J. L. Russo. 1981. A revision of the scaly toadfishes, genus *Batrachoides*, with descriptions of two new species from eastern Pacific. Bull. Mar. Sci. 31(2):197–233.

Collette, B. B., and J. L. Russo. 1984. Morphology, systematics, and biology of the Spanish mackerels (*Scomberomorus*, Scombridae). Fish Bull. U. S. 82:545–692.

Collette, B. B., and J. L. Russo. 1986. Systematic status of the suborder Scombroidei (Abstr.), p. 938. *In*: Indo-Pacific fish biology. Proc. 2nd Internat. Conf. on Indo-Pacific Fishes, T. Uyeno, R. Arai, T. Taniuchi and K. Matsuura (eds.). Ichthyol. Soc. Japan, Tokyo.

Collette, B. B., N. V. Parin, and M. S. Nizinsky. 1992. Catalog of type specimens of Recent fishes in the National Museum of Natural History, Smithsonian Institution, 3: Beloniformes (Teleostei). Smithson. Contr. to Zool. No. 525.

Collette, B. B., G. E. McGowen, N. V. Parin, and S. Mito. 1984. Beloniformes: Development and relationships, p. 335–354. *In*: Ontogeny and systematics of fishes, H. G. Moser, W. J. Richards,

D. M. Cohen, M. P. Fahay, A. W. Kendall, Jr. and S. L. Richardson (eds.). Am. Soc. Ichthyol. and Herpetol. Sp. Pub. No. 1.

Collette, B. B., T. Potthoff, W. J. Richards, S. Ueyanagi, J. L. Russo, and Y. Nishikawa. 1984. Scombroidei: Development and relationships, p. 591–619. *In*: Ontogeny and systematics of fishes, H. G. Moser, W. J. Richards, D. M. Cohen, M. P. Fahay, A. W. Kendall, Jr. and S. L. Richardson (eds.). Am. Soc. Ichthyol. and Herpetol. Sp. Pub. No. 1.

Collins, T. M. In Press. Molecular comparison of transisthmian species pairs: Rates and patterns of evolution. *In*: Environmental and biological change in Neogene and Quaternary tropical America. J. B. C. Jackson, A. G. Coates, and A. Budd (eds.). University of Chicago Press, Chicago.

Colombo, L., and D. W. Johnson. 1972. Corticosteroidogenesis in vitro by the head kidney of *Tilapia mossambica* (Cichlidae, Teleostei). Endocrinology 91(2):450–462.

Colombo, L., H. Bern, and J. Pieprzyk. 1971. Steroid transformations by the corpuscles of Stannius and the body kidney of *Salmo gairdneri* (Teleostei). Gen. Comp. Endocrinol. 16(1):74–84.

Colombo, L., P. Colombo Belvedere, and A. Marconato. 1979. Biochemical and functional aspects of gonadal biosynthesis of steroid hormones in teleost fishes, p. 443–451. *In*: Symposium on hormonal steroids in fish. India Nat. Acad. Science, New Delhi.

Colombo, L., P. Colombo Belvedere, A. Marconato, and F. Bentivegna. 1982. Pheromones in teleost fish, p. 84–94. *In*: Reproductive physiology of fish, C. J. J. Richter and H. J. Th. Goos (compilers). Cent. for Agric. Pub. and Documentation (Pudoc)., Wageningen, Netherlands.

Colville, T. P., R. H. Richards, and J. W. Dobbie. 1983. Variations in renal corpuscular morphology with adaptation to sea water in the rainbow trout, *Salmo gairdneri* Richardson. J. Fish. Biol. 23:451–456.

Committee on Animal Nutrition. 1993. Nutrient requirements of fish. Board Agric., Natl. Res. Council. National Academy Press, Washington, DC.

Compagno, L. J. V. 1973. Interrelationships of living elasmobranchs, p. 15–61. *In*: Interrelationships of fishes, P. H. Greenwood, R. S. Miles and C. Patterson (eds.). Academic Press, New York.

Compagno, L. J. V. 1977. Phyletic relationships of living sharks and rays. Am. Zool. 17:303–322.

Compagno, L. J. V. 1988. Sharks of the order Carchariniformes. Princeton University Press, Princeton, NJ.

Compagno, L. J. V. 1990. Relationships of the megamouth shark, *Megachasma pelagios* (Lamniformes: Megachasmidae) with comments on its feeding habits, p. 357–380. *In*: Elasmobranchs as living resources, H. L. Pratt, Jr., S. H. Fisheries Service, NOAA Tech. Dept. 90.

Compagno, L. J. V. 1991. The evolution and diversity of sharks, p. 15–22. *In*: Discovering sharks, S. H. Gruber (ed.). Am. Littoral Soc., Highlands, NJ.

Conover, D. O., and B. E. Kynard. 1981. Environmental sex determination: Interaction of temperature and genotype in a fish. Science 213:577–579.

Conroy, D. A. 1975. An evaluation of the present state of world trade in ornamental fish. FAO Tech. Pap. 146:1–128.

Constantz, G. D. 1981. Life history patterns of desert fishes, p. 237–290. *In*: Fishes in North American deserts, R. J. Naiman and D. L. Soltz (eds.). John Wiley and Sons Inc., New York.

Conte, F. P. 1969. Salt secretion, p. 241–292. *In*: Fish physiology, Vol. 1, W. S. Hoar and D. J. Randall (eds.). Academic Press, New York.

Conte, F. P., and H. H. Wagner. 1965. Development of osmotic and ionic regulation in juvenile steelhead trout, *Salmo gairdneri*. Comp. Biochem. Physiol. 14:603–620.

Cook, J. A., K. R. Bestgen, D. L. Propst, and T. L. Yates. 1992. Allozymic divergence and systematics of the Rio Grande silvery minnow. *Hybognathus amarus* (Teleostei: Cyprinidae). Copeia 1992:36–44.

Cook, S. (ed.) 1987. Sharks, and inquiry into biology, behavior, fisheries and use. EM 833D, Oregon State University Extension Service, Corvallis, OR.

Coombs, S. 1981. Interspecific differences in hearing capabilities for select teleost species, p. 173–178. *In*: Hearing and sound communication in fishes, W. N. Tavolga, A. N. Popper and R. R. Fay (eds.), Springer-Verlag, New York.

Coombs, S., and J. Janssen. 1989. Peripheral processing by the lateral line system of the mottled sculpin (*Cottus bairdi*), p. 299–319. *In*: The mechanosensory lateral line, S. Coombs, P. Görner and H. Münz (eds.). Springer-Verlag, New York.

Coombs, S., J. Janssen, and J. F. Webb. 1988. Diversity of lateral line systems: Evolutionary and functional considerations, p. 553–593. *In*: Sensory biology of aquatic animals. J. Atema, R. R. Fay, A. N. Popper and W. N. Tavolga (eds.). Springer-Verlag, New York.

Cooper, J. C., and P. J. Hirsch. 1982. The role of chemoreception in salmonid homing, p. 343–362. *In*: Chemoreception in fishes, T. J. Hara (ed.). Elsevier Scientific Publishing Co., Amsterdam, Oxford, New York.

Corwin, J. T. 1981. Audition in elasmobranchs, p. 81–105. *In*: Hearing and sound communication in fishes. W. N. Tavolga, A. N. Popper and R. R. Fay (eds.). Springer-Verlag, New York.

Cott, H. B. 1940. Adaptive coloration in animals. Methuen, London.

Coull, J. R. 1933. World fisheries resources. Rutledge, London and New York.

Courtenay, W. R., Jr., D. P. Jennings, and J. D. Williams. 1991. Exotic fishes, p. 97–107. *In*: Common and scientific names of fishes from the United States and Canada. C. R. Robins, R. M. Bailey, C. E. Bond, J. R. Brooker, E. A. Lachner, R. N. Lea and W. B. Scott. Am. Fish. Soc. Sp. Pub. 20.

Cowey, C. B., and J. R. Sargent. 1979. Nutrition, p. 1–69. *In*: Fish physiology, Vol. VIII, Bioenergetics and growth, W. S. Hoar, D. J. Randall, and J. R. Brett (eds.). Academic Press, New York.

Cowey, C. B., A. M. Mackie, and J. G. Bell. 1985. Nutrition and feeding in fish. Academic Press, London.

Cowx, I. G. (ed.). 1991. Catch effort sampling strategies. Their application in freshwater fisheries management. Fishing News Books, Oxford.

Crane, J. M. 1981. Feeding and growth by the sessile larvae of the teleost *Porichthys notatus*. Copeia 1981:895–897.

Crescitelli, F. 1991. The scotopic photoreceptors and their visual pigments of fishes: Functions and adaptations. Vision Res. 31(3):339–348.

Crim, L. W. 1981. Control of gonadotropic hormone secretion (GTH) by the rainbow trout pituitary gland. Evidence of GTH inhibition by catecholamine and stimulation of GTH release by some other neuroregulatory factors, p. 442. *In*: Neurosecretion, D. S. Farner and K. Lederis (eds.). Plenum, New York.

Crim, L. W. 1982. Environmental modulation of annual and daily rhythms associated with reproduction in teleost fishes. Can. J. Fish. Aquat. Sci. 39(1):17–21.

Crossman, E. J., and D. E. McAllister. 1986. Zoogeography of freshwater fishes of the Hudson Bay drainage, Ungava Bay and the Arctic archipelago, p. 53–104. *In*: Zoogeography of North American freshwater fishes, C. H. Hocutt and E. O. Wiley (eds.). John Wiley and Sons, New York.

Crowder, L. B. 1985. Optimal foraging and feeding mode shifts in fishes. Env. Biol. Fish 12(1):57–62.

Culp, J. M. 1989. Nocturnally constrained foraging of a lotic minnow (*Rhinichthys cataractae*). Can. J. Zool. 67:2008–2012.

Culver, D. C. 1982. Cave life: Evolution and ecology. Harvard University Press, Cambridge.

Curry, E. 1939. The histology of the digestive tube of the carp (*Cyprinus carpio communis*). J. Morph. 65:53–78.

Cushing, D. H. 1968. Fisheries biology, a study on population dynamics. University of Wisconsin Press, Madison, WI.

Cushing, D. H. 1988. The provident sea. Cambridge University Press, Cambridge.

Cushing, D. H., and F. R. Harden-Jones. 1968. Why do fish school? Nature 218:918–920.

Daniel, J. F. 1934. The elasmobranch fishes. University of California Press, Berkeley.

Daniel, T. L. 1981. Fish mucus: *in situ* measurements of polymer drag reduction. Biol. Bull. 160:376–382.

Daniels, R. A. 1987. Comparative life histories and microhabitat use in three sympatric sculpins (Cottidae: *Cottus*) in northeastern California. Env. Biol. Fishes 19:93–110.

Darnell, R. N., and Meierotto, R. R. 1962. Determination of feeding chronology in fishes. Trans. Am. Fish. Soc. 91:313–320.

Darlington, P. J., Jr. 1957. Zoogeography. John Wiley & Sons, London.

Darwin, C. 1859. On the origin of species by means of natural selection. Facsimile, 1967. Harvard University Press, Cambridge, MA.

Davies, D. H., R. Lawson, S. J. Burch, and J. Hanson. 1986. Relationships of cartilaginous fishes. 1. An immunological study of holocephalans and elasmobranchs based on the antigenic characteristics of isolated serum transferrin, p. 173–182. *In*: Indo-Pacific fish biology, T. Uyeno, R. Arai, T. Taniuchi and K. Matsuura (eds.). Proc. Second Int. Conf. on Indo-Pacific Fishes. The Ichthyol. Soc. of Japan, Tokyo.

Davis, G. E. 1968. Estimation of food consumption rates, p. 204–225. *In*: Methods for assessment of fish production in fresh waters, W. E. Ricker (ed.). Blackwell Scientific Publications, Oxford.

Davis, G. E., and C. E. Warren. 1965. Trophic relations of a sculpin in laboratory stream communities. J. Wildl. Mgt. 29:845–871.

Davis, P. S., M. E. Forster, B. Davison, and G. H. Satchell. 1987. Cardiac function in the New Zealand hagfish, *Eptatretus cirrhatus*. Physiol. Zool. 60:233–240.

Day, M. H. (ed.). 1981. Vertebrate locomotion. Symp. Zool. Soc. London 48. Academic Press, New York.

Dean, B. 1895. Fishes, living and fossil. MacMillan, New York.

Dean, B. D. 1906. Chimaeroid fishes and their development. Pub. Carnegie Inst., Washington 32:1–195.

de Beaufort, L. F. 1951. Zoogeography of the land and inland waters. Sidgwick and Jackson, London.

DeBeer, G. R. 1937. The development of the vertebrate skull. Clarendon Press, Oxford.

Decker, D. J., and G. R. Goff (eds.). 1987. Valuing wildlife: Economic and social perspectives. Westview Press, Boulder and London.

De Groot, S. J. 1971. On the relationships between morphology of the alimentary tract, food and feeding behavior in flatfishes (Pisces: Pleuronectiformes). Netherl. J. Sea Res. 5:121–196.

de Jager, S., and W. J. Dekkers. 1975. Relations between gill structure and activity in fish. Netherl. J. Zool. 25(3):276–308.

Demski, L. S., and M. Schwanzel-Fukuda. 1987. The terminal nerve (nervus terminalis) structure, function and evolution. Ann. New York Acad. Sci. 519.

Denison, R. 1978. Placodermi, p. 1–128. *In*: Handbook of paleoichthyology, Vol. 2, H.-P. Schultze (ed.). Gustav Fischer Verlag, Stuttgart, New York.

Denison, R. 1979. Acanthodii, p. 1–62. *In*: Handbook of paleoichthyology, Vol. 5, H.-P. Schultze (ed.). Gustav Fischer Verlag, Stuttgart, New York.

Denton, E. J. 1970. On the organization of reflecting surfaces in some marine animals. Phil. Trans. Roy. Soc. Lond. B 258:285–313.

Denton, E. J., and J. A. B. Gray. 1988. Mechanical factors in the excitation of the lateral lines of fishes, p. 595–617. *In*: Sensory biology of aquatic animals, J. Atema, R. R. Fay, A. N. Popper and W. N. Tavolga (eds.). Springer-Verlag, New York.

Denton, E. J., and J. A. B. Gray. 1989. Some observations on the forces acting on neuromasts in fish lateral line canals, p. 229–263. *In*: The mechanosensory lateral line, S. Coombs, P. Görner and H. Münz (eds.). Springer-Verlag, New York.

Denton, E. J., and P. J. Herring. 1976. Filter pigments in photophores. Cruise Report No. 46, R.R.S. Discovery cruise 77. Institute of Oceanographic Sciences.

Denton, E. J., and J. A. C. Nicol. 1964. The choroidal tapeta of some cartilaginous fishes (Chondrichthyes). J. Mar. Biol. Assoc. U.K. 44:219–258.

Denton, E. J., and J. A. C. Nicol. 1966. A survey of reflectivity in silvery teleosts. J. Mar. Biol. Assoc. U.K. 46:655–722.

Depêche, J., and R. Billard. 1994. Embryology in fish, a review. Soc. Francaise d'Ichtyologie.

de Sylva, D. P. 1976. Attacks by bluefish (*Pomatomus saltatrix*) on humans in South Florida. Copeia 1976(1):196–198.

de Sylva, D. P. 1984a. Mugiloidei: Development and relationships, p. 530–533. *In*: Ontogeny and systematics of fishes, H. G. Moser, W. J. Richards, D. M. Cohen, M. P. Fahay, A. W. Kendall, Jr. and S. L. Richardson (eds.). Am. Soc. Ichthyol. and Herpetol. Sp. Pub. 1.

de Sylva, D. P. 1984b. Sphyraenoidei: Development and relationships, p. 534–540. *In*: Ontogeny and systematics of fishes, H. G. Moser, W. J. Richards, D. M. Cohen, M. P. Fahay, A. W. Kendall,

Jr. and S. L. Richardson (eds.). Am. Soc. Ichthyol. and Herpetol. Sp. Pub. 1.

de Sylva, D. P. 1984c. Polynemoidei: Development and relationships, p. 540–541. *In*: Ontogeny and systematics of fishes, H. G. Moser, W. J. Richards, D. M. Cohen, M. P. Fahay, A. W. Kendall, Jr. and S. L. Richardson (eds.). Am. Soc. Ichthyol. and Herpetol. Sp. Pub. 1.

Devlin, R. H., T. Y. Yesaki, C. A. Biagi, E. M. Donaldson, P. Swanson, and W.-K. Chan. 1994. Extraordinary salmon growth. Nature 371:209–210.

DeVries, A.L. 1971. Freezing resistance in fishes, p. 157–190. *In*: Fish physiology, Vol. VI, W. S. Hoar and D. J. Randall (eds.). Academic Press, New York.

Dey, D. B., and D. M. Damkaer. 1990. Effects of specteral irradiance on the early development of chinook salmon. Prog. Fish.-Cult. 53(3):141–154.

Dharmamba, M. 1979. Corticosteroids and osmoregulation in fishes. Proc. Indian Nat. Sci. Acad. B45(5):515–525.

Diamond, J. 1971. The Mauthner cell, p. 265–346. *In*: Fish physiology, Vol. V, W. S. Hoar and D. J. Randall (eds.). Academic Press, New York.

Diana, J. S. 1980. Diel activity pattern and swimming speeds of northern pike (*Esox lucius*) in Lac Ste. Anne, Alberta. Can. J. Fish. Aquat. Sci. 37:1454–1458.

DiJulio, D. H., and G. W. Brown. 1975. Urea in the coelacanth. 1974 Research in Fisheries; Annual Report of College of Fisheries, University of Washington (Cont. No. 415), Seattle.

Dill, P. A. 1971. Perception of polarized light by yearling sockeye salmon. J. Fish. Res. Bd. Can. 28:1319–1322.

Dipper, F. A., C. B. Bridges, and A. Menz. 1977. Age, growth and feeding in the ballan wrasse *Labrus bergylta* Ascanius 1767. J. Fish Biol. 11(2):105–120.

DiSilvestro, R. L. 1987. The endangered kingdom: The struggle to save America's wildlife. Wiley, New York.

Dobbs, G. H., III, and A. L. DeVries. 1975. The aglomerular nephron of Antarctic teleosts: A light and electron microscopic study. Tissue and Cell. 7(1):159–170.

Dodd, J. M. 1983. Reproduction in cartilaginous fishes (Chondrichthyes), p. 31–95. *In*: Fish physiology, Vol. IXA, W. S. Hoar, D. J. Randall and E. M. Donaldson (eds.). Academic Press, New York, London.

Dodd, J. M., M. H. I. Dodd, and N. Jenkins. 1980. Presence of a gonadotropin in the rachendach hypophyse of the pituitary gland of the rabbit fish *Hydrolagus colliei* (Chondrichthyes; Holocephali). Gen. Comp. Endocrinol. 40:342–343. (abstr.)

Doherty, P., and T. Fowler. 1994. An empirical test of recruitment limitation in a coral reef fish. Science 263:935–939.

Dominey, W. J. 1984. Effects of sexual selection and life history on speciation: Species flocks in African cichlids and Hawaiian *Drosophila*, p. 231–249. *In*: Evolution of fish species flocks, A. A. Echelle and I. Kornfield (eds.). University of Maine Press, Orono, Maine.

Donaldson, E. M. 1973. Reproductive endocrinology of fishes. Am. Zool. 13:909–927.

Donaldson, E. M., and G. A. Hunter. 1983. Induced final maturation, ovulation,and spermiation in cultured fish, p. 351–403. *In*:

Fish physiology, Vol. IXB, W. S. Hoar, D. J. Randall and E. M. Donaldson (eds.). Academic Press, New York, London.

Donaldson, E. M., and J. R. McBride. 1974. Effect of ACTH and salmon gonadotropin on interrenal and thyroid activity of the gonadectomized adult sockeye salmon, *Oncorhynchus nerka*. J. Fish. Res. Bd. Can. 31:1211–1214.

Donaldson, E. M., U. H. M. Fagerlund, D. A. Higgs, and J. R. McBride. 1979. Hormonal enhancement of growth, p. 455–597. *In*: Fish physiology, Vol. VIII, Bioenergetics and growth, W. S. Hoar, D. J. Randall and J. R. Brett (eds.). Academic Press, New York.

Dooley, J. K. 1972. Fishes associated with the pelagic *Sargassum* complex, with a description of the sargassum community. Cont. Mar. Sci. Univ. Tex. 16:1–32.

Dooley, J. K., and J. Collura. 1988. Homing behavior in tidepool fishes. Underwater Nat. 17:3–6.

Douglas, R., and M. Djamgoz (eds.). 1990. The visual system of fish. Chapman and Hall, London.

Douglas, R. H., J. Eva, and N. Guttridge. 1988. Size constancy in goldfish (*Carassius auratus*). Behav., Brain Res. 30:37–42.

Douglas, R. H., and C. W. Hawryshyn. 1990. Behavioral studies of fish vision: Analysis of visual capabilities, p. 373–418. *In*: The visual system of fish, R. Douglas and M. Djamgoz (eds.). Chapman and Hall, London.

Döving, K. B., and K. Holmberg. 1974. A note on the function of the olfactory organ of the hagfish *Myxine glutinosa*. Acta Physiol. Scand. 91:430–432.

Döving, K. B., R. Selset, and G. Thommesen. 1980. Olfactory sensitivity to bile acids in salmonid fishes. Acta Physiol. Scand. 108: 123–131.

Dowling, T. E., and W. S. Moore. 1984. Level of reproductive isolation between two cyprinid fishes, *Notropis cornutus* and *N. chrysocephalus*. Copeia 1984:617–628.

Dowling, T. E., and W. S. Moore. 1985. Evidence for selection against hybrids in the family Cyprinidae (genus *Notropis*). Evolution 39:152–158.

Dowling, T. E., W. R. Hoeh, G. R. Smith, and W. M. Brown. 1992. Evolutionary relationships of shiners in the genus *Luxilus* (Cyprinidae) as determined by analysis of mitochondrial DNA. Copeia 1992:306–322.

Drenner, R. W., G. L. Vinyard, M. Gophen, and S. R. McComas. 1982. Feeding behavior of the cichlid, *Sarotherodon galilaeum*: Selective predation on Lake Kenneret plankton. Hydrobiologia 87:117–120.

Drenner, R. W., G. L. Vinyard, K. D. Hambright, and M. Gophen. 1987. Particle ingestion by *Tilapia galilea* is not affected by the removal of gill rakers and microbranchiospines. Trans. Am. Fish. Soc. 116:272–276.

Dugatkin, L. A., and D. S. Wilson. 1992. The prerequisites for strategic behavior in bluegill sunfish, *Lepomis macrocheirus*. Anim. Behav. 44:223–230.

Dugatkin, L. A., G. J. Fitzgerald, and J. Lavorie. 1994. Juvenile three-spined sticklebacks avoid parasitized conspecifics. Env. Biol. Fishes 39:215–218.

Duman, J. G., and A. L. DeVries. 1974a. Freezing resistance in winter flounder *Pseudopleuronectes americanus*. Nature 247: 237–238.

Duman, J. G., and A. L. DeVries. 1974b. The effects of temperature and photo-period on antifreeze production in cold water fishes. J. Exp. Zool. 190(1):89–98.

Dunham, R. A. 1986. Selection and crossbreeding responses for cultured fish, p. 391–400. *In*: X Breeding programs for swine, poultry, and fish, G. E. Dickerson and R. K. Johnson (eds.). University of Nebraska Bd. Regents, Lincoln.

Dunn, J. R. 1983. Development and distribution of the young of northern smoothtongue, *Leuroglossus schmidti* (Bathylagidae), in the northeast Pacific, with comments on the systematics of the genus *Leuroglossus* Gilbert. Fish. Bull. U. S. 81:23–40.

Dunn, J. R. and H. C. Matarese. 1984. Gadidae: Development and relationships, p. 283–299. *In*: Ontogeny and systematics of fishes, H. G. Moser, W. J. Richards, D. M. Cohen, M. P. Fahay, A. W. Kendall, Jr. and S. L. Richardson (eds.). Am. Soc. Ichthyol. and Herpetol. Spec. Pub. 1.

Dupree, H. K. 1976. Studies on nutrition and feeds of warm water fishes. Proc. 1st Int. Conf. Aquaculture Nutr., University of Delaware, p. 65–84.

Durbin, A. G., E. G. Durbin, P. G. Verity, and T. J. Smayda. 1981. Voluntary swimming speeds and respiration rates of a filter-feeding planktivore, the Atlantic menhaden, *Brevoortia tyrannus*, (Pisces: Clupeidae). Fish Bull U. S. 78(4):877–886.

Duthie, G. G. 1982. The respiratory metabolism of temperature-adapted flatfish at rest and during swimming activity and the use of anaerobic metabolism at moderate swimming speeds. J. Exp. Biol. 97:359–373.

Dye, J. C., and J. H. Meyer. 1986. Central control of the electric organ discharge in weakly electric fish, p. 71–102. *In*: Electroreception, T. H. Bullock and W. Heiligenberg (eds.). Wiley, New York.

Dymond, J. R. 1964. A history of ichthyology in Canada. Copeia 1964:33–41.

Eastman, J. T. 1993. Antarctic fish biology. Academic Press, San Diego.

Eaton, R. C. (ed.) 1991. Neuroethology of the Mauthner system. Brain, Behav. and Evol. 37(5):250–332.

Ebeling, A. W. 1967. Zoogeography of tropical deep-sea animals. Stud. Trop. Oceanogr., Miami 5:593–613.

Ebeling, A. W., and W. H. Weed, III. 1973. Fishes of the western North Atlantic. Order Xenoberyces (Stephanoberyciformes). Sears Found. Mar. Res., Memoir (Yale University) 1(6):397–478.

Ebeling, A. W., P. Bernal, and A. Zuleta. 1970. Emersion of the amphibious Chilean clingfish *Sicyases sanguinus*. Biol. Bull. 139: 115–137.

Ebert, D. J. and S. P. Filipek. 1988. Fish community structure and zonation related to stream habitat. Proc. Ann. Conf. Southeast. Assoc. Fish and Wildl. Agencies 42:234–242.

Echelle, A. A., and D. T. Mosier. 1981. All-female fish: A cryptic species of *Menidia* (Atherinidae). Science 212(4501):1411–1413.

Echelle, A. A., A. F. Echelle, and C. D. Crozier. 1983. Evolution of an all-female fish, *Menida clarkhubbsi* (Atherinidae). Evolution 37:772–784.

Ehrlich, P. R. 1975. The population biology of coral reef fishes. Ann. Rev. Ecol. Syst. 6:211–247.

Eigenmann, C. H. 1894. On the viviparous fishes of the Pacific coast of North America. Bull. U. S. Fish. Comm. 12:381–478.

Eilertson, C. D., P. K. O'Connor, and M. A. Sheridan. 1991. Somatostatin-14 and somatostatin-25 stimulate glycogenolysis in rainbow trout, *Oncorhynchus mykiss*, liver incubated in vitro: A systemic role for somatostatins. Gen. Comp. Endo. 82(2):192–205.

Ekman, S. 1953. Zoogeography of the sea. Sidgwick and Jackson. London.

Elgar, M., and B. Crespi (eds.). 1992. Cannibalism: Ecology and evolution among diverse taxa. Oxford University Press, New York.

Elger, M., and M. Hentschel. 1981. The glomerulas of a stenohaline freshwater teleost, *Carassius auratus gibilio*, adapted to saline water. Cell Tissue Res. 220:73–85.

Elliot, D. K. 1987. A reassessment of *Astraspis desiderata*, the oldest North American vertebrate. Science 237:190–192.

Elliot, D. K., and E. J. Loeffler. 1989. A new agnathan from the Lower Devonian of arctic Canada, and review of the tessellated heterostracans. Palaeontology 32, part 2:883–391.

Elshoud-Oldenhave, M. J. W., and J. W. M. Osse. 1976. Functional morphology of the feeding system in the ruff—*Gymnocephalus cernua* (L. 1758)—(Teleostei, Percidae). J. Morph. 150:399–422.

Engana, A. C., and J. E. McCosker. 1984. Attacks on divers by white sharks in Chile. Calif. Fish and Game 70:173–179.

Engstrom, K. 1963. Cone types and cone arrangements in teleost retinae. Acta Zool. 44(1–2):179–243.

Ercolini, A., and R. Berti. 1975. Light sensitivity experiments and morphology studies of the blind phreatic fish *Phreatichthys* Vinciguerra from Somalia. Monit. Zool. Ital., Suppl. 6(2):29–43.

Erikson, L. O. 1978. Nocturnalism versus diurnalism—dualism within fish individuals, p. 68–89. *In*: Rhythmic activity in fishes, J. E. Thorpe (ed.). Academic Press, New York.

Eschmeyer, W. N. 1990. Genera in a classification, p. 435–495. *In*: Catalog of the genera of Recent fishes, W. N. Eschmeyer (ed.). Calif. Acad. Sci., San Francisco.

Eschmeyer, W. N., and R. M. Bailey. 1990. Genera of Recent fishes, p. 7–433. *In*: Catalog of the genera of Recent fishes, W. N. Eschmeyer (ed.). Calif. Acad. Sci., San Francisco.

Evans, D. H. 1980. Osmotic and ionic regulation by freshwater and marine fishes, p. 93–122. *In*: Environmental physiology of fishes, M. A. Ali (ed.). Plenum Press, New York.

Evans, D. H. 1993. The physiology of fishes. CRC Press, Boca Raton, FL.

Fahay, M. P., and D. F. Markle. 1984. Gadiformes: development and relationships, p. 265–283. *In*: Ontogeny and systematics of fishes, H. G. Moser, W. J. Richards, D. M. Cohen, M. P. Fahay, A. W. Kendall, Jr. and S. L. Richardson (eds.). Am. Soc. Ichthyol. and Herpetol. Sp. Pub. 1.

Fairbairn, D. J. 1981. Which witch is which? A study of the stock structure of witch flounder (*Glyptocephalus cynoglossus*) in the Newfoundland region. Can. J. Fish. Aquat. Sci. 38:782–794.

Fänge, R. 1963. Structure and function of the excretory organs of myxinoids, p. 516–529. *In*: The biology of *Myxine*, A. Brodal and R. Fänge (eds.). Universitetsforlaget, Oslo.

Fänge, R. 1992. Fish blood cells, p. 1–54. *In*: Fish physiology, Vol. XII, Part B, W. S. Hoar, D. J. Randall, and A. P. Farrell (eds.). Academic Press, San Diego.

Fänge, R., and K. Fugelli. 1962. Osmoregulation in chimaeroid fishes. Nature 196:689.

Fänge, R., and K. Fugelli. 1965. The osmotic adjustment in the euryhaline teleosts, the flounder, *Pleuronectes flesus* L., and the three-spined stickleback, *Gasterosteus aculeatus* L. Comp. Biochem. Physiol. 15(3):283–292.

Fänge, R., and D. Grove. 1979. Digestion, p. 161–260. *In*: Fish physiology, Vol VIII, Bioenergetics and growth, W. S. Hoar, D. J. Randall and J. R. Brett (eds.). Academic Press, New York.

FAO. 1975. Review of the state of world fishery resources. FAO Fish. Circ. 710 Rev. 1:41.

FAO. 1979. Yearbook of fishery statistics, Vol. 48. Food and Agric. Org. of the United Nations, Rome.

FAO. 1990. Yearbook of fishery statistics, Vol. 66. Food and Agric. Org. of the United Nations, Rome.

FAO. 1991. Yearbook of fishery statistics, Vol. 68. Food and Agric. Org. of the United Nations, Rome.

FAO. 1995. Yearbook of fishery statistics, Vol. 77 for 1993. Food and Agric. Org. of the United Nations, Rome.

Farquhar, G. B. (ed.). 1970. Proceedings of an international symposium on biological sound scattering in the ocean. Maury Center for Ocean Science, Dept. of the Navy, Washington, DC.

Farrell, A. P. 1991. Circulation of body fluids, p. 509–558. *In*: Comparative animal physiology (Fourth Edition), Environmental and metabolic animal physiology, C. L. Prosser (ed.). Wiley-Liss, New York.

Farris, J. S. 1990. Haeckel, history, and Hull. Syst. Zool. 39(1):81–88.

Fay, R. R., and A. N. Popper. 1980. Structure and function in teleost auditory systems, p. 3–42. *In*: Comparative studies of hearing in vertebrates, A. N. Popper and R. R. Fay (eds.). Springer-Verlag, New York, Heidelberg.

Feder, M. E., and W. W. Burggren. 1985a. Skin breathing in vertebrates. Sci. Am. 253(5):126–142.

Feder, M. E., and W. W. Burggren. 1985b. Cutaneous gas exchange in vertebrates: Design, patterns, control and implications. Biol. Rev. Cambridge Philos. Soc. 601:1–45.

Feeney, R. E., and R. Hofman. 1973. Depression of freezing point by glycoproteins from an antarctic fish. Nature 243:357–359.

Feng, A. S. 1991. Electric organs and electroreceptors, p. 319–334. *In*: Comparative animal physiology (Fourth Edition). Neural and integrative animal physiology, C. L. Prosser (ed.). Wiley, New York.

Feng, A. S., and J. C. Hall. 1991. Mechanoreception and phonoreception, p. 247–316. *In*: Comparative animal physiology (Fourth Edition), Neural and integrative animal physiology, C. L. Prosser (ed.). Wiley-Liss, New York.

Fenwick, J. C. 1970. The pineal organ, p. 91–108. *In*: Fish physiology, Vol. III, W. S. Hoar and D. J. Randall (eds.). Academic Press, New York.

Fernald, R. D. 1990. The optical system of fishes, p. 45–62. *In*: The visual system of fish, R. H. Douglas and M. B.A. Djamgoz (eds.). Chapman and Hall, London.

Fernald, R. D., and S. E. Wright. 1983. Maintenance of optical quality during crystalline lens growth. Nature 301:618–620.

Fernald, R. D., and S. E. Wright. 1985. Growth of the visual system in the African cichlid fish, *Haplochromis burtoni*: Accommodation. Vision Res. 25:163–170.

Fernando, M. M., and D. J. Grove. 1974. Melanophore aggregation in the plaice (*Pleuronectes platessa* L.). I. Changes in *in vivo* sensitivity to sympathomimetic amines. Comp. Biochem. Physiol. 48A(4):711–721.

Ferris, S. D. 1984. Tetraploidy and the evolution of catostomid fishes, p. 55–93. *In*: Evolutionary genetics of fishes, B. J. Turner (ed.). Plenum Press, New York.

Fierstine, H. L., and V. Walters. 1968. Studies in locomotion and anatomy of scombroid fishes. Mem. So. Calif. Acad. Sci. 6:1–31.

Filho, O. M., L. A. C. Bertollo, and P. M. G. Junior. 1980. Evidences for a multiple sex chromosome system with female heterogamety in *Apareiodon affinis* (Pisces, Parodontidae). Caryologia 33:83–89.

Fine, M. L. 1981. Mismatch between sound production and hearing in the oyster toadfish, p. 257–263. *In*: Hearing and sound communication in fishes, W. N. Tavolga, A. N. Popper and R. R. Fay (eds.). Springer-Verlag, New York.

Fineran, B. A., and J. A. C. Nicol. 1974. Studies on the eyes of New Zealand parrot fishes (*Labridae*). Proc. R. Soc. Lond. B186 (1084):217–247.

Finger, T. E. 1982. Somatotopy in the representation of the pectoral fin and free fin rays in the spinal cord of the sea robin, *Prionotus carolinus*. Biol. Bull. Mar. Biol. Lab., Woods Hole. 163:154–161.

Finger, T. R. 1982. Interactive segregation among three species of sculpins (*Cottus*). Copeia 1982:680–694.

Fingerman, M. 1965. Chromatophores. Physiol. Rev. 45:296–339.

Fink, S. V., and W. L. Fink. 1981. Interrelationships of the ostariophysan fishes (Teleostei). J. Linn. Soc. (Zool.) 72(4):297–353.

Fink, W. L. 1981. Ontogeny and phylogeny of tooth attachment modes in teleost fishes. J. Morphol. 167:167–184.

Fink, W. L. 1984a. Basal euteleosts: Relationships, p. 202–206. *In*: Ontogeny and systematics of fishes, H. G. Moser, W. J. Richards, D. M. Cohen, M. P. Fahay, A. W. Kendall, Jr. and S. L. Richardson (eds.). Am. Soc. Ichthyol. and Herpetol. Sp. Publ. 1.

Fink, W. L. 1984b. Stomiiforms: Relationships, p. 181–184. *In*: Ontogeny and systematics of fishes, H. G. Moser, W. J. Richards, D. M. Cohen, M. P. Fahay, A. W. Kendall, Jr. and S. L. Richardson (eds.). Am. Soc. Ichthyol. and Herpetol. Sp. Publ. 1.

Fink, W. L. 1984c. Salmoniforms: Introduction, p. 139. *In*: Ontogeny and systematics of fishes, H. G. Moser, W. J. Richards, D. M. Cohen, M. P. Fahay, A. W. Kendall, Jr. and S. L. Richardson (eds.). Am. Soc. Ichthyol. and Herpetol. Sp. Publ. 1.

Fink, W. L. 1985. Phylogenetic relationships of the stomiid fishes (Teleostei: Stomiiformes). Misc. Pub. Mus. Zool., The University of Michigan No. 171.

Fink, W. L., and S. V. Fink. 1979. Central Amazonia and its fishes. Comp. Biochem. Physiol. 63A:13–29.

Fink, W. L., and S. V. Fink. 1986. A phylogenetic analysis of the genus *Stomias,* including the synonymization of *Macrostomias*. Copeia 1986:494–503.

Fink, W. L., and S. H. Weitzman. 1982. Relationships of the stomiiform fishes (Teleostei) with a description of *Diplophos*. Bull. Mus. Comp. Zool. 150(2):31–93.

Fish, M. P., and W. H. Mowbray. 1970. Sounds of western North Atlantic fishes. Johns Hopkins Press, Baltimore and London.

Fishelson, L. 1974. Histology and ultrastructure of the recently found toxic gland in the fish *Meiacanthus nigrolineatus* (Blenniidae). Copeia 1974(2):386–392.

Fisher, W. L., and W. D. Pearson. 1987. Patterns of resource utilization among four species of darters in three central Kentucky streams, p. 69–76. *In*: Community and evolutionary ecology of North American stream fishes, W. J. Matthews and D. C. Heins (eds.) University of Oklahoma Press, Norman.

Fletcher, G. I. 1977. Circannual cycles of blood plasma freezing point and NA$^+$ and Cl$^-$ concentrations in Newfoundland winter flounder (*Pseudopleuronectes americanus*): Correlation with water temperature and photoperiod. Can. J. Zool. 55:789– 795.

Fletcher, G. L., D. R. Idler, A Vaisius, and C. L. Hew. 1989. Hormonal regulation of antifreeze protein gene expression in winter flounder. Fish Physiol. and Biochem. 7(1–4):387–393.

Foran, J. A., M. Cox, and D. Croxton. 1989. Sport fish consumption advisories and projected cancer risks in the Great Lakes basin. Am. J. Publ. Health 79(3):322–325.

Forey, P. L. 1973a. Relationships of elopomorphs, p. 351–368. *In*: Interrelationships of fishes, P. H. Greenwood, R. S. Miles and C. Patterson (eds.). Zool. J. Linn. Soc. 53 (Suppl. 1). Academic Press, New York.

Forey, P. L. 1973b. A revision of the elopiform fishes, fossil and Recent. Bull. Br. Mus. (Nat. Hist.) Geol. Suppl. 10:1–222.

Forey, P. L. 1980. *Latimeria*: A paradoxical fish. Proc. R. Soc. Lond. B208:369–384.

Forey, P. L. 1987. The Downtonian ostracoderm *Sclerodus* Agassiz (Osteostraci: Tremataspididae). Bull. Br. Mus. (Nat. Hist.) Geol. 41(1):1–30.

Forey, P. L. 1988. Golden jubilee for the coelacanth, *Latimeria chalumnae*. Nature 336:727–732.

Forey, P. L. 1991. *Latimeria chalumnae* and its pedigree, p. 75–97. *In*: The biology of *Latimeria chalumnae* and the evolution of coelacanths, J. A. Musick, M. N. Bruton and E. K. Balon (eds.). Env. Biol. Fish. 32(1–4).

Forey, P. L., and P. Janvier. 1993. Agnathans and the origin of jawed vertebrates. Nature 361:129–134.

Forey, P. L., and P. Janvier. 1994. Evolution of the early vertebrates. Am. Scientist 82:554–565.

Forster, R. P., and L. Goldstein. 1969. Formation of excretory products, p. 313–350. *In*: Fish physiology, Vol. I, W. S. Hoar and D. J. Randall (eds.). Academic Press, New York and London.

Foster, K. W. 1933. Color changes in *Fundulus* with special reference to the color changes of the iridosomes. Proc. Nat. Acad. Sci. 19: 535–540.

Fouchereau-Peron, M., Y. Arlot-Bonnemains, M. S. Mouktar, and G. Milhaud. 1986. Adaptation of rainbow trout (*Salmo gairdnerii*) to sea water: Changes in calcitonin levels. Comp. Biochem. Physiol. 82A(1):83–87.

Fouz, B., R. F. Conchas, A. E. Toranzo, and C. Amaro. 1992. *Vibrio damsela* strain virulence for fish and mammals. Fish Health Sect. Am. Fish. Soc. Newsletter 20(1):3–4.

Fowler, H. W. 1964–1973. A catalog of world fishes. Q. J. Taiwan Mus., Vols. 1–19.

Fox, D. L. 1957. The pigments of fishes, p. 367–385. *In*: The physiology of fishes, M. E. Brown (ed.). Academic Press, New York.

Fox, H. M., and H. G. Vevers. 1960. The nature of animal colours. Macmillan, New York.

Fox, P. J. 1978. Preliminary observations on different reproduction strategies in the bullhead (*Cottus gobio* L.) in northern and southern England. J. Fish. Biol. 12(1):5–11.

Franck, J. P. C., J. M. Wright, and B. J. Andrews. 1992. Genetic variability in a family of microsatellite DNAs in tilapia (Pisces; Cichlidae). Genome 35:719–725.

Frankel, O. H., and M. E. Soulé. 1981. Conservation and evolution. Cambridge University Press, Cambridge.

Fraser, T. H. 1972. Some thoughts about the teleostean fish concept—the Paracanthopterygii. Jpn. J. Ichthyol. 19:232–242.

Frazzetta, T. H. 1994. Feeding mechanisms in sharks and other elasmobranchs, p. 31–57. *In*: Advanced comparative environmental physiology 18: Biomechanics of feeding in vertebrates, V. L. Bels, M. Chardon, and P. Vandewalle (eds.). Springer-Verlag, Berlin.

Freihofer, W. C. 1970. Some nerve patterns and their systematic significance in paracanthopterygian, salmoniform, gobioid, and apogonid fishes. Proc. Cal. Acad. Sci. 38:215–264.

Freihofer, W. C. 1973. Trunk lateral line nerves, hyoid arch, gill rakers, and olfactory bulb location in atheriniform, mugiloid and percoid fishes. Occas. Pap. Cal. Acad. Sci. 95:1–31.

Fricke, H., O. Reinicke, H. Hofer, and W. Nachtigall. 1987. Locomotion of the coelacanth *Latimeria chalumnae* in its natural habitat. Nature 329:331–333.

Fries, E. F. B. 1958. Iridescent white reflecting chromatophores (Antaugophores) iridoleucophores in certain teleost fishes, particularly in *Bathygobius*. J. Morphol. 103:203–254.

Fritsche, T. R., R. L. Eastburn, L. H. Wiggins, and C. A. Terhune, Jr. 1989. Praziquantel for treatment of human *Nanophyetus salmincola* (*Troglotrema salmincola*) infection. J. Infect. Dis. 160(5):896–899.

Fritzsche, R. A. 1984. Gasterosteiformes: development and relationships, p. 398–405. *In*: Ontogeny and systematics of fishes, H. G. Moser, W. J. Richards, D. M. Cohen, M. P. Fahay, A. W. Kendall, Jr. and S. L. Richardson (eds.). Am. Soc. Ichthyol. and Herpetol. Sp. Publ. 1.

Fry, F. E. J. 1947. Effect of the environment on animal activity. University of Toronto, Biology Series, SS/Publ., Ontario Fish. Res. Lab. 68:1–62.

Fry, F. E. J. 1971. The effect of environmental factors on the physiology of fish, p. 1–98. *In*: Fish physiology, Vol. VI, W. S. Hoar and D. J. Randall (eds.). Academic Press, New York.

Fryer, G., and T. D. Iles. 1972. The cichlid fishes of the great lakes of Africa: Their biology and evolution. T. F. H. Publications, Neptune City, NJ.

Fujii, R. 1969. Chromatophores and pigments, p. 307–353. *In*: Fish physiology, Vol. III, W. S. Hoar and D. J. Randall (eds.). Academic Press, New York.

Fujii, R. 1993. Coloration and chromatophores, p. 535–562. *In*: The physiology of fishes, D. H. Evans (ed.). CRC Press, Boca Raton, FL.

Fujita, I., P. W. Sorenson, N. E. Stacey, and T. Hara. 1991. The olfactory system, not the terminal nerve, functions as the primary chemosensory pathway mediating responses to sex pheromones in male goldfish. Brain, Behav. Evol. 38:313–321.

Fukasawa, S., T. Suda, and S. Kubota. 1988. Identification of luminous bacteria isolated from the light organ of the fish *Acropoma japonicum*. Agric. Biol. Chem. 52(1):285–286.

Furukawa, A. 1976. Diet in yellowtail culture. Proc. 1st Int. Conf. Aquaculture Nutr., University of Delaware, p. 85–104.

Gall, G. 1983. Genetics of fish: A summary of conclusions. Aquaculture 33:383–394.

Gandolfi, G. 1969. A chemical sex attractant in the guppy *Poecilia reticulata* Peters (Pisces, Poeciliidae). Monitore Zool. Ital. 3:89–98.

Gans, C. 1993. Evolutionary origin of the vertebrate skull, p. 1–35. *In*: The skull, Vol. 2, J. Hanken and B. Hall (eds.). University of Chicago Press, Chicago and London.

Garcia-Fernandez, J., and P. W. H. Holland. 1994. Archetypal organization of the amphioxus *Hox* gene cluster. Nature 370:563–566.

Gardiner, B. G. 1970. Osteichthyes. McGraw-Hill Yearbook Science and Technology 1970:284–286.

Gardiner, B. G. 1973. Interrelationships of teleostomes, p. 105–135. *In*: Interrelationships of fishes, P. H. Greenwood, R. S. Miles and C. Patterson (eds.). Academic Press, New York.

Gardiner, B. G. 1984a. The relationships of the palaeoniscoid fishes, a review based on new specimens of *Mimia* and *Moythomasia* from the Upper Devonian of Western Australia. Bull. Br. Mus. (Nat. Hist.), Geol. 37:173–427.

Gardiner, B. G. 1984b. The relationships of placoderms. J. Vert. Paleont. 4(3):379–395.

Garrett, G. P. 1982. Variation in the reproductive traits of the Pecos pupfish *Cyprinodon pecoensis*. Amer. Midl. Nat. 108:355–363.

Gaudet, J-L. 1971. Report of the 1970 workshop on fish feed technology and nutrition. U. S. Fish and Wildl. Serv., Bur. of Sport Fish and Wildl., Resource Publ. 102.

Gee, J. H. 1968. Adjustment of buoyancy by longnose dace (*Rhinichthys cataractae*) in relation to velocity of water. J. Fish. Res. Bd. Can. 25:1485–1496.

Gee, J. H. 1972. Adaptive variation in swimbladder length and volume in dace, genus *Rhinichthys*. J. Fish. Res. Bd. Can. 29:119–127.

Gee, J. H. 1974. Behavioral and developmental plasticity of buoyancy in the longnose, *Rhinichthys cataractae,* and blacknose, *R. atratulus* (Cyprinidae) dace. J. Fish. Res. Bd. Can. 31:35–41.

Gemne, G., and K. B. Döving. 1969. Ultrastructural properties of primary olfactory neurons in fish (*Lota lota* L.). Am. J. Anat. 126(4): 457–476.

Geraudie, J., and F.-J. Meunier. 1980. Elastoidin actinotrichia in coelacanth fins: A comparison with teleosts. Tissue and Cell 12:637–645.

Gerber, G. P., and J. M. Haynes. 1988. Movements and behavior of smallmouth bass, *Micropterus dolomieui,* and rock bass, *Ambloplites rupestris*, in southcentral Lake Ontario and two tributaries. J. Freshwater Ecol. 4:425–440.

Gerking, S. D. 1958. The restricted movement of fish populations. Biol. Rev. 34:221–242.

Gerking, S. D. 1994. Feeding ecology of fishes. Academic Press, San Diego.

Gery, J. 1969. The fresh-water fishes of South America, p. 828–847. *In*: Biogeography and ecology in South America, Vol. 2, E. J. Fittkau, J.Illies, H. Klinge, G. H. Schwabe, and H. Sioli (eds.). Monographiae Biologicae, Vol. 19. Dr. W. Junk, The Hague.

Gery, J. 1977. Characoids of the world. T. F. H. Publications, Neptune City, NJ.

Gharrett, A. J., and J. E. Seeb. 1990. Practical and theoretical guidelines for genetically marking fish populations. Am. Fish. Soc. Symp. 7:407–417.

Gharrett, A. J., and F. M. Utter. 1982. Scientists detect genetic differences. Sea Grant Today 12(2):3–4.

Gharrett, A. J., S. M. Shirley, and G. R. Tromble. 1987. Genetic relationships among populations of Alaskan chinook salmon (*Oncorhynchus tshawytscha*). Can. J. Fish. Aquat. Sci. 43:765–774.

Ghiselin, M. T. 1974. A radical solution to the species problem. Syst. Zool. 23:536–544.

Ghittino, P. 1972. Aquaculture and associated diseases of fish of public health importance. J. Am. Vet. Med. Assoc. 161(11): 1476–1485.

Gibson, R. N. 1969. The biology and behavior of littoral fish. Oceanogr. Mar. Biol. Ann. Rev. 7:367–410.

Gibson, R. N. 1982. Recent studies on the biology of intertidal fishes. Oceanogr. Mar. Biol. Ann. Rev. 20:363–414.

Gibson, R. N., and I. A. Ezzi. 1985. Effect of particle concentration on filter- and particulate-feeding in the herring *Clupea harengus*. Mar. Biol. 88(2):109–116.

Gilbert, P. W. 1963. The visual apparatus of sharks, p. 283–326. *In*: Sharks and survival, P. W. Gilbert (ed.). D. C. Heath and Co., Boston.

Gill, H. S., A. H. Weatherly and T. Bhesania. 1982. Histochemical characterization of myotomal muscle in the bluntnose minnow, *Pimephales notatus* Rafinesque. J. Fish. Biol. 21:205–214.

Gill, H. S., A. H. Weatherly, R. Lee, and D. Legere. 1989. Histochemical characterization of myotomal muscle of five teleost species. J. Fish. Biol. 34(3):375–386.

Gill, T. N. 1891. The characteristics of the Dactylopteroidea. Proc. U.S. Nat. Mus. 13:243–248.

Ginberg, A. S. 1972. Fertilization in fishes and the problem of polyspermy. Acad. of Sciences of the USSR. Translated from Russian by Israel Program for Scientific Translations, Jerusalem.

Gjerde, B. 1993. Breeding and selection, p. 187–208. *In*: Salmon aquaculture, K. Heen, R. L. Monohan and F. Utter (eds.). Fishing News Books, Blackwell, Oxford.

Gjedrem, T. 1983. Genetic variation in quantitative traits and selective breeding in fish and shellfish. Aquaculture 33:51–72.

Gjosaeter, J., and K. Kawaguchi. 1980. A review of the world resources of mesopelagic fish. FAO Fish. Tech. Pap. 193:1–151.

Glass, C. W., C. S. Wardle, and W. R. Mojsiewicz. 1986. A light intensity threshold for schooling in the Atlantic mackerel *Scomber scombrus*. J. Fish. Biol. 29:71–82.

Glass, H. J., N. L. MacDonald, and J. R. Stark. 1987. Carbohydrate and protein digestion in Atlantic halibut. Comp. Biochem. Physiol. B 86B:281–289.

Gnyubkin, V. F. 1989. Response of pigmented cornea of whitespotted greenling to changes in illumination. Biol. Morya Vladivostok (1):25-32 (Engl. Abstr.)

Gnyubkin, V. F., and A. G. Gamburtseva. 1981. Morphological variation in the coloration of the cornea of the eye of a fish. J. Ichthyol. 21(1):175–181.

Gold, J. R. 1979. Cytogenetics, p. 353–405. *In*: Fish physiology, Vol. VII, W. S. Hoar, D. J. Randall and J. R. Brett (eds.). Academic Press, New York.

Gold, J., W. J. Karel, and M. R. Strand. 1980. Chromosome formulae of North American fishes. Prog. Fish-Cult. 42:10–23.

Goldman, B., and F. H. Talbot. 1976. Aspects of the ecology of coral reef fishes, p. 125–154. *In*: Biology and geology of coral reefs, Vol. IV, Biology 2, O. A. Jones and R. Endean (eds.). Academic Press, New York.

Goldschmidt, T. 1991. Egg mimics in haplochromine cichlids (Pisces, Perciformes) from Lake Victoria. Ethol. 88:177–190.

Goldspink, G. 1981. The use of muscles during flying, swimming, and running from the point of view of energy saving, p. 219–238. *In*: Vertebrate locomotion, M. H. Day (ed.). Symp. Zool. Soc. London. Academic Press, New York.

Goldstein, L., R. P. Forster, and G. M. Fanelli, Jr. 1964. Gill blood flow and ammonia excretion in the marine teleost, *Myxocephalus scorpius*. Comp. Biochem. Physiol. 12(4):489–499.

Gomahr, A., M. Palzenberger, and K. Kotrschal. 1992. Density and distribution of external taste buds in cyprinids. Env. Biol. Fish. 33(1):125–134.

Gon, O. and P. C. Heemstra (eds.). 1990. Fishes of the southern ocean. J. L. B. Smith Institute of Ichthyology, Grahamstown.

Goodman, M. L. 1990. Preserving the genetic diversity of salmonid stocks: A call for federal regulation of hatchery programs. Environ. Law 20:111–116.

Goodrich, E. S. 1909. Cyclostomes and fishes. *In*: A treatise on zoology, Part 9, First Fascicle, R. Lankester (ed.). Adam and Charles Black, London. Reprint ed. 1964, A. Asher, Amsterdam.

Goodrich, E. S. 1930. Studies on the structure and development of vertebrates. Constable and Co. London. Reprint ed. 1958, Dover Publ., New York.

Goody, P. C. 1969. The relationships of certain Upper Cretaceus teleosts with special reference to the myctophoids. Bull. Br. Mus. (Nat. Hist.) Geol. Suppl. 7:1–255.

Goodyear, C. P., and D. H. Bennett. 1979. Sun compass orientation of immature bluegill. Trans. Am. Fish. Soc. 108:555–559.

Gorbman, A., and K. Davey. 1991. Endocrines, p. 693–754. *In*: Comparative animal physiology (Fourth Edition). Neural and integrative animal physiology, C. L. Prosser (ed.). Wiley-Liss, New York.

Gorbman, A., and A. Tamarin. 1985. Early development of oral, olfactory, and adenohypophyseal structures of agnathans and its evolutionary implications, p. 165–185. *In*: Evolutionary biology of primitive fishes. NATO ASI Series A: Life Sciences 103.

Gorbman, A., W. W. Dickhoff, S. R. Vigna, N. B. Clark, and C. L. Ralph. 1983. Comparative endocrinology. Wiley-Interscience, New York.

Gordon, D. J., D. F. Markle, and J. E. Olney. 1984. Ophidiiformes: Development and relationships, p. 308–319. *In*: Ontogeny and systematics of fishes, H. G. Moser, W. J. Richards, D. M. Cohen, M. P. Fahay, A. W. Kendall, Jr. and S. L. Richardson (eds.). Am. Soc. Ichthyol. and Herpetol. Sp. Publ. 1.

Gordon, M. 1946. Interchanging genetic mechanisms for sex determination in fishes under domestication. J. Hered. 37:307–320.

Gordon, M. S., I. Boetius, D. H. Evans, R. McCarthy, and L. C. Oglesby. 1969. Aspects of the physiology of terrestrial life in amphibious fishes. 1. The mudskipper, *Periophthalmus sobrinus*. J. Exp. Biol. 50:141–149.

Gorlick, D. L., P. D. Atkins, and G. S. Losey. 1987. Effect of clean- ing by *Labroides dimidiatus* (Labridae) on an ectoparasite population infecting *Pomacentrus vaiuli* (Pomacentridae) at Enewetok Atoll. Copeia 1987:41–45.

Gorr, T., and T. Kleinschmidt. 1993. Evolutionary relationships of the coelacanth. Am. Scientist 81(1):72–82.

Gosline, W. A. 1959. Four new species, a new genus, and a new suborder of Hawaiian fishes. Pacific Sci. 13:67–77.

Gosline, W. A. 1960a. Hawaiian lava-flow fishes. Part IV. *Snyderidia canina* Gilbert, with notes on the osteology of ophidioid families. Pacific Sci. 14:373–381.

Gosline, W. A. 1960b. Contributions toward a classification of modern isospondylous fishes. Bull. Br. Mus. (Nat. Hist.) Zool. 6:325–365.

Gosline, W. A. 1961. Some osteological features of modern lower teleostean fishes. Smithsonian Misc. Coll. 142(3):1–42.

Gosline, W. A. 1963. Considerations regarding the relationships of the percopsiform, cyprinodontiform, and gadiform fishes. Occas. Pap. Mus. Zool., Univ. Mich. (629):1–38.

Gosline, W. A. 1965. Teleostean phylogeny. Copeia 1965:186–194.

Gosline, W. A. 1968. The suborders of perciform fishes. Proc. U. S. Nat. Mus. 124:1–78.

Gosline, W. A. 1969. The Morphology and systematic position of the alepocephaloid fishes. Bull Br. Mus. (Nat. Hist.) Zool. 18(6):183–218.

Gosline, W. A. 1970. A reinterpretation of the teleostean fish order Gobiesociformes. Proc. Calif. Acad. Sci., Ser. 4. 38(19):363–382.

Gosline, W. A. 1971. Functional morphology and classification of teleostean fishes. University Press of Hawaii, Honolulu.

Gosline, W. A. 1973. Considerations regarding the phylogeny of cypriniform fishes, with special reference to structures associated with feeding. Copeia 1973:761–776.

Gosline, W. A. 1980. The evolution of some structural systems with reference to the interrelationships of modern lower teleostean fish groups. Jpn. J. Ichthyol. 27:1–28.

Gosline, W. A. 1983. The relationships of the mastacembelid and synbranchid fishes. Jpn. J. Ichthyol. 29:323–328.

Gosline, W. A. 1985. A possible relationship between aspects of dentition and feeding in the centrarchid and anabantoid fishes. Env. Biol. Fish. 12(3):161–168.

Gosline, W. A. 1987. Jaw structures and movement in higher teleostean fishes. Jpn. J. Ichthyol. 34(1):21–32.

Gosline, W. A., N. B. Marshall, and G. W. Mead. 1966. Order Iniomi. Characters and synopsis of families, p. 1–18. *In*: Fishes of the western North Atlantic. Mem. Sears Found. Mar. Res. 1(5).

Goulding, M. 1980. The fishes and the forest: Explorations in Amazonian natural history. University of California Press, Berkeley.

Grady, J. M., R. C. Cashner, and J. S. Rogers. 1990. Evolutionary and biogeographic relationships of *Fundulus catenatus* (Fundulidae). Copeia 1990:315–323.

Graham, J. B., and R. H. Rosenblatt. 1970. Aerial vision: Unique adaptation in an intertidal fish. Science 168:586–588.

Grande, L. 1982. A revision of the fossil genus *Diplomystus* with comments on the interrelationships of clupeomorph fishes. Amer. Mus. Nov. 2728:1–34.

Grande, L., and W. E. Bemis. 1991. Osteology and phylogenetic relationships of fossil and recent paddlefishes (Polyodontidae) with comments on the interrelationships of Acipenseriformes. Soc. Vert. Paleo. Mem. No. 1.

Grant, W. S., and F. M. Utter. 1984. Biochemical population genetics of Pacific herring (*Clupea pallasi*). Can. J. Fish. Aquat. Sci. 41:856–864.

Grasse, P. P. (ed.). 1958. Agnathes et poissons: Anatomie, ethologie, systematique. *In*: Traite de zoologie, Vol. 13, 3 parts. Masson et Cie, Paris.

Graves, J. E., S. D. Ferris, and A. E. Dizon. 1984. Close genetic similarity of Atlantic and Pacific skipjack tuna (*Katsuwonus pelamis*) demonstrated with restriction endonuclease analysis of mitochondrial DNA. Mar. Biol. 79:315–319.

Green, J. M. 1971. High tide movements and homing behavior of the tidepool sculpin, *Oligocottus maculosus*. J. Fish Res. Bd. Can. 28:383–389.

Green, J. M. 1975. Restricted movements and homing of the cunner, *Tautogolabrus adspersus* (Walbaum) (Pisces: Labridae). Can. J. Zool. 53:1427–1431.

Greene, C. W., and C. H. Greene. 1914. The skeletal musculature of the king salmon. Bur. Comm. Fish. Doc. 796 (issued Aug. 9, 1914; republished 1915 in Bur. Comm. Fish. Vol. 33 for 1913).

Greenwood, P. H. 1968. The osteology and relationships of the Denticipitidae, a family of clupeomorph fishes. Bull. Br. Mus. (Nat. Hist.) Zool. 16:213–273.

Greenwood, P. H. 1974. The cichlid fishes of Lake Victoria, East Africa: The biology and evolution of a species flock. Bull Br. Mus. (Nat. Hist.) Zool. Supp. 6:1–134.

Greenwood, P. H. 1977. Notes on the anatomy and classification of elopomorph fishes. Bull. Br. Mus. (Nat. Hist.) Zool. 32:65–102.

Greenwood, P. H. 1983. The zoogeography of African freshwater fishes: Bioaccountancy or biogeography? p. 179–199. *In*: Evolution, time and space: The emergence of the biosphere, R. W. Sims, J. H. Price and P. E. S. Whalley (eds.). Academic Press, London, New York.

Greenwood, P. H., 1987. The natural history of lungfishes, p. 163-179. *In*: The biology and evolution of lungfishes, W. E. Bemis, W. W. Burggren, and N. E. Kemp (eds.). Alan R. Liss, Inc., New York.

Greenwood, P. H., and G. V. Lauder. 1981. The protractor pectoralis muscle and the classification of teleost fishes. Bull. Br. Mus (Nat. Hist.) Zool. 41(4):213–234.

Greenwood, P. H., and D. E. Rosen. 1971. Notes on the structure and relationships of the alepocephaloid fishes. Am. Mus. Nov. 2473:1–41.

Greenwood, P. H., and K. S. Thomson. 1960. The pectoral anatomy of *Pantodon buchholzi* Peters (a freshwater flying fish) and the related Osteoglossidae. Proc. Zool. Soc. London 35:283–301.

Greenwood, P. H., R. S. Miles, and C. Patterson (eds.). 1973. Interrelationships of fishes. Academic Press, New York.

Greenwood, P. H., G. S. Myers, D. E. Rosen, and S. H. Weitzman. 1967. Named main divisions of teleostean fishes. Proc. Biol. Soc. Wash. 80:227–228.

Greenwood, P. H., D. E. Rosen, S. H. Weitzman, and G. S. Myers. 1966. Phyletic studies of teleostean fishes, with a provisional classification of living forms. Bull. Am. Mus. Nat. Hist. 131: 339–456.

Greer-Walker, M., and G. A. Pull. 1975. A survey of red and white muscle in marine fish. J. Fish Biol. 7:295–300.

Gregory, W. K. 1933. Fish skulls. Trans Am. Phil. Soc. 23:75–481 (offset reprinted. 1959. Eric Lundberg Publ. Co., Laurel, FL.)

Grey, M. 1956. The distribution of fishes found below a depth of 2000 meters. Fieldiana (Zool.) 36(2):75–337.

Grier, H. J. 1981. Cellular organization of the testis and spermatogenesis in fishes. Am. Zool. 21:345–357.

Griffith, R. W., and P. K. T. Pang. 1979. Mechanisms of osmoregulation in the coelacanth: Evolutionary implications, p. 79–93. *In*: The biology and physiology of the living coelacanth, J. E. McCosker and M. D. Lagios (eds.). Occ. Pap. Calif. Acad. Sci. (134).

Griffith, R. W., B. L. Umminger, B. F. Grant, P. K. T. Pang, and G. E. Pickford. 1974. Serum composition of the coelacanth, *Latimeria chalumnae* Smith. J. Exp. Zool. 187:87–102.

Grizzle, J. M., and W. A. Rogers. 1976. Anatomy and histology of the channel catfish. Auburn University Agri. Exper. Sta., Auburn, AL.

Grobecker, D. B., and T. W. Pietsch. 1979. High-speed cinematographic evidence for ultrafast feeding in antennariid angler fishes. Science 205:1161–1162.

Groot, C., and L. Margolis (eds.). 1991. Pacific salmon life histories. UBC Press, Vancouver.

Grossman, G. D. 1982. Dynamics and organization of a rocky intertidal fish assemblage: The persistence and resilience of taxocene structure. Am. Nat. 119:611–637.

Grossman, G. D., J. R. Dowd, and M. Crawford. 1990. Assemblage stability in stream fishes: A review. Environ. Manage. 14(5): 661–671.

Gruber, J. B., D. K, Hamasaki, and B. L. Davis. 1975. Window to the epiphysis in sharks. Copeia 1975:378–380.

Gruber, S. H., and J. F. Cohen. 1978. Visual system of the elasmobranchs: State of the art 1960–1975, p. 11–105. *In*: Sensory biology of sharks, skates and rays, E. S. Hodgson and R. F. Mathewson (eds.). U. S. Navy, Arlington, VA.

Gubanov, Y. P. 1972. On the biology of the thresher shark (*Alopias vulpinus* Bonnaterre) in the northwest Indian Ocean. J. Ichthyol. 12(4):591–600.

Gudger, E. W. 1930. The candiru: the only vertebrate parasite of man. P. B. Hoeber, Inc., New York.

Guerrero, R. D. 1979. Use of hormonal steroids for artificial sex reversal of *Tilapia*, p. 512–514. *In*: Symposium on hormonal steroids in fish, B. I. Sundararaj and S. V. Goswami (eds.). New Delhi, Ind. Nat. Acad. Sci. (also listed as Proc. Ind. Nat. Acad. Sci. B45[5]).

Guillette, L. J., Jr. 1987. The evolution of viviparity in fishes, amphibians and reptiles: An endocrine approach, p. 523–562. *In*: Hormones and reproduction in fishes, amphibians, and reptiles, D. O. Norris and R. E. Jones (eds.). Plenum Press, New York and London.

Guillory, V., and W. E. Johnson. 1986. Habitat, conservation status, and zoogeography of the cyprinodont fish *Cyprinodon variegatus hubbsi* (Carr). Southwestern Nat. 31:95–100.

Gulland, J. A. 1983. Fish stock assessment, Vol. 1. John Wiley and Sons, New York.

Gulland, J. A. (ed.) 1988. Fish population dynamics: Implications for management. Wiley & Sons, New York.

Gunderson, D. R. 1993. Surveys of fisheries resources. John Wiley & Sons, Inc., New York, Chichester, Brisbane, Toronto, Singapore.

Gunther, A. C. L. G. 1880. An introduction to the study of fishes. A. and C. Black, Edinburgh.

Gupta, S. 1974. Observations on the reproductive biology of *Mastacembelus armatus* (Lacepede). J. Fish Biol. 6:13–21.

Guthrie, D. M. 1986. Role of vision in fish behavior, p. 75–113. *In*: The behavior of teleost fishes, T. J. Pitcher (ed.). Johns Hopkins Univ. Press, Baltimore.

Haedrich, R. L. 1967. The stromateoid fishes: Systematics and a classification. Bull. Mus. Comp. Zool., Harv. Univ. 135:31–319.

Haedrich, R. L. 1983. Estuarine fishes, p. 183–207. *In*: Ecosystems of the world 26, Estuaries and enclosed seas, B. H. Ketchum (ed.). Elsevier, New York.

Hafeez, M. 1971. Light microscopic studies on the pineal organ in teleost fishes with special regard to its function. J. Morphol. 134:281–313.

Hagedorn, M., and R. Zelick. 1989. Relative dominance among males is expressed in the electric organ discharge of a weakly electric fish. Anim. Behav. 38:520–525.

Hagedorn, M., M. Womble, and T. E. Finger. 1990. Synodontid catfish: A new group of weakly electric fish. Brain, Behav. Evolution. 35(5):268–277.

Haglund, T. R., D. G. Buth, and R. Lawson. 1993. Allozyme variation and phylogenetic relationships of Asian, North American, and European populations of the ninespine stickleback, *Pungitius pungitius*, p. 438–452. *In*: Systematics, historical ecology, and North American freshwater fishes, R. S. Mayden (ed.). Stanford University Press, Stanford.

Hale, R. C., J. Greaves, J. L. Gunderson, and R. F. Mothershead II. 1991. Occurrence of organochlorine contaminants in tissues of the coelacanth *Latimeria chalumnae*, p. 361–367. *In*: The biology of *Latimeria chalumnae* and the evolution of coelacanths, J. A. Musick, M. N. Bruton and E. K. Balon (eds.). Environ. Biol. Fish. 32(1–4).

Halstead, B. W. 1970. Poisonous and venemous marine animals of the world, Vol. 3. U.S. Government Printing Office, Washington, DC.

Halstead, B. W. 1978. Poisonous and venomous marine animals of the world (Revised Edition). Darwin Press, Princeton, NJ.

Halstead, B. W. 1988. Poisonous and venomous marine animals of the world (Second Revised Edition). Darwin Press, Princeton, NJ.

Halstead, B. W. 1992. Dangerous aquatic animals of the world. A color atlas. (In collaboration with P. S. Auerbach.) Darwin Press, Princeton, NJ.

Halstead, B. W., and J. M. Vinci. 1988. Biology of poisonous and venomous marine animals, p. 1–30. *In*: Handbook of natural toxins, vol. 3, A. Tu (ed.). Marcel Dekker Inc., New York and Basel.

Halstead, B. W., P. S. Auerbach, and D. Campbell. 1990. A colour atlas of dangerous marine animals. Wolfe Medical Publications Ltd., London.

Halstead, L. B. 1982. Evolutionary trends and the phylogeny of the Agnatha, p. 159–196. *In*: Problems of phylogenetic reconstruction, K. A. Joysey and A. E. Friday (eds.). Systematics Assoc. Sp. Vol. 21.

Halver, J. E. (ed.). 1972. Fish nutrition. Academic Press, New York.

Halver, J. E. 1976. Formulating practical diets of fish. J. Fish. Res. Bd. Can. 33:1032–1039.

Halver, J. E. 1989. The vitamins, p. 31–109. *In*: Fish nutrition (Second Edition), J. E. Halver (ed.). Academic Press, San Diego, London.

Hamilton, W. J., III, and R. M. Peterman. 1971. Countershading in the colourful reef fish, *Chaetodon lunula*: Concealment, communication or both. Anim. Behav. 19(2):357–364.

Hamlett, W. C. 1989. Evolution and morphogenesis of the placenta in sharks, p. 35–52. *In*: Evolutionary and contemporary biology of elasmobranchs, W. C. Hamlett and B. Toa (eds.). J. Expt. Zool. Suppl. 2. Liss, New York.

Haneda, Y. 1986. On a new type of luminous fishes and squids, ingested luminescence, p. 838–839. *In*: Indo-Pacific fish biology, T. Uyeno, R. Arai, T. Taniuchi, and K. Matsuura (eds.). Proc. 2nd Internat. Congr. on Indo-Pac. Fishes. Ichthyol. Soc. Japan, Tokyo.

Hanken, J., and B. Hall (eds.) 1993. The skull, Vols. 1, 2, and 3. University of Chicago Press, Chicago and London.

Hansen, L. P., W. C. Clarke, R. L. Saunders, and J. E. Thorpe. 1989. Salmonids smoltification III: Special issue, Aquaculture 82:1–390.

Hanson, M. 1993. Chemoreception, p. 191–218. *In*: The physiology of fishes, D. H. Evans (ed.). CRC Press, Boca Raton.

Hanson, M., H. Westerberg, and M. Oblad. 1990. The role of magnetic statoconia in dogfish (*Squalus acanthias*). J. Exp. Biol. 151:205–218.

Hara, T. J. 1971. Chemoreception, p. 79–120. *In*: Fish physiology, Vol. V, W. S. Hoar and D. J. Randall (eds.). Academic Press, New York.

Hara, T. J. 1975. Olfaction in fish. Prog. Neurobiol. 5:271–335.

Hara, T. J. 1992a. Fish chemoreception. Chapman and Hall, London.

Hara, T. J. 1992b. Overview and introduction, p. 1–12. *In*: Fish chemoreception, T. J. Hara (ed.). Chapman and Hall, London.

Hara, T. J. 1992c. Mechanisms of olfaction, p. 150–170. *In*: Fish chemoreception, T. J. Hara (ed.). Chapman and Hall, London.

Hara, T. J. 1993. Chemoreception, p. 191–218. *In*: The physiology of fishes, D. H. Evans (ed.). CRC Press, Boca Raton.

Hara, T. J., and B. Zielinski. 1989. Structural and functional development of the olfactory organ in teleosts. Trans. Am. Fish. Soc. 118(2): 183–194.

Harden–Jones, F. R. 1963. The reaction of fish to moving backgrounds. J. Exp. Biol. 40(3):437–446.

Hardisty, M. W. 1971. Gonadogenesis, sex differentiation and gametogenesis, p. 295–359. *In*: The biology of lampreys, Vol. 1, M. W. Hardisty and I. C. Potter (eds.). Academic Press, London, New York.

Hardisty, M. W. 1979. Biology of the cyclostomes. Chapman and Hall, London.

Hardisty, M. W., and I. C. Potter (eds.). 1971a. The biology of lampreys, Vol. 1. Academic Press, London.

Hardisty, M. W., and I. C. Potter. 1971b. The behaviour, ecology and growth of larval lampreys, p. 85–125. *In*: The biology of lampreys, M. W. Hardisty and I. C. Potter (eds.). Academic Press, London, New York.

Hardisty, M. W., and I. C. Potter. 1971c. The general biology of adult lampreys, p. 127–206. *In*: The biology of lampreys, Vol. 1, M. W. Hardisty and I. C. Potter (eds.). Academic Press, London, New York.

Hardisty, M. W., and I. C. Potter. 1971d. Paired species, p. 249–277. *In*: The biology of lampreys, Vol. 1. Academic Press, London and New York.

Hardisty, M. W., and I. C. Potter. 1972. The biology of lampreys, Vol. 2. Academic Press, London.

Hardisty, M. W., and I. C. Potter. 1981. The biology of lampreys, Vol. 3. Academic Press, London.

Harkness, L., and H. C. Bennet-Clark. 1978. The deep fovea as a focus indicator. Nature 272(5656):814–816.

Harless, J., R. S. Nairn, R. Svensson, K. D. Kallman, and D. C. Morizot. 1991. Mapping of two thyroid hormone receptor-related (*erb*A-like) DNA sequences to linkage groups U4 and XII of *Xiphophorus* fishes (Poeciliidae). J. Hered. 82:256–259.

Harless, J., R. Svensson, K. D. Kallman, D. C. Morizot, and R. S. Nairn. 1990. Assignment of an *erb*B-like DNA sequence to linkage group VI in fishes of the the the genus *Xiphophorus* (Poeciliidae). Cancer Genet. Cytogenet. 50:45–51.

Harosi, F. I., and Y. Hashimoto. 1983. Ultraviolet visual pigment in a vertebrate: A tetrachromatic cone system in the dace. Science, Wash. 222(4627):1021–1023.

Harrell, L. W., and T. L. Deardorf. 1990. Human nanophyetiasis: Transmission by handling naturally infected coho salmon (*Oncorhynchus kisutch*). J. Infect. Dis. 16(1):146–148.

Harrington, R. W., Jr. 1971. How ecological and genetic factors interact to determine when self-fertilizing hermaphrodites of *Rivulus marmoratus* change into functional secondary males with reappraisal of the modes of intersexuality among fishes. Copeia 1971:389–432.

Harris, A. S., S. Bieger, R. W. Doyle, and J. M. Wright. 1991. DNA fingerprinting of tilapia, *Oreochromis niloticus*, and its application to aquaculture genetics. Aquacult. 92:157–163.

Harris, G. W., and B. T. Donovan. 1966. The pituitary gland. University of California Press, Berkeley and Los Angeles.

Harrison, C. M. H., and G. Palmer. 1968. On the neotype of *Radiicephalus elongatus* Osorio with remarks on its biology. Bull. Br. Mus. (Nat. Hist.) Zool. 16(5):187–211.

Hart, J. L. 1973. Pacific fishes of Canada. Bull. 180. Fish. Res. Bd. Can., Ottawa.

Harvey, B. 1993. Cryopreservation of fish spermatozoa, p. 175–179. *In*: Genetic conservation of salmonoid fishes, J. G. Cloud and G. H. Thorgaard (eds.). Plenum Press, New York, London.

Harvey, E. N. 1940. Living light. Princeton University Press, Princeton, NJ.

Hashimoto, Y. 1979. Marine toxins and other bioactive marine metabolites. Jpn. Scient. Soc. Press, Tokyo.

Hasler, A. D., and A. T. Scholz. 1978. Olfactory imprinting in coho salmon, p. 356–369. *In*: Animal migration, navigation, and homing, K. Schmidt-Koenig and W. T. Keeton (eds.). Springer-Verlag, Berlin.

Hasler, A. D., and A. T. Scholz. 1983. Olfactory imprinting and homing in salmon. Investigations into the mechanism of the imprinting process. Springer-Verlag, Berlin.

Hasler, A. D., and W. J. Wisby. 1951. Discrimination of stream odors by fishes and relation to parent stream behavior. Am. Nat. 85:223–238.

Hasler, A. D., and W. J. Wisby. 1958.The return of displaced largemouth bass and green sunfish to a "home" area. Ecol. 39:289–293.

Hastings, J. W., and J. G. Morin. 1991. Bioluminescence, p. 131–170. *In*: Neural and integrative animal physiology: Comparative animal physiology (Fourth Edition), C. L. Prosser (ed.). Wiley-Liss, New York.

Hattori, S. 1970. Reproductive aspects of fish resources, p. 209–222. *In*: Ocean developments. 4. Exploitation of fisheries resources, T. Sasaki (ed.). Ocean Dev. Centy. Press, Tokyo. (In Japanese.)

Hawkes, J. W. 1974. The structure of fish skin. 2. The chromatophore unit. Cell. Tissue Res. 149(2):159–172.

Hawkins, A. D. 1981. The hearing abilities of fish, p. 109–133. *In*: Hearing and sound communication in fish, W. N. Tavolga, A. N. Popper and R. R. Fay (eds.). Springer-Verlag, New York.

Hawkins, A. D. 1986. Underwater sound and fish behavior, p. 129–169. *In*: The behavior of teleost fishes, T. J. Pitcher (ed.). Johns Hopkins Press, Baltimore.

Hawryshyn, C. W. 1991. Light adaptation properties of the ultraviolet-sensitive cone mechanism in comparison to the other receptor mechanisms of gold fish. Visual Neurosci. 6:293–301.

Hawryshyn, C. W. 1992. Polarization vision in fish. Amer. Sci. 80(2):164–175.

Hawryshyn, C. W., and W. N. McFarland. 1987. Cone photoreceptor mechanisms and the detection of polarized light in fish. J. Comp. Physiol. A 160:459–465.

Hayes, F. R. 1948. The growth, general chemistry, and temperature relations of salmonid eggs. Q. Rev. Biol. 24:281–308.

Healey, M. C. 1991. Life history of chinook salmon, p. 311–393. *In*: Pacific salmon life histories, C. Groot and L. Margolis (eds.). UBC Press, Vancouver.

Healey, M. C., and R. Prieston. 1973. The interrelationships among individuals in a fish school. Fish. Res. Bd. Can. Tech. Rep. No. 389.

Healy, E. G. 1972. The central nervous system, p. 307–372. *In*: The biology of lampreys, Vol. 2, M. W. Hardisty and I. C. Potter (eds.). Academic Press, London.

Heath, M. 1989. Transport of larval herring (*Clupea harengus* L.) by the Scottish coastal current, p. 85–91. *In*: The early life history of fish, J. H. S. Blaxter, J. C. Gamble and H. v. Westernhagen (eds.). Rapp. P.-v. Reun. Cons. Int. Explor. Mer 191.

Heczko, E. J., and B. H. Seghers. 1981. Effects of alarm substance on schooling in the common shiner (*Notropis cornutus* Cyprinidae), p. 25–29. *In*: Ecology and ethology of fishes, D. L. G. Noakes and J. A. Ward (eds.). Junk, The Hague.

Heemstra, P. C. 1980. A revision of the zeid fishes (Zeiformes: Zeidae) of South Africa. J. L. B. Smith Inst., Ichthyol. Bull. 41.

Heen, K., R. L. Monahan, and F. Utter. 1993. Salmon aquaculture. Fishing News Books, Oxford.

Heierhorst, J., K. Leideris, and D. Richter. 1993. Vasotocin neuropeptide precursors and genes of teleost and jawless fish, p. 339–356. *In*: Biochemistry and molecular biology of fish, Vol. 2: Molecular biology frontiers, P. W. Hochachka and T. P. Mommsen (eds.). Elsevier, Amsterdam.

Heiligenberg, W. 1993. Electrosensation, p. 137–160. *In*: The physiology of fishes, D. H. Evans (ed.). CRC Press, Boca Raton.

Heiligenberg, W., and G. Rose. 1985. Neural correlates of the jamming avoidance response (JAR) in the weakly electric fish *Eigenmannia*. Trends-Neurosci. 8(10):442–449.

Heins, D. C. 1979. A comparative life history of a closely related group of minnows (*Notropis*: Cyprinidae) inhabiting streams of the Gulf Coastal plain. Ph.D. Dissertation, Tulane University, New Orleans.

Heins, D. C., and J. A. Baker. 1987. Analysis of factors associated with intraspecific variation in propagule size of a stream-dwelling fish, p. 223–231. *In*: Community and evolutionary ecology of North American stream fishes, W. J. Matthews and D. C. Heins (eds.). University of Oklahoma Press, Norman.

Helfman, G. S. 1986. Fish behavior by day, night and twilight, p. 366–387. *In*: The behavior of teleost fishes, T. Pitcher (ed.). Johns Hopkins University Press, Baltimore.

Helm, T. 1976. Dangerous sea creatures. Funk and Wagnalls, New York.

Hemmingson, E. A. 1991. Respiratory and cardiovascular adaptations in hemoglobin-free fish: Resolved and unresolved problems, p. 191–203. *In*: Biology of antarctic fish, G. di Prisco, B. Maresca and B. Tota (eds.). Springer-Verlag, Berlin, Heidelberg.

Hemmingson, E. A., and E. L. Douglas. 1972. Respiratory and circulatory responses in a hemoglobin-free fish, *Chaenocephalus aceratus*, to changes in temperature and oxygen tension. Comp. Biochem. Physiol. 43A(4A):1031–1043.

Henderson, I. W., and I. C. Jones. 1973. Hormones and osmoregulation in fishes. Ann. Inst. Michel Pacha 5(2):69–235.

Hennig, W. 1967. Phylogenetic systematics. Illinois University Press, Urbana, Illinois. Translated by D. D. Davis and R. R. Zangerl.

Hensley, D. A. 1986. Current research on Indo-Pacific bothids, [Abstract] p. 941. *In*: Indo-Pacific fish biology, T. Uyeno, R. Arai, T. Taniuchi, and K. Matsubara (eds.). The Ichthyol. Soc. of Japan, Tokyo.

Hensley, D. A., and E. H. Ahlstrom. 1984. Pleuronectiformes: relationships, p. 670–687. *In*: Ontogeny and systematics of fishes, H. G. Moser, W. J. Richards, D. M. Cohen, M. P. Fahay, A. W. Kendall, Jr. and S. L. Richardson (eds.). Am. Soc. Ichthyol. and Herpetol. Sp. Pub. 1.

Herald, E. S. 1961. Living fishes of the world. Doubleday, Garden City, NY.

Herbold, B. 1984. Structure of an Indiana stream fish association: Choosing an appropriate model. Amer. Nat. 124:561–572.

Herke, S. W., I. Kornfield, P. Moran, and J. R. Moring. 1990. Molecular confirmation of hybridization between northern pike (*Esox lucius*) and chain pickerel (*E. niger*). Copeia 1990:846–850.

Herring, P. J. 1967. The pigments of plankton at the sea surface. Aspects of marine zoology, Symp. Zool. Soc. Lond. (19):215–235.

Herring, P. J. 1977. Bioluminescence of marine organisms. Nature 276:788–793.

Herring, P. J. 1982. Aspects of the bioluminescence of fishes. Oceanogr. Mar. Biol. Ann. Rev. 20:415–470.

Herring, P. J. 1990. Bioluminescent communication in the sea, p. 245–264. *In*: Light and life in the sea, P. J. Herring, A. K. Campbell, M. Whitfield and L. Maddox (eds.). Cambridge University Press, Cambridge.

Herring, P. J. 1992. Bioluminescence of the oceanic apogonid fishes *Howella brodiei* and *Florenciella lugubris*. J. Mar. Biol. Assoc. U.K. 72(1):139–149.

Herring, P. J., and J. G. Morin. 1978. Bioluminescence in fishes, p. 273–329. *In*: Bioluminescence in action, P. J. Herring (ed.). Academic Press, London.

Herring, P. J., A. K. Campbell, M. Whitfield, and L. Maddock. 1990. Light and life in the sea. Cambridge University Press, Cambridge.

Hettler, W. F. 1984. Description of eggs, larvae, and early juveniles of gulf menhaden, *Brevoortia patronus*, and comparisons with Atlantic menhaden, *B. tyrannus*, and yellowfin menhaden, *B. smithi*. Fish. Bull. U. S. 82:85–95.

Heuter, R. E., and P. W. Gilbert. 1991. The sensory world of sharks, p. 48–55. *In*: Discovering sharks, S. H. Gruber (ed.). Am. Litt. Soc., Highlands, NJ.

Hickman, C. P., and B. F. Trump. 1969. The kidney, p. 91–239. *In*: Fish physiology, Vol. I, W. S. Hoar and D. J. Randall (eds.). Academic Press, New York.

Higashi, G. I. 1985. Foodborne parasites transmitted to man from fish and other aquatic foods. Food Technology, March 1985: 69–74; 111–112.

Hilborn, R., and C. J. Walters. 1992. Quantitative fisheries stock assessment: Choice, dynamics, and uncertainty. Chapman and Hall, New York.

Hildebrand, M. 1982. Analyses of vertebrate structure, (Second Edition). Wiley, New York.

Hildenberg, M. 1988. Analysis of vertebrate structure, (Third Edition). Wiley, New York.

Hile, R., P. H. Eschmeyer, and G. F. Langer. 1951. Decline of the lake trout fishery in Lake Michigan. U. S. Fish and Wildl. Serv. Fish. Bull. 52:77–95.

Hilgerd, T. 1983. Ciguatera food poisoning: A circum-tropical fisheries problem, p. 1–7. *In*: Natural toxins and human pathogens in the marine environment, R. Colwell (ed.). Maryland Sea Grant Pub.

Hillyard, S. D. 1981. Energy metabolism and osmoregulation in desert fishes, p. 385–409. *In*: Fishes in North American deserts, R. J. Naiman and D. L. Soltz (eds.). John Wiley and Sons, New York.

Hindar, K., N. Ryman, and F. Utter. 1991. Genetic effects of cultured fish on natural fish populations. Can. J. Fish. Aquat. Sci. 48:945–957.

Hinegardner, R., and D. E. Rosen. 1972. Cellular DNA content and the evolution of teleostean fishes. Am. Nat. 106:621–644.

Hlohowskyj, I., and T. E. Wissing. 1986. Substrate selection by fantail (*Etheostoma flabellare*), greenside (*E. blennioides*) and rainbow (*E. caeruleum*) darters. Ohio J. Sci. 86:124–129.

Hoar, W. S. 1969. Reproduction, p. 1–72. *In*: Fish physiology, Vol. III. W. S. Hoar and D. J. Randall (eds.). Academic Press, New York.

Hoar, W. S. 1976. Smolt transformation: Evolution, behavior, and physiology. J. Fish. Res. Bd. Can. 33:1234–1252.

Hoar, W. S., and D. J. Randall (eds.). 1969. Fish physiology, Vol. II: The endocrine system. Academic Press, New York.

Hoar, W. S., and D. J. Randall (eds.). 1970. Fish physiology, Vol. IV: The nervous system, circulation, and respiration. Academic Press, New York.

Hoar, W. S., and D. J. Randall (eds.). 1978. Fish physiology, Vol. VII: Locomotion. Academic Press, New York.

Hoar, W. S., and D. J. Randall (eds.). 1984. Fish physiology, Vol. X: Gills, Parts A and B. Academic Press, Orlando.

Hoar, W. S., and D. J. Randall (eds.). 1988. Fish physiology, Vol. XI. Academic Press, San Diego.

Hoar, W. S., D. J. Randall, and A. P. Farrell (eds.). 1988. Fish physiology, Vol. XII. Academic Press, San Diego.

Hoar, W. S., D. J. Randall, and J. R. Brett (eds.). 1979. Fish physiology, Vol. VIII: Bioenergetics and growth. Academic Press, New York.

Hoar, W. S., D. J. Randall, and E. M. Donaldson (eds.). 1983. Fish physiology, Vol. IX, Part A. Academic Press, New York and London.

Hobson, E. S. 1971. Cleaning symbiosis among California inshore fishes. Fish. Bull. 69:491–523.

Hochachka, P. W., and T. P. Mommson (eds.) 1991. Biochemistry and molecular biology of fish I. Elsevier, Amsterdam, Oxford, New York.

Hocutt, C. H., and E. O. Wiley. 1986. The zoogeography of Northern American freshwater fishes. John Wiley and Sons, New York.

Hoese, H. D., and R. H. Moore. 1977. Fishes of the Gulf of Mexico. Texas A&M University Press, College Station, TX, and London.

Hoekstra, D., and J. Janssen. 1985. Non-visual feeding behavior of the mottled sculpin, *Cottus bairdi*, in Lake Michigan. Env. Biol. Fish. 12(2):111–117.

Holliday, F. G. T. 1969. The effects of salinity on the eggs and larvae of teleosts, p. 293–311. *In*: Fish physiology, Vol. I, W. S. Hoar and D. J. Randall (eds.). Academic Press, New York.

Holmes, E. B. 1985. Are lungfishes the sister group of tetrapods? Biol. J. Linn. Soc. 25(4):379–397.

Holmes, R. L., and J. N. Ball. 1974. The pituitary gland. A comparative account. Cambridge University Press, Cambridge.

Holmes, W. N., and E. M. Donaldson. 1969. The body compartments and the distribution of electrolytes, Part 1, p. 1–89. *In*: Fish physiology, Vol. I, W. S. Hoar and D. J. Randall (eds.). Academic Press, New York.

Holmgren, S. 1985. Substance P in the gastrointestinal tract of *Squalus acanthias*. Molec. Physiol. 8:119.

Holmgren, S., D. J. Grove, and S. Nilsson. 1985. Substance P acts by releasing 5-hydroxytryptamine from enteric neurons in the stomach of the rainbow trout, *Salmo gairdnerii*. Neuroscience 14:683.

Homewood, B. 1994. Vampire fish show their teeth. New Scientist 1957:7.

Honore, E. K., and P. H. Klopfer. 1990. A concise survey of animal behavior. Academic Press, San Diego.

Hoogenboezem, W., J. G. M. van den Boogart, F. A. Sibbing, E. H. R. R. Lammens, A. Terlouw, and J. W. M. Osse. 1991. A new model of particle retention and branchial sieve adjustment in filter-feeding bream (*Abramis brama*, Cyprinidae). Can. J. Fish. Aquat. Sci. 48(1):7–18.

Hopkins, C. D. 1983. Functions and mechanisms in electrorecep-tion, p. 215–259. *In*: Fish neurology, Vol. I, R. G. Northcutt and R. E. Davis (eds.). University of Michigan Press, Ann Arbor.

Hora, S. L. 1938. Notes on the biology of the fresh-water grey mullett *Mugil corsula* H., with observations on the probable mode of origin of aerial vision in fishes. J. Bombay Nat. Hist. Soc. 40:62–68.

Hori, M. 1991. Feeding relationships among cichlid fishes in Lake Tanganyika: Effects of intra- and interspecific variations of feeding behavior on their coexistence. Ecol. Int. Bull. 19:89–101.

Horn, M. H. 1983. Optimal diets in complex environments: Feeding strategies of two herbivorous fishes from a temperate rocky intertidal zone. Oecologia 58(3):345–350.

Horn, M. H. Symp. 1983; publ. 1984. Stromateoidei: Development and relationships, p. 620–628. *In*: Ontogeny and systematics of fishes, H. G. Moser, W. J. Richards, D. M. Cohen, M. P. Fahay, A. W. Kendall, Jr. and S. L. Richardson (eds.). Am. Soc. Ichthyol. and Herpetol. Spec. Pub. 1.

Horn, M. H. 1989. Biology of marine herbivorous fishes. Oceanogr. and Mar. Biol. Ann. Rev. 27:167–272.

Horn, M. H., and L. G. Allen. 1978. A distributional analysis of California coastal marine fishes. J. Biogeog. 5:23–42.

Horn, M. H., and R. N. Gibson. 1988. Intertidal fishes. Sci. Am. 258(1):64–70.

Horton, R. E. 1945. Erosional development of streams: Quantitative physiography factors. Bull. Geol. Soc. Amer. 56:275–370.

Houde, A. E., and J. A. Endler. 1990. Correlated evolution of female mating preference and male color pattern in *Poecilia reticulata*. Science 248:1405–1408.

Howe, K. M., D. L. Stein, and C. E. Bond. 1980. First records off Oregon of the pelagic fishes *Paralepis atlantica, Gonostoma atlanticum,* and *Aphanopus carbo*, with notes on the anatomy of *Aphanopus carbo*. Fish. Bull. 77(3):700–703.

Howes, G. J. 1985. Cranial muscles of gonorhynchiform fishes, with comments on generic relationships. Bull. Br. Mus. (Nat. Hist.) 49(2): 273–303.

Howes, G. J. 1989. Phylogenetic relationships of macrouroid and gadoid fishes based on cranial myology and arthrology, p. 113– 128. *In*: Papers on the systematics of the gadiform fishes, D. M. Cohen (ed.). Los Angeles Co., Mus. Sci. Ser. No. 32.

Howes, G. J., and O. A. Crimmen. 1990. A review of the Bathygadidae (Teleostei: Gadiformes). Bull. Br. Mus. (Nat. Hist.) Zool. 56(2): 155–203.

Howes, G. J., and C. P. J. Sanford. 1987. Oral ontology of the ayu, *Plecoglossus altivelis* and comparisons with the jaws of other salmoniform fishes. Zool. J. Linn. Soc. 89(2):133–169.

Hoyt, J. W. 1975. Hydrodynamic drag reduction due to fish slimes, p. 653–672. *In*: Swimming and flying in nature, Vol. 2, T. Y.-T. Wu, C. J. Brokaw and C. Brennan (eds.). Plenum Press, New York.

Hubbs, C. L. 1943. Terminology of early stages of fishes. Copeia 1943:260.

Hubbs, C. L. 1961. The marine vertebrates of the outer coast, p. 137–147. *In*: Symposium: The biogeography of Baja California and adjacent seas, Pt. 2, Marine biotas. Syst. Zool. 9(3 and 4): 134–147.

Hubbs, C. L. 1964a. History of ichthyology in the United States after 1850. Copeia 1964:42–60.

Hubbs, C. L. 1964b. David Starr Jordan. Systematic Zoology 13(4):195–200.

Hubbs, C. L., and T. E. B. Pope. 1937. The spread of the sea lamprey through the Great Lakes. Trans. Am. Fish. Soc. 66:172–176.

Hubbs, C. L., and I. C. Potter. 1971. Distribution, phylogeny and taxonomy, p. 1–65. *In*: The biology of lampreys, M. W. Hardisty and I. C. Potter (eds.). Academic Press, London.

Hubbs, C. L., and L. P. Schultz. 1931. The scientific name of the Columbia River chub. Occ. Pap. Mus. Zool., University of Michigan. 232:1–6.

Hudson, R. C. L. 1969. Polyneural innervation of the fast muscles of the marine teleost (*Cottus scorpius*) L. J. Exp. Biol. 50:47–67.

Huet, M. 1959. Profiles and biology of western European streams as related to fish management. Trans. Am. Fish. Soc. 88:153–163.

Hughes, G. M. 1966. The dimensions of fish gills in relation to their functions. J. Exp. Biol. 45:177–195.

Hughes, G. M. 1984. General anatomy of the gills, p. 1–72. In: Fish physiology, Vol. X. Part A, W. S. Hoar and D. J. Randall (eds.). Academic Press, Orlando.

Hughes, G. M., B. R. Singh, R. N. Thakur, and J. S. D. Munshi. 1974. Areas of the air breathing surfaces of *Amphipnous cuchia* (Ham.). Proc. Ind. Nat. Sci. Acad., Part B 40(4):379–392.

Hughes, J. M., and M. E. Potter. 1991. Scombroid-fish poisoning. New Engl. J. Med. 324(11):766–768.

Hulley, P. A. 1972. The family Gurgesiellidae (Chondrichthyes, Batoidei) with reference to *Pseudoraja atlantica* Bigelow and Schroeder. Copeia 1972:356–359.

Hunt, B. P. 1960. Digestion rate and food consumption of Florida gar, warmouth and largemouth bass. Trans. Am. Fish. Soc. 89:206–211.

Hunter, J. R. 1971. Sustained speed of jack mackerel, *Trachurus symmetricus*. Fish. Bull. U. S. 69(2):267–271.

Hunter, J. R. 1981. Feeding ecology and predation of fish larvae, p. 34–77. *In*: Marine fish larvae, R. Lasker (ed.). Washington Sea Grant Program. University of Washington Press, Seattle.

Hunter, J. R., and J. R. Zweifel. 1971. Swimming speed, tail beat frequency, tail beat amplitude, and size in jack mackerel, *Trachurus symmetricus,* and other fishes. Fish. Bull. 69:253–267.

Hutchinson, G. 1957. A treatise on limnology. John Wiley and Sons, Inc., New York.

Hyatt, K. D. 1979. Feeding strategy, p. 71–119. *In*: Fish physiology, Vol. VIII, Bioenergetics and growth, W. S. Hoar, D. J. Randall, and J. R. Brett (eds.). Academic Press, New York.

Hyman, L. H. 1942. Comparative vertebrate anatomy (Second Edition). University of Chicago Press, Chicago.

Hyman, L. H. 1979. Hyman's comparative anatomy (Third Edition), M. H. Wake (ed.). University of Chicago Press, Chicago and London.

Hynes, H. B. N. 1970. The ecology of running waters. University of Toronto Press, Toronto.

Ibarra, M., and D. J. Stewart. 1989. Longitudinal zonation of sandy beach fishes in the Napo River Basin, eastern Ecuador. Copeia 1989:364–381.

Ida, H. 1976. Removal of the family Hypoptychidae from the sub-order Ammodytoidei, order Perciformes, to the suborder Gasterosteoidei, order Syngnathiformes. Jpn. J. Ichthyol. 23:33–42.

Igarashi, S., and T. Kamiya. 1972. Atlas of the vertebrate brain. University Park Press, Baltimore, London, Tokyo.

Ihssen, P. E., L. R. McKay, I. McMillan, and R. B. Phillips. 1990. Ploidy manipulation and gynogenesis in fishes: Cytogenetic fisheries applications. Trans. Am. Fish. Soc. 119:698–717.

Ingmanson, D. E., and W. J. Wallace. 1989. Oceanography: An introduction (Fourth Edition). Wadsworth Publ. Co., Belmont, CA.

Inoue, K., S. Yamashita, J. Hata, S. Kabeno, S. Asada, E. Nagahisa, and T. Fujita. 1990. Electroporation as a new technique for producing transgenic fish. Cell Differ. Develop. 29:123–128.

International Union for Conservation of Nature and Natural Resources (IUCN). 1988. International Union for Conservation of nature red list of threatened animals. Gland, Switzerland and Cambridge, U. K.

Ivantsoff, W., B. Said, and A. Williams. 1987. Systematic position of the family Dentatherinidae in relationship to Phallostethidae and Atherinidae. Copeia 1987:649–658.

Ivlev, V. S. 1945. The biological productivity of waters. (translated by W. E. Ricker.) J. Fish. Res. Bd. Can. 23:1707–1759. (Translated 1966).

Ivoylov, A. A. 1986. Classification and nomenclature of tilapias of the tribe Tilapiinii (Cichlidae). New commercial fishes in warm waters of the USSR. J. Ichthyol. 26(3):97–109.

Iwamoto, R. N., A. M. Saxton, and W. K. Hershberger. 1982. Genetic estimates for length and weight of coho salmon during freshwater rearing. J. Heredity 73:187–191.

Jackson, J. B. C., A. G. Coates, and A. Budd. 1995. Environmental and biological change in Neogene and Quaternary tropical America. University of Chicago Press, Chicago.

Jacöbowski, M., and M. Whitear. 1990. Comparative morphology and cytology of taste buds in teleosts. Z. mikrosk.-anat. Forsch. 104: 529–560.

Jaiswal, A. G., and D. K. Belsare. 1973. Comparative anatomy and histology of the caudal neurosecretory system in teleosts. Z. Mikrosk, Anat. Forsch. 87(5/6):589–609.

Jamieson, B. G. M. 1991. Fish evolution and systematics: Evidence from spermatozoa. Cambridge University Press, Cambridge, New York, Port Chester, Melbourne, Sydney.

Janssen, J. 1978. Feeding-behaviour repertoire of the alewife, *Alosa pseudo-harengus*, and the ciscoes, *Coregonus hoyi* and *C. artedii*. J. Fish. Res. Bd. Can. 35:249–253.

Janssen, J. 1990. Localization of substrate vibrations by the mottled sculpin (*Cottus bairdi*). Copeia 1990:349–355.

Janssen, W. A. 1970. Fish as potential vectors of human bacterial disease, p. 284–290. *In*: A symposium on diseases of fish and shellfishes, S. F. Snieszko (ed.). Am. Fish. Soc. Sp. Pub. 5.

Janvier, P. 1981. The phylogeny of the Craniata, with particular reference to the significance of fossil "agnathans." J. Vert. Paleont. 1(2):121–159.

Janvier, P. 1984. The relationships of the Osteostraci and Galeaspida. J. Vert. Paleont. 4(3):344–358.

Janvier, P. 1985. Ces e'tranges betes du Montana. La Recherche 16:98–100.

Janvier, P. 1986. Le nouvelles conceptions de la phylogenie de la classification des "Agnathes" et des Sarcopterygiens, p. 123–138.

In: Les poissons: Classification et phylogenese, Y. François and M. L. Beuchot (eds.). Oceanis 12(3).

Janvier, P. 1993. Patterns of diversity in the skull of jawless fishes, p. 131–188. *In*: The skull, Vol. 2, J. Hanken and B. Hall (eds.). University of Chicago Press, Chicago.

Janvier, P., and R. Lund. 1983. *Hardistiella montanensis* n. gen. et sp. (Petromyzonida) from the Lower Carboniferous of Montana, with remarks on the affinities of lamprey. J. Vert. Paleon. 2(4):407–413.

Jarvik, E. 1968. The systematic position of the Dipnoi, p. 223–245. *In*: Current problems of lower vertebrate phylogeny, T. Orvig (ed.). Wiley-Interscience, New York.

Jarvik, E. 1980. Basic structure and evolution of vertebrates, Vols. 1 and 2. Academic Press, New York.

Jenkins, D. B. 1989. The utricle in *Ictalurus punctatus,* p. 73–78. *In*: Hearing and sound communication in fishes, W. N. Tavolga, A. N. Hopper, and R. R. Fay (eds.). Springer-Verlag, New York.

Jobling, M. 1994. Fish energetics. Chapman and Hall, London.

Johansen, K. 1970. Air breathing in fishes, p. 361–411. *In*: Fish physiology, Vol. IV, W. S. Hoar and D. J. Randall (eds.). Academic Press, New York.

Johansen, K., and H. Gesser. 1986. Fish cardiology: Structural, haemodynamic, electromechanical and metabolic aspects, p. 71–85. *In*: Fish physiology: Recent advances, S. Nilsson and S. Holmgren (eds.). Croom Helm, London.

Johansen, K., and R. Strahan. 1963. The respiratory system of *Myxine glutinosa* L., p. 352–371. *In*: The biology of *Myxine*, A. Brodal and R. Fänge (eds.). Universitetsforlaget, Oslo.

Johansen, S., P. H. Guddal, and T. Johansen. 1990. Organization of the mitochondrial DNA of Atlantic cod, *Gadus morhua.* Nucl. Acids Res. 18:411–419.

John, G., P. U. G. K. Reddy, and S. D. Gupta. 1984. Artificial gynogenesis in two Indian major carps, *Labeo rohita* (Ham.) and *Catla catla* (Ham.). Aquaculture 42(2):161–169.

Johnson, D. W. 1973. Endocrine control of hydromineral balance in teleosts. Am. Zool. 13:799–818.

Johnson, F. H., and Y. Haneda. 1966. Bioluminescence in progress. Princeton University Press, Princeton, NJ.

Johnson, G. D. 1975. The procurrent spur: An undescribed perciform caudal character and its phylogenetic implications. Occ. Pap. Calif. Acad. Sci. (121):1–23.

Johnson, G. D. 1984. Percoidei: Development and relationships, p. 464–498. *In*: Ontogeny and systematics of fishes, H. G. Moser, W. J. Richards, D. M. Cohen, M. P. Fahay, A. W. Kendall, Jr. and S. L. Richardson (eds.). Am. Soc. Ichthyol. and Herpetol. Spec. Pub. 1.

Johnson, G. D. 1986. Scombroid phylogeny: An alternative hypothesis. Bull. Mar. Sci. 39:1–41.

Johnson, G. D. 1992. Monophyly of the euteleostean clades—Neoteleostei, Eurypterygii and Ctenosquamata. Copeia 1992: 8–25.

Johnson, G. D. 1993. Percomorph phylogeny: Progress and problems. Bull. Mar. Sci. 52(1):3–28.

Johnson, G. D., and E. B. Brothers. 1993. Schindleria: A paedomorphic goby (Teleostei: Gobioidei). Bull. Mar. Sci. 52(1): 441–471.

Johnson, G. D., and C. Patterson. 1993. Percomorph phylogeny: A survey of acanthomorphs and a new proposal. Bull. Mar. Sci. 52(1):554–626.

Johnson, G. D., and C. Patterson. 1995. Interrelationships of lower Euteleostei. Abst. of ASIH Meeting, p. 124–125, Edmonton, Alta., Canada.

Johnson, P. B., H. Zhou, and M. A. Adams. 1990. Gustatory sensitivity of the herbivore *Tilapia zilli* to amino acids. J. Fish. Biol. 35(4): 387–393.

Johnson, R. K. 1974. A revision of the alepisaurid family Scopelarchidae (Pisces: Myctophiformes). Fieldiana (Zool.) 66: 1–249.

Johnson, R. K. 1982. Fishes of the families Evermannellidae and Scopelarchiidae: Systematics, morphology, interrelationships, and zoogeography. Fieldiana (Zool.) N.S. 12:1–252.

Johnston, I. A. 1981. Structure and function of fish muscles. Symp. Zool. Soc. Lond. 48:71–113.

Johnston, I. A. 1983. Dynamic properties of fish muscle, p. 36–67. *In*: Fish biomechanics, P. W. Webb and D. Weihs (eds.). Praeger, New York.

Jollie, M. 1962. Chordate morphology. Reinhold Publishing, New York.

Jones, D. R., J. W. Kiceniuk, and O. S. Bamford. 1974. Evaluation of the swimming performance of several fish species from the MacKenzie River. J. Fish. Res. Bd. Can. 31:1641–1647.

Jordan, D. S. 1901. The fish fauna of Japan, with observation on the geographical distribution of fishes. Science, New Series XIV: 545–567.

Jordan, D. S. 1905. A guide to the study of fishes, Vols. 1 and 2. Henry Holt, New York.

Jordan, D. S. 1923. A classification of fishes, including families and genera as far as known. Stanford University Publ., Biol. Sci. 3.

Jordan, D. S. 1928. The distribution of fresh-water fishes. Annual Report of the Smithsonian Institution, 1927:335–385.

Jorgensen, J. M. 1989. Evolution of octavolateralis sensory cells, p. 115–149. *In*: The mechanosensory lateral line, S. Coombs, P. Görner and H. Münz (eds.). Springer-Verlag, New York.

Jorstad, K. E., O. Skaala, and G. Dahle. 1991. The development of biochemical and visible genetic markers and their potential use in evaluating interaction between cultured and wild fish populations. ICES Marine Sci. Symp. 192:200–205.

Joyce, J. E., W. W. Smoker, R. Heintz, and A. J. Gharrett. 1994. Survival to fry and seawater tolerance of diploid and triploid hybrids between chinook (*Oncorhynchus tshawytscha*), chum (*O. keta*), and pink salmon (*O. gorbuscha*). Can. J. Fish. Aquatic. Sci. 51 (supp.1):25–30.

Junger, H., K. Kotrschal, and A. Goldschmid. 1989. Comparative morphology and ecomorphology of the gut in European cyprinids (Teleostei). J. Fish Biol. 34:315–336.

Kalmijn, A. J. 1971. The electric sense of sharks and rays. J. Exp. Biol. 55:371–383.

Kalmijn, A. J. 1977. The electric and magnetic sense of sharks, skates, and rays. Oceanus 20(3):45–52.

Kalmijn, A. J. 1978. Electric and magnetic sensory world of sharks, skates, and rays, p. 507–528. *In*: Sensory biology of sharks, skates and rays, E. S. Hodgson and R. F. Mathewson (eds.). Office of Naval Research, Arlington, VA.

Kalmijn, A. J. 1985. Theory of electromagnetic orientation: A further analysis, p. 525–563. *In*: Comparative physiology of sensory systems, L. Bolis and R. D. Keynes (eds.). Press Syndicate, University of Cambridge, Cambridge.

Kalmijn, A. J. 1989. Functional evolution of lateral line and inner ear sensory systems, p. 187–215. *In*: The mechanosensory lateral line, S. Coombs, P. Görner and H. Münz (eds.). Springer-Verlag, New York.

Kamler, E. 1992. Early life history of fish: An energetics approach. Chapman and Hall, London.

Kan, T. T., and C. E. Bond. 1981. Notes on the biology of the Miller Lake lamprey *Lampetra* (*Entosphenus*) *minima*. Northwest Science 55(1):70–74.

Kanazawa, A. 1985. Essential fatty acid and lipid requirement of fish, p. 281–298. *In*: Nutrition and feeding in fish, C. B. Cowey, A. M. Mackie, and J. G. Bell (eds.). Int. Symp. on Feeding and Nutrition in Fish, Aberdeen, U.K.

Kapoor, B. G., H. Smit, and I. A. Verighina. 1975. The alimentary canal and digestion in teleosts. Adv. in Mar. Biol. 13:109–239.

Kapuscinski, A. R., and E. M. Hallerman. 1990. Transgenic fish and public policy: Anticipating environmental impacts of transgenic fish. Fisheries 15(1):2–11.

Karplus, I. 1978. A feeding association between the grouper *Epinephelus fasciatus* and the moray eel *Gymnothorax griseus*. Copeia 1978:164.

Kashiwagi, M., and R. Sato. 1969. Studies on the osmoregulation of the chum salmon, *Oncorhynchus keta* (Walbaum). I. The tolerance of eyed period eggs, alevins and fry of the chum salmon to seawater. Tohoku J. Agr. Res. 20(1):41–47.

Katsuki, Y., and K. Yanagisawa. 1982. Chemoreception in the lateral line, p. 227–242. *In*: Chemoreception in fishes, T. J. Hara (ed.). Elsevier, Amsterdam.

Kaufman, L. S., and K. F. Liem. 1982. Fishes of the suborder Labroidei (Pisces: Perciformes): phylogeny, ecology, and evolutionary significance. Breviora 472.

Kavaliers, M. 1980. The pineal organ and circadian rhythms of fishes, p. 631–643. *In*: Environmental physiology of fishes, M. A. Ali (ed.). Plenum Press, New York.

Keast, A., and D. Webb. 1966. Mouth and body form relative to feeding ecology in the fish fauna of a small lake, Lake Opinicon, Ontario. J. Fish. Res. Bd. Can. 23:1845–1874.

Keene, M. J., and K. A. Tighe. 1984. Beryciformes: Development and relationships, p. 383–392. *In*: Ontogeny and systematics of fishes, H. G. Moser, W. J. Richards, D. M. Cohen, M. P. Fahay, A. W. Kendall, Jr. and S. L. Richardson (eds.). Am. Soc. Ichthyol. and Herpetol. Spec. Pub. 1.

Keenleyside, M. H. A. 1979. Diversity and adaptation in fish behavior. Springer-Verlag, Berlin.

Keenleyside, M. H. A. 1991. Parental care, p. 191–308. *In*: Cichlid fishes: behaviour, ecology and evolution, M. H. A. Keenleyside (ed.). Chapman and Hall, London, New York.

Keller, C. H., M. Kawasaki, and W. Heiligenberg. 1991. The control of pacemaker modulations for social communication in the weakly electric fish *Sternopygus*. J. Comp. Physiol., A. 169(4): 441–450.

Kemp, N. E. 1987. The biology of the Australian lungfish, *Neoceratodus forsteri*, p. 181–198. *In*: The biology and evolution of lungfishes, W. E. Bemis, W. W. Burggren, and N. E. Kemp (eds.). Alan R. Liss, Inc., NY.

Kendall, A. W., Jr. E. H. Ahlstrom, and H. G. Moser. 1984. Early life history stages of fishes and their characters, p. 11–22. *In*: Ontogeny and systematics of fishes, H. G. Moser, W. J. Richards, D. M. Cohen, M. P. Fahay, A. W. Kendall, Jr. and S. L. Richardson (eds.). Am. Soc. Ichthyol. and Herpetol. Spec. Pub. 1.

Kent, G. C. 1992. Comparative anatomy of the vertebrates. Mosby, St. Louis, MO.

Kido, K. 1988. Phylogeny of the family Liparididae, with the taxonomy of the species found around Japan. Mem. Fac. Fish., Hokkaido University 35(1):125–256.

Kieffer, J. D., and P. W. Colgan. 1992. The role of learning in fish behavior. Rev. Fish Biol. Fish. 2:125–143.

Kille, A. 1960. Fertilization of the lamprey egg. Exp. Cell. Res. 20:12–27.

Kimura, K., K. Tsukamoto, and T. Sugimoto. 1993. Migration of the Japanese eel larvae in the subtropical gyre-effect of the tradewind. Paper presented at the Fourth Indo-Pacific Fish Conf. (Bangkok). Abstr. p. 73 of program.

Kincaid, H. L. 1983. Inbreeding in fish populations used for aquaculture. Aquaculture 33:215–227.

King, M.-C., and A. C. Wilson. 1975. Evolution at two levels in humans and chimpanzees. Science 188:107–116.

Kirpichnikov, V. S. 1981. Genetic bases of fish selection. Springer-Verlag, New York, Berlin, Heidelberg.

Kirschner, L. E. 1991. Water and ions, p. 13–107. *In*: Environmental and metabolic animal physiology: Comparative animal physiology (Fourth Edition), C. L. Prosser (ed.). Wiley-Liss, New York.

Kiyohara, S., I. Hidaka, and T. Tamura. 1975. Gustatory response in the puffer. 2. Single fiber analyses. Bull. Jpn. Soc. Sci. Fish 41(4):383–391.

Klauswitz, W., J. E. McCosker, J. E. Randall, and H. Zetsche. 1978. *Hoplolatilus chlupatyi* n sp., un nouveau poisson marin des Philippines (Pisces, Perciformes, Percoidei, Branchiostegidae). Rev. Fr. Aquariol. 2:41–48.

Kleerekoper, H. 1969. Olfaction in fishes. Indiana University Press, Bloomington.

Klimley, A. P. 1994. The predatory behavior of the white shark. Am. Sci. 82:122–133.

Knutson, S., and T. Grav. 1976. Seawater adaptation in Atlantic salmon (*Salmo salar* L.) at different experimental temperatures and photoperiods. Aquaculture 8:169–187.

Kocher, T. D., and K. R. McKaye. 1983. Defense of heterospecific cichlids by *Cyrtocara moorii* in Lake Malawi, Africa. Copeia 1983: 544–547.

Kodric-Brown, A. 1983. Determinants of male reproductive success in pupfish (*Cyprinodon pecoensis*). Anim. Behav. 31(1): 128–137.

Koelz, W. 1929. Coregonid fishes of the Great Lakes. Bull. U.S. Bur. of Fish. 43:297–643.

Koenig, C. C., and R. J. Livingston. 1976. The embryological development of the diamond killifish. Copeia 1976:435–449.

Konishi, J., and Y. Zotterman. 1963. Taste functions in the carp. Acta Physiol. Scand. 52:150–161.

Kornfield, I. L. 1978. Evidence for rapid speciation in African cichlid fishes. Experientia 34:335–336.

Kortmulder, K. 1987. Ecology and behavior in tropical freshwater fish communities. Arch. Hydrobiol. Beih. 28:503–513.

Kosswig, C. 1964. Polygenic sex determination. Experientia 20: 190–199.

Kotrschal, K. 1991. Solitary chemosensory cells: Taste, common chemical sense or what? Rev. Fish Biol. Fish. 1:3–22.

Kotrschal, K., and M. Palzenberger. 1992. Neuroecology of cyprinids: Comparative, quantitative histology reveals diverse brain patterns. Env. Biol. Fish. 37(1):135–152.

Kottcamp, G. M., and P. B. Moyle. 1972. Use of disposable beverage cans by fish in the San Joaquin Valley. Trans. Am. Fish. Soc. 101:566.

Kottelat, M., and X. L. Chu. 1988. Revision of *Yunnanitus* with descriptions of a miniature species flock and six new species from China (Cypriniformes: Homalopteisidae). Env. Biol. Fishes 23: 65–93.

Kowarsky, J. 1973. Extra-branchial pathways of salt exchange in a teleost fish. Comp. Biochem. Physiol. 46A:477–486.

Kozhov, M. 1963. Lake Baikal and its life. Monographiae Biologiae, Vol. 11. W. Junk, The Hague.

Kramer, B. 1990. Electrocommunication in teleost fishes. Springer-Verlag, Berlin, Heidelberg.

Krejsa, R. J., and H. C. Slavkin. 1987. Earliest craniate teeth identified: The hagfish-conodont connection. J. Dent. Res. 66 (Spec. Issue):144.

Krejsa, R. J., H. C. Slavkin, P. Bringas, Jr. and M. Nakamura. 1987. Hagfish tooth development and morphology compared with conodonts: Agnathan ancestors identified. Proc. Finn. Dent. Soc. 83:227. (Abstract only seen.)

Krishnamurthy, V. G., and H. A. Bern. 1971. Innervation of the corpuscles of Stannius. Gen. Comp. Endocrinol. 16(1):162–165.

Krogh, A. 1939. Osmotic regulation in aquatic animals. Cambridge University Press, Cambridge.

Krygier, E. E., and W. G. Pearcy. 1986. The role of estuarine and offshore nursery areas for young English sole, *Parophrys vetula* Girard, of Oregon. Fish. Bull. U. S. 84(1):119–132.

Kuehne, R. A. 1962. A classification of streams, illustrated by fish distribution in an eastern Kentucky creek. Ecol. 43:608–614.

Kuhlenbeck, H. 1975. The central nervous system of vertebrates, Vol. 4: Spinal cord and deuterencephalon. S. Karger, Basel.

Lack, D. 1954. The natural regulation of animal numbers. Oxford University Press, Oxford.

Ladich, F. 1988. Sound production by the gudgeon, *Gobio gobio* L., a common European freshwater fish (Cyprinidae, Teleostei). J. Fish Biol. 32(5):707–715.

Lagios, M. D. 1975. The pituitary gland of the coelacanth *Latimeria chalumnae* Smith. Gen. Comp. Endocrinol. 25(2):126–146.

Lagios, M. D. 1979. The coelacanth and the Chondrichthyes as sister groups: A review of shared apomorph characters and a cladistic analysis and reinterpretation, p. 25–44. *In*: The biology and physiology of the living coelacanth, J. E. McCosker and M. D. Lagios (eds.). Occ. Papers Calif. Acad. Sciences (134).

Lagios, M. D. 1982. *Latimeria* and the Chondrichthyes as sister taxa: A rebuttal to recent attempts at refutation. Copeia 1982:942–948.

Lagler, K. F. 1956. Freshwater fishery biology (Second Edition). Wm. C. Brown, Dubuque, IA.

Lagler, K. F., J. W. Bardach, and R. R. Miller, 1962. Ichthyology. John Wiley and Sons, New York.

Lagler, K. F., J. W. Bardach, R. R. Miller, and D. R. M. Passino. 1977. Ichthyology (Second Edition). John Wiley and Sons, New York.

Lahlou, B. 1980. Les hormones dans l'osmoregulation des poissons, p. 201–204. *In*: Environmental physiology of fishes, M. A. Ali (ed.). Plenum Press, New York.

Lahlou, B., I. W. Henderson, and W. H. Sawyer. 1969. Renal adaptations by *Opsanus tau*, a euryhaline aglomerular teleost, to dilute media. Am. J. Physiol. 216:1266–1272.

Lall, S. P. 1989. The minerals, p. 219–257. *In*: Fish nutrition (Second Edition), J. E. Halver (ed.). Academic Press, San Diego.

Lam, T. J. 1982. Applications of endocrinology to fish culture. Can. J. Fish. Aquat. Sci. 39(1):111–137.

Lam, T. J. 1983. Environmental influences on gonadal activity in fish, p. 65–116. *In*: Fish physiology, Vol. IXB, W. S. Hoar, D. J. Randall and E. M. Donaldson (eds.). Academic Press, New York, London.

Land, M. F. 1991. Polarizing the world of fish. Nature 353:118–119.

Lane, S., A. J. McGregor, S. G. Taylor, and A. J. Gharrett. 1990. Genetic marking of an Alaskan pink salmon population, with an evaluation of the mark and marking process. Am. Fish. Soc. Symp. 7:395–406.

Langille, L., E. D. Stevens, and A. Anantaraman. 1983. Cardiovascular and respiratory flow dynamics, p. 92–137. *In*: Fish biomechanics, P. W. Webb and D. Weihs (eds.). Praeger, New York.

Lanzing, W. J. R. 1974. Sound production in the cichlid *Tilapia mossambica* Peters. J. Fish. Biol. 6:341–347.

Lasker, R. 1981a. Marine fish larvae. University of Washington Press, Seattle.

Lasker, R. 1981b. The role of a stable ocean in larval fish survival and subsequent recruitment, p. 80–87. *In*: Marine fish larvae, R. Lasker (ed.). University of Washington Press, Seattle.

Lauder, G. V. 1979. Feeding mechanics in primitive teleosts and in the halecomorph fish *Amia calva*. J. Zool. Lond. 187:543–578.

Lauder, G. V. 1980a. The role of the hyoid apparatus in the feeding mechanism of the coelacanth *Latimeria chalumnae*. Copeia 1980:1–9.

Lauder, G. V. 1980b. Evolution of the feeding mechanism in primitive actinopterygian fishes: A functional anatomical analysis of *Polypterus, Lepisosteus* and *Amia*. J. Morph. 163:283–317.

Lauder, G. V. 1980c. Hydrodynamics of prey capture by teleost fishes. Biofluid Mech. 2:161–181.

Lauder, G. V. 1982a. Patterns of evolution in the feeding mechanism of actinopterygian fishes. Amer. Zool. 22(2):275–285.

Lauder, G. V. 1982b. Structure and function in the tail of the pumpkinseed sunfish (*Lepomis gibbosus*). J. Zool. London 97:483–495.

Lauder, G. V. 1983. Food capture, p. 280–311. *In*: Fish biomechanics, P. W. Webb and D. Weihs (eds.). Praeger, New York.

Lauder, G. V. 1989. Caudal fin locomotion in ray-finned fishes: historical and functional analysis. Am. Zool. 29(1):85–102.

Lauder, G. V., and L. E. Lanyon. 1980. Functional anatomy of feeding in the blue gill sunfish, *Lepomis macrochirus*: *in vivo* measurement of bone strain. J. Exp. Biol. 84:33–55.

Lauder, G. V., and K. F. Liem. 1980. The feeding mechanism and cephalic myology of *Salvelinus fontinalis*: Form, function, and evolutionary significance, p. 365–390. *In*: Charrs, salmonid fishes of the genus *Salvelinus,* E. K. Balon (ed.). Junk, The Hague.

Lauder, G. V., and K. F. Liem. 1981. Prey capture by *Luciocephalus pulcher*: Implication for models of jaw protrusion in teleost fishes. Environ. Biol. Fish. 6(3/4):257–268.

Lauder, G. V., and K. F. Liem. 1983a. The evolution and interrelationships of the actinopterygian fishes. Bull. Mus. Comp. Zool. 150(3): 95–197.

Lauder, G. V., and K. F. Liem. 1983b. Patterns of diversity and evolution in ray-finned fishes, p. 1–24. *In*: Fish neurobiology, Vol. 1, R. G. Northcutt and R. E. Davis (eds.). University of Michigan Press, Ann Arbor.

Lauman, J., U. Pern, and V. Blum. 1974. Investigations on the function and hormonal regulation in the anal appendages in *Blennius parvo* Riso. J. Exp. Zool. 190(1):47–56.

Laurent, P., and S. Dunel-Erb. 1984. The pseudobranch morphology and function, p. 285–323. *In*: Fish physiology, Vol. X, Part B, W. S. Hoar and D. J. Randall (eds.). Academic Press, Orlando.

Lawry, J. V. 1974. Lantern fish compare downwelling light and bioluminescence. Nature 247:155–157.

Lazier, C. B., and M. E. MacKay. 1993. Vitellogenin gene expression in teleost fish, p. 391–405. *In*: Biochemistry and molecular biology of fish, Vol. 2: Molecular biology frontiers, P. W. Hochacha and T. P. Mommsen (eds.). Elsevier, Amsterdam.

Lea, R. N., and D. J. Miller. 1985. Shark attacks off the California and Oregon coasts: An update, 1980–84, p. 136–150. *In*: Biology of the white shark, J. A. Siegel and C. C. Swift (eds.). Mem. So. Calif. Acad. Sci. 9.

Leatherland, J. F. 1982. Environmental physiology of the teleostean thyroid gland: A review. Environ. Biol. Fish. 7(1):83–110.

LeBoeuf, B. J., J. E. McCosker, and J. Hewitt. 1987. Crater wounds on northern elephant seals: the cookie cutter shark strikes again. Fish. Bull. U. S. 85(2):387–392.

LeBoeuf, B. J., M. Reidman, and R. S. Keyes. 1982. White shark predation on pinnipeds in California coastal waters. Fish. Bull. U. S. 80(4):881–895.

Lee, D. J., and G. B. Putnam. 1973. The response of rainbow trout to varying protein/energy ratios in a test diet. J. Nutr. 103:916–922.

Lee, D. S., C. R. Gilbert, C. H. Hocutt, R. E. Jenkins, D. E. McAllister, and J. R. Stauffer, Jr. 1980. Atlas of North American freshwater fishes. North Carolina State University, Raleigh.

Leis, J. M. 1984. Tetradontiformes: Relationships, p. 459–463. *In*: Ontogeny and systematics of fishes, H. G. Moser, W. J. Richards, D. M. Cohen, M. P. Fahay, A. W. Kendall, Jr. and S. L. Richardson (eds.). Am. Soc. Ichthyol. and Herpetol. Spec. Pub. 1.

Leitritz, E., and R. C. Lewis. 1976. Trout and salmon culture. California Fish Bull. 164.

Leu, M. R. 1989. A late Permian freshwater shark from eastern Australia. Paleontology 32, part 2:265–286.

Levine, J. S., and E. F. MacNichol, Jr. 1979. Visual pigments in teleost fishes: Effects of habitat, microhabitat, and behavior on visual system evolution. Sens. Process 3(2):95–131.

Lewis, E. R., E. L. Leverenz, and W. S. Bialek. 1985. The vertebrate inner ear. CRC Press, Boca Raton, FL.

Li, H. W., C. B. Schreck, C. E. Bond, and E. Rexstad. 1987. Factors influencing changes in fish assemblages of Pacific northwest streams, p. 193–202. *In*: Community and evolutionary ecology of North American stream fishes, W. J. Matthews and D. C. Heins (eds.). University of Oklahoma Press, Norman.

Licht, J. H., and W. S. Harris. 1973. The structure, composition and elastic properties of the teleost bulbus arteriosus in the carp *Cyprinus carpio*. Comp. Biochem. Physiol. 46A:699–708.

Liem, K. F. 1963. The comparative osteology and phylogeny of the Anabantoidei (Teleostei, Pisces). Ill. Biol. Monograph No. 30.

Liem, K. F. 1967a. Functional morphology of the integumentary, respiratory and digestive systems of the synbranchoid fish *Monopterus albus*. Copeia 1967:375–388.

Liem, K. F. 1967b. A morphological study of *Luciocephalus pulcher,* with notes on gular elements in other recent teleosts. J. Morph. 121(2):103–134.

Liem, K. F. 1978. Modulatory multiplicity in the functional repertoire of the feeding mechanism in cichlid fishes. I. Piscivores. J. Morph. 158:323–360.

Liem, K. F. 1980a. Air ventilation in advanced teleosts: Biochemical and evolutionary aspects, p. 57–91. *In*: Environmental physiology of fishes, M. A. Ali (ed.). Plenum Press, New York.

Liem, K. F. 1980b. Adaptive significance of intra- and interspecific differences in the feeding repertories of cichlid fishes. Amer. Zool. 20:295–314.

Liem, K. F. 1980c. Acquisition of energy by teleosts: Adaptive mechanisms and evolutionary patterns, p. 299–334. *In*: Environmental physiology of fishes, M. A. Ali (ed.). Plenum Press, New York.

Liem, K. F. 1981. Larvae of air-breathing fishes as countercurrent flow devices in hypoxic environments. Science 211(4487):1177–1179.

Liem, K. F. 1990. Aquatic versus terrestrial feeding modes: Possible impacts on the trophic ecology of vertebrates. Amer. Zool. 30:209–221.

Liem, K. F. 1991. Functional morphology, p. 129–150. *In*: Cichlid fishes—behaviour, ecology and evolution, M. H. A. Keenleyside (ed.). Chapman and Hall, London.

Lighthill, M. J. 1983. Epilogue: Toward a more fully integrated fish biomechanics, p. 372–375. *In*: Fish biomechanics, P. W. Webb and D. Weihs (eds.). Praeger, New York.

Lighthill, M. J., and R. Blake. 1990. Biofluid dynamics of balistiform and gymnotiform location. Part I. Biological background and analysis of elongated body theory. J. Fluid Dynamics 212: 183–207.

Liley, N. R. 1982. Hormones and reproductive behavior in fishes, p. 73–116. *In*: Fish physiology, Vol. III, W. S. Hoar and D. J. Randall (eds.). Academic Press, New York.

Liley, N. R. 1969. Chemical communication in fish. Can. J. Fish. Aquat. Sci. 39(1):22–35.

Lim, T. M. 1981. Effects of schooling prey on the hunting behavior of a fish predator, p. 132. *In*: Ecology and ethology of fishes, D. L. G. Noakes and J. A. Ward (eds.). Junk, The Hague.

Limbaugh, C. 1961. Cleaning symbiosis. Sci. Am. 205(2):42–49.

Lin, H. R., and R. E. Peter. 1986. Induction of gonadotropin secretion and ovulation in teleosts using LHRH analogs and catecholaminergic drugs: A review, p. 667–670. *In*: The First Asian fisheries forum, J. L. Maclean, L. B. Dizon and L. V. Hosillos (eds.). Asian Fish. Soc., Manila, Philippines.

Lindberg, G. U. 1974. Fishes of the world: A key to families and a checklist. John Wiley and Sons, New York. (English translation by Israel Program for Scientific Translations Ltd.)

Lindsay, G. J. H. 1984. Distribution and function of digestive tract chitinolytic enzymes in fish. J. Fish. Biol. 24:529–536.

Lindsey, C. C. 1975. Pleomersism, the widespread tendency among related fish species for vertebral number to be correlated with maximum body length. J. Fish. Res. Bd. Can. 32:2453–2469.

Lindsey, C. C. 1978. Form, function, and locomotory habits in fish, p. 1–100. *In*: Fish physiology, Vol. VII, W. S. Hoar and D. J. Randall (eds.). Academic Press, New York.

Lineaweaver, T. H., III, and R. H. Backus. 1969. The natural history of sharks. J. B. Lippincott, Philadelphia, PA.

Linhart, O., R. Billard, and J. P. Proteau. 1993. Cryopreservation of European catfish (*Silurus glanis* L.) spermatozoa. Aquaculture 115(3-4):347–359.

Linnaeus, C. 1758. Systema naturae per regna tria naturae, secundum classes, ordines, genera, species, cum characteribus, differentiis, synonymis, locis. Tomus I. Holmiae:1–823.

Lippe, C., and C. Ardizzone. 1989. Urea transport and its regulation across the gall bladder of *Torpedo marmorata*, p. 143–145. *In*: Evolutionary and contemporary biology of elasmobranchs, W. C. Hamlett and B. Tota (eds.). J. Exp. Zool. Suppl. 2. Liss, New York.

Litman, G. W., C. T. Amemiya, R. N. Haire, and M. J. Shamblott. 1990. Antibody and immunoglobulin diversity. BioScience 40:751–757.

Little, E. E. 1983. Behavioral function of olfaction and taste in fish, p. 351–376. *In*: Fish neurobiology, Vol. 1, R. G. Northcutt and R. E. Davis (eds.). University of Michigan Press, Ann Arbor.

Littler, M. M., D. E. Littler, S. M. Blair, and J. N. Norris. 1985. Deepest known plant life discovered on an uncharted seamount. Science 227:57–59.

Livingston, M. E. 1987. Morphological and sensory specializations of five New Zealand flatfish species, in relation to feeding behavior. J. Fish. Biol. 31:775–795.

Lobel, P. S. 1991. Mating strategies of coastal marine fishes. Oceanus 34:19–26.

Lobel, P. S. 1992. Sounds produced by spawning fishes. Environ. Biol. Fish. 33:351–358.

Locket, N. A. 1974. The choroidal tapetum lucidum of *Latimeria*. Proc. R. Soc. Lond. B186(1084):281–290.

Locket, N. A. 1980a. Some advances in coelacanth biology. Proc. R. Soc. Lond. B208:265–307.

Locket, N. A. 1980b. Variation of architecture with size in the multiple-bank retina of a deep-sea teleost, *Chauliodus sloani*. Proc. R. Soc. Lond. B208:223–242.

Long, J. A. 1989 (Listed as 1986 in text). A new rhizodontiform fish from the early Carboniferous of Victoria, Australia, with remarks on the phylogenetic position of the group. J. Vert. Paleont. 9(1):1–17.

Long, J. A. 1990. Heterochrony and the origin of tetrapods. Lethaia 23(2):157–166.

Long, J. A. 1995. The rise of fishes. Johns Hopkins University Press, Baltimore and London.

Long, W. L., and W. W. Ballard. 1976. Normal embryonic stages of

the white sucker, *Catostomus commersoni*. Copeia 1976:342–351.

Losey, G. S. 1972. The ecological importance of cleaning symbiosis. Copeia 1972:820–833.

Losey, G. S. 1978. The symbiotic behavior of fishes, p. 1–31. *In*: The behavior of fish and other aquatic organisms, D. I. Mostofsky (ed.). Academic Press, New York.

Lovell, R. T. 1989. Diet and fish husbandry, p. 549–604. *In*: Fish nutrition (Second Edition), J. E. Halver (ed.). Academic Press, San Diego, London.

Lowe-McConnell, R. H. 1969. Speciation in tropical freshwater fishes. Biol. J. Linn. Soc. 1:51–75.

Lowe-McConnell, R. H. 1975. Fish communities in tropical freshwater. Longmans, London and New York.

Lowe-McConnell, R. H. 1987. Ecological studies in tropical fish communities. Cambridge University Press, Cambridge.

Lowenstein, D. 1971. The labyrinth, p. 207–240. *In*: Fish physiology, Vol. V, W. S. Hoar and D. J. Randall (eds.). Academic Press, New York, London.

Lubbock, R. 1980. Why are clownfishes not stung by sea anemones? Proc. R. Soc. Lond. B207:35–61.

Lubzens, E., S. Rothbard, and A. Hadani. 1993. Cryopreservation and viability of spermatozoa from the ornamental Japanese carp (nishikigoi). Isr. J. Aquacult. Bamidgeh 45(4):169–174.

Lund, R. 1977. New information on the evolution of the bradyodont Chondrichthyes. Fieldiana Geol. 33(28):521–539.

Lund, R. 1986. The diversity and relationships of the Holocephali, p. 97–106. *In*: Indo-Pacific fishes, T. Uyeno, R. Arai, T. Tarniuchi and K. Matsuura (eds.). Ichthyol. Soc. of Japan, Tokyo.

Lund, R. 1989. New petalodonts (Chondrichthyes) from the Upper Mississippian Bear Gulch limestone (Naimurian E_2b) of Montana. J. Vert. Paleontology 9(3):350–368.

Lund, R. 1990. Chondrichthyan life history styles as revealed by the 320 million years old Mississippian of Montana. Env. Biol. Fishes 27(1):1–19.

Lund, R. 1991. Shadows in time—a capsule history of sharks, p. 23–28. *In*: Discovering sharks, S. H. Gruber (ed.). Sp. Pub. 14, Am. Littoral Soc., Highlands, NJ.

Lundberg, J. G. 1993. Freshwater fishes, p. 156–190. *In*: Biological relationships between Africa and South America, P. Goldblatt (ed.). Yale University Press, New Haven.

Lundberg, J. G., and L. A. McDade. 1990. Systematics, p. 65–108. *In*: Methods for fish biology, C. B. Schreck and P. B. Moyle (eds.). Amer. Fish. Soc., Bethesda, MD.

Lundberg, J. G., A. Machado–Allison, and R. F. Kay. 1986. Miocene characoid fishes from Colombia: Evolution or extirpation. Science 224:208–209.

Lutz, P. L. 1975a. Osmotic and ionic composition of the polypteroid *Erpetoichthys calabaricus*. Copeia 1975:119–123.

Lutz, P. L. 1975b. Adaptive and evolutionary aspects of the ionic content of fishes. Copeia 1975:369–373.

Lyon, E. P. 1904. On rheotropism. I. Rheotropism in fishes. Am. J. Physiol. 12:149–161.

Lythgoe, J. N. 1975a. The structure and phylogeny of iridescent corneas in fishes, p. 253–262. *In*: Vision in fishes: New approaches in research, M. A. Ali (ed.). Plenum Press, New York and London.

Lythgoe, J. N. 1975b. The iridescent cornea of the sand goby *Pomatoschistus minutus* (Pallas), p. 263–277. *In*: Vision in fishes: New approaches in research, M. A. Ali (ed.). Plenum Press, New York.

Lythgoe, J. N. 1979. The ecology of vision. Clarendon Press, Oxford.

Lythgoe, J. N., and J. Shand. 1982. Change in spectral reflexions from the iridophores of the neon tetra. J. Physiol. 325:23–34.

Lythgoe, J. N., and J. Shand. 1983. Diel color changes in the neon tetra *Paracheirodon innesi*. Env. Biol. Fish. 8(4):249–254.

Macey, M. J., G. E. Pickford, and R. E. Peter. 1974. Forebrain localization of the spawning reflex response to exogenous neurohypophyseal hormones in the killifish, *Fundulus heteroclitus*. J. Exp. Zool. 190(3):269–280.

Maetz, J. 1969. Seawater teleosts: Evidence for a sodium-potassium exchange in the branchial sodium-excreting pump. Science 166:613–615.

Maetz, J. 1974. Aspects of adaptation to hypo-osmotic hyper-osmotic environments, p. 1–167. *In*: Biochemical and biophysical perspectives in marine biology, Vol. 1, D. C. Malins and J. R. Sargent (eds.). Academic Press, New York.

Magid, A. M. A., and M. M. Babiker. 1975. Oxygen consumption and respiratory behavior of three Nile fishes. Hydrobiologia 46(4): 359–367.

Magnuson, J. J. 1962. An analysis of aggressive behavior, growth, and competition for food and space in medake (*Oryzias latipes,* Pisces, Cyprinodontidae). Can. J. Zool. 40:313–363.

Magnuson, J. J. 1973. Comparative study of adaptations for continuous swimming and hydrostatic equilibrium of scombroid and xiphoid fishes. U. S. Fish and Wildl. Serv. Fish. Bull. 71:337–356.

Magurran, A. E. 1990. The adaptive significance of schooling as an anti-predator defence in fish. Ann. Zool. Fennici 27:51–66.

Magurran, A. E., and J. A. Bendelow. 1990. Conflict and cooperation in White Cloud Mountain minnow school. J. Fish Biol. 37: 77–83.

Maisey, J. G. 1980. An evaluation of jaw suspension in sharks. Am. Mus. Novit. 706:1–19.

Maisey, J. G. 1982. The anatomy and relationships of Mesozoic hybodont sharks. Am. Mus. Novit. 2724:1–48.

Maisey, J. G. 1984. Higher elasmobranch phylogeny and biostratigraphy. Zool. J. Linn. Soc. 82:33–54.

Maisey, J. G. 1985. Relationships of the megamouth shark, *Megachasma*. Copeia. 1985:228–231.

Maisey, J. G. 1986. Heads and tails: A chordate phylogeny. Cladistics 2(3):201–256.

Major, P. F. 1978. Predator-prey interaction in two schooling fishes, *Caranx ignobilis* and *Stolephorus purpureus*. Anim. Behav. 26: 760–777.

Manner, H. W. 1975. Vertebrate development. Kendall Hunt, Dubuque, IA.

Manooch III, C. S. 1984. Fisherman's guide: Fishes of the southeastern United States. North Carolina State Mus. of Nat. Hist., Raleigh.

Mansueti, R. 1963. Symbiotic behavior between small fishes and jellyfishes with new data on that between the stromateid *Peprilis alepidotus* and the scyphomedusa *Chrysaora quinquecirrha*. Copeia 1963:40–80.

Marchalonis, J., and S. F. Schluter. 1990. Origins of immunoglobins and immune recognition molecules. BioScience 40:758–768.

Maretic, Z. 1988. Fish venoms, p. 445–476. *In*: Handbook of natural toxins, Vol. 3, Marine toxins and venoms, A. T. Tu (ed.). Marcel Dekker Inc., New York and Basel.

Margulis, L. 1981. Symbiosis in cell evolution. W. H. Freeman and Co., San Francisco.

Markle, D. F. 1989. Aspects of character homology and phylogeny of the Gadiformes, p. 59–88. *In*: Papers on the systematics of the gadiform fishes, D. M. Cohen (ed.). Los Angeles Co. Mus. Sci. Ser. No. 32.

Markle, D. F., and J. E. Olney. 1980. A description of the vexillifer larvae of *Pyramodon ventralis* and *Snyderidia canina* (Pisces, Carapidae) with comments on classification. Pac. Sci. 34(2): 173–180.

Markle, D. F., and J. E. Olney. 1990. Systematics of the pearlfishes (Pisces: Carapidae). Bull. Mar. Sci. 47(2):269–410.

Marshall, C. R. 1987. A list of fossil and extant dipnoans, p. 15–23. *In*: The biology and evolution of lungfishes, W. E. Bemis, W. W. Burggren and N. E. Kemp (eds.). Alan R. Liss, Inc., New York.

Marshall, N. B. 1954. Aspects of deep sea biology. Hutchinsons, London.

Marshall, N. B. 1962. Observations on the Heteromi, an order of teleost fishes. Bull. Br. Mus. (Nat. Hist.) Zool. 9(6):249–270.

Marshall, N. B. 1966. The life of fishes. Universe Books, New York.

Marshall, N. B. 1967. The olfactory organs of bathypelagic fishes, p. 57–70. *In*: Aspects of marine zoology, N. B. Marshall (ed.). Academic Press, New York.

Marshall, N. B. 1970. The life of fishes. Universe Books, New York.

Marshall, N. B. 1971. Explorations in the life of fishes. Harvard University Press, Cambridge.

Marshall, N. B. 1979. Developments in deep-sea biology. Blandford Press, Poole, Dorset.

Marshall, N. B. 1980. Deep sea biology: Developments and perspectives. Garland S.T.P.M. Press, New York.

Marshall, N. B., and T. Iwamoto. 1973. Fishes of the western North Atlantic. Family Macrouridae. Sears Found. Mar. Res., Memoir (Yale University) 1(6):496–665.

Marshall, N. B., and G. W. Mead. 1966. Order Iniomi. Characters and synopsis of families, p. 1–18. *In*: Fishes of the western North Atlantic, D. M. Cohen (ed.-in-chief). Mem. Sears Found. Mar. Res., New Haven.

Martin, F. D. 1972. Factors influencing the local distribution of *Cyprinodon variegatus* (Pisces: Cyprinodontidae). Trans. Am. Fish. Soc. 101:89–93.

Martin, W. R. 1949. The mechanism of environmental control of body form in fishes. University of Toronto Stud. Biol. Ser. 58: 1–81.

Marui, T., and J. Caprio. 1992. Teleost gustation, p. 171–198. *In*: Fish chemoreception, T. J. Hara (ed.). Chapman and Hall, London.

Marx, J. L. 1988. Gene-watcher's feast served up in Toronto. Science 242:32–33.

Masuda, H., K. Amaoka, C. Araga, T. Uyeno, and T. Yoshsino. 1984. The fishes of the Japanese Archipelago. Tokai U. Press, Tokyo.

Matarese, A. C. 1989. Phylogenetic relationships of the ronquils (Perciforms: Bathymasteridae). (Abstr.) Program and Abstr. 65th Ann. Meeting ASIH, p. 115.

Matsubara, K., K. Arai, and R. Suzuki. 1995. Survival potential and chromosomes of progeny of triploid and pentaploid females in the loach (*Misgurnus anguillicaudatas*). Aquacult. 131:37–48.

Matsuoka, M., and T. Iwai. 1983. Adipose fin cartilage found in some teleostean fishes. Jpn. J. Ichthyol. 30(1):37–46.

Matsuyama, M., M. Hamada, T. Ashitani, M. Kashiwagi, T. Iwai, K. Okuzawa, H. Tanaka, and H. Kagawa. 1993. Development of LHRH a copolymer pellet polymerized by ultraviolet and its application for maturation in red sea bream *Pagrus major* during the non-spawning season. Bull. Jpn. Soc. Sci. Fish. 59(8):1361–1369.

Matsuura, K. 1929. Phylogeny of the superfamily Balistoidea (Pisces: Tetradontiformes). Mem. Fac. Fish., Hokkaido University 26(1/2):49–169.

Matthews, L. H. 1955. The evolution of viviparity in vertebrates. Mem. Soc. Endocrinol. 4:129–148.

Matthews, W. J. 1985. Critical current speeds and microhabitats of the benthic fishes *Percina roanoka* and *Etheostoma flabellare*. Env. Biol. Fish. 12:303–308.

Matthews, W. J., and D. C. Heins (eds.). 1987. Community and evolutionary ecology of North American stream fishes. University of Oklahoma Press, Norman.

Matty, A. J. 1985. Fish endocrinology. Croom Helm, London and Sydney.

Matty, A. J. 1986. Nutrition, hormones and growth. Fish Physiol. Biochem. 2:141–150.

May, B., and K. R. Johnson. 1990. Composite linkage map of salmonid fishes (*Salvelinus, Salmo,* and *Oncorhynchus*), p. 4.151–4.159. *In*: Genetic maps, locus maps of complex genomes (Fifth Edition), S. J. O'Brien (ed.). Cold Spring Laboratory Press, New York.

May, R. C. 1974. Larval mortality in marine fishes and the critical period concept p. 3–19. *In*: The early life history of fish, J. H. S. Blaxter (ed.). Springer-Verlag, New York, Heidelberg.

Mayden, R. L. 1989. Phylogenetic studies of North American minnows, with emphasis on the genus *Cyprinella* (Teleostei: Cypriniformes). University of Kansas, Mus. Nat. Hist. Misc. Pub. No. 80.

Mayden, R. L. (ed.). 1992. Systematics, historical ecology, and North American freshwater fishes. Stanford University Press, Palo Alto.

Mayden, R. L., and R. H. Matson. 1992. Systematics and biogeography of the Tennessee shiner, *Notropis leuciodus* (Cope) (Teleostei: Cyprinidae). Copeia 1992:954–968.

Mayden, R. L., and E. O. Wiley. 1992. The fundamentals of phylogenetic systematics, p. 14–185. *In*: Systematics, historical ecology, and North American freshwater fishes, R. L. Mayden (ed.). Stanford, University Press, Palo Alto.

Mayden, R. L., W. J. Rainboth, and D. G. Buth. 1991. Phylogenetic

systematics of the cyprinid genera *Mylopharodon* and *Ptychocheilus:* Comparative morphometry. Copeia 1991:819–834.

Maynard, D. J. 1988. Status signalling and the social structure of juvenile coho salmon. Diss. Abst. Int. Pt. Biol. Sci. and Eng. 48(12).

Mayr, E. 1963. Animal species and evolution. Harvard University Press, Cambridge.

McAllister, D. E. 1968. Evolution of branchiostegals and classification of teleostome fishes. Bull. Nat. Mus. Can. 221:1–239.

McArdle, J., and A. M. Bullock. 1987. Solar ultraviolet radiation as a causal factor of "summer syndrome" in cage-reared Atlantic salmon, *Salmo salar* L.: A clinical and histopathological study. J. Fish. Dis. 10(4):255–264.

McCormick, C. A. 1983. Organization and evolution of the octavolateralis area of fishes, p. 179–213. *In*: Fish neurobiology, Vol. I, R. G. Northcutt and R. E. Davis (eds.). University of Michigan Press, Ann Arbor.

McCormick, S. D. 1990. Cortisol directly stimulates differentiation of chloride cells in tilapia opercular membrane. Am. J. Physiol. 259(4):R857–R863.

McCosker, J. E. 1985. White shark attack behavior: Observations of and speculations about predator and prey strategies. Mem. So. Calif. Acad. Sci. 9:123–135.

McCune, A. R., and B. Schaffer. 1986. Triassic and Jurassic fishes: Patterns and diversity, p. 171–181. *In*: The beginning of the age of dinosaurs. Cambridge University Press, Cambridge.

McCune, A. R., K. S. Thomson, and P. E. Olsen. 1984. Semionotid fishes from the mesozoic Great Lakes of North America, p. 27–44. In: Evolution of fish species flocks, A. A. Echelle and I. Kornfield (eds.). University of Maine Press, Orono.

McDonald, P., R. A. Edwards, and J. F. D. Greenhalgh. 1966. Animal nutrition. Oliver and Boyd, Edinburgh.

McDowall, R. M., 1969. Relationships of galaxioid fishes with a further discussion of salmoniform classification. Copeia 1969: 769–824.

McDowall, R. M. 1988. Diadromy in fishes. Timber Press.

McDowall, R. M. 1990. New Zealand freshwater fishes: A natural history and guide (Second Edition). Heinemann-Reed, Auckland.

McDowall, R. M., and B. J. Pusey. 1983. *Lepidogalaxias salamandroides* Mees—a redescription, with natural history notes. Rec. West. Aus. Mus. 11(1):11–23.

McDowall, R. M., B. M. Clark, G. J. Wright, and T. G. Northcote. 1993. Trans-2-cis-6-nonadienal: The cause of cucumber odor in osmerid and retropinnid smelts. Trans. Am. Fish. Soc. 122(1): 144–147.

McDowell, S. B. 1973. Order Heteromi (Notacanthiformes), p. 1–228. *In*: Fishes of the western North Atlantic, D. M. Cohen (ed.-in-chief). Mem. Sears Found. Mar. Res., New Haven.

McFall-Ngai, M. J. 1990. Crypsis in the pelagic environment. Am. Zool. 30(1):175–188.

McFall-Ngai, M. J., and P. V. Dunlap. 1982. Three new modes of luminescence in the leiognathid fish *Gazza minuta*: Discrete luminescence. Mar. Biol. 73(3):227–237.

McFall-Ngai, M., and J. G. Morin. 1991. Camouflage by disruptive illumination in leiognathids, a family of shallow-water bioluminescent fishes. J. Exp. Biol. 156:119–137.

McFall-Ngai, M. J., F. Cressitelli, J. Childress, and J. Horwitz.

1986. Patterns of pigmentation in the eye lens of the deep-sea hatchet fish *Argyropelecus affinis* Garman. J. Comp. Physiol. A159:791–800.

McHugh, J. L. 1967. Estuarine nekton, p. 581–620. *In*: Estuaries, G. Lauff (ed.). Amer. Assoc. Adv. Sci. Publ. No. 83, Washington, DC.

McKaye, K. R. 1977. Defense of a predator's young by a herbivorous fish: An unusual strategy. Am. Nat. 111:301–315.

McKaye, K. R. 1984. Behavioral aspects of cichlid reproductive strategies: Patterns of territoriality and brood defence in Central American substratum spawners and African mouth brooders, p. 245–273. *In*: Fish reproduction: Strategies and tactics, G. W. Potts and R. J. Wooton (eds.). Academic Press, London, Orlando.

McKaye, K. R., and T. Kocher. 1983. Head ramming behavior by three paedophagous cichlids in Lake Malawi, Africa. Anim. Behav. 31(1):206–210.

McKaye, K. R., and A. Marsh. 1983. Food switching by two specialized algae-scraping cichlid fishes in Lake Malawi, Africa. Oecologia 56(2–3):245–248.

McKaye, K. R., and N. M. McKaye. 1977. Communal care and kidnapping of young by parental cichlids. Evol. 31(3):674–681.

McKaye, K. R., and M. K. Oliver. 1980. Geometry of a selfish school: Defence of cichlid young by bagrid catfish in Lake Malawi, Africa. Anim. Behav. 28(4):1287.

McKeown, B. 1984. Fish migration. Croom Helm, London.

McNulty, J. A. 1981. A quantitative morphological study of the pineal organ in the goldfish, *Carassius auratus*. Can. J. Zool. 59(7):1312–1325.

McVay, J. A., and H. W. Kaan. 1940. The digestive tract of *Carassius auratus*. Biol. Bull. 78(1):53–67.

Mead, G. W., and R. M. Bailey. 1963. The poeciliid fishes (Cyprinodontiformes), their structure, zoogeography, and systematics. Bull. Am. Mus. Nat. Hist. 126:1–176.

Mead, G. W., E. Bertelsen, and D. M. Cohen. 1964. Reproduction among deep sea fishes. Deep-Sea Res. 11:569–596.

Meador, M. R., and W. E. Kelso. 1990. Physiological responses of largemouth bass, *Micropterus salmoides,* exposed to salinity. Can. J. Fish. Aquat. Sci. 47:2358–2363.

Meffe, G. K., and F. F. Snelson, Jr. (eds.). 1989. Ecology and evolution of livebearing fishes (Poeciliidae). Prentice Hall, Englewood Cliffs, NJ.

Meagher, S., and T. E. Dowling. 1991. Hybridization between the cyprinid fishes *Luxilus albeolus, L. cornutus,* and *L. cerasinus* with comments on the proposed hybrid origin of *L. albeolus*. Copeia 1991:979–991.

Meinke, D. K. 1982. A light and scanning electron microscope study of microstructure, growth and development of the dermal skeleton of *Polypterus* (Pisces: Actinopterygii). J. Zool. 197(3):355–382.

Meisami, E. 1991. Chemoreception, p. 335–434. *In*: Neural and integrative animal physiology. C. L. Prosser (ed.). Wiley, New York.

Meredith, M., and J. White. 1987. Interactions between the olfactory system and the terminal nerve, p. 349–368. *In*: The terminal nerve (nervus terminalis) structure function and evolution, L. Danski and M. Schwanzel-Fukuda (eds.). Ann. New York Acad. Sci. Vol. 519.

Merrick, J. R., and G. E. Schmida. 1984. Australia freshwater fishes: Biology and management. Griffin Press, Netley, Australia.

Mertz, E. T. 1969. Amino acid and protein requirements of fish, p. 233–244. *In*: Fish in research, O. W. Neuhaus and J. E. Halver (eds.). Academic Press, New York.

Meyer, A. 1993. Evolution of mitochondrial DNA in fishes, p. 1–38. *In*: Biochemistry and molecular biology of fishes, Vol. 2: Molecular biology frontiers, P. W. Hochachka and T. P. Mommsen (eds.). Elsevier, Amsterdam.

Meyer, A., and A. C. Wilson. 1990. Origin of tetrapods inferred from their mitochondrial DNA affiliation to lungfish. J. Mol. Evol. 31:359–364.

Meyer, A., T. D. Kocher, P. Basasibwaki, and A. C. Wilson. 1990. Monophyletic origin of Lake Victoria cichlid fishes suggested by mitochondrial DNA sequence. Nature 347:550–553.

Michael, J. H. 1984. Additional notes on the repeat spawning by Pacific lamprey. Calif. Fish and Game 70(3):186–188.

Miles, H. M. 1971. Renal function in migrating adult coho salmon. Comp. Biochem. Physiol. 38A:787–926.

Miles, H. M., and L. S. Smith-Vaniz. 1968. Ionic regulation in migrating juvenile coho salmon, *Oncorhynchus kisutch*. Comp. Biochem. Physiol. 26:381–398.

Miles, R. S. 1973. Relationships of acanthodians, p. 63–104. *In*: Interrelationships of fishes, P. H. Greenwood, R. S. Miles and C. Patterson (eds.). J. Linn. Soc., Vol. 53, Suppl No. 1. Academic Press, London.

Miller, D. J., and R. N. Lea. 1972. Guide to the coastal marine fishes of California. Calif. Fish Bull. 157.

Miller, R. R., J. D. Williams, and J. E. Williams. 1989. Extinctions of North American fishes during the past century. Fisheries 14(6):22–38.

Miller, R. V. 1964. The morphology and function of the pharyngeal organs in the clupeid, *Dorosoma petenense* (Gunther). Chesapeake Sci. 5:194–199.

Miller, T. J., L. B. Crowder, J. A. Rice, and E. A. Marshall. 1988. Larval size and recruitment mechanisms in fishes: Toward a conceptual framework. Can. J. Fish Aquat. Sci. 45:1657–1670.

Mills, D. H. 1989. Ecology and management of Atlantic salmon. Chapman and Hall, London.

Milne, L., and M. Milne. 1972. The senses of animals and men. Atheneum, New York.

Minckley, W. L., and J. E. Deacon (eds.). 1991. Battle against extinction: Native fish management in the American west. University of Arizona Press, Tucson and London.

Minckley, W. L., D. A. Hendrickson, and C. E. Bond. 1986. Zoogeography of freshwater fishes in western North America. John Wiley and Sons, New York.

Mochioka, N., and M. Iwazuma. 1993. Function of peculiar large teeth of leptocephalus eel larvae. Paper presented at Fourth Indo-Pacific Fish Conf. (Bangkok). Abstr. p. 73 of program.

Mok, H. K. 1981. Sound production in the naked goby, *Gobiosoma bosci* (Pisces, Gobiidae)—a preliminary study, p. 447–455. *In*: Hearing and sound communication in fishes, W. N. Tavolga, A. N. Popper, and R. R. Fay (eds.). Springer-Verlag, New York.

Moller, P. 1976. Electric signals and schooling behavior in a weakly electric fish, *Marcusenius cyprinoides* L. (Mormyriformes). Science 193:697–699.

Molnar, G., E. Tomassy, and I. Tolg. 1967. The gastric digestion of living predatory fish, p. 135–149. *In*: The biological basis of freshwater fish production, S. D. Gerking (ed.). Blackwell Scientific Publications, Oxford.

Montgomery, J. C. 1988. Sensory physiology, p. 79–98. *In*: Physiology of elasmobranch fishes, T. J. Shuttleworth (ed.). Springer-Verlag, Berlin, Heidelberg.

Montgomery, J. C., N. W. Pinkhurst, and B. A. Foster. 1989. Limitations on visual food-location in the planctivorous antarctic fish *Pagothenia borchgrevinki*. Experientia 45:395–397.

Montgomery, W. L. 1975. Interspecific associations of sea basses (Serranidae) in the Gulf of California. Copeia 1975:785–787.

Moore, A., S. M. Freake, and I. M. Thomas. 1990. Magnetic particles in the lateral line of the Atlantic salmon (*Salmo salar* L.). Phil. Trans. Roy. Soc. Lond. B 329:11–15.

Moore, J. A. 1990. A reanalysis of the osteology and phylogeny of "Beryciformes." Abstract, p. 192, 70th Ann. Meeting. Am. Soc. Ichthyol. and Herpetol.

Moore, J. A. 1993. Phylogeny of the Trachichthyiformes (Teleostei: Percomorpha). Bull. Mar. Sci. 52(1):114–136.

Moore, K. S., S. Wehrli, H. Roder, M. Rogers, J. N. Forest, Jr. D. McGrimmon, and M. Zazloff. 1993. Squalamine: An ammosterol antibiotic from the shark. Proc. Nat. Acad. Sci. 90(4):1354–1358.

Moran, P., and I. Kornfield. 1993. Retention of an ancestral polymorphism in the mbuna species flock (Teleostei: Cichlidae) of Lake Malawi. Mol. Biol. Evol. 10:1015–1029.

Morgan, J. D., and G. K. Iwama. 1990. The energetics of ion regulation in freshwater resident and anadromous juvenile rainbow trout (*Oncorhynchus mykiss*). Bull. Aquacult. Assoc. Canada 90–4:57–60.

Morgan, M., and P. W. A. Tovell. 1973. The structure of the gill of the trout, *Salmo gairdneri* (Richardson). Z. Zellforsch. Mikrosk. Anat. 142:147–162.

Morin, J. G. 1981. Bioluminescent patterns in shallow tropical marine fishes, p. 569–574. *In*: The reef and man, E. D. Gomez, C. E. Birkeland, R. W. Buddenmeier, R. E. Johannes, J. A. Marsh, Jr. and R. T. Tsuda (eds.). Proceedings of the Fourth International Symposium, Vol. 2. Marine Science Center, Quezon City, Philippines.

Morizot, D. C. 1990. Use of fish gene maps to predict ancestral vertebrate genome organization, p. 207–234. *In*: Isozymes: structure, function, and use in biology and medicine. Z. I. Ogita and C. L. Markert (eds.). Wiley-Liss, New York.

Morizot, D. C., S. A. Slaugenhaupt, K. D. Kallerman, and A. Chakravarti. 1991. Genetic map of fishes of the genus *Xiphophorus* (Teleostei: Poeciliidae). Genetics 127:399–410.

Morizot, D. C., S. W. Calhoun, L. L. Clepper, M. E. Schmidt, J. H. Williamson, and G. J. Carmichael. 1991. Multispecies hybridization among native and introduced centrarchid basses in central Texas. Trans. Am. Fish. Soc. 120:283–289.

Morris, R. 1960. General problems of osmoregulation with special reference to cyclostomes, p. 1–16. *In*: Hormones in fish, I. C. Jones (ed.). Sympos. Zool. Soc. Lond., London.

Morris, R. 1972. Osmoregulation, p. 193–239. *In*: The biology of lampreys, Vol. 2, M. W. Hardisty and I. C. Potter (eds.). Academic Press, London.

Morris, S. L., and A. J. Gaudin. 1975. The cranial osteology of *Amphistichus argenteus* (Pisces: Embiotocidae). Bull. So. Cal. Acad. Sci. 75:29–38.

Morrow, J. D., G. R. Margolies, J. Rowland, and L. J. Roberts, II. 1991. Evidence that histamine is the causative toxin of scombroid-fish poisoning. New Engl. J. Med. 324(11):716–720.

Morrow, J. E. 1964. General discussion and key to families, Suborder Stomiatoidea, p. 71–76. *In*: Fishes of the western North Atlantic, part 4. Mem. Sears Found. Mar. Res.

Moser, H. G. 1981. Morphological and functional aspects of marine fish larvae, p. 90–131. *In*: Marine fish larvae, R. Lasker (ed.). University of Washington Press, Seattle.

Moser, H. G., and E. H. Ahlstrom. 1974. Role of larval stages in systematic investigations of marine teleosts: The Myctophidae, a case study. Fish. Bull. U. S. 72:391–413.

Moser, H. G., W. J. Richards, D. M. Cohen, M. P. Fahay, A. W. Kendall, Jr. and S. L. Richardson (eds.). 1984. Ontogeny and systematics of fishes. Am. Soc. Ichthyol. and Herpetol. Sp. Publ. 1.

Mosse, P. R. L., and R. C. L. Hudson. 1977. The functional roles of different muscle fibre types identified in the myotomes of marine teleosts: A behavioral, anatomical and histochemical study. J. Fish. Biol. 11:417–430.

Mossman, H. W. 1987. Vertebrate fetal membranes. Rutgers University Press, New Brunswick, NJ.

Motta, P. J. 1982. Functional morphology of the head of the inertial suction feeding butterflyfish, *Chaetodon miliaris* (Perciformes, Chaetodontidae). J. Morph. 174(3):283–312.

Motta, P. J. 1984a. Tooth attachment, replacement and growth in the butterflyfish, *Chaetodon miliaris* (Chaetodontidae, Perciformes). Can. J. Zool. 62(2):183–189.

Motta, P. J. 1984b. Mechanics and functions of jaw protrusion in teleost fishes: A review. Copeia 1984:1–18.

Motta, P. J. 1985. Functional morphology of the head of Hawaiian and mid-Pacific butterfly-fishes (Perciformes, Chaetodontidae). Env. Biol. of Fishes 13(3):253–276.

Motta, P. J. 1987. A quantitative analysis of ferric iron in butterflyfish teeth (Chaetodontidae, Perciformes) and the relationship to feeding ecology. Can. J. Zool. 65(1):106–112.

Moulton, J. M. 1960. Swimming sounds and the schooling of fishes. Biol. Bull. Woods Hole 119:210.

Moulton, J. M. 1963. Acoustic behaviour of fishes, p. 665–693. *In*: Acoustic behaviour of animals, R.-G. Busnel (ed.). Elsevier, Amsterdam.

Moyle, P. B. 1976. Inland fishes of California. University of California Press, Berkeley.

Moyle, P. B., and J. J. Cech, Jr. 1982. Fishes: An introduction to ichthyology. Prentice-Hall, Englewood Cliffs, NJ.

Moyle, P. B., and B. Herbold. 1987. Life-history patterns and community structure in stream fishes of western North America: Comparisons with eastern North America and Europe, p. 25–32. *In*: Community and evolutionary ecology of North American stream fishes, W. J. Matthews and D. C. Heins (eds.). University of Oklahoma Press, Norman, OK.

Moyle, P. B., and R. M. Leidy. 1992. Loss of biodiversity in aquatic ecosystems: Evidence from fish faunas, p. 127–169. *In*: Conservation biology. The theory and practice of nature conservation, preservation, and management, P. L. Fiedler and S. K. Jain (eds.). Chapman and Hall, New York.

Moyle, P. B., and J. E. Williams. 1990. Biodiversity loss in the temperate zone: Decline of the native fish fauna of California. Conser. Biol. 4:275–284.

Moy-Thomas, J. A., and R. S. Miles. 1971. Palaeozoic fishes. W. B. Saunders, Philadelphia, PA.

Müller, K. 1978a. The flexibility of the circadian system of fish at different latitudes, p. 91–104. In: Rhythmic activity of fishes, J. E. Thorpe (ed.). Academic Press, New York.

Müller, K. 1978b. Locomotor activity of fish and environmental oscillations, p. 1–19. In: Rhythmic activity of fishes, J. E. Thorpe (ed.). Academic Press, New York.

Munford, J., and L. Greenwald. 1974. The hypoglycemic effects of external insulin in fish and frogs. J. Exp. Zool. 90(3):341–345.

Munk, O. 1982. Cones in the eye of the deep-sea teleost *Diretmus argenteus.* Vision Res. 22:179–181.

Munk, O. 1984. Non-spherical lenses in the eyes of some deep-sea teleosts. Arch. Fischereiwiss 34(2–3):145–153.

Munk, O., and R. D. Frederickson. 1974. On the function of aphakic apertures in teleosts. Vidensk. Medd. Dan. Naturhist. Foren. Khb. 137:65–94.

Muntz, W. R. A. 1976. On yellow lenses in mesopelagic animals. J. Mar. Biol. Assn. U. K. 56:963–976.

Muntz, W. R. A. 1983. Bioluminescence and vision, p. 217–238. In: Experimental biology at sea, A. G. MacDonald and I. G. Priede (eds.). Academic Press, London, New York.

Muntz, W. R. A. 1990. Stimulus, environment and vision in fishes, p. 491–511. In: The visual system of fish, R. Douglas and M. Djamgos (eds.). Chapman and Hall, London, New York, Tokyo, Melbourne, Madras.

Muntz, W. R. A., and G. S. V. Mouat. 1984. Annual variations in the visual pigments of brown trout inhabiting lochs providing different light environments. Visual Pigment Biochem. 24:1575–1580.

Munz, F. W. 1971. Vision: Visual pigments, p. 1–32. In: Fish physiology, Vol. V., W. S. Hoar and D. J. Randall (eds.). Academic Press, New York and London.

Münz, H. 1989. Functional organization of the lateral line periphery, p. 285–297. In: The mechanosensory lateral line, S. Coombs, P. Görner and H. Münz (eds.). Springer-Verlag, New York.

Murata, M., K. Miyagawa–Kohshima, K. Nakanishi, and Y. Naya. 1986. Characterization of compounds that induce symbiosis between sea anemone and fish. Science 234:585–587.

Murphy, C. J., and H. C. Howland. 1990 (1991). The functional significance of crescent-shaped pupils and multiple pupillary apertures. J. Exp. Zool. Suppl. 5:22–28.

Murphy, R. C. 1971. The structure of the pineal organ of the bluefin tuna, *Thunnus thynnus.* J. Morph. 133:1–16.

Murphy, R. W., J. W. Sites, Jr. D. G. Buth, and C. H. Haufler. 1990. Proteins I: Isozyme electrophoresis, p. 45–126. In: Molecular systematics, D. H. Hillis and C. Moritz (eds.). Sinauer Associates, Sunderland, MA.

Musick, J. A., M. W. Bruton, and E. K. Balon (eds.). 1991. The biology of *Latimeria chalumnae* and the evolutionary coelacanths. Environ. Biol. of Fish. 32(1–4).

Myers, G. S. 1938. Freshwater fishes and West Indian zoogeography. Smithsonian Report for 1937, Pub. 3645:339–364.

Myers, G. S. 1941. The fish fauna of the Pacific ocean, with special reference to zoogeographical regions and distribution as they affect the international aspects of the fisheries. Proc. 6th Pacific Sci. Cong. (3):201–210.

Myers, G. S. 1949. Salt tolerance of fresh-water fish groups in relation to zoogeographical problems. Bijdr. Dierk 28:315–322.

Myers, G. S. 1950. Flying of the half beak (*Euleptorhamphus*). Copeia 1950:320.

Myers, G. S. 1951. Freshwater fishes and East Indian zoogeography. Stanford Ichthyol. Bull. 4:11–21.

Myers, G. S. 1958. Trends in the evolution of teleostean fishes. Stanford Ichthyol. Bull. 7(3):27–30.

Myers, G. S. 1964. A brief sketch of the history of ichthyology in America to the year 1850. Copeia 1964:33–41.

Myrberg, A. A., Jr. 1981. Sound communication and interception in fishes, p. 395–425. In: Hearing and sound communication in fishes, W. N. Tavolga, A. N. Popper and R. R. Fay (eds.). Springer-Verlag, New York.

Myrberg, A. A., Jr. and D. R. Nelson. 1991. The behavior of sharks: What have we learned? p. 92–100. In: Discovering sharks, S. H. Gruber (ed.). Spec. Pub. 14, Am. Littoral Soc., Highlands, NJ.

Myrberg, A. A., Jr. M. Mohler, and J. D. Catala. 1986. Sound production by males of a coralreef fish (*Pomacentrus partitus*): Its significance to females. Anim. Behav. 34(3):913–923.

Nachtigall, W. (ed.). 1977. The physiology of movement: Biomechanics. Gustav Fischer Verlag, Stuttgart.

Nagahama, Y., R. S. Nishioka, H. A. Bern, and R. L. Gunther. 1975. Control of prolactin secretion in teleosts, with special reference to *Gillichthys mirabilis* and *Tilapia mossambica.* Gen. Comp. Endocrinol. 25(2):166–188.

Naiman, R. J., and D. L. Soltz (eds.). 1981. Fishes in North American deserts. John Wiley and Sons, New York.

Naitoh, T., A. Morikawa, and Y. Omura. 1985. Adaptation of a common freshwater goby, yoshinobori, *Rhinogobius branneus* Temminck and Schlegel to various backgrounds including those containing different sizes of black and white checkerboard pattern. Zool. Sci. 2(1):59–63. (Abstract seen.)

Nakabo, T. (ed.). 1993. Fishes of Japan with pictorial keys to the species. Tokai University Press, Tokyo.

Nakamura, H. 1969. Tuna distribution and migration. Fishing News (Books), London.

Nakano, H., and M. Tabuchi. 1990. Occurrence of the cookiecutter shark *Isistius brasiliensis* in surface waters of the North Pacific Ocean. Jpn. J. Ichthyol. 37(1):60–63.

Nash, C. E. 1988. A global overview of aquaculture production. J. World Aquacult. Soc. 19(2):51–58.

Nash, J. 1931. The number and size of glomeruli in the kidneys of fishes, with observations on the morphology of the renal tubules of fishes. Am. J. Anat. 47(2):425–446.

National Academy of Sciences–National Research Council. 1973. Nutrient requirements of trout, salmon, and catfish. Nutrient Requirements of Domestic Animal Series. National Academy of Sciences No. 11.

National Research Council. 1981. Nutrient requirements of coldwater fishes. National Academy Press, Washington, DC.

National Research Council. 1983. Nutrient requirements of warmwater fishes. National Academy Press, Washington, DC.

Natochin, Y., and G. P. Gusev. 1970. The coupling of magnesium secretion and sodium reabsorption in the kidney of teleosts. Comp. Biochem. Physiol. 37(1):107–111.

Neilsen, J. G., and O. Munk. 1964. A hadal fish (*Bassogigas profundissimus*) with a functional swim bladder. Nature 204(4958): 594–595.

Neilson, J. D., and R. I. Perry. 1990. Diel vertical migrations in young fish: A facultative or obligate phenomenon, p. 115–168. *In*: Advances in marine biology, Vol. 26, R. S. Russell (ed.). Academic Press, London.

Nellen, W. 1986. A hypothesis on the fecundity of bony fish. Meeresforsch. 31:75–89.

Nelson, D. R. 1991. Shark repellants: How effective, how needed, p. 106–108. *In*: Discovering sharks, S. H. Gruber (ed.). American Littoral Soc., Highlands, NJ.

Nelson, G. J. 1969a. Gill arches and the phylogeny of fishes, with notes on the classification of vertebrates. Bull. Am. Mus. Nat. Hist. 141(4):475–552.

Nelson, G. J. 1969b. Infraorbital bones and their bearing on the phylogeny and geography of osteoglossomorph fishes. Am. Mus. Nov. 2394:1–37.

Nelson, G. J. 1969c. Origin and diversification of teleostean fishes. Ann. N.Y. Acad. Sci. 167:18–30.

Nelson, G. J. 1970. Pharyngeal denticles (placoid scales) of sharks with notes on the dermal skeleton of vertebrates. Am. Mus. Nov. 2415:1–26.

Nelson, G. J. 1972. Cephalic sensory canals, pitlines, and the classification of esocoid fishes, with notes on galaxiids and other teleosts. Am. Mus. Nov. 2492:1–49.

Nelson, G. J. 1973. Relationships of clupeomorphs, with remarks on the structure of the lower jaw in fishes, p. 333–349. *In*: Interrelationships of fishes, P. H. Greenwood, R. S. Miles and C. Patterson (eds.). Academic Press, New York.

Nelson, J. S. 1976. Fishes of the world. Wiley-Interscience, New York.

Nelson, J. S. 1982. Two new South Pacific fishes of the genus *Ebinania* and contributions to the systematics of Psychrolutidae (Scorpaeniformes). Can. J. Zool. 60(6):1470–1504.

Nelson, J. S. 1984. Fishes of the world (Second Edition). John Wiley and Sons, New York.

Nelson, J. S. 1994. Fishes of the world (Third Edition). John Wiley and Sons, New York.

Nelson, K., and M. Soulé. 1987. Genetical conservation of exploited fishes, p. 345–368. *In*: Population genetics in fishery management, N. Ryman and F. M. Utter (eds.). University of Washington Press, Seattle.

Neudecker, S. 1989. Eye camouflage and false eyespots: Chaetodontid responses to predators. Env. Biol. Fishes 25(1–3): 143–157.

Neumeyer, C. 1986. Wavelength discrimination in goldfish. J. Comp. Physiol. A 158:203–213.

Ngan, P. V., K. Hamamori, I. Hanyu, and T. Hibiya. 1974. Measurement of blood pressure of carp. Jpn. J. Ichthyol. 21(1):1–8.

Nicol, J. A. C. 1963. Some aspects of photoreception and vision in fishes. Adv. Mar. Biol. 1:171–208.

Nicol, J. A. C. 1969. Bioluminescence, p. 355–400. *In*: Fish physiology, Vol. III, W. S. Hoar and D. J. Randal. (eds.). Academic Press, New York and London.

Nicol, J. A. C. 1989. The eyes of fishes. Clarendon Press, Oxford.

Nicoletta, P. F. 1991. The relationship between male ornamentation and swimming performance in the guppy, *Poecilia reticulata*. Behav. Ecol. Sociobiol. 28(5):365–370.

Nicoll, C. S., S. W. Wilson, R. Nishioka, and H. A. Bern. 1981. Blood and pituitary prolactin levels in tilapia (*Sarotherodon mossambicus*; Teleostei) from different salinities as measured by a homologous radioimmunoassay. Gen. Comp. Endocrinol. 44:365–373.

Nieuwenhuys, R. 1959. The structure of the telencephalon of the teleost *Gasterosteus aculeatus*. I. Proc. Kon Ned. Akad. Wet., Series C. 62:341–362.

Nieuwenhuys, R. 1962. Trends in the evolution of the actinopterygian forebrain. J. Morphol. 111:69–88.

Nieuwenhuys, R., and E. Pouwels. 1983. The brain stem of actinopterygian fishes, p. 25–87. *In*: Fish neurobiology, Vol. I, R. G. Northcutt and R. E. Davis (eds.). University of Michigan Press, Ann Arbor.

Nikolsky, G. V. 1961. Special ichthyology. Washington, DC. Department of Commerce, Israel program for Scientific Translation.

Nikolsky, G. V. 1963. The ecology of fishes. Academic Press, New York.

Nilsson, S. 1986. Control of gill blood flow, p. 86–101. *In*: Fish physiology: Recent advances, S. Nilsson and S. Holmgren (eds.). Croom Helm, London, Sydney, Dover, NH.

Noakes, D. L. G. 1978. Ontogeny of behavior in fishes: A survey and suggestions, p. 103–125. *In*: The development of behavior: Comparative and evolutionary aspects, G. M. Burghardt and M. Bekoff (eds.). Garland STPM Press, New York.

Noakes, D. L. G. 1986. Genetic basis of behavior, p. 3–22. *In*: The behavior of teleost fishes, T. J. Pitcher (ed.). Johns Hopkins University Press, Baltimore.

Noback, C. R. 1977. The nervous system, p. 465–519. *In*: Chordate structure and function, A. G. Kluge (ed.). Macmillan, New York.

Noller, H. F., and C. R. Woese. 1981. Secondary structure of 16S ribosomal RNA. Science 212:403–411.

Noller, H. F., V. Hoffarth, and L. Zimniak. 1992. Unusual resistance of peptidyl transferase to protein extraction procedures. Science 258(5062):1416–1419.

Nordlie, F. G., S. J. Walsh, D. C. Haney, and T. F. Nordlie. 1991. The influence of ambient salinity on routine metabolism in the teleost *Cyprinodon variegatus* Lacepede. J. Fish Biol. 38:115–122.

Norman, J. R. 1934. A systematic monograph of the flatfishes (Heterosomata). Br. Mus. (Nat. Hist.) 1:1–459.

Norman, J. R. 1957. A draft synopsis of the orders, families and genera of Recent fishes and fish-like vertebrates. Br. Mus. (Nat. Hist.), London. Unpublished photo offset copies.

Norman, J. R., and P. H. Greenwood. 1975. A history of fishes (Third Edition). Halsted Press, New York.

Normark, B. B., A. R. McCune, and R. G. Harrison. 1991. Phylogenetic relationships of neopterygian fishes, inferred from mitochondrial DNA sequences. Mol. Biol. Evol. 8(6):819–834.

Norris, D. O., and R. E. Jones. 1987. Hormones and reproduction in fishes, amphibians, and reptiles. Plenum Press, New York and London.

Norris, K. S., and B. Mohr. 1983. Can odontocoetes debilitate prey with sound? Am Nat. 122(1):85–104.

Northcutt, R. G. 1986. Electroreception in nonteleost bony fishes, p. 257–285. *In*: Electroreception, T. H. Bullock and W. Heiligenberg (eds.). Wiley, New York.

Northcutt, R. G. 1989a. Phylogeny and innervation of lateral lines, p. 17–78. *In*: The mechanosensory lateral line, S. Coombs, P. Görner and H. Münz (eds.). Springer-Verlag, New York, Berlin, Heidelberg.

Northcutt, R. G. 1989b. Brain variation and phylogenetic trends in elasmobranch fishes, p. 83–100. *In*: Evolutionary and contemporary biology of elasmobranchs, W. C. Hamlett and B. Tota (eds.). Alan R. Liss, New York.

Northcutt, R. G., and W. E. Bemis. 1993. Cranial nerves of the coelacanth, *Latimeria chalumnae* (Osteichthyes: Sarcopterygii: Actinistia), and comparisons with other Craniata. Brain, Behav. Evol. 42(S1):v–x,1–76.

Northcutt, R. G., and R. E. Davis. 1983. Telencephalic organization in ray-finned fishes, p. 203–236. *In*: Fish neurobiology, R. G. Northcutt and R. E. Davis (eds.). University of Michigan Press, Ann Arbor.

Northcutt, R. G., and C. Gans. 1983. The genesis of neural crest and epidermal placodes: A reinterpretation of vertebrate origins. Quart. Rev. Biol. 58(1):1–28.

Nursall, J. R. 1962. Swimming and the origin of paired appendages. Am. Zool. 2:127–141.

O, W., and T. H. Chan. 1974. A cytological study on the structure of the pituitary gland of *Monopterus albus* (Zuiew). Gen. Comp. Endocrinol. 24(2):208–222.

Obruchev, D. V. 1967. Class Placodermi, p. 168–259. *In*: Fundamentals of paleontology, Vol. XI, Agnatha, Pisces. Jerusalem. Israel Program for Scientific Translations.

O'Connell, C. P. 1981. Development of organ systems in the northern anchovy, *Engraulis mordax*, and other teleosts. Am. Zool. 21:429–446.

O'Day, W. T., and H. R. Fernandez. 1974. *Aristostomias scintillans* (Malacosteidae): A deep-sea fish with visual pigment apparently adapted to its own bioluminescence. Vision Res. 14:545–550.

Odiorne, J. M. 1959. Color changes, p. 387–401. *In*: The physiology of fishes, Vol. 2, M. E. Brown (ed.). Academic Press, New York.

Ogawa, M. 1961. Comparative study of the external shape of the teleostean kidney with relation to phylogeny. Sci. Rep. Tokyo Kyoiku Daigaku B10:61–68.

Ogawa, M. 1962. Comparative study on the internal structure of the teleostean kidney. Sci. Rep. Saitama Univ. B4(2):107–131.

Oguri, M. 1964. Rectal glands of marine and freshwater sharks: Comparative histology. Science 144:1151–1152.

Ohizumi, Y., and M. Kobayashi. 1990. Co-dependent excitatory effects of maitotoxin on smooth and cardiac muscle, p. 133–143. *In*: Marine toxins: Origin, structure and molecular pharmacology, S. Hall and G. Strichartz (eds.). ACS Symp. Ser. 418.

Ohno, S. 1970. Evolution by gene duplication. Springer-Verlag, Berlin and New York.

Oikari, A. 1975. Seasonal changes in plasma and muscle hydromineral balance in three Baltic teleosts, with special reference to the thermal response. Ann. Zool. Fenn. 12(3):230–236.

Okiyama, M. 1984. Myctophiformes: Relationships, p. 254–259. *In*: Ontogeny and systematics of fishes, H. G. Moser, W. J. Richards, D. M. Cohen, M. P. Fahay, A. W. Kendall, Jr. and S. L. Richardson (eds.). Am. Soc. Ichthyol. and Herpetol. Sp. Pub. 1.

Olla, B. L., and M. W. Davis. 1990. Effects of physical factors on the vertical distribution of larval walleye pollock *Theragra chalcogramma* under controlled laboratory conditions. Mar. Ecol. Prog. Ser. 63:105–112.

Olney, J. E. 1984. Lampriformes: Development and relationships, p. 368–379. *In*: Ontogeny and systematics of fishes, H. G. Moser, W. J. Richards, D. M. Cohen, M. P. Fahay, A. W. Kendall, Jr. and S. L. Richardson (eds.). Am. Soc. Ichthyol. and Herpetol. Sp. Pub. 1.

Olney, J. E., G. D. Johnson, and C. G. Baldwin. 1993. Phylogeny of lampridiform fishes. Bull. Mar. Sci. 52(1):137–169.

Olsèn, K. H. 1992. Kin recognition in fish mediated by chemical senses, p. 229–248. *In*: Fish chemoreception, T. J. Hara (ed.). Chapman and Hall, London.

Olsen, P. E., and A. R. McCune. 1991. Morphology of the *Semionotus* species group from the early Jurassic part of the Newark Supergroup of eastern North America with comments on the family Semionotidae (Neopterygii). J. Vertebr. Paleont. 11(3):269–292.

Omura, Y., and M. A. Ali. 1981. Ultrastructure of the pineal organ of the killifish, *Fundulus heteroclitus,* with special reference to the secretory function. Cell Tissue Res. 219(2):355–369.

Ono, R. D. 1983. Dual motor innervation in the axial musculature of fishes. J. Fish. Biol. 22(4): 395–408.

Ono, R. D., and S. G. Poss. 1981. Structure and innervation of the swimbladder musculature in the weakfish *Cynoscion regalis* (Teleostei: Sciaenidae). Can. J. Zool. 60:1955–1967.

Oppenheimer, J. R. Mouthbreeding in fishes. Anim. Behav. 18: 493–503.

Orlov, Yu. A., and D. V. Obruchev (eds.). 1967. Fundamentals of paleontology, Vol. XI, Agnatha, Pisces. Jerusalem. Israel Program for Scientific Translations.

Orr, J. W., and R. A. Fritzche. 1993. Revision of the pipefishes family Solenostomidae. Copeia 1993:168–182.

Orvig, T. 1968a. The dermal skeleton; general considerations, p. 373–397. *In*: Current problems of lower vertebrate phylogeny. Proc. 4th Nobel Symp., T. Orvig (ed.). John Wiley & Sons, New York.

Orvig, T. (ed.). 1968b. Current problems of lower vertebrate phylogeny. Wiley-Interscience, New York.

Oshima, K., W. E. Hahn, and A. Gorbman. 1969. Electroencephalographic olfactory responses in adult salmon to waters traversed in the homing migration. J. Fish. Res. Bd. Can. 26:2123–2133.

Osse, J. W. M. 1969. Functional morphology of the head of the perch (*Perca fluviatilis* L.): An electromyographic study. Neth. J. Zool. 19:289–392.

Osse, J. W. M., and M. Muller. 1980. A model of suction feeding in teleostean fishes with some implications for ventilation, p. 335–352. *In*: Environmental physiology of fishes, M. A. Ali (ed.). Plenum Press, New York.

Owen, H. G. 1981. Constant dimensions or an expanding earth? p. 179–192. *In*: Chance, change, and challenge. The evolving earth, L. R. M. Cocks (ed.). Br. Mus. (Nat. Hist.), London.

Packard, G. C. 1974. The evolution of air-breathing in Paleozoic gnathostome fishes. Evol. 28:320–325.

Page, L. M. 1983. Handbook of the darters. T. F. H. Publications, Neptune, NJ.

Paine, M. D. 1990. Life history tactics of darters (Percidae: Etheostomatiini) and their relationship with body size, reproductive behavior, latitude and rarity. J. Fish Biol. 37:473–488.

Palmer, A. R. 1983. The decade of North American geology—1983 time scale. Geology 11:503–504.

Palmer, J. D. 1974. Biological clocks in marine organisms. Wiley-Interscience, New York.

Pan, J. 1984. The phylogenetic position of the Eugaleaspida in China. Proc. Linn. Soc. N. S. W. 107(3):309–319.

Panchen, A. L., and T. R. Smithson. 1988. The relations of the earliest tetrapods, p. 1–32. In: The phylogeny and classification of the tetrapods, Vol. 1: Amphibians, reptiles, birds, M. J. Benton (ed.). Clarenden Press, Oxford Press, Oxford.

Pandian, T. J. 1967. Intake, digestion, absorption and conversion of food in the fishes Megalops cyprinoides and Ophiocephalus striatus. Mar. Biol. 1:16–32.

Pang, P. K. T. 1977. Osmoregulation in elasmobranchs. Amer. Zool. 17:365–377.

Pang, P. K. T., and R. K. Pang. 1986. Hormone and calcium regulation in Fundulus heteraclitus. Am. Zool. 26(1):225–234.

Pankhurst, N. W. 1989. The relationship of ocular morphology to feeding modes and activity periods in shallow marine teleosts from New Zealand. Environ. Biol. Fish. 26(3):201–211.

Pankhurst, N. W., and J. M. Lythgoe. 1982. Structure and colour of the integument of the European eel Anguilla anguilla (L.). J. Fish. Biol. 21(3):279–296.

Parenti, L. R. 1981. A phylogenetic and biogeographic analysis of cyprinodontiform fishes (Teleostei, Atherinomorpha). Bull. Am. Mus. Nat. Hist. 168:335–557.

Parenti, L. R. 1984a. On the relationships of phallostethid fishes (Atherinomorpha) with notes on the anatomy of Phallostethus dunckeri Regan, 1913. Am. Mus. Novit. 2779.

Parenti, L. R. 1984b. A taxonomic revision of the Andean killifish genus Orestias (Cyprinidontiformes, Cyprinodontidae). Bull. Am. Mus. Nat. Hist. 178(2):107–214.

Parenti, L. R. 1993. Relationships of atherinomorph fishes (Teleostei). Bull. Mar. Sci. 52:170–196.

Parenti, L. R., and M. Rauchenberger. 1989. Systematic overview of the poeciliine, p. 3–12. In: Ecology and evolution of livebearing fishes (Poeciliidae), G. K. Meffe and F. F. Snelson, Jr. (eds.). Prentice-Hall, Englewood Cliffs, NJ.

Parker, J. B. 1942. Some observations on the reproductive system of the yellow perch (Perca flavescens). Copeia 1942:223–226.

Parker, T. J., and W. A. Haswell. 1962. A text-book of zoology, Vol. 2 (Seventh Edition). (Revised by A. J. Marshall.) Macmillan, London.

Parks, A. M. 1969. The neurohypophysis, p. 111–205. In: Fish physiology, Vol. II, W. S. Hoar and D. J. Randall (eds.). Academic Press, New York.

Parrish, R. D., D. L. Mallicote, and R. A. Klingbiel. 1986. Age dependent fecundity, number of spawnings per year, sex ratio, and maturation stages in northern anchovy. Fish. Bull. U. S. 84:503–517.

Parry, G. 1966. Osmotic adaptation in fishes. Biol. Rev. 41:392–444.

Pasha, S. M. K. 1964. The anatomy and histology of the alimentary canal of an herbivorous fish, Tilapia mossambica (Peters). Proc. Ind. Acad. Sci., Sec. B 59(6):340–349.

Patterson, C. 1964. A review of Mesozoic acanthopterygian fishes with special reference to those from the English Chalk. Phil. Trans Roy. Soc. Lond., B 247:213–482.

Patterson, C. 1965. The phylogeny of the chimaeroids. Philos. Trans. R. Soc. Lond. (Biol. Sci.) B 249:101–219.

Patterson, C. 1967a. Are the teleosts a polyphyletic group? Coloq. Int. Cent. Nat. Res. Scient. 163:93–109.

Patterson, C. 1967b. Classes Selachii and Holocephali, p. 666–675. In: The fossil record, W. B. Harland et al. (eds.). London Geological Society, London.

Patterson, C. 1967c. Menaspis and the bradyodonts, p. 171–205. In: Current problems of lower vertebrate phylogeny, T. Orvig (ed.). Wiley-Interscience, New York.

Patterson, C. 1967d. A second specimen of the Cretaceous teleost Protobrama and the relationships of the suborder Tselfatioidei. Ark. Zool. 19:215–234.

Patterson, C. 1968. The caudal skeleton in Lower Liassic pholidophoroid fishes. Bull. Br. Mus. (Nat. Hist.) Geol. 16(5):201–239.

Patterson, C. 1973. Interrelationships of holosteans, p. 233–305. In: Interrelationships of fishes, P. H. Greenwood, R. S. Miles, and C. Patterson (eds.). J. Linn. Soc. (Zool.) 53., Supp. 1. Academic Press, New York.

Patterson, C. 1975a. The braincase of pholidophorid and leptolepid fishes, with a review of the actinopterygian braincase. Phil. Trans. Roy. Soc. London B 269:275–579.

Patterson, C. 1975b. The distribution of mesozoic freshwater fishes. Mem. Mus. Nat. Hist. Nat. (Paris), (Ser. A, Zool.) 87:156–173.

Patterson, C. 1977. The contribution of paleontology to teleostean phylogeny, p. 579–643. In: Major patterns in vertebrate evolution, M. K. Hecht, P. C. Goody and B. M. Hecht (eds.). Plenum Press, New York.

Patterson, C. 1981. The development of the North American fish fauna—a problem of historical biogeography, Chapter 20. In: The evolving biosphere, P. L. Forey (ed.). Br. Mus. (Nat. Hist.), London.

Patterson, C. 1982. Morphology and interrelationships of primitive actinopterygian fishes. Amer. Zool. 22:241–259.

Patterson, C., and G. D. Johnson. 1995. The intermuscular bones and ligaments of teleostean fishes. Smithsonian Contr. to Zool., No. 559. Smithsonian Institution Press, Washington, DC.

Patterson, C., and D. E. Rosen. 1977. Review of ichthyodectiform and other Mesozoic teleost fishes and the theory and practice of classifying fossils. Bull. Am. Mus. Nat. Hist. 158:81–172.

Patterson, C., and D. E. Rosen. 1989. The Paracanthopterygii revisited: Order and disorder, p. 5–36. In: Papers on the systematics of gadiform fishes, D. M. Cohen (ed.). Nat. Hist. Mus., Los Angeles Co. Sci. Ser. 32.

Patterson, C., and A. B. Smith. 1987. Is the periodicity of extinction a taxonomic artifact? Nature 330:248–251.

Patzner, R. A. 1978. Experimental studies on the light sense in the hagfish Eptatretus burgeri and Paramyxine atami. Helgolander Wiss. Meeresunters. 31(1–2):180–190.

Pearson, D. M. 1982. Primitive bony fishes, with especial reference to *Cheirolepis* and palaeonisciform actinopterygians. Zool. J. Linn. Soc. 74–35–67.

Pella, J. J., and G. B. Milner. 1987. Use of genetic marks in stock composition analysis, p. 247–276. *In*: Population genetics in fishery management, N. Ryman and F. M. Utter (eds.). University of Washington Press, Seattle.

Penrith, M. J. 1972. Earliest description and name for the whale shark. Copeia 1972:362.

Peter, R. E. 1981. Gonadotropin secretion during reproductive cycles in teleosts: Influence of environmental factors. Gen. Comp. Endocrinol. 45:294–305.

Peterson, G. L., and H. Shehadeh. 1971. Changes in blood components of the mullet, *Mugil cephalus* L. following treatment with salmon gonadotropin and methyltestosterone. Comp. Biochem. Physiol. 38B:451–457.

Pezold, F. 1984. Evidence for multiple sex chromosomes in the freshwater goby, *Gobionellus shufeldti* (Pisces: Gobiidae). Copeia 1984:235–238.

Pfeiffer, W. 1962. The fright reaction of fish. Biol. Rev. Cambridge Phil. Soc. 37:475–511.

Pfeiffer, W. 1977. The distribution of fright reaction and alarm substances in fishes. Copeia 1977:517–539.

Pflieger, W. L. 1975. The fishes of Missouri. Missouri Department of Conservation, Jefferson City.

Phillips, A. M., Jr. 1969. Nutrition, digestion and energy utilization, p. 391–432. *In*: Fish physiology, Vol. I, W. S. Hoar and D. J. Randall (eds.). Academic Press, New York.

Phillips, M. J. 1989. The feeding sounds of rainbow trout, *Salmo gairdneri* Richardson. J. Fish. Biol. 35(4):589–592.

Phleger, C. F., and M. R. Grigor. 1990. Role of wax esters in determining buoyancy in *Hoplostethus atlanticus* (Beryciformes: Trachichthyidae). Mar. Biol. 105(2):229–233.

Picirrili, J. A., T. S. McConnell, A. J. Zaug, H. F. Noller, and T. R. Cech. 1992. Aminoacyl esterase activity of the *Tetrahymena* ribozyme. Science 256:1420–1424.

Pickens, P. E., and W. N. McFarland. 1964. Electric discharge and associated behaviour in the stargazer. Anim. Behav. 12:362–367.

Pickford, G., and J. W. Atz. 1957. The physiology of the pituitary gland of fishes. New York Zool. Soc., New York.

Pietsch, T. W. 1976. Dimorphism, parasitism and sex: Reproductive strategies among deepsea ceratioid anglerfishes. Copeia 1976(4): 781–793.

Pietsch, T. W. 1978a. The feeding mechanism of *Stylephorus chordatus* (Teleostei: Lampridiformes): Functional and ecological implications. Copeia 1978 (2):255–262.

Pietsch, T. W. 1978b. Evolutionary relationships of the sea moths (Teleostei: Pegasidae) with a classification of gasterosteiform families. Copeia 1978(3):517–529.

Pietsch, T. W. 1979. Systematics and distribution of ceratioid anglerfishes of the family Caulophrynidae with the description of a new genus and species from the Banda Sea. Cont. in Science. Nat. Hist. Mus., Los Angeles Co. (310):1–25.

Pietsch, T. W. 1981. The osteology and relationships of the anglerfish genus *Tetrabrachium* with comments on lophiiform classification. Fish. Bull. U.S. 79:387–419.

Pietsch, T. W. 1984. Lophiiformes: Development and relationships, p. 320–325. *In*: Ontogeny and systematics of fishes, H. G. Moser, W. J. Richards, D. M. Cohen, M. P. Fahay, A. W. Kendall, Jr. and S. L. Richardson (eds.). Am. Soc. Ichthyol. and Herpetol. Sp. Pub. 1.

Pietsch, T. W., and D. B. Grobecker. 1978. The compleat angler: Aggressive mimicry in an antennariid anglerfish. Science 201: 369–370.

Pietsch, T. W., and D. B. Grobecker. 1987. Frogfishes of the world. Stanford University Press, Stanford.

Pike, C. S., III. 1990. Uncovering the ages of sharks and its importance in fisheries management, p. 109–111. *In*: Discovering sharks, S. H. Gruber (ed.). Spec. Publ. 14, Am. Littoral Soc., Highlands, NJ.

Piper, R. G., I. B. McElwain, L. E. Orme, J. P. McCraren, L. G. Fowler, and J. R. Leonard. 1982. Fish hatchery management. U. S. Fish and Wildlife Service, Washington, DC.

Pister, E. P. 1981. The conservation of desert fishes, p. 411–445. *In*: Fishes in North American deserts, R. J. Naiman and D. L. Soltz (eds.). Wiley, New York.

Pitcher, T. J. (ed.). 1986. The behavior of teleost fishes. Johns Hopkins Press, Baltimore.

Pitcher, T. J., B. L. Partridge, and C. S. Wardle. 1976. A blind fish can school. Science 194:963–965.

Platt, C. 1983. The peripheral vestibular system of fishes, p. 89–123. *In*: Fish neurobiology, R. G. Northcutt and R. E. Davis (eds.). University of Michigan Press, Ann Arbor.

Platt, C., and A. N. Popper. 1981. Fine structure and function of the ear, p. 3–38. *In*: Hearing and sound communication in fishes, W. N. Tavolga, A. N. Popper and R. R. Fay (eds.). Springer-Verlag, New York.

Platt, C., A. N. Popper, and R. R. Fay. 1989. The ear as part of the octavolateralis system, p. 633–651. *In*: The mechanosensory lateral line, S. Coombs, P. Görner and H. Münz (eds.). Springer-Verlag, New York.

Poggendorf, D. 1976. The absolute threshold of learning in the bullhead (*Ameiurus nebulosus*) and contributions to the physics of the Weberian apparatus of the Ostariophysi, p. 147–181. *In*: Sound reception in fishes, W. N. Tavolga (ed.). Dowdin, Hutchinson, and Ross, Inc., Stroudsburg, PA.

Pohajdak, B., B. Dixon, and G. R. Stuart. 1993. Immune system, p. 191–205. *In*: Biochemistry and molecular biology of fish, Vol. 2: Molecular biology frontiers, P. W. Hochachka and T. P. Mommsen (eds.). Elsevier, Amsterdam.

Policansky, D. 1982. The asymmetry of flounders. Sci. Am. 246(5): 116–122.

Popper, A. N. 1983. Organization of the inner ear and auditory processing, p. 125–178. *In*: Fish neurobiology, R. G. Northcutt and R. E. Davis (eds.). University of Michigan Press, Ann Arbor.

Popper, A. N., and S. Coombs. 1980. Auditory mechanisms in teleost fishes. Am. Scient. 68(4):429–440.

Popper, A. N., P. H. Rogers, W. M. Saidel, and M. Cox. 1988. Role of the fish ear in sound processing, p. 687–710. *In*: Sensory biology of aquatic animals, J. Atema, R. R. Fay, A.N. Popper and W. N. Tavolga (eds.). Springer-Verlag, New York.

Postlethwait, J., and J. L. Hopson. 1989. The nature of life. Random House, New York.

Potter, I. C. 1980. The Petromyzoniformes with particular reference to paired species. Can. J. Fish. Aquat. Sci. 37:1595–1615.

Potter, I. C., and F. W. H. Beamish. 1977. The freshwater biology of adult anadromous sea lampreys *Petromyzon marinus*. J. Zool., London (1977)181:113–130.

Potter, I. C., Lord Richard Percy, D. L. Barber, and D. J. Macey. 1982. The morphology, development and physiology of blood cells, p. 233–292. *In*: The biology of lampreys, M. W. Hardisty and I. C. Potter (eds.). Academic Press, London, New York.

Potthoff, T., and S. Kelley. 1982. Development of the vertebral column, fins and fin supports, branchiostegal rays, and squamation in the swordfish, *Xiphias gladius*. Fish. Bull. U. S. 80:161–186.

Potthoff, T., W. J. Richards, and S. Ueyanagi. 1980. Development of *Scombrolabrax heterolepis* (Pisces, Scombrolabracidae) and comments on familial relationships. Bull. Mar. Sci. 30:329–357.

Potts, G. W. 1970. The schooling ethology of *Lutjanus monostigma* (Pisces) in the shallow reef environment of Aldabra. J. Zool. Lond. 161:223–235.

Potts, W. T. W., M. A. Foster, P. P. Rudy and G. P. Howells. 1967. Sodium and water balance in the cichlid teleost, *Tilapia mossambica*. J. Expt. Biol. 47(3):461–470.

Power, M. E. 1987. Predator avoidance by grazing fishes in temperate and tropical streams: Importance of stream depth and prey size, p. 333–351. *In*: Predation: Direct and indirect impacts in aquatic communities, W. C. Kerfoot and A. Sih (eds.). University Press of New England, Dartmouth.

Powers, D. A. 1991. Evolutionary genetics of fish. Advances in Genetics 29:119–228.

Prazdnikova, N. V. 1967. Peculiarities of the distinction of visual images by fish, p. 79–86. *In*: Behavior and reception in fish, G. S. Karzinkin and G. A. Maliukina (eds.). Moscow, "Nauka," Akademiya Nauk SSR, Ministerstvo Rybnogo Khozyaistva SSR, Ikhtiologicheskaya Komissiya. Transl. by Robert M. Howland.

Prescott, B. D. 1984. "Scombroid poisoning" and bluefish: the Connecticut connection. Conn. Med. 48:110.

Prosser, C. L. 1973. Water: Osmotic balance; hormonal regulation, p. 1–78. *In*: Comparative animal physiology, C. L. Prosser (ed.). W. B. Saunders, Philadelphia.

Prosser, C. L., and E. J. DeVillez. 1991. Feeding and digestion, p. 205–229. *In*: Environmental and metabolic animal physiology. C. L. Prosser (ed.). Wiley-Liss, New York.

Prosser, C. L., and L. B. Kirschner. 1973. Inorganic ions, p. 79–110. In: Comparative animal physiology, C. L. Prosser (ed.). W. B. Saunders, Philadelphia.

Prunet, P., and G. Boeuf. 1989. Plasma prolactin levels during smoltification in Atlantic salmon, *Salmo salar*. Aquaculture 82:297–305.

Prunet, P., G. Boeuf, and L. M. Houde. 1985. Plasma and pituitary prolactin levels in rainbow trout during adaptation to different salinities. J. Exp. Zool. 235:187–196.

Pusey, B. J. 1989. Aestivation in the teleost fish *Lepidogalaxias salamandroides* Mees. Comp. Biochem. Physiol. 92A(1):137–138.

Pusey, B. J. 1990. Seasonality, aestivation and the life history of the salamanderfish *Lepidogalaxias salamandroides*. Env. Biol. Fish 29(1):15–26.

Quinn, T. P. 1988. Estimated swimming speeds of migrating adult sockeye salmon. Can. J. Zool. 66:2160–2163.

Quinn, T. P., B. A. ter Hart, and C. Groot. 1989. Migratory orientation and vertical movements of a homing adult sockeye salmon, *Oncorhynchus nerka*, in coastal waters. Anim. Behav. 37:587–599.

Radakov, D. V. 1973. Schooling in the ecology of fish. Halstead Press, New York.

Radinsky, L. B. 1987. The evolution of vertebrate design. University of Chicago Press, Chicago.

Raleigh, R. F. 1967. Genetic control in the lakeward migrations of sockeye salmon (*Oncorhynchus nerka*) fry. J. Fish. Res. Bd. Can, 24:2613–2622.

Raleigh, R. F. 1971. Innate control of migration of salmon and trout fry from natal gravels to rearing areas. Ecol. 52:291–297.

Ralston, S. L., and M. H. Horn. 1986. High tide movements of the temperate-zone herbivorous fish *Cebidichthys violaceus* (Girard) as determined by ultrasonic telemetry. J. Exp. Mar. Biol. Ecol. 98:35–50.

Rand, D. M., and G. V. Lauder. 1981. Prey capture in the chain pickerel, *Esox niger*: Correlations between feeding and locomotor behavior. Can. J. Zool. 59(6):1072-1078.

Randall, J. E. 1967. Food habits of reef fishes of the West Indies. Proc. Int. Conf. Trop. Oceanogr., Univ. Miami Stud. in Trop. Oceanogr. 5:665–847.

Randall, J. E. 1980. A survey of ciguatera at Enewetak and Bikini, Marshall Islands, with notes on the systematics and food habits of ciguatoxic fishes. Fish. Bull. U.S. 78:201–249.

Randall, J. E. 1982. Examples of antitropical and antiequatorial distribution of Indo-West Pacific fishes. Pacific Sci. 35(3):197–209.

Randall, J. E. 1985. Shunts in fish gills, p. 71–82. *In*: Cardiovascular shunts: Phylogenetic, ontogenetic and clinical aspects, K. Johansen and W. Burggren (eds.). Munksgaard, Copenhagen.

Randall, J. E., and J. T. Millington. 1990. Triggerfish bite—a little-known marine hazard. J. Wild. Med. 1:79–85.

Randall, J. E., G. R. Allen, and R. C. Steene. 1990. Fishes of the Great Barrier Reef and the Coral Sea. University of Hawaii Press, Honolulu.

Rangeley, R. W. 1994. The effects of seaweed harvesting on fishes: A critique. Env. Biol. Fishes 39:319–323.

Rankin, H. C., and P. Moller. 1986. Social behavior of the African electric catfish, *Malapterurus electricas*, during intra- and interspecific encounters. Ethology 73(3):177–190.

Rankin, J. C., I. W. Henderson, and J. A. Brown. 1983. Osmoregulation and the control of kidney function, p. 66–88. *In*: Control processes in fish physiology, J. C. Rankin, T. J. Pitcher and R. T. Duggan (eds.). Wiley, New York.

Raschi, W., and L. A. Mackanos. 1989. The structure of the ampullae of Lorenzini in *Dasyatis garouaensis* and its implications on the evolution of freshwater electroreceptive systems, p. 101–111. *In*: Eighth Int. Symp. on Morphological Sciences, W. C. Hamlett and B. Tota (eds.). Rome, Italy.

Rass, T. S., and G. U. Lindberg. 1972. Modern concepts of the classification of living fishes. J. Ichthyol. 11:302–319.

Rauchenberger, M. 1989. Systematics and biogeography of the genus *Gambusia* (Cyprinodontiformes: Poeciliidae). Am. Mus. Novit. 1951.

Ray, D. L. 1950. The peripheral nervous system of *Lampanyctus leucopsaurus*. J. Morphol. 87:61–178.

Rayner, J. M. V. 1981. Flight adaptations in vertebrates, p.

137–172. *In*: Vertebrate locomotion, Symp. Zool. Soc. Lond. (48), M. H. Day (ed.). Academic Press, New York.

Read, L. J. 1971. Body fluids and urine of the holocephalan, *Hydrolagus colliei*. Comp. Biochem. Physiol. 39A:185–192.

Redding, J. M., and R. Patiño. 1993. Reproductive physiology, p. 503–534. *In*: The physiology of fishes, D. H. Evans (ed.). CRC Press, Boca Raton.

Redding, J. M., R. Patiño, and C. B. Schreck. 1991. Cortisol effects on plasma electrolytes and thyroid hormones during smoltification in coho salmon, *Oncorhynchus kisutch*. Gen. Comp. Endocrinol. 81:373–382.

Reebs, S. G., and P. W. Colgan. 1992. Proximal cues for nocturnal egg care in convict cichlids, *Cichlasoma nigrofasciatum*. Anim. Behav. 43:209–214.

Refstie, T. 1983. Induction of diploid gynogenesis in Atlantic salmon and rainbow trout using irradiated sperm and heatshock. Can. J. Zool. 61(11):211–216.

Regan, C. T. 1906–1908. Pisces, p. 1–203. *In*: Biologia Centrali-Americana. Porter, London.

Regan, C. T. 1929. Fishes, Encyclopaedia Britannica. 14th ed. 9: 305–329.

Rehnberg, B. G., and C. B. Schreck. 1986. The olfactory L.-serine receptor in coho salmon: Biochemical specificity and behavioral response. J. Comp. Physiol. 159:61–67.

Reimers, N. 1979. A history of a stunted brook trout population in an alpine lake: A lifespan of 24 years. Cal. Fish and Game 65:196–215.

Reinert, R. E., B. A. Knuth, N. A. Kamrin, and Q. J. Stober. 1991. Risk assessment, risk management, and fish consumption advisories in the United States. Fisheries 16(6):5–12.

Reutter, K. 1982. Taste organ in the barbel of the bullhead, p. 77–91. *In*: Chemoreception in fishes, T. J. Hara (ed.). Elsevier, Amsterdam.

Reutter, K. 1992. Structure of the peripheral gustatory organ, represented by the siluroid fish *Plotosus lineatus* (Thunberg), p. 60–78. *In*: Fish chemoreception, T. J. Hara (ed.). Chapman and Hall, London.

Reznick, D., and J. A. Endler. 1982. The impact of predation on life history evolution in Trinidadian guppies (*Poecilia reticulata*). Evol. 36:160–177.

Rhyther, J. H. 1959. Potential productivity of the sea. Science 130:602–608.

Richards, W. J. 1984. Elopiformes: Development, p. 60–62. *In*: Ontogeny and systematics of fishes, H. G. Moser, W. J. Richards, D. M. Cohen, M. P. Fahay, A. W. Kendall, Jr. and S. L. Richardson (eds.). Am. Soc. Ichthyol. and Herpetol. Sp. Pub. 1.

Richards, W. J., and J. M. Leis. 1984. Labroidei: Development and relationships, p. 542–547. *In*: Ontogeny and systematics of fishes, H. G. Moser, W. J. Richards, D. M. Cohen, M. P. Fahay, A. W. Kendall, Jr. and S. L. Richardson (eds.). Am. Soc. Ichthyol. and Herpetol. Sp. Pub. 1

Ricker, W. E. 1954. Stock and recruitment. J. Fish. Res. Bd. of Canada 11:559–623.

Ricker, W. E. (ed.). 1968. Methods for assessment of fish production in fresh waters. IBP handbook No. 3. Blackwell Scientific, Oxford, Edinburgh.

Riddell, B. E. 1993. Spatial organization of Pacific salmon: What to conserve?, p. 23–41. *In*: Genetic conservation of salmonid fishes,

J. G. Cloud and G. H. Thorgaard (eds.). Plenum Press, New York.

Riggs, A. 1970. Properties of fish hemoglobins, p. 208–252. *In*: Fish physiology, Vol. IV, W. S. Hoar and D. J. Randall (eds.). Academic Press, New York and London.

Rimmer, D. W., and W. J. Wiebe. 1987. Fermentative microbial digestion in herbivorous fishes. J. Fish Biol. 31:229–236.

Rivas, L. R. 1953. The pineal apparatus of tunas and relative scombroid fishes as a possible light receptor controlling phototactic movements. Bull. Mar. Sci. Gulf Carib. 3:168–180.

Roberts, C. D. 1993. Comparative morphology of spined scales and their phylogenetic significance in the Teleostei. Bull. Mar. Sci. 52(1):60–113.

Roberts, J. L. 1975. Active branchial and ram gill ventilation in fishes. Biol. Bull. 148(1):85–105.

Roberts, R. J., and A. M. Bullock. 1989. Nutritional pathology, p. 423–473. *In*: Fish nutrition (Second Edition). J. E. Halver (ed.). Academic Press, San Diego.

Roberts, T. R. 1969. Osteology and relationships of characoid fishes, particularly the genera *Hepsetus, Salminus, Hoplias, Ctenolucius,* and *Acestrorhynchus*. Proc. Cal. Acad. Sci., Ser. 4, 36 (15):391–500.

Roberts, T. R. 1971. The fishes of the Malaysian family Phallostethidae (Atheriniformes). Breviora 374:1–27.

Roberts, T. R. 1973. Interrelationships of ostariophysans, p. 373–395. *In*: Interrelationships of fishes, P. H. Greenwood, R. S. Miles and C. Patterson (eds.). Academic Press, New York.

Roberts, T. R. 1975. Geographical distribution of African freshwater fishes. Zool. J. Linn. Soc. 57(4):249–319.

Roberts, T. R. 1981. Sundasalangidae, a new family of minute freshwater salmoniform fishes from southeast Asia. Proc. Calif. Acad. Sci. 42(9):295–302.

Roberts, T. R. 1982. Unculi (horny projections arising from single cells), an adaptive feature of the epidermis of ostariophysan fishes. Zool. Scripts 11(1):55–76.

Roberts, T. R. 1984a. *Amazonsprattus scintilla*, new genus and species from the Rio Negro, Brazil, the smallest known clupeomorph fish. Proc. Calif. Acad. Sci 43(2):317–321.1

Roberts, T. R. 1984b. Skeletal anatomy and classification of the neotenic Asian salmoniform superfamily Salangoidea (icefishes or noodlefishes). Proc. Calif. Acad. Sci. 43(13):179–220.

Roberts, T. R. 1989. The freshwater fishes of western Borneo (Kalimantan Barat, Indonesia). Mem. Calif. Acad. Sci. 14.

Roberts, T. R. 1992. Systematic revision of the Old World freshwater fish family Notopteridae. Icththyol. Explor. Freshwater 2(4): 311–383.

Robertson, D. R., H. P. A. Sweatman, E. A. Fletcher, and M. G. Cleland. 1976. Schooling as a mechanism for circumventing the territoriality of competitors. Ecol. 57:1208–1220.

Robertson, J. D. 1963. Osmoregulation and ionic composition of cells and tissues, p. 504–515. *In*: The biology of *Myxine,* A. Brodal and R. Fänge (eds.). Universitetsforlaget, Oslo.

Robins, C. H., and C. R. Robins. 1976. New genera and species of dysommine and synaphobranchine eels (Synaphobranchidae) with an analysis of the Dysomminae. Proc. Acad. Nat. Sci. Philad. 127(18):249–280.

Robins, C. R. 1971. Distributional patterns of fishes from coastal and shelf waters of the tropical western Atlantic. FAO Fish Rep. 71(2):249–255.

Robins, C. R. 1974. Billfish biology: Facts for the fisherman. Addendum to The International Marine Angler 36(5):1–4.

Robins, C. R. 1989. The phylogenetic relationships of the anguilliform fishes, p. 9–33. *In*: Fishes of the western North Atlantic, E. B. Böhlke (ed.), Part 9, Vol. 1: Orders Anguilliformes and Saccopharygiformes. Mem. Sears Found. Mar. Res.

Robins, C. R., D. M. Cohen, and C. H. Robins. 1979. The eels *Anguilla* and *Histiobranchus* photographed on the floor of the deep Atlantic in the Bahamas. Bull. Mar. Sci. 29(3):401–405.

Robins, C. R., R. M. Bailey, C. E. Bond, J. R. Brooker, E. A. Lachner, R. N. Lea, and W. B. Scott. 1980. A list of common and scientific names of fishes from the United States and Canada (Fourth Edition). Am. Fish. Soc. Sp. Pub. 12.

Robins, C. R., R. M. Bailey, C. E. Bond, J. R. Brooker, E. A. Lachner, R. N. Lea, and W. B. Scott. 1991a. Common and scientific names of fishes from the United States and Canada (Fifth Edition). Am. Fish. Soc. Sp. Pub. 20.

Robins, C. R., R. M. Bailey, C. E. Bond, J. R. Brooker, E. A. Lachner, R. N. Lea and W. B. Scott. 1991b. World fishes important to North Americans. Am. Fish. Soc. Sp. Pub. 21.

Robinson, B. W., and D. S. Wilson. 1994. Character release and displacement in fishes: A neglected literature. Am. Nat. 144:596–627.

Roche, E. T., and B. W. Halstead. 1972. The venom apparatus of California rockfishes (family Scorpaenidae). Calif. Dept. Fish and Game Bull. 156.

Rohde, F. C., R. G. Arndt, and J. C. S. Wang. 1976. Life history of the freshwater lampreys, *Okkelbergia aepyptera* and *Lampetra lamottenii* (Pisces: Petromyzonidae) on the Delmarva Peninsula (East Coast, United States). Bull. So. Calif. Acad. Sci. 75(2):99–111.

Rojo, A. L. 1991. Dictionary of evolutionary fish osteology. CRC Press, Boca Raton.

Romer, A. S. 1970 The vertebrate body (Fourth Edition). W. B. Saunders, Philadelphia.

Romer, A. S., and T. S. Parsons. 1978. The vertebrate body: Shorter version. W. B. Saunders, Philadelphia.

Ronan, M. 1986. Electroreception in cyclostromes, p. 209–224. *In*: Electroreception, T. H. Bullock and W. Heiligenberg (eds.). Wiley, New York.

Ronan, M. C., and D. Bodznick. 1986. End buds: Non-ampullary electroreceptors in adult lampreys. J. Comp. Physiol. 158(1):9–15.

Rosa, I. L., and R. S. Rosa. 1987. *Pinquipes* Cuvier and Valenciennes and Pinguipedidae Gunther, the valid names for the fish taxa usually known as *Mugiloides* and Mugiloididae. Copeia 1987:1048–1051.

Rosen, D. E. 1962. Comments on the relationships of the North American cave fishes of the family Amblyopsidae. Am. Mus. Novit. 2109.

Rosen, D. E. 1964. The relationships and taxonomic position of the halfbeaks, killifishes, silversides, and their relatives. Bull. Am. Mus. Nat. Hist. 127:217–268.

Rosen, D. E. 1973. Interrelationships of higher euteleostean fishes, p. 397–513. *In*: Interrelationships of fishes, P. H. Greenwood, R.

S. Miles and C. Patterson (eds.). J. Linn. Soc. (Zool.) 53, Suppl.1. Academic Press, New York.

Rosen, D. E. 1974. Phylogeny and zoogeography of salmoniform fishes and relationships of *Lepidogalaxias salamandroides*. Bull. Am. Mus. Nat. Hist. 153(2):265–326.

Rosen, D. E. 1975. A vicariance model of Caribbean biogeography. Syst. Zool. 24:431–464.

Rosen, D. E. 1978. Vicariant patterns and historical explanation in biogeography. Syst. Zool. 27(2):159–188.

Rosen, D. E. 1984. Zeiforms as primitive plectognath fishes. Am. Mus. Novit. 2782.

Rosen, D. E. 1985. An essay on euteleostean classification. Am. Mus. Novit. 2827.

Rosen, D. E., and L. R. Parenti. 1981. Relationships of *Oryzias,* and the groups of atherinomorph fishes. Am. Mus. Novit. 2719:1–25.

Rosen, D. E., and C. Patterson. 1969. The structure and relationships of the paracanthopterygian fishes. Bull. Am. Mus. Nat. Hist. 141(3):357–474.

Rosen, D. E., P. L. Forey, B. G. Gardiner, and C. Patterson. 1981. Lungfishes, tetrapods, paleontology, and plesiomorphy. Bull. Am. Mus. Nat. Hist. 167(4):159–276.

Rosen, M. W., and N. E. Cornford. 1970. Publication no. 193 of the Naval Undersea Research and Development Center, San Diego.

Rosen, M. W., and N. E. Cornford. 1971. Fluid friction of fish slimes. Nature 234:49–51.

Rosenblatt, R. H. 1967. The zoogeographic relationships of the marine shore fishes of tropical America. Stud. trop. Oceanogr., Miami (5):579–592.

Rosenblatt, R. H. 1984. Blennioidei: Introduction, p. 551. *In*: Ontogeny and systematics of fishes, H. G. Moser, W. J. Richards, D. M. Cohen, M. P. Fahay, A. W. Kendall, Jr. and S. L. Richardson (eds.). Am. Soc. Ichthol. and Herpet. Sp. Pub. 1.

Ross, D. M. 1963. The sense organs of *Myxine glutinosa* L., p. 150–160. *In*: The biology of *Myxine*, A. Brodal and R. Fänge (eds.). Universitetsforlaget, Oslo.

Ross, R. M., G. S. Losey, and M. Diamond. 1983. Sex change in a coral-reef fish: Dependence of stimulation and inhibition on relative size. Science 221:574–575.

Ross, R. M., T. F. Hourigan, M. M. F. Lutnesky, and I. Singh. 1990. Multiple simultaneous sex changes in social groups of a coral-reef fish. Copeia 1990:427–433.

Rothschild, B. J. 1981. More food from the sea? BioScience 31(3):216–222.

Rothschild, B. J. (ed.). 1983. Global fisheries. Perspectives for the 1980s. Springer, New York.

Roule, L. 1922. Description de *Scombrolabrax heterolepis* nov. gen. nov. sp., poisson abyssal nouveau de l'Ile Madere. Bull. Inst. Oceanogr. (Monaco) 408:1–8.

Rounsefell, G. A. 1975. Ecology, utilization and management of marine fisheries. C. V. Mosby, St. Louis.

Rovainen, C. M. 1978. Muller cells, "Mauthner" cells and other identified reticulospinal neurons in the lamprey, p. 245–269. *In*: Neurobiology of the Mauthner cell, D. Faber and H. Korn (eds.). Raven Press, New York.

Royce, W. F. 1972. Introduction to the fishery sciences. Academic Press, New York.

Royce, W. F. 1987. Fishery development. Academic Press, Orlando.

Ruetter, K. 1982. Taste organ in the barbel of the bullhead, p. 77–92. *In*: Chemoreception in fishes, T. J. Hara (ed.). Elsevier Scientific Publishing Amsterdam.

Russell, F. E. 1969. Poisons and venoms, p. 401–449. *In*: Fish physiology, Vol. III, W. S. Hoar and D. J. Randall (eds.). Academic Press, New York.

Russell, F. S. 1976. The eggs and planktonic stages of British marine fishes. Academic Press, London and New York.

Ruzzanti, D. E., and R. W. Doyle. 1991. Rapid behavioral changes in medaka (*Oryzias latipes*) caused by selection for competitive and noncompetitive growth. Evol. 45:1936–1946.

Ryther, J. H. 1969. Photosynthesis and fish production in the sea. Science 166:72–76.

Ryther, J. H. 1981. Mariculture, ocean ranching, and other culture-based fisheries. BioScience 31(3):223.

Sacca, R., and W. W. Burggren. 1982. Oxygen uptake in air and water in the air-breathing reedfish *Calamoichthys calabaricus*: Role of skin, gills and lungs. J. Exp. Biol. 97:179–186.

Sagua, V. O. 1979. Observations on the food and feeding habits of the African electric catfish *Malapterurus electricus* (Gmelin). J. Fish Biol. 15:61–69.

Saheki, K., S. Kobayashi, and T. Kawanishi. 1989. "*Salmonella* contamination of eel culture ponds" (in Japanese, abstract in English). Jpn. Soc. Sci. Fisheries 55(4):675–679.

Saint-Paul, U., and G. Bernardinho. 1988. Behavioral and ecomorphological responses of the neotropical pacu *Piaractus mesopotamicus* (Teleostei, Serrasalmidae) to oxygen-deficient waters. Exp. Biol. 48:19–26.

Sakamoto, K. 1984. Interrelationships of the family Pleuronectidae (Pisces: Pleuronectiformes). Mem. Fac. Fish. Hokkaido Univ. 31(1/2):95–215.

Sale, P. F. 1991. The ecology of fishes on coral reefs. Academic Press, New York.

Sanford, C. P. J. 1990. The phylogenetic relationships of salmonoid fishes. Bull. Br. Mus. (Nat. Hist.) Zool. 56(2):145–153.

Sargent, J., R. J. Henderson, and D. H. Tocher. 1989. The lipids, p. 153–218. *In*: Fish nutrition (Second Edition), J. E. Halver (ed.). Academic Press, San Diego.

Satchell, G. H. 1991. Physiology and form of fish circulation. Cambridge University Press, Cambridge.

Saunders, R. L. 1961. The irrigation of the gills in fishes. I. Studies of the mechanism of branchial irrigation. Can. J. Zool. 39(5): 677–683.

Saunders, R. L. 1962. The irrigation of the gills in fishes. II. Efficiency of oxygen uptake in relation to respiratory flow activity and concentrations of oxygen and carbon dioxide. Can. J. Zool. 40(5):817–862.

Sazima, I. 1977. Possible case of aggressive mimicry in a neotropical scale-eating fish. Nature 170(5637):510–512.

Schaefer, K. M. 1985. Body temperatures in troll-caught frigate tuna, *Auxis thazard*. Copeia 1985:231–233.

Schaeffer, B. 1967. Comments on elasmobranch evolution, p. 3–35. *In*: Sharks, skates and rays. P. W. Gilbert, R. F. Mathewson and D. P. Rall (eds.). Johns Hopkins Press, Baltimore.

Schaeffer, B. 1968. The origin and basic radiation of the Osteichthyes, p. 207–222. *In*: Current problems of the lower vertebrate phylogeny, T. Orvig (ed.). Wiley-Interscience, New York.

Schaeffer, B. 1969. Adaptive radiation of the fishes and the fish-amphibian transition. Ann. N.Y. Acad. Sci. 167:5–17.

Schaeffer, B. 1973. Interrelationships of chondrosteans, p. 207–226. *In*: Interrelationships of fishes, P. H. Greenwood, R. S. Miles and C. Patterson (eds.). Academic Press, New York.

Schaeffer, B. 1981. The xenacanth shark neurocranium, with comments on elasmobranch monophyly. Bull. Amer. Mus. Nat. Hist. 169:3–66.

Schaeffer, B., and D. E. Rosen. 1961. Major adaptive levels in the evolution of the actinopterygian feeding mechanism. Am. Zool. 1(2):187–204.

Schaeffer, B., and M. Williams. 1977. Relationships of fossil and living elasmobranchs. Amer. Zool. 17:293–302.

Schemmel, C. 1980. Studies on the genetics of feeding behavior in the cave fish *Astyanax mexicanus* f. *anoptichthys*. Z. Tierpsychol. 53:9–22.

Schindler, D. W., S. E. M. Kasian, and R. H. Hesslein. 1989. Losses of biota from American aquatic communities due to acid rain. Env. Monit. and Assess. 12:269–285.

Schlichting, H. 1960. Boundary layer theory (Fourth Edition). McGraw-Hill, New York.

Schliewen, U. K., D. Tautz, and S. Paabo. 1994. Sympatric speciation suggested by monophyly of crater lake cichlids. Nature 368:629–632.

Schlupp, I., J. Parzefall, and M. Schott. 1991. Male mate choice in mixed bisexual/unisexual breeding complexes of *Poecilia* (Teleostei: Poeciliidae). Ethology 88(3):215–222.

Schmitt, R. J., and S. J. Holbrook. 1984. Ontogeny of prey selection by black surfperch *Embiotoca jacksoni* (Pisces: Embiotocidae): The roles of fish morphology, foraging behavior, and patch selection. Mar. Ecol. Prog. Ser. 18(3):225–239.

Schneider, H., and A. D. Hasler. 1960. Lauate and lauterzeugung beim Susswassertrommler *Aplodinotus grunniens* Rafinesque (Sciaenidae: Pisces). Z. Vergl. Physiol. 4:499–517.

Scholander, P. F. 1954. Secretion of gases against high pressures in the swim bladder of deep sea fishes. II. The rete mirabile. Biol. Bull., Woods Hole 107:260–277.

Scholander, P. F. 1957. The wonderful net. Scient. Amer. 196(4):96–107.

Scholander, P. F., and L. Van Dam. 1954. Secretion of gases against high pressures in the swim bladder of deep sea fishes. I. Oxygen dissociation in blood. Biol. Bull., Woods Hole 107:247–259.

Scholes, J. H. 1975. Colour receptors, and their synaptic connexions in the retina of a cyprinid fish. Philos. Trans. R. Soc. Lond. (Biol. Sci.) 270(902):61–118.

Schreck, C. B. 1974. Control of sex in fishes. Virginia Polytechnic Institute, Blacksburg.

Schreck, C. B. 1981. Stress and compensation in teleostean fishes: Response to social and physical factors, p. 295–321. *In*: Stress and fish, A. D. Pickering (ed.). Academic Press, London and New York.

Schreck, C. B., and P. F. Scanlon. 1977. Endocrinology in fisheries and wildlife. Fisheries 2(3):20–27.

Schreck, C. B., C. Bradford, M. S. Fitzpatrick, and R. Patino. 1989. Regulation of the interrenal of fishes: non-classical control mechanisms. Fish Physiol. Biochem. 7:259–265.

Schuijf, A., and R. J. A. Buwalda. 1980. Underwater localization—

A major problem in fish acoustics, p. 43–78. *In*: Comparative studies in hearing in vertebrates, A. N. Popper and R. R. Fay (eds.). Springer-Verlag, New York.

Schultz, L. P., and E. M. Stern. 1948. The ways of fishes. Van Nostrand Co. Inc., New York.

Schultz, R. J. 1973. Origin and synthesis of a unisexual fish, p. 207–211. *In*: Genetics and mutagenesis of fish. J. H. Schroder (ed.). Springer-Verlag, New York, Heidelberg, Berlin.

Schultz, R. J. 1989. Origins and relationships of unisexual poeciliids, p. 69–87. *In*: Ecology and evolution of livebearing fishes, G. K. Meffe and F. F. Snelson, Jr. (eds.). Prentice Hall, Englewood Cliffs, NJ.

Schultz, R. J. 1993. Genetic regulation of temperature-mediated sex ratios in the livebearing fish *Poeciliopsis lucida*. Copeia 1993: 1148–1151.

Schultze, H.-P. 1993. Patterns of the diversity in the skills of jawed fishes, p. 189–254. *In*: The skull, Vol. 2, J. Hanken and B. Hall (eds.). University of Chicago Press, Chicago, London.

Schultze, H.-P., and M. Arsenault. 1985. The panderichthyid fish Elpistostege: A close relative of tetrapods. Palaeontology 28: 293–310.

Schwanzara, S. A. 1967. The visual pigments of freshwater fishes. Vision Res. 7:121–148.

Schwartz, F. J. 1972. World literature to fish hybrids with an analysis by family, species, and hybrid. Gulf Coast Res. Lab. Publ. 3:1–328.

Schwartz, F. J. 1981. World literature to fish hybrids with an analysis by family, species, and hybrid: Supplement I. NOAA Tech. Rep., NMFS SSRF-750:1–507.

Schwassman, H. O. 1975. Refractive state, accommodation, and resolving power of the fish eye, p. 279–288. *In*: Vision in fishes: New approaches in research, M. A. Ali (ed.). Plenum Press, New York.

Schwassman, H. O., and L. Kruger. 1965. Experimental analysis of the visual system of the four-eyed fish (*Anableps microlepis*). Vision Res. 5:269–281.

Schweigert, J. F., F. J. Ward, and J. W. Clayton. 1977. Effects of fry and fingerling introductions on walleye (*Stizostedion vitreum vitreum*) production in West Blue Lake, Manitoba. J. Fish. Res. Bd. Can. 34:2142–2150.

Scott, W. B., and E. L. Crossman. 1972. Freshwater fishes of Canada. Bull. 184, Fish. Res. Bd., Canada. Ottawa.

Seaberg, K. G., and J. B. Moyle. 1964. Feeding habits, digestion rates and growth of some Minnesota warmwater fishes. Trans. Am. Fish. Soc. 93:269–285.

Segaar, J., J. P. C. de Bruin, and M. E. van der Meche-Jacobi. 1983. Influence of chemical receptivity on reproductive behavior of the male three-spined stickleback (*Gasterosteus aculeatus* L.) Behavior 86:100–166.

Seikei, T. 1989. Albinism of hatchery reared flounder (*Paralichthys olivaceous*) as a result of deformation of asymmetrical development of skin structure, p. 489 (abstr.). *In*: The early life history of fish, 3rd ICES Symposium, J. H. S. Blaxter, J. C. Gamble and H. v. Westernhagen (eds.). Cons. Int. Explor. Mer 191.

Seliger, H. H. 1962. Direct action of light in naturally pigmented muscle fibers. I. Action spectrum for contraction in eel iris sphincter. J. Gen. Physiol. 46:333–342.

Selset, R., and K. B. Doving. 1980. Behavior of mature anadromous char (*Salmo alpinus* L.) towards odorants produced by smolts of their population. Acta Physiol. Scand. 108:113–122.

Seret, B. 1986. Classification et phylogenese des chondrichthyes. Oceanis 11(3):161–180.

Sette, O. E. 1943. Biology of the Atlantic mackerel (*Scomber scombrus*) of North America: Part I. Early life history, including growth, drift and mortality of the egg and larval populations. U.S. Fish Wildl. Serv. Fish. Bull. 50:149–237.

Shaklee, J. B., and C. S. Tamaru. 1981. Biochemical and morphological evolution of Hawaiian bonefishes (*Albula*). Syst. Zool. 30:125–146.

Shaklee, J. B., C. Busack, A. Marshall, and S. R. Phelps. 1990. The electrophoretic analysis of mixed-stock fisheries of Pacific salmon, p. 235–265. *In*: Isozymes: Structure, function, and use in biology and medicine, Z.-I. Ogita and C. L. Markert (eds.). Wiley-Liss, New York.

Shand, J., J. C. Partridge, S. N. Archer, G. W. Potts, and J. N. Lythgoe. 1988. Spectral absorbance changes in the violet/blue sensitive cones of the juvenile pollack, *Pollachius pollachius*. J. Comp. Physiol. 163(5):699–703.

Sharma, S. 1971. Homology of the so-called "head-kidney" in certain Indian teleosts. Ann. Zool., Agra 7(2):20–40.

Sharma, S., and A. Sharma. 1975. A note on the caudal neurosecretory system and seasonal changes observed in the urophysis of *Rita rita* (Bleeker). Can. J. Zool. 53(3):357–360.

Shaw, E. 1978. Schooling fishes. Am. Sci. 66(2):166–175.

Shaw, E., and A. Tucker. 1965. The optomotor response of schooling carangid fishes. Anim. Beh. 8:330–336.

Shedlock, A. M., J. D. Parker, D. A. Crispin, T. W. Pietsch, and G. C. Burmer. 1992. Evolution of the salmonid mitochondrial control region. Mol. Phylogen. Evol. 1:179–192.

Shelton, G. 1970. The regulation of breathing, p. 293–359. *In*: Fish physiology, Vol. IV, W. S. Hoar and D. J. Randall (eds.). Academic Press, New York and London.

Shetter, D. 1949. A brief history of the sea lamprey problem in Michigan waters. Trans. Am. Fish. Soc. 76:160–176.

Shirai, S. 1992. Squalean phylogeny. University of Hokkaido Press.

Shulman, G. E. 1974. Life cycles of fish: Physiology and biochemistry. Halstead Press, New York.

Shuttleworth, T. J. 1988. Salt and water balance—extrarenal mechanisms, p. 171–199. *In*: Physiology of elasmobranch fishes, T. J. Shuttleworth (ed.). Springer-Verlag, Berlin.

Sibbing, F. A. 1982. Pharyngeal mastication and food transport in the carp (*Cyprinus carpio* L.): A cineradiographic and electromyographic study. J. Morphol. 172:223–258.

Sideleva, V. G. 1982. Seismosensory systems and ecology of the Baikalian sculpins (Cottoidei). Novisibirisk: Isv. Nauka, Akad. Nauk. SSSR. (*In* Russian)

Siebert, D. J. 1990. Book review: Papers on the systematics of gadiform fishes by D. M. Cohen (ed.), 1989. Copeia 1990(3):889–893.

Sigler, W. F., and J. W. Sigler. 1987. Fishes of the Great Basin: A natural history. University of Nevada Press, Reno.

Sigler, W. F., and J. W. Sigler. 1990. Recreational fisheries: Management, theory and application. University of Nevada Press, Reno and Las Vegas.

Siitonen, L., and G. A. E. Gall. 1989. Response to selection for early spawn date in rainbow trout., *Salmo gairdneri*. Aquaculture 78:153–161.

Sikkel, P. C. 1989. Egg presence and developmental stage influence spawning site choice by female garibaldi. Anim. Behav. 38:447–456.

Sikkel, P. C. 1994. Filial cannibalism in a paternal-caring marine fish: The influence of egg develompental stage and position in the nest. Anim. Behav. 47:1149–1158.

Silver, W. L., and T. E. Finger. 1984. Electrophysiological examination of a non-olfactory, non-gustatory chemosense in the searobin, *Prionotus carolinus*. J. Comp. Physiol. A 154(2A):167–174.

Simon, H. 1971. The splendor of iridescence; structural colors in the animal world. Dodd and Mead, New York.

Simons, J. R. 1970. The direction of the thrust produced by the heterocercal tails of two dissimilar elasmobranchs: The Port Jackson shark, *Heterodontus portusjacksoni* (Meyer), and the piked dogfish, *Squalus megalops* (Macleay). J. Exp. Biol. 52:95–107.

Sinclair, M. 1988. Marine populations: An essay on population regulation and speciation. Wash. Sea Grant Program. University of Washington Press, Seattle.

Sivak, J. G. 1975. Accommodative mechanisms in aquatic vertebrates, p. 289–297. *In*: Vision in fishes: New approaches in research, M. A. Ali (ed.). Plenum Press, New York.

Sivak, J. G. 1976. Optics of the eye of the 'four-eyed fish' (*Anableps anableps*). Vision Res. 16:531–534.

Sivak, J. G. 1980. Accommodation in vertebrates: A contemporary survey, p. 281–330. *In*: Current topics in eye research, Vol. 3., J. A. Zadunaicky and H. Dawson (eds.). Academic Press, New York.

Sivak, J. G. 1990. Optical variability of the fish lens, p. 63–80. *In*: The visual system of fish, R. H. Douglas and M. B. A. Djamgoz (eds.). Chapman and Hall, London.

Skaala, O., and K. E. Jorstad. 1988. Inheritance of the fine-spotted pigmentation pattern of brown trout, *Salmo trutta* L., p. 3–4. *In*: Trouts in streams and lakes. Polish Arch. Hydrobiol. 35.

Skaala, O., G. Dahle, K. E. Jorstad, and G. Naevdal. 1990. Interactions between natural and farmed fish populations: Information from genetic markers. J. Fish Biol. 36:449–460.

Skelton, P. 1993. The complete guide to the freshwater fishes of southern Africa. Southern Book Publishers.

Smeets, W. J., A. J. R . Nieuwenhuys, and B. L. Roberts. 1983. The central nervous system of cartilaginous fishes. Springer-Verlag, Berlin, Heidelberg, New York.

Smith, D. G. 1984. Elopiformes, Notacanthiformes and Anguilliformes: Relationships, p. 94–102. *In*: Ontogeny and systematics of fishes, H. G. Moser, W. J. Richards, D. M. Cohen, M. P. Fahay, A. W. Kendall, Jr. and S. L. Richardson (eds.). Am. Soc. Ichthyol. and Herpetol. Sp. Pub. 1.

Smith, D. G., and P. H. J. Castle. 1982. Larvae of the nettastomid eels: Systematics and distribution. Dana Report 90:1–44.

Smith, G. R. 1981a. Late Cenozoic freshwater fishes of North America. Ann. Rev. Ecol. Syst. 12:163–193.

Smith, G. R. 1981b. Effects of habitat size on species richness and adult body size of desert fishes, p. 125–172 . *In*: Fishes in North American deserts, R. J. Naiman and D. J. Soltz (eds.). John Wiley and Sons, New York.

Smith, G. R. 1987. Fish speciation in a western North American Pliocene rift lake. Palaios 2:436–445.

Smith, G. R. 1992. Phylogeny and biogeography of the Catostomidae, freshwater fishes of North America and Asia, p. 778–826. *In*: Systematics, historical ecology, and North American freshwater fishes, R. L. Mayden (ed.). Stanford University Press, Palo Alto.

Smith, G. R., and R. F. Stearley. 1989. The classification and scientific names of rainbow and cutthroat trouts. Fisheries 14(1):4–10.

Smith, G. R., J. Rosenfield, and J. Porterfield. 1995. Processes of origin and criteria for preservation of fish species. *In*: Evolution and the aquatic ecosystem, J. Neilsen (ed.). Am. Fish. Soc. Spec. Publ., Bethesda, MD. (In Press)

Smith, H. W. 1930. Metabolism of the lungfish, *Protopterus aethiopicus*. J. Biol. Chem. 88:97–130.

Smith, J. L. B. 1939. A living coelacanthid fish from South Africa. Trans. R. Soc. So. Afr. 28:1–106.

Smith, L. S. 1966. Blood volumes of three salmonids. J. Fish. Res. Bd. Can. 23(9):1439–1446.

Smith, L. S. 1973. Introductory anatomy and biology of selected fish and shellfish. College of Fisheries, University of Washington, Seattle.

Smith, L. S. 1982. Introduction to fish physiology. T. F. H. Publications, Neptune, NJ.

Smith, L. S. 1989. Digestive functions in teleost fishes, p. 332–423. *In*: Fish nutrition (Second Edition), J. E. Halver (ed.). Academic Press, San Diego, London.

Smith, M. M., and M.-M. Chang. 1990. The dentition of *Diabolepis speratus* Chang and Yu, with further consideration of its relationships and the primitive dipnoan dentition. J. Vert. Paleont. 10(4):420–433.

Smith, M. M., and P. C. Heemstra. 1986. Family No. 263: Balistidae, p. 876–882. *In*: Smith's sea fishes, M. M. Smith and P. C. Heemstra (eds.). Springer-Verlag, Berlin, Heidelberg, New York, London, Paris, Tokyo.

Smith, R. J. F. 1982. Reaction of *Percina nigrofasciata, Ammocrypta beani,* and *Etheostoma swaini* (Percidae, Pisces) to conspecific and intergeneric skin extracts. Can. J. Zool. 60(5):1067–1072.

Smith, R. J. F. 1985. The control of fish migration. (Zoophysiology, Vol. 17). Springer-Verlag, Berlin.

Smith, R. J. F. 1992. Alarm signals in fishes. Rev. Fish. Biol. Fish 2(1):33–63.

Smith, R. L. 1980. Ecology and field biology (Third Edition). Harper and Row, New York.

Smith, R. S. 1989. Nutritional energetics, p. 1–29. *In*: Fish nutrition (Second Edition), J. E. Halver (ed.). Academic Press, San Diego.

Sneath, P., and R. R. Sokal. 1973. Numerical taxonomy. W. H. Freeman, San Francisco.

Snyder, D. E. 1976. Terminologies for intervals of larval fish development, p. 41–58. *In*: Great Lakes fish egg and larvae identification: Proceedings of a workshop, J. Boreman (ed.). U.S. Fish Wildl. Serv. Biol. Serv. Prog. FWS/OBS-76/23.

Snyder, D. E. 1981. Contributions to a guide to the cypriniform fish larvae of the upper Colorado River system. U.S. Dep. Inter., Bur. Land Manage., Colo. Off., Biol. Sci. Ser. 3.

Soehn, K. L., and M. V. H. Wilson. 1990. A complete, articulated heterostracan from Wenlockian (Silurian) beds of the Delorme group, Mackenzie Mountains, Northwest Territories, Canada. J. Vert. Paleont. 10(4):405–419.

Soin, S. G. 1968. Adaptational features in fish ontogeny. Jerusalem, Israel Program for Scientific Translations, translated from Russian 1971.

Solar, I. I., E. M. Donaldson, and D. Douville. 1991. A bibliography of gynogenesis and androgenesis in fish. Can. Tech. Rep. Fish. Aquat. Sci. 1788.

Somiya, H. 1979. 'Yellow lens' eyes and luminous organs of *Echinostoma barbatum* (Stomiatoidei, Melanostomiatidae). Jpn. J. Ichthyol. 25(4):269–272.

Somiya, H., and T. Tamura. 1973. Studies on the visual accommodation in fishes. Jpn. J. Ichthyol. 20(4):193–206.

Soltz, D. L., and R. J. Naiman. 1981. Fishes in deserts: Symposium rationale, p. 1–9. *In*: Fishes in North American deserts, R. J. Naiman and D. L. Soltz (eds.). John Wiley and Sons Inc., New York.

Song, J., and R. G. Northcutt. 1991. Morphology, distribution and innervation of the lateral line receptors of the Florida gar, *Lepisosteus platyrhincus*. Brain, Behav. and Evol. 37(1):10–37.

Sonnenworth, A. C., Z. A. McGee, and B. D. Davis. 1980. Other pathogenic microorganisms: L-phase variants, p. 789–805. *In*: Microbiology (Third Edition), B. D. Davis, R. Dulbecco, S. N. Eisen and H. S. Ginsberg (eds.). Harper and Row, Hagerstown.

Sorenson, P. W. 1992. Hormones, pheromones and chemoreception, p. 199–228. *In*: Fish chemoreception, T. J. Hara (ed.). Chapman and Hall, London.

Sower, S. A. 1990. Neuroendocrine control of reproduction in lampreys. Fish Physiol. Biochem. 8(5):365–374.

Sower, S. A., C. B. Schreck, and E. M. Donaldson. 1982. Hormone induced ovulation of coho salmon (*Oncorhynchus kisutch*) held in seawater and freshwater. Can. Jour. of Fish. and Aquat. Sci. 39(4):627–632.

Spangler, G. R., D. S. Robson, and H. A. Regier. 1980. Estimates of lamprey-induced mortality in whitefish, *Coregonus clupeiformis*. Can. J. Fish. and Aquat. Sci. 39(11):2146–2158.

Springer, V. G. 1978. Synonymization of the family Oxudercidae, with comments on the identity of *Apocryptes cantoris* Day (Pisces: Gobiidae). Smithson. Contr. Zool. 270.

Springer, V. G. 1982. Pacific plate biogeography with special reference to shorefishes. Smithson. Contr. Zool. (367):1–182.

Springer, V. G. 1983. *Tyson belos*, new genus and species of western Pacific fish (Gobiidae, Xenesthminae), with discussion of gobioid osteology and classification. Smithson. Contr. Zool. 390.

Springer, V. G., and T. H. Fraser. 1976. Synonymy of the fish families Cheilobranchidae (=Alabetidae) and Gobiesocidae, with descriptions of two new species of *Alabes*. Smithson. Contr. Zoo. 234.

Springer, V. G., and W. F. Smith-Vaniz. 1972. Mimetic relationships involving fishes of the family Blenniidae. Smithson. Contr. Zool. No. 112.

Srivastava, C. B. L., A. Gopesh, and M. Singh. 1981. A new neurosecretory system in fish, located in the gill region. Experientia 37(8):850–851.

Srivastava, G., and C. B. L. Srivastava. 1991. A lens-like specialization for photic input in the pineal window of an Indian catfish, *Heteropneustes fossilis*. Experientia 47(7):698–700.

Stabell, O. B. 1992. Olfactory control of homing behaviour in salmonids, p. 249–270. *In*: Fish chemoreception, T. J. Hara (ed.). Chapman and Hall, London.

Stacey, N. E. 1981. Hormonal regulation of female reproductive behavior in fish. Am. Zool. 21:305–316.

Stahl, B. 1967. Morphology and relationships of the Holocephali with special reference to the venous system. Bull. Mus. Comp. Zool. Harvard 135(3):141–213.

Stanley, H. P. 1961. Studies on the genital systems and reproduction in the chimaeroid fish *Hydrolagus colliei* (Lay and Bennett). Ph.D. thesis, Oregon State University.

Stanley, J. G., and K. E. Snead. 1974. Artificial gynogenesis and its application in genetics and selective breeding of fish, p. 527–536. *In*: The early life history of fish. J. H. S. Blaxter (ed.). Springer-Verlag, New York, Heidelberg.

Stanley, S. M. 1989. Earth and life through time. Freeman, New York.

Stearley, R. F., and G. R. Smith. 1993. Phylogeny of the Pacific trouts and salmons (*Oncorhynchus*) and genera of the family Salmonidae. Trans. Am. Fish. Soc. 122:1–33.

Stearns, S. C. 1976. Life-history tactics: A review of ideas. Quart. Rev. Biol. 51:3–47.

Stearns, S. C. 1992. The evolution of life histories. Oxford University Press, New York.

Stearns, S. C., and R. E. Crandall. 1984. Plasticity for age and size at sexual maturity: A life-history response to unavoidable stress, p. 13–33. *In*: Fish reproduction: Strategies and tactics, G. W. Potts and R. J. Wootton (eds.). Academic Press, London.

Steen, J. B. 1970. The swim bladder as a hydrostatic organ, p. 441–443. *In*: Fish physiology, Vol. IV, W. S. Hoar and D. J. Randall (eds.). Academic Press, New York and London.

Steen, J. B., and A. Kruysse. 1964. The respiratory function of teleostean gills. Comp. Biochem. Physiol. 12:127–142.

Steigenberger, L. W., and P. A. Larkin. 1974. Feeding activity and rates of digestion of northern squawfish. J. Fish. Res. Bd. Can. 31(4):411–420.

Stein, D. L., and C. E. Bond. 1985. Observations on the morphology, ecology, and behaviour of *Bathylychnops exilis* Cohen. J. Fish Biol. 27:215–228.

Stein, D. L., and W. G. Pearcy. 1982. Aspects of reproduction, early life history, and biology of macrourid fishes off Oregon U.S.A. Deep-Sea Res. 29(11A):1313–1329.

Stensio, E. A. 1968. The cyclostomes, with special reference to the diphyletic origin of the Petromyzontida and Myxinoidea, p. 13–71. *In*: Current problems of lower vertebrate phylogeny, T. Orvig (ed.). Wiley-Interscience, New York.

Stephenson, R. L., and M. S. Power. 1989. Observations on herring larvae retained in the Bay of Fundy: Variability in vertical movement and position of the patch edge, p. 177–183. *In*: The early life history of fish, J. H. S. Blaxter, J. C. Gamble and H. v. Westernhagen (eds.). Rapp. P.-v. Reun. Cons. int. Explor. Mer, 191.

Stepien, C. A., and R. H. Rosenblatt. 1991. Patterns of gene flow and genetic divergence in the northeastern Pacific Clinidae (Teleostei: Blennioidei), based on allozyme and morphological data. Copeia 1991:875–896.

Stevens, C. E. 1988. Comparative physiology of the vertebrate digestive system. Cambridge University Press, London and New York.

Stevens, E. G., W. Watson, and A. C. Matarese. 1984. Notothenoidea: development and relationships, p. 561–564. *In*: Ontogeny and systematics of fishes, H. G. Moser, W. J. Richards, D. M. Cohen, M. P. Fahey, A. W. Kendall, Jr. and S. L. Richardson (eds.). Am. Soc. Ichthyol. and Herpetol. Sp. Pub. 1.

Stevens, J. K., and K. E. Parsons. 1980. A fish with double vision. Nat. Hist. 89(1):62–67.

Stiassny, M. L. J. 1986. The limits and relationships of the acanthomorph teleosts. J. Zool. (Lond.) B 1:411–460.

Stiassny, M. L. J. 1990. Notes on the anatomy and relationships of the bedotiid fishes of Madagascar, with a taxonomic revision of the genus *Rheocles* (Atherinomorpha: Bedotiidae). Am. Mus. Novit. 2979.

Stiassny, M. L. J. 1991. Phylogenetic intrarelationships of the family Cichlidae: An overview, p. 1–35. *In*: Cichlid fishes—behaviour, ecology and evolution, M. H. A. Keenleyside (ed.). Chapman and Hall, London.

Stiassny, M. L. J. 1993. What are grey mullets? Bull. Mar. Sci. 52(1):197–219.

Stiassny, M. L. J., and M. C. C. de Pinna. 1994. Basal taxa and the role of cladistic patterns in the evaluation of conservation priorities: A view from freshwater, p. 235–249. *In*: Systematics and conservation evaluation. P. I. Forey and C. J. Humphries (eds.). Syst. Assoc. Spec. Vol. No. 50 (London).

Stiassny, M. L. J., and J. S. Jensen. 1987. Labroid interrelationships revisited: Morphological complexity, key innovations, and the study of comparative diversity. Bull. Mus. Comp. Zool. 151(5): 269–319.

Stickney, R. R. 1979. Principles of warmwater aquaculture. Wiley, New York.

Stock, D. W., and G. S. Whitt. 1992. Evidence from 18S ribosomal RNA sequences that lampreys and hagfishes form a natural group. Science 257:787–789.

Stoss, J. 1983. Fish gamete preservation and spermatozoan physiology, p. 305–350. *In*: Fish physiology, Vol. IXB, W. S. Hoar, D. J. Randall and E. M. Donaldson (eds.). Academic Press, New York, London.

Stouder, D. J., K. L. Fresh, and R. J. Feller. 1994. Theory and application of fish feeding ecology. University of South Carolina Press, Columbia.

Streisinger, G., C. Walker, N. Dower, D. Knauber, and F. Singer. 1981. Production of clones of homozygous zebrafish *Brachidanio rerio*. Nature 291:293–296.

Subdehar, N., and P. D. Prasado Ras. 1974. Effects of some corticosteroids and metopirone on the corpuscles of Stannius and interrenal gland of the catfish, *Heteropneustes fossilis* (Bloch). Gen. Comp. Endocrinol. 23(4):403–414.

Sulak, K. J. 1975. Cleaning behavior in the centrarchid fishes, *Lepomis macrochirus* and *Micropterus salmoides*. Anim. Behav. 23:331–334.

Sullivan, J. A., and R. J. Schultz. 1986. Genetic and environmental basis of variable sex ratios in laboratory strains of *Poeciliopsis lucida*. Evolution 40:152–158.

Summerfelt, R. C., and G. E. Hall (eds.). 1987. Age and growth of fish. Iowa State University Press, Ames.

Sundararaj, B. I. 1981. Reproductive physiology of teleost fishes: A review of present knowledge and needs for future research. FAO. ADCP/Rep. 381/16, Rome.

Sundararaj, B. I., and S. Vasali. 1976. Photoperiod and temperature control in the regulation of reproduction in the female catfish. J. Fish. Res. Bd. Can. 33:959–971.

Sundness, G., T. Enns, and P. F. Scholander. 1958. Gas secretion in fishes lacking a rete mirabile. J. Exp. Biol. 35(3):671–676.

Suyehiro, Y. 1942. A study of the digestive system and feeding habits of fish. Jpn. J. Zool. 10:1–303.

Suzuki, R., M. Kishida, and T. Hirano. 1990. Growth hormone secretion during long term incubation of the pituitary in the Japanese eel, *Anguilla japonica*. Fish Physiol. Biochem. 8(2):159–165.

Svardson, G. 1949. Natural selection and egg number in fish. Rep. Inst. Freshwater Res. (Drottningholm) 29:115–122.

Sverdrup, H., M. W. Johnson, and R. H. Fleming. 1947. The oceans: Their physics, chemistry, and general biology. Prentice Hall, Englewood Cliffs.

Svetovidov, A. N. 1948. Gadiformes, p. 1–304. *In*: Fauna of the U.S.S.R., Fishes, 9(4), E. N. Pavlovskii and A. A. Shtakel'berg (eds.). Zoological Institute, Akademii Nauk SSSP. (Translated for the National Science Foundation and Smithsonian Institution, Washington, DC, 1962.)

Swanson, B. L., and D. V. Swedberg. 1980. Decline and recovery of the Lake Superior Gull Island Reef lake trout (*Salvelinus namaycush*) and the role of sea lamprey (*Petromyzon marinus*) predation. Can. J. Fish. and Aquat. Sci. 39(11):2074–2080.

Swift, C. C., T. R. Haglund, M. Ruiz, and R. N. Fisher. 1993. The status and distribution of the freshwater fishes of southern California. Bull. S. Calif. Acad. Sci. 92(3):101–167.

Swift, C. C., J. L. Nelson, C. Maslow, and T. Stein. 1989. Biology and distribution of the tidewater goby, *Eucyclogobius newberryi* (Pisces: Gobiidae) of California. Contr. Sci. Los Angeles Cty. Mus. Nat. Hist. 404:19.

Swift, C. C., C. R. Gilbert, S. A. Bortone, G. H. Burgess, and R. W. Yerger. 1986. Zoogeography of the freshwater fishes of the southeastern United States: Savannah River to Lake Pontchartrain, p. 213–265. *In*: The zoogeography of North American freshwater fishes, C. H. Hocutt and E. O. Wiley (eds.). John Wiley and Sons, New York.

Szabo, T. 1974. Anatomy of the specialized lateral line organs of electroreception, p. 13–58. *In*: Handbook of sensory physiology, Vol. III, pt. 3, A. Fessard (ed.). Springer-Verlag, Berlin.

Tabata, M., T. Tamura, and H. Niwa. 1975. Origin of the slow potentials in the pineal organ of the rainbow trout. Vision Res. 15(6):737–740.

Taggart, J. B., and A. Ferguson. 1990. Hypervariable minisatellite DNA single locus probes for the Atlantic salmon, *Salmo salar* L. J. Fish Biol. 37:991–993.

Takemura, A. 1984. Acoustical behavior of the freshwater go by *Odontobutis obscura*. Bull. Jpn. Soc. Sci. Fish. 50(4):561–564.

Talbot, L. M. 1987. The ecological value of wildlife to the well-being of human society, p. 179–186. *In*: Valuing wildlife: Economic and social perspectives, D. J. Decker and G. R. Goff (eds.). Westview Press, Boulder and London.

Talwar, P. K., and R. K. Kacker. 1984. Commercial sea fishes of India. Pooran Press, Calcutta (pub. by Director, Zoological Survey of India).

Tamura, T., and I. Hanyu. 1980. Pineal sensitivity in fishes, p. 477–496. *In*: Environmental physiology of fishes, M. A. Ali (ed.). Plenum Press, New York.

Tamura, T., and W. J. Wisby. 1963. The visual sense of pelagic fishes, especially the visual axis and accommodation. Bull. Mar. Sci. Gulf Caribb. 13(3):433–448.

Tave, D. 1986. Genetics for fish hatchery managers. AVI Publishing, Westport, CT.

Tavolga, W. N. 1971. Sound production and detection, p. 135–205. *In*: Fish physiology, Vol. V, W. S. Hoar and D. J. Randall (eds.). Academic Press, New York.

Taylor, E. B., T. D. Beacham, amd M. Kaeriyama. 1994. Population structure and identification of North Pacific Ocean chum salmon (*Oncorhynchus keta*) revealed by an analysis of minisatellite DNA variation. Can. J. Fish. Aquat . Sci. 51:1430–1442.

Taylor, L. R., L. J. V. Compagno, and P. J. Struhsaker. 1983. Megamouth—a new species, genus and family of lamnoid shark (*Megachasma pelagios,* Family Megachasmidae) from the Hawaiian Islands. Proc. Cal. Acad. Sci. 43(8):87–110.

Taylor, S. L., and R. K. Bush. 1988. Allergy by ingestion of seafoods, p. 149–183. *In*: Marine toxins and venoms, A. T. Tu (ed.). Marcel Dekker, Inc., New York, Basel.

Teal, J. M., and F. G. Carey. 1967. Skin respiration and oxygen debt in the mudskipper *Periophthalmus sobrinus*. Copeia 1967:677–679.

Tesch, F. W. 1977. The eel. Chapman and Hall, London.

Tesch, F. W. 1978. Horizontal and vertical swimming of eels during the spawning migration at the edge of the continental shelf, p. 378–391. *In*: Animal migration, navigation and homing, K. Schmidt-Koenig and W. T. Keeton (eds.). Springer-Verlag, Berlin.

Tester, A. L. 1963. Olfaction, gustation, and the common chemical sense in sharks, p. 255–282. *In*: Sharks and survival, P. W. Gilbert (ed.). D. C. Heath, Boston.

Tester, A. L., J. I. Kendall, and W. B. Milisen. 1972. Morphology of the ear of the shark genus *Carcharinus* with particular reference to the macula neglecta. Pacific Sci. 26:264–274.

Theis, D., and W.-E. Reif. 1985. Phylogeny and evolutionary ecology of Mesozoic Neoselachii. N. Jb. Geol. Palaont. Abh. 169: 333–361.

Thibault, R. E., and R. J. Schultz. 1979. Reproductive adaptations among viviparous fishes (Cyprinodontiformes: Poeciliidae). Evolution 32:320–333.

Thomas, W. K., and A. T. Beckenbach. 1989. Variations in salmonid mitochondrial DNA: Evolutionary constraints and mechanisms of substitution. J. Mol. Evol. 24:218–227.

Thompson, E. M., and F. I. Tsuji. 1989. Two populations of the marine fish *Porichthys notatus,* one lacking in luciferin essential for bioluminescence. Mar. Biol. 102:161–165.

Thompson, E. M., B. G. Nafpaktitis, and F. I. Tsuji. 1988. Dietary uptake and blood transport of *Vargula* (crustacean) luciferin in the bioluminescent fish, *Porichthys notatus*. Comp. Biochem. Physiol. 89A(2):203–209.

Thomson, D. A., L. T. Findley, and A. N. Kerstitch. 1979. Reef fishes of the Sea of Cortez. John Wiley and Sons, New York.

Thomson, K. S. 1966. Intracranial mobility in the coelacanth. Science 153:999–1000.

Thomson, K. S. 1967. Mechanisms of intracranial kinetics in fossil rhipidistian fishes (Crossopterygii) and their relatives. J. Linn. Soc., Zool. 46:223–253.

Thomson, K. S. 1976. On the heterocercal tail in sharks. Paleobiol. 2(1):19–38.

Thomson, K. S. 1991a. Where did tetrapods come from? Am. Scientist 79(6):488–490.

Thomson, K. S. 1991b. Living fossil: The story of the coelacanth. W. W. Norton, New York.

Thomson, K. S., and D. F. Simanek. 1977. Body form and locomotion in sharks. Amer. Zool. 17:343–354.

Thorgaard, G. H. 1976. Robertsonian polymorphism and constitutive heterochromatin distribution in chromosomes of the rainbow trout (*Salmo gairdneri*). Cytogenet. Cell Genet. 17:174–184.

Thorgaard, G. H. 1977. Heteromorphic sex chromosomes in male rainbow trout. Science 196:900–902.

Thorgaard, G. H. 1978. Sex chromosomes in the sockeye salmon: A Y-autosome fusion. Can. J. Genet. Cytol. 20:349–354.

Thorgaard, G. H. 1983. Chromosomal differences among rainbow trout populations. Copeia 1983: 650–662.

Thorgaard, G. H., and S. K. Allen. 1987. Chromosome manipulation and markers in fishery management, p. 319–331. *In*: Population genetics in fishery management, N. Ryman and F. M. Utter (eds.). University of Washington Press, Seattle.

Thorpe, J. E. 1993. Impacts of fishing on genetic structure of salmonid populations, p. 67–80. *In*: Genetic conservation of salmonid fishes, J. G. Cloud and G. H. Thorgaard (eds.). Plenum Press, New York.

Thorson, T. B. 1961. Partitioning of body water in Osteichthyes: Phylogenetic implications in aquatic vertebrates. Biol. Bull. 120:238–254.

Thorson, T. B. 1967. Osmoregulation in freshwater elasmobranchs, p. 265–270. *In*: Sharks, skates and rays, P. W. Gilbert, R. F. Mathewson and O. P. Rall (eds.). Johns Hopkins, Baltimore.

Thresher, R. E. 1984. Reproduction in reef fishes. T. F. H. Publications, Neptune City, NJ.

Todd, J. H. 1971. The chemical language of fishes. Sci. Am. 244(5):98–108.

Todd, J. H., J. Atema, and J. E. Bardach. 1968. Chemical communication in social behavior of a fish, the yellow bullhead (*Ictalurus natalis*). Science 158:672–673.

Torrey, T. W. 1962. Morphogenesis of the vertebrates. John Wiley & Sons, New York.

Tota, B. 1989. Myoarchitecture and vascularization of the elasmobranch heart ventricle, p. 122–135. *In*: Evolutionary and contemporary biology of elasmobranchs, W. C. Hamlett and B. Tota (eds.). J. Exp. Zool. Supp. 2.

Trautman, M. B. 1981. The fishes of Ohio. Revised edition. Ohio State University Press.

Travers, R. A. 1981. The interarcual cartilage: A review of its development, distribution and value as an indicator of phyletic relationships in euteleostean fishes. J. Nat. Hist. 15:853–871.

Travers, R. A. 1984. A review of the Mastacembeloidei, a suborder of synbranchiform teleost fishes. Part II: Phylogenetic analysis. Bull. Br. Mus. (Nat. Hist.) Zool. 47:83–150.

Trewavas, E. 1973. On the cichlid fishes of the genus *Pelmatochromis* with proposal of a new genus for *P. congicus*; on the relationship between *Pelmatochromis* and *Tilapia* and the recognition of *Sarotherodon* as a distinct genus. Bull. Br. Mus. (Nat. Hist.) Zool. 25(1):1–26.

Trewavas, E. 1981. Nomenclature of tilapias of southern Africa. J. Limnol. Soc. South Africa 7(1):42.

Tricas, T. C., and J. E. McCosker. 1984. Predatory behavior of the white shark (*Carcharodon carcharias*), with notes on its biology. Proc. Calif. Acad. Sci. 43(14):221–238.

Trnski, T., J. M. Leis, and P. Wirtz. 1989. Pholidichthyidae—convict blennies, engineerfishes, p. 259–261. *In*: The larvae of Indo-Pacific shorefishes, J. M. Leis and T. Trnski (eds.). University of Hawaii Press, Honolulu.

Tsukamoto, K. 1992. Discovery of the spawning area for Japanese eel. Nature 356(6372):789–791.

Tsukamoto, K., A. Umezawa, O. Taketa, N. Mochioka, and T. Kayihara. 1989. Age and birth date of *Anquilla japonica* leptocephali collected in western North Pacific in September 1986. Bull. Jpn. Soc. Sci. Fish. 55(6):1023–1028.

Tu, A. T. (ed.). 1988. Handbook of natural toxins: Vol. 3, Marine toxins and venoms. Marcel Dekker Inc., New York, Basel.

Tucker, D. 1983. Fish chemoreception: Peripheral anatomy and physiology, p. 311–349. *In*: Fish neurobiology, Vol. 1, R. G. Northcutt and R. E. Davis (eds.). University of Michigan Press, Ann Arbor.

Tuge, H., K. Uchinashi, and H. Shimamura. 1968. An atlas of the brains of fishes of Japan. Tzukiji Shokan, Tokyo.

Turner, C. L. 1936. The absorptive processes in the embryos of *Parabrotula dentiens,* a viviparous, deep-sea brotulid fish. J. Morphol. 59:313–325.

Turner, C. L. 1946. Male secondary sexual characters of *Dinematichthys illuocoetoides.* Copeia 1946:92–96.

Turner, C. L. 1947. Viviparity in teleost fishes. Sci. Monthly 65: 508–518.

Turner, S. 1986. *Thelodus macintoshi* Stetson 1928, the largest known thelodont (Agnatha: Thelodonti). Breviora 486.

Turner, S. 1991. Monophyly and interrelationships of the Thelodonti, p. 87–120. *In*: Early vertebrates and related probems of evolutionary biology, M. M. Chang, Y. H. Liu and G. R. Zhang (eds.). Science Press, Beijing.

Turner, S., and G. C. Young. 1992. Thelodont scales from the Middle-Late Devonian Aztec siltstone, southern Victoria Land, Antarctica. Antarct. Sci. 4(1):89–105.

Tyler, A. V. 1970. Rates of gastric emptying in young cod. J. Fish. Res. Bd. Can. 27:1177–1189.

Tyler, J. C. 1980. Osteology, phylogeny, and higher classification of the fishes of the order Plectognathi (Tetraodontiformes). NOAA Tech. Rep. NMFS Circ. 434.

Tyler, J. C., G. D. Johnson, I. Nakamura, and B. B. Collette. 1989. Morphology of *Luvarus imperialis* (Luvaridae), with a phylogenetic analysis of the Acanthuroidei (Pisces). Smiths. Cont. Zool. 485.

Ultsch, G. R., H. Boschung, and M. J. Ross. 1978. Metabolism, critical oxygen tension, and habitat selection in darters (*Etheostoma*). Ecol. 59:99–107.

Underhill, J. C. 1986. The fish fauna of the Laurentian Great Lakes, the St. Lawrence lowlands, Newfoundland and Labrador, p. 105–137. *In*: The zoogeography of the freshwater fishes of North America, C. H. Hocutt and E. O. Wiley (eds.). John Wiley and Sons, New York.

Ursin, E. 1979. Growth in fishes, p. 63–87. *In*: Fish phenology: Anabolic adaptiveness in teleosts, P. J. Miller (ed.). Sym. Zool. Soc. Lond. 44. Academic Press, London.

USDA. 1991. Aquaculture situation and outlook. U. S. Dept. Agric., Washington, DC.

USDA. 1992. Aquaculture situation and outlook. U. S. Dept. Agric., Washington, DC.

Usher, M. L., C. Talbot, and F. B. Eddy. 1991. Effects of transfer to seawater on growth and feeding of Atlantic salmon smolts. Aquaculture 94(4):309–326.

Utter, F., G. Milner, G. Stahl, and D. Teel. 1989. Genetic structure of chinook salmon, *Oncorhynchus tshawytscha,* in the Pacific Northwest. Fish. Bull., U.S. 87:239–264.

Uyeno, T. 1991. Observations on locomotion and feeding of released coelacanths, *Latimeria chalumnae.* Env. Biol. Fishes 32:267–173.

Uyeno, T., and G. R. Smith. 1972. Tetraploid origin of the karyotype of catostomid fishes. Science 175:644–646.

Uyeno, T., and T. Tsutsumi. 1991. Stomach contents of *Latimeria chalumnae* and further notes on its feeding habits. Env. Biol. Fishes 32:275–279.

Van Beneden, R. J. 1993. Oncogenes, p. 113–136. *In*: Biochemistry and molecular biology of fish, Vol. 2: Molecular biology frontiers, P. W. Hochachka and T. P. Mommsen (eds.). Elsevier, Amsterdam.

Vandepitte, J., P. Lemmens, and L. De Swert. 1983. Human edwarsiellosis traced to ornamental fish. J. Clin. Microbiol. 17(1): 165–167.

Van Oosten, J. 1957. The skin and scales, p. 207–244. *In*: The physiology of fishes, M. E. Brown (ed.). Academic Press, New York.

van der Waal, B. C. W. 1974. Observations on the breeding habits of *Clarias gariepinus* (Burchell). J. Fish Biol. 6:23–27.

van Weerd, J. H., and C. J. J. Richter. 1991. Sex pheromones and ovarian development in teleost fish. Comp. Biochem. Physiol. 100A:517–527.

Vari, R. P. 1979. Anatomy, relationships, and classification of the families of Citharinidae and Distichodontidae (Pisces, Characoidea). Bull. Br. Mus. (Nat. Hist.) Zoology 36:261–344.

Vari, R. P. 1983. Phylogenetic relationships of the families Curimatidae, Prochilodontidae, Anostomidae, and Chilodontidae (Pisces: Characiformes). Smithson. Contr. Zool. 378.

Vari, R. P. 1988. The Curimatidae, a lowland neotropical fish family (Pisces: Characiformes); distribution, endomism, and phylogenetic biogeography, p. 343–377. *In*: Proceedings of a workshop on neotropical distribution patterns, W. R. Heyer and P. E. Vanzolini (eds.). Academia Brasiliera de Ciencias, Rio de Janeiro.

Vari, R. P. 1989. A phylogenetic study of the neotropical characiform family Curimatidae (Pisces: Ostariophysi). Smithson. Contr. Zool. 507.

Vari, R. P., and H. Ortega. 1986. The catfishes of the neotropical family Helogenidae (Ostariophysi: Siluroidei). Smithson. Contr. Zool. 442.

Veith, W. J. 1980. Viviparity and embryonic adaptations in the teleost *Clinus superciliosus.* Can. J. Zool. 58:1–12.

Verheijen, F. J., and J. H. Reuter. 1969. The effect of alarm substance on predation among minnows. Anim. Behav. 17:551–554.

Vickers, T. 1961. A study of the so-called "chloride-secretory" cells of the gills of teleosts. Q. J. Micros. Sci. 60:507–518.

Videler, J. J. 1977. Mechanical properties of fish tail joints, p. 183–194. *In*: Physiology of movement—biomechanics, W. Nachtigall (ed.). Gustave Fischer Verlag, Stuttgart.

Videler, J. J. 1981. Swimming movements, body structure and propulsion in cod, *Gadus morhua*, p. 1–27. *In*: Vertebrate locomotion, Symp. Zool. Soc. Lond. (48), M. E. Day (ed.). Academic Press, New York.

Videler, J. J. 1993. Fish swimming. Chapman and Hall, London, Glasgow, New York, Tokyo, Melbourne, Madras.

Videler, J. J., and D. Weihs. 1982. Energetic advantages of burst-and-coast swimming of fish at high speeds. J. Exp. Biol. 97:169–178.

Vielkind, J. R., K. D. Kallman, and D. C. Morizot. 1989. Genetics of melanomas in *Xiphophorus* fishes. J. Aquatic. Anim. Health 1:69–77.

Viete, S. and W. Heiligenberg. 1991. The development of the jamming avoidance response (JAR) in *Eigenmannia*: An innate behavior indeed. J. Comp. Physiol. A 169(1):15–33.

Vilches-Troya, J., R. F. Dunn, and D. P. O'Leary. 1984. Relationship of the vestibular hair cells to magnetic particles in the otolith of the guitarfish (*Rhinobatos productus*) sacculus. J. Comp. Neurol. 236(4):489–494.

Vine, P. J. 1974. Effects of algal grazing and aggressive behavior of the fishes *Pomacentrus lividus* and *Acanthurus sohal* on coral reef ecology. Mar. Biol. 24:131–136.

Vinyard, G. L. 1982. Variable kinematics of Sacramento perch (*Archoplites interruptus*) capturing evasive and non-evasive prey. Can. J. Fish. and Aq. Sciences 39:208–211.

Vladykov, V. D., and E. Kott. 1979. Satellite species among the holarctic lampreys (Petromyzonidae). Can. J. Zool. 57(4):860–867.

Vodicnik, M. J., R. E. Kral, V. L. de Vlaming, and L. W. Crim. 1978. The effects of pinealectomy on pituitary and plasma gonadotropin levels in *Carassius auratus* exposed to various photoperiod-temperature regimes. J. Fish. Biol. 12:187–196.

Vodicnik, M. J., J. Olcese, G. Delahunty, and V. de Vlaming. 1979. The effects of blinding, pinealectomy and exposure to constant dark conditions on gonadal activity in the female goldfish, *Carassius auratus*. Environ. Biol. Fish. 4:173–177.

Vogel, W. O. P. 1985. The caudal heart of fishes: Not a lymph heart. Act. Anat. 121:41–45.

von der Emde, G. 1990. Discrimination of objects through electrolocation in the weakly electric fish *Gnathonemus petersii*. J. Comp. Physiol. A:167:413–422.

Wagner, H. H. 1974. Seawater adaptation independent of photoperiod in steelhead trout. Can. J. Zool. 52:805–812.

Wagner, H. H., F. P. Conte, and J. L. Fessler. 1969. Development of osmotic and ionic regulation in two races of chinook salmon, *Oncorhynchus tshawytscha*. Comp. Biochem. Physiol. 19:325–341.

Wagner, H. J. 1990. Retinal structure of fishes, p. 109–158. *In*: The visual system of fish, R. H. Douglas and M. B. A. Djamgoz (eds.). Chapman and Hall, London.

Wahlqvist, I. 1980. Effects of catecholamines on isolated systemic and branchial vascular beds of the cod, *Gadus morhua*. J. Comp. Physiol. 137:139–143.

Wahlqvist, I., and S. Nilsson. 1980. Adrenergic control of the cardio-vascular system of the Atlantic cod, *Gadus morhua*, during "stress." J. Comp. Physiol. 137:145–150.

Wainwright, S. A. 1983. To bend a fish, p. 68–90. *In*: Fish biomechanics, P. W. Webb and D. Weihs (eds.). Praeger, New York.

Waiwood, K., and J. Majkowski. 1984. Food consumption and diet composition of cod, *Gadus morhua*, inhabiting the southwestern Gulf of St. Lawrence. Env. Biol. Fish. 11(1):63–78.

Wake, M. H. (ed.). 1979. Hyman's comparative anatomy (Third Edition). University of Chicago Press, Chicago.

Wake, M. H. 1987. Urogenital morphology of dipnoans, with comparisons to other fishes and to amphibians, p. 199–216. *In*: The biology and evolution of lungfishes, W. E. Bemis, W. W. Borggren and M. E. Kemp (eds.). Alan Liss Inc., New York.

Wales, J. H. 1950. Swimming speed of the western sucker *Catastomus occidentalis* Ayres. Calif. Fish and Game 36:433–434.

Walker, B. W. 1959. The timely grunion. Nat. Hist. 68(6):302–307.

Walker, M. M. 1984. Learned magnetic field discrimination in yellowfin tuna, *Thunnus albacares*. J. Comp. Physiol. 155:673–679.

Walker, M. M., A. E. Dizon, and J. L. Kirschvink. 1982. Geomagnetic field detection by yellowfin tuna, p. 755–758. *In*: Oceans 82 Conf. Record: Industry, government, education. Partners in Progress, Washington, DC.

Walker, M. M., J. L. Kirschvink, S. R. Chang, and A. E. Dizon. 1984. A candidate magnetic sense organ in the yellowfin tuna, *Thunnus albacares*. Science 224:751–753.

Wallace, R. A., and K. Selman. 1981. Cellular and dynamic aspects of oocyte growth in teleosts. Am. Zool. 21:325–343.

Wallbrunn, H. M. 1958. Genetics of the Siamese fighting fish, *Betta splendens*. Genetics 43:289–298.

Walls, G. L. 1963. The vertebrate eye and its adaptive radiation. Hafner, New York.

Walsh, S. J., and B. B. Burr. 1981. Distribution, morphology and life history of the least brook lamprey, *Lampetra aepyptera* (Pisces: Petromyzontidae), in Kentucky. Brimleyana 6:83–100.

Walters, C. 1986. Adaptive management of renewable resources. MacMillan, New York.

Walters, L. H., and V. Walters. 1965. Laboratory observations on the cavernicolous poeciliid from Tabasco, Mexico. Copeia 1965:214–223.

Walters, V. 1963. The trachypterid integument and an hypothesis on its hydrodynamic function. Copeia 1963:260–270.

Walvig, F. 1963. The gonads and the formation of the sexual cells, p. 530–580. *In*: The biology of *Mxyine*, A. Brodal and R. Fänge (eds.). Universitetesforlaget, Oslo.

Wang, R. J., and J. A. C. Nicol. 1974. The tapetum lucidum of gars (Lepisosteidae) and its role as a reflector. Can. J. Zool. 52(12):1523–1530.

Waples, R. S. 1987. A multispecies approach to the analysis of gene flow in marine shore fishes. Evolution 41:385–400.

Ward, P. I. 1988. Sexual dichromatism and parasitism in British and Irish freshwater fish. Anim. Behav. 36(4):1210–1215.

Wardle, C. S. 1975. Limit of fish swimming speed. Nature, Lond. 255:725–727.

Wardle, C. S., and P. He. 1988. Burst swimming speeds of mackerel, *Scomber scombrus* L. J. Fish Biol. 32(3):471–478.

Wardle, C. S., J. J. Videler, T. Arimoto, J. M. Franco, and P. He. 1989. The muscle twitch and the maximum swimming speed of giant bluefin tuna, *Thunnus thynnus* L. J. Fish Biol. 35:129–137.

Warner, R. R. 1984. Mating behavior and hermaphroditism in coral reef fishes. Am. Scient. 72:128–136.

Warner, R. R., and D. R. Robertson. 1978. Sexual patterns in the labroid fishes of the western Caribbean. I: The wrasses (Labridae). Smithson. Contr. Zool. 254:1–27.

Warren, C. E. 1971. Biology and water pollution control. W. B. Saunders, Philadelphia.

Warren, C. E., and G. E. Davis. 1967. Laboratory studies on the feeding, bioenergetics, and growth of fish, p. 175–214. *In*: The biological basis of fish production, S. D. Gerking (ed.). Blackwell Scientific, Oxford.

Washington, B. B., W. N. Eschmeyer, and K. M. Howe. 1984. Scorpaeniformes: Relationships, p. 438–447. *In*: Ontogeny and systematics of fishes, H. G. Moser, W. J. Richards, D. M. Cohen, M. P. Fahay, A. W. Kendall, Jr. and S. L. Richardson (eds.). Am. Soc. Ichthyol. and Herptol. Sp. Pub. 1.

Washington, B. B., H. G. Moser, W. A. Laroche, and W. J. Richards. 1984. Scorpaeniformes: Development, p. 405–477. *In*: Ontogeny and systematics of fishes, H. G. Moser, W. J. Richards, D. M. Cohen, M. P. Fahay, A. W. Kendall, Jr. and S. L. Richardson (eds.). Am. Soc. Ichthyol. and Herpetol. Sp. Pub. 1.

Watiporn, P. 1988. Effects of carotenoid pigments from different sources on color changes of fancy carp, *Cyprinus carpio*. Abstract Master of Science theses (Fish. Sci.). Notes Fac. Fish. Kasetsart Univ. 10:1.

Watson, D. M. S. 1937. The acanthodian fishes. Philos. Trans. R. Soc. (Biol. Sci.) 228:49–146.

Watson, W. 1982. Development of eggs and larvae of the white croaker, *Genyonemus lineatus* Ayres (Pisces: Sciaenidae), off the southern California coast. Fish. Bull. U.S. 80:403–417.

Watson, W., A. C. Matarese, and E. G. Stevens. 1984. Trachinoidea: development and relationships, p. 554–561. *In*: Ontogeny and systematics of fishes, H. G. Moser, W. J. Richards, D. M. Cohen, M. P. Fahay, A. W. Kendall, Jr. and S. L. Richardson (eds.) Am. Soc. Ichthyol. and Herpetol. Sp. Pub. 1.

Watson, W., E. G. Stevens, and A. C. Matarese. 1984. Trachinoidea: development and relationships, p. 552–554. *In*: Ontogeny and systematics of fishes, H. G. Moser, W. J. Richards, D. M. Cohen, M. P. Fahay, A. W. Kendall, Jr. and S. L. Richardson (eds.) Am. Soc. Ichthyol. and Herpetol. Sp. Pub. 1.

Watts, E. H. 1960. The relationship of fish locomotion to the design of ships, p. 27–40. *In*: Vertebrate locomotion, Symp. Zool. Soc. Lond. (5).

Weatherly, A. H. 1972. Growth and ecology of fish populations. Academic Press, London.

Weatherly, A. H. 1976. Factors affecting maximization of fish growth. J. Fish. Res. Bd. Can. 33:1046–1058.

Webb, J. F. 1989. Gross morphology and evolution of the mechanosensory lateral-line system in teleost fishes. Brain, Behav. Evol. 3(1):34–53.

Webb, P. W. 1971a. The swimming energetics of trout. I. Thrust and power output at cruising speeds. J. Exp. Biol. 55:489–520.

Webb, P. W. 1971b. The swimming energetics of trout. II. Oxygen consumption and swimming efficiency. J. Exp. Biol. 55:521–540.

Webb, P. W. 1973. Kinematics of pectoral fin propulsion in *Cymatogaster aggregata*. J. Exp. Biol. 59:697–710.

Webb, P. W. 1975. Hydrodynamics and energetics of fish propulsion. Bulletin 190, Fish. Res. Bd. Can.

Webb, P. W. 1975. Efficiency of pectoral-fin propulsion of *Cymatogaster aggregata*, p. 573–584. *In*: Swimming and flying in nature, Vol. 2, T. Y.-T. Wu, C. J. Brokaw and C. Brennan (eds.). Plenum Press, New York.

Webb, P. W. 1978. Hydrodynamics: nonscombroid fish, p. 189–237. *In*: Fish physiology, Vol. VII, W. S. Hoar and D. J. Randall (eds.). Academic Press, New York.

Webb, P. W. 1982. Locomotor patterns in the evolution of the actinopterygian fishes. Amer. Zool. 22:329–342.

Webb, P. W. 1984. Form and function in fish swimming. Sci. Am. 251(1):72–82.

Webb, P. W., and R. S. Keyes. 1981. Division of labor between median fins in swimming dolphin (Pisces: Coryphaenidae). Copeia 1981:901–904.

Webb, P. W., and R. S. Keyes. 1982. Swimming kinetics of sharks. Fish. Bull. U.S. 80(4):803–812.

Webster, D., and M. Webster. 1974. Comparative vertebrate morphology. Academic Press, New York.

Weibe, J. P. 1968. The reproductive cycle of the viviparous seaperch, *Cymatogaster aggregatus* Gibbons. Can. J. Zool. 46:1221–1234.

Weichert, C. K. 1965. Anatomy of the chordates (Third Edition). McGraw-Hill, New York.

Weihs, D. 1973. Hydromechanics of fish schooling. Nature 245:48–50.

Weihs, D. 1974. Energetic advantages of burst swimming in fish. Jour. Theoret. Biol. 48:215–229.

Weihs, D. 1981. Body section variations in sharks—an adaptation for efficient swimming. Copeia 1981:217–219.

Weihs, D. 1989. Design features and mechanics of axial locomotion in fish. Am. Zool. 29:151–160.

Weihs, D., and P. W. Webb. 1983. Optimization of locomotion, p. 339–371. *In*: Fish biomechanics, P. W. Webb and D. Weihs (eds.). Praeger, New York.

Weihs, D., R. S. Keyes, and D. M. Stalls. 1981. Voluntary swimming speeds of two species of large carcharhinid sharks. Copeia 1981:219–222.

Weintraub, B. 1992. How to fool a pirahana: Two heads are better. Nat. Geog. 182(5).

Weisel, G. F. 1962. Comparative study of the digestive tract of a sucker, *Castostomus catostomus,* and a predaceous minnow, *Ptychocheilus oregonense*. Am. Midl. Nat. 68:334–346.

Weisel, G. F. 1967. Early ossification in the skeleton of the sucker (*Catostomus macrocheilus*) and the guppy (*Poecilia reticulata*). J. Morphol. 121:1–8.

Weisel, G. F. 1973. Anatomy and histology of the digestive system of the paddlefish (*Polyodon spathula*). J. Morphol. 140:243–256.

Weithman, A. S. 1993. Socioeconomic benefits of fisheries, p. 159–177. *In*: Inland fisheries in North America, C. C. Kohler and W. A. Hubert (eds.). Amer. Fish. Soc., Bethesda, MD.

Weitzman, S. H. 1967. The origin of the stomiatoid fishes with comments on the classification of salmoniform fishes. Copeia 1967:507–540.

Weitzman, S. H. 1974. Osteology and evolutionary relationships of the Sternoptychidae, with a new classification of stomiatoid families. Bull. Am. Mus. Nat. Hist. 153:329–478.

Welcomme, R. L. 1979. Fisheries ecology of floodplain rivers. Longman, Ltd., London and New York.

Wells, L. 1980. Lake trout (*Salvelinus namaycush*) and sea lamprey (*Petromyzon marinus*) populations in Lake Michigan, 1971–78. Can. J. Fish. and Aquat. Sci. 37(11):2047–2051.

Wendelaar Bonga, S. E. 1993. Endocrinology, p. 469–502. *In*: The physiology of fishes, D. H. Evans (ed.). CRC Press, Boca Raton.

Westneat, M. W., W. Hoese, C. A. Pell, and S. A. Wainwright. 1993. The horizontal septum: Mechanisms of force transfer in locomotion of scombrid fishes (Scombridae, Perciformes). J. Morphol. 217(2):183–204.

Wetherbee, B. 1990. Feeding biology of sharks, p. 74–76. *In*: Discovering sharks, S. H. Gruber (ed.). Spec. Publ. 14, Am. Littoral Soc., Highlands, NJ.

Wetherington, J. D., R. A. Schenck, and R. C. Vrijenhoek. 1989. The origins and ecological success of unisexual *Poeciliopsis*: The frozen niche-variation model, p. 259–275. *In*: Ecology and evolution of livebearing fishes, G. K. Meffe and F. F. Snelson, Jr. (eds.). Prentice Hall, Englewood Cliffs, NJ.

Wheeler, A. 1975. Fishes of the world: An illustrated dictionary. MacMillan, New York.

Wheeler, P. A., and G. H. Thorgaard. 1991. Cryopreservation of rainbow trout semen in large straws. Aquaculture 93:95–100.

White, B. N. 1986. The isthmian link, antitropicality and American biogeography: Distributional history of the Atherinopsinae (Pisces: Atherinidae). Syst. Zool. 35:176–194.

White, B. N. 1994. Vicariance biogeography of the open-ocean Pacific. Prog. Oceanog. 34:257–284.

White, B. N., R. J. Lavenberg, and G. E. McGowen. 1984. Atheriniformes: development and relationships, p. 356–362. *In*: Ontogeny and systematics of fishes, H. G. Moser, W. J. Richards, D. M. Cohen, M. P. Fahay, A. W. Kendall, Jr. and S. L. Richardson (eds.) Am. Soc. Ichthyol. and Herpetol. Sp. Pub. 1.

Whitear, M. 1986a. Epidermis, p. 8–38. *In*: Biology of the integument. 2. Vertebrates, J. Berester-Hahn, A. G. Matoltsy and K. S. Richards (eds.). Springer-Verlag, Berlin, Heidelberg, New York, Tokyo.

Whitear, M. 1986b. Dermis, p. 39–64. *In*: Biology of the integument. 2. Vertebrates, J. Berester-Hahn, A. G. Matoltsy and K. S. Richards (eds.). Springer-Verlag, Berlin, Heidelberg, New York, Tokyo.

Whitear, M. 1992. Solitary chemosensory cells, p. 103–125. *In*: Fish chemoreception, T. J. Hara (ed.). Chapman and Hall, London.

Whitear, M., and K. Kotrschal. 1988. The chemosensory anterior dorsal fin in rocklings (*Galdropsarus* and *Ciliata,* Teleostei, Gadidae): activity, fine structure and innervation. J. Zool. London 216:339–366.

Whitehead, P. J. P. 1985. FAO species catalog. Clupeoid fishes of the world (suborder Clupeoidei), Part 1—Chirocentridae, Clupeidae and Pristigasteridae. FAO Fish. Synop. No. 125, Vol. 7 (pt. 1).

Whitehead, P. J. P., and J. H. S. Blaxter. 1989. Swimbladder form in clupeoid fishes. Zool. J. Linn. Soc. 97(4):299–372.

Whitney, R. R. 1969. Schooling of fishes relative to available light. Trans. Am. Fish. Soc. 98:497–504.

Wicander, R., and J. S. Monroe. 1989. Historical geology. West Publishing Co., St. Paul.

Wicht, H., and R. G. Northcutt. 1992. The forebrain of the Pacific hagfish: A cladistic reconstruction of the ancestral craniate forebrains. Brain, Behav. Evol. 40(1):25–64.

Wiley, E. O. 1976. The phylogeny and biogeography of fossil and recent gars (Actinopterygii: Lepisosteidae). Mus. Nat. Hist., University of Kansas Misc. Publ. No. 64:1–111.

Wiley, E. O. 1992. Phylogenetic relationships of the Percidae (Teleostei; Perciformes): A preliminary hypothesis, p. 247–267. *In*: Systematics, historical ecology, the North American freshwater fishes, R. L. Mayden (ed.). Stanford University Press, Palo Alto.

Wiley, M. L., and B. B. Collette. 1970. Breeding tubercles and contact organs in fishes: Their occurrence, structure, and significance. Bull. Am. Mus. Nat. Hist. 143:145–216.

Wilhelm, W. 1980. The disputed feeding behavior of a paedophagous Haplochromine cichlid (Pisces) observed and discussed. Behavior 74:310–323.

Williams, C. D., and J. E. Deacon. 1991. Ethics, federal legislation, and litigation, p. 109–121. *In*: Battle against extinction: Native fish management in the American West, W. L. Minckley and J. E. Deacon (eds.). University of Arizona Press, Tucson and London.

Williams, J. E., and C. E. Bond. 1983. Status and life history notes on the native fishes of the Alvord Basin, Oregon and Nevada. Great Basin Naturalist 43(3):409–420.

Williams, J. E., D. B. Bowman, J. E. Brooks, A. A. Echelle, R. J. Edwards, D. A. Hendereson, and J. J. Landye. 1985. Endangered aquatic ecosystems in North American deserts with a list of vanishing species of the region. J. Arizona-Nevada Acad. of Sci. 20(1):1–61.

Williams, J. E., J. E. Johnson, D. A. Hendrickson, S. Contreras-Balderas, J. D. Williams, M. Navarro-Mendoza, D. E. McAllister, and J. E. Deacon. 1989. Fishes of North America endangered, threatened, or of special concern: 1989. Fisheries 14(6): 2–20.

Williamson, D. S. 1843. The statistical account of Tongland. Cited in D. Mills, 1989. Ecology and management of Atlantic salmon. Chapman and Hall, London.

Wilson, E. O. 1975. Sociobiology: The new synthesis. Harvard University Press, Cambridge.

Wilson, J. G., and O. Munk. 1964 A hadal fish (*Bassogigas profondissimus*) with a functional swim bladder. Nature 204(4958): 594–595.

Wilson, M. V. H., and M. W. Caldwell. 1993. New Silurian and Devonian fork-tailed thelodonts are jawless vertebrates with stomachs and deep bodies. Nature 361(6411):442–444.

Wilson, M. V. H., and P. Veilleux. 1982. Comparative osteology and relationships of the Umbridae (Pisces: Salmoniformes). Zool. J. Linnaean Society 76(4):321–352.

Wilson, M. V. H., and R. R. G. Williams. 1992. Phylogenetic, biogeographic, and ecological significance of early fossil records of North American teleostean fishes, p. 224–246. *In*: Systematics, historical ecology, and North American freshwater fishes, R. L. Mayden (ed.). Stanford University Press, Palo Alto.

Wilson, R. P. 1985. Amino acid and protein requirements of fish, p. 1–16. *In*: Nutrition and feeding in fish, C. B. Cowey, A. M. Mackie and J. J. Bell (eds.). Academic Press, London, Orlando.

Wilson, R. P. 1989. Amino acids and proteins, p. 111–151. *In*: Fish nutrition (Second Edition), J. E. Halver (ed.). Academic Press, San Diego, London.

Wilson, R. P., P. G. Ryan, A. James, and M. P.-T. Wilson. 1987. Conspicuous coloration may enhance prey capture in some piscivores. Anim. Behav. 35(5):1558–1560.

Winberg, G. G. 1956. Rate of metabolism and food requirements of fishes. Lenina, Minsk, Nauchnye Trudy Belorusskovo Gosudarstrennovo Universiteta im, V. I. Trans. Ser. 194, Fish. Res. Bd. Can.

Windell, J. T. 1967. Rates of digestion in fishes, p. 150–173. *In*: The biological basis of freshwater fish production, S. D. Gerking (ed.). Blackwell Scientific, Oxford.

Windell, J. T., and D. O. Norris. 1969, Gastric digestion and evacuation in rainbow trout. Progr. Fish-Cult. 31:20–26.

Winemiller, K. O. 1989. Development of dermal lip protuberances or aquatic surface respiration in South American characid fishes. Copeia 1989:382–390.

Winemiller, K. O. 1990. Caudal eyespots as deterrents against fin predation in the neotropical cichlid *Astronotus ocellatus*. Copeia 1990:665–673.

Winemiller, K. O., and L. C. Kelso-Winemiller. 1993. Fin-nipping piranhas: Predatory response of piranhas to alternative prey. Nat. Geog. Res. and Explor. 9:344–357.

Winemiller, K. O., and K. A. Rose. 1993. Why do most fish produce so many tiny offspring. Am. Nat. 142:585–603.

Winfield, I. J., and J. S. Nelson (Eds.). 1991. Cyprinid fishes: Systematics, biology and exploitation. Chapman and Hall, London.

Winterbottom, R. 1974. The familial phylogeny of the Tetraodontiformes (Acanthopterygii: Pisces) as evidenced by their comparative myology. Smithson. Contr. Zool. 155.

Wisby, W. J., and A. D. Hasler. 1954. Effect of olfactory occlusion in migrating silver salmon (*Oncorhynchus kisutch*). J. Fish. Res. Bd. Can. 11:472–478.

Withers, N. W. 1988. Ciguatera fish toxins and poisoning, p. 31–61. *In*: Marine toxins and venoms, A. T. Tu (ed.). Marcel Dekker, Inc., New York, Basel.

Witte, F. 1981. Initial results of the ecological survey of the haplochromine cichlid fishes from the Mwanza Gulf of Lake Victoria. Neth. J. Zool. 31:175–202.

Wittenberg, J. B., and R. L. Haedrich. 1974. The choroid rete mirabile of the fish eye. II. Distribution and relation to the pseudobranch and to the swimbladder rete mirabile. Biol. Bull. 146:137–156.

Wolf, N. G. 1983. The behavioral ecology of herbivorous fishes in mixed species groups. PhD. Thesis, Cornell University, Ithaca, NY.

Wolf, N. G. 1985. Odd fish abandon mixed-species groups when threatened. Behav. Ecol. and Sociobiol. 17:47–52.

Wolinsky, E. 1980. Mycobacteria, p. 724–742. *In*: Microbiology (Third Edition), B. D. Davis, R. Dulbecco, S. N. Eisen and H. S. Ginsberg (eds.). Harper and Row, Hagerstown.

Wood, E. 1985. Exploitation of coral reef fishes for the aquarium trade. Marine Conservation Society, Ross-On-Wye, U.K. (Abstract only seen.)

Wood, R. M., and R. L. Mayden. 1992. Systematics, evolution, and biogeography of *Notropis chlorocephalus* and *N. lutpinnis*. Copeia 1992:68–81.

Wooton, R. J. 1984. Introduction: Strategies and tactics in fish reproduction, p. 1–12. *In*: Fish reproduction: Strategies and tactics, G. W. Potts and R. J. Wootton (eds.). Academic Press, London, Orlando.

Wooton, R. J. 1984. A functional biology of sticklebacks. University of California Press, Berkeley.

Wooton, R. J. 1990. Ecology of teleost fishes. Chapman and Hall, London.

Wooton, R. J. 1992. Fish ecology. Chapman and Hall, New York.

Wourms, J. P. 1977. Reproduction and development in chondrichthyan fishes. Amer. Zool. 17:379–410.

Wourms, J. P. 1981. Viviparity: The maternal-fetal relationship in fishes. Amer. Zool. 21:473–515.

Wright, J. M. 1993. DNA fingerprinting of fishes, p. 57–91. *In*: Biochemistry and molecular biology of fish. Vol. 2: Molecular biology frontiers, P. W. Hochachka and T. P. Mommsen (eds.). Elsevier, Amsterdam.

Wu, T.Y.-T, C. J. Brokaw, and C. Brennan (eds.). 1975. Swimming and flying in nature, Vol. 2. Plenum Press, New York.

Wunder, W. 1936. Physiologie der Susswasserfische Mitteleuropas. E. Schweizerbart, Stuttgart.

Wydoski, R. S., and R. R. Whitney. 1979. Inland fishes of Washington. University of Washington Press, Seattle.

Yabe, M. 1981. Osteological review of the family Icelidae Berg, 1940 (Pisces: Scorpaeniformes), with comment on the validity of this family. Bull. Fac. Fish. Hokkaido University 32(4):293–315.

Yalden, D. W. 1985. Feeding mechanisms as evidence for cyclostome monophyly. Zool. J. Linn. Soc. 84:291–300.

Yamagami, K. 1981. Mechanisms of hatching in fish: Secretion of hatching enzyme and enzymatic choriolysis. Am. Zool. 21(2): 459–471.

Yamamoto, M. 1982. Comparative morphology of the peripheral olfactory organ in teleosts, p. 39–59. *In*: Chemoreception in fishes. T. J. Hara (ed.). Elsevier, Amsterdam.

Yamamoto, T. 1969a. Inheritance of albinism in the medaka, *Oryzias latipes*, with special reference to gene interaction. Genetics 62:797–809.

Yamamoto, T. 1969b. Sex differentiation, p. 117–175. *In*: Fish physiology, Vol. III, W. S. Hoar and D. J. Randall (eds.). Academic Press, New York.

Yamauchi, A. 1980. Fine structure of the fish heart, p. 119–148. *In*: Hearts and heart-like organs, Vol. 1, G. H. Bourne (ed.). Academic Press, London.

Yamazaki, F. 1974. On the so-called "cobalt" variant of rainbow trout. Bull. Jpn. Soc. Sci. Fish. 40(1):17–25.

Yamazaki, F. 1976. Applications of hormones in fish culture. J. Fish. Res. Bd. Can. 33:948–958.

Yancey, P. H., M. E. Clark, S. C. Hand, and G. N. Somero. 1982. Living with water stress: Evolution of osmolyte systems. Science 217:1214–1222.

Yasumoto, T., N. Seino, Y. Murakami, and M. Murata. 1987. Toxins produced by benthic dinoflagellates. Biol. Bull.,Woods Hole 172(1):128–131.

Yates, G. T. 1983. Hydromechanics of body and caudal fin propulsion, p. 177–213. *In*: Fish biomechanics, P. W. Webb and D. Weihs (eds.). Praeger, New York.

Yone, Y. 1976. Nutritional studies of red sea bream. Proc. 1st Int. Conf. Aquaculture Nutr., University of Delaware, p. 39–64.

Young, G., B. T. Bjornsson, P. Prunet, R. Lin, and H. A. Bern. 1989. Smoltification and seawater adaptation in coho salmon (*Oncorhynchus kisutch*): Plasma prolactin, growth hormone, thyroid hormones and cortisol. Gen. Comp. Endocrinol. 74:335–345.

Young, J. Z. 1933. Comparative studies on the physiology of the iris. I. Selachians. Proc. Roy. Soc. B112:228–241.

Young, R. E., and C. E. Ropper. 1977. Intensity regulation of bioluminescence during countershading in living midwater animals. Fish. Bull. U.S. 75:239–252.

Youson, J. H. 1970. Observations on the opisthonephric kidney of the sea lamprey of the Great Lakes, *Petromyzon marinus* L., at various stages during the life cycle. Can. J. Zool. 48(6):1313–1316.

Youson, J. H. 1981. The kidneys, p. 191–261. *In*: The biology of lampreys, M. W. Hardisty and I. C. Potter (eds.). Academic Press. London.

Youson, J. H. 1988. First metamorphosis, p. 135–196. *In*: Fish physiology, Vol. IXB, W. S. Hoar, D. J. Randall and E. A. Donaldson (eds.). Academic Press, New York.

Youson, J. H., and D. B. McMillan. 1970. The opisthonephric kidney of the sea lamprey of the Great Lakes, *Petromyzon marinus* L. I. The renal corpuscle. Am. J. Anat. 127:207–232.

Yu, X., and X.-Y. Yu. 1990. A schizothoracine fish species *Diptychus dipogon* with a very high number of chromosomes. Chromosome Inf. Serv. 48:17–18.

Zadunaisky, J. A. 1984. The chloride cell: The active transport of chloride and the paracellular pathways, p. 129–176. *In*: Fish physiology, XB, W. S. Hoar and D. J. Randall (eds.). Academic Press, Orlando.

Zahuranec, B. J. (ed.). 1983. Shark repellents from the sea: New perspectives. AAAS Sel. Sympos. Westview Press, Boulder.

Zakon, H. H. 1986. The electroreceptive periphery, p. 103–156. *In*: Electroreception, T. H. Bullock and W. Heiligenberg (eds.). John Wiley, New York.

Zangerl, R. 1973. Interrelationships of early chondrichthyans, p. 1–14. *In*: Interrelationships of fishes, P. H. Greenwood, R. S. Miles and C. Patterson (eds.). Academic Press, New York.

Zangerl, R. 1984. On the microscopic anatomy and possible function of the "spine-brush" complex of *Stethacanthus* (Elasmobranchii: Symmoriida). J. Vert. Paleont. 14(3):372–378.

Zangerl, R., and G. R. Case. 1973. Iniopterygia, a new order of chondrichthyan fishes from the Pennsylvanian of North America. Fieldiana (Geol.) 6:1–67.

Zehren, S. J. 1979. The comparative osteology and phylogeny of the Beryciformes (Pisces:Teleostei). Evol. Monog. University of Chicago 1.

Zeiske, E., J. Caprio, and S. H. Gruber. 1986. Morphological and electrophysiological studies on the olfactory organ of the lemon shark, *Negaprion brevirostris* (Poey), p. 381–391. *In*: Indo-Pacific fish biology. Proc. Second Internat. Conf. on Indo-Pacific Fishes, T. Uyeno, R. Arai, T. Taniuchi, and K. Matsuura (eds.). Ichthyol. Soc. Japan, Tokyo.

Zeiske, E., B. Theisen, and H. Breuker. 1992. Structure, development and evolutionary aspects of the peripheral olfactory system, p. 13–39. *In*: Fish chemoreception, T. J. Hara (ed.). Chapman and Hall, London.

Zenkevitch, L. 1963. Biology of the seas of the U.S.S.R. George Allen and Unwin Ltd., London.

Zottoli, S. J. 1978. Comparative morphology of the Mauthner cell in fish and amphibians, p. 13–45. *In*: Neurobiology of the Mauthner cell, D. S. Faber and H. Korn (eds.). Raven Press, New York.

Zottoli, S. J. 1981. Electrophysiological and morphological characterization of the winter flounder Mauthner cell. J. Comp. Physiol. 144:541–553.

Zuyev, G. V., and V. v. Belyayev. 1970. An experimental study of the swimming of fish in groups as exemplified by the horsemackerel (*Trachurus mediterraneus ponticus* Aleev.). J. Ichthyol. 10:545–548.

Greek and Latin Word Roots and Terms

a- (Gr)   without, not, absence of something

acanth (Gr)   thorn, spine

acipenser (L)   sturgeon

actin (Gr)   ray or beam

ala (L)   wing

alb (L)   white

ali (L)   wing, other

-alis (L)   pertaining to

alope (Gr)   fox

ambly- (Gr)   blunt

amia (Gr)   a kind of fish

ammo (Gr)   sand

amphi (G)   double, on both sides

an- (Gr)   without, not

anabas (Gr)   gone up

anguilla (L)   eel

anoplo (Gr)   unarmed

antenna (NL)   feeler

anti (Gr)   opposed, against

aphritis (Gr)   a fish

aplo (or haplo) (Gr)   single, simple

-arch (Gr)   anus, rectum

arch-, archi- (Gr)   ancient, first, primitive; important, chief

argent- (L)   silvery

arthro (Gr)   a joint

arti (Gr)   entire, complete (in Hyperoartia, not pierced), even numbered

aspis, aspidos (Gr)   a shield

aster (Gr)   a star

atherin (Gr)   a kind of smelt

aulo (Gr)   a tube or pipe

aur, -at, -ic (L)   gold, golden

aur, -is (L)   ear

bagr (L)   a kind of fish

balist (L)   catapult

barb (L)   a beard

bathy (Gr)   deep

batrach (Gr)   a frog

belon (Gr)   a needle, dart

beryc (L)   a kind of fish

boös (G)   an ox or bull

bov (L)   cow, ox

brachi (Gr)   the arm

brad (Gr)   slow

branch (Gr)   gill

bun (Gr)   a mound or hill

calam (L)   a reed

callo (Gr)   beautiful, (L) hard

calva (L)   bald or smooth

camp (Gr)   a sea creature, a bending, (L) a field or plain

carchar (Gr)   jagged, sharp

carin (L)   keel

cato (Gr)   downwards, inferior

caud (L)   tail

caul (L <Gr)   cabbage, a stalk

cent (Gr)   puncture, point, center

cephal (Gr)   the head

ceps (L)   the head

cerat (Gr)   a horn

cerd (Gr)   a fox

cet (Gr)   whale

chano (Gr)   the open mouth, yawn

charac (Gr)   a kind of (sea) fish

chauliod (Gr)   having protruding teeth

cheimarr (Gr)   a winter torrent

chiasm (Gr)   diagonal, marked with a cross

chil (Gr)   lip

chimaera (Gr)   a mythical monster

chir (Gr)   a hand

chit -on, -in (Gr)   outer coat or covering

chlamyd (Gr)   cloak or mantle

chondr (Gr)   cartilage, also grain

cipit (L)   the head

cirr (Gr)   yellow; (L) curl

clad (Gr)   branch

clistic (Gr)   enclosed

cobit (Gr)   gudgeon-like fish

cochl (Gr)   spiral-shelled mollusk

coelo (Gr)   a hollow

coelum (L)   the heavens

coet (Gr)   bed

cotto (Gr)   head

cottus (Gr)   sculpin

cranio (Gr)   head

crosso (Gr)   fringe

cteno (Gr)   comb

cybium (Gr)   tuna

cyclo (Gr)   circle

cyprin (Gr)   carp

dasy (Gr)   with much hair

dent (L)   tooth

derm (Gr)   skin

di- (Gr)   two

distal (Eng.)   at a point away from the center

echene (Gr)   holding ships, remora

echino (Gr)   like a hedgehog; (L) prickly

elasm (Gr)   a plate

ele (Gr)   a swamp

eleuthero (Gr)   free

elop (Gr)   name of a marine fish

en- (Gr)   in

endo (Gr)   within

engraulis (Gr)   a small fish

epi- (Gr)   over or upon

erp (Gr)   creeper

esox (L)   pike

ethmos (Gr)   a strainer

eu- (Gr)   true, good

exo- (Gr)   external

flavi (L)   yellow

fontinalis (L)   of a spring

formes (L)   in the shape of

gado (Gr)   cod

galax (Gr)   milky

galeo (Gr)   shark-like

gaster (Gr)   stomach

gempyl (NL)   a mackerel-like fish

genyp (Gr)   jaw

giga (Gr)   giant

gladius (L)   sword

glanid (L, Gr)   sheatfish

glosso (Gr)   tongue

gnatho (Gr)   jaw

gobi (L)   a fish of small value

gono (Gr)   seed or offspring

grammus (Gr)   writing

gymno (<Gr)   naked

haemal (Gr)   blood red
halo (Gr)   salt
helo (Gr)   marsh
hept (Gr)   seven
heteros (Gr)   different
hex (Gr)   six
hippo (Gr)   horse
holo (Gr)   whole
hybo (Gr)   hump-backed
hyo- (<Gr)   hog
hyper (<Gr)   beyond
hypnos (Gr)   sleep

icost (Gr)   twenty
ict- (L)   fish
-icus (L)   belonging to
inio (Gr)   back of the head
intercalary (L)   inserted
iso (Gr)   equal
istio (Gr)   a sail

korso (Gr)   side of the forehead; the temple
kurt (Gr)   curved

lachrymos (L)   tearful
laemus (Gr)   throat
lamni (Gr)   a voracious fish
lamprid (Gr)   bright
lati (L)   the side
lepido (<Gr)   scale or peel
lepto (Gr)   thin or delicate
leucas (<Gr)   white
levator (Gr)   a lifter
lobi (Gr)   a lobe
loph (Gr)   a crest
lucio (L)   light
lupus (L)   wolf
lys (Gr)   loose

macro (Gr)   long (or large)
mala (L)   jaw (or cheekbone)
mandibul (L)   jaw
mantellum (L)   a cloak
masta (Gr)   a breast
mega (Gr)   great
melano (Gr)   black
mere (<Gr)   part
meso (Gr)   middle
meta (Gr)   between
mimi (L)   mimicry

mira (L)   marvelous and strange
mixi (Gr)   mingling
momo (Gr)   ridicule
mono (Gr)   one
mormyro (L)   a sea fish
morph (Gr)   form or shape
muraeno (L)   a moray
mycto (Gr)   nose
mylio (<Gr)   grinding
myo (<Gr)   muscle
myri (Gr)   infinite
myxin (Gr)   mucus
myzon (Gr)   suck

nark (<Gr)   torpid
nect (<Gr)   swimming
nema (<Gr)   thread
neo (Gr)   new
nesthid (Gr)   hungry
neuro (Gr)   nerve
not (Gr)   the back or (L) not
nym, onym (Gr)   a name

ocell (L)   a little eye
odons (Gr)   tooth
odus (Gr)   tooth
ogco (Gr)   a protruberance
oidei (Gr)   form of or type of
omo (Gr)   the shoulder
onco (Gr)   hook
ophido (Gr)   serpent (actually, the root is probably ophio)
opistho (<Gr)   behind
opleg (Gr)   armor or tool (the root may be opl)
ops (Gr)   aspect or late
orbito (L)   a circle or ring
orecto (Gr)   stretched out
osmer (Gr)   smell or odor
osphro (Gr)   to smell
ostei (Gr)   a door
osteo (Gr)   bone
otic (Gr)   of the ear
oxy (Gr)   sharp

paedia (Gr)   child
paleao (Gr)   ancient
panto (Gr)   all
para (Gr)   near or beside
parietal (Gr)   walls
pegas (Gr)   solid or strong

pemph(igo) (Gr)   a bubble
peri (Gr)   near
petro (Gr)   a rock
phallo (Gr)   penis
pharyngo (Gr)   the pharynx
phidi (<Gr)   thrifty or stingy
phili (Gr)   love
pholi (Gr)   a hole
pholido (Gr)   a scale or spot
phor (Gr)   a thief
phore (Gr)   weaving
phracto (Gr)   fenced in or protected
phthalmo (Gr)   the eye
phyllo (Gr)   a leaf
physi (Gr)   a bladder
pimelod (Gr)   fat and soft
pinna (L)   a feather or wing
placo (Gr)   flat and wide
platy (Gr)   broad and flat
pleco (Gr)   to weave
plect (Gr)   plaited or twisted
pleuro (Gr)   a rib or the side
plio (Gr)   more
pnoi (Gr)   breath
pnous (Gr)   breathing
pogon (Gr)   bearded
poly (Gr)   many
pomatom (Gr)   a cover
pomi (L)   a fruit tree
post (L)   after
pre (L)   before
pria (Gr)   a saw
pristio (Gr)   a file
pro (L)   before
proximal (L)   nearest
pseph (Gr)   a pebble
psett (Gr)   a flatfish
pseud (Gr)   false
pter (Gr)   fin
pterygo (Gr)   fin or wing
ptycho (Gr)   a fold or layer
ptyct (Gr)   folded
pungi (L)   puncture
pycno (Gr)   dense
pylo (Gr)   grate or orifice

quadrate (Gr)   squared

raj (L)   a flat fish or ray
ramph (Gr)   a crooked beak
retro (L)   backward

rhina (Gr)   a file or rasp
rhino (Gr)   nose
rhyac (Gr)   a brook
rhyncho (Gr)   a beak
rostrat (L)   a snout or beak

sacco (L)   sack-like
salang (Gr)   a kind of fish
salmo (L)   salmon
sapid (L)   savory
saur (Gr)   lizard
scaen (Gr)   clumsy, crooked
scapano (Gr)   a spade
scombro (Gr)   mackerel
scopus (Gr)   a watcher
scorp (Gr)   scorpion
scylio (Gr)   dogfish
selachi (Gr)   a cartilaginous fish
serri (L)   a saw
siluri (L)   a kind of river fish
siren (L)   a mermaid-like creature
    who lured sailors to their deaths
    with singing
soleo (L)   the bottom of the foot,
    sandal
spatula (L)   a spoon
sphyrae (Gr)   hammers

sphyrno (L)   hammer-like
spondyl (Gr)   vertebra
squalo (L)   a kind of sea fish
squat (L)   skate
stego (Gr)   roof
sten (Gr)   a narrow, confining space
stephano (Gr)   a crown
stetho (Gr)   breast or chest
stichae (Gr)   rows
stomato (Gr)   mouth
stroma (Gr)   a mattress
stygnos (Gr)   hated
sub- (L)   under
supra- (L)   above
sym- (Gr)   together
syn- (Gr)   together

taenio (Gr)   ribbon
tecto (Gr)   molten
teleo (Gr)   perfect
tera (Gr)   a monster
tetra (Gr)   four
thelo (Gr)   nipple
thenoid (Gr)   palm of the hand
tho (Gr)   quick
thriss (Gr)   a kind of anchovy
tome (L)   part or book

torp (L)   numb
trabecular (L)   marked with cross
    bars
trachino (L)   horse mackerel
trans- (L)   across
treti (Gr)   pierced
triakis (Gr)   three-pointed
tricho (Gr)   hair
troctes (Gr)   gnawer
trypao (Gr)   to bore
typhlos (Gr)   blind

uro (Gr)   the tail

velifer (L)   veiled
vulpes (L)   fox

xena (Gr)   a stranger
xiph (Gr)   sword

zanc (Gr)   sickle
zei (Gr)   a kind of fish
zoarco- (Gr)   life-supporting
zoön (Gr)   animal
zygon (Gr)   a pair, yoke

Glossary

Adaptation   A structure, function, or behavior that makes the fish more fit, or a process (some form of natural selection) that makes a structure, function, or behavior more fit.

Acanthopterygian   Refers to spiny-rayed teleosts.

ACTH   Adrenocorticotropic hormone.

Aestivation   The dormant state of certain animals during periods of drought or high temperatures.

Allele   One of two or more alternative forms of a gene at a chromosome locus; one use of the term *gene*.

Allele frequency   The relative number of a particular allele at a locus in a population. A population of diploid (2N) organisms would have 2N alleles at a locus. The frequency of a particular allele is its relative proportion of that total. Often referred to as *gene frequency*.

Allopatric   Distributions that do not overlap geographically.

Allopolyploid   A polyploid derived from complete chromosomal set additions (e.g., an increase from 2N to 4N) from chromosome sets of two different species. See *autopolyploid*.

Allozymes   Alleles for a protein encoded by a locus that can be separated electrophoretically. Because both alleles are ordinarily expressed in a heterozygote, the expression is usually codominant.

Amino acid   The fundamental building block of proteins specified by nucleotide codons in DNA (or messenger RNA).

Arcualia   Bow-shaped components of vertebral arches.

Artificial propagation   Fish cultural activity, generally involving modification of natural spawning, incubation, or rearing environments.

Artificial selection   Fish cultural activity in which breeders are chosen on the basis of heritable traits (whether purposeful or inadvertent).

Autonomic nervous system   Those efferent motor fibers and their ganglia that regulate bodily functions not under voluntary (conscious) control. The system is composed of two antagonistic parts: parasympathetic and sympathetic.

Autopolyploid   A polyploid derived from complete chromosomal set additions (e.g., an increase from 2N to 4N) involving two species. See *allopolyploid*.

Backcross   A mating in which a hybrid is crossed with a parental type.

Batesian mimicry   Mimicry in which an uncommon but palatable species resembles a more common but unpalatable species.

Biomass   The total weight of all members of a species (taxon) or group of species (taxa) in a given area at an instant in time. ("Total weight of all organisms in a particular habitat or area; the term is also used to designate the total weight of a particular species or group of species.")

Bowman's capsule   The expanded proximal end of the kidney tubule surrounding the glomerulus.

Broodstock   Adult fish used for artificial propagation.

Chromosomal complement   The entire set of chromosomes in a cell or that characterize a species. Normally the complement is diploid (2N), but polyploids may have additional haploid (N) sets and be triploid (3N), tetraploid (4N), etc.

Chromosome   A structure comprised mostly of DNA and protein and found in nuclei of cells that organizes the genetic information. Genes are arranged linearly along chromosomes. During meiosis and mitosis, chromosomes condense to facilitate apportioning of the genetic material between daughter cells.

Circadian   Pertaining to 24-hour biological cycles.

Clade   A branch of the tree of life (see *monophyletic*); a natural group.

Cladistic tree   A tree diagram showing the sequence of evolutionary branching of a study group.

Cladistics   A method of phylogenetic analysis formalized by Willi Hennig (1967), in which monophyletic (natural) groups are diagnosed by shared, derived characters called synapomorphies.

Cladogram   A cladistic tree diagram, showing branching sequence.

Codominant   Both alleles are expressed in a heterozygous (Aa) individual so the individual is distinguishable from both homozygous (AA and aa) types. Most allozyme electrophoretic banding patterns reflect codominant inheritance.

Codon   Three contiguous nucleotides in a nucleotide sequence (gene) that specify a particular amino acid in the protein encoded.

Commensalism   A form of symbiosis in which one species benefits and the other neither suffers nor benefits.

Commissure   A linking or connecting of parts of the nervous or lateral-line systems, usually from one side to the other.

Congruence   The consistency of change in character states of different characters with each other over the branching sequence of a clado-gram.

Convergence   The attainment of functionally similar structures by distantly related taxa.

Cristae   Patches of sensory cells (neuromasts) at the juncture of the semicircular canals and the utriculus.

Cross   A mating.

Cryophylic   Refers to organisms that thrive at relatively low temperatures.

Dam   Female parent in a cross. See *sire*.

Defining characters   These are assigned without benefit of a cladis-tic attempt to discover the mono-phyly or individuality of the group being defined.

Demersal   Refers to aquatic organ-isms living on or in close associa-tion with the substrate (bottom).

Derived   Modified relative to the primitive condition.

Diagnostic characters   These are discovered in a cladistic analysis.

Diapause   The state of suspended development.

Diastole   The dilation phase of the heart action.

Diploid (2N)   Having a chromosomal complement consisting of two homologous chromosomes of each type—that is, two haploid (N) com-plements.

Diverticulum   Any blind sac or pouch connected to a larger cavity.

DNA (deoxyribonucleic acid)   Lin-ear, double-stranded molecule con-sisting of (deoxyribo) nucleotides, which encode the genetic informa-tion of an organism. The code lies in the specific order of the nucleotides. The strands are held together by complementary **A-T** and **G-C** nucleotide pairing. See *nucleotide* and *replication*.

Dominant   An allele that is expressed in the phenotype of both homozy-gous (AA) and heterozygous (Aa)

individuals. It can also refer to the phenotype expressed. The other allele (phenotype) is recessive.

Ecophenotypic   A trait whose varia-tion is controlled by environmental conditions (e.g., high meristic counts caused by lower develop-mental temperatures at high lati-tudes).

Electrophoresis   A technique for separating molecules on the basis of their intrinsic charge. It is applied to proteins to obtain allozyme data and to DNA frag-ments to estimate their size.

Emmetropic   Refers to normal ocu-lar vision (i.e., not near- or far-sighted).

Endemic   Native or confined (restricted) to a particular geo-graphical region.

Endogenous   Originating within. Produced from within the body, an organ, or a geographical area.

Endolymph   Fluid contained within the inner ear (membranous labyrinth).

Epistasis   Differential expression of a phenotype as a result of interaction between alleles at more than one locus.

Epithelium   The thin layer of tissue covering internal and external body surfaces.

Exaptation   Restrictive term for adaptations that began in a different adaptive context. For example, if the swimbladder originated as a respiratory organ, one might say it is an exaptation for buoyancy and hearing.

Extensive aquaculture   Cultured organisms released into the natural environment for part of their life cycles. See *intensive aquaculture*.

Fecundity   An organism's capacity to produce offspring. (In fishes, often expressed as the number of eggs produced per female.)

Fenestra   An aperture in a bone or a transparent portion of a membrane.

Fimbria   A bordering fringe.

Fitness   The productivity of a partic-ular phenotype in a population rel-ative to the most productive phenotype in that population, in that environment. If there is a genetic basis for the phenotype, fitness-based selection can result in evolutionary change.

Follicle   Any small sac or pit.

Fontanelle   A membrane-covered opening in a bone.

Foramen   A small opening or perfo-ration in any body structure.

Gametes   Mature, haploid, male or female reproductive cells.

Ganglion   A concentration of nerve cell bodies located outside the cen-tral nervous system.

Gene   The basic unit of inheritance, the nucleotide sequence that carries encoded information for a particu-lar trait. Genes are linearly arranged on chromosomes. *Gene* is often used in a broader context to refer to loci or alleles.

Gene flow   Exchange of alleles between populations (in one or both directions). See *migration*.

Gene pool   The aggregate genetic composition of a population. Often quantified in terms of allele frequencies.

Genetic diversity   The genetic varia-tion that exists within a population or unit of interest.

Genetic drift   Random variation of allele frequency in a population from one generation to another that results from random errors in sam-pling gametes. Because such errors result from a finite sample of gametes of their parents, the genetic composition of progeny may differ from that of their parents.

Genetic integrity   The degree to which the genetic composition of a population resembles its natural

Iteroparous Spawning more than once before dying. See *semel-parous.*

Lineage A genetically continuous population or group.

Linkage Genes physically connected on the same chromosome may be transmitted together in a gamete and are, therefore, linked and do not follow Mendel's law of random assortment.

Locus (plural, **loci**) The site on a chromosome of the nucleotide sequence that encodes information for a particular trait. A gene.

Maculae Patches of sensory cells (neuromasts) in the utriculus, sacculus, and lagena.

Meiosis Cell division involved in reducing the chromosomal complement from diploid (2N) to haploid (N) for gamete production.

Mendelian trait A trait resulting from the expression of alleles at a single locus. The phenotypes of these traits usually have a discrete distribution.

Meninges Protective membrane enclosing (surrounding) the brain and spinal cord.

Metamorphosis The stage in development during which an animal undergoes a radical change in form and function.

Microphagous Refers to organisms that feed on relatively small food items.

Micropyle An aperture in the vitelline membrane of an egg, through which the sperm enters.

Migration Movement of individuals or populations from one geographic location to another. In genetics, the movement of genes from one gene pool to another (gene flow).

Mimicry Imitation of another organism or object in the environment (in form, color, and/or behavior).

Mitochondrion A subcellular organelle involved in aerobic metabolism. Mitochondria possess DNA that carries information for some of their functions. Mitochondria are haploid and are believed to be maternally inherited in fish.

Mitosis Proliferative cell division during growth or development in which resultant cells have the same diploid (2N) complement as the parental cells from which they are derived.

Monophyletic A natural group including all descendants of a common ancestor (i.e., a clade). Also called *holophyletic.*

MSH Melanophore-stimulating hormone; can cause either dispersion or aggregation of pigment.

Müllerian mimicry Mimicry in which several unrelated animal species distasteful to their predators resemble each other.

Mutation An alteration in the nucleotide sequence of a gene. Allelic differences result from mutation. Mutation is the ultimate source of all genetic variation.

Myelinate Refers to nerves ensheathed by a fatty membrane.

Myoid Contractile segment of visual cell.

Native Fish populations historically indigenous to an area.

Natural selection As a result of different fitnesses, genetically based phenotypes contribute differentially to the following generation. These genetic changes in the population reflect natural selection.

Neuromasts Mechanoreceptors of the acousticolateralis system.

Nictitating membrane The third "eyelid" present in many vertebrates (aids in cleaning and protecting the eye).

Nucleotide The fundamental subunit of DNA (and RNA) structure. Nucleotides are comprised of the nitrogenous ring structures adenine [A], thymine [T] (uracil [U] in RNA), cytosine [C], and guanine [G] attached to a deoxyribose sugar (ribose in RNA), which is attached to a phosphate. Alternating (deoxy)ribose and phosphates form the scaffolding of DNA and RNA polymers.

Nuptial Refers to breeding.

Ocellus An eyelike marking.

Oligotrophic Refers to bodies of water, especially lakes, with low biological productivity.

Ontogeny Development through the life cycle.

Otoliths Calcareous nodules in the utriculus, sacculus, and lagena of the membranous labyrinth.

Outgroups The relatives, outside the study group, included in a cladistic analysis as a source of information about the primitive starting point of character states.

Oviparous Refers to egg-laying animals.

Ovoviviparity Condition of retention and incubation of eggs within the ovary or oviduct. The young receive no (or little) nourishment from the female.

Paedophagous Refers to a fish that eats larvae or juveniles.

Parabranchial cavity Cavity bounded by the operculum and branchiostegal membrane; receives water that has passed through gills.

Paraphyletic An unnatural assemblage defined in such a way as to exclude part of a clade—for example, Pisces (usually excludes Sarcopterygii) or Perciformes (usually excludes Scorpaeniformes, Pleuronectiformes, etc.).

Parr That stage in anadromous salmonids between yolk sac absorption and transformation to the smolt stage prior to seaward migration.

Parsimony In cladistics, the princi-

state, uninfluenced by anthropogenic causes.

Genetic marker   An allele that may characterize a population or group of populations. The marker may be qualitative (there or not) or quantitative (present at greater or lesser frequency in the marked group). Genetic markers are the basis of genetic stock identification and have been used to determine parentage.

Genetic stock identification (GSI)   Use of genetic differences occurring among populations or aggregates of populations to estimate proportionate contributions to mixtures, such as mixed stock fisheries.

Genome   The entire set of genetic information (nucleotide sequences) of an individual.

Genotype   The allelic composition of an individual. *Genotype* can refer to a single locus, to several loci, or to the total genomic content.

Glomerulus   A knot of small blood vessels contained within Bowman's capsule at the proximal end of a kidney tubule.

Glycoproteins   Organic molecules composed of carbohydrates and proteins.

Gnathostomes   Jawed vertebrates.

Gonadotropic hormones   Hormones, secreted by the anterior lobe of the pituitary, that induce development of gametes.

Gonopodium   A modified anal fin that functions as a copulatory organ.

Gynogen   An offspring possessing only maternal chromosomes. If males are the heterogametic sex, gynogens are exclusively female.

Haploid (N)   Having a chromosomal complement consisting of a single chromosome of each type. Gametes of normal diploid (N) organisms are haploid.

Hemopoietic   Refers to the production of blood cells.

Heritability   The degree to which offspring resemble their parents as a result of their shared genetic background. Heritability can be in the narrow sense (h^2), which takes into account only the genetic determinants of similarity that can be bred for predictability (the additive genetic variation V_A), or in the broad sense (H^2), which takes into account all the genetic determinants (both additive, V_A, and dominance, V_D, variation).

Heritable variation   That portion of an organism's variation that is controlled by genetic pathways rather than environment.

Hermaphrodite   An organism with both male and female sex organs. Sequential hermaphrodites do not simultaneously possess both sets. See *protandry* and *protogyny*.

Heterogametic   The sex that has two different sex chromosomes (e.g., XY or WZ) can produce two different gametic types, X or W carrying and Y or Z carrying.

Heterozygous   Refers to a state in which two alleles at a locus in a diploid (2N) individual differ (e.g., Aa).

Holarctic   A biogeographic region that includes the arctic and north temperate zones. It includes mostly the northern parts of Eurasia and North America.

Homologous   (1) Refers to characters, in different taxa, that are structurally similar due to common evolutionary origin. (2) Refers to chromosomes that carry information for the same set of traits and pair during meiosis. Although a particular gene that they carry may specify the same trait, they may be alternatives (different alleles) for that trait.

Homology   Similarity due to common ancestry. Usually hypothesized by similarity of position, composition, and embryological origin. A synapomorphy may be confirmed to be a homology by demonstration that it is uniquely diagnostic of a clade, as shown by congruence with other characters in a cladistic tree analysis (e.g., the Weberian apparatus of Otophysi).

Homoplasy   Similarity due to convergent, parallel, or reversed evolution (e.g., spines of catfish, goldfish, and sunfish), in contrast homology.

Homozygous   Refers to a state in which two alleles at a locus in a diploid (2N) individual are the same (e.g., AA or aa).

Hormones   Chemical substances that are released from endocrine glands, are transported via the circulatory system, and regulate a wide range of physiological functions.

Hybrid   Offspring of a cross between two genetically dissimilar individuals. Intraspecies hybrids result from crossing individuals of two different strains. Interspecies hybrids may result from crossing individuals of two species.

Illicium   Modified first dorsal fin ray of angler fishes, used to attract prey.

Individual lineage or clade   In the restricted philosophical sense, a taxon meeting strict criteria of genetic continuity within spatial and temporal boundaries, irreplaceable for purposes of conservation.

Intensive aquaculture   Cultured organisms maintained in captivity during their life cycle. See *extensive aquaculture*.

Intergrades   As used in this book, individuals that are, for whatever reason, intermediate between two species (or other taxa).

Introgression   The introduction of genes from one population (or species) into the gene pool of another. Introgression may compromise genetic integrity.

Intromittent organ   Male copulatory organ.

ple dictating the choice of cladistic hypotheses requiring the fewest ad hoc assumptions about character convergence, parallelism, and reversal.

Parthenogenesis Production of off-spring by a female with no male contribution.

Pelagic Refers to organisms that inhabit open waters of the oceans (or large lakes).

Peritoneum Membrane lining the coelom.

Phagocytosis The process by which certain cells engulf other cells or foreign particles.

Phenetics The study of patterns of similarity among organisms. Also, the estimation of relationship by overall similarity rather than special (shared, derived) similarities.

Phenotype The observed expression of a trait or characters of an organism, determined by interaction of its genes and the environment; *Phenotype* can refer to a single Mendelian character or to the entire multilocus organism. Species diagnoses must be based on parts of the phenotype that are distinctive due to inheritance rather than environmental effects.

Photophores Light-producing organs.

Phylogenetic tree A cladistic branching diagram with added information about times of branching and morphological change.

Phylogeny Evolutionary relationships among species based on their descent from a common ancestor.

Placenta An intimate association of embryonic and maternal membranes through which gases, nutrients, and wastes are exchanged.

Pleiotropy The effect of a single locus on more than one trait. See *polygenic trait.*

Polyandrous Refers to species in which one female mates with more than one male.

Polyculture The use of two or more species to utilize two or more trophic levels within a culture system. (The same effect can be obtained in some instances by the use of distinct life stages of one species.)

Polygenic trait A trait resulting from the combined expression of alleles at numerous loci. The phenotype of these traits generally has a continuous distribution and may be referred to as quantitative or metric traits.

Polyphyletic An unnatural group defined to include two or more groups that are not closely related (not of the same immediate line of descent)—for example, Apodes (for unrelated eel-shaped animals).

Polyploid Possessing three or more haploid sets of chromosomes. See *allopolyploid, autopolyploid, triploid,* and *tetraploid.*

Population A group of organisms belonging to the same species that occupy the same locality at the time of reproduction, have a reasonable chance of interbreeding, and produce progeny that will generally interbreed with progeny of other members of the population.

Primary production The rate at which plants and other autotrophs accrue biomass or energy and nutrients. Expressed as weight/area/time or calories/area/time.

Productivity The relative contribution of offspring to the next generation by a unit (e.g., an individual, phenotype, population, etc.). If differential productivity (whether it is attributable to increased fecundity, survival, or other factors) is heritable, it reflects fitness and, through natural selection, may lead to evolutionary changes.

Protandry A condition in which sequential hermaphrodites that are first males become females later in life.

Protogyny A condition in which sequential hermaphrodites that are first females become males later in life.

Recessive An allele that is expressed only when homozygous (aa). It can also refer to the trait expressed by homozygous recessive alleles.

Recombination The occurrence of combinations of genes in an organism that were in neither parental type but that can be transmitted to the organism's progeny. Recombination is a result of random assortment of unlinked loci or physical exchange between chromosomes of linked loci during meiosis.

Replication Synthesis of DNA based on the complementarity of A-T and G-C nucleotide pairs. At the end of replication, there are two complete copies of the original DNA molecule.

Ribosome The site of protein synthesis (translation) in cells. Constructed of protein and RNA.

RNA (ribonucleic acid) Linear molecule consisting of (ribo) nucleotides transcribed from the DNA sequences that encode the genetic information of an organism. The code lies in the specific order of the nucleotides. The information specifies the amino acid sequence in a protein (messenger RNAs) or other RNAs involved in cell operations.

Semelparous Spawning once and then dying, as do Pacific salmon. See *iteroparous.*

Sexual dimorphism Morphological variation within a species correlated with the sex of the individual.

Sire Male parent in a cross. See *dam.*

Sister chromatids At cell division following DNA replication, each chromosome has been duplicated. Identical copies of the same DNA sequence are sister chromatids.

Each homolog produces a pair of sister chromatids.

Smolt The seaward migrating stage of anadromous salmonids.

Stegural A bone flanking the uroneurals in the dorsal tier of caudal fin supports. The family Salmonidae is diagnosed by a unique fan shape of this bone.

Stenophagy A narrow range of preferred foods.

Steroids A class of organic compounds composed of four interlocking carbon rings. Cholesterol and male and female sex hormones are included.

Stock A term that varies with context and ranges in meaning from a discrete, largely reproductively isolated subpopulation to an aggregation of populations managed as a unit. Alternatively, a genetic strain.

Strain A group of individuals derived from a common genetic origin; a lineage.

Symbiosis An intimate living arrangement (relationship) between two species. (One or both species may benefit from the association.)

Sympatric Two or more species that occupy the same geographic range.

Sympatry Temporal and spatial overlap of the ranges of two or more species.

Symplesiomorphy Shared, primitive character.

Synapomorphy Shared, derived character.

Systole The contraction phase of the heart.

Telolecithal Refers to eggs in which the yolk is concentrated at the vegetal pole.

Tetraploid (4N) A polyploid that has four complete haploid (N) sets of chromosomes.

Trait Manifestation of a genetically determined character. A trait may be Mendelian or polygenic.

Transcription Synthesis of RNA from DNA based on the complementarity of A-(T or U) and G-C nucleotide pairs. The RNA product is complementary to the DNA strand that served as a template.

Transgenic A genetically manipulated organism that received a portion of its genetic information from another species. The genetic transfer often involves molecular genetic or recombinant DNA technology.

Translation Protein is synthesized at the ribosome from information encoded in the nucleotide sequence of the messenger RNA molecule under direction of the amino acid–specifying codons.

Triploid (3N) A polyploid that has three complete haploid (N) sets of chromosomes.

Trophic Refers to nutrition or to ecological levels or mode of feeding.

Triturate To grind.

Velum In hagfishes, a scroll-like pharyngeal membrane that acts as a respiratory pump. In adult lampreys, the fleshy, tentacled flap guarding the opening to the respiratory tube.

Viviparity The condition in which fertilized ova (eggs) are retained within the female and derive nourishment from the female via placenta or from secretions.

Wild Fishes naturally produced, not artificially propagated or cultured. Often refers to native, self-sustaining populations.

Zymogen The inactive precursor of an enzyme.

Index

Order Anguilliformes — eels
Order Saccopharyngiformes — gulpers
Subdivision Clupeomorpha
Order Clupeiformes — herrings
Subdivision Euteleostei
Superorder Ostariophysi
Order Gonorhynchiformes — milkfish
Order Cypriniformes — carps
Order Characiformes — characins
Order Siluriformes — catfishes
Order Gymnotiformes — knifefishes
Superorder Protacanthopterygii
Order Esociformes — pikes
Order Osmeriformes — smelts
Order Salmoniformes — salmons
Superorder Stenopterygii
Order Stomiiformes — stomiiforms
Order Ateleopodiformes — jellynose fishes
Superorder Cyclosquamata
Order Aulopiformes — aulopiforms
Superorder Scopelomorpha
Order Myctophiformes — lanternfishes
Superorder Lampridiomorpha
Order Lampridiformes — opah, oarfishes
Superorder Polymixiomorpha
Order Polymixiiformes — beardfishes
Superorder Paracanthopterygii
Order Percopsiformes — troutperch
Order Ophidiiformes — cusk eels, brotulas
Order Gadiformes — cods
Order Batrachoidiformes — toadfishes
Order Lophiiformes — anglerfishes
Superorder Acanthopterygii
Series Mugilomorpha
Order Mugiliformes — mullets
Series Atherinomorpha
Order Atheriniformes — silversides
Order Beloniformes — flyingfishes
Order Cyprinodontiformes — cyprinodonts
Series Percomorpha
Order Stephanoberyciformes — stephanoberycoids
Order Beryciformes — berycoids
Order Zeiformes — dories
Order Gasterosteiformes — sticklebacks, seahorses
Order Synbranchiformes — swamp eels
Order Scorpaeniformes — scorpionfishes
Order Perciformes — perches
Order Pleuronectiformes — flatfishes
Order Tetraodontiformes — trunkfishes, porcupinefishes